D1457673
CO-BRH-771

PROCEEDINGS OF THE 1980 GUANGZHOU CONFERENCE ON THEORETICAL PARTICLE PHYSICS

VOLUME ONE OF TWO VOLUMES

SCIENCE PRESS

Beijing, China, 1980

Distributed by

VAN NOSTRAND REINHOLD COMPANY

New York Cincinnati Toronto London Melbourne

Published by Science Press, Beijing
Distributed by Van Nostrand Reinhold Company
New York, Cincinnati, Toronto, London, Melbourne
Printed in Hong Kong

First published 1980
ISBN 0-442-20273-3
Science Press Book No. 1894-2

前 言

在中国科学院、教育部、中国科学技术协会等单位的赞助下，广州粒子物理理论讨论会于一九八〇年一月五日至十日在广东从化举行。为了避免翻译所带来的困难，采用中国普通话作为讨论会的正式语言。国内粒子物理理论学家和海外华裔同行举行这样的盛会，这还是第一次。会议中的报告和讨论以及会议外的交流非常成功。经邀请在讨论会上所做的三个粒子物理实验的报告，也引起与会的粒子物理理论学家很大的兴趣。

在新中国诞生的时候，国内的粒子物理理论工作者屈指可数。在以后十年中参加这方面研究工作的人数略有增加，但是研究工作主要是沿着当时国际上流行的研究方向进行的。一九五五年一月上旬，毛主席在一次会议上向我提出"质子和中子又是什么东西组成的?"这样一个问题的时候，我和当时国际上绝大部分物理学家一样，认为质子和中子是"基本粒子"，是物质的最基本的单元。毛主席微笑着说："我看不见得。质子、中子、电子也仍然是可分的。现在实验上虽然还没有证明，将来实验条件发展了，将会证明它们是可分的。你们不信？反正我信。"这个问题的深度显然超出了当时粒子物理理论研究的深度，因此，自然也超出当时国内粒子物理理论工作者研究和探讨的深度。

一九六四年北京科学讨论会在北京举行。在这期间毛主席会见了日本学者坂田昌一教授。毛主席很重视坂田运用唯物的观点和辩证的方法进行科学研究而取得的重要成就。一九六五年《红旗》杂志翻译和刊载了坂田的文章：《新基本粒子观的对话》。这鼓励了一些国内粒子物理理论研究工作者去学习方法论，并用之于分析粒子物理的实验结果和探讨粒子物理理论发展的方向。他们这才意识到，十年前毛主席的预言正在开始实现。于是他们发挥集体智慧，经过九个月的工作，提出了强子结构的层子模型。之所以将组成强子的成分称为"层子"，是想强调这样的观点：即层子也不是物质的最基本的单元，只是物质结构的无限层次中的一个层次而已。这是国内粒子物理理论工作者第一次自己开辟新的研究方向并沿着这个方向进行工作。工作的成果曾经在北京科学讨论会一九六六年暑期物理讨论会上报告。

可惜紧接着，国内粒子物理理论研究工作和自然科学其它领域中的基础科学研究一样，中断了许多年。从一九七一年开始，杨振宁教授多次来华讲学，对于国内开展规范场的研究工作起了推动作用。从一九七二年开始，李政道教授也多次来华讲学，对于在国内开展关于孤粒子和口袋模型的研究产生了重要的影响。很多国外的粒子物理学家也陆续来中国讲学，其中既有理论学家，也有实验学家；既有华裔学者，也有外国学者。所有这些都有利于国内的粒子物理理论研究工作的恢复。一九七六年阻碍国内科学发展的障碍终于被拔除，这样才有可能在一九七七年夏在黄山召开粒子物理理论座谈会。一九七八年夏在庐山举行中国物理学会第三次大会，全国粒子物理理论研究工作者再次得到机会进行讨论和交流。同年秋，又在桂林举行了粒子物理理论座谈会。所有这一切都表明，国内的粒子物理理论研究工作开始恢复。

一九七九年四月到六月，李政道教授再次来华讲学，受到热烈欢迎。我们和李政道教授探讨举行一次国内粒子物理理论学家和海外华裔同行参加的讨论会的可能性，立即得到李教授的热情赞成和支持。我们又写信征求杨振宁教授的意见，也同样得到杨教授的热情赞成和支持。于是我们向其他海外华裔粒子物理理论学家发出邀请，也得到热情的响应。几位海外华裔粒子物理实验学家也欣然接受邀请，派代表在讨论会上提出报告。正是由于海外华裔粒子物理学家的普遍的、热情的支持，广州粒子物理理论讨论会才开得如此成功。在此，我谨向所有来广州参加讨论会的海外同行表示衷心的感谢。

这次会议之所以能够成功地举行，也是由于国内多方面的支持和帮助。特别是广东省对我们的支持，使得这次会议能在风景优美的从化举行。这次讨论会是在顾问委员会直接领导下召开的，讨论会的科学秘书和论文集的编辑，做了大量的、复杂的科学组织工作和讨论会论文集的编辑工作。正是由于这些同志的不懈努力，才能使讨论会论文集早日和读者见面。我谨代表组织委员会对这些同志表示深切的感谢。

组织委员会主任委员　**钱三强**

一九八〇年三月

FOREWORD

Sponsored by the Academia Sinica, the Ministry of Education of China, the Chinese Association of Science and Technology, etc., the Guangzhou Conference on Theoretical Particle Physics was held in Conghua of Guangdong Province from 5th to 10th January of 1980. To avoid the need of translation, the common speech of the Chinese language was adopted as the official language of the Conference. This was the first time for Chinese theorists of particle physics and their oversea colleagues of Chinese origin to hold such a fruitful meeting. Reports and the ensuing discussions during the session and the exchange of views after the sessions were very successful. Three invited talks on recent experiments of particle physics also attracted great interest from those attending the meeting.

There were very few theorists of particle physics in this country when New China was born. In the following decade their number increased slowly, while their work still followed closely the prevalent world trend of research in this field. At a meeting in early January of 1955, Chairman Mao Zedong raised the question: "What are then the proton and the neutron composed of?" and asked for my view. Just like the vast majority of physicists of the world at that time, I regarded the proton and the neutron as "elementary particles", which meant the most fundamental building blocks of matter. Smiling, Chairman Mao said, "This seems unlikely to me. The proton, the neutron and the electron may still be further divided. Though there is no experimental proof now, conditions in experiments will continue to develop. Future experiments are going to prove that they are divisible. Do you believe it? I believe it anyway, even if you scientists don't believe it." The problem raised penetrated much deeper than most research problems which engaged the attention of the theoretical physicists of the world at that time. Naturally it also penetrated far beyond the depth of the investigations and explorations of theoretical physicists in this country.

The Peking Symposium was held in 1964 in Peking. Chairman Mao met then the Japanese physicist, Prof. S. Sakata. Chairman Mao valued highly the important scientific contribution of Prof. Sakata who adopted the materialistic viewpoint and the method of dialectics in his research. Prof. Sakata's article "Dialogues Concerning a New View of Elementary Particles" was translated into Chinese and published in the periodical "Red Flag" in 1965. It encouraged some of our theorists of particle physics to embark upon a study of methodology, and to try to use it in analysizing the experimental results and discussing the direction of the development of the theory of particle physics. They came to realize only then that the prediction of Chairman Mao's ten years ago began to be fulfilled. After nine months of hard work by collective effort, they put forward the Straton Model of the structure of hadrons. The constituents of the hadrons were named "Straton" to stress the following viewpoint: Neither are the stratons the most fundamental building blocks of matter, they constitute also only one of an infinite number of strata of the structure of matter. This was the first time for theorists of particle physics of this country to initiate a new direction of research by themselves and work steadily along it. Part of the research results were presented at the 1966 Summer Physics Colloquium of the Peking Symposium.

Immediately afterwards, unfortunately, however, research in theoretical particle physics in China was interrupted for many years, so was fundamental research in other branches of natural sciences. Starting from 1971, Prof. C. N. Yang frequently visited China and accepted invitations to talk on important developments in particle physics. As a consequence many theoretical particle physicists in this country since then have become interested in and began to work on the gauge field theory. Prof. T.D. Lee also came to China from time to time since 1972 and lectured on main theoretical topics of particle physics, which had important influence on our theoreticians who started working on the solition and the bag model. Besides, many other particle physicists from abroad have come to China on lecture tours, among whom there were theorists as well as experimentalists, colleagues of Chinese origin as well as foreign friends. All this has been beneficial to the recovery of research activity of theoretical particle physics in this country. The obstacles to the progress of science in China were finally removed in 1976. It was then possible to hold the Huangshan discussion meeting on the theory of particle physics in the summer of 1977. In the summer of 1978, when the Third Congress of the Chinese Physical Society took place in Lushan, theoretical particle physicists of the whole country had the opportunity to meet again for discussions and exchange of views. In the autumn of the same year, a further discussion meeting on particle physics was held in Guilin. All this was good indication that the research of theoretical particle physics in China began to recover.

From April to June of 1979, Prof. T.D. Lee came to China again to deliver a series of lectures which won enthusiastic acclaim. During this time we consulted with Prof. Lee concerning the possibility of convening a conference participated by theoretical particle physicists from this country and those of Chinese origin from abroad. Prof. Lee gave approval and support immediately. We then wrote to Prof. C.N. Yang asking for this opinion and also received enthusiastic approval and support. Then we sent invitations to other theoretical particle physicists of Chinese origin working abroad. Their responses were very favourable. We also invited some experimental particle physicists of Chinese origin to send their representatives to the Conference to give reports on their recent achievements. It was the universal enthusiastic support of the particle physicists of Chinese origin from abroad which made the Guangzhou Conference on Theoretical Particle Physics such a success. I would like to take this opportunity to extend my sincere thanks to all our colleagues from abroad, who came to Guangzhou to contribute to this Conference.

The success of this Conference was also due to the support and help we received from many sources in this country. In particular, it was the hospitality of Guangdong Province which enabled us to hold the Conference in such beautiful surroundings in Conghua. The Conference went smoothly under the efficient direction of the Advisory Committee. The large amount of work involved in organizing the scientific activities and in preparing the Proceedings of this Conference was undertaken by the scientific secretaries and the editors of the Proceedings. The early publication of the Proceedings is mainly due to their tireless efforts. On behalf of the Organizing Committee, I would like to offer them my deep gratitude.

Qian San-qiang (Tsien San-tsiang)
Chairman of the Organizing Committee
March, 1980

CONTENTS

VOLUME ONE

II. REPORTS AT THE PARALLEL SESSIONS

A

VOLUME TWO

B

D

III. REPORTS ON RECENT EXPERIMENTAL RESULTS PRESENTED AT THE PLENARY SESSION

OPENING ADDRESS

Zhou Pei-yuan (Chou Pei-yuan)
Vice President, Chinese Academy of Sciences

Ladies and Gentlemen,
Friends and Comrades,

The 1980 Guangzhou Conference on Theoretical Particle Physics sponsored by the Academia Sinica, the Ministry of Education of China and the Chinese Association of Science and Technology is now open. On behalf of the Academia Sinica, I would like to extend our warmest welcome and cordial greetings to the oversea Chinese sciennists, to the physicist of Chinese descent from abroad, to our colleagues from Hongkong and Macao, to the participants of institutes of scientific research and colleges and universities from different parts of this country, to the representatives of the host province Guangdong attending this opening ceremony, and to all other friends and comrades present at this meeting.

In the course of preparation for this conference, we received active cooperation and generous assistance from many quarters, especially from Guangdong Province. We also got enthusiastic support from colleagues both at home and abroad, especially from Professors T.D.Lee and C.N.Yang, who made many constructive suggestions, provided us with the necessary information and contributed in many other ways to the successful convention of this Conference. To all of them I would like to express our sincere thanks.

Up to now, 50 overseas scientists and physicists of Chinese descent from Australia, Germany, Malaysia, Singapore, the United Kingdom and the United States, together with scientists from Hongkong and Macao, have already arrived and submitted to the Conference about 40 contributions. Over 100 participants from various research institutes and institutions of higher learning in China have submitted about another 80 ones. Most of these contributions will be presented and discussed at plenary sessions or at the paralleled sessions. These reports and the subsequent discussions will contribute to the understanding of the phenomena of high energy physics. They will also have important influence on the development of science in this country.

Friends and comrades! Important achievements have been made during the last decades in the investigation of particle physics. It is still in the course of vigorous development, these investigations also raised problems, the fundamental solution of which will require still greater efforts. In China, the study of theoretical particle physics began early in the fifties. The Straton Model put forward in 1965-1966 pioneered the theoretical investigation of the structure of hadrons. But in the ensuing decade, basic researches in natural science in China were interrupted, bringing serious damage to science in this country. In recent years, the Academia Sinica and the institutions of higher learning have been endeavouring to promote and support renewed efforts in basic researches. Up to now, two symposia on a nationwide scale on the theory of particle physics and a number of discussion meetings on specific topics have been held.

The main aim of this Conference is to review the important progress and present the new results in various areas of theoretical particle physics, including the hadron structure and the strong interaction phenomenology, the gauge field theory and QCD, the grand unified theories and gravitation, cosmology, etc. The contacts between scientists and the exchange of ideas during the Conference will be of great benefit to the raise of the level of research in theoretical particle physics in this country.

This Conference is also a reunion of theoretical particle physicists of China with their overseas colleagues of Chinese descent. It offers an excellent opportunity for exchange of views, for mutual helping and learning from each other. I ask all of you to put forward your views freely and thoroughly. I hope there will be extensive and lively discussions. It is to be expected that by the collective effort of all of us present here, the Conference will fulfil the task it has set out to accomplish.

To conclude, I wish this Conference a great success; wish all of you good health and a happy new year!

REMINISCENCES OF THE STRATON MODEL

Hung-yuan Tzu
Institute of High Energy Physics
Academia Sinica

Abstract

A brief account of the straton model of the structure of the hadrons proposed 14 years ago is given, including the titles of 42 papers and notes published in Chinese in 1966 and the names of their authors.

The straton model of the structure of the hadrons was proposed 14 years ago. From September of 1965 to June of 1966, there were 39 people taking part in the research work, most of them were young graduates from the universities. The results were presented in 42 papers and notes in "Atomic Energy" and "Acta Scientiarum Naturalium Universitatis Pekinensis" published in 1966. The main results were reported at the 1966 Summer Physics Colloquium of the Peking Symposium. Shortly afterwards, all periodicals of natural sciences in China ceased to publish. These works were thus never published in foreign languages. Colleagues outside China have heard about these works, but know little about their contents.

It is very rare to have works completed more than 10 years ago talked upon at today's international scientific conference. The viewpoint maintained by the straton model that the hadrons have internal structure is now generally accepted by the international community of high energy physicists. Works on the structure of the hadrons which began to appear 14 years ago were exploratory and preliminary investigations. Today, the structure of hadrons becomes one of the important areas of research in high energy physics. Since 1972, many colleagues of Chinese origin visiting China enquired about the actual contents of the straton model. Owing to the particular historical circumstances, it was only possible to give very brief account on very few occassions. Even the names of the author's were not mentioned. We would like thus to ask your kind permission to use this tribune to give colleagues abroad a simple account of the historical facts concerning the straton model.

In the following are listed the names of the authors from various institutes and universities:

C.H. Chang, S. Chen , T.H. Ho , T.C. Hsien,
H.C. Huang, C.S. Ju , B.A. Li , T.N. Ruan ,
T.Z. Ruan , T.S. Tu , H.Y. Tzu , R. Wang ,
D.Z. Xu , X.C. Yang , H.Y. Yu , R.Y. Zhou ,

Institute of Atomic Energy, Academia Sinica

C.Y. Chao , K.T. Chao , J. Chen* , C.S. Gao,
N. Hu , C.S. Huang , L.S. Liu , C.T. Qian,
T.H. Qin , X.C. Song , K.C. Yang, ,
Peking University
I. An , T.S. Chang , W.Y. Chao , T.K. Chen,
C.Y. Chu , Y.B. Dai , B.Y. Hou , S.W. Hu ,
L.S. Zhou ,
Institute of Mathematics, Academia Sinica
B.H. Chao , Y.Y. Liu , B.R. Zhou
The Chinese University of Science and Technology

The following list contains articles in "Atomic Energy" published in March of 1966:

1. A relativistic model of the structure of particles with strong interaction.
2. The three body semi-leptonic decay of charged mesons.
3. The electro-magnetic vertices of mesons in the structure model.
4. The electro-magnetic decays of π^0 and η mesons in the relativistic structure model.
5. The electro-magnetic vertices and decay of $1^- \longrightarrow \gamma$, $1^- \longrightarrow 0^- + \gamma$ in the relativistic structure model.
6. The electro-magnetic decays of the vector mesons in

---

* On leave of absence from Jiling University

the relativistic structure model.

7. The electro-magnetic form factors of the nucleons in the relativistic structure model and its comparison with experiments.
8. The magnetic moments and the electro-magnetic form factors of the baryons.
9. The weak vertices of $\frac{1}{2}^+$ baryons.
10. The electro-magnetic and weak vertices involving $\frac{1}{2}^+$ – $\frac{3}{2}^+$ baryons in the relativistic structure model.
11. The electro-magnetic vertices of the baryon resonances in the relativistic structure model.
12. The SU(3) symmetric and SU(6) symmetric wave functions of the mesons and baryons.
13. The coupling constants, magnetic moments and the weak magnetic moments of the "quarks" and mesons.
14. The rotational model of the baryons.
15. The static electro-magnetic properties of the hadrons and the structure model.
16. A possible model of the elementary particles.

The following list contains the articles in "Acta Scientiarum Naturalium Universitatis Pekinensis" published in May of 1966.

1. The covariant field theoretical method for the structure model of hadrons.
2. The $\frac{1}{2}^+$ baryon currents in the structure model.

3. The electro-magnetic form factors of the $\frac{1}{2}^+$ baryons according to the structure model.
4. The weak interaction with leptons of $\frac{1}{2}^+$ baryons in the structure model.
5. The $\frac{3}{2}^+ - \frac{1}{2}^+$ baryon currents in the structure model.
6. The decay of the $N^{*+} \longrightarrow \gamma + p$ in the structure model.
7. The decay of $\Omega^- \longrightarrow \Xi^0 + \ell^- + \tilde{\nu}_\ell$ in the structure model.
8. On the connection of the inner movement and outer symmetry with the independent structure form factors of baryon currents in the structure model.
9. The meson currents in the structure model.
10. The electro-magnetic and weak decays of mesons in the structure model.
11. The charge of ur-hadrons and the electro-magnetic decays of vector mesons.
12. The space-time symmetry properties of the structure wave functions of the mesons.
13. The structure wave functions of mesons and the Bethe-Salpeter Equation.
14. The effective mass of the basic particles in bound states.
15. The rotational excited states of the elementary particles.
16. The individual-particle approximation of the higher

excited states.

The following list contains articles in "Atomic Energy" published in July 1966.

1. A preliminary investigation on the internal wave functions of the 35-plet mesons.
2. On the para-statistics and the structure model.
3. A discussion on the universal weak interaction in the structure model.
4. The problem of the Cabbibo Angle in the relativistic structure model.
5. On the sub-strong interaction.
6. A kind of wave functions of the excited states of the baryons and their application.
7. The wave functions of excited states of mesons and their electro-magnetic vertices.
8. The radiative corrections to $\frac{\pi \rightarrow e+\nu}{\pi \rightarrow \mu+\nu}$, $\frac{K \rightarrow e+\nu}{K \rightarrow \mu+\nu}$ in the structure model.
9. The electro-magnetic 3 body decay of the vector meson ω^0.
10. The Bethe-Salpeter wave functions in the structure model of hadrons and the super-strong interactions.

In as early as 1949, Fermi and Yang[1] proposed a model of the structure of the π mesons as the bound states of a nucleon and an anti-nucleon. Then Sakata proposed in 1955 the Sakata scheme of the structure of the hadrons, and called the three kinds of constituents of the hadrons the "Fundamental Particles in the True Sense".[2] In 1964, Gell-Mann[3] proposed a modification of the Sakata scheme, the quark scheme, and called the three kinds of the constituents of the hadrons the "quarks". In 1965, the number of the so-called "elementary particles" discovered was already comparable with the number of chemical elements on the periodic table, including baryon resonances with the spin as high as 11/2. The symmetries among the properties and the transmutations among the hadrons indicated that they might have an unified basis. The measurement of the electro-magnetic form factors of the nucleons not only proved that the nucleons are not geometrical points but have a definite size, it also showed that these form factors are of the dipole form. All these were indications that the hadrons have internal structure. The birth and the development of the theory of the hadron structure were to become a natural trend of the development of the theory of the particle physics.

Of course, there were difficulties in developing a theory of the hadron structure at that time. It was not sure whether there are new unknown laws of dynamics operating inside the hadron. The detailed form of the strong interaction was not known. No reliable mathematical method was available to handle the strong interaction. However, we don't know the exact form of the nuclear forces, neither do we know any reliable mathematical method to deal with the strong interaction even now. But these have not prevented the birth and the development of the theory of nuclear structure.

Judging from the main features of the experimental results of high energy physics, it seems safe to assume that concepts of quantum numbers and eigenvalues, the concepts of probability waves are still good concepts in the high energy physics. So that even inside the hadron, we might be justified to use the wave functions to represent the physical states and the operators to represent the dynamical variables. As a first step in developing a theory of the hadron structure, we tried to introduce the internal wave functions to describe the internal structure of the hadrons. The transition matrix elements for processes involving hadrons can then be expressed in terms of the values of these wave functions at the origin, the overlapping integral of these wave functions and the interactions involving stratons. These can be checked by the experiments, the results of which can in turn be used to improve the theory. The dynamical laws and the equations of motion which determine the internal wave functions of the hadrons were left for future investigation.

As we did not know, whether the quark scheme or other variety of modified Sakata scheme were going to describe the constituents of the hadrons correctly, we called these constituents "stratons" in general. The problems whether there are only three kinds of stratons, what are their properties, ... etc., are to be solved in the course of the development of the experiments and the theories. There were also scientific reasons for proposing such a name. The history of science reminded us that the notion of "fundamental constituents of matter " remains an illusion up to now. The search after an absolute rest system of reference represented by the ether was futile, as it simply does not exist. It might very well be, that even the constituents of the hadrons are not the fundamental constituents of matter either. They might be just one of the many strata of the structure of matter which remain to be discovered. The name "straton" was proposed to express such an idea.

At a meeting in early January of 1955, Chairman Mao Tse-tung asked Tsien San-tsiang: "Atomic nuclei are composed of proton and neutron. What are then the proton and the neutron composed of?" Tsien San-tsiang answered: "Judging from our present knowledge, the proton and the neutron are elementary particles." Smiling, Chairman Mao said: "It seems unlikely to me. The proton, the neutron and the electron are still divisible. Though there is no experimental proof now, conditions in experiments will continue to develop. Future experiments are going to prove that they are divisible. Do you believe it? I believe it anyway, even if you scientists don't believe it." According to our preliminary comprehension, the prediction was based upon the following viewpoint: the law of the unity of opposites is the fundamental law of the universe. All things are unities of opposites, contain inner contradictions. Elementary particles are no exception. Neither are the constituents inside the hadrons. The structure of the matter have infinite strata. After 25 years, experiments prove conclusively, that the proton and the neutron are not the fundamental constituents of matter , they do have internal structures. The correct scientific viewpoint and the correct philosophical viewpoint are both true reflections of the physical world. As a matter of course, they should agree with each other.

From the symmetries of the space-time, it follows that the wave function of the pseudo-scalar meson contains 4 Lorentz invariant functions, while that of the vector meson includes 8 Lorentz invariant functions. The wave functions of the $\frac{1}{2}^+$ and $\frac{3}{2}^+$ baryons are even more complicated. The problem then was, what kind of zero-th order approximation of the wave function should be used. A process, about which there was accurate experimental data and which was the simplest to treat theoretically, was $\pi^+ \longrightarrow \mu^+ + \nu_\mu$. Only one unknown parameter a appears in its transition amplitude

$$\langle 0 \left| J^W(0) \right| \pi \rangle = a \sqrt{\frac{m}{E}} \cdot P_\mu \qquad (1)$$

J^W_μ is here the charged weak current; m, E, P are the mass, energy and 4 - momentum of the π -meson respectively. It was convenient to start with this process in the discussion of the internal wave functions of the hadrons. In the structure model, this parameter turned out to be

$$|a|^2 = \frac{2}{m^2} \left| \psi(o) \right|^2 (1 - \bar{v}^2) \qquad (2)$$

It is proportional to the probability $|\psi(o)|^2$ for the straton and the anti-straton to meet each other. It also depends on the average velocity square $\bar{v}^2$, and thus depends in turn on the effective mass of the straton inside the meson. This latter feature is peculiar to the decay of 0^- meson through a V - A interaction. As a matter of fact,

$$\frac{\Gamma \; \pi^+ \longrightarrow e^+ + \nu_e}{\Gamma \; \pi^+ \longrightarrow \mu^+ + \nu_\mu} = \frac{m_e^2}{m_\mu^2} \left(\frac{m_\pi^2 - m_e^2}{m_\pi^2 - m_\mu^2} \right)^2 = 1.3 \times 10^{-4} \qquad (3)$$

The main factor $\frac{m_e^2}{m_\mu^2}$ is due to this origin.

The average momentum of the straton inside the meson can be estimated from the radius of the meson. The average velocity of the straton inside the meson can be then estimated, if an effective mass of the straton inside the meson is assumed. Since there exists approximate SU(6) symmetry for the hadrons, the contribution from the kinetic energy term, which breaks the SU(6) symmetry should not be very important. It might be justified to assume that $v^2 \ll c^2$, and can thus be neglected. It is then possible to derive the value of $|\psi(o)|^2$ from the decay rate of $\pi^+ \longrightarrow \mu^+ + \nu_\mu$. The radius of π - mesons can in turn be estimated. It turns out, that if the wave function is of the exponential type, the radius of the π-meson derived is 1.1 x 10^{-13} cm. If the wave function is of the gaussian type, the radius of the π-meson turns out to be 0.63 x 10^{-13} cm. It appears to be of the correct order of magnitude. Though there was no reliable experimental data for the radius of the π-meson, the radius of the nucleon had already been measured and turned out to be 0.8 x 10^{-13} cm.

We could make the other extreme assumption: The effective mass of the straton inside the meson is very small, its average velocity approaches the velocity of the light. The radius of the π-meson derived therefrom is then very small. For example, if the effective mass of the straton inside the meson is 10 MeV/c^2, the radius of the π-meson derived is then only of the order 10^{-15} cm, evidently too small. As the zero-th order approximation, we assumed, that $v^2 \ll c^2$ and can then be neglected. We assumed further, that in the rest system of the hadrons, the spin and the unitary spin wave functions have SU(6) symmetry and can then be written down directly. The wave function of a free-moving hadron can then be derived by a Lorentz transformation. The wave function of the 0^- meson is then of the form

$$\Phi^{Ps}_{P,\zeta}(x_1, x_2) = \frac{i}{2\sqrt{2}} \sqrt{\frac{M}{E}} \mathcal{A}(\zeta)(1 - \frac{i\hat{p}}{M})\gamma_5 \Psi^{Ps}_{\zeta}(x) e^{iPX} \tag{4}$$

The corresponding expression for the 1^- meson is:

$$\Phi^{V}_{P,\zeta,\lambda}(x_1, x_2) = \frac{1}{2\sqrt{2}} \cdot \sqrt{\frac{M}{E}} \mathcal{A}(\zeta)(1 - \frac{i\hat{p}}{M}) \hat{f}^{\lambda} \Psi^{V}_{\zeta}(x) e^{iPX} \tag{5}$$

The indices Ps and V specify the pseudoscalar and the vector mesons respectively, x_1 and x_2 denote the coordinates of the straton and anti-straton. X is the coordinate of the center of mass, while x is the relative coordinate between the straton and the anti-straton. P,M,E are respectively the 4-momentum, the mass and the energy of the meson. ζ denotes the unitary spin state, λ the state of polarization. $\mathcal{A}(\zeta)$ is the unitary spin wave function, while $\Psi(x)$ is the internal space-time wave function of the meson.

The wave function of $\frac{1}{2}^+$ baryon is

$$\Phi_{lm}(x_1, x_2, x_3, i, j, k, \alpha, \beta, \gamma) = \frac{1}{\sqrt{12}} \sqrt{\frac{M}{E}} \left\{ \epsilon_{ijl} \left[\left(1 - \frac{i\hat{p}}{M}\right) \gamma_5 C \right]_{\alpha\beta} \times \right.$$

$$\left. \times U_{\gamma}^{(\lambda)}(P)\, \delta_{mk} + \begin{pmatrix} i & j & k \\ \alpha & \beta & \gamma \end{pmatrix} \right\} \Psi_{\zeta}^{(\frac{1}{2})}(x, x') e^{iPX} \tag{6}$$

The corresponding expression for the $\frac{3}{2}^{+}$ baryon is

$$\Phi_{lmn}(x_1, x_2, x_3, i, j, k, \alpha, \beta, \gamma) = \sqrt{\frac{3}{4}} \sqrt{\frac{M}{E}}\, d_{lmn}^{ijk} \times$$

$$\times \left[\left(1 - \frac{i\hat{p}}{M}\right) \gamma_{\mu} C \right]_{\alpha\beta} \psi_{\gamma\mu}^{(\lambda)}(P)\, \Psi_{\zeta}^{(\frac{3}{2})}(x, x') e^{iPX} \tag{7}$$

x_1, x_2, x_3 are the coordinates of the three stratons, X the coordinate of the center of mass, x,x' the internal relative coordinates. i,j,k are unitary spin coordinates, α, β, γ the spin coordinates. l,m,n label the unitary spin state ζ , while λ represents the polarization state of the baryon. $\begin{pmatrix} i & j & k \\ \alpha & \beta & \gamma \end{pmatrix}$ represents terms cyclic in (i,α) (j,β), (k,γ). $d_{\ell mn}^{ijk}$ are a set of numbers completely symmetrical with respect to the upper and lower indices separately. C is the charge conjugation matrix. $U^{(\lambda)}$ (P) is the Dirac' wave function of a free, spin $\frac{1}{2}$ particle with the momentum P and the polarization λ . $\psi_{\mu}^{(\lambda)}(p)$ is the solution of the Rarita - Schwinger equation for a free spin $\frac{3}{2}$ particle with the momentum P and the polarization λ . Thus for each hadron there remains only one unknown internal space-time wave function $\psi(x,x')$

Wave functions (6) and (7) are completely symmetric . To ensure Fermi statistics for the stratons, the existence of an additional $SU^{(2)}(3)$ space has to be postulated, which is now called the colour space. Then an additional factor of $SU^{(2)}(3)$ singlet has to be attached to the wave functions (4), (5), (6) and (7), as was shown in the paper. No. 16 published in March of 1966. The only practical consequence for the following discussions is that the value of $|\psi(o)|^2$ derived is smaller by a factor 3.

These wave functions are the consequences of the following hypothesis: The external motion of the hadron might be highly relativistic, but its internal motion is of low velocities. Simultaneously with us, both Dalitz[4] and Morpurgo[5] put forward the quark model of the structure of the hadrons. The wave functions they used are non-relativistic, and are therefore different from ours. In certain processes, the velocities of the hadrons involved might be quite large. For example, during $\omega \rightarrow \pi^0 + \gamma$ the velocity of π^0 in the center of mass system of ω is 94% of the light velocity. Furthermore, some properties of the hadrons, such as the electro-magnetic multipole moments and the form factors, are intimately connected with relativistic effects.

If it is assumed further, that hadrons belonging to the same SU(6) multiplet all have the same internal space-time wave function, and are of the same mass, the results derived would then reduce to those of the theory of SU(6) symmetry. Of course, SU(6) is not a good symmetry. The mass of ϕ is 7.2 times that of π , while the mass of Ω^- is 1.8 times that of p. Under such an approximation, the results for processes involving mesons might differ from the actual values by one order of magnitude, while the results for those involving baryons might differ by a factor of 2. However, it might still be useful to start the investigation with such an approximation and introduce improvements after systematic comparisons between the theoretical and the experimental results.

The stratons involved in various processes are physical stratons, their interactions should include the effect of the renormalization. The coefficients of renormalization are parameters of the theory, the values of which are to be determined from the electro-magnetic and and weak properties of the nucleons. Once the forms of the wave functions and the interactions are available, we can proceed to calculate the various properties of the hadrons and the transition probabilities of various processes.

The transition rate for $K^+ \longrightarrow \mu^+ + \nu_\mu$ was calculated first. If the internal wave function of K^+ was taken to be same as that of the π^+, if $\sin\theta_c = 0.25$, then the result obtained is $1.3 \times 10^7 \sec^{-1}$, only $\frac{1}{4}$ of the experimental value. Recently, the measured value of the radius of K^+, r_k, is different from that of π^+, r_π . Thus their $|\psi(o)|^2$ should differ from each other. If we assume

$$\frac{r_k}{r_\pi} = 0.63 \tag{8}$$

the theoretical value obtained would agree with the experimental value of the decay rate. Experimentally, there are:

$$r_k = 0.51 \pm 0.07 \times 10^{-13} \text{ cm}$$

$$r_\pi = 0.711 \pm 0.018 \times 10^{-13} \text{ cm}$$

$$\frac{r_k}{r_\pi} = 0.72 \pm 0.10 \tag{9}$$

It is interesting to note, that the experimental ratio of the decay rates of $D \longrightarrow K^+ + K^-$ and $D \longrightarrow \pi^+ + \pi^-$ is[6]

$$\frac{D \longrightarrow K^+ + K^-}{D \longrightarrow \pi^+ + \pi^-} = 3.4 \begin{matrix} +2.8 \\ -1.2 \end{matrix} \tag{10}$$

If the strong interaction is neglected, these two processes can be represented by the following diagramms

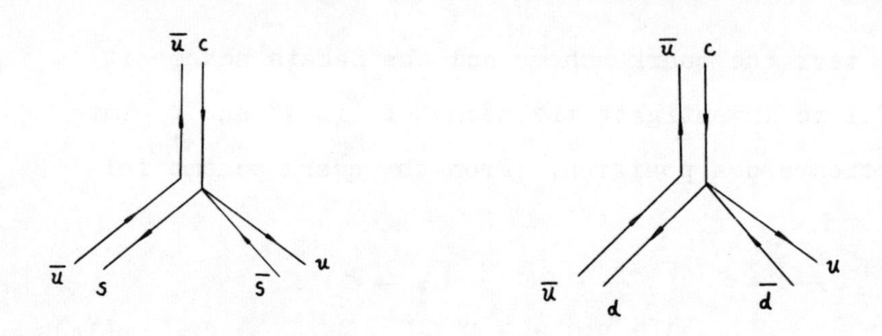

Evidently, the decay rate should include a factor $|\psi(o)|^2$. It should also include a factor representing the phase space. If (8) is assumed, which is derived from the decay rates of $K^+ \rightarrow \mu^+ + \nu_\mu$ and $\pi^+ \rightarrow \mu^+ + \nu_\mu$, there is then

$$\frac{|\psi_K(o)|^2}{|\psi_\pi(o)|^2} = 4 \qquad (11)$$

Further there is

$$\frac{p_K}{p_\pi} = 0.86 \qquad (12)$$

The contribution of these two effects to the decay rate ratio is

$$\frac{|\psi_K(o)|^2 \cdot p_K}{|\psi_\pi(o)|^2 \cdot p_\pi} = 3.43 \qquad (13)$$

Strong interaction effects make the theoretical treatment of these nonleptonic decay processes very difficult. However, it seems, that the above-mentioned effects should be included under any circumstances.

To test the quark scheme and the Sakata scheme it is useful to investigate the decay of ω, ρ^0 and ϕ into an electron and a positron. From the quark scheme follows:

$$\Gamma_{\omega \to e^+ + e^-} : \Gamma_{\rho^0 \to e^+ + e^-} : \Gamma_{\phi \to e^+ + e^-}$$

$$= 1 : 9.1 : 1.5 \qquad (14)$$

If the Sakata scheme is used and the $\omega - \phi$ mixing is taken into account, the corresponding ratios turn out to be

$$1 : 7.5 : 0.10 \qquad (15)$$

The experimental values are

$$1 : 8.7 \pm 2.2 : 1.6 \pm 0.2 \qquad (16)$$

which agree with the result of the quark scheme but disagree with the result of the Sakata scheme.

To test the quark scheme and the Sakata scheme further, it is useful to investigate the γ - decays of the hadrons turning into another hadron. To calculate the transition amplitude for the process involving one hadron in the initial state and one hadron in the final state, it is necessary to evaluate the overlapping integral between the wave functions of the hadrons in the initial and final state. If the internal wave functions of the two hadrons are different from each other, the overlapping integral is smaller than 1. Even if the internal wave functions of the two hadrons are identical in their rest system, their overlapping integral becomes smaller than 1, if there is relative motion between them, as the wave functions are subjected to the effect of the Lorentz contraction. With these effects taken into account, the following decay widths were obtained in the quark scheme. The corresponding experimental values are also given in the following table:

Decay process	Theoretical decay width in KeV.	Experimental decay width in KeV.
$\omega \longrightarrow \pi^0 + \gamma$	930	890 ± 50
$\omega \longrightarrow \eta + \gamma$	9	9 ± 40
$\rho^0 \longrightarrow \pi^0 + \gamma$	100	37 ± 11
$\rho^0 \longrightarrow \eta + \gamma$	110	
$\phi \longrightarrow \pi^0 + \gamma$	10	6 ± 2
$\phi \longrightarrow \eta + \gamma$	250	66 ± 8
$\phi \longrightarrow \eta' + \gamma$	1	
$K^{+*} \longrightarrow K^+ + \gamma$	80	74 ± 35
$K^{0*} \longrightarrow K^0 + \gamma$	330	
$\eta' \longrightarrow \rho + \gamma$	150	< 380
$\eta' \longrightarrow \omega + \gamma$	20	< 25
$\Delta^+ \longrightarrow p + \gamma$	470	600 ± 40
$\Sigma^0 \longrightarrow \Lambda + \gamma$	8	11 ± 3

There are serious contradictions between the results of the Sakata scheme and the experiments. For example, the theoretical value of $\Gamma_{\omega \longrightarrow \pi^0 + \gamma}$ in the Sakata scheme is as high as 2500 KeV. They are therefore not given in the table. With the preliminary and approximate nature of the theory kept in the mind, no serious contradictions exist between the quark scheme results and the present experimental results for 8 processes. However, there exist serious contradictions for 2 processes. But consistent and reliable experimental values exist only for $\Gamma_{\omega \longrightarrow \pi^0 + \gamma}$. Most other experimental values are preliminary. Some results are changing from one experiment to another. "Review of Particle Properties" of 1978 is of the opinion that to give an average experimental value to $\Gamma_{\Delta^+ \longrightarrow p + \gamma}$ is meaningless.

Similar methods can be used to calculate the electromagnetic properties of the hadrons, including their magnetic moments, electric quadrupole moments and magnetic octapole moments. Only the results from the quark scheme will be given. The theoretical and the experimental values of the magnetic moments of $\frac{1}{2}^+$ baryons are listed in the following table. The unit used is $\mu_o = \frac{e}{2m_p}$, where m_p is the mass of the proton.

	Theoretical value	Experimental value
p	2.793 (input)	2.793
n	- 1.86	- 1.91
Λ	- 0.88	- 0.606 ± 0.034
Σ^+	2.58	2.83 ± 0.25

Σ^0	0.86	
Σ^-	- 0.86	- 1.48 ± 0.37
Ξ^0	- 1.67	
Ξ^-	- 0.84	- 1.85 ± 0.75

There are serious disagreements between the theoretical and the experimental values for the magnetic moments of Σ^- and Ξ^-, which are disquieting. However, cases of new and old experimental data differing by more than two standard deviations are not extremely rare in experiments of high energy physics.

The magnetic moments of $\frac{3}{2}^+$ baryons and the 1^- mesons are given by one single formula:

$$\frac{Q}{2M} + \frac{\mu_p Q}{2m_p} \qquad (17)$$

Q and M represent here the charge and the mass of the hadron respectively. $\mu_p = 1.793$ is the coefficient for the anomalous magnetic moment of the proton. It is interesting to note that since the velocity of the internal motion of the stratons inside the hadrons has been neglected, the magnetic moments of the hadrons depend no more on the effective mass of the straton. The magnetic moments have contributions from two sources. The first part comes from relativistic effects, determined by the recoil of the hadron as a whole. Its value depends on the mass of the hadron. The second part comes from the anomalous magnetic moments of the stratons.

The electric quadrupole moments of the 1^- mesons are given by the formula:

$$- \frac{2Q}{3M^2} \left(1 - \frac{\pi^2}{15} \right) \tag{18}$$

while those of the $\frac{3}{2}^+$ baryons are given by the formula:

$$- \frac{Q}{M^2} \tag{19}$$

The magnetic octapole moments of the $\frac{3}{2}^+$ baryons are given by the formula:

$$- \frac{3\sqrt{3}}{20} \cdot \frac{Q}{M^3} \left(1 + \frac{\mu_p M}{m_p} \right) \tag{20}$$

To be precise, in a magnetic field of the form

$$\vec{H} = r^2 \vec{Y}_{J,L,M}(\theta, \varphi) A \tag{21}$$

$$J = 3, \quad L = 2, \quad M = 0.$$

the energy of the magnetic octapole is:

$$E = \sqrt{\frac{3}{\pi}} \cdot \frac{Q}{8M^3} \left(1 + \frac{\mu_p M}{m_p} \right) A \tag{22}$$

The decay width of $0^- \longrightarrow \gamma + \gamma$ has also been calculated. The values given by the quark scheme and the corresponding experimental values are given in the following table.

	Theoretical value	Experimental value
$\pi^0 \longrightarrow \gamma + \gamma$	6.6. e.v.	7.9 ± 0.5 e.v.
$\eta^0 \longrightarrow \gamma + \gamma$	400 e.v.(no η-η' mixing) 760 e.v.(mixing angle 10°)	323 ± 9 e.v.
$\eta' \longrightarrow \gamma + \gamma$	9 kev.(no η-η' mixing) 6 kev.(mixing angle 10°)	5.8 ± 1.1 kev.

It is to be noted that the experimental value of Γ_{tot} of η changed by a factor of 3 around 1976.

The above results were obtained by assuming that the velocity of the internal motion of the stratons inside the hadrons can be neglected. The actual situation is certainly not that simple. If the effects of the internal velocity are to be taken into account in the higher approximation, the theory would become very much more complicated. A preliminary investigation has been carried out in this direction. It was found that these effects lead to a non-zero electric form factor of the neutron and a non-zero E2 transition amplitude in $\Delta^+ \longrightarrow p + \gamma$.

The semi-leptonic weak decay of the hadron has also been calculated. The simplest process is then $\pi^+ \longrightarrow \pi^0 + e^+ + \nu_e$. The calculation gives a transition probability of 0.412 sec^{-1}, while the corresponding experimental value is 0.392 ± 0.027 sec^{-1}. The theoretical results about $K^+ \longrightarrow \pi^0 + e^+ + \nu_e$ and $K^+ \longrightarrow \pi^0 + \mu^+ + \nu_\mu$ are given together with the experimental results in the following table.

	Theoretical results	Experimental results
$\sin\theta_c$	0.25 (input)	
$\Gamma_{K\mu 3}/\Gamma_{Ke3}$	0.65	0.663 ± 0.018
ξ	-0.92	$-0.72 +\pm 0.21$ (1971)
		-1.0 ± 0.3 (1972)
		-0.36 ± 0.40 (1978)
		(From spectrum analysis)
		-0.20 ± 0.15 (1978)
		(From branching ratio)
λ_+	-0.074	-0.029 ± 0.004

The experimental value of ξ changed considerably during the recent years. But the experimental value of λ_+ remained fairly constant. The discrepancy between the theoretical and the experimental values of λ_+ seems to be real.

The theoretical and the experimental values of the branching ratios of the semi-leptonic weak decays of the baryons are listed in the following table.

	Theoretical value	Experimental value
$n \longrightarrow p + e^- + \tilde{\nu}_e$	(input)	
$\Lambda \longrightarrow p + e^- + \tilde{\nu}_e$	9.35×10^{-4}	$(8.07 \pm 0.28) \times 10^{-4}$
$\Sigma^- \longrightarrow n + e^- + \tilde{\nu}_e$	1.07×10^{-3}	$(1.08 \pm 0.04) \times 10^{-3}$
$\Sigma^0 \longrightarrow p + e^- + \tilde{\nu}_e$	3.02×10^{-8}	

Decay	Theory	Experiment
$\Sigma^+ \longrightarrow \Lambda + e^+ + \nu_e$	1.65×10^{-5}	$(2.02 \pm 0.47) \times 10^{-5}$
$\Xi^- \longrightarrow \Lambda + e^- + \tilde{\nu}_e$	0.62×10^{-3}	$(0.69 \pm 0.18) \times 10^{-3}$
$\Xi^- \longrightarrow \Sigma^0 + e^- + \tilde{\nu}_e$	9.2×10^{-5}	$< 5 \times 10^{-4}$
$\Xi^0 \longrightarrow \Sigma^+ + e^- + \tilde{\nu}_e$	2.9×10^{-4}	$< 1.1 \times 10^{-3}$
$\Sigma^- \longrightarrow \Lambda + e^- + \tilde{\nu}_e$	5.4×10^{-5}	$(6 \pm 0.6) \times 10^{-5}$
$\Sigma^0 \longrightarrow \Sigma^+ + e^- + \tilde{\nu}_e$	4.0×10^{-21}	
$\Sigma^- \longrightarrow \Sigma^0 + e^- + \tilde{\nu}_e$	1.5×10^{-10}	
$\Xi^- \longrightarrow \Xi^0 + e^- + \tilde{\nu}_e$	2.4×10^{-10}	
$\Omega^- \longrightarrow \Xi^0 + e^- + \tilde{\nu}_e$	7×10^{-3}	
$\Omega^- \longrightarrow \Xi^0 + \mu^- + \tilde{\nu}_\mu$	7×10^{-3}	

It is interesting to note that experimental data for the semi-leptonic decays of Ω^- might be available in the not too distant future. The theory shows that the main contribution to these processes comes from the pseudovector interaction.

The theory of symmetry gives relations among the masses of the mesons. It also gives relations among the masses of the baryons. The straton model gives besides a relation between the masses of the mesons and the baryons.

$$\frac{m_K^2 - m_\pi^2}{m_\rho^2 - m_\pi^2} = \frac{M_\Omega - M_\Delta}{4(M_\Delta - M_N)} \qquad (23)$$

The experimental value of the left-hand side is 0.39, while that of the right-hand side is 0.38.

Above is a brief account of the part of the results

obtained 14 years ago. Since then, the investigation on the structure of the hadrons has made very important progresses, which are opening the path leading to a fundamental theory of the strong interaction. Simultaneously, these will also lead to a fundamental theory of the structure of the hadrons. Of course, it would take a long time to reach these goals.

On the other hand, the number of stratons and anti-stratons discovered is already comparable to the number of hadrons discovered up to 1955, when Sakata proposed his composite model of the hadrons. Ten kinds of leptons and anti-leptons have already been discovered. There exists already indication of symmetry between the stratons and the leptons. The investigations on the mass spectra of the hadrons and the symmetries of their properties in the first half of the sixties paved the way for exploring and understanding the internal structure of the hadrons. The current investigations on the mass spectra of the stratons and the leptons and the symmetries of their properties might also pave the way for the exploration and understanding the internal structures of the stratons and leptons. It seems that the stratons might indeed be just one stratum of the many strata of the structure of matter. Looked at in this way, the progress of the high energy physics is very rapid indeed. If we regard

Sakata's proposal 25 years ago as the beginning of the exploration of the internal structure of the hadrons, we are rapidly approaching the threshold for the exploration of a still deeper stratum of the structure of matter. It seems that we are now facing the starting point of a new stage of the development of physics.

References

1. E. Fermi and C.N. Yang, Phys. Rev. 76, 1739 (1949).
2. S. Sakata, Progr. Theoret. Phys., 16, 686, (1956).
3. M. Gell-Mann, Phys. Letters, 8, 214 (1964)
   G. Zweig, CERN preprints TH.401 and TH.412 (1964).
4. R.H. Dalitz and D.G. Sutherland, Phys. Rev., 146, 1180 (1966).
5. C. Becci and G. Morpurgo, Phys. Rev. 140B, 687 (1965).
6. G.S. Abrams et al, Phys. Rev. Letters, 43, 477 (1979).

QCD AND THE BAG MODEL OF HADRONS

T. D. Lee

Columbia University, New York, N. Y. 10027

Abstract

By regarding the QCD vacuum as a perfect, or nearly perfect, dia-electric medium, we show how the quark confinement mechanism can be understood. The color dielectric constant κ characterizes the long-range order effect of QCD. The hadrons are viewed as soliton solutions with $\kappa = 1$ inside and $\kappa = 0$, or $\cong 0$, outside. The hadron masses and other dynamic properties can then be calculated in terms of the power series expansion of $\alpha = g^2/4\pi$ around these soliton solutions, where g is a coupling inside the hadron.

1. Vacuum As a Color Dia-electric

In this talk I would like to discuss some of my recent efforts to link QCD as a fundamental theory to the various bag models, such as the MIT bag[1] and the SLAC bag[2]. My approach will essentially be phenomenological.

To give QCD a well-defined meaning, we first contain the whole system within a volume of size L^3 to avoid infrared difficulties. In QCD there are quark fields ψ_q^a where q, the flavor index, $= u, d, s, \cdots$, and a, denoting the color, $= 1, 2, 3$. In addition, we have the gauge fields V^i where $i = 1, 2, \cdots, 8$ representing the color index of the gauge-field quanta. Let g_L be the renormalized coupling constant in the long wavelength limit, momentum $k \sim L^{-1}$. It is possible to prove under rather general assumptions and valid to all orders in $(\text{coupling})^2$

$$g_L > g_\ell \qquad \text{if} \quad L > \ell \quad . \tag{1.1}$$

This result is connected with the familiar "asymptotic freedom" property[3] of the theory: When $\ell \to 0$, g_ℓ decreases to 0. As there is no scale in the theory (assuming that all quarks are of zero mass), this implies that when ℓ increases, g_ℓ must also increase, which leads to (1.1). The difficulty lies in the infrared limit. When $L \to \infty$, it seems likely that g_L may $\to \infty$, or at least $\gg 1$. Since the true physical system is one with $L = \infty$, we may always be in the ultra-strong coupling limit. As we shall see, this difficulty can be resolved by regarding the vacuum as a color dia-electric medium.

Let us introduce κ_L which is called the color dielectric constant of the vacuum in a volume L^3. As a convention, we shall adopt a standard renormalized coupling constant g, defined by

$$g = g_\ell \quad \text{when } \ell = \text{some arbitrarily chosen length, say the proton radius.} \tag{1.2}$$

The color dielectric constant κ_L will then be defined as

$$g_L^2 = \frac{g^2}{\kappa_L} . \tag{1.3}$$

Consequently, in accordance with (1.2)

$$\kappa_\ell = 1 \quad \text{when } \ell = \text{proton radius}. \tag{1.4}$$

Equation (1.1) now implies

$$\kappa_L < \kappa_\ell \quad \text{if } L > \ell . \tag{1.5}$$

To the second order[3] we have,

$$\frac{\kappa_L}{\kappa_\ell} = \frac{1}{1 + \frac{1}{2\pi}\frac{g^2}{4\pi}(11 - \frac{2}{3} n_f) \ln \frac{L}{\ell} + O(g^4)} \tag{1.6}$$

where n_f is the number of quark flavor varieties (assumed to be < 17). With the convention (1.4) we find that the color dielectric constant of the vacuum in an infinite volume $L = \infty$ is

$$\kappa_{\infty} < 1 \tag{1.7}$$

which is valid to all orders in g^2. The ultra-strong coupling difficulty mentioned before corresponds to

$$\kappa_{\infty} = 0 , \tag{1.8}$$

or

$$\kappa_{\infty} \ll 1 . \tag{1.9}$$

In the former we call the vacuum in QCD a perfect color dia-electric medium, in the latter a nearly perfect color dia-electric medium.

2. A Hypothetical Problem

To see how this will lead to quark confinement and bag models, let us consider a hypothetical problem in classical electromagnetism. In quantum electrodynamics, our usual convention is to set the dielectric constant of the vacuum state $\kappa_{vac} = 1$. It is then possible to prove that all physical media have their dielectric constants $\kappa > 1$. This can be seen most easily by using the familiar formula

$$\vec{D} = \vec{E} + 4\pi\vec{P} \tag{2.1}$$

where $\vec{D}$ is the displacement vector, $\vec{E}$ the electric field, and $\vec{P}$ the polarization vector. Since under $\vec{E}$, all atoms have their polarization $\vec{P}$ in the same direction as $\vec{E}$, so as to produce a screening effect, we have $\kappa > 1$.

In this section, we shall consider a hypothetical problem. Let us imagine that in classical electromagnetism, but without the quantum theory of atoms, there could be a medium with its dielectric constant

$$\kappa \equiv \kappa_{med} \ll 1 , \quad \text{or} \quad \cong 0 , \tag{2.2}$$

i.e., this hypothetical medium is anti-screening. Now, suppose we place a small charge distribution ϵ in the medium. As we shall show, no matter how small ϵ is, the medium will develop a hole surrounding the charge. Inside the hole, we have $\kappa = 1$, but outside $\kappa = \kappa_{med}$, as shown in Fig. 1. To see this, let us assume that such a hole is formed. Because of the anti-screening nature of the medium, the induced charge on the inner surface of the hole is of the same sign as ϵ. Consequently, if we want to reduce the size of the hole, we must do work to overcome the repulsion between ϵ and the induced charge. That work is infinite if the hole is to be eliminated in toto. Hence the hole will not disappear. [This

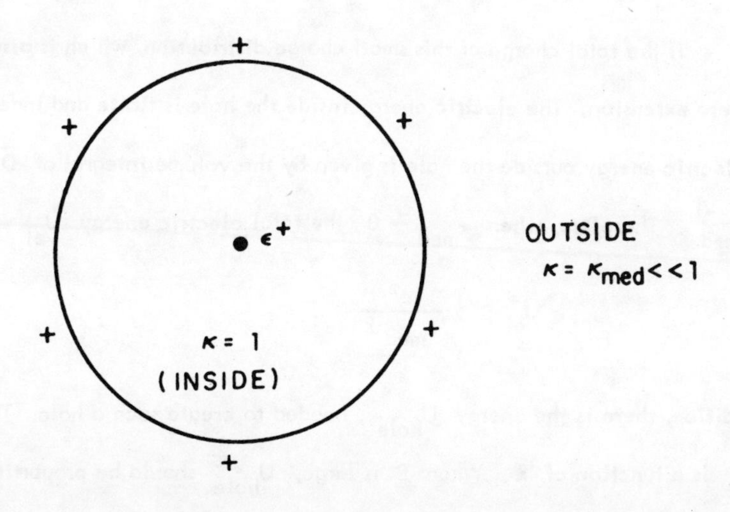

Fig. 1

situation is completely different for a normal medium whose dielectric constant is > 1. Imagine that a similar hole were created. The corresponding induced charge would be of the opposite sign to ϵ. The hole would automatically shrink to 0, resulting in the usual Coulomb field distribution around the charge. The medium will remain homogeneous if ϵ is sufficiently small.] Here, no matter how small ϵ is, when $\kappa_{med} \to 0$ inhomogeneity in the medium in the form of holes must occur.

We may estimate the radius R of such a hole. Let D_{out} and E_{out} be, respectively, the normal components of $\vec{D}$ and $\vec{E}$ outside the hole when $r = R$. Similarly, let D_{in} and E_{in} be the corresponding components inside the hole when $r = R$. For a spherical hole we have

$$D_{in} = E_{in} = D_{out} = \frac{\epsilon}{R^2}$$

and

$$E_{out} = \frac{\epsilon}{\kappa_{med} R^2} ,$$

where ϵ is the total charge of this small charge distribution, which is assumed to have a non-zero extension. The electric energy inside the hole is finite and independent of κ_{med}. The electric energy outside the hole is given by the volume integral of $\vec{D} \cdot \vec{E}$, which is $\propto \kappa_{med}^{-1}$. Therefore, when $\kappa_{med} \to 0$, the total electric energy U_{el} is

$$U_{el} \sim \frac{1}{2} \frac{\epsilon^2}{\kappa_{med} R} . \tag{2.3}$$

In addition, there is the energy U_{hole}, needed to create such a hole. The amount U_{hole} is a function of R. When R is large, U_{hole} should be proportional to the volume plus a term proportional to the surface, etc. We may write

$$U_{hole} = \frac{4\pi}{3} R^3 p + 4\pi R^2 s + \cdots \tag{2.4}$$

where $p, s, \cdots$ are positive constants. The total energy of the system is given by

$$M = U_{el} + U_{hole} , \tag{2.5}$$

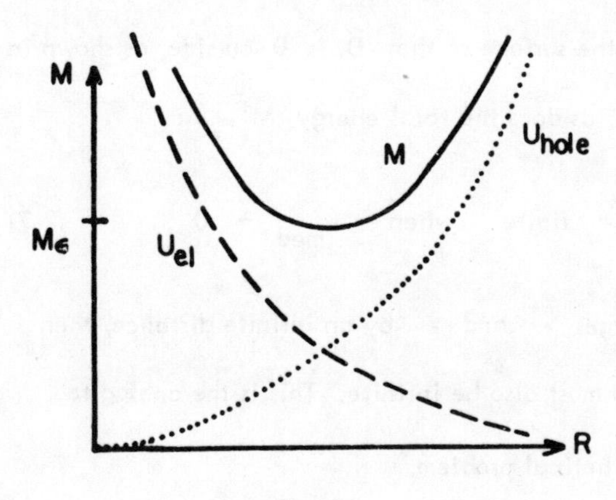

Fig. 2

which is drawn schematically in Fig. 2. Its minimum, as determined by $dM/dR = 0$, gives (in the approximation $U_{hole} \cong \frac{4\pi}{3} R^3 p$)

$$M \equiv M_\epsilon \sim \frac{4}{3} \left(\frac{\epsilon^2}{2\kappa_{med}} \right)^{3/4} (4\pi p)^{1/4}$$

from which we can draw the conclusion that, if the total charge $\epsilon \neq 0$, then $R \neq 0$. Furthermore, the total energy

$$M_\epsilon \to \infty \quad \text{when} \quad \kappa_{med} \to 0 \,. \tag{2.6}$$

It is not difficult to show that the same conclusion can also be reached without the approximation $U_{hole} \cong \frac{4\pi}{3} R^3 p$.

Next, we replace the single charge distribution by a dipole distribution; i.e. two small charge distributions of total charge ϵ^+ and ϵ^-. It is easy to see that when κ_{med} is sufficiently small, the minimal energy state again requires the formation of a hole around these two charges. As before, inside the hole $\kappa = 1$, and outside $\kappa = \kappa_{med}$. When $\kappa_{med} \to 0$, at the surface of the hole the electric field inside should be parallel to the surface so that $\vec{D}$ is 0 outside, as shown in Fig. 3. Thus, U_{el} remains finite, as does the total energy $M_{\epsilon^+\epsilon^-}$:

$$M_{\epsilon^+\epsilon^-} = \text{finite} \quad \text{when} \quad \kappa_{med} \to 0 \ . \tag{2.7}$$

If we try to separate the two charges ϵ^+ and ϵ^- by an infinite distance, then because of (2.6) the work required must also be infinite. This is the analog to "quark confinement" in our hypothetical problem.

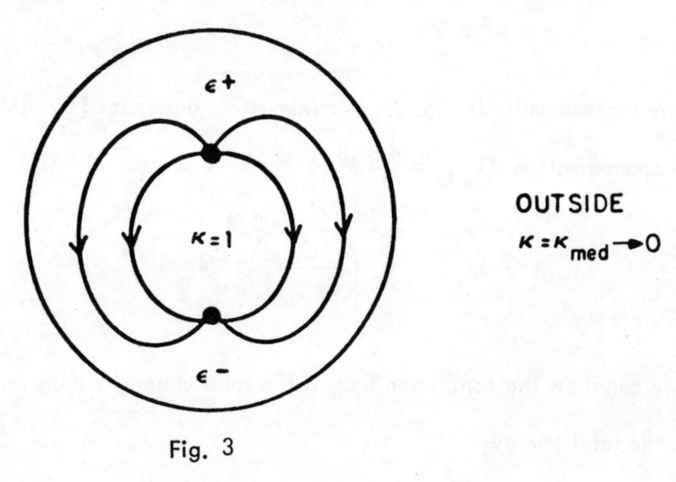

Fig. 3

3. Quark Confinement

We now return to the corresponding problem in QCD, discussed in Section 1. By following the same argument given in the above section, we see that whenever there are quarks or antiquarks there must be inhomogeneity in the space surrounding these particles. We may call these either bags[1,2], or domain structures, or nontopological soliton solutions[4,5], or (if the total color is zero) simply "hadrons". The vacuum is assumed to be a perfect (or nearly perfect) color dia-electric medium. κ_∞ is 0, or $\ll 1$. If the total "color" is non-zero, the mass of such a bag would be infinite when $\kappa_\infty \to 0$, in analogy with (2.6). However, just as in (2.7), the bag mass remains finite when $\kappa_\infty \to 0$, if inside the bag one has a color singlet such as

$$\text{meson: } \bar{q}^a q^a \text{ , or baryon: } \epsilon_{abc} q^a q^b q^c$$

where a, b, c are color indices which can vary from 1 to 3. (See Fig. 4.)

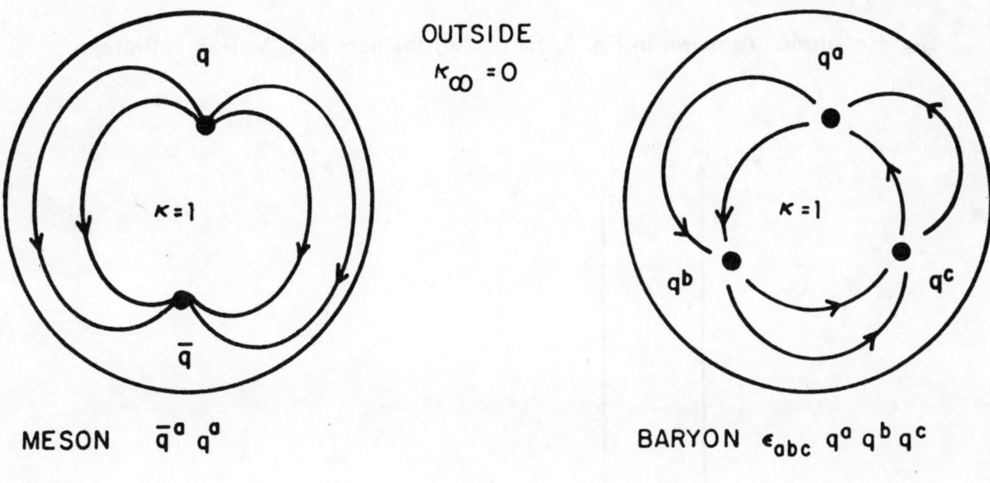

Fig. 4

From this one can derive that the work required to separate the quarks to a large distance r is approximately proportional to r. The quark confinement is then "explained" by the assumption that the vacuum in QCD is a perfect (or nearly perfect) color dia-electric.

Of course, at present it is not known that the mass M_q of a truly free quark is indeed infinite. However, the fact that no free quark has been observed so far in the final state of any high energy collision sets a lower bound $M_q > 5$ GeV for $q = u, d$ and s. It can be shown that this lower bound implies[6]

$$\kappa_\infty < .013 \left(\frac{g^2}{4\pi}\right) \tag{3.1}$$

where g is defined by (1.2).

4. A New Power Series Expansion

One may wonder in what sense we have by-passed the difficulty of ultra-strong coupling, mentioned previously. To understand this, let us consider other problems in physics in which interactions are also ultra-strong and yet there are no mathematical difficulties. A good example is the interaction potential $V(r)$ between two He atoms. As shown in Fig. 5, for $r < a$, the potential V is $\cong$ infinite;

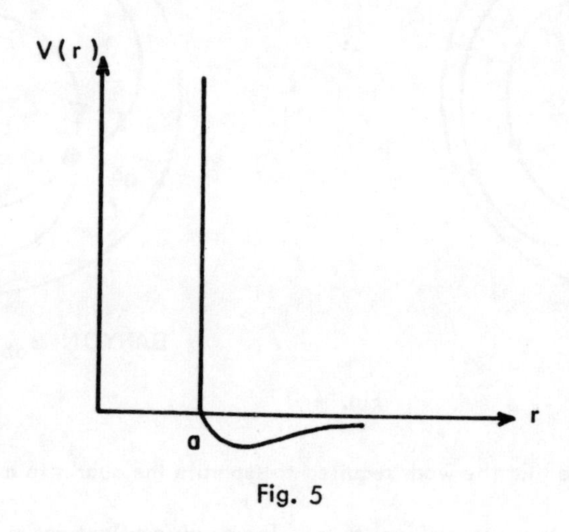

Fig. 5

for $r > a$, V is small. Because the strong potential is of a repulsive nature, the two He atoms will avoid the strong interaction region, and therefore this strong interaction does not present any difficulty. The usual method is to replace the strong repulsive part of V by a hard-sphere potential which leads to a new boundary value problem: The distance r between the two He atoms is restricted to a region $r \geqslant a$ in which the potential $V(r)$ is relatively weak, and therefore may be regarded as a perturbation.

From Fig. 4, we see that when $\kappa_\infty \to 0$, i.e. $g_\infty^2 = g^2/\kappa_\infty \to \infty$, the quarks are confined to a domain in which $\kappa = 1$. Thus the ultra-strong region, $g_\infty \to \infty$, exerts a repulsion against the quarks and antiquarks. Just as in the above problem of He-He interaction, the particles automatically stay away from the ultra-strong interaction region. Because of Lorentz invariance, the corresponding boundary value problem in the present case requires the relativistic soliton solution, as has been given in the literature.[5,6] When one expands around these soliton solutions, only the coupling g inside the soliton is the relevant parameter, and that resolves the difficulty. [It is only when one attempts to make a simple-minded plane wave expansion around a homogeneous background, that one has to use g_∞ as the expansion parameter. Exactly the same kind of difficulty would appear if one were to force a similar plane-wave expansion in the aforementioned He - He problem.] Because QCD is asymptotically free, the quarks inside the hadron behave approximately like free particles; furthermore, their effective masses can be relatively small.

In the following we shall discuss how to obtain such a power series expansion in terms of the g^2 inside the hadron (not the g_∞^2 outside). For example, the hadron mass M can be expanded

$$M = M_0 + \alpha M_1 + \alpha^2 M_2 + \cdots \tag{4.1}$$

where $\alpha = g^2/4\pi$ and g^2 is defined by (1.2).

Since our theory is relativistic invariant, the velocity of light $c/(\kappa\mu)^{\frac{1}{2}}$ in the vacuum must remain c. Hence

$$\kappa\mu = 1 \tag{4.2}$$

where κ is the color dielectric constant and μ the corresponding "magnetic" susceptibility. When $\kappa = \kappa_\infty \to 0$ one must have $\mu = \mu_\infty \to \infty$.

With (4.2), we can consider κ to be a Lorentz-invariant quantity. It may be useful to introduce a phenomenological spin-0 field σ which depends linearly on κ:

$$\sigma \propto 1 - \kappa \ . \tag{4.3}$$

Inside the hadron we have $\kappa = 1$ and, therefore, $\sigma = 0$. Outside the hadron, κ is κ_∞ and σ is σ_{vac}. In the case that $\kappa_\infty = 0$, (4.3) becomes simply

$$\sigma = \sigma_{vac}(1 - \kappa) \ .$$

The phenomenological Lagrangian density is assumed to be

$$\mathcal{L} = -\frac{1}{4}\kappa V^a_{\mu\nu} V^a_{\mu\nu} - \psi^\dagger \gamma_4(\gamma_\mu D_\mu + f\sigma + m)\psi - \frac{1}{2}\left(\frac{\partial\sigma}{\partial x_\mu}\right)^2 - U(\sigma) + \text{counter terms} \tag{4.4}$$

where $\dagger$ denotes the Hermitian conjugation, $x_\mu = (\vec{r}, it)$,

$$V^a_{\mu\nu} = \frac{\partial}{\partial x_\mu} V^a_\nu + \frac{\partial}{\partial x_\nu} V^a_\mu + g\, C^{abc} V^b_\mu V^c_\nu \ ,$$

$$D_\mu = \frac{\partial}{\partial x_\mu} - \frac{1}{2} i g \lambda^a V^a_\mu \ , \tag{4.5}$$

the λ^a's are the standard Gell-Mann matrices related to C^{abc} by the commutation relation

$$[\lambda^a, \lambda^b] = 2i\, C^{abc} \lambda_c \ ,$$

the γ_μ's are the usual Hermitian Dirac matrices and the function $U(\sigma)$ has an absolute minimum at $\sigma = \sigma_{vac}$ and a local minimum at $\sigma = 0$ with

$$\text{and} \qquad \begin{aligned} U(\sigma_{vac}) &= 0 \\ U(0) &\equiv p > 0 \ , \end{aligned} \tag{4.6}$$

as shown in Fig. 6.

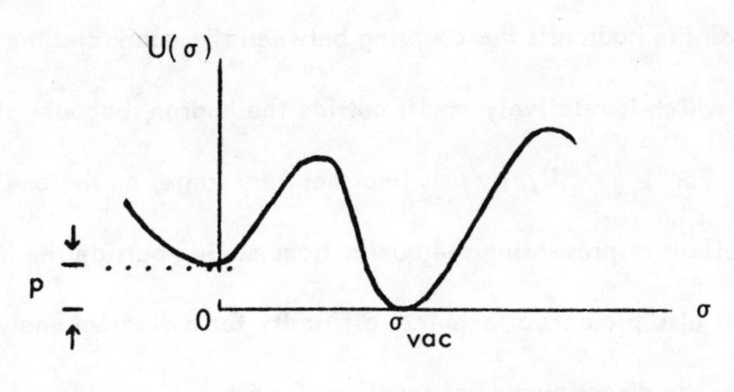

Fig. 6

As we shall see, the detailed form of $U(\sigma)$ is not important. If one wishes, one may assume $U(\sigma)$ to be simply a quartic function of σ. In (4.4) ψ denotes the quark field which besides being a color triplet also has F flavors, m is the mass matrix for quarks inside the hadron, and g and f are both renormalizable coupling constants. Since σ is only a phenomenological field, describing the long-range collective effects of QCD, its short wavelength components do not exist in reality. The counter terms in $\mathcal{L}$ are for renormalization; they consist <u>only</u> of those due to loop diagrams of the vector gauge field V_μ^a and the quark field ψ. In the following, we shall ignore all σ-loops; i.e., σ will be approximated by a classical field.

In our phenomenological Lagrangian density, the σ-field is coupled to the quarks in two ways: one is through the vector field via the quark - V_μ^4 coupling g and $-\frac{1}{4}\kappa V_{\mu\nu}^a V_{\mu\nu}^a$ which, because of (4.3), couples V_μ^a with σ, the other is a direct quark - σ coupling f. So far as the quark confinement problem is concerned, there is no need to have the direct quark - σ coupling; it suffices to have the vacuum be a perfect dia-electric $(\kappa_\infty = 0)$. The origin of the f-coupling lies in the dichotomy that only inside the hadron is the coupling between the quark and the vector gauge field really g, which is relatively small; outside the hadron, because of (1.3) it is actually $g/\kappa_\infty^{\frac{1}{2}}$. For $\kappa_\infty \ll 1$, $g/\kappa_\infty^{\frac{1}{2}}$ becomes very large; on the one hand this has the desirable effect of preventing the quarks from moving outside the hadron, but on the other hand it also presents a technical difficulty for a diagram analysis in powers of g^2. In (4.4) the direct quark - σ coupling f with

$$f\,\sigma_{vac} \gg \text{hadron mass} \tag{4.7}$$

is introduced phenomenologically to give a convenient alternative formulation of the same physical effect, but bypassing the above difficulty. As we shall see, because of (4.7) the f-coupling restricts the quarks to staying always inside the hadron, and that enables us to take full advantage of the relative smallness of the quark - V_μ^a coupling g inside the hadron. With the f - coupling we may now expand any physical observable, say the hadron mass M, in a power series of the form (4.1). Phenomenologically, if f were zero, it would be difficult to derive the zeroth-order term M_0, since when $a = 0$ both g and $g/\kappa_\infty^{\frac{1}{2}}$ should be zero; hence, without the f-coupling quarks would be unconfined.

5. Zeroth Order

With the direct quark - σ coupling f , we can now in the zeroth-order calculation neglect the exchange of V_μ^a . The description of the hadron reduces to that of a simple soliton model, consisting of the scalar σ field and the quark ψ field. The relevant part of the Lagrangian density (4.4) now consists simply of

$$-\psi^\dagger \gamma_4 \left(\gamma_\mu \frac{\partial}{\partial x_\mu} + f\sigma + m \right) \psi - \tfrac{1}{2}\left(\frac{\partial \sigma}{\partial x_\mu} \right)^2 - U(\sigma) \quad . \tag{5.1}$$

As mentioned before, since σ is only a phenomenological field that has no short wavelength components, we shall neglect all σ - loop diagrams. The remaining σ - diagrams are all tree diagrams, which correspond to the quasi-classical approximation that has been extensively studied in the literature[5]. Here we give only a summary of the results for light quark hadrons, with $m = 0$. In this case, ψ and σ can be reduced to c. no. functions which satisfy

$$(-i\vec{\alpha}\cdot\vec{\nabla} + f\beta\sigma)\psi = \epsilon\psi$$

and

$$-\nabla^2\sigma + U'(\sigma) = -fN\psi^\dagger\beta\psi \tag{5.2}$$

where $U'(\sigma) = dU/d\sigma$, $\vec{\alpha}$ and β are the usual Dirac matrices, $\epsilon > 0$, ψ satisfies $\int \psi^\dagger \psi \, d^3r = 1$ and N is the total number of quarks and antiquarks,

$N = 2$ for mesons

and

$N = 3$ for baryons.

As shown in Ref. 5 for $f\sigma_{vac}$ >> hadron mass and under very general assumptions, the hadron acquires a well-defined surface $\mathcal{S}$. One has

$$\psi = 0$$

and

$$\sigma = \sigma_{vac} \qquad \text{outside} \qquad ; \tag{5.3}$$

on the inner side of the surface $\mathcal{S}$,

$$-i\,\beta\,\vec{\alpha}\cdot\hat{n}\,\psi = \psi \tag{5.4}$$

where $\hat{n}$ is the unit vector normal to $\mathcal{S}$, and inside $\mathcal{S}$ the field σ is $\cong 0$, with its deviation from zero proportional to $\psi^\dagger \beta\, \psi$. For the $s_{\frac{1}{2}}$ orbit, $\mathcal{S}$ is a spherical surface of radius R. Inside $\mathcal{S}$, (5.2) can be further reduced, through scaling, to a simple system of two coupled first-order differential equations:

$$\frac{du}{d\rho} = (-1 + u^2 - v^2)\,v$$

and

$$\frac{dv}{d\rho} + \frac{2v}{\rho} = (1 + u^2 - v^2)\,u \tag{5.5}$$

where ρ is related to the radius r by $\rho = \epsilon r$,

$$\psi \propto \begin{pmatrix} u \\ i(\vec{\sigma}\cdot\vec{r}/r)\,v \end{pmatrix} \zeta \ ,$$

$$\zeta = \begin{pmatrix} 0 \\ 1 \end{pmatrix} \quad \text{or} \quad \begin{pmatrix} 1 \\ 0 \end{pmatrix} \ ,$$

$\vec{\sigma}$ is the usual Pauli spin matrix, and the variables ρ, u and v are all dimensionless. Although (5.5) does not contain any explicit parameters, its solutions form a one-parameter family. As $\rho \to 0$, one has $v \to 0$ and $u \to u(0)$. For every $u(0)$ between 0 and a critical value $u_c = 1.7419$, there is a solution of (5.5). The solution can be obtained by direct integration from $\rho = 0$ to $\rho = \rho_0$. At $\rho = \rho_0$, one has $u(\rho_0) = v(\rho_0)$ and therefore the boundary condition (5.4) is satisfied. The radius of the hadron is given by

$$R = \rho_0/\epsilon \ .$$

A convenient parameter to label these solutions can be either $u(0)$ or the integral

$$n \equiv \int (u^2 + v^2)\, d^3\rho \tag{5.6}$$

over the region $\rho \leqq \rho_0$. As the initial value $u(0) \to 0$, one has $n \to 0$, but as $u(0) \to u_c = 1.7419$, $n \to \infty$. Examples of these two limiting solutions are given in Fig. 7.

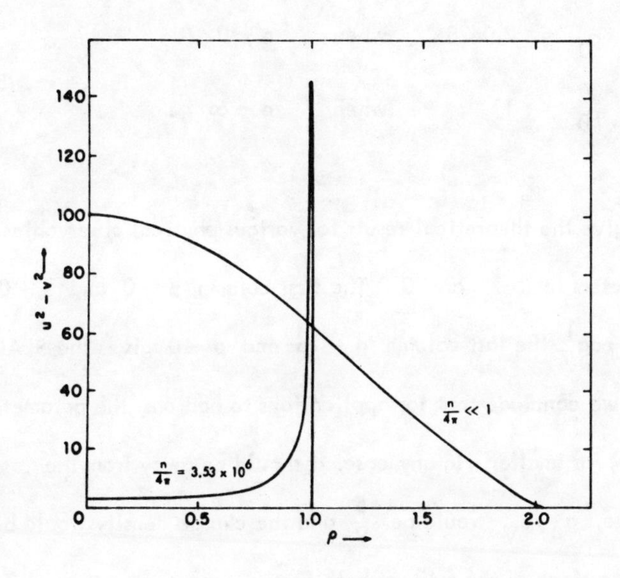

Fig. 7

The physical description of the soliton solution then resembles that of a gas bubble (i.e., the hadron) inside a medium (i.e., the vacuum). The hadron mass is determined by three parameters:

$$p\,, \quad s \quad \text{and} \quad n \tag{5.7}$$

where p is defined by (4.6) which represents the pressure of the medium on the bubble,

s is the surface tension which arises because σ changes from 0 to σ_{vac} across the soliton surface, and n determines the gas pressure inside the bubble which is due to both the kinetic energy of quarks and the excitation energy of σ. In either of the limits $n \to 0$ or $n \to \infty$ the hadron mass has the form, in the notation of (4.1)

$$M_0 = \frac{N \rho_0}{R} + \frac{1}{3} 4\pi R^3 p + 4\pi R^2 s \qquad (5.8)$$

where

$$\rho_0 = 2.0428 \quad \text{when} \quad n \to 0$$

and

$$\rho_0 = 1 \quad \text{when} \quad n \to \infty \quad . \qquad (5.9)$$

In Table 1 we give the theoretical result for various physical observables when two of the parameters in (5.7) are 0. The first column $n \to 0$ and $s \to 0$ corresponds to the MIT bag[1], the last column $n \to \infty$ and $p \to 0$ gives the SLAC bag[2]. From Table 1 we conclude that for applications to hadrons, the parameter n could be either O(1) or smaller. In any case, it should be away from the $n \to \infty$ limit. Otherwise, g_A/g_V would be $\frac{5}{9}$ and the charge density would be distributed only on the surface of the soliton; both features seem to be quite different from those of the physical nucleon.

The low-lying hadron states are made of light quarks, u, d and s, and their antiparticles. All hadrons are color singlets. In the zeroth approximation there are no spin-dependent interactions, and for convenience we may set the masses of these three light quarks to be 0 inside the hadron. The mesons are $q\bar{q}$ compounds; their lowest energy states have a 36-fold degeneracy consisting of the pseudoscalar nonet, π, K, η and η', and the vector nonet, ρ, K_V, ω, ϕ (each having a 3-fold

Physical observable	Experimental value	Theoretical value in some limiting cases			
		$n \to 0$		$n \to \infty$	
		$s \to 0$	$p \to 0$	$s \to 0$	$p \to 0$
r_p	$3.86/m_N$	$4.25/m_N$	$4.78/m_N$	$2.86/m_N$	$3.21/m_N$
μ_p	$2.79/(2m_N)$	$2.36/(2m_N)$	$2.66/(2m_N)$	$1.90/(2m_N)$	$2.14/(2m_N)$
μ_n	$-0.685\,\mu_p$	$-\frac{2}{3}\mu_p$	$-\frac{2}{3}\mu_p$	$-\frac{2}{3}\mu_p$	$-\frac{2}{3}\mu_p$
g_A/g_V	1.25	1.09	1.09	5/9	5/9

Table 1. Root mean-squared charge radius r_N, magnetic moment μ_N and g_A/g_V of the nucleon N. The parameter n is defined by (5.6). The first column $n \to 0$, $s \to 0$ corresponds to the MIT bag, the last column $n \to \infty$, $p \to 0$ to the SLAC bag.

degeneracy due to spin). The baryons are qqq compounds; their lowest energy states have a 56-fold degeneracy consisting of the spin-$\frac{1}{2}$ octet N, Λ, Σ and Ξ, and the spin-$\frac{3}{2}$ decuplet Δ, $\Sigma^*(1385)$, $\Xi^*(1530)$ and Ω. These states exhibit SU(6) symmetry. The 36-meson states consist of two SU(6) representations, $\underline{35}$ and $\underline{1}$; the 56-baryon states form a single SU(6) representation. In order to split the SU(6) degeneracy, we must include the spin-dependent interaction due to the exchanges of gauge quanta, which are the QCD corrections.

6. Gauge Field Propagator

To carry out the QCD corrections we need the propagator $D_{\mu\nu}(x, x')$ of the gauge field inside a hadron. Let $\mathcal{S}$ be the surface of the hadron; inside $\mathcal{S}$ the color dielectric constant $\kappa = 1$ and outside $\kappa = \kappa_\infty \to 0$. In the Coulomb gauge we can decompose $D_{\mu\nu}$ into transverse and longitudinal components:

$$D_{\mu\nu} = \begin{cases} (D_{ij})_{tr} & \begin{matrix}\mu = i \neq 4 \\ \nu = j \neq 4\end{matrix} \\ (D_{44})_{long} & \mu = \nu = 4 \\ 0 & \text{otherwise} \end{cases}$$

where, when expressed in powers of κ_∞

$$(D_{44})_{long} = O(\frac{1}{\kappa_\infty}) + O(1) + O(\kappa_\infty) + \cdots ,$$

$$(D_{ij})_{tr} = O(1) + O(\kappa_\infty) + \cdots .$$

For example, if $\mathcal{S}$ is a sphere of radius R then

$$(D_{44}(x, x'))_{long} = \delta(t-t') \frac{1}{4\pi} \Big[(\frac{1}{\kappa_\infty} - 1) \frac{1}{R} + \frac{1}{|\vec{r} - \vec{r}\,'|} + \sum_{\ell=1}^{\infty} \frac{\ell+1}{\ell} (\frac{rr'}{R^2})^{\ell} \frac{1}{R} P_\ell(\cos\theta) \Big] + O(\kappa_\infty) \qquad (6.1)$$

where $x_\mu = (\vec{r}, it)$, $x'_\mu = (\vec{r}\,', it')$ and $\theta = \angle(\vec{r}, \vec{r}\,')$. From (6.1) one sees that as $\kappa_\infty \to 0$, $D_{44} \to \infty$. Because of this, the matrix element of the S-matrix between any two non-color-singlet states $|a\rangle$ and $|b\rangle$ diverges when $\kappa_\infty \to 0$. Nevertheless, it can be shown that the matrix element between any two color singlets satisfies

$$\langle b | S | a \rangle = O(1)$$

when $\kappa_\infty \to 0$. Furthermore, between color singlets (6.1) can be replaced by

$\delta(t-t')\, G(\vec{r}, \vec{r}\,')$ where

$$G(\vec{r}, \vec{r}\,') = \frac{1}{4\pi}\left[\frac{1}{|\vec{r}-\vec{r}\,'|} + \sum_{\ell=1}^{\infty} \frac{\ell+1}{\ell}\left(\frac{rr'}{R^2}\right)^{\ell} \frac{P_\ell(\cos\theta)}{R}\right] + O(\kappa_\infty) \ . \tag{6.2}$$

Likewise $(D_{ij})_{tr}$ can be worked out explicitly.

Out gauge field propagator $D_{\mu\nu}$ depends on κ_∞ and the shape of the surface $\mathcal{S}$. Only for color singlet states and in the limits $\kappa_\infty \to 0$ and 4-momentum transfer $\to \infty$, does the propagator $D_{\mu\nu}$ reduce to the usual one used in the literature.

7. First Order

We now turn to the spectroscopy of hadron levels, such as those listed in Table 2.

π (140)	η (549)	η' (958)
ρ (770)	ω (783)	ϕ (1020)
N (940)	Δ (1232)	etc.

Table 2.

As mentioned in Section 5, the mesons have a 36-fold degeneracy and the baryons a 56-fold degeneracy. Part of the lifting of this degeneracy is due to the fact that the s quark "mass" is heavier than those of u and d by ~ 100 MeV. That breaks the flavor-SU(3) symmetry and leads to the familiar Gell-Mann-Okubo mass formula. The exchange of gauge-field quanta brings in a spin-dependent interaction and that lifts the degeneracy of states with different spins, such as those between π and ρ and between N and Δ. For simplicity, we shall concentrate only on these four states.

To evaluate the first-order correction in α, we include only the diagram (1) in Fig. 8 for mesons, and the corresponding one in Fig. 9 for baryons. For low-lying hadron states, the quarks are all in the $s_{\frac{1}{2}}$ state. In the approximation of

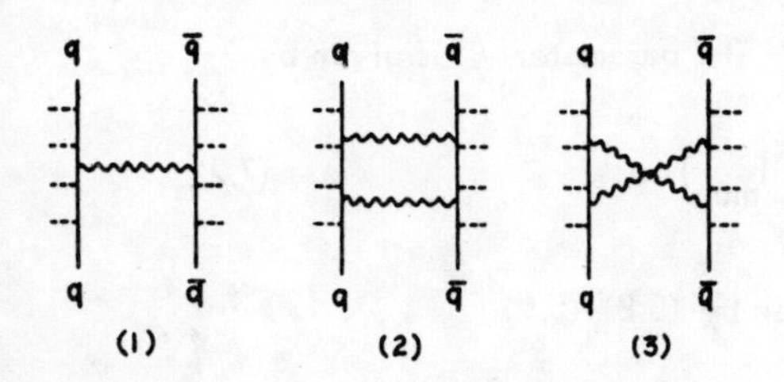

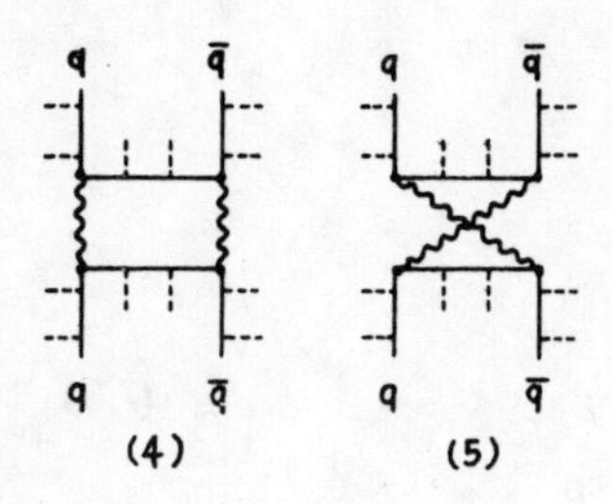

--- σ line
〰 V_μ^a line

Fig. 8

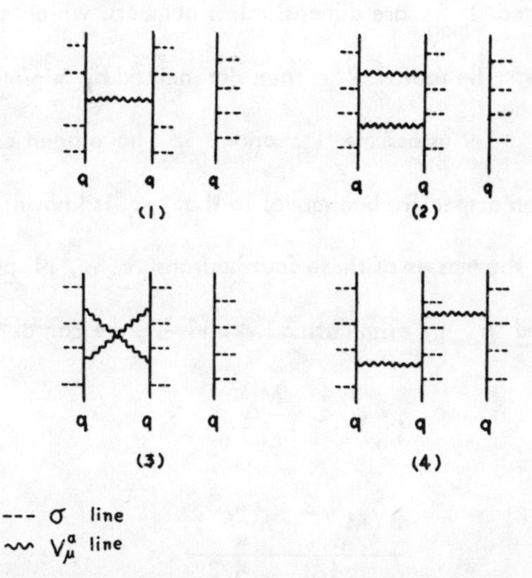

Fig. 9

--- σ line
〰 V_μ^a line

zero quark mass, the meson and the baryon masses, with the first-order diagrams included, can always be written in the form

$$M = \frac{N\xi}{R} + \frac{4\pi}{3} R^3 p \qquad (7.1)$$

where $N = 2$ for mesons and 3 for baryons. The parameter ξ is given by

$$\xi = \rho_0 - \alpha (I_{el} - \lambda I_{mag}) \tag{7.2}$$

where ρ_0 is the zeroth-order parameter given by (5.8)-(5.9)

$$\rho_0 = \begin{cases} 2.0428 & \text{MIT Bag} \\ 1 & \text{SLAC Bag} \end{cases} ,$$

$$\lambda = \begin{cases} -3 & \text{for } \pi \\ -1 & \text{for } N \\ 1 & \text{for } \Delta \text{ and } \rho \end{cases} ; \tag{7.3}$$

I_{el} and I_{mag} are dimensionless numbers, which can be computed from Figures 8 and 9. The radius R is then determined by minimizing M.

Let us assume I_{el} and I_{mag} have been calculated. Furthermore we have chosen a specific bag model so that ρ_0 is known. By using (7.1)-(7.2), we can express the masses of these four hadrons π, ρ, N and Δ in terms of two parameters α and p. By eliminating α and p, we can derive two mass formulas:

$$\left(\frac{2}{3}\right)^{3/4} \frac{M_\Delta}{M_\rho} = 1 , \tag{7.4}$$

$$\frac{3\,(M_\rho^{4/3} - M_\pi^{4/3})}{4\,(M_\Delta^{4/3} - M_N^{4/3})} = 1 . \tag{7.5}$$

As we shall see, exactly the same two mass formulas emerge when I_{el}, I_{mag} and ρ_0 are assigned different numbers. This is quite fortunate since the calculations of I_{el} and I_{mag} are complicated and (7.4)-(7.5) are the maximum information that can be gotten from (7.1)-(7.2) for these four hadron masses. The experimental values of the lefthand sides of (7.4) and (7.5) are 1.180 and 1.187 respectively, which agrees reasonably well with the theoretical predictions.

To see how these two remarkable mass formulas can be derived without first determining I_{el}, I_{mag} and ρ_0, we note that according to (7.3) the λ values for Δ and ρ are the same. Hence, because of (7.2), the ξ values for Δ and ρ must also be the same. From (7.1), we see that the only difference is $N = 2$ for ρ and $N = 3$ for Δ. By minimizing M with respect to R, we find M to be $\propto N^{3/4}$, and that leads to (7.4).

In (7.2) ρ_0 and αI_{el} only appear in the combination $(\rho_0 - \alpha I_{el})$. Thus, even if we treat I_{el}, I_{mag}, ρ_0, α and p all as unknown parameters, (7.1) actually depends only on three combinations $(\rho_0 - \alpha I_{el})$, αI_{mag} and p. By eliminating these three from the four hadron masses, we derive (7.5).

8. Remarks

Quark confinement is a large-scale phenomena. Therefore, at least on the phenomenological level, it should be understandable through a quasi-classical macroscopic theory, much like the London-Landau theory for superfluidity. In this talk, we suggest that the use of the color dielectric constant of the vacuum serves such a purpose. The hadrons are color-singlet composites of q and $\bar{q}$, describable by domain structures in the vacuum. In this way we are able to connect the bag models with QCD.

References

1. A. Chodos, R. L. Jaffe, K. Johnson, C. B. Thorn and V. F. Weisskopf, Phys. Rev. D9, 3471 (1974).

2. W. A. Bardeen, M. S. Chanowitz, S. D. Drell, M. Weinstein and T.-M. Yan, Phys.Rev. D11, 1094 (1975).

3. H. D. Politzer, Phys.Rev.Lett. 30, 1346 (1973); D. Gross and F. Wilczek, Phys. Rev.Lett. 30, 1343 (1973); G. 't Hooft, Phys.Lett. B61, 455, B62, 444 (1973), first presented in some brief remarks at the Marseilles meeting, 1972 (unpublished).

4. R. Friedberg, T. D. Lee and A. Sirlin, Phys.Rev. D13, 2739 (1976), Nucl.Phys. B115, 1, 32 (1976).

5. R. Friedberg and T. D. Lee, Phys.Rev. D15, 1694, D16, 1096 (1977), D18, 2623 (1978).

6. T. D. Lee, Phys.Rev. D19, 1802 (1979).

QUANTUM MECHANICAL TREATMENT OF A DAMPED HARMONIC OSCILLATOR

Peng Huan-wu

Institute of Theoretical Physics

Academia Sinica

Abstract

Given a damping force proportional to the velocity, the quantum condition between the physical variables, the displacement and the momentum, has to be modified by the damping. It is possible, however, to change to new canonical variables by using the so-called canonicalization transformation, and then we can apply the usual quantum mechanics, including the usual physical interpretation, with the help of statistical average values. As an example, the case of a one-dimensional damped harmonic oscillator under the action of an arbitrary external force has been worked out. No difficulty arises, and the results are reasonable.

It is not clear, in quantum mechanics, whether a description involving frictional force is truly significant. Assuming that such a description is meaningful, we shall treat in what follows the case of a damped harmonic oscillator driven by an external force quantum mechanically.

From the equations of motion

$$\frac{dx}{dt} = \frac{p}{M} , \quad \frac{dp}{dt} = - \frac{C}{M} t - Kx + f(t) , \tag{1}$$

follows

$$\frac{d}{dt} (xp - px) = - \frac{C}{M} (xp - px) . \tag{2}$$

This depends only on the form of the frictional force which varies as the velocity with a proportional constant C. It thus appears that the usual quantum condition between the displacement x and the momentum p has to be modified and that probably

$$(xp - px)\, e^{Ct/M} = i\hbar . \tag{3}$$

This is compatible with the equations of motion (1) and coincides with the usual quantum condition when C vanishes. For a non-vanishing frictional constant, the necessary modification mentioned above originates from the fact that in both classica and quantum mechanics the physical variables x and p no longer form a pair of canonicalization conjugate variables.

We now try to make a change of variables from the origina non-canonical pair x and p to a new pair X and P_1 which shall form a pair of canonically conjugate variables. This is indee possible for a damped harmonic oscillator, because the transformed equations of motion obtained from (1) can be written in the canonical form with a suitable Hamiltonian. If one such canonicalization transformation $x, p \rightarrow X_1, P_1$ can be found, then there exist other canonicalization transformations $x, p \rightarrow X_2, P_2$, where X_2, P_2 and X_1, P_1 are related to each other by a canonical transformation; and we can make use of this freedom to simplify the Hamiltonian. We take, as canonical variables,

$$X = x\, e^{ct/2M} \quad , \quad P = (p - \tfrac{1}{2}Cx)\, e^{ct/2M} \tag{4}$$

and obtain for the equations of motion (1) the canonical form

$$\frac{dX}{dt} = \frac{\partial H}{\partial P} = \frac{XH - HX}{i\hbar}, \quad \frac{dP}{dt} = -\frac{\partial H}{\partial X} = \frac{PH - HP}{i\hbar}, \tag{5}$$

with the following simple expression for the Hamiltonian

$$H = \frac{1}{2M} P^2 + \tfrac{1}{2}\left(K - \frac{C^2}{4M}\right) X^2 - f(t)\, e^{ct/2M}\, X. \tag{6}$$

The usual canonical quantum condition

$$XP - PX = i\hbar \tag{7}$$

agrees with (3). With a change of variables (4) we have thus transformed the equations of motion (1) and the quantum condition (3) into the usual form (5) to (7) and can treat the latter in the usual way. We can change from the above Heisenberg picture to the Schroedinger picture

$$-\frac{\hbar}{i}\frac{\partial \psi}{\partial t} = H\psi \tag{8}$$

and adopt the usual statistical interpretation of the wave

function ψ at time t.

We must remember that most physical quantities of interest are given in terms of the physical variables x and p. For example, the displacement at time t is given by x itself, the driving power P_f, the power dissipated P_c and the total energy of the oscillator E at time t are given bv

$$P_f = f(t)\frac{p}{M}, \quad P_c = C\frac{p^2}{M^2}, \quad E = \frac{p^2}{2M} + \frac{K}{2}x^2 \quad (9)$$

With the help of the inverse transformation of (4) these quantities can be expressed in terms of the canonical variables X and P at time t. On changing to the Schroedinger picture we can evaluate the average values of these quantities at time t with the help of the wave function $\psi(t)$, thus for example we have for the total energy

$$\bar{E}(t) = \frac{\int \psi^*(t)\, e^{-Ct/M}\left[\frac{1}{2M}\left(P-\frac{C}{2}X\right)^2+\frac{K}{2}X^2\right]\psi(t)}{\int \psi^*(t)\,\psi(t)} \quad (10)$$

Here on the right hand side we deliberately leave the representation unspecified. If we use the X -representation then $\psi(t)$ depends on X and P is interpreted to denote $(\hbar/i)\partial/\partial X$. The integral sign is to be completed by dX, the range of integration being from minus to plus infinity.

If there is of no driving force, f(t) = 0, the Hamiltonian (6) does not involve t explicitly. Then even for a stationary state where the time factor of $\psi(t)$ is compensated by that of $\psi^*(t)$, the total energy $\bar{E}(t)$ decays with time just as in classical mechanics.

In order to examine whether the quantum mechanical treatment indicated above will give reasonable results for physical interpretation, we shall solve the Schroedinger equation in the case of a general driving force and then evaluate the average values of the physical quantities of interest mentioned above. For simplicity, we consider the case of underdamping where we can write

$$K - C^2/(4M) = M\Omega^2 \tag{11}$$

with a real Ω of the dimension of frequency. We choose the units $\hbar = M = \Omega = 1$, so that $K = 1 + C^2/4$.

It is convenient to introduce the Fock variables

$$\xi = \frac{X - iP}{\sqrt{2}}, \qquad \xi = \frac{X + iP}{\sqrt{2}} \tag{12}$$

in terms of which the Hamiltonian (6) becomes

$$H = \xi\xi^* + 1/2 - f(t)\, e^{ct/2} (\xi + \xi^*)/\sqrt{2} \tag{13}$$

We use the same asterisk to denote the Hermitian conjugate of a non-commuting quantity or the ordinary conjugate of a complex number. The quantum condition (7) becomes

$$\xi^*\xi - \xi\xi^* = 1 \tag{14}$$

The Schroedinger equation in the ξ-representation, in which ξ^* is represented by $\partial/\partial\xi$, is a linear partial differential equation of the first order and therefore easy to solve:

$$-\frac{\partial \psi}{i\partial t}=\left[\xi \frac{\partial}{\partial \xi}+\frac{1}{2}-f(t)e^{ct/2}\frac{1}{\sqrt{2}}\left(\xi+\frac{\partial}{\partial \xi}\right)\right]\psi \tag{15}$$

Let us assume that the driving force starts to act from a certain instant $t = 0$, while, prior to this instant, we have for the general solution a superposition of stationary-state solutions with constants given arbitrarily

$$\psi(\xi,t)=\sum_{n=0}^{\infty} a_n \psi_n(\xi,t), \quad (t\leq 0),$$

where

$$\psi_n(\xi,t)=\frac{1}{\sqrt{n!}}\xi^n e^{-i(n+1/2)t} \quad . \tag{16}$$

The general solution of (15) for $t \geq 0$ is obtained from the integrals of the associated system of ordinary equations

$$\frac{d\xi}{\xi-\frac{1}{\sqrt{2}}f(t)e^{ct/2}}=\frac{dt}{\frac{1}{i}}=\frac{d\psi}{\left(\frac{1}{\sqrt{2}}f(t)e^{ct/2}\xi-\frac{1}{2}\right)\psi},$$

namely

$$\xi e^{-it}+i\beta=c_1, \quad \ln\psi+\frac{it}{2}-i\beta^* c_1-\alpha^*=\ln c_2 .$$

Expressing the second constant of integration c_2 as an arbitrary function of the first constant of integration c_1, we obtain

$$\psi=\psi(\xi,t)=e^{-\alpha-\frac{it}{2}+i\beta^*\xi e^{-it}}\phi(\xi e^{-it}+i\beta), (t\geq 0). \tag{17}$$

Here we have introduced the abbreviations

$$\beta=\int_0^t e^{-iv}\frac{1}{\sqrt{2}}f(v)e^{cv/2}dv, \tag{18}$$

$$\alpha=\int_0^t e^{-iu}\frac{1}{\sqrt{2}}f(u)e^{cu/2}du\int_0^u e^{iv}\frac{1}{\sqrt{2}}f(v)e^{cv/2}dv,$$

which together with their complex conjugate are related by the identity $\alpha + \alpha^* = \beta^* \beta$. (18a)

On joining the solutions (16) and (17) at $t = 0$, we determine the arbitrary function in terms of the constants a_n :

$$\phi(\xi) = \sum_{n=0}^{\infty} a_n (\xi^n / \sqrt{n!}) \tag{19}$$

The final wave function, if expanded into a double series, is

$$\psi = \sum_{n=0}^{\infty} \sum_{m=0}^{\infty} \psi_m G_{mn} a_n$$

where

$$G_{mn} = \sum_{r=0}^{\infty} \frac{\sqrt{n!}\,\sqrt{m!}}{r!\,(n-r)!\,(m-r)!} (i\beta)^{n-r} (i\beta^*)^{m-r} e^{-\alpha}$$

agrees with the kernel obtained for a forced harmonic oscillator by the method of path integral [1] .

The complex conjugate wave function $\psi^*(\xi^*, t)$ is obtained from (17) and (19) in the ξ^* -representation.

In evaluating an integral of the form

$$\int \psi^* O \psi = \int \psi^*(\xi^*, t)\, O(\xi, \xi^*; t)\, \psi(\xi, t) \tag{20}$$

we may proceed as follows. For non-commuting quantities ξ and ξ^* obeying (14) we can read the equation from left to right as an operator equation in the ξ -representation, interpreting ξ^* as a differential operator to the right. We can read the same equation from right to left as an operator equation in the ξ^* -representation, interpreting ξ as a differential operator to the left. The wave function (17) can then be replaced by the same operator to the right of which a factor unity which is to be operated on is understood. Similarly the complex conjugate wave function $\psi^*(\xi^*, t)$ in (20) can be replaced by the Hermitian conjugate operator of the above operator, and to the left of which a factor unity is similarly understood. We then make use of (14) and bring the exponential factor of ξ^* from left to right in accordance with the operator identity

$$e^{-ib\xi^*} g(\xi, \xi^*) = g(\xi - ib, \xi^*) e^{-ib\xi^*} \tag{21}$$

and similarly bring the exponential factor of ξ from right to left in accordance with the Hermitian conjugate identity of (21). The factor $e^{-ib\xi^*}$, when brought to the right end, operates on the unity factor to give again a unity factor and hence can be dropped. Similarly, the factor $e^{ib^*\xi}$, when brought to the left end, can also be dropped. Thus, remembering (18a), we have

$$\int \psi^* O \psi = \int \phi^*(\xi^* e^{it}) O(\xi - i\beta e^{it}, \xi^* + i\beta^* e^{-it}; t) \phi(\xi e^{-it}). \tag{22}$$

Further we note from the series form (19), the identity

$$\xi e^{-it} \phi(\xi e^{-it}) = \sum_{n=0}^{\infty} (\sqrt{n}\, a_{n-1}) \frac{1}{\sqrt{n!}} (\xi e^{-it})^n = A\, \phi(\xi e^{-it}),$$

where A denotes an operator acting only on the coefficients a_n defined by

$$A\, a_n = \sqrt{n}\; a_{n-1}. \tag{23}$$

Together with similar relations for the conjugate we write (22) as

$$\int \psi^* O \psi = \int \phi^*(\xi^* e^{it}) O((A-i\beta)e^{it}, (A^*+i\beta^*)e^{-it}; t)\phi(\xi e^{-it}). \tag{24}$$

We shall not need the explicit specification of the integral sign in the complex Fock representation discussed elsewhere, but we do need it to note the following orthonormal relations [2]

$$\int \psi_m^* \psi_n = \int \frac{1}{\sqrt{m!}}(\xi^* e^{it})^m \frac{1}{\sqrt{n!}}(\xi e^{-it})^n = \delta_{mn} \,. \tag{25}$$

Finally we obtain the simple result

$$\int \psi^* O \psi = \sum_{n=0}^{\infty} a_n^* O((A-i\beta)e^{it}, (A^*+i\beta^*)e^{-it}; t)\, a_n \,. \tag{26}$$

In particular, by putting O = 1 we obtain

$$\int \psi^* \psi = \sum a_n^* a_n \tag{27}$$

and, by putting

$$O(\xi, \xi^*; t) = x = e^{-ct/2}(\xi+\xi^*)/\sqrt{2} \tag{28}$$

we obtain for the average value of displacement

$$\bar{x}(t) = e^{-ct/2}(-i\beta e^{it} + i\beta^* e^{-it})/\sqrt{2} +$$

$$+\frac{1}{\sqrt{2}} e^{-ct/2}\left(\frac{\sum a_n^* \sqrt{n}\, a_{n-1}}{\sum a_n^* a_n} e^{it} + \frac{\sum a_{n-1}^* \sqrt{n}\, a_n}{\sum a_n^* a_n} e^{-it}\right). \tag{29}$$

Here we may distinguish a part involving the driving force and another part involving the constants a_n occuring in the initial value of the wave function. The latter is damped, while the former, after substitution from (18), is given by

$$e^{-ct/2}(-i\beta e^{it}+i\beta^* e^{-it})/\sqrt{2}$$

$$=\int_0^t \{e^{-c(t-v)/2}\sin(t-v)\} f(v)dv \tag{30}$$

which agrees with the solution of the classical equation of motion for the damped harmonic oscillator with the initial conditions $x = dx/dt = 0$ at $t = 0$.

We can calculate similarly the average values of the other physical quantities of interest, as given by (9). After a time when those contributions involving the constants a_n are damped out, we are left with contributions involving the driving force only, which can be more simply obtained from (9) and the above classical (30).

Our Hamiltonian is simpler than that used by other authors [3][4], so that results needed for physical interpretation are easier to obtain. The above example shows that, provided a canonicalization transformation can be found, there is no difficulty to apply quantum mechanics to dissipative systems.

References

1. Feynmann, R.P. and A.R. Hibbs, Quantum Mechanics and Path Integrals, McGraw-Hill, 1965, p.234.
2. Dirac, P.R.S. London, 180 (1942) 1.
   Peng H.W., Proc. Roy. Irish Acad., 51 A 8 (1947) 113.
3. Feshbach, H and Y. Tikochinsky, Festschrift for I.I. Rabi, ed. by L. Motx (New York Acad. of Sci., New York, 1977).
4. Greenberger, D.M., J. Math. Phys., 20 (1979) 762.

   cf. also Khandekar, D.C. and S.V. Iawande, J. Math. Phys. 20 (1979), 1870 where an exact solution of a time-dependent quantal harmonic oscillator with damping and a perturbing force is obtained. (Note added in proof)

Quantum Field Theory and Statistical Mechanics

Barry M. McCoy*

Institute for Theoretical Physics, State University of New York,
Stony Brook, New York 11794, U.S.A

and

Tai Tsun Wu†

Gordon McKay Laboratory, Harvard University, Cambridge, Mass. 02138, U.S.A.

ABSTRACT

Quantum field theories may be treated by (1) perturbation series, (2) the summation of the perturbation series, and (3) non-perturbative methods. These three treatments are of very different nature. Since the first one is well known, we review here some aspects of the second and the third methods. For the second approach, we illustrate with the theory that has led, nearly ten years ago, to the prediction of rising total cross sections for hadrons together with a number of related aspects such as the movement of the dips. Recent advances in the third approach rely heavily on calculations from statis-

---

*Work supported in part by the United States National Science Foundation under No. DMR-79-08556.

†Work supported in part by the United States Department of Energy under contract No. DE-AS02-76ER03227.

tical mechanics, and we illustrate them in some detail with two examples. The first example is the Thirring model, and the lessons that we learn from it are the dynamic generation of mass and the corresponding dynamic determination of the coupling constant. Such dynamic determination of the coupling constant has been the dream of physicists for a long time. The second example is the Ising field theory in two dimensions, and the lessons that we learn from it are the existence of multiple phases coming from the same Lagrangian and the corresponding lack of relevance of classical paths. In particular, it is found that, for the spontaneously broken phase, the mass spectrum for the two-point Green's function is purely continuum, and the three and higher point Green's functions have string structures. If both features persist in four dimensions, then it is natural to postulate the existence of indefinite-mass particles, or imps for short. We discuss in detail the phenomenology of such novel objects, which may be quarks.

1. INTRODUCTION

Quantum field theory is the physical theory born, just over half a century ago, of the marriage between quantum mechanics of a finite number of degrees of freedom and classical field theory of an infinite number of degrees of freedom. Typical of most new theories, it was first treated by perturbation methods. In the very interesting case of quantum electrodynamics, perturbation methods seem to be especially appropriate because of the smallness of the fine structure constant. Unlike most other theories, however, it was soon found that beyond the lowest order the perturbation series is divergent term by term.

Approximately twenty years after its conception this difficulty of quantum field theory was brilliantly overcome with the invention of renormalization.[1] It may be said that renormalization is what distinguishes quantum field theory from classical field theories. Thus renormalizable quantum field theories can be studied on three different levels:[2]

1. by the renormalized perturbation series,

2. by the summation of the renormalized perturbation series, and

3. by non-perturbative methods with renormalization.

The second method, the summation of the renormalized perturbation series, is sometimes referred to as non-perturbative. This terminology is very confusing, and shall not be used here. We emphasize that the second method and the third method may well produce different results. One of the many mathematical examples is given by the function

$$f(g) = \begin{cases} e^{-(g-g_0)^{-1}} & \text{for } g > g_0 \\ 0 & \text{for } g \leqslant g_0 \end{cases}$$

where g_0 may be zero or positive. In either case, if we expand f(g) through the Taylor series about the point g = 0 and then sum the series, we get 0 instead of f(g). While this mathematical example is of ancient origin, it is a rather recent realization that this does happen in the case of the Thirring model.[3]

A great deal is known about the renormalized perturbation series; we can in principle calculate as many terms of the series as we wish. Accordingly. this first method will not be discussed any further; instead, we review here some aspects of the second and the third methods. Admittedly, our knowledge about these methods is still quite primitive, and a great deal of further work is needed.

In Sec. 2 we illustrate the second method by an example that has been treated in great deal in the literature, namely the high-energy behavior of quantum field theory. The rest of this paper deals with the third one, non-perturbative methods with renormalization. Since the original development of renormalization is concerned exclusively with the perturbation series, clearly the first question to be answered is what is renormalization in the context of non-perturbative methods? To obtain a natural answer to this question, we go back to classical field theory, and remember that there an elementary method to treat the field equations (partial differential equations) is to replace them by partial difference equations. The reason is that the process of shrinking the lattice distance to zero is surely the most obvious way of giving meaning to the concept of a continuum. This reason holds equally well for quantum and classical field theories. Accordingly, we may put the quantum field theory on a lattice - lattice renormalization.

The following major difference between the classical and quantum field theories must be emphasized. While in both cases the basic idea is put in the field theory on the lattice, lattice renormalization in the quantum case is much more fundamental than the method of finite differences in the classical case. The reason is that, while the classical field theory is defined by the Lagrangian, the quantum field theory is defined by the Lagrangian <u>and</u> the renormalization procedure. This is most clearly seen in the Ising field theory,[4] i.e., the quantum field theory obtained from the two-dimensional Ising model.[5] In this case, for the same Lagrangian there are two field theories of very different nature, corresponding respectively to the phases with and without spontaneous symmetry breaking. The phase with spontaneous symmetry breaking has rather peculiar properties, of which one possible interpretation is the presence of particles whose masses are not well defined.[6]

In Sec. 3, we discuss lattice renormalization in some detail. Two examples are invoked as illustrations: the Thirring model in Sec. 4 and the Ising field theory in Sec. 5. In Sec. 6 we present finally, on a more speculative level, some phenomenological considerations in connection with the indeterminate-mass particle just mentioned.

2. SUMMATION OF PERTURBATION SERIES

In this section, we present the method of summing the renormalized perturbation theory and use it to discuss the prediction of the rising total cross section made a decade ago.[7]

"Summation of the renormalized perturbation series" can be interpreted in two entirely different ways.

(1) Calculate the perturbation series exactly and then sum this exact series. While this interpretation is perfectly reasonable, it is not especially useful because it can be carried out in fact for very few cases, those where the problem is exactly solvable. For example, this procedure has been carried out[8] for the Schwinger model.[9]

(2) Calculate order by order the <u>leading terms</u> of the perturbation series and then sum these leading terms. This interpretation is much more useful, but it must be admitted that the correctness of this procedure of summing the leading terms is a pure assumption. In this paper and indeed in this entire report from here on, we shall use only this interpretation.

This procedure of summing the leading terms has been used, over two decades ago, by Lee, Huang, and Yang[10] in the context of statistical mechanics with a great deal of success. In order for the concept of the leading term to be meaningful, we must have a small parameter in the problem. In other words, we are summing the terms, which are order by order the largest with respect to this small parameter.

Let us consider for definiteness the case of massive quantum electrodynamics, where a fermion f interacts with a vector particle V. In the massless limit, this f and V correspond respectively to the electron and the proton in the usual quantum electrodynamics. But a much more profound difference is that we shall not take the coupling constant g to be small, as in the case of the usual quantum electrodynamics. Consequently, we must look for another small parameter as required by the procedure of summing the leading terms. We consider the case of the high-energy limit where the center-of-mass energy $\sqrt{s}$ is very large and hence s^{-1} is the small parameter. It can of course be rendered dimensionless by using the f mass, the V mass, and/or some momentum transfer.

Physically, why are we interested in such a study? The purpose is of course to have some understanding of strong interactions at extremely high energies. While it seems to us only natural to turn to quantum field theory in attempting to gain this understanding, this thinking has not always been accepted. Indeed, ten years ago quantum field theory was unfashionable in the high-energy physics community. Consequently, talks which are based on a field-theory calculation would usually be prefaced by an apology which brought in phenomenology and physical interpretation. The fashion has changed since then, and today we may present the field-theory calculation first and its physical interpretation second. Accordingly, while in 1970 we would have called this section "The Rising Total Cross Section" and have studied this question by the summation of the perturbation series, we now in 1980 call this section "Summation of Perturbation Series" and will obtain the rising total cross section as a result.

When this study of the high-energy behavior of quantum field theory was initiated over a decade ago,[11] there was actually another motivation. At that time, as well as now, there were two schools of thought on high-energy processes. The first is due to Yang, Byers, Chou, Benecke and Yen[12] and is going to be reported later in this Conference by Chou. In this point of view, unitarity in the s channel plays a central role, while that in the t channel or the cross channel is considered to be less important. In the opposite school of thought due to the Regge-pole phenomenologists,[13] unitarity in the t channel is of primary importance, while that in the s channel is relegated to a secondary role. Since both points of view have led to a great deal of success in the sense of understanding the experimental data, it is natural to inquire whether a model can be constructed to incorporate both points of view at the same time in the sense of permitting both unitarities to play central roles simultaneously. Such a model is indeed provided by quantum field theory in general and quantum electrodynamics in particular. After the rising total cross section was predicted theoretically in 1970[7] and experimentally verified three years later,[14] it was found that there is no difficulty in incorporating this rising total cross section in the droplet model of Yang et al.[15] This aspect of more recent developments will also be reported in the summary talk of Chou.

Let us describe very briefly the computational procedure in the case of Compton scattering, i. e., the elastic scattering process

$$f + V \to f + V$$

in the limit of very large s but fixed t (the square of momentum transfer). For this process, the matrix element is of the order of s^0 in the second order, and is larger by some factors of $\ell n\ s$ in the fourth order. In other words, to these orders the differential cross section $d\sigma/dt$ decreases very rapidly at high energies. In order to have a finite $d\sigma/dt$ in this limit, we need a matrix element of the order of s. Thus the first question is: Is there such a large matrix element in some order? The answer is yes, and it occurs in the sixth order. With this information, the next question is: Are there even larger matrix elements in higher orders? The answer is again yes, and in the tenth order the matrix element is found to be of the order of $s\ \ell n\ s$ With this additional information, the situation becomes clear: the orders of the terms to be summed together with their orders of magnitudes are as follows

6th order	s
10th order	$s \ln s$
14th order	$s(\ln s)^2$
18th order	$s(\ln s)^3$

etc. The sum of these leading terms yields,[16] give or take a few $\ln s$,

$$i\, s^{1+Ag^4}$$

where $A = \dfrac{11}{512\pi}$.

That the exponent is larger than unity is not to be interpreted as a violation of the Froissart bound. Rather, it ought to be regarded as a realization of the strongly absorptive "potential" with a coupling constant increasing with energy, as conceived in Froissart's original paper.[17] Thus the Froissart bound on the total cross section is saturated. Of course, Froissart's other bounds in the same paper cannot be saturated.

The following extremely simple picture[7] therefore emerges for scattering processes at extremely high energies. Since this picture applies not only to massive quantum electrodynamics as discussed here, but also to a number of other quantum field theories, it is expected to hold also for hadronic processes. For such extremely high energies and for fixed impact parameter, complete absorption occurs through the production of relatively low-energy particles. The radius of this region, referred to as the black disk, increases as $\ln s$. This black disk is surrounded by a gray region (i. e., partially absorbing region) of a radial size of order 1. Inelastic diffraction dissociation occurs in this region.

This physical picture, while simple, already contains a number of deep consequences for scattering processes at extremely high energies. We list a few.[7, 18]

(A) The total cross section σ_{tot} increases without bound as $(\ln s)^2$. This is the saturation of the Froissart bound.

(B) The integrated elastic cross section σ_{el} also increases without bound as $(\ln s)^2$.

(C) The differential cross section for elastic scattering has many dips located approximately at

$$\sqrt{t} \sim \text{constant} \frac{j_1^{(n)}}{\ln s}$$

where $j_1^{(n)}$ are the zeros of the Bessel function J_1.

(D) The locations of the maxima for the differential cross section between the dips also move in; furthermore, the values of $d\sigma/dt$ at these maxima increase as s increases.

(E) Consider an inelastic diffractive process. The integrated cross section for such a process increases without bound as $\ln s$. When there is no exchange of spin and parity, the differential cross section for such a process has many dips located approximately at

$$\sqrt{t} \sim \text{constant} \frac{j_o^{(n)}}{\ln s}$$

where $j_o^{(n)}$ are the zeros of the Bessel function J_o. It should be emphasized that these results are supposed to hold only at extremely high energies. Unfortunately, by extremely high energies we mean not only that s/M^2 is large but also that $\ln(s/M^2)$ is large, where M is some typical mass scale, presumably about 1 GeV for hadronic processes.

When these theoretical predictions where first obtained ten years ago, there was not the slightest bit of evidence even for the first item, the increasing total cross section. In order to encourage experimentalists to look for this increase, it was necessary to produce some rough quantitative estimate for the magnitude of this phenomenon. This was essentially an impossible task, and Cheng, Walker and Wu[17] used the Serpukhov data and the assumption of exchange degeneracy for Kp channel. When eventually experimental data were obtained in 1973 from the ISR of CERN, the observed increase in proton-proton total cross section[14] turned out to be about 50% higher than the predicted value.[18] In view of the assumptions involved, this must be considered to be remarkable agreement. Predictions in such a complicated situation about a new phenomenon cannot be as accurate as postdictions with the experimental data in hand.

Detailed comparison between theory and experiment has been summarized at various times, such as the 18th International Conference on High Energy Physics at Tbilisi.[19] Some of the verified predictions include the existence of the pionization plateau,[20] ratio of the real to imaginary parts for forward scattering,[21] rising total elastic cross section,[7,14] movement of the dip in elastic scattering to smaller momentum transfer,[22] and the rising pionization plateau.[23] It is to be emphasized that in each of these cases, the predictions were given on field-theoretic grounds before the experimental verification.

3. LATTICE RENORMALIZATION

We now turn our attention to lattice renormalization so that we will be able to go beyond the physics of perturbation theory.[24] Actually there are two somewhat different ways to proceed. We can either use a Minkowski metric and put only the N-dimensional coordinate space components on a lattice, or we can carry out a Wick rotation to imaginary times and put the entire resulting (N + 1)Euclidean space on a lattice. Both schemes are useful in practice and we discuss them separately.

A. Euclidean Renormalization

(1) Take the classical continuum action S and put it on an (N+1) dimensional Euclidean lattice. In general the action depends on several numerical parameters: bare masses, bare coupling constants, and an overall scale.

(2) Calculate Green's functions by means of the general formula[25]

$$<\phi(x_1) \dots \phi(x_n)> = Z^{-1}\int\{\prod d\,\phi_i\}\,\phi\,(x_1)\dots\,\phi(x_n)e^S \qquad (3.1)$$

with

$$Z = \int\{\prod d\phi_i\}\;e^S\;. \qquad (3.2)$$

In (3.2) there is an independent field ϕ_i for each site of the lattice, and we integrate these ϕ_i over all allowed values. The ϕ_i commute (anticommute) for bosons (fermions). At this stage we take the limit of an infinite lattice; thus all infrared cutoffs are removed.

(3) Consider the two-point function on the lattice. In general, as the distance between the two points $|x_1 - x_2|$ approaches infinity, the two-point function approaches its limiting value as $e^{-\kappa|x_1 - x_2|}$. The inverse

correlation length κ is a function of the parameters of the Lagrangian. To construct a field theory we must find those values of the parameters where $\kappa = 0$ on the lattice.

(4) Renormalization is carried out by letting the parameters approach (or equal) those special values where $\kappa = 0$ as the lattice spacing $a \to 0$ while keeping quantities such as the physical mass fixed. Renormalized Green's functions are defined by

$$<\phi(x_1 \ldots \phi(x_n)>_{ren} = \lim_{a\to 0} Z_3(a)^{-n/2} <\phi(x_1)\ldots\phi(x_n)> , \tag{3.3}$$

where $Z_3(a)$, called the wave-function renormalization constant, is chosen so that the limit in (3.3) exists and is not zero. If the renormalized Green's functions so defined are rotationally invariant, then after a Wick rotation a relativistic quantum field theory has been constructed from the classical action.

B. Minkowski Renormalization

(1') Calculate the Hamiltonian which corresponds to the Lagrangian, and put the N space coordinates on a lattice.

(2') The many-body wave function for the ground state is calculated by solving the Schrödinger equation, and Green's functions are obtained by carrying out integrations over this ground-state wave function.

(3') The excitation spectrum of the Hamiltonian is calculated and the values of the parameters are found which make the mass gap vanish.

(4') Renormalization is now carried out as in (3.3) directly in Minkowski space.

There are at least four ways in which the above procedure, either version A.) or version B.), can fail to give a non-trivial field theory.

(a) There may be no values of the Lagrangian parameters which make the inverse correlation length (or mass gap) vanish. Recent Monte Carlo calculations[26] indicate that this may indeed be the case for the Z_2 lattice gauge theory.

(b) There may not exist a $Z_3(a)$ which renormalizes all the n-point functions.

(c) The resulting Green's function may not be Lorentz invariant.

(d) Even if the Green's functions are Lorentz invariant, they still may not be different from those of a free field.

We say that the quantum field theory is trivial if the difficulty (d) occurs. Several decades ago, Landau[27] conjectured that quantum electrodynamics is trivial, because summation of leading logarithms gives an unacceptable singularity, the Landau pole. Since then, there has been a great deal of work done on possible triviality of field theories[28], especially in connection with the ϕ^4 theory in four dimensions. Last year, Baker and Kincaid[29] gave a most convincing numerical demonstration that this ϕ^4 theory in four dimensions is indeed trivial.

4. THIRRING MODEL

We wish to illustrate, with one example each, the two versions of lattice renormalization given in the last section. These examples are the Thirring model [3] for the Minkowski version, and the Ising field theory[4] for the Euclidean version. In both cases, N = 1 and the field theory, after being placed on a lattice, corresponds to a solvable problem of statistical mechanics.

In this section we consider the Thirring model with no base mass. The non-perturbative renormalization effects to be illustrated are the dynamic generation of mass and the corresponding dynamic determination of the coupling constant.[30]

The Lagrangian density for the massless Thirring model is

$$L(x) = i \bar{\psi} \gamma_\mu \partial^\mu \psi - g(\bar{\psi} \gamma_\mu \psi)(\bar{\psi}\gamma^\mu \psi) . \tag{4.1}$$

In order to treat this model non-perturbatively, it is most convenient to use a special case of the lattice version of Luther[31] and Lüscher[32]. Their Hamiltonian is

$$H = \sum_n \left[\frac{i}{2a} V(\psi_n^\dagger \psi_{n+1} - \psi_{n+1}^\dagger \psi_n) - \frac{G}{2a}(\psi_n^\dagger \psi_n - \tfrac{1}{2})(\psi_{n+1}^\dagger \psi_{n+1} - \tfrac{1}{2}) \right], \tag{4.2}$$

where a is the lattice spacing. When the g defined in the Sommerfield sense is larger than $-\pi/2$, then $G = -\frac{4\varepsilon}{\pi} \cot\varepsilon$ and $V = \frac{2\varepsilon}{\pi \sin\varepsilon} > 0$, where $\varepsilon = \pi(g + \pi/2)/(g + \pi)$.

By means of the Jordan-Wigner[33] transformation

$$\psi_n^\dagger = (\sigma_n^x + i\sigma_n^y) \prod^{n-1} (i\sigma_j^z) \tag{4.3}$$

we find

$$H = -\frac{V}{4a} \left\{ \sum_n (\sigma_n^x \sigma_{n+1}^x + \sigma_n^y \sigma_{n+1}^y) + \Delta \sigma_n^z \sigma_{n+1}^z \right\} \tag{4.4}$$

when $\Delta = G/2V$.

We restrict ourselves to $\Delta < 1$, for otherwise the spin model is frozen in the ground state. It is most important that Yang and Yang[34] demonstrated that this ground state energy per spin, in the limit of an infinite chain, is an analytic function of Δ except at $\Delta = -1$. This lack of analyticity is the underlying reason for dynamic mass generation.

For $\Delta > -1$, there is no gap between the ground state and the excited states, and thus the massless Thirring model remains massless. For $\Delta < -1$, the situation is completely different. In particular, the spectrum of the lowest lying excited states[35] is

$$E(q) = \frac{V}{\pi a} K(k_1) \sinh\lambda \, (1-k_1^2 \cos^2 q)^{\frac{1}{2}}, \tag{4.5}$$

where q is a "momentum" variable, $\lambda = \cosh^{-1}(-\Delta)$, and $K(k_1)$ is the complete elliptic integral of the second kind of modulus k_1 determined by

$$\pi K[(1 - k_1^2)^{\frac{1}{2}}] = \lambda K(k_1). \tag{4.6}$$

The spectrum (4.5) has a gap for $\Delta < -1$; as $\Delta \to -1$, the gap vanishes. Therefore, no relativistic field theory can be constructed for fixed $\Delta < -1$, but <u>two</u> distinct ones can be obtained from $\Delta = -1$, which corresponds to $g = -\pi/2$:

(1) Keep g fixed at $-\pi/2$ and then let $a \to 0$. The absence of the gap implies that the result is a massless theory, which is the limiting case of those obtained from $\Delta > -1$.

(2) Let $a \to 0$ and $\Delta \to -1-$ simultaneously in such a fashion that the mass gap

$$M = \frac{1}{32} V a^{-1} \lambda e^{-\pi^2/\lambda} \tag{4.7}$$

is fixed. With a suitable scaling of the q, the dispersion relation is

$$E(p) = (M^2 + p^2)^{1/2}. \tag{4.8}$$

Therefore, there is dynamic mass generation. Furthermore, the coupling constant g can take on only one value, namely $-\pi/2$.

5. ISING FIELD THEORY

We next consider the Ising field theory.[3,24] The non-perturbative renormalization effects to be illustrated are the existence of multiple phases coming from the same Lagrangian and the corresponding lack of relevance of classical paths.

The action for the two-dimensional Ising model is[5]

$$S = \frac{1}{2} K \sum_{\vec{x}} \sum_{\{\vec{\delta}\}} \sigma(\vec{x})\, \sigma(\vec{x} + \vec{\delta}) \tag{5.1}$$

where $\vec{x}$ takes values on a square lattice of spacing a, the sum $\{\vec{\delta}\}$ is over the nearest neighbors, and $\sigma(\vec{x})$ is constrained to take only the values +1 and -1. This action (5.1) may be rewritten in the form

$$S = -\frac{1}{4} K a^2 \sum_{\vec{x}} \sum_{\{\vec{\delta}\}} \left[\frac{\sigma(\vec{x}) - \sigma(\vec{x}+\vec{\delta})}{a}\right]^2 + \text{constant} \tag{5.2}$$

which, as $a \to 0$, formally approaches $-\frac{1}{2} K \int (\partial_\mu \sigma)^2 \, d^2x$. Accordingly, the Ising field theory may be formally regarded as either the N = 1 non-linear σ model or the Euclidean field theory with the Lagrangian density

$$-\frac{1}{2}K(\partial_\mu \sigma)^2 - \lambda\sigma^4 + m_0\sigma^2 \tag{5.3}$$

in the limit where $m_0 = 2\lambda$ and $\lambda \to \infty$.

The important feature of this example is the following. It has been known for a long time that, as a function of K, a second-order phase transition[5] occurs in the two-dimensional Ising model when

$$K = K_c = \frac{1}{2} \ln(1 + \sqrt{2}). \tag{5.4}$$

Because of this second-order phase transition, we can construct not one but two quantum field theories, one by taking the limit $K(a) \to K_c+$ and a second one by $K(a) \to K_c-$. This construction was carried out for the two-point function in 1973[36], and for the general n-point function in 1977.[4] Since in both cases K(a) takes on the value K_c in the limit, these two field theories have the same Lagrangian, including the same bare coupling constants. This phenomenon that two different lattice renormalizations of the same Lagrangian can give rise to different field theories is referred to as the existence of different phases of the same quantum field.

Because of the importance of the Ising field theory, we shall now discuss in some detail the n-point function for the case $K(a) \to K_c+$, where there is spontaneous symmetry breaking.

α) Two-point function

In momentum space the two-point Greens function (i.e. the propagator) has the spectral representation

$$G_2^c(k) = \int_0^\infty dm \frac{\rho(m)}{k^2+m^2} \tag{5.5}$$

where

$$\rho(m) = 0 \qquad \text{for } 0 \leqslant m \leqslant m_0 \tag{5.6}$$

$= \text{const}(m^2-m_0^2)^{\frac{1}{2}} / m^2$ for $m_0 \leqslant m \leqslant 2m_0$ and for $m \geqslant 2m_0$, $\rho(m)$ has threshold singularities wherever m/m_0 is an integer. This weight function contains no delta function and therefore $G_2^c(k)$ contains no poles.

β) Three-point function

The three-point function of the Ising model for $T<T_c$ has a profoundly different behavior for large separation if the three points are collinear or not.[37] Away from collinearity

$$G_3^c(\vec{r}_1,\vec{r}_2,\vec{r}_3) \sim -\tan\tfrac{1}{2}\theta_1 \tan\tfrac{1}{2}\theta_2 \tan\tfrac{1}{2}\theta_3 \, (r_{12}r_{23}r_{13})^{-\frac{1}{2}} e^{-(r_{12}+r_{23}+r_{13})} \tag{5.7}$$

where $r_{ij} = |r_i - r_j|$ and θ_i are the 3 interior angles of the triangle. Near the collinearity with point 2 between the points 1 and 3 we define the variable

$$\tau^2 = r_{12} + r_{23} - r_{13} = O(1) \tag{5.8}$$

Then we have for $r_{13} \to \infty$

$$G_3^c(\vec{r}_1,\vec{r}_2,\vec{r}_3) \sim - \text{const } r_{13}^{-2} e^{-2r_{13}} F(\tau) \tag{5.9}$$

where

$$F(\tau) = (2\pi^2)^{-1} e^{-2\tau^2} + \frac{1}{2}\pi^{-3/2} \tau e^{-\tau^2} \text{Erf}\tau$$

$$+ (4\pi)^{-1} [1-(\text{Erf}\tau)^2] \tag{5.10}$$

where

$$\text{Erf}\tau = 2\pi^{-\frac{1}{2}} \int_0^\tau dx\, e^{-x^2} \tag{5.11}$$

(Here we take $m_0 = 2$)

γ) Cut breakup

The string structure is also present in the higher n-point functions. The form of G_n^c changes when all r_i become nearly collinear.[38] From this we may calculate how the two-point function behaves when a small magnetic field h is put on the system. We find that the cut breaks up into a series of poles[39] in such a way that both the spacing and the residue vanish as $h \to 0$. By calculating the four-point function for positive h one can study the S-matrix which describes the scattering of the particles represented by these poles. However, when $h \to 0$, this S-matrix description eventually ceases to be adequate for the physics of the field theory.

δ) Absence of classical path

From the same Lagrangian, two field theories of very different nature are obtained. For the Ising field theory, the field variables can take on only the values +1 and -1, and hence the only possible "classical" configurations are $\sigma(\vec{x}) = +1$ for all $\vec{x}$ and $\sigma(\vec{x}) = -1$ for all x. Clearly neither of these "classical" configurations has any relevance to either phase of the Ising field theory. This point can be seen very clearly from the point of view of statistical mechanics. Both of the above "classical" configurations above correspond to $T = 0$, while the two field theories come from $T \to T_c\pm$. Since T_c is non-zero, fluctuations are of paramount importance for both field theories. We may say that, while field theories that arise from $T \to 0+$ are somewhat classical, those with different phases from $T_c \neq 0$ are truly quantum mechanical in nature.

Since the Ising field theory is obtained by combining quantum field theory with the two-dimensional Ising model, it is of great interest in itself. However, it must be remembered that the real purpose of studying models is to learn about our actual four-dimensional world. For this reason, it is necessary to give physical interpretations to these results from the Ising field theory.

As mentioned above, the Ising field theory is a limiting case of the ϕ^4 field theory with the unrenormalized coupling constant approaching infinity. Thus let us consider a ϕ^4 classical field. This classical theory possesses "kink" solutions where the energy is localized. If periodic boundary conditions are imposed, then only states that consist of an equal number of kinks and anti-kinks are allowed. If there are particles in the quantum theory that correspond to classical kinks and anti-kinks, they may resemble their classical ancestors.[40] Thus one argues that the cut in the two-point function of the Ising field theory merely comes from two-particle states.

There are two difficulties with this interpretation. First, while the argument may perhaps hold for weak couplings, the coupling in the Ising field theory is surely strong; indeed it is as strong as possible in the sense that the bare coupling constant is infinite. Secondly, the starting point of this interpretation is the classical theory while the Ising field theory is truly quantum mechanical in nature. In other words, semi-classical arguments depend on stationary phase integration around a classical path, while for the Ising field theory there is no classical path at all.

Nevertheless, because of the simplicity of this interpretation of the Ising field theory for $T \to T_c-$, let us inquire what tests it must pass before it can be considered to be viable. Since one of the most striking properties of this theory is the string property of the n-point function, the most basic test is clearly the derivation of the Green's function strings for three and higher point functions on the basis of the above semi-classical argument with kinks and anti-kinks. We find it difficult to believe that G-strings can be obtained this way. If they can be, however, it means that the $T<T_c$ Ising model is prototypical for the broken phase of all ϕ^4 models.

Because of this difficulty, we find it more attractive to pursue an alternative interpretation where the cut itself is considered to be a representation of the physical object directly. Since such an object has no well-defined mass, we shall call it an indeterminate-mass particle, or imp for short.[6] This point of view is extremely quantum mechanical in outlook. The phenomenology of imps is to be discussed in detail in the following section.

6. PHENOMENOLOGY OF THE IMP

An introduction to non-S-matrix physics.

Following the introduction of quarks[41] in 1964, a vast amount of effort has been expended in attempts to detect them as free particles. All of these searches have failed. However, the concept of quark is so useful in explaining experimental data that it seems impossible to abandon it. Therefore the physics community is being compelled, much against its will, to entertain the possibility that fields exist which do not possess a particle interpretation. The lack of observation of free quarks compels us to go beyond the bounds of the physics which is contained in the S-matrix. This non-S-matrix physics is a completely new realm of endeavor which requires new concepts. In this section we will present an attempt to construct a phenomenology of non-S-matrix physics. Our point of departure will be the properties of the spontaneously broken Ising field theory of the previous section. This is a natural place to begin to learn about non-S-matrix physics because, indeed, it has no S-matrix description at zero magnetic field. However, we will rapidly progress beyond the point where any Ising analogy can guide us and we will conclude with a discussion of a few of the many questions which arise.

i) In the Ising field theory we saw that an imp can be described by a collection of poles in the limit where the spacing and residue go to zero. Since in atomic physics systems with many closely spaced levels are large in spatial extent we conclude that the particles described by these poles will be large. Moreover, in the limit of zero spacing this size goes to infinity. On the other hand, it is difficult to see how in a production process by particles an imp could be produced which was much larger than its parents. Therefore the simplest picture is that the size of an imp will increase indefinitely with time. This will guarantee that there are no asymptotic states.

This phenomenology surely depends on the rate of expansion. However, without a compelling mechanism which would depress the expansion rate the reasonable assumption is that the radius grows as a reasonable fraction of the speed of light.

ii) Since the size of an imp increases with time, its internal structures, such as mass and charge distributions, must also vary with time, possibly in a complicated manner. What can we say about the variation of such internal structures? In the case of the quark-imps, all that we know is that they are not detected by ionization devices located far away from the points of production. These ionization devices are sensitive to a point particle of charge $\pm e$, but not to $\pm\frac{1}{3}e$. Thus, the charge distribution of a large quark-imp may be smooth or granular with localized total charges of no more than $\frac{1}{3}e$. In either case, when the radius of the quark-imp reaches the size of optical wavelengths it will cease to couple effectively to optical photons and cannot be detected in ionization devices such as bubble chambers or spark chambers. Such large particles could carry off unobserved charge and would give rise to reactions which appear to violate charge conservation.

iii) For a short time after its production the imp will ionize normally and can be seen in nuclear emulsions. These tracks will have the characteristic that the terminus does not show increasing ionization or star but just fizzles out. If the expansion rate were $\frac{1}{10}$ c, and if a 400GeV proton fragmented into 3 quark imps with their branch points located at $\sim \frac{1}{3} m_p$ then a track length of a fraction of a millimeter could be observed in silver bromide.

iv) Far away from the production vertex imps can be detected by their possible electromagnetic, strong and weak interaction. In the first case a gas

Cerenkov counter just above threshold may be used. In the other two cases, by a suitable combination of bending magnets and shields, charged imps may be transported to regions of space inaccessible to neutral particles. Possible inelastic interactions of imps could then be observed with, for example,a hadron calorimeter.

v) In the above discussion we have tacitly assumed a linear rate of growth for imp radii. In practice there will be a distribution of growth rates related to the time dependence of the internal structure. This will lead, for example, to a distribution of track lengths in nuclear emulsions.

vi) With a growth rate of $\frac{1}{10}$ c it is unlikely that imps can be observed in a magnetic levitometer experiment. However, if in nuclei or in macroscopic matter the growth of imps can be arrested and /or the imps can be bound then the levitometer experiments can succeed.

The phenomenology outlined above has made many rather arbitrary assumptions which are certainly open to modification. We will discuss several modifications explicitly.

1) Do imps have to be smooth? As already mentioned there seems to be nothing to force us to assume that an imp cannot have one or more expanding or non-expanding granular centers (or even strings or discs) in addition to the expanding cloud. Such centers will facilitate experimental detection. It is possible that a granualr center could even leave a track in a sensitive ionization device. It is therefore important to emphasize that it takes more than the observation of a track to demonstrate the existence of a particle. For a track to be called a particle it must respond to classical external forces according to F = ma, with a well defined mass. Because of the mass indeterminacy and because of the coherence, a non-expanding center in an imp cannot respond to a classical field independently of the rest of the cloud of the

imp. Such a track from a granular center in an imp is expected to have a non-conventional kinematics.

2) Can imps oscillate in time: It is possible to avoid asymptotic states and also indefinite expansion if the size of the imp oscillates. Such an object could have the property of leaving a track consisting of isolated parts separated by regions of invisibility.

3) Is the rate of imp expansion independent of the production process? One can argue that if imps have strong interactions their mass uncertainty may be of the order of the pion mass. If this is the case, then, because of the smallness of the pion mass, the rate of expansion will be roughly independent of the production process.

4) Can imps be focussed? If the growth of imps can be arrested in matter there may be other ways to stop the growth, or even to make the imp reduce in size. For incoherent beams, such a reduction in beam size is achieved through, for example, strong focussing using quadrupole magnets. It may be interesting to inquire whether the coherence of the imp prevents focussing. On the other hand, if imps can be focussed we can envisage the possibility of a matter transmitter.

5) Can large imps interact? We have already argued that large imps will not interact with photons. On the other hand, because of their coherence properties it may be that large imps can interact strongly with each other.

6) Can imps be produced singly? Our discussion of imp production has tacitly assumed that they be produced in pairs or more. For quarks such pair production is to be expected. However for bosons it is not clear to us if single imps can be produced.

7) What role can imps play in cosmology? In particular there appears to be much more matter than anti-matter in the universe. Could the missing

anti-matter have been carried off by imps?

8) Can charge density oscillate in space in an imp? More generally one can ask the relation of the distribution of mass, charge, and strong and weak coupling constants.

9) Is there possibly a finite level spacing and hence a maximum size of imps?

All of these questions, and many more, are open to speculation. Roughly speaking, the major feature of this novel physics is that the experimental results depend in an interesting way on the distance to the production vertex. If quarks are indeed imps, there seem to be many possibilities for new and dramatic physical effects to occur.

Acknowledgments

One of us (TTW) thanks Professor Chen Ning Yang and Professor Tsien San-tsiang for the kind invitation to the 1980 Guangzhou Conference on Theoretical Particle Physics, and Professor Hung Cheng for collaboration on the material presented in Sec. 2.

REFERENCES

[1] S. Tomonaga, Prog. Theor. Phys. 1, 27 (1946); Phys. Rev. 74, 224 (1948); J. Schwinger, Phys. Rev. 73, 416 (1948); 74, 1439 (1948); R. P. Feynman, Rev. Mod. Phys. 20, 367 (1948); Phys. Rev. 74, 939 and 1430 (1948); F. J. Dyson, Phys. Rev. 75, 486 (1949).

[2] K. Symanzik, in New Developments in Quantum Field Theory and Statistical Mechanics (ed. M. Levy and P. Mitter, Cargese 1976) p. 265.

[3] W. E. Thirring, Ann. Phys. 3, 91 (1958); Nuovo Cim. 9, 1007 (1958); V. Glaser, Nuovo Cim. 9, 990 (1958); K. Johnson, Nuovo Cim. 20, 773 (1961); C. M. Sommerfield, Ann. Phys. 26, 1 (1964); B. Klaiber, Helv. Phys. Acta 37, 554 (1964); and in Lectures in Theoretical Physics, Vol. XA, (ed. A. Barut and W. Britten, Gordon and Breach, New York, 1968) p. 141.

[4] B. M. McCoy, C. A. Tracy, and T. T. Wu, Phys. Rev. Lett. 38, 793 (1977); M. Sato, T. Miwa, and M. Jimbo, Proc. Japan Acad. 53A, 6 (1977).

[5] E. Ising, Z. Physik 31, 253 (1925); L. Onsager, Phys. Rev. 65, 117 (1944); Chen Ning Yang, Phys. Rev. 85, 808 (1952); B. M. McCoy and T. T. Wu, The Two-Dimensional Ising Model (Harvard University Press, Cambridge, Mass.,1973).

[6] B. M. McCoy and T. T. Wu, Phys. Lett. 72B, 219 (1977); Phys. Rep. 49C, 193 (1979).

[7] H. Cheng and T. T. Wu, Phys. Rev. Lett. 24, 1456 (1970).

[8] I. O. Stamatescu and T. T. Wu, Nuclear Phys. B143, 503 (1978).

[9] I. Bialynicki-Birula, Nuovo Cimento 10, 1150 (1958); J. Schwinger, Phys. Rev. 128, 2425 (1962).

[10] T. D. Lee, Kerson Huang, and C. N. Yang, Phys. Rev. 106, 1135 (1957).

[11] H. Cheng and T. T. Wu, Phys. Rev. Letters 22, 666 (1969); Phys. Rev. 182, 1852, 1868, 1873, and 1899 (1969).

[12] N. Byers and Chen Ning Yang, Phys. Rev. 142, 976 (1966); T. T. Chou and Chen Ning Yang, in Proceedings of the Second International Conference on High Energy Physics and Nuclear Structure, Rehovoth, Israel, 1967 (ed. G. Alexander, North-Holland, Amsterdam, The Netherlands, 1967); and Phys. Rev. 170, 1591 (1968); J. Benecke, T. T. Chou, Chen Ning Yang, and E. Yen, Phys. Rev. 188, 2159 (1969).

[13] T. Regge, Nuovo Cim. 14, 951 (1959) and 18, 947 (1960); For a review of Regge phenomenology up to 1968, see V. D. Barger and D. B. Cline, Phenomenological Theories of High Energy Scattering - An Experimental Evaluation (W. A. Benjamin, Inc., New York, 1969).

[14] U. Amaldi et al., Phys. Lett. 44B, 112 (1973); S. R. Amendolia et al., Phys. Lett. 44B, 119 (1973).

[15] T. T. Chou and Chen Ning Yang, Nuclear Phys. B107, 1 (1976).

[16] H. Cheng amd T. T. Wu, Phys. Rev. D1, 467 (1970).

[17] M. Froissart, Phys. Rev. 123, 1053 (1961).

[18] H. Cheng, J. K. Walker, and T. T. Wu, contribution to the 16th International Conference on High Energy Physics (1972); H. Cheng and T. T. Wu in Fifth International Conference on High Energy Collisions, Stony Brook (ed. C. Quigg, AIP Conference Proceedings No. 15, 1973).

[19] T. T. Wu, in the Proceedings of the 18th International Conference on High Energy Physics, Tbilisi (1976), p. A5-23.

[20] H. Cheng and T. T. Wu, Phys. Rev. Lett. 23, 1311 (1969).

[21] Theory: H. Cheng, J. K. Walker and T. T. Wu, Phys. Lett. 44B, 283 (1973); Experiment: Bartenev et al., Phys. Rev. Lett. 31, 1367 (1973).

[22] Theory: Reference [6];C. Rubbia et al., contribution to the 16th International Conference on High Energy Physics (1972).

[23] Theory: H. Cheng and T.T.Wu, Phys. Lett. 45B, 367 (1973); Experiment: K. Guettler et al., Phys. Lett. 64B, 111 (1976).

[24] B. M. McCoy and T. T. Wu, Scientia Sinica 22, 1021 (1979); and in Bifurcation Phenomena in Mathematical Physics and Related Topics , Proceedings of the NATO ASI at Cargese (ed. C. Bardos and D. Bessis, 1979).

[25] R. P. Feynman, Doctoral Dissertation, Princeton University (1942); Phys. Rev. 80, 440 (1950); I. M. Gel'fand and A. M. Yaglom, Uspekhi Math. Nauk 11, 77 (1956) English translation; J. Math. Phys. 1, 48 (1960); L. D. Faddeev in Methods in Field Theory (ed. R. Balian and J. Zinn-Justin, North-Holland Publishing Co., 1976) p. 1; V. N. Popov, Functional Integrals in Quantum Field Theory and Statistical Physics (Atomicdat, Moscow, 1976).

[26] M. Creutz, L. Jacobs, and C. Rebbi, Phys. Rev. Lett. 42, 1390 (1979).

[27] L. D. Landau in Niels Bohr and the Development of Physics (ed. W. Pauli, Pergamon Press, London, 1955) p. 52.

[28] H. Lehmann, K. Symanzik, and W. Zimmermann, Nuovo Cimento 2, 425 (1955); K. G. Wilson and J. Kogut, Phys. Reports 12C, 75 (1974).

[29] G. A. Baker and J. M. Kincaid, Phys. Rev. Lett. 42, 1431 (1974).

[30] B. M. McCoy and T. T. Wu, Phys. Lett. 87B, 50 (1979).

[31] A. Luther, Phys. Rev. B14, 2153 (1976).

[32] M. Lüscher, Nucl. Phys. B117, 475 (1976).

[33] P. Jordan and E. Wigner, Z. Physik 47, 631 (1928).

[34] Chen Ning Yang and Chen Ping Yang, Phys. Rev. 150, 321 and 327 (1966); and 151, 258 (1966).

[35] J. des Cloizeaux and M. Gaudin, J. Math. Phys. 7, 1384 (1966); J. D. Johnson, S. Krinsky, and B. M. McCoy, Phys. Rev. A8, 2526 (1973).

[36] E. Barouch, B. M. McCoy, and T. T. Wu, Phys. Rev. Lett. 31, 1409 (1973); C. A. Tracy and B. M. McCoy, Phys. Rev. Lett. 31, 1500 (1973); T. T. Wu, B. M. McCoy, C. A. Tracy, and E. Barouch, Phys. Rev. B13, 316 (1976).

[37] B. M. McCoy and T. T. Wu, Phys. Rev. D18, 1243 (1978).

[38] B. M. McCoy and T. T. Wu, Phys. Rev. D18, 1253 (1978).

[39] B. M. McCoy and T. T. Wu, Phys. Rev. D18, 1259 (1978).

[40] S. Coleman in New Phenomena in Subnuclear Physics (ed. A. Zichichi, Plenum Press, 1977) Part A, p. 351.

[41] M. Gell-Mann, Phys. Lett. 8, 214 (1964); G. Zweig, CERN preprints Th.401 and TH.412 (1964).

A REVIEW ON THE RESEARCH IN CLASSICAL GAUGE FIELD THEORY IN CHINA DURING THE RECENT YEARS

Hua-chung Lee
Zhongshan University
Yong-shi Wu
Institute of Theoretical Physics, Academia Sinica.
Ting-chang Hsien
Institute of High Energy Physics, Academia Sinica.
Bo-yu Hou
Northwest University
I-shi Duan
Lanzhou University
Han-ying Guo
Institute of Theoretical Physcis, Academia Sinica.

This is a review whose main purpose is to introduce part of the results of the research in classical gauge field theory done in China during the recent years. Being limited by both the level of us the reviewers and the time we had in preparing this review, we have not been able to review all the works in this field thoroughly and exactly -- even some important works might have been missed.

This is what we must apologize to the authors beforehand, in particular to the mathematicians of Fudan University, whose works in cooperation with Prof. Yang had been familiar to scientists in China and abroad, and some of them had been reviewed on the 19th International Conference on High Energy Physics at Tokyo, 1978, hence, we take the liberty of not repeating them here. The mathematicians of Fudan University have done much on the mathematical structure of the gauge field theory, on which Prof. Gu and Hu have reported during their recent visit abroad, therefore, we only quote these results briefly from their review article*. There are also some investigations on the monopole atom, however, as they apparently do not belong to the gauge field theory of the monopole, hence we have not included them in the present review. Results of the research on classical gauge field theory in China have been mainly published on the four main Chinese jounals:"Scientia Sinica", "Acta Physica", "Physica Energiae Fortis et Physica Nuclearis" and "Kexue Tongbao"**, they have numbered about 50 up to October, 1979. Besides, there have been also 17 papers on the gauge theory of gravitation published on these journals. As the gauge theory of

---

* "Mathematical Structure of Gauge Fields", Preprint of the Institute of Mathematics of Fudan University, 1978, Mar.

** Papers published in other journals of various Universities have not been listed here. Of course, apart from the classical gauge field theory, there are other directions of research in the gauge field theory, they are not included in the present review.

gravitation is an independent and interesting field, we have listed them separately. Finally, as a supplement to the present review, we have listed five review articles published earlier in China in the reference*.

At times we will point out some concepts and methods as well as some conclusions made independently by Chinese physicists. Some of them were either published at about the same time, or at a later earlier or latter time, as the papers containing the analogous results published abroad. Most of the conclusions reached by the Chinese authors are more or less in accordance with those published abroad. We will give a list of comparison in what follows. This review will cover the following topics:

1. The dual charge, the topological properties of gauge fields,
2. Basic quantities of gauge fields,
3. Classical solutions of the gauge field equations,

---

* The notations used for the references quoted in the present review are as follows: [A] - papers published in China; [AG] - papers on the gauge theory of gravitation published in China; [R] -review articles published in China; [C] -papers published abroad which are quoted to compare with the results obtained by Chinese authors. Papers presented to the present conference are not reviewed in this article.

4. Fibre bundle treatment of the topological properties of gauge fields,
5. The gauge theory of gravitation,
6. Other topics: the angular momentum operator and the wave function of the charged particle around a magnetic monopole; gauge field on the coset space; etc...

I. DUAL CHARGE AND THE TOPOLOGICAL PROPERTITIES OF GAUGE FIELDS

The terminology of "dual charge" had been widely adopted in the Chinese literature, which corresponds to what is called by "topological charge" nowadays. The concept of "dual charge" was introduced in China [A2], [A3], [A9] to describe the magnetic charge-- which is the dual of the electric charge- and the "magnetic monopole charge" -- which is the dual of the self interacting charge of the non-abelian gauge fields. Because the concept of topology in mathematics was not known to the authors then, hence the starting point was some sort of intuitively simple and rather naive imagination. Lee, Hsien and Kuo[A2] considered a two dimensional spherical surface S^2 and associated a frame at each point on the surface, representing the coordinate system of the SO(3) group. Now Carry out a parallel displacement of a vector along the equator around the surface once, the vector might change by an angle. Imagine the sphere is divided by the equator into two regions-- the upper and the lower halves --, if the angle changed by parallely displacing a vector along the equator round the upper half sphere and along the equator round the lower half sphere are different, then they can differ only by an integer factor of 2π, this integer defines a charge -- the so called dual charge. This case corresponds to the "non-trivial topology" discussed later. In the language of gauge field theory, let the vector on the spherical surface be the basic vector of the frame of the SO(3) gauge group, then according to the correspondence between the parallel displacement of the vector and the gauge potential:

$$\Gamma^{i}_{j\mu} \longleftrightarrow e\, W^{a}_{\mu} (T_a)^{i}_{j} \tag{1}$$

where $\Gamma^{i}_{j\mu}$ represents the Christoffel symbol on the surface, characterizing the parallel displacement, T_a the generators of the gauge group. $\Gamma^{i}_{j\mu}$ can be calculated in the differential geometry, and hence the gauge potential, and one can prove that the gauge potential determined in this way is just the one corresponding to a dual charge g, moreover

$$ge = n \tag{2}$$

where e is the self coupling constant and n an integer.

In fact the gauge potential thus obtained is just the static sourceless solution found by Wu and Yang[C2]. From this we see that the Wu-Yang solution is related to the dual charge and is equivalent to the U(1) monopole gauge: under certain gauge transformations it can be changed into U(1) magnetic monopole solutions — the single singular string electromagnetic potential of Dirac or the double singular string potential of Schwinger. When such gauge transformations exist, the non-abelian group is called abelianizable. Hou[A4] pointed out the abelianizability of a gauge group is the existence of a vector $n_\mu (x)$ which is invariant under parallel displacement

$$D_\mu n_\mu = 0$$

where D_μ represents the covariant differentiation. This invariant vector is called generally the Higgs field. Thus, from the view point of non-abelian gauge field theory, the singular string is superficial.

These concepts and methods, in particular the correspondence method between the parallel displacement on the spherical surface and the gauge potential, were extended to the case of the N dimensional space with an SO(N) gauge symmetry group to study the dual charge, the synchrosphericosymmetric gauge potentials with or without singular strings, etc, therein. Also, the SU(2) magnetic monopole with the O(5) symmetry, the instanton solution in the E^4 space, etc, were derived[A7], [A9], [A37] in this way, which will be mentioned later in Sec. III.

At the same time, Duan, Ge and Hou [A3] discussed the topological properties of the SU(2) gauge fields in the apparent existence of the Higgs field. They proposed a method of decomposing the gauge potential and came to the conclusion that the dual charge (magnetic monopole charge) is related only to the topology of the Higgs field. This important conclusion was also made by Arafune, Goebel and Freund [C3]. [A3] was done independently. In the existence of the Higgs field, the SU(2) gauge potential can be decomposed as

$$
\begin{aligned}
W_\mu(x) &\equiv A_\mu(x) + B_\mu(x) \\
A_\mu(x) &\equiv (W_\mu \cdot n)\; n - \frac{1}{e}[n, \partial_\mu n] \qquad (3) \\
B(x) &\equiv \frac{1}{e}[n, D_\mu n] \\
n &\equiv \frac{\phi(x)}{|\phi(x)|} \qquad (4) \\
\phi(x) &= \phi^a T^a \qquad (5)
\end{aligned}
$$

The physical interpretation is : A_μ represents the electromagnetic potential which produces the electromagnetic field $f_{\mu\nu}$; B_μ the gauge covariant potential of the charged vector field. The electromagnetic tensor $f_{\mu\nu}$ can be written in the form

$$
\begin{aligned}
f_{\mu\nu} &= n\cdot(\partial_\mu A_\nu - \partial_\nu A_\mu + e[A_\mu, A_\nu]) \qquad (6) \\
&= \partial_\mu(A_\nu\cdot n) - \partial_\nu(A_\mu\cdot n) - \frac{1}{e} n\cdot[\partial_\mu n, \partial_\nu n]
\end{aligned}
$$

and the complete gauge field strength $G_{\mu\nu}$ has the form

$$
\begin{aligned}
G_{\mu\nu}(x) = f_{\mu\nu}(x)n(x) + e[B_\mu, B_\nu] + \{\partial_\mu B_\nu + e[A_\mu, B_\nu] \\
- \partial_\nu B_\mu - e[A_\nu, B_\mu]\} \qquad (7)
\end{aligned}
$$

Noticing that the third term is perpendicular to n and one finds

$$f_{\mu\nu} = \frac{\phi}{|\phi|} G_{\mu\nu} - \frac{\phi}{|\phi|^3}[D_\mu \phi, D_\nu \phi] \tag{8}$$

which is the expression for the electromagnetic component in the SU(2) gauge field theory proposed by 't Hooft [C5]. It is clear that the direction of n(x) defines the electromagnetic subgroup U(1), W_μ and $G_{\mu\nu}$ are locally decomposed with regard to this subgroup, and the unit Higgs field n (x) gives the Kronecker mapping from Minkowski space to the coset space $SU(2)/U(1) \sim S^2$. Integrating $f_{\mu\nu}$ over various closed space-like surfaces, it is easily seen that the mapping degree and the homotopic property of the mapping determine the magnetic charge of the monopole, and one thus comes to the conclusion that the magnetic charge is related only to the topology of the Higgs field. Duan and Ge [A39] pointed out that the existence of the magnetic monopoles and their motion in space is determined by the zeros of the Higgs field $\phi(x)$ and their trajectories and proved that the magnetic charge density is of the form of $\delta(\phi(x))$, i.e., point sources of magnetic charge with δ-function singularity are produced at points where $\phi(x) = 0$. The topological property of the mapping of the point set of the ordinary space manifold to the isospace is related directly to the covering number and the number and distribution of the zeros of $\phi(x)$. They also wrote down the differential form of the conserved topological magnetic charge current. Hou [A6] derived the gauge independent conserved Noether current. Later, the above mentioned decomposition was extended to the cases of SU(N) and SO(N) [A35].

Gu [A27] studied the U(1) gauge field in the SU(N) theory in the presence of the Higgs field and the quantized values of its dual charge in the theory. Later, his discussion was extended to the case of general compact gauge groups. Thus the investigation initiated by 't Hooft was extended to very general cases and the problem of the values of the dual charge and their geometrical meaning is completely resolved.

II. BASIC QUANTITIES OF GAUGE FIELDS*

The basic quantitites in the gauge field theory are the field strength, the gague potential and the field source. In the abelian gauge field theory, the basic physical quantity is the field strength, as to the gauge potential, there exists the gauge uncertainty. Such uncertainty exists both for the field strength and the gauge potential in the non-abelian gauge field theory. Morevoer, as Wu and Yang [C6] had pointed out that there exist two gauge non-equivalent potentials corresponding to the same field strength in the SU(2) case, hence the gauge field is not completely determined by the field strength. The problem raised is that what are the most appropriate quantities to describe the gauge field in order to minimized the gauge uncertainty? and to what extent does the field strength determine the gauge field? Xia [A29],

* Many presentation in this section are quoted from the review article [R2].

Hu[R2], Yan[A42] and Shen[R2] made a thorough study of these problems and concluded that for the SU(2) synchrospherico-symmetric gauge field, the field strength and the source can determine the gauge field. In many cases, the field strength and the source cannot determine the gauge field. For an arbitrary non-abelian group G, there always exist an infinite number of non-equivalent sourceless potentials corresponding to the same field strength. Gu and Yang[A28] proved that for a general group, the field strength and its gauge derivatives up to the p-th order can always determine the gauge field, where p = 2 for the cases of SU(2) and SU(3), and p assumes different values for different groups. Hou[A12] analysed the common physical characteristics and the geometrical interpretation of a theory with non-equivalent gauge fields corresponding to the same field strengths and external sources.

The Bohm-Aharonove experiment showed that even in the abelian case the field strength in a multi-connected space cannot determine the electromagnetic field, the same field strength might correspond to different physical circumstances. Wu and Yang[C4] pointed out that the phase factors of closed path loops are the appropriate basic quantity to describe the electromagnetic field. Gu[A14] proposed another basic quantity to describe general gauge fields: the phase factors of closed path loops passing a fixed point. They defined an equivalent class of the gauge fields, and the gauge uncertainty reduces to the adjoint transformations of the gauge group.

For the electromagnetic field in the multi-connected region, the field strength together with a set of complete and independent phase factors of closed loops just determine the electromagnetic field. For the non-abelian cases, it can be proved that the gauge field strength and their gauge derivatives up to a certain order, together with a set of independent basic phase factors of closed loops can determine the gauge field.

Wu[A24] called the phase factors the holonomy functionals, which are the elements of the holonomy group. Starting from Yang's integral formalism of gauge fields, he was able to reformulate the equations of motion for gauge fields as functional differential equations of the phase factors of closed loops. As a preliminary application of these equations, he proved that there exists a set of infinite number of conservation laws in the functional differential form for the gauge field in a two dimensional space-time.

The physical significance of the phase factors of closed loops as the criteria of the phenomena of the confinement of quarks was emphasized by Wilson [C7]. Gu's work gave an exact and more general definition in the mathematical sense to the phase factors of closed loops and discussed their significance in the view of the basic quantity to describe the gauge fields.

III. CLASSICAL SOLUTIONS OF THE GAUGE FIELD EQUATIONS

The research in the classical solutions for gauge field equations in China was started in 1974. Since then a number of papers have been published in this field, including the synchrospherico-symmetric solutions, the instanton solution, multi-instanton solutions and other solutions in various forms, etc. In 1975, in discussing the problem of the dual charge, Lee, Hsien and Kuo developed a method to acquire static synchrospherico-symmetric solutions of the gauge field theory in an N dimensional space-time with SO(N) gauge symmetry [A2], [A9]. This was known later as the "factor space method" or the "coset space method", in which the spherico-symmetric gauge field is written as $W_\mu^a(x) = f(r)\,\overline{W}_\mu^a(\theta,\varphi)$, where the "kinematic" factor $\overline{W}_\mu^a(\theta,\varphi)$ of the gauge potential is completely determined by the invariance under the synchronous rotation of the configuration space and the isospace, in a way just as the kinematic factor of the S-matrix is completely determined by the Lorentz invariance. Another factor f(r) is determined by a simplified equation of motion, and from which one obtains solutions more general than those obtained by starting from the self-dual condition. To determine the "kinematic" factor one first uses the Riemanian geometry on an S^{N-1} surface to find out the connection in the factor space $SO(N)/SO(N-1) \sim S^{N-1}$, and then by the correspondence relation of the gauge potential and the connection derives the potential. This method was applied to the gauge groups SO(3), SO(4), SO(5) and SU(N) to obtain the Wu-Yang

SU(2) static sourceless solution, the SU(2) magnetic monopole with SO(5)symmetry similar to what was obtained by Yang [C8], the relation between various solutions with single singular string, double singular string as well as without singular string and thus the understanding of the superficial nature of the singular string from the view point of non-abelian gauge groups. The BPST instanton solution for the SO(4) case was also derived by using this method [A7], and was argued [A19] to coincide with the result obtained by Jakiw and Rebbi [C9]. It was shown that only in the cases of N = 4 and 10 (four and ten dimensional space-time) can the solution be of finite action. After the completion or publication of these works, the authors found some similar or identical conclusions from papers of Wilczek [C10], Cremmer and Scherk [C11], Deser [C12], and Arafune et al [C3].

Gu and Hu [A5], [R2] discussed the general spherico-symmetric gauge field in the most general mathematical sense and gave a method to determine the general structure of the gauge fields when the gauge group G contains a symmetric group H of space-time. This method was applied to determine various types of spherically symmetric gauge fields, in particular it gave the complete classification of the SO(2) spherical gauge fields into three types:

1. strict-spherically symmetric,
2. synchro-spherically symmetric and
3. spherically symmetric reducible to U(1).

They also wrote down the SU(N) spherically symmetric gauge fields, including the SU(3) spherically symmetric gauge field given in [C13] . Xia [A29] , [A40] and Yan [A42] discussed the general expression of the gauge potential determined from the field strength and field source in more detail. They analyzed all SU(2) solutions in the sourceless case and in the cases with source, gave the general expression for the solution if ever it exists as well as many interesting examples. These works were done earlier than the works of analogous content published abroad [C27], [C28], [C29].

By using the general form of the synchrospherically symmetric gauge fields given by Hu [R2], Peng [A30] [A41] derived the self-dual and anti-self-dual SU(2) instanton solutions with arbitrary integer topological number q, which is also a kind of synchrospherically symmetric gauge field. After the completion of this work, the autor found that his result was similar to that in Witten's paper [C14]. The $q > 1$ instanton solutions were first found in these papers.

The dual condition were reduced to three nonlinear harmonic-like equations under appropriate gauge by Yang [C15]. Based on [C15] Wu and Peng [A16] reformulated these equations to simplify the ansatz for solution and provided a more explicit view of the relation between the existing solutions and the Yang equations.

Wang, Chang and Hou [A18] obtained the multi-magnetic monopole solution in the SU(2) gauge field theory. Wang and Hou [A25] obtained the multi-synchrospherically symmetric solutions. Wang[A44] obtained the SU(4) instanton solution with $q = \pm 2$. Wang, Hsien and Zheng [A20] obtained the SU(4) magnetic monopole and dyon solutions, and in another paper [A15] showed that in a curved space-time, the coupled non-abelian gauge field equations and the Einstein equations allow an abelianizable point magnetic monopole solution. Hou and Hou[A23] obtained the zero energy solutions of a fermion with various isospin in the background field of point magnetic monopole and extended magnetic monopole. Wang, Zheng, Wang, Zhang and Hsien [A10] discussed the bound state solution of the monopole and the spinor field. In contrary to the conclusion made in [C 16], they proved that for the scalar coupling of the spinor field and the Higgs field, and under a certain restraint on the magnitude of the coupling constant, the bound state exists. Zhu [A13] discussed the magnetic monopole potential of the meridian form, which is different from the lattitude form given by Dirac, though free from singular string, but is multi-valued in the azimuthal angle φ. Tu and Ruan [A11] discussed the potential of a magnetic monopole moving with arbitrary acceleration and pointed out the difference between the potentials of a magnetic monopole and an electric charge moving with varying velocity.

IV. FIBRE BUNDLE TREATMENT OF THE TOPOLOGICAL PROPERTIES OF GAUGE FIELDS

The geometrical concept of the fibre bundle in mathematics and the physical concept of gauge fields were originally developed independently one of the other. However, now it becomes clear that the gauge fields are none other than the connections on the fibre bundles, and the mathematical theory of fibre bundles is a very appropriate and powerful tool for dealing with the topological complexity of gauge fields. With deeper understanding of the importance of the topological properties of gauge fields, people believe more strongly that not only the gauge fields can be expressed in the language of fibre bundles, but very probable all the properties (especially the global topological properties) of gauge fields cannot be clarified without the help of the theory of fibre bundles.

Chinese physicists' attention to the theory of fibre bundles was drawn by a talk on the integral formalism of gauge fields given by Prof. C.N.Yang in the summer of 1972, in which he metioned that a mathematician had remarked to him that gauge theories might have a kinship to the theory of fibre bundles. Afterwards mathematician Look worked out the correspondence relation between the mathematical theory of connections on the principal bundles and the physical idea of gauge fields [A1]. In particular, he showed the equivalence of Yang's integral formalism of gauge fields [C17] and the parallel displacement of (or the horizontal lift to) the principla bundle, thus arriving at a unified treatment of the differential and integral formalism of the gauge field theory by means of the theory of connections, In his lectures [R1], instead of the abstract mathematical formulation, Look presented a concrete description of fibre bundles in terms of local coordinates, which is much more acceptable to the physicists. In 1975, Wu and Yang [C4] demonstrated the relation between the topological properties of the fibre bundles and the monopoles in physics, and thus further revealed the usefulness of the fibre bundle theory in the theory of gauge fields. Based on these works, a series of efforts were made in studying the fibre bundle treatment of the topological properties of gauge fields.

When the topology of a fibre bundle is nontrivial, it is insufficient to describe the bundle by using only one coordinate patch, just like a two dimensional sphere in the theory of manifolds has to be described by using at least two coordinate patches to avoid singularities. Wu and Yang [C4] introduced the idea of using several coordinate patches to describe a global gauge, and gave a description of magnetic monopoles without the Dirac singular strings. This demonstrates the interplay of the nontrivial topology of a fibre bundle and physics. Wu, Chen, Tu and Guo [A8], [A31], [A33], [A46], [A48] discussed not only the Dirac monopole, but also the BPST instanton [C20] and the 't Hooft monopole [C5] with the help of the fibre bundle theory. They first made a systematic topological classification of gauge fields by using a classification theorem in the theory of fibre bundles. According to this theorem, the equivalent classes of the principal bundles $P(S^n, G)$ can be characterized by the homotopic classification of the characteristic maps of these bundles. The physical significance is that the gauge types of the gauge fields on S^n with G as the gauge group are in one-to-one correspondence to the elements of the homotopic groups $\pi_{n-1}(G)$

In some cases, the element of $\pi_{n-1}(G)$ can be characterized by the mapping degree D of the map from S^{n-1} into G. For example, D = 2eg = the first Chern class for the principal bundles $P(S^2, U_1)$ [C4], [A33]; D = no. of instantons = the second Chern class [A33] for the principal bundles $P(S^4, SU_2)$. Thus one obtains the topological classification of electromagnetic fields characterized by the magnetic charge satisfying the Dirac quantization condition, and that of the SU(2) gauge fields on E^4 (which becomes S^4 after one-point compactification) by the number of instantons. The topological structures of some principal bundles were also discussed in these papers. Wu and Guo [A31] showed that the principal bundle $P_D(S^2, U_1)$ corresponding to the magnetic charge g ($g = D/2e$) is actually isomorphic to the Hopf bundle S^3/Z_D (Z_D being the cyclic group of order D), and if one assigns properly a metric to this bundle for a static monopole, then the bundle $P_D(S^2, U_1)$ can also be viewed as S^3/Z_D in the sense of a Riemannian manifold [A31]. Generalizing to the one instanton case, the corresponding principal bundle $P_1(S^4, SU_2)$ with $D = 1$ is isomorphic to the Hopf bundle S^7, and, by assigning a proper metric, it can be viewed as an S^7 directly[A31].

The physical significance of studying the structure of the principal bundles corresponding to the monopole and instanton is that the expressions for the angular momentum operator of a particle moving in the field of a monopole or an instanton can be derived thereby [A33], [A48]. In quantum mechanics, the expression for the angular momentum operator of a particle is derived from the infinitesimal generators of the rotation group SO_3, which leaves the 3-space(or S^2) invariant. In the case of a static monopole, the principal bundle $P(S^2, U_1)$ constitutes an S^3 itself, so the symmetry group leaving not only S^3 but also the bundle structure of S^3 invariant is a subgroup $U_2 \approx SU_2 \times U_1$ of SO_4, which has four infinitesimal generators. The generator corresponding to U_1 is shown to be the ordinary charge operator, and the other three corresponding to SU_2 are identified with the angular momentum operators of a charged particle around the monopole:

$$\vec{L} = \vec{r} \times (\vec{p} - e\vec{A}) - eg\,\frac{\vec{r}}{r} \tag{9}$$

where $\vec{A}$ is the Wu-Yang potential for the static monopole. Eq. (9)had been conjectured for a long time in order to obtain a correct commutation relation of the angular momentum operators. Now it is derived directly from the invariance of the structure of the principal bundle corresponding to a monopole, and the last term in eq.(9)appears naturally.In addition,the Wu-Yang monopole harmonics[C18]and their relationship to the wavefunctions of a symmetric top[A43]are also derived by the bundle technique. Similarly,the symmetry group leaving both the metric and the bundle structure of the one-instanton principal bundle S^7 invariant is Sp(2),which is locally isomorphic to SO(5). In [A48] the expression for the angular momentum operators of a particle in the one-instanton background field was derived to be

$$L_{\mu\nu} = -i\{x_\mu(\partial_\nu + A_\nu) - x_\nu(\partial_\mu + A_\mu)\} - ier^2 F_{\mu\nu} \qquad (10)$$

just of the same form as conjectured by Yang [C19].

For the spontaneously broken gauge theories with Higgs fields, the analysis based on the theory of fibre bundles is very useful too. In [A13] the connection between an SU(2) bundle and its U(1) subbundles was discussed with the help of the concept of the holonomy group, and SU(2) gauge fields were classified on the basis of this connection. It was pointed out in [A8] that according to theorems in the theory of fibre bundles, not only the relation between the value of the magnetic charge and the topolgy of the Higgs fields, but also a systematic procedure for obtaining the induced SO(2) gauge potentials (the electromagnetic potentials in the Georgi-Glashow model) from the SO(3) gauge potentials $\vec{W}_\mu$ and the isotriplet Higgs field $\vec{\phi}(x)$ can be derived. The result reads

$$A_\mu = \vec{\phi} \cdot \vec{W}_\mu - \frac{\phi_1 \partial_\mu \phi_2 - \phi_2 \partial_\mu \phi_1}{1 + \phi_3} \tag{11}$$

and the electromagnetic field-strength corresponding to it is exactly the gauge invariant one defined by 't Hooft:

$$F_{\mu\nu} = \partial_\mu A_\nu - \partial_\nu A_\mu = \vec{\phi} \cdot \vec{F}_{\mu\nu} - \vec{\phi} \cdot D_\mu \vec{\phi} \times D_\nu \vec{\phi} \tag{12}$$

where $\vec{\phi}(x)$ is normalized to $\vec{\phi}(x) \cdot \vec{\phi}(x) = 1$ Furthermore, it was concluded according to the theory of fibre bundles that the usual conclusion claiming the value of the magnetic charge depends only on the topology of Higgs fields is valid only for the trivial SO(3) bundle. Additional considerations are needed for the case of nontrivial SO(3) bundle.

The relation between gauge fields and vector bundles was also discussed [A44]. With the SO(N) broken gauge theories as an example, the Riemannian structure of the vector bundle was given, and the 't Hooft's definition of the electromagnetic field strength was obtained in this way. In broken gauge theories, the Higgs field belongs to some vector spaces on which the gauge group acts. Let the set of the corresponding vector spaces V_x at every point x in space-time M be denoted by E. According to the theory of fibre bundles, it is a vector bundle E(M,V,G,P,) with M as the base, V the fibre, G the structure group, and associated with the principal bundle P; the Higgs field is a cross section on E, $\varphi : M \to E$. The results obtained in this way are identical to those obtained in [A8] and [A46] by means of principal bundles.

V. THE GAUGE THEORY OF GRAVITATION

Since 1972, a series of researches on the gauge theory of gravitation have been made in China and about 20 papers have been published, in which various schemes with the Lorentz group[AG1, 3-5, 6-8, 9], the Poincaré group[AG 12] the de Sitter group[AG 7, 10, 11, 13] or the conformal group[AG 15] as the gauge group have been investigated. Some authors have also studied the supergravity and proposed a Yang-Mills type theory of conformal supergravity[AG 19].

One of the specialities in these works is that not only the curvature of space-time and the energy-momentum tensor of matter, but also the torsion of space-time and the spin current of the matter are considered, and Yang-Mills type gauge theories of gravitation are established by referring to Einstein's basic consideration in proposing his field equations.

It is well known that Einstein proposed that the geometric property of space-time should be determined by the mass of matter and he named this idea "the Mach principle" and regarded it as one of the basic principles in general relativity for estabishing his gravitational field equations. However, in deriving his equations, Einstein required that the conservation properties of the geometric quantities determined by the energy-momentum tensor of matter should be the same as those of the energy-momentum tensor itself. Since conservation properties are always related to certain symmetries, the above basic consideration of Einstein can be summarized as : The geometric properties of space-time is determined by the physical properties of matter corresponding to the same symmetries.

In a report on the gauge theory of gravitation [AG2], the authors named the above principle "the Einstein principle" and proposed that it be taken as one of the basic principles of the gauge theory of gravitation. According to this principle and considering the role of the energy-momentum tensor as well as the role of the spin current, which can also affect the geometric properties of space-time, and by including the torsion into the geometry in addition to the curvature and analyzing the symmetries corresponding to these quantities, the gravitational field equations should imply the following logical relations:

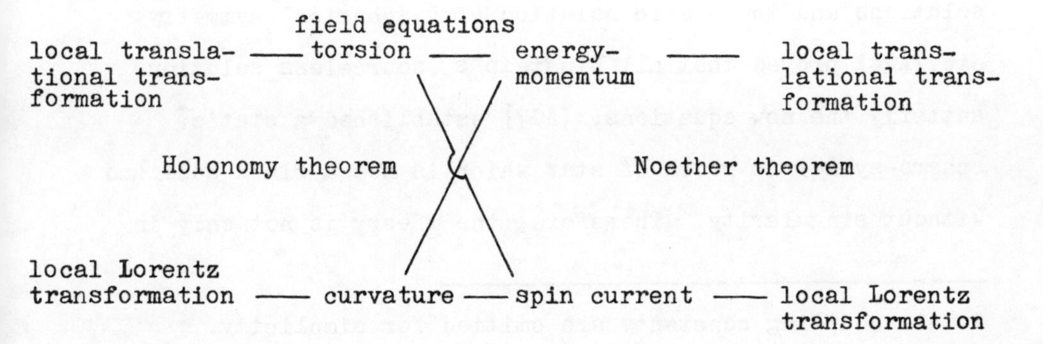

It is not difficult to show that these relations correspond to that of the usual Yang-Mills theory.

In 1973, based on the analysis of general relativity and Yang's sourceless gravitational field equations, Kuo, Wu and Zhang [AG3] proposed the following gravitational Lagrangian*

$$L_L = F - \frac{1}{4} F^{ab}_{\mu\nu} F^{\mu\nu}_{ab} \tag{13}$$

where the first term represents the Einstein-Cartan Lagrangian, $F = F^{ab}_{\mu\nu} e^{\mu}_{a} e^{\nu}_{b}$, $F^{ab}_{\mu\nu}$ the SO(3.1) gauge field strength, e^{μ}_{a} the Lorentz tetrad, and the second term the kinematic energy of the gauge field.The gravitational variables are the tetrad e^{μ}_{a} and the SO(3,1)gauge potential B^{ab}_{μ}. The field equations can be derived by means of the variational principle. Afterwards, Wu, Chen and Tsou [AG4], Zhang and Kuo [AG5], Huang and Kuo [AG8] Huang [AG9], Jiang [AG6], Tsou, Chen, Ho and Kuo [AG 1, 17] investigated the problems of the sourceless and torsion-free exact solutions of these field equations, the propagation of light in the gravitational field, cosmological solutions and the static solution with spherical symmetry, etc. [AG4] proved that all Einstein's sourceless solutions satisfiy the new equations. [AG7] established a static, sphero-symmetric model of star which is not a black hole and without singularity. Therefore, the theory is not only in

---

* All coupling constants are omitted for simplicity.

agreement with the existing experriments and observations, but also may lead to some results concerning the problems of the black holes and singularity which differ from those obtained from general relativity. Recently, some foreign authors have also proposed this theory and obtained some similar results [C23], [C24].

Based on the above scheme, the kinematic term of the torsion fields is introduced, and the Lagrangian has the form

$$L_p = F - \frac{1}{4} F^{ab}_{\mu\nu} F^{\mu\nu}_{ab} - \frac{1}{4} T^{a}_{\mu\nu} T^{\mu\nu}_{a} \tag{14}$$

(In a recent preprint, Nieh [C25] has also discussed this Lagrangian.)

Wu, Lee and Kuo [AG7] found that Lagrangian (14) with the cosmological constant Λ could be obtained from a Yang-Mills type Lagrangian with a local de Sitter gauge invariance. Kuo [AG10] presented a geometric interpretation to this local de Sitter invariance. An, Chen, Tsou and Kuo [AG11] proposed another de Sitter gravity. Guo [AG13] explained some problems in these two schemes. According to these works about the de Sitter gravity under a special de Sitter gauge condition, the de Sitter gauge potentials can be expressed as

$$B_\mu = \begin{pmatrix} B^{a}_{b\mu} & \theta e^{a}_{\mu} \\ e_{b\mu} & 0 \end{pmatrix} \in \text{de Sitter algebra} \quad \theta = \pm 1 \tag{15}$$

and the corresponding de Sitter curvature tensor is

$$F_{\mu\nu} = \begin{pmatrix} F^{a}_{\nu\mu b} + \theta e^{a}_{b\mu\nu} & \theta T^{a}_{jk} \\ T_{bjk} & 0 \end{pmatrix} \quad (16)$$

where $e^{a}_{b\mu\nu} = e^{a}_{\mu} e_{b\nu} - e^{a}_{\nu} e_{b\mu}$. At the same time, the Yang-Mills type Lagrangian of the de Sitter curvature can be decomposed as

$$\begin{aligned} L_{d} &= -\frac{1}{4} \operatorname{Tr} (F_{\mu\nu} F^{\mu\nu}) \\ &= F + 2\Lambda - \frac{1}{4} F^{ab}_{\mu\nu} F^{\mu\nu}_{ab} \end{aligned} \quad (17)$$

One can prove that the corresponding field equations also contain the Einstein's sourceless solutions with the cosmological constant. Moreover, due to the properties of the de Sitter group, the cosmological constant Λ comes from the global de Sitter invariance, since there is only one dimensionless coupling constant to be introduced to describe gravity, and thus the Newton gravitational constant can be composed from this dimensionless coupling constant and the cosmological constant. Recently, Townsend[C23] also discussed a de Sitter gravity with the same Lagrangian. Hsu[C26] also discussed this problem and obtained some interesting results.

Based on the de Sitter gravity, Chen, Guo, Li and Zhang [AG 15, 16] established a theory of conformal gravity and a conformal supergravity. The Lagrangian has been taken as a Yang-Mills type trace-invariant or super-trace-invariant respectively. [AG 19] pointed out that the Rarita-Schwinger Lagrangian can be obtained form this supertrace-invariant.

Recently, Guo and Zhang [AG12] gave a geometric interpretation of the L_L and L_P type Lagrangians from the viewpoint of Riemannian geometry and Riemann-Cartan geometry of the fibre bundles with corresponding gauge groups.

They pointed out that from a Riemannian scalar curvature of a Lorentz frame bundle follows the Lagrangian:

$$R_{LB} = R - \frac{1}{4} F^{ab}_{\mu\nu} \tag{18}$$

and L_L follows from a special Riemann-Cartan scalar curvature. Likewise, from the Riemannian scalar curvature of a Poincaré frame bundle follows the Poincaré gravitational Lagrangian:

$$R_{PB} = R - \frac{1}{4} F^{ab}_{\mu\nu} F_{ab} - \frac{1}{4} T^{a}_{\mu\nu} T_{a} \tag{19}$$

and L_P in (14) follows from a special Riemann-Cartan scalar curvature of the bundle. In fact, their work extended the theory proposed by Chang et al [C21] to the case in which torsion is included.

Based on the above research on the classical aspects of the gauge theory of gravitation, Guo, Zhao and Yan [AG 14] discussed the problem of quantum gravity. It seems possible to establish a unitary and renormalizable quantum theory of gravitation.

VI. OTHER TOPICS

1. The Angular Momentum Operator and the Wave Function of a Charged Particle Around a Monopole

The study of the angular momentum operator and the wave function of a charged particle around a magnetic monopole is a very old topic. In the earlier discussions, the variables were chosen to be the two degrees of freedom on the spherical surface, hence the problem of singular string exists. Hsien and Hou [A42] considered a hidden degree of freedom in the monopole-charged particle system, i.e., the rotation around the axis connecting the magnetic monopole and the charged particle as well as the corresponding synchronous or multi-synchronous gauge transformations, and introduced the angular momentum operator containing the three Euler's angles-- the so-called symmetric top operator -- and the corresponding wave function $D^J_{m,-q}(\alpha, \beta, \gamma)$ such that apart from the angular quantum number m along the z-axis, there exists another quantum number q, which is the spin quantum number along the axis connecting the two particles and is the product of the magnetic and electric charge $q = eg$. Starting from the top operator and the wave function $D^J_{m,-q}(\alpha, \beta, \gamma)$ it is very natural to obtain the angular momentum operators and the wave functions in the ordinarily adopted gauges. Hou [A49] discussed the structure of the angular momentum operator of the magnetic monopole in an N-dimensional apace-time with an SO(N) symmetry and the problem of reducing it to an expression containing explicitly the field strengths. As it has been mentioned earlier in the discussion of the fibre bundle treatment of the topological properties of gauge fields, Wu, Chen and Guo [A33] constructed a Hopf bundle from the U(1) group and the S^2 basis and introduced a metric for the bundle and found out the expression for the connection. They were able to prove that the operator leaving the metric on the bundle and the fibre invariant is the angular momentum operator of the natural connection. Wu and Wu [A48] applied this method to deduce the angular momentum operator of the SU(2) megnetic monopole with O(5) symmetry.

2. The Pure Gauge Fields on a Coset Space [A36].

In the discussion of the SU(3) x SU(3) chiral symmetry model, the non-linear σ-model is un-renormalizable, but in the lowest order perturbation calculation, it yields results in agreement with experiments. The linear σ - model is renormalizable, but it cannot yield results in agreement with experiments. In the light of the gauge invariant theory, Chou, Tu and Ruan [A36] proposed a unified treatment of the linear and non-linear σ -models: just like what has been done in the non-abelian gauge field theory, introduce a gauge freedom into the theory such as to make the theory invariant under the gauge transformations, then in a certain gauge the theory is renormalizable, though the particle picture is not explicit; while in another gauge, the theory is non-renormalizable, but with an explicit particle picture and could yield physical results by perturbation calculation. The introduction of this gauge freedom is just to introduce the so called"pure scalar gauge fields on a coset" in the theory. It is obvious that the scalar property of these "gauge fields" indicates that they are different from those in the ordinary sense, but still this theory is constructed from the requirement of local gauge invariance. Let the symmetry group of the physical system be G, H be an arbitrary subgroup of it, G/H the coset of G with regard to H, and their elements be **g**, h, and ϕ respectively. There exists the following discomposition:

$$g \quad \phi_0 = \phi(g, \phi_0)h(g, \phi_0) \tag{20}$$

i.e., h(**g**, ϕ_0) is the mapping of **g** into the subgroup H through the coset element ϕ_0 ,

$$g \xrightarrow{\phi_0} h(g, \phi_0) \tag{21}$$

where ϕ_0 is called the pure scalar gauge field on the coset space. For cases when the global property of the group G is trivial, ϕ_0 is merely a gauge freedom, which can be gauged away by choosing an appropriate gauge. By choosing different ϕ_0 one can obtain the " renormalizable gauge " or the " physical gauge ". If the subgroup H is a U(1), then its gauge field is the electromagnetic field, and if the system contains the magnetic monopole, then the monopole and its electromagnetic fields can be described by ϕ_0 . No singular string exists in such electromagnetic field, and the equations of motion of both the monopole and the electromagnetic field are determined by the same Lagrangian. By utilizing ϕ_0 from a system which is invariant locally with respect to the subgroup and globally with repect to G, one can construct a Lagrangian which is invariant locally with respect to G to describe the system. Take the SU(2) × SU(2) chiral group as an example, the locally invariant theory constructed in this way is equivalent to the linear σ -model in the renormalizable gauge. In another gauge, it looks rather like the non-linear σ -model, but with some abundant terms, and it is just these terms which guarantee the renormalizability of the theory.

Above, we have reviewed the works completed in recent years on the classical gauge field theory in China. From this we can see that a number of works have been done to discuss the global property of gauge fields; the dual charge or the topological number; and the topological properties of gauge fields starting from simple, concrete examples, from the differential geometry of Higgs fields, from various classical solutions of the field equations, and from the theory of fibre bundles. In the beginning of these efforts, several groups started their activities almost simultaneously in different places in China with very little exchanges and discussions among them, not to say with their foreign colleagues. The situation was changed since 1977. A series of conferences have been held since then: the Second Colloquiumon Theoretical Particla Physics at Peking in Mar. 1977; the Topical Meeting on Gauge Field Theory at Guangzhou in May, 1979; the Parrallel Session on Theoretical Particle Physics of the Annual Meeting of the Physical Society at Lushan in Aug.1979; and the Meeting on Ideas in Microscopic Physics at Kweilin in Oct. 1979. There were vivid discussions and debates in these conferences, which greatly stimulated the studies in this field, and as a by-product, initiated the collaboration between physicists and mathematicians. However, we are still in a rather secluded situation, and the developments abroad in this field come to us mainly via published materials. Several years ago it was only through the visits of scholars such as Profs. T.D.Lee, C.N.Yang and others could we learn about the latest developments in particle physics and grasp rapidly the mainstream of the development abroad. The situation of scientific exchanges has been improved during the past two years, and we are looking forward to discussions, exchanges and criticisms in this field from our colleagues from abroad.

References

A Papers on general classical gauge field theory published in China (up to October, 1979).

A1 Acta Physics 23, 249 (1974) K.H. Look: Gauge field and the connection on the principal fibre bundle.

A2 ibid 25, 514 (1976) H.C. Lee, T.C. Hsien and S.H. Kuo: On the problem of the dual charge in non-abelian gauge theory - a discussion on the problem of magnetic monopole. (Journal of Sun Yatsen University (Natural Science) 1975, No. 3)

A3 ibid 25, 514 (1976) Y.S. Duan, M.L. Ge and B.Y. Hou: Dual Charge of a Non-Abelian Gauge Field (Journal of Lanzhou University (Natural Science) 1975, No. 2)

A4 ibid 26, 83 (1977) B.Y. Hou: On the Structure of SU(2) Gauge Field.

A5 ibid 26, 155 (1977) C.H. Gu and H.S. Hu: On Spherically Symmetric Gauge Fields and Monopoles.

A6 ibid 26, 433 (1977) Gauge Invariant Conserved Current of a Non-Abelian Gauge Field.

A7 ibid 26, 544 (1977) H.C. Lee, T.C. Hsien and S.H. Kuo: Soliton Solutions of Non-Abelian Gauge Fields.

A8	Physica Energiae Fortis et Physica Nuclearis	1, 52 (1977) Y.S. Chen, Y.S. Wu, T.S. Tu and H.Y. Kuo: Dual Charges of a Sub-group in Gauge Theories -- Magnetic Mo-nopoles of SU(2) and SO(3) Gauge Fields.
A9	ibid	2, 23 (1978) H.C. Lee, T.C. Hsien and S.H. Kuo: On the Problem of the Dual Charge (Magnetic Monopole) of Non-Abelian Groups (2).
A10	ibid	2, 35 (1978) M.C. Wang, H.T. Cheng, K.L. Wang , C.K. Chang and T.C. Hsien: Spinor Monopole is Possible.
A11	ibid	2, 49 (1978) T.S. Tu and T.N. Ruan: Various Expressions for the Field Strengths of a Monopole with Different Velocities and the Internal Form of the Potential of a Monopole.
A12	ibid	2, 61 (1978) B.Y. Hou: Non-Equivalent SU(2) Gauge Potentials Under Equal Field Strengths and Sources.
A13	ibid	2, 93 (1978) T.P. Tzu: The Nonsingular Vector Potential of a Magnetic Monopole in the direction of Meridian.
A14	ibid	2, 97 (1978) C.H. Gu: The Loop Phse-Factor Approach to Gauge Fields (Journal of Fudan University (Natural Science) 1976, No. 2)

A15 ibid 2, 275 (1978) K.L. Wang, T.C. Hsien and H.T. Cheng: On the Magnetic Monopole and Dipole Solutions of the SU(N) Gauge Field Theory in a Curved Space.

A16 ibid 2, 87 (1978) Y.S. Wu and C.K. Peng: A Note on the Equations of SU(2) Instantons.

A17 ibid 2, 295 (1978) C.H. Gu: The U_1 Gauge Field in a SU_N Gauge Field and the Quantized Values of Dual Charges.

A18 ibid 2, 368 (1978) Y.K. Wang, K.Y. Chang and B.Y. Hou: A Series of Sourceless Solutions of SU(2) Gauge Fields -- the SU(2) and U(1) Potentials of Static Monopole System. (Journal of Northwest University (Natural Science), 1976, No. 2-3).

A19 ibid 2, 371 (1978) Y.S. Wu, H.C. Lee, T.C. Hsien and S.H. Kuo: On a Pseudo-Particle Solution of the Yang-Mills Field. (Journal of Sun Yatsen University (Natural Science), 1977, No. 1).

A20 ibid 2, 403 (1978) K.L. Wang, T.C. Hsien and H.T. Cheng: On the SU(4) Monopole and Dyon Solutions.

A21 ibid 2, 481 (1978) T.N. Ruan and T.S. Tu: Strong Interaction and Chiral Gauge Field.

A22 ibid 2, 511 (1978) Y.S. Duan and M.L. Ge: Geometrical Description of Decomposibility of a Non-Abelian Gauge Potential. (Journal of Lanzhou University (Natural Science), 1977, No. 3).

A23 ibid 3, 255 (1979) B.Y. Hou and B.Y. Hou: Static Sphero-symmetrical Sourceless Solutions of SU(2) Gauge Field - The Self-duality Uniqueness and the Self-Induced Current.

A24 ibid 2, 382 (1979) Y.S. Wu: The Holonomy Functional Formulation of Yang-Mills Field Equations.

A25 ibid 3, 555 (1979) P. Wang and B. Y. Hou: The Multisynchrono-Spherical Solution of the SU(N) Gauge Theory.

A26 Scientia Sinica 18, 483 (1975) C.H. Gu and C.N. Yang: Some Problems on the Gauge Fields (Journal of Fudan University (Natural Science), 1975, No. 1).

A27 ibid 19, 320 (1976) C.H. Gu: Electromagnetic Field and the Global Gauge Field of the U(1) Group. (Journal of Fudan University (Natural Science), 1975, No. 3) (The volume and page numbers of this article refers to the Chinese Edition of Scientia

Sinica).

A28 ibid 20, 47, 177 (1975) C.H. Gu and C.N. Yang: Some Problems on the Gauge Field Theories II, III (Journal of Fudan University (Natural Science), 1976, No. 1).

A29 ibid 20, 145 (1977) D.X. Xia: On Field Strengths and Gauge Potentials of Yang-Mills Fields. (Journal of Fudan University (Natural Science), 1976, No. 1)

A30 ibid 20, 345 (1977) C.K. Peng: Yang-Mills Pseudo Particle With Any Integer Dual Charge.

A31 ibid 20, 186 (1977) Y.S. Wu and H.Y. Kuo: The Structure of Principal bundle for Dual Charges in Gauge Fields.

A32 ibid 21, 767 (1978) C.H. Gu, C.L. Shen, H.S. Hu and C.N. Yang: A Geometrical Interpretation of Instanton Solutions in Euclidean Space.

A33 ibid 21, 193 (1978) Y.S. Wu, S. Chen, H.Y. Kuo: The $U_2 \simeq SU(2) \times U(1)$ Invariance of the Monopole Principal Fibre Bundle.

A34 ibid 21, 317 (1978) T.S. Tu, T.T. Wu and C.N. Yang: Interaction of Electron, Magnetic Monopole and Photon.

A35 ibid 21, 446 (1979) B.Y. Hou and Y.S. Duan: The Decomposition and Reduction of Gauge Field and Dual charged solution of Abelianizable Field.

A36 ibid 22, 37 (1979) K.C. Chou, T.S. Tu and T.N. Ruan: The Pure Gauge Fields on the Coset Space.

A37 ibid 22, 169 (1979) H.C. Lee, T.C. Hsien and S.H. Kuo: On the Problem of Dual Charge (Magnetic Monopole) in Non-Abelian Gauge Groups.

A38 Kexue Tongbao 20, 273 (1975) B.Y. Hou: Electro-Magneto Dual Double-Covariant Gauge Fields. (Journal of Northwest University (Natural Science), 1975, No. 2)

A39 ibid 21, 282 (1976) Y.S. Duan and M.L. Ge: Magnetic Monopoles and Higgs Fields. (Journal of Lanzhou University (Natural Science) , 1975, No. 2)

A40 ibid 21, 330 (1976) D.X. Xia: On Field Strengths and Gauge Potential of Yang-Mills Fields.

A41 ibid 22, 255 (1977) C.K. Peng: Yang-Mills Pseudo Particles with Any Integer Dual Charge.

A42 ibid 22, 427 (1977) S.Z. Yan and D.X. Xia: On the Strength, Source and Gauge Potential of Yang-Mills Fields.

A43 ibid 22, 204 (1977) T.C. Hsien and B.Y. Hou: The Wave Function of a Charged Particle in the Field Around a Magnetic Monopole.

A44 ibid 22, 433 (1977) H.Y. Kuo: The Theory of Principal Vector Bundle of Gauge Fields.

A45 ibid 22, 527 (1977) P. Wang: Pseudo Particle Solution of the SU(4) Gauge Field.

A46 ibid 23, 150 (1978) Y.S. Wu, S. Chen, T.S. Tu and H.Y. Kuo: The Monopoles in the Broken SO_3 and SU_2 Gauge Theories.

A47 ibid 23, 598 (1978) C.H. Gu: Breaking of Symmetry and Dual Charges in Gauge Fields.

A48 ibid 23, 598 (1978) Y.S. Wu and C.S. Wu: $U_2(Q)$ Invariance of the Principal Bundles for SU_2 Dual Charges of the Corresponding Angular Momentum Operators.

A49 ibid 24, 16 (1979) B.Y. Hou: Angular Momentum Operator of Non-Abelian Gauge Field Monopole.

A50 ibid 24, 492 (1979) C.H. Gu and H.S. Hu: On the Gauge Condition for Gauge Fields.

AG Papers on the gauge theory of gravitation published in China.

AG1 Scientia Sinica 19, 199 (1976) S. Chen, T.H. Ho, C.L. Tsou and H.Y. Kuo: On Yang's Spherically Symmetric Static Gravitational Field.

AG2 ibid 22, 628 (1979) C.L. Tsou, P. Huang, Y.C. Chang, G.D. Lee, Y. An, S. Chen, L.N. Chang, T.H. Ho and H.Y. Kuo: Some Researches on Gauge Theories of Gravitation.

AG3 Kexue Tongbao 18, 72 (1973) H.Y. Kuo, Y.S. Wu and Y.C. Chang: A Scheme for the Gauge Theory of Gravitation.

AG4 ibid 18, 119 (1973) Y.S. Wu, C.L. Tsou and S. Chen: Some Vacuum Solutions of the Gauge Theory of Gravitation and the Tests by the Three Known Experiments.

AG5 ibid 18, 122 (1973) Y.C. Chang and H.Y. Kuo: The Electromagnetic Field in the Gauge Theory of Gravitation.

AG6 ibid 19, 320 (1974) S. Chang: The Twist Free Vacuum Solutions of the Gauge Theory of Gravitation.

AG7 ibid 19, 509 (1974) Y.S. Wu, K.D. Lee and H.Y. Kuo: Gravitational Lagrangian and Local de Sitter Invariance.

AG8 ibid 19, 512 (1974) P. Huang and H.Y. Kuo: An Interior Solution for the Einstein-Yang Equation.

AG9 ibid 20, 561 (1975) P. Huang: Qualitative Analysis for the Solution of the Einstein-Yang Equation.

AG10 ibid 21, 31 (1976) H.Y. Kuo: The Local de Sitter Invariance.

AG11 ibid 21, 379 (1976) I. An, S. Chen, C.L. Tsou and H. Y. Kuo: A New Theory of Gravitation -- Gravitational Theory with Local de Sitter Invariance.

AG12 ibid 24, 103 (1979) H.Y. Kuo and L.N. Chang: The Actions of Gauge Fields for Gravitation and Their Principal Bundle Formulation.

AG13 ibid 24, 202 (1979) On the de Sitter Gauge Theory for Gravitation.

AG14 ibid 24, 587 (1979) M.L. Yan, B.H. Chao and H.Y. Kuo: Renormalization of Gravitational Field with Torsion.

AG15 ibid 24, 596 (1979) S. Chen, K.D. Lee and H.Y. Kuo: A Theory of Conformal Supergravity.

AG16 Acta Astronomica Sinica 17, 147 (1976) C.L. Tsou, S. Chen, T.H. Ho and H.Y. Kuo: A Spherically Symmetric Star Model in the Gravitational Gauge Theory.

R Review articles published in China.

R1 K.H. Look: Lecture Notes on the Theory of Gauge Fields and Fibre Bundles. (Monograph of the Institute of Mathematics, Academia Sinica, 1975).

R2 C.H. Gu: Mathematical Structure of Gauge Fields (Preprint of the Institute of Mathematics of Fudan University, 1978).

R3 S.H. Kuo and H.C. Lee: Pseudoparticle Physics (Collection of Review Articles Presented to the Canton Colloquium on the Theory of Gauge Fields, 1979).

R4 H.Y. Kuo: Progress of the Gauge Theory of Gravitation (Presented at the 1978 Canton Conference on Gravitation and Relativistic Celetial Physics)

C Paper published abroad, which are quoted to compare with the results obtained by Chinese authors.

C1 C.N. Yang, R.L. Mills, Phys. Rev. 96, 191 (1954).

C2 T.T. Wu, C.N. Yang in "Properties of Matter Under Unusual Conditions 349 (1968)."

C3 J. Arafune, F.G.O. Freund, C.J. Goobel, J.M. Math, Phys. 16, 433 (1975).

C4 T.T. Wu, C.N. Yang, Phys. Rev. D12, 3845 (1975).

C5 G. 't Hooft, Nucl. Phys. B79 276 (1974).

C6 T.T. Wu, C.N. Yang, Phys. Rev. D12, 3943 (1975).

C7 K. Wilson, Phys. Rev. D10, 2455 (1974).

C8 C.N. Yang: SU(2) Magnetic Monopole with an O_5 Symmetry (Lecture given in Peking 1976).

C9 R. Jackiw, C. Robbi, Phys. Rev. D14, 517 (1976).

C10 F. Wilczek: In "Quark Confinement and Field Theory" (1976).

C11 E. Cremer, J. Scherk, Nucl. Phys. B118, 61 (1977).

C12 S. Doser, Phys. Lett. B64, 464 (1976).

C13 A. Chakrabarti, Ann. Inst. Henri Poincare A23, 239 (1975).

C14 E. Witten, Phys. Rev. Lett. 38, 121 (1977).

C15 C.N. Yang, Phys. Rev. Lett. 38, 1377 (1977).

C16 J. Swank, L. Swank, T. Deralli, Phys. Rev. D12, 1096 (1975).

C17 C.N. Yang, Phys. Rev. Lett. 33, 445 (1974).

C18 T.T. Wu, C.N. Yang, Nucl. Phys. B107, 365 (1976).

C19 C.N. Yang, J. Math, Phys. 19, 320 (1978).

C20 A.A. Belavin, A.M. Polykov, A.S. Sohwartz, V.S. Tyupkin, Phys. Lett. B59, 85 (1975).

C21 F. Mansouri, L.N. Chang, Phys. Rev. D13, 3192 (1976).

C22 P.K. Townsend, Phys. Rev. D15, 2795 (1977).

C23 E.E. Fairchild Jr., Phys. Rev. D16, 2438 (1977).

C24 S. Ramaswamy, P.B. Yasskin, Phys. Rev. D19, 2264 (1979).

C25 H.T. Nieh, ITP-SB-79-76.

C26 J.P. Hsu, Cosmic Matter-Antimatter Asymmetry and 4-dimensional gauge symmetry of Gravity (Paper presented to the present conference).

C27 P. Roskies, Phys. Rev. D15, 1731 (1977).

C28 M. Calvo, Phys. Rev. D15, 1733 (1977).

C29 S. Solomon, Nucl. Phys. B147, 174 (1979).

New Particles*

Tung-Mow Yan
Newman Laboratory of Nuclear Studies
Cornell University, Ithaca, New York 14853
U.S.A.

The theoretical ideas for the new particles, the ψ and Υ families, are reviewed. The predictive power and the shortcomings of the nonrelativistic quark model are described. Predictions for hadronic transitions in the multipole expansion of QCD are presented.

*Supported in part by the National Science Foundation.

I. Introduction

Since November 1974 many new particles have been discovered.[1] They fall into different categories:

1. ψ-particles: J/ψ, ψ', ψ'', χ_J, U, ...
2. Charmed particles: D, D^*, ...
3. T-particles: T, T', T".
4. Heavy lepton: τ.

In today's talk I will concentrate on ψ and T particles. Practically all properties of the heavy lepton τ were anticipated by Prof. Y. S. Tsai[2] before its discovery. He is going to give a talk on the subject in the parallel session. As for the charmed particles, they involve light quarks so they cannot be treated by the nonrelativistic quark model.

There are quite a few of ψ and T particles. Clearly, the most interesting questions are their spectra and their decays. We will see that many gross features of these particles can be understood by ancient ideas of the nonrelativistic quark model and the so-called OZI rule. In the quark model all mesons are interpreted as bound states of a quark and an antiquark. The OZI rule states that a decay is allowed (large) if the corresponding quark diagram is connected, as in $\rho \to \pi\pi$; a decay is forbidden (small) if the corresponding quark diagram is disconnected, as in $\phi \to 3\pi$.

For heavy quarks these ideas can be justified to some extent. The relative motion of heavy quarks in a bound state is slow and a non-relativistic potential description should suffice. Also the bound states of heavy constituents are small in size; this may explain why the OZI forbidden decays are small. For example, consider the decay of a heavy quarkonium into ordinary hadrons. In QCD this decay occurs through annihilation into gluons. The annihilation takes place at short distances and the effective coupling constant is small according to asymptotic freedom.

Let's apply these ideas to the new particles. These particles are very heavy and their total decay widths are unusually narrow. It is then natural to postulate that their constituents are very heavy. The narrow widths are then "explained" by assuming that the heavy quarks carry new flavors and these new particles lie below the production threshold of the new flavors; they can only have OZI forbidden decays.

Aside from the obvious that anything new is interesting, we may ask why these new particles have attracted so much attention. In their original paper on charmonium, Appelquist and Politzer[3] argued that the large mass of the charmed quark c implies that the $c\bar{c}$ states are small compared to the familiar hadrons, and that, because of asymptotic freedom, the $c\bar{c}$ system would be well described by a Coulombic interaction. The $c\bar{c}$ system would be "the hydrogen atom of hadrons". Later development did not support this picture, but there are several reasons for the wide interest in the new particles:

1. They are very different from ordinary hadrons. They have unusually narrow widths and there are many of them. To characterize the relationship of a particle's width Γ to its mass m we introduce the parameter

$$\rho = \frac{\Gamma}{m} \tag{1.1}$$

For ordinary hadrons we have

$$\rho = 0.01 - 0.1 \ . \tag{1.2}$$

For J/ψ and ψ' we have

$$\begin{aligned} \rho(\psi) &= 2\times10^{-5} \\ \rho(\psi') &= 5\times10^{-5} \end{aligned} \tag{1.3}$$

despite the huge phase space available for decays to various channels. To appreciate how small these numbers are, the $W^{\pm}$ particle in the Weinberg-Salam model, a particle without strong interactions, will have

$$\rho(W) \simeq 10^{-2} \quad . \tag{1.4}$$

The contrast between $\rho(\psi)$, $\rho(\psi')$ and $\rho(W)$ blurs the distinction between a hadron and a non-hadron.

For the first time in particle physics, there are so many narrow states of the same $Q\overline{Q}$ system. So far, there are J/ψ, ψ', ψ'', χ_J and U(2983) in the charmonium family, and Υ, Υ' and Υ'' in the upsilon family. Certainly there are more to come in the near future.

2. New particles imply existence of new flavors. According to the OZI rule, the narrow widths of J/ψ and Υ imply that the constituents of these particles carry new flavors. In fact, the name charmonium was suggested on the assumption that J/ψ, ψ', ..., are bound states of the long-awaited charmed quark[4] and its antiquark. Although many people didn't accept this interpretation until the discovery of charmed particles several years later, it is now generally taken for granted that Υ's are bound states of yet another new heavy quark b and its antiquark $\overline{b}$.

An important issue raised by these new particles is how many flavors there are in the subnuclear world. If we assume that the b-quark is the charge $-\frac{1}{3}$ partner of a third Cabbibo doublet and the heavy lepton τ has its own neutrino, then there is a symmetry between leptons and quarks: each has three generations.

Leptons: $\begin{pmatrix} e \\ \nu_e \end{pmatrix}$, $\begin{pmatrix} \mu \\ \nu_\mu \end{pmatrix}$, $\begin{pmatrix} \tau \\ \nu_\tau \end{pmatrix}$;

Quarks: $\begin{pmatrix} u \\ d \end{pmatrix}$, $\begin{pmatrix} c \\ s \end{pmatrix}$, $\begin{pmatrix} t \\ b \end{pmatrix}$.

Is this symmetry accidental? Or has it deeper significance? Are there a finite or infinite number of generations? Are quarks and leptons made of even smaller constituents?[5]

3. New particles give strong support to the quark model. Many narrow states exist in one family below the OZI threshold. This offers the opportunity to compare the spectrum predicted by the quark model with the observed particle masses. In contrast with many of other quark model results which are consequences of SU(6) symmetry, prediction of spectrum requires knowledge of the nature of the potential. The non-relativistic quark model is tremendously successful here. This success lends strong support to the notion that quarks are the constituents of hadrons.
4. New particles act as sources of gluons. In QCD J/ψ and Τ decay by annihilation into gluons and the subsequent conversion of gluons into ordinary hadrons. According to Sterman and Weinberg[6] if one is only interested in the global features of the decay such as energy flow, angular distributions and correlations one can ignore how gluons materialize as hadrons. The problem is then the same as if they decay directly into gluons. Thus, study of the decays of J/ψ, Τ and other heavy quarkonia will reveal the properties of the gluons. Since gluons

do not carry any flavor, heavy quarkonia are perhaps the most convenient sources of gluons.

It is impossible to cite all the contributions to the subject in my talk. Fortunately, there are a number of excellent review articles available. They are listed in Ref. 7. My own work on the subject has been in collaboration with my colleagues at Cornell: E. Eichten, K. Gottfried, T. Kinoshita, and K. Lane. J. Kogut also participated in the early stage of our effort. Our recent work appears in two comprehensive articles.[8]

II. Qualitative Description

The quark model is only a model, not a real theory. But it possesses predictive power. To appreciate how much one can learn from such a simple model, it is instructive to repeat the arguments our group at Cornell actually went through almost exactly five years ago.[9] Let's imagine that the time is December 1974 and we are given the experimental information:

i) $M(J/\psi) = 3.1$ GeV, $\Gamma_e = 4.8\pm0.6$ keV,

ii) $M(\psi') = 3.7$ GeV, $\Gamma_e = 2.1\pm0.3$ keV,

iii) The decay $\psi' \to \psi + 2\pi$ had a large branching ratio.

From iii) we infer that ψ and ψ' were different states of a $c\bar{c}$ system described by a nonrelativistic Hamiltonian,

$$H \approx \frac{p^2}{2m} + V(r) \qquad (2.1)$$

The leptonic width is given by the Van Royen-Weisskopf formula,

$$\Gamma_e = \frac{16\pi\alpha^2 e_Q^2}{M^2} |\psi_n(0)|^2 \qquad (2.2)$$

Given the sizeable electronic widths of ψ and ψ', it is natural to assign them to the 1^3S and 2^3S states of the $c\bar{c}$ system respectively. The data and Eq. (2.2) combine to give the ratio

$$\eta = \frac{|\psi_{2S}(0)|^2}{|\psi_{1S}(0)|^2} = 0.62 \ . \tag{2.3}$$

The ratio η is sensitive to the nature of the potential. For example, the values of η for a few models are

$$\begin{aligned} &\text{Coulomb potential:} \quad && \eta = \tfrac{1}{8} \ , \\ &\text{Linear potential:} \quad && \eta = 1 \ , \\ &\text{Harmonic oscillator:} \quad && \eta = 1.5 \ . \end{aligned} \tag{2.4}$$

The experimental value of η corresponds to a potential that is more confining than Coulombic but less so than a harmonic oscillator. The qualitative features of the spectrum can be surmised by visual interpolation between these two cases [see Fig. 1]. It shows that there should be a P-multiplet at 3.4-3.5 GeV. This immediately brings with it the exciting prediction of E1 photon cascades $\psi' \to {}^3P_J + \psi$. There should also be pseudoscalar partners (the ${}^1S_0$ states) near and below J/ψ and ψ'. M1 transitions between the ${}^3S_1$ and ${}^1S_0$ states are also to be expected. Furthermore, the charmed quark mass is naively $m_c \sim \frac{1}{2} m_\psi \sim 1.6$ GeV. So the charmed meson mass is expected to be approximately $m_D \sim \frac{1}{2} m_\psi + m_q \sim 1.9$ GeV.

Although it has been five years since these predictions were made some of them still await confirmation. The following is a brief summary of the experimental situation:

1. The P states were first seen in 1975. There are now three well-established states at energies 3.41, 3.51 and 3.55 GeV. All the evidence is consistent with the assignment of J = 0,1,2 to the three states, respectively. All the transitions $\psi' \to P_J + \gamma$ and $P_J \to \gamma + \psi$

have been observed except $P_o(3.41) \to \gamma + J/\psi$.

2. The charmed particles were first found in 1976. They have the properties of charm as predicted by Glashow, Iliopoulos and Maiani.[4]

3. In 1979 a narrow peak was observed[10] in the inclusive photon spectra of two processes,

$$\psi' \to \gamma + \text{anything},$$

$$\psi \to \gamma + \text{anything},$$

corresponding to a missing mass of

$$m = 2.983 \pm 0.015 \text{ GeV}. \tag{2.5}$$

This is the obvious candidate for the charmonium ground state η_c.

So far, the 2^1S_0 state and the photon transition $^3P_0 \to \gamma + \psi$ have not yet been seen.

To go beyond this qualitative description, we will have to choose a specific potential. We will see that this approach can lead to new predictions. The most interesting one is the position of a 1^3D state. Without a potential, its position is uncertain except that it should be somewhere between 2S and 3S. The model we used predicted its position to be about 70 MeV above ψ'. This was later confirmed by experiments[11] in 1977.

III. The Charmonium Model

A. Choice of Potential

The choice of potential is not a trivial problem. There are some theoretical conjectures about its asymptotic behavior as $r \to \infty$ and $r \to 0$. According to the flux tube picture abstracted from lattice gauge theories, the potential $V(r)$ should behave linearly at large distances,

$$V(r) \xrightarrow[r \to \infty]{} \frac{1}{a^2} r . \quad (3.1)$$

Asymptotic freedom of QCD suggests that

$$V(r) \xrightarrow[r \to 0]{} -\frac{4}{3} \frac{\alpha_s}{r} . \quad (3.2)$$

Many potentials can have the same limiting behavior given by Eqs. (3.1) and (3.2), but give very different interpolation between the two limits. However, in view of the discussion on the relation of η to the potential, it is simplest to use

$$V(r) = \frac{1}{a^2} r - \frac{\kappa}{r} . \quad (3.3)$$

Here we have relaxed the identification of the strength of the Coulomb potential with the strong interaction coupling constant α_s.[12]

Eq. (3.3) is extensively used by our group at Cornell and others. But many alternatives have also appeared in literature. Here we only mention a few of them. Because of the approximate equality of the spacings $M_{\Upsilon'} - M_{\Upsilon} \simeq M_{\psi'} - M_{\psi}$, Quigg and Rosner[13] proposed a logarithmic potential which has the unique feature that spacings are independent of quark masses. However, a pure logarithmic potential has no theoretical justification. But one can approximate the intermediate $Q\overline{Q}$ potential by a logarithmic one. This was done by Bhanot and Rudaz[14]:

$$\begin{aligned} V &= -\frac{4}{3}\frac{\alpha_s}{R} & R &< R_1 \\ &= \ln (R/R_0) & R_1 &\le R \le R_2 \qquad (3.4) \\ &= \frac{1}{a^2} R & R &> R_2 \end{aligned}$$

Another potential has been proposed by Richardson[15] which incorporates both (3.1) and (3.2) with a single parameter. The potential is most simply expressed in momentum space

$$V(q) = -\frac{4}{3}\frac{12\pi}{33-2n_f}\frac{1}{q^2\,\ell n(1+q^2/\Lambda^2)} \tag{3.5}$$

where n_f is the number of flavor.

Instead of obtaining the spectrum by first guessing a potential, Quigg, Rosner and Thacker[16] have applied the technique of inverse scattering to construct the potential from data on the masses of the charmonium and upsilon states and their wave functions at the origin as deduced from the leptonic widths. The potential so obtained is then used to predict new states.

All these models give similar results for the charmonium system. Their predictions for the $b\bar{b}$ system will be presented after we have discussed the Cornell model.

B. The Cornell Model

When the model was proposed[9], only ψ and ψ' were known. Consequently one had to fix the parameters from the ψ'-ψ mass difference and the leptonic widths, subject to the general requirement that $(\frac{v}{c})^2$ remains "small". While this strategy led to many interesting predictions, it was also clear that the model can't give a quantitative fit to the data. To be precise, today we have a far richer body of data; these include (below OZI threshold):

1. Positions of spin triplet states ψ, $\chi_{J=0,1,2}$ and ψ'.
2. Leptonic widths $\Gamma_e(\psi)$ and $\Gamma_e(\psi')$.
3. E1 transition rates $\psi' \to \chi_J + \gamma$

None of the simple potential models proposed so far can fit all of the data. We will illustrate the situation by the specific Coulomb plus linear model. We define the model by

1. The potential is given by (3.3),

2. the leptonic width is given by (2.2), and

3. El rates are given by the nonrelativistic dipole approximation.

In this model[17] there are three parameters m_c, a, and κ that characterize the ψ family. Let's use the spacings ψ'-$P_{c.o.g.}$ and ψ'-ψ as two conditions to restrict the three parameters. We choose the third combination of these parameters to be v_ψ^2, the mean square velocity in the 1S state. The various transition rates can then be computed as functions of v_ψ^2. If the model were actually capable of giving a fit to all the data, these quantities would all lead to the same small value of v_ψ^2.

Unfortunately, we are not in this happy situation. This is illustrated by Fig. 2, which shows the ground state leptonic width, $\Gamma(\psi \to e^+e^-)$, and one of the El rates, $\Gamma(\psi' \to \gamma\, ^3P_0)$, as functions of v_ψ^2. Both of these curves are computed from the naive nonrelativistic rate fomrulas. The measured values of $\Gamma(\psi \to e^+e^-)$ and $\Gamma(\psi' \to \gamma\, ^3P_0)$ are also shown in Fig. 2. We note that there is no value of v_ψ^2 that yields a fit to both rates: the El rate is small, so to say, and requires low velocity motions, but such low velocities give far too large an annihilation probability. But there are theoretical reasons why it is difficult to fit Γ_e and El rates by the simple-minded approach. Celmaster[18] has observed that a direct translation of the one-loop contribution in QED to QCD gives a large correction to the Van Royen-Weisskopf formula,

$$\Gamma_e = \Gamma_0(1 - \frac{16}{3\pi}\alpha_s) \qquad (3.6)$$

where Γ_0 is given by Eq. (2.2). When other higher-order effects are included, we would have more generally

$$\Gamma_e = \Gamma_0\, f(\alpha_s) \quad . \qquad (3.7)$$

The correction factor $f(\alpha_s)$ is insensitive to the particular $Q\overline{Q}$ state in question. Thus, the ratios of leptonic widths with one $Q\overline{Q}$ family should still be well described by the model, even if the absolute widths are not.

As for the E1 rates, they are very sensitive to wave functions. Even in atomic physics[19], Hartree-Fock calculation of E1 rates in atoms with $Z \lesssim 20$ and only <u>one</u> valence electrons are frequently off by as much as 50%. If nonrelativistic approximation is good, the ratios of rates should be independent of wave functions. If we denote by Γ_J the rate for $\psi' \to P_J + \gamma$ then theoretically

$$\Gamma_J \sim (2J+1)k_J^3 \ . \qquad (3.8)$$

Using the experimental values for k we find

$$\Gamma_0:\Gamma_1:\Gamma_2 = 1:0.85:0.66 \ . \qquad (3.9)$$

Experimentally the ratios are[10]

$$\Gamma_0:\Gamma_1:\Gamma_2 = (7\pm1.5):(7\pm1.5):(7\pm1.5) \ . \qquad (3.10)$$

Eqs. (3.9) and (3.10) are not inconsistent. The discrepancy between (3.9) and (3.10) is an indication that relativistic and other corrections may be of order of 20-30%. This is not unexpected since $v_\psi^2 = 0.2$ in our model.

We must now choose a value of v_ψ^2 to fix our parameters. This choice is, of necessity, somewhat arbitrary and subjective. Our choice is $v_\psi^2 = 0.2$. Our parameters are[8]

$$m_c = 1.84 \text{ GeV}$$
$$a = 2.34 \text{ GeV}^{-1} \qquad (3.11)$$
$$\kappa = 0.52 \ .$$

Detailed predictions are given in ref. 8. The spectrum is listed in Table I as well as some properties of these states. The spectrum is compared with the structures observed at SPEAR and DORIS. There is a rather satisfying correspondence between the model and observation. The computed E1 rates are about a factor of 2-3 too large. There is no improvement here over the predictions that preceded the discovery of the P-states.

C. Spin-singlet ψ-states

The experimental search for spin-singlet ψ states has proved to be much more difficult than the search for the low-lying spin-triplets which all have been found.

Earlier there were candidates for both 1^1S [X(2830) observed at DORIS] and 2^1S [χ(3450) and χ(3600) possibly seen in $\psi' \to \gamma\gamma\psi$]; their properties disagreed so badly with the model that it has been argued[20] that they were not the hyperfine partners of ψ and ψ'. These states have not been seen by the crystal ball collaboration at SPEAR in their detailed study of inclusive single photon spectrum.[10] But they have found what appeared to be the charmonium ground state η_c. Its essential properties are[10]

$$m = 2983 \pm 15 \text{ MeV}$$

$$\Gamma = 20^{+20}_{-10} \text{ MeV} \tag{3.12}$$

$$\Gamma(\psi' \to \eta_c + \gamma) = 0.7 \text{ keV}$$

$$\Gamma(\psi \to \eta_c + \gamma) = 0.7 \text{ keV}$$

No error is quoted for the M1 transition rates. The ψ'-ψ mass difference is

$$m_\psi - m_{\eta_c} = 112 \text{ MeV} \tag{3.13}$$

It agrees very well with the prediction of ref. 21, which is based on dispersion relation sum rules for heavy quarks. Our model gives[8]

$$\Gamma(\psi' \to \eta_c + \gamma) = 1.0 \text{ keV} \tag{3.14}$$

$$\Gamma(\psi \to \eta_c + \gamma) = 2 \text{ keV} . \tag{3.15}$$

It is puzzling that the theory agrees so well with the preliminary data for the hindered transition $\psi' \to \eta_c + \gamma$, which is very sensitive to wave function details, whereas it disagrees significantly with the model-insensitive allowed transition $\psi \to \eta_c + \gamma$.

D. QCD Predictions

We now turn to a brief discussion of QCD predictions for heavy quarkonia. We already mentioned that the naive Van Royen-Weisskopf formula for Γ_e does not work because of large high order QCD correction. We will see that none of the QCD predictions work too well either.

The total width of η_c given in (3.12) is much bigger than expected. Lowest order QCD calculations give

$$\frac{\Gamma_{had}(n^3S_1)}{\Gamma_{had}(n^1S_0)} = \frac{5(\pi^2-9)}{27\pi} \alpha_s \tag{3.16}$$

$$\frac{\Gamma_{had}(n^3S_1)}{\Gamma_e(n^3S_1)} = \frac{5}{18} \frac{\pi^2-9}{\pi} \frac{\alpha_s^3}{\alpha^2} \tag{3.17}$$

From the measured total width and leptonic width of ψ we have

$$\alpha_s \sim 0.2 . \tag{3.18}$$

That in turn gives a much smaller value for $\Gamma_{had}(\eta_c)$:

$$\Gamma_{had}(\eta_c) \cong 6 \text{ MeV} . \tag{3.19}$$

But the lowest order QCD formulas may be unreliable. Indeed, a surprisingly large correction is found for the ratio[22]

$$R = \frac{\Gamma(n^3S_0 \to gg)}{\Gamma(n^3S_0 \to \gamma\gamma)} = R_o(1 + 22.14 \frac{\alpha_s}{\pi}) \tag{3.20}$$

It is clearly of great interest to know whether the high order correction to the $n^3S_1 \to ggg$ is large or small. In QED, it is known that the radiative corrections to the three-photon decay rate of orthopositronium has a very large coefficient[23]:

$$\Gamma_{3\gamma} \cong \Gamma_0(1 - 10\,\frac{\alpha}{\pi}) \tag{3.21}$$

Other QCD related predictions do not fare well with experiments:

1. Naive E1 rate calculation gives the ratio

$$\frac{\Gamma(P_2 \to \gamma\psi)}{\Gamma(P_0 \to \gamma\psi)} \simeq 2.76 \tag{3.22}$$

and lowest order QCD predicts

$$\frac{\Gamma(P_2 \to gg)}{\Gamma(P_0 \to gg)} = \frac{4}{15} \tag{3.23}$$

We may combine Eqs. (3.22) and (3.23) to obtain

$$\frac{B(P_2 \to \gamma\psi)}{B(P_0 \to \gamma\psi)} \simeq 10.4 \tag{3.24}$$

Experimentally[10] the ratio (3.24) is larger than 22.

2. Lowest QCD formula and the potential model wave function give

$$\Gamma(P_0 \to gg) = 2\text{-}3 \text{ MeV} \tag{3.25}$$

But the measured width[10] is

$$\Gamma(P_0) = 7\pm3 \text{ MeV} \tag{3.26}$$

3. Lowest order QCD predicts

$$\frac{B(P_2 \to \gamma\gamma)}{B(P_0 \to \gamma\gamma)} = 1 \tag{3.27}$$

Experimentally[10] $P_2 \to \gamma\gamma$ has been seen but the other decay mode has not.

Until we have a better understanding of the high order effects, lowest order QCD predictions appear not to be very reliable.

E. Spin Dependent Forces

The model described above did not incorporate the spin dependent part of the potential. There have been many attempts[24] to relate the spin dependent part to the spin-independent potential. The most popular approach has been to assume that the heavy quark binding can be summarized by an instantaneous Bethe-Salpeter kernel consisting of vector and scalar interaction terms

$$K = V_{Coul}(\vec{k}^2)\gamma_1^{\mu}\gamma_{2\mu} + V_v(\vec{k}^2)\Gamma_1^{\mu}\Gamma_{\mu 2} + V_s(\vec{k}^2)1_1 1_2$$

$$\Gamma_{\mu} = \gamma_{\mu} - \frac{i\lambda}{2m_c}\sigma_{\mu\nu}k^{\nu} \qquad (3.28)$$

In the nonrelativistic limit, the spin-independent potential is

$$V_o = V_{Coul} + V_v + V_s \qquad (3.29)$$

It is customary to identify the sum $V_v + V_s$ with the linear confining potential.

Recently Eichten and Feinberg[25] have made a general analysis of the problem. They studied the gauge invariant 4-point function of a $Q\overline{Q}$ pair separated by a spatial distance $\vec{R}$ and propagating for a time interval T. The spin dependent potential appears as a correction to the static energy in the expansion of the inverse powers of heavy quark masses. Their essential conclusions are

1. Only the classical spin-orbit term and the Thomas precession effect are related to the spin-independent static potential.
2. Other spin-dependent terms involve expectation values over the Wilson loop of products of color magnetic field with itself or with color electric field.

3. To obtain an explicit expression for the spin-dependent potential, they assume that effects involving color magnetic field are short range, calculable in perturbation theory. In this approximation the spin-dependent potential is completely determined.

4. If the result is applied to the Cornell model, it gives a reasonably good description of the P state splittings. But the mass difference ψ-η_c is only 65 MeV which is too low compared with the experimental value (3.13).

Clearly our understanding of the spin-dependence force is still quite limited. Since the much richer spectroscopy of the T family will be mapped out at CESR in the near future, it will require a good theory of the spin-dependent forces to understand the fine and h.f. structures of these levels.

F. Coupling to Decay Channels

When the total energy W exceeds the charm threshold, at 3.726 GeV, $c\bar{c}$ states can undergo OZI-allowed decays into charmed mesons. At that point the naive model has broken down. The Cornell group[8,26,27] has incorporated OZI-allowed decay phenomena into the model. The mathematical details of how this is done were described in Ref. 8.
The most important results of this coupled channel calculation are as follows:

1. Below charm threshold, mixing between $c\bar{c}$ states and the charmed meson sector is rather small, and leads to modest (~20%) reductions in the E1 rates predicted by the naive model. The leptonic width of ψ' that emerges from the coupled channel calculation is in complete agreement with experiment.

2. The input experimental mass differences between D and D^* is the only source of spin dependence in our model. It induces an S-D mixing.

This S-D mixing is adequate to provide us a quite complete understanding of $\psi(3772)$ as the $^3D_1$ $c\bar{c}$ state mixed with 2^3S by both open and closed decay channels.

3. The decay amplitudes for $c\bar{c} \to \bar{c}q + q\bar{c}$ are oscillatory functions of momentum, with a node structure that is determined by the radial nodes in the $c\bar{c}$ wave function. This provides a qualitative understanding of the peculiar branching ratios into various charmed meson channels observed at 4.03 GeV. The prediction of a zero in $\sigma(e^+e^- \to D\bar{D})$ near 4.0 GeV should be tested experimentally because it would, for the first time, confirm in detail the structure of a quark-antiquark radial wave function.

4. There is a complex structure in the measured $R(e^+e^- \to \text{hadrons})$ in the 4 GeV region. More careful measurements in this region are needed. They may reveal degrees of freedom not accommodated in the charmonium model. These include string vibrational states[28] and/or $Q\bar{Q}q\bar{q}$ exotic states.[29]

G. Υ System

Just before I left Cornell for this conference, a third narrow Υ resonance had been discovered at CESR (Cornell Electron Storage Ring) by the CLEO[30] and Columbia-Stony Brook collaborations.[31] Called Υ", it is presumably the $3^3S(b\bar{b})$. The measured properties of the three states Υ, Υ' and Υ" are summarized in Table II. Predictions of various models for the $b\bar{b}$ system are given in Table III. It is seen that predictions of all these models are quite similar.

In the Cornell model, the mass of b-quark is fixed at[32]

$$m_b = 5.17 \text{ GeV} \tag{3.30}$$

There are three possible choices for the Coulomb parameter κ in this model.

i) It is fixed at the value determined for ψ-family $\kappa = 0.52$. It then gives

$$m(\Upsilon') - m(\Upsilon) = 591 \text{ MeV} \tag{3.31}$$

which is about 5% higher than the experimental value.

ii) If κ is identified with the running coupling constant of QCD at the mass of ψ, then the renormalization group equation gives

$$\kappa(\Upsilon) = 0.33 \tag{3.32}$$

which leads to a value much too low for the mass difference $\Upsilon' - \Upsilon$:

$$m(\Upsilon') - m(\Upsilon) = 454 \text{ MeV} \tag{3.33}$$

iii) We may fix $\kappa(\Upsilon)$ by requiring the model to give the observed 2S-1S spacing. In this way we find

$$\kappa(\Upsilon) = 0.48 \tag{3.34}$$

The predictions given in Table III correspond to this choice of κ with a and m_b kept at

$$a = 2.34 \text{ GeV}^{-1}$$
$$m_b = 5.17 \text{ GeV} \tag{3.35}$$

An important question in the Υ system is the mass of the B meson (the lightest $b\bar{u}$ meson). Let's estimate it by the quark model formula

$$m_B = m_D + m_b - m_c + \frac{3}{4}\left(1 - \frac{m_c}{m_b}\right)(m_{D^*} - m_D) \tag{3.36}$$

In the Cornell model this gives

$$m_B = 5.26 \text{ GeV} . \tag{3.37}$$

The vector meson B* has a mass given by

$$m_{B^*} - m_B \cong \frac{m_c}{m_b}(m_{D^*} - m_D) \simeq 50 \text{ MeV} \tag{3.38}$$

Thus, according to Table III, the $4S(b\bar{b})$ is expected to lie above the $B\bar{B}$

threshold. A crucial question is the width of the S state. In principle, this can be studied by a coupled channel calculation if the positions of the thresholds are known. Since this is not the case, it is not practical to embark on such a project. Instead, let's try some empirical method.

For the known vector mesons ρ, ϕ, and ψ'' their decay widths into two pseudoscalar mesons appear to satisfy the empirical rule

$$\Gamma(V \to P\overline{P}) = 0.2 \left(1 - \frac{4m^2}{M^2}\right)^{3/2} M \tag{3.39}$$

where M and m are the masses of the vector and pseudoscalar mesons. We will estimate the width $\Gamma(T''' \to B\overline{B})$ from (3.39). The other width $\Gamma(T''' \to B\overline{B}^* + \overline{B}B^*)$ can be estimated by the phase space corrected relation

$$\Gamma(B\overline{B}^* + \overline{B}B^*) = 4 \left(\frac{p^*}{p}\right)^3 \Gamma(B\overline{B}) \tag{3.40}$$

where p^* and p are the momenta of the decay products in the two decay modes.

It should be emphasized that there is a basic difference between the $4S(b\overline{b})$ and all other vector mesons mentioned above. The 4S state is a radial excitation; therefore, its decay amplitudes have nodes. This is not the case for other vector mesons since their wave functions are not oscillatory. It may happen that one of its decays is suppressed because of the presence of a node at the decay momentum. The estimates made by (3.39) and (3.40) must be regarded as rough guides.

We will fix m_B = 5.26 GeV and m_{B^*} = 5.31 GeV. But we will compute the widths for several values of $M(T''')$:

$M(T''')$ = 10.63 GeV

$$\begin{aligned} \Gamma(B\overline{B}) &= 6.3 \text{ MeV} \\ \Gamma(B\overline{B}^* + B^*\overline{B}) &= 10.2 \text{ MeV} \\ \Gamma_{tot} &= 16.5 \text{ MeV} \end{aligned} \tag{3.41}$$

$$M(\Upsilon''') = 10.60 \text{ GeV}$$

$$\Gamma(B\overline{B}) = 3.9 \text{ MeV}$$

$$\Gamma(B\overline{B}^* + B^*\overline{B}) = 3.6 \text{ MeV} \tag{3.42}$$

$$\Gamma_{tot} = 7.5 \text{ MeV}$$

$$M(\Upsilon''') = 10.58 \text{ GeV}$$

$$\Gamma(B\overline{B}) = 2.5 \text{ MeV}$$

$$\Gamma(B\overline{B}^* + B^*\overline{B}) = 1.4 \text{ MeV} \tag{3.43}$$

$$\Gamma_{tot} = 4 \text{ MeV}$$

Thus, our guess for the width of the 4S is

$$\Gamma(\Upsilon''') \sim 5\text{-}20 \text{ MeV} . \tag{3.44}$$

It certainly should be carefully looked for at CESR.

IV. Hadronic Transitions and Multipole Expansion of QCD[33]

For an excited state of a $Q\overline{Q}$ system that lies below the OZI threshold, in addition to the gluonic annihilations it can also undergo another type of hadronic decays--the hadronic transitions:

$$\Phi' \to \Phi + \text{light hadrons}$$

where Φ' and Φ are members of the same $Q\overline{Q}$ system. In QCD this process proceeds through emission of gluons by the heavy quarks and subsequent conversion of gluons into light hadrons. Since the energy available to the light hadrons is small compared with the masses of Φ' and Φ, the emitted gluons are predominantly soft. Gottfried[34] pointed out that the heavy quark system moves slowly and has dimensions small compared to those of the emitted light quark system. Therefore the heavy quarks can be treated nonrelativistically, and a multipole expansion of the changing gauge field converges rapidly for sufficiently large heavy quark mass.

The situation is reminiscent of soft photon emission in atomic or nuclear physics. There, an extended distribution of charge and current are replaced by one or a few multipole moments. Without detailed knowledge of the bound states, it is possible to extract a lot of information. For example,

1. selection rules,
2. angular distributions and correlations,
3. relations among transitions between two multiplets that follow from the Wigner-Eckart theorem,
4. rate estimates from the order of multipole moments.

We will see that multipole expansion of QCD will enable us to obtain similar information for hadronic transitions.

There are, of course, differences between photon transitions and hadronic transitions. Gluons are confined and at least two or more gluons have to be emitted. Because of these differences, it is difficult to give a straightforward estimate of the expansion parameter ka, the product of the effective gluon momentum and the size of the heavy auark system. Nevertheless, let us be most simple-minded, and use two-gluon emission for our estimate. Then

$$k \simeq \frac{1}{2}\,\Delta E = 300\text{ MeV} \tag{4.1}$$

for both transitions $\psi' \to \psi$ and $\Upsilon' \to \Upsilon$. The sizes of these states can be estimated by a potential model. In the Cornell model we find

$$\begin{aligned}(ka)_\psi &\sim 0.7\\ (ka)_\Upsilon &\sim 0.2\end{aligned} \tag{4.2}$$

So the multipole expansion should be rather good for the $b\bar{b}$ system, but it is only marginal for the $c\bar{c}$ system. In the following we will assume that it works for both systems.

For a $Q\overline{Q}$ system the effective Hamiltonian including the coupling to the gauge field is

$$H' = H + H_I \tag{4.3}$$

$$H = \frac{p^2}{m_Q} + V(r) \tag{4.4}$$

$$H_I = -Q_a A_{a0} + \vec{d}_a \cdot \vec{E}_a + \vec{m}_a \cdot \vec{B}_a + \ldots \tag{4.5}$$

where Q_a, d_a, and $\vec{m}_a$ are the color charge, color electric dipole, and color magnetic dipole, respectively. The gauge field operators A_{a0}, $\vec{E}_a$, and $\vec{B}_a$ are evaluated at the center of the $Q\overline{Q}$ system. In (4.4) V(r) is the effective $Q\overline{Q}$ potential resulting from complicated gluon exchanges. The gauge fields in (4.5) will only couple to the emitted light quarks; they should not be used to produce further $Q\overline{Q}$ interaction since those effects are already incorporated in V(r).

For the transitions

$$\Phi' \rightarrow \Phi + 2\pi$$

the matrix element is given by second order perturbation

$$M = \langle \Phi 2\pi | \vec{x}\cdot\vec{E} \frac{1}{E' - H_8 + iD_o} \vec{x}\cdot\vec{E} | \Phi' \rangle \tag{4.6}$$

where $\vec{x}$ is the relative coordinate betw-en Q and $\overline{Q}$. Other notations are

$$\begin{aligned} D_o &= \partial_o - gA_o \\ A_o &= \sum_a F_a A_{ao} \\ (F_a)_{bc} &= f_{bac} \\ E_i O E_j &= \sum_{i,j} E_{ia} O_{ab} E_{jb} \end{aligned} \tag{4.7}$$

and H_8 is the projection of H in (4.4) into the $Q\overline{Q}$ octet sector.

Although $\vec{E}_i$ and D_o are very complicated objects, they behave as c-numbers to the $Q\overline{Q}$ system. Therefore, M is a reducible second rank tensor in the quark variables. We can decompose it into irreducible tensors $T^{(0)}$, $T_{ij}^{(1)}$ and $T_{ij}^{(2)}$:

$$T_{ij} = x_i \frac{1}{E' - H_s + iD_o} x_j = \delta_{ij} T^{(0)} + T_{ij}^{(1)} + T_{ij}^{(2)} \tag{4.8}$$

We will assume that V(r) is spin-independent. The Wigner-Eckart theorem implies that

$$\langle \Phi 2\pi | E^i T^{ij} E^j | \Phi' \rangle$$

$$= \sum_{k=0,1,2} c_{ij} \begin{Bmatrix} k & \ell & \ell' \\ S & J' & J \end{Bmatrix} \langle \ell\alpha 2\pi \| E^i T^{(k)} E^j \| \ell' \alpha' \rangle \tag{4.9}$$

where $\begin{Bmatrix} k & \ell & \ell' \\ S & J' & J \end{Bmatrix}$ is a 6-J symbol, and c_{ij} is a Clebsch-Gordan coefficient. The total angular momentum, orbital angular momentum, spin and other quantum numbers are labelled by J,ℓ,S, and α for Φ and with a prime for Φ'.

Generally, there are three reduced matrix elements. The interference can be eliminated after angular integrations. For example,

$$\frac{d\Gamma(J' \to J)}{dM_{\pi\pi}} = \sum_{k=0,1,2} C_k(M_{\pi\pi}) \begin{Bmatrix} k & \ell & \ell' \\ S & J' & J \end{Bmatrix}^2 \tag{4.10}$$

We now draw some simple conclusions from (4.10):

1. When $\ell = \ell' = 0$, only $T^{(0)}$ contributes and we have

$$\begin{aligned} d\Gamma(\psi' \to \psi + 2\pi) &= d\Gamma(\eta_c' \to \eta_c + 2\pi), \\ d\Gamma(\Upsilon' \to \Upsilon + 2\pi) &= d\Gamma(\eta_b' \to \eta_b + 2\pi), \\ d\Gamma(\Upsilon'' \to \Upsilon' + 2\pi) &= d\Gamma(\eta_b'' \to \eta_b' + 2\pi) \ , \end{aligned} \tag{4.11}$$

where η_c, η_c', η_b, η_b' and η_b'' are the spin singlet partners of ψ, ψ', Υ, Υ' and Υ'', respectively. From the known rate for $\psi' \to \psi + 2\pi$, we predict

$$\Gamma(\eta_c' \to \eta_c + 2\pi) \simeq 110 \text{ keV} \tag{4.12}$$

Furthermore, the angular and mass distributions of the pions should be identical in both $\psi' \to \psi$ and $\eta_c' \to \eta_c$ transitions.

2. For the Υ system the first two P-multiplets should lie below the OZI threshold. Hadronic transitions between them should be observable. There are nine transitions between the two spin triplet P-multiplets. Eq. (4.10) gives six relations. Three are contained in the reciprocity relation

$$(2J+1)d\Gamma(J' \to J) = (2J'+1)d\Gamma(J \to J') \tag{4.13}$$

The other three are

$$\begin{aligned} d\Gamma(1 \to 1) &= d\Gamma(0 \to 0) + \frac{1}{4}\, d\Gamma(0 \to 1) + \frac{1}{4}\, d\Gamma(0 \to 2), \\ d\Gamma(1 \to 2) &= \frac{5}{12}\, d\Gamma(0 \to 1) + \frac{3}{4}\, d\Gamma(0 \to 2), \\ d\Gamma(2 \to 2) &= d\Gamma(0 \to 0) + \frac{3}{4}\, d\Gamma(0 \to 1) + \frac{7}{20}\, d\Gamma(0 \to 2). \end{aligned} \tag{4.14}$$

In both (4.13) and (4.14) we have used the short-hand notation $d\Gamma(J' \to J)$ for $\frac{d\Gamma}{dM_{\pi\pi}}(\Phi_{J'}' \to \Phi_J + 2\pi)$.

We turn to another example to illustrate the predictive power of multipole expansion. Let's compare the two transitions

$$\psi' \to \psi + 2\pi, \qquad \Upsilon' \to \Upsilon + 2\pi$$
$$\psi' \to \psi + \eta, \qquad \Upsilon' \to \Upsilon + \eta\,.$$

Since η is a pseudoscalar, the transition goes through second order M1 transitions rather than E1 transitions. Thus

$$M(\Phi' \to \Phi + 2\pi) \sim \langle r^2 \rangle_{\Phi'} \tag{4.15}$$

$$M(\Phi' \to \Phi + \eta) \sim \left(\frac{1}{m_Q}\right)^2 \tag{4.16}$$

So we expect[34]

$$\frac{\Gamma(\psi' \to \psi + 2\pi)}{\Gamma(T' \to T + 2\pi)} \sim \frac{\langle r^2\rangle^2_{\psi'}}{\langle r^2\rangle^2_{T'}} \sim 10 . \qquad (4.17)$$

But

$$\frac{\Gamma(\psi' \to \psi + \eta)}{\Gamma(T' \to T + \eta)} \sim \left(\frac{m_b}{m_c}\right)^4 \left|\frac{P_\eta(\psi')}{P_\eta(T')}\right|^3 \sim 400 \qquad (4.18)$$

where we have included a p-wave phase space correction since $T' \to T + \eta$ has a very small available phase space. Thus $T' \to T + \eta$ will be much more suppressed than the transition $T' \to T + 2\pi$.

So far we have only made use of the information from the heavy quarks. The dependence of the transition amplitude on the light hadrons' momenta is in general unknown. But in the case of soft pion emission, current algebra and PCAC provide some information. Thus we would like to combine the multipole expansion and current algebra. We therefore assume that multipole expansion and PCAC are compatible. We will see that this assumption leads to interesting predictions.

Brown and Cahn[35] have studied $\psi' \to \psi + 2\pi$ using current algebra. They showed that if the σ-term $[\sim m_\pi^2]$ is neglected, then the matrix element $M(\psi' \to \psi + \pi_1 + \pi_2)$ is bilinear in q_1 and q_2. We will use this result to write

$$\begin{aligned}
\langle \ell\alpha 2\pi \| E^i T^{(2)} E^j \| \ell'\alpha'\rangle &= A(q_1^i q_2^j + q_2^i q_1^j - \tfrac{2}{3}\delta^{ij}\, \vec{q}_1 \cdot \vec{q}_2) \\
\langle \ell\alpha 2\pi \| E^i T^{(1)} E^j \| \ell'\alpha'\rangle &= 0 \qquad (4.19)\\
\langle \ell\alpha 2\pi \| E^i T^{(0)} E^j \| \ell'\alpha'\rangle &= \delta^{ij}(B q_1^\mu q_{2\mu} + C q_{10} q_{20})
\end{aligned}$$

where A, B and C are constants.

For $\psi' \to \psi\pi\pi$ and $T' \to T\pi\pi$, only $T^{(0)}$ and therefore B and C contribute. Brown and Cahn by appealing to data conclude that[36]

$$C(\psi' \to \psi\pi\pi) \not\approx 0. \tag{4.20}$$

For $\psi'' \to \psi\pi\pi$ only $T^{(2)}$, and hence A, contributes. In this case, the shapes of the mass distribution and angular distributions are uniquely determined. We plot the mass distributions to show that they are very different (Fig. 3). Although $\psi'' \to \psi\pi\pi$ is not a major decay mode, the same shape will also apply to transitions such as $2^3P_{J'} \to 1^3P_J + 2\pi$ $(J' \pm J)$. Hopefully experiments at CESR will test these distributions.

You may ask "Do these predictions depend on the vector nature of the gauge field?" The answer is "Yes". There is strong evidence from quark model spectroscopy for the existence of color and that all hadrons are color singlets. Let's accept this. Then in QCD the "monopole term" $Q_a A_{a0}$ can't produce a transition since

$$Q_a |\text{hadrons}\rangle = 0 .$$

But for the scalar gluons, the "monopole term" is <u>not</u> proportional to the total color charge. Therefore the monopole term can produce a transition. The leading operator for $\Delta L = 0$ transitions is a rotational scalar and is order 1 (in powers of size). This difference has the important consequence that the ratio of the rates for $\psi' \to \psi + 2\pi$ and $\Upsilon' \to \Upsilon + 2\pi$ is very different:

$$\text{Scalar gluon:} \qquad \frac{\Gamma(\psi' \to \psi\pi\pi)}{\Gamma(\Upsilon' \to \Upsilon\pi\pi)} \simeq 1 \tag{4.21}$$

$$\text{Vector gluon:} \qquad \frac{\Gamma(\psi' \to \psi\pi\pi)}{\Gamma(\Upsilon' \to \Upsilon\pi\pi)} \cong 10 \tag{4.22}$$

Gottfried[34] has argued that (4.21) is already ruled out by experiments at Fermilab.

V. Conclusions

The new particles have offered a testing ground for various ideas of quark dynamics. In particular, the nonrelativistic quark model has provided a simple framework for semiquantitative understanding of the systematics of these particles. Nevertheless, a model of this type cannot fit all the data, and the data does not support a particular model. This should not be surprising since the quark model is a very crude approximation to the complex dynamics of quarks and gluons. An infinite number of degrees of freedom have been truncated. For this reason it is important to search for evidence of degrees of freedom associated with gluon and/or color. They can manifest as extra states in e^+e^- annihilation.[37]

The expected richer spectrum of the Υ system will certainly reveal more information about the interquark potential, in particular, the spin-dependent forces. It should be recognized, however, that the existence of heavier quark systems will not shed more light on the nature of the quark confinement, since only the short distance part of the potential will be felt by these small systems.

For the Υ system many hadronic transitions will be accessible for experimental study. The multipole expansion of QCD gives many predictions for these transitions. I am looking forward to the day when these predictions will be tested.

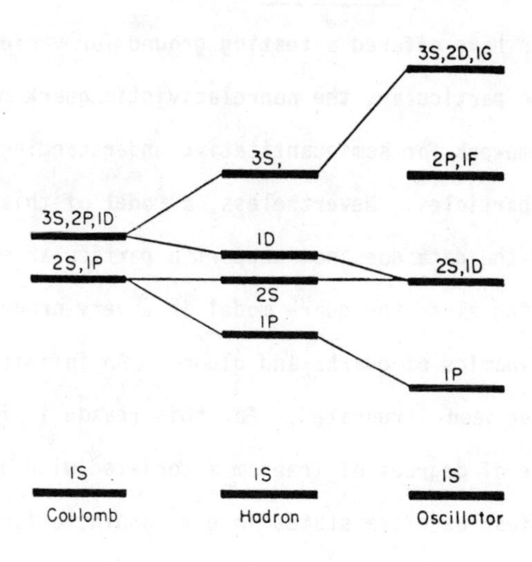

Fig. 1.

The qualitative features of the quark-antiquark excitation spectrum can be inferred by visual interpolation, as shown in this figure. Imagine a continuous deformation of an oscillator potential into a Coulomb field, and the associated change of the bound states. The potential of interest in mesonic spectroscopy is between the two extremes, whence the spectrum labelled "hadron". The parameters of the potentials are chosen so that the 2S-1S difference is fixed. In the case of the "hadron", not all levels are shown for the sake of clarity. Note that we label states by (n+l), where n is the number of radial nodes.

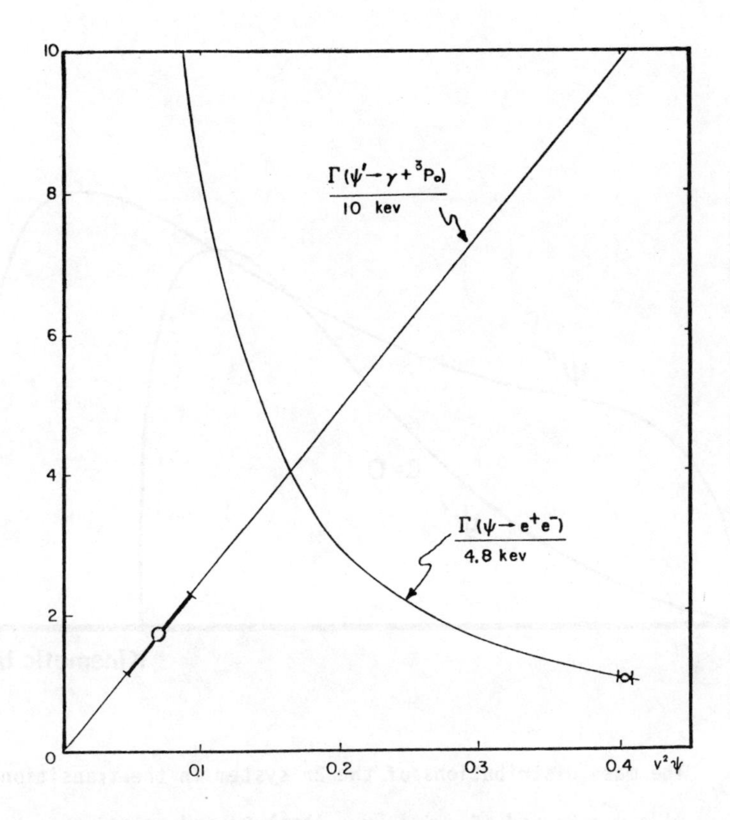

Fig. 2.

The E1 transition rate $\Gamma(\psi' \to \gamma + {}^3P_0)$ and the electronic width $\Gamma(\psi \to e^+e^-)$ as functions v_ψ^2. The energy differences E(2S) - E(1S) and E(1P) - E(1S) are held fixed at their experimental values. The measured values, together with their errors, are indicated.

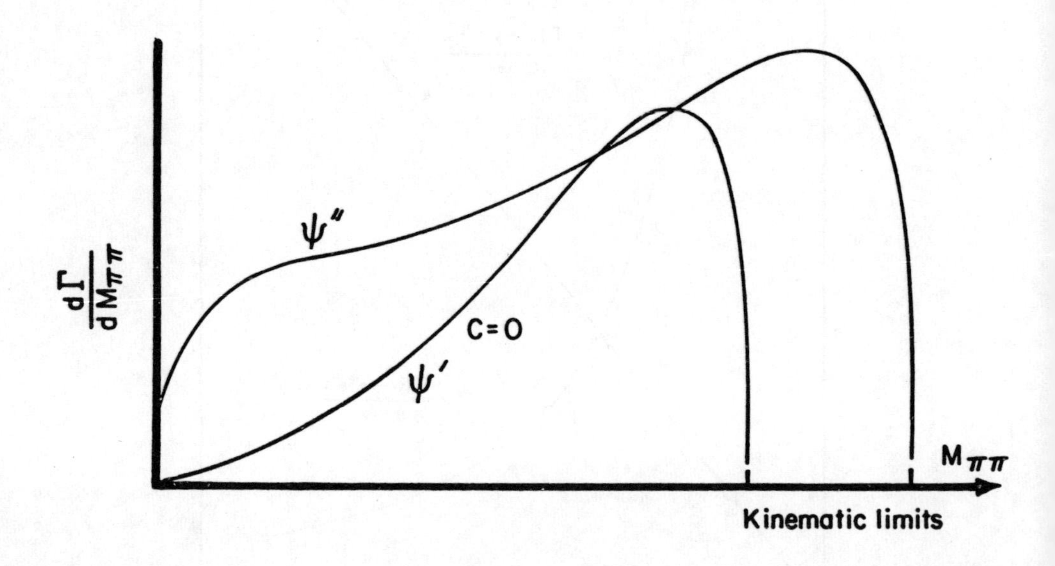

Fig. 3. The mass distributions of the 2π system in the transitions $\psi' \to \psi + 2\pi$ and $\psi'' \to \psi + 2\pi$. Absolute and relative normalization are arbitrary.

Table I. $c\bar{c}$ bound states in naive model, and their properties. Parameters used are m_c = 1.84 GeV, a = 2.34 GeV^{-1} and κ = 0.52.

State	Mass (GeV)	Γ_{ee} (keV)[b]	$\langle \frac{v^2}{c^2} \rangle$	$\langle r^2 \rangle^{1/2}$ (fm)	Candidate
1S	3.095[a]	4.8	0.20	0.47	ψ(3095)
1P	3.522[a]	---	0.20	0.74	$\chi_{0,1,2}$(3522 5)
2S	3.684[a]	2.1	0.24	0.96	ψ'(3684)
1D	3.81	---	0.23	1.0	ψ'(3772)[c]
3S	4.11	1.5	0.30	1.3	ψ(4028)
2D	4.19	---	0.29	1.35	ψ(4160)[d]
4S	4.46	1.1	0.35	1.7	ψ(4414)
5S	4.79	0.8	0.40	2.0	

[a]Input

[b]Correction factor $(1 - \frac{4\kappa}{\pi})$ = 0.341 is included.

[c]See Ref. 11.

[d]See Flügge's review of Ref. 1.

Table II. Measured properties of upsilon states. The first error is statistical, the second systematic.

	DORIS[a]	DORIS[b]	CLEO[c]	CLEO[d]
$M(\Upsilon')-M(\Upsilon)$ (MeV)	555±11	560±10	560.7±0.8±3.0	559±1±3
$\Gamma_e(\Upsilon')/\Gamma_e(\Upsilon)$	0.23±0.08	0.31±0.09	0.44±0.06±0.04	0.39±0.06
$M(\Upsilon'')-M(\Upsilon)$ (MeV)	---	---	891.1±0.7±3.0	889±1±5
$\Gamma_e(\Upsilon'')/\Gamma_e(\Upsilon)$	---	---	0.35±0.04±0.03	0.32±0.04

[a]Data from C. W. Dardeen et al., Phys. Lett. 76B, 246 (1978) and 78B, 364 (1978) and G. Flugge, Tokyo Conference Proceedings, Tokyo 1978, p. 807.

[b]Data from J. K. Bienlein et al., Phys. Lett. 78B, 360 (1978).

[c]Data from CLEO collaboration, Ref. 30.

[d]Data from Columbia-Stony Brook collaboration, Ref. 31.

Table III. Predictions for upsilon states in different models. Quantities bearing an asterisk are inputs.

	Coulomb plus linear[a]	Bhanot and Rudaz[b]	Richardson[c]	Logarithmic[d]	Inverse scattering[e]
$M(\Upsilon')-M(\Upsilon)$ (MeV)	560*	560	555	590	560*
$\Gamma_e(\Upsilon')/\Gamma_e(\Upsilon)$	0.39	0.44	0.42	0.45	0.26
$M(\Upsilon'')-M(\Upsilon)$ (MeV)	900	880	886	910	900
$\Gamma_e(\Upsilon'')/\Gamma_e(\Upsilon)$	0.27	0.30	--	0.29	0.27
$M(\Upsilon''')-M(\Upsilon)$ (MeV)	1170	1140	1150	1140	1170
$\Gamma_e(\Upsilon''')/\Gamma_e(\Upsilon)$	0.23	0.26	--	0.21	--

[a] Ref. 8.

[b] Ref. 14.

[c] Ref. 15.

[d] Ref. 13.

[e] This is taken from a talk by J. Rosner at the 1979 Meeting of the Division of Particles and Fields of the APS, Montreal, Quebec, October 1979. The author gave several predictions depending on the assumed correction factor to the Van Royen-Weisskopf formula. The one listed fits the mass difference $\Upsilon''-\Upsilon$ and it here corresponds to a correction factor of unity.

References

1. For a review of the experimental situation of the new particles, see G. Feldman and M. Perl, Phys. Reports 33C, 285 (1977); G. Flügge, in 1979 Tokyo Proceedings, pp. 793-810; Report on Crystal Ball Collaboration at Irvine Conference, December 1979. Unless stated otherwise, data are taken from these reviews.
2. Y. S. Tsai, Phys. Rev. D4, 2821 (1971).
3. T. Appelquist and H. D. Politzer, Phys. Rev. Lett. 34, 43 (1975).
4. S. L. Glashow, J. Iliopoulos, and L. Maiani, Phys. Rev. D2, 1285 (1970).
5. For a speculation along this line, see H. Harari, Phys. Lett. 86B, 83 (1979).
6. G. Sterman and S. Weinberg, Phys. Rev. Lett. 39, 1436 (1977).
7. For recent theoretical reviews, see: K. Gottfried, in Proc. of the 1977 International Symposium on Lepton and Photon Interactions at High Energies, Hamburg, F. Gutbrod, ed. (DESY, Hamburg, 1978); V. A. Novikov et al., Phys. Reports 41C, 1 (1978); T. Appelquist, R. M. Barnett, and K. Lane, Ann. Rev. Nucl. Part. Sci. 28, 387 (1978); J. D. Jackson, C. Quigg, and J. L. Rosner, in Proc. XIX Int. Conf. on High Energy Physics, Tokyo, 1978, edited by S. Homma, M. Kawaguchi, and H. Miyazawa (Phys. Soc. Japan, Tokyo, 1978), p. 381; M. Krammer and H. Krasemann, DESY 79/20 (Lectures at 18th Int. Universitatswochen fur Kernphysik, Schladming, Austria, Feb. 28-Mar. 10, 1979, unpublished); and C. Quigg, FERMILAB-CONF-79/74-THY, Sept. 1979, to be published in Proc. Int. Symp. on Lepton and Photon Interactions at High Energies, Batavia, Ill., Aug. 23-27, 1979.
8. E. Eichten, K. Gottfried, T. Kinoshita, K. D. Lane and T. M. Yan, Phys.

Rev. D17, 3090 (1978), ibid. D21, 203 (1980), and ibid., D21, 313 (1980).

9. E. Eichten, K. Gottfried, T. Kinoshita, J. Kogut, K. D. Lane and T. M. Yan, Phys. Rev. Lett. 34, 369 (1975).

10. C. Peck, report on Crystal Ball data at the 1979 Meeting of the Division of Particles and Fields of the APS, October 1979, Montreal, Quebec, and report at the Irvine Conference, Irvine, California, December 1979.

11. P. A. Rapidis et al., Phys. Rev. Lett. 39, 529 (1977).

12. In the second article of ref. 8, arguments are given to support this viewpoint.

13. C. Quigg and J. L. Rosner, Phys. Lett. 71B, 153 (1977).

14. G. Bhanot and S. Rudaz, Phys. Lett. 78B, 114 (1978).

15. J. L. Richardson, Phys. Lett. 82B, 272 (1979).

16. H. B. Thacker, C. Quigg, and Jonathan L. Rosner, Phys. Rev. D18, 274, 287 (1978); C. Quigg, H. B. Thacker, and Jonathan L. Rosner, Phys. Rev. D21, 234 (1980); Jonathan F. Schonfeld, Waikwok Kwong, Jonathan L. Rosner, C. Quigg, and H. B. Thacker, FERMILAB-PUB-79/77-THY, to be submitted to Ann. Phys. (N.Y.). See also H. Grosse and A. Martin, Nucl. Phys. B148, 413 (1979; Bruce McWilliams, Phys. Rev. D20, 1227 (1979).

17. The discussion here is presented in more detail in ref. 8.

18. R. Barbieri et al., Nucl. Phys. B105, 125 (1976); W. Celmaster, preprint SLAC-PUB-2151 (July 1978); E. C. Poggio and H. J. Schnitzer, Brandeis preprint (January 1979); L. Bergström, H. Snellman and G. Tengstrand, Stockholm preprint TRITA-TFY-78-3.

19. W. L. Wiese, M. W. Smith and B. M. Miles, Atomic Transition Probabilities, U.S. National Bureau of Standards, National Standards Ref. Data Series-22 (U.S. GPO, 1969).

20. See, for example, K. Gottfried, ref. 7.

21. M. A. Shifman et al., Phys. Lett. 77B, 80 (1978).

22. R. Barbieri et al., Nucl. Phys. B154, 535 (1979).

23. W. E. Caswell, G. P. Lepage, and J. Sapirstein, Phys. Rev. Lett. 38, 388 (1977).

24. For a review of these attempts, see the article by Appelquist, Barnett and Lane in ref. 7 and works cited in ref. 10.

25. E. Eichten and F. Feinberg, Phys. Rev. Lett. 43, 1205 (1979).

26. K. D. Lane and E. Eichten, Phys. Rev. Lett. 37, 477 (1976).

27. See also A. Le Yaouanc et al., Phys. Lett. B71, 397 (1977).

28. S.-H. H. Tye, Phys. Rev. D13, 3416 (1976); R. C. Giles and S.-H. H. Tye, ibid. D16, 1079 (1977).

29. K. T. Chao, Nucl. Phys. B, to be published.

30. D. Andrews et al., Cornell preprint CLNS 80/445 (1980).

31. T. Böhringer et al., to be published.

32. In ref. 8 it is shown that this value of m_b is insensitive to the other model parameters. For an earlier discussion of heavy quark spectroscopy, see E. Eichten and K. Gottfried, Phys. Lett. 66B, 286 (1977).

33. Material in this section is drawn from a forthcoming paper being prepared by T. M. Yan.

34. K. Gottfried, Phys. Rev. Lett. 40, 598 (1978); M. B. Voloshin, ITEP preprint, Moscow, 1978; G. Bhanot, W. Fischler and S. Rudaz, Nucl. Phys. B155, 208 (1979); M. E. Peskin, ibid. B156, 365 (1979); G. Bhanot and M. E. Peskin, ibid. B156, 391 (1979).

35. L. Brown and R. Cahn, Phys. Rev. Lett. 35, 1 (1975).

36. G. S. Abrams et al., Phys. Rev. Lett. 33, 1453 (1974); G. S. Abrams et al., Phys. Rev. Lett. 34, 1181 (1975).

37. W. Buchmüller and S.-H. H. Tye, Cornell preprint CLNS-438 (1979), to appear in Phys. Rev. Lett.; Kuang-Ta Chao and S. F. Tuan, these Proceedings.

Large Scale Effects via Factorization and Renormalization Group*

York-Peng Yao

Randall Laboratory of Physics
University of Michigan
Ann Arbor, MI 48109

In certain physical processes where one of the energy, momentum, or mass scales is much larger than the rest and where the flow of this large scale can be properly ordered, then the relevant matrix elements can be evaluated by the factorized formula

$$\sum_i C_i \, \Gamma^n(0_i)$$

where C_i's are process independent universal coefficient functions which contain all the relevant behavior of the largc scale. They can be calculated via renormalization group technique. 0_i's are effective vertices, the matrix elements of which $(\Gamma^n(0_i))$ may contain long distance effects, such as due to binding. The neglected terms are smaller by order $(1/\text{large scale})^2$.

Deeply inelastic production, inclusive annihilation processes, etc, are some examples. Another class of problems where factorization occurs pertain to effects of heavy particles in low energy light particle phenomena.

The tools for analysis of this sort are Zimmermann's algebraic renormalization and gauge invariance.

---

* Supported in part by the U.S. Department of Energy.

I. INTRODUCTION

There are quite a few processes in nature whose interaction at short distance can be identified and isolated. For example, in deeply inelastic production $e + p \to e + X$,[1] the momentum of the virtual photon is hignly space-like. The focus here is on the commutator

$$\int d^4x \cdot e^{-iq\cdot x} \langle p|[J_\mu(x), J_\nu(0)]|p\rangle$$

which, in the rest system of the proton, can be written as

$$\int d^4x \exp\left\{i[x_-((\nu^2+q^2)^{1/2}+\nu) - x_+((\nu^2+q^2)^{1/2}-\nu)]\right\} \cdot [J_\mu(x), J_\nu(0)]$$

where

$$x_\pm = \frac{1}{2}(x^\circ \pm x_3), \qquad \nu = -p\cdot q/m$$

As $q^2 \to \infty$, with $\omega = -2p\cdot q/q^2 \geq 1$ fixed, the region being studied is $x_- \cong 0$ and $x_+ = \text{finite}$. Because of the causal requirement

$$x^2 = x_\perp^2 - 4\,x_+x_- \cong x_\perp^2 \leq 0$$

we have

$$x_\perp^2 \cong 0$$

or

$$x^2 \cong 0$$

We are investigating physics near the light cone. This can be made more explicit when we expand[2]

$$[J_\mu(x), J_\nu(0)] = \sum_{n,i} x_{\mu_1} \cdots x_{\mu_n} C_n^i(x^2) O_{\mu\nu}^{i\,\mu_1\cdots\mu_n}$$

Here, $C_n(x^2)$ contain all the physics near $x^2 = 0$, which can be projected out by taking moments.[3]

In a simple kinematical argument due to Feynman,[4] one can interpret results of such a process as taking an instant snapshot of the distribution of constituents in the proton. This suggestion has provided the most intriguing stimulus and put short distance physics in the limelight.

Many people have extended the short distance analysis to other processes, especially in the past two years. This has broadened the tests of quantum chromodynamics considerably. These include

(a) moments of structure functions of some inclusive processes,[5] such as

$$e + p \to e + \text{anything}$$

$$e^+ + e^- \to h + \text{anything}$$

$$e^+ + e^- \to h_1 + h_2 + \ldots + h_n + \text{anything}$$

$$p + p \to \mu^+ + \mu^- + \text{anything}$$

(b) exclusive processes,[6] such as form factor of

$$e + \pi \to e + \pi$$

or the structure function of

$$e + \pi \to e + \text{anything}$$

in the $x = -q^2/2p\cdot q \to 1$ limit.

(c) There is a class of problems which also test short distance physics. This has to do with effects of heavy particles in processes involving only light external particles at low energy.[7]

The role here is equivalent to using (g-2)/2 to test QED renormalization program. A by-product of this is to give precise meaning to the notion of effective interaction.

There are two characteristic features which are particularly worth noting in the results obtained. (1)Factorization:[8] one is looking for processes whose measured quantities can be written as

T = S(short distance) · L(long distance)

where the division into short and long distance is determined in principle by the sizes of the participating composite hadronic systems. In the deeply inelastic production, the size is presumably the proton or pion radius, although this issue is not well understood. (2) Improved perturbation calculation: S is calculable through the renormalization group technique.[9] This becomes even more meaningful in QCD because of asymptotic freedom,[10] i.e. the coupling constant becomes weak in the large scale limit.

As for L, in some cases, it can be eliminated either by taking ratios at different short distance points or by abstracting its parameterization from other processes. Or, one can apply semi-phenomenological methods to give an estimate. We shall not go into this aspect here.

Now, there is one single technique which is universally applicable to all these processes. Because the underlying physics of the renormalization group is the freedom to change scales, there should exist certain 'orderedness' in the response of a system to the change of scale. The machinery which helps us to organize the ordered flow is the Zimmermann's algebraic identity.[11] The central idea here is to shuffle the integrand in such a way that one shall be able to systematically short-circuit the flow of the large scale and extract out the short distance physics. The complementary component in such an analysis will be certain matrix elements which may be plagued with long distance effects, due to binding and mass singularities.

II. GENERAL TECHNIQUE FOR FACTORIZATION

As noted earlier, the technique which is useful in short distance analysis is the Zimmermann's algebraic identity. I shall in the following give two examples to show how it works.

Example 1: [7]

Consider the problem of extracting $0(1/M^2)$ effects of the muon loop on the electron self energy.

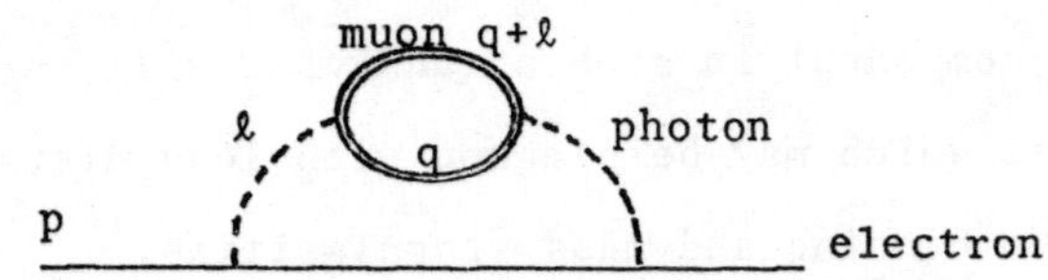

The renormalized electron self energy is given by

$$\sum(p) = \frac{e^2}{(2\pi)^4 i} \int d^4\ell \; (1-t_1^{\,p}) \; I(1-t_o^{\,\ell}) \; \bar{\pi}\,(\ell^2)$$

where

$$(g_{\lambda\kappa} - \frac{\ell_\lambda \ell_\kappa}{\ell^2}) \; \ell^2 \; \bar{\pi}(\ell^2) = \frac{e^2}{(2\pi)^4 i} \int d^4q \; \mathrm{Tr} \left\{ \gamma_\lambda \; \frac{1}{M+\gamma\cdot q} \; \gamma_\kappa \cdot \frac{1}{M+\gamma\cdot(q+\ell)} \right\}$$

is the muon contribution to the photon polarization. M is the muon mass. t_o^ℓ is a Taylor operator, i.e.

$$(1-t_o^{\,\ell}) \; \bar{\pi}(\ell^2) = \bar{\pi}(\ell^2) - \bar{\pi}(0)$$

and $t_1^{\,p}$ is another Taylor operator

$$(1-t_1{}^p)\ I(p,\ell) = I(p,\ell)-I(o,\ell)-\gamma\cdot p\ \frac{\partial}{\partial\gamma\cdot p}\ I(o,\ell)$$

where

$$I(p,\ell) = \gamma_\mu\ \frac{1}{\mu+\gamma\cdot(p-\ell)}\ \gamma_\nu\ (g^{\mu\nu}-\frac{\ell^\mu\ell^\nu}{\ell^2})\ \frac{1}{\ell^2}$$

To extract the $O(1/M^2)$ effects, we do it by the following rearrangement

$$(1-t_1{}^p)(1-t_1{}^\ell) = (1-t_3{}^p)(1-t_2{}^\ell) + (1-t_3{}^p)(t_2{}^\ell-t_o{}^\ell)$$

$$+ (t_3{}^p-t_1{}^p)(1-t_o{}^\ell)$$

Then

$$\sum(p) = \frac{e^2}{(2\pi)^4 i}\int d^4\ell\ [(1-t_3{}^p)I(1-t_2{}^\ell)\ \overline{\pi}$$

$$+ (1-t_3{}^p)I(t_2{}^\ell - t_o{}^\ell)\ \overline{\pi}$$

$$+ (t_3{}^p-t_1{}^p)I(1-t_o{}^\ell)\ \overline{\pi}] \qquad (2.1)$$

Let us consider the first term in the above expression. There,

$$(1-t_2{}^\ell)\overline{\pi}(\ell) = \overline{\pi}(\ell)-\overline{\pi}\ (0) - \ell^2\ \frac{\partial}{\partial\ell^2}\ \overline{\pi}(0)$$

Its asymptotic behavior is certainly better than that in the region $0 \lesssim |\ell^2| \lesssim M^2$, which is

$$\sim (\frac{\ell^2}{M^2})^2$$

Also

$$(1-t_3{}^p)\ I(p,\ell) = I(p,\ell)-I(o,\ell)-\gamma\cdot p\ \frac{\partial}{\partial\gamma\cdot p}\ I(o,\ell)$$

$$-\ \frac{(\partial\cdot p)^2}{2!}\ (\frac{\partial}{\partial\gamma\cdot p})^2\ I(o,\ell)-\ \frac{(\partial\cdot p)^3}{3!}(\frac{\partial}{\partial\gamma\cdot p})^3\ I(o,\ell)$$

$$\sim \quad (\frac{1}{\ell^2})^4 \quad \text{as } \ell^2 \to \infty$$

Thus, putting these two behaviors together, we conclude that the first term of Eq. (2.1) goes as $1/M^4$ and can be neglected.

We look at the second term of Eq. (2.1), where

$$(t_2{}^{\ell}-t_o{}^{\ell})\bar{\pi}(\ell) = \ell^2 \frac{\partial}{\partial \ell^2} \bar{\pi}(0) = - \frac{\alpha}{15\pi} \frac{\ell^2}{M^2}$$

Because of the $(1-t_3{}^{p})$ operation, the remaining integration is finite. We can regard the contribution of $(t_2{}^{\ell}-t_o{}^{\ell})\bar{\pi}$ as

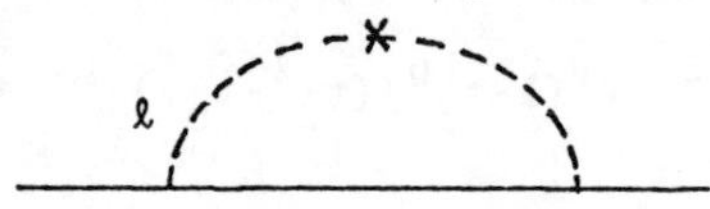

where the cross (x) denotes an insertion in the photon propagator with a vertex $-\ell^2$, multiplied by a certain coefficient. In fact, we have a local operator $-\frac{i}{4}\int d^4x\, F^{\mu\nu}\, \partial^2\, F_{\mu\nu}$ inserted into the photon propagator, with a coefficient $\frac{\alpha}{15\pi}\frac{1}{M^2}$. Note that the coefficient has all the muon mass effects, while the operator is evaluated in the theory with electrons and photons only.

We come to the last term of Eq. (2.1), in which

$$(t_3{}^{p}-t_1{}^{p})\ I(p,\ell) = \frac{(\gamma\cdot p)^2}{2!}\left(\frac{\partial}{\partial \gamma\cdot p}\right)^2 I(p=o,\ell)$$
$$+ \frac{(\gamma\cdot p)^3}{3!}\left(\frac{\partial}{\partial \gamma\cdot p}\right)^3 I(p=o,\ell)$$

Note that this is a polynomial in p. For large ℓ, it goes like $1/(\ell^2)^3$. Thus, up to logarithm factors, we know that the integral is of order $1/M^2$. We can summarize their contributions as

i.e. some operators inserted into the electron propagator with vertices p^2 and p^3 and certain coefficients. These operators are $-i\int d^4x\ \bar{\psi}(\gamma\cdot\frac{1}{i}\partial)^2\psi$ and $-i\int d^4x\ \bar{\psi}(\gamma\cdot\frac{1}{i}\partial)^3\psi$ with coefficients, which are respectively $-\frac{\alpha^2}{40\pi^2}\frac{1}{M^2}$ and $-\frac{\alpha^2}{4\pi^2M^2}(\frac{1}{30}\ell n\frac{M^2}{m^2}+\frac{47}{900})$. Again, we notice that the operators and their matrix elements make no reference to muons, while the coefficient functions contain all the short distance (muon mass) dependence. This is factorization.

Since this example is a prototype to illustrate factorization, it is worth rephrasing the procedure.

There are two paths for the flow of the large scale M. The first is within the muon loop. To confine it in this region, we apply the operation $(t_2^{\ell}-t_o^{\ell})\bar{\pi}$. Let us make a cut to denote it

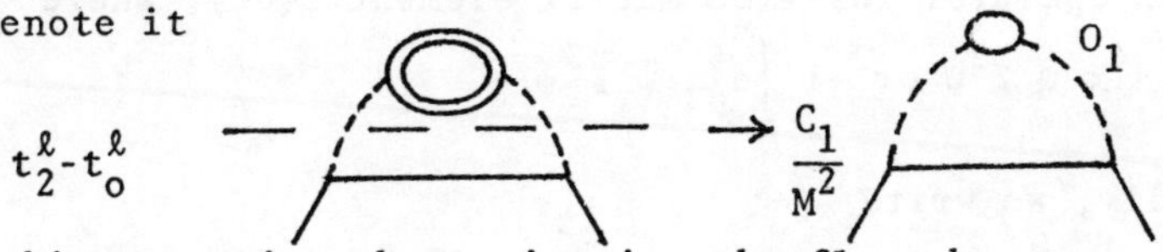

Note that this operation short circuits the flow, because the momentum ℓ into the upper part is evaluated at null value. Thus, the upper part turns into a coefficient function C_1, while the lower part becomes an operator inserted matrix element $\Sigma(O_1)$, with

$$O_1 = -i\int d^4x\ F^{\mu\nu}\ \partial^2\ F_{\mu\nu}$$

Power counting in momentum integration dictates that the renormalization of this matrix element requires two extra subtraction over the naive degree of divergence (d=1). This is why $(1-t_p^{\ 3})$ is applied.

Let us rewrite the diagram as

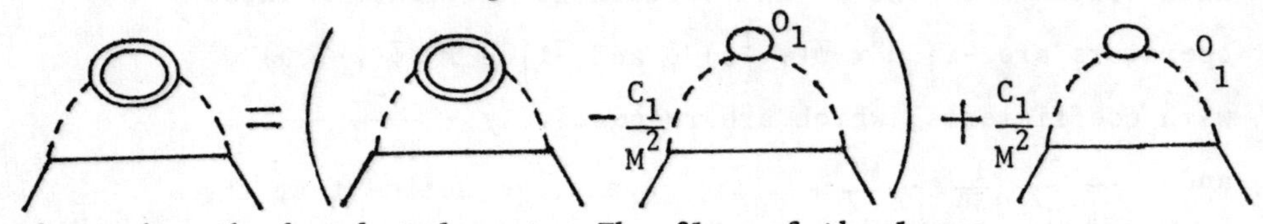

and examine the bracketed part. The flow of the large scale now moves into the photon line. We extract its contribution by the short circuiting operation $t_p^{\ 3}-t_p^{\ 1}$. This is again denoted by a cut

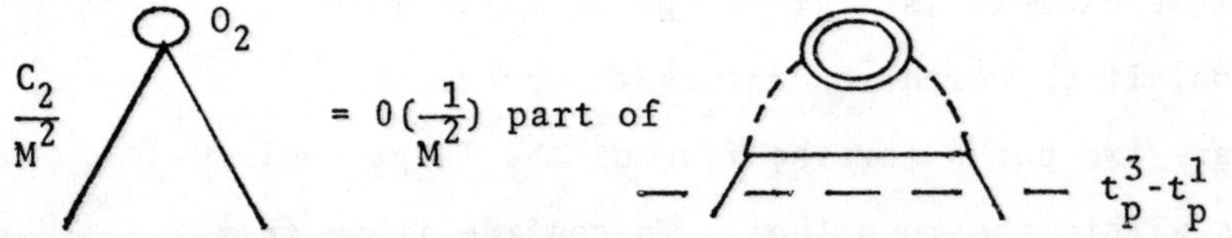

$\frac{C_2}{M^2}$ O_2 $= O(\frac{1}{M^2})$ part of $t_p^3-t_p^1$

The upper part becomes a coefficient C_2, while the lower part is an operator inserted matrix element $\Sigma(O_2)$, where $O_2 = -i\int d^4x\, \overline{\psi}\, \not{\partial}^2\psi$ or $-i\int d^4x\, \overline{\psi}\, \not{\partial}^3\psi$.

Finally, we write

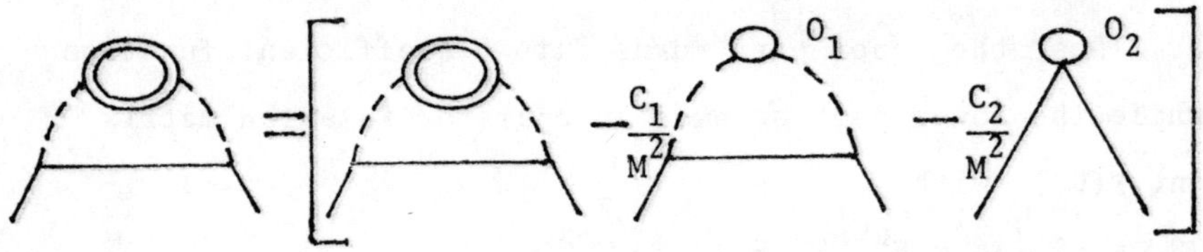

$= \Big[\; - \frac{C_1}{M^2} \, O_1 \; - \frac{C_2}{M^2} \, O_2 \Big]$

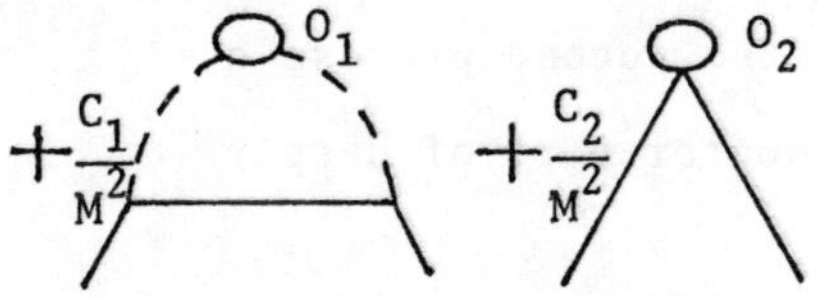

$$+\frac{C_1}{M^2}\ \ O_1 \qquad +\frac{C_2}{M^2}\ \ O_2$$

the combination in square brackets has no $1/M^2$ contribution and we have accomplished factorization to this order.

To summarize, what we need in the general case are two main ingredients to extract large scale contributions, viz. clever algebraic rearrangement and a set of rules for power counting. The former is provided by Zimmermann. The latter may differ from process to process and the justification of it usually is tenuous.

Example 2:[12]

Here we shall give an example to show how counting should be done in an inclusive process. To simplify kinematics, let us consider $(\phi^3)_6$.

The basic inclusive processes here are

$\phi(q) \rightarrow \phi(p)$ + anything

and

$\phi(q) + \phi(p) \rightarrow$ anything

The first corresponds to the decay of a time-like ($q^2<0$) scalar particle. The momentum p of one of the final particles is analyzed. This processes is equivalent to $e^+e^- \to h+x$, where the 'photon' is a scalar. In the second process q is space-like ($q^2>0$) and it is the counter part of deeply inelastic production.

In the following treatment, there is no essential difference in the two situations and we shall for concreteness use kinematics of the decay process.

Consider the diagram

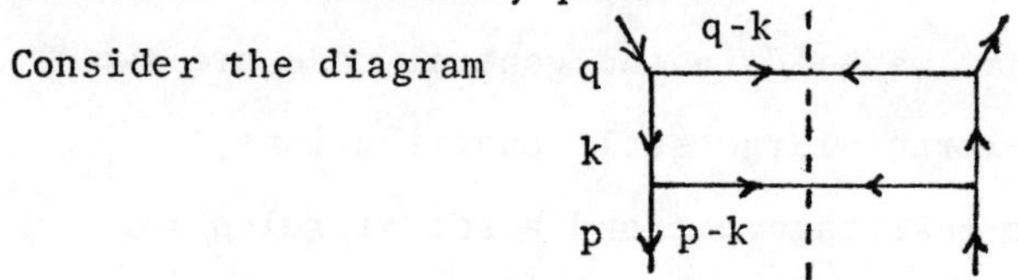

of which the relevant integral is

$$I = \int d^6k \; \delta((q-k)^2+m^2) \frac{1}{(k^2+m^2)^2} \delta((k-p)^2+m^2)$$

$$\theta(q^\circ-k^\circ)\theta(k^\circ-p^\circ)$$

The short distance aspect manifests itself in the accumulation of powers of $\ell n q^2$, which originate from the lack of sharp transverse momentum cutoff. We must develope a set of counting rules to facilitate isolating such factors.

Let us work in a system where q is along the 5th direction. The light cone variables are

$$a_\pm = \frac{1}{\sqrt{2}} (a^\circ \pm a_5) \qquad a_\perp = a_{1,2,3,4}.$$

The scalar product of two vectors becomes

$$a\cdot b = \vec{a}_\perp\cdot\vec{b}_\perp - a_+b_- - a_-b_+.$$

Among such systems, we choose one in which q_+ is very large, but q_- is finite. Also p^μ is finite. Thus, it follows that

$\omega \equiv 2p\cdot q/q^2 \cong p_-/q \quad \leq 1$

One of the δ-function has as its argument

$$0 = (q-k)^2+m^2 = -2(q_+-k_+)(q_--k_-)+k_\perp^2+m^2$$

which gives

$$q_\pm \geq k_\pm$$

In particular, k_- is finite. The other δ-functions gives

$$(k-p)^2+m^2 = 0$$

or

$$k_+ = \frac{(k_\perp-p_\perp)^2+m^2}{2(k_--p_-)} + p_+$$

Taken together, these suggest that we should assign 'degrees' as follows

$k_\perp$	1
k_+	2
k_-	0

Now, for the integral at hand, there are two routings of large q^2. One is through the first line

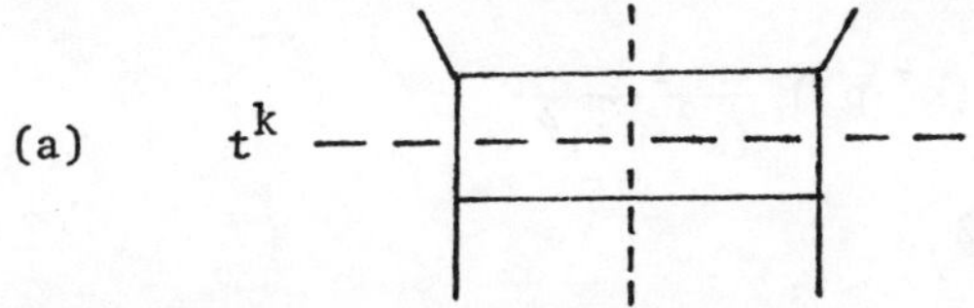

and the other is throughout the whole graph

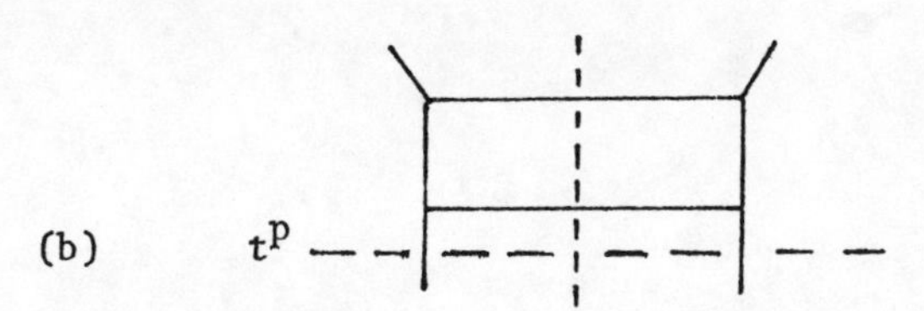

The integral is of course perfectly finite, since it is a tree. However to extract the $\ell n\ q^2$ factor, we shall pretend that it had divergences and count with the rules

d' = 2+6 x number of integration
+ number of $k_\perp$ in the numerator
+ 2 x number of k_+ in the numerator
- contribution from δ-functions and denominators.

We write

$$1 = (1-t^p)(1-t^k) + (1-t^p)\ t^k + t^p$$

with

$$t^p\ f(p) = f(p=\hat{p})$$

$$t^k\ g(k) = g(k=\hat{k})$$

and

$$\hat{p} = (o, p_-, \vec{o}_\perp)$$

$$\hat{k} = (o, k_-, \vec{o}_\perp)$$

Note that t^k acts only on the upper horizontal line and

$$\left.\frac{2k\cdot q}{q^2}\right|_{k=\hat{k}} = \frac{k_-}{q_-} = \omega'$$

The contribution of diagram (a) according to this arrangement is

$$I^{(a)} = \int d^6k\ \frac{1}{q^2}\ \delta(1-\omega')\theta(k^- - p^-)\ \frac{1}{(k^2+m^2)^2}$$

$$\cdot\ (1-t^p)\delta((k-p)^2+m^2)$$

which is factorized by taking moments

$$\int_0^1 d\omega\, \omega^{\sigma+2}\, I^{(a)}$$

$$= \int d(\frac{p_-}{q_-})\, (\frac{p_-}{q_-})^{\sigma+2}\, I^{(a)}$$

$$= \int d(\frac{k_-}{q_-})(\frac{k_-}{q_-})^{\sigma+2}\, p_-^{\sigma+3}\, k_-^{-\sigma-3}\, I^{(a)}$$

$$= \int d\omega'\, \omega'^{\sigma+2}\, \frac{1}{q^2}\, \delta(1-\omega')$$

$$\cdot \int d^6k\, (k_-/p_-)^{-\sigma-3}\, \theta(\frac{k_-}{p_-} - 1)\frac{1}{(k^2+m^2)^2}\, (1-t^p)\,\delta((k-p)^2+m^2)$$

which is of the form $A_\sigma(q^2)\, \Gamma_\sigma(p\ )$, where

$$A_\sigma(q^2) \sim \int d\omega'\, \omega'^{\sigma+2}\, \frac{1}{q^2}\, \delta(1-\omega')$$

$\Gamma_\sigma(p)$ is called the cut vertex and it is quite suggestive that it satisfies the equation

$$\Gamma_\sigma(p^2)=1+\int d^6k\, (k_-/p_-)^{-\sigma-3}\, \theta(\frac{k_-}{p_-} - 1)\, |\Delta(k^2)|^2\, T(k,p)$$

where $T(k,p)$ is a two particle irreducible scattering kernel.

We can do the same for the flow in diagram (b), which is short circuited by t^p in the algebraic identity. Factorization is obtained and a cut vertex (which is trivial in this case) is introduced.

General Case:

Given an integrand I_Γ corresponding to a diagram Γ, we can identify the one particle irreducible divergent parts (γ). These are called renormalization parts and their degrees of divergence are $d(\gamma) \geq 0$. Let us indicate by $t^{d(\gamma)}$ the Taylor operation which renders γ finite. We construct all the forests U, each of which is made up of a set of non-overlapping renormalization parts. Note that a null set is also considered as a forest. The BPH Z scheme states that I_Γ is made finite after the indicated operations:[11]

$$R_\Gamma = \sum_U \sum_{\gamma \varepsilon U} (-t)^{d(\gamma)} \, I_\Gamma .$$

Now, we make a cut to divide the diagrams Γ into two parts. For example, in the problem of the effects of heavy particles in light particle low energy physics (Example 1),[7] the upper part contains at least all the heavy particle lines and must be one particle irreducible. The rest is the lower part.

In the decaying problem (Example 2).[12] The upper part contains at least both of the decaying time-like scalars and the lines which have been cut should be one particle irreducible The rest is the lower part.

We shall describe briefly how factorization is proven in general. We assume that a suitable new set of divergence counting rules has been chosen, which are distinct from the set used for renormalization.

Let us call the upper part τ and denote the corresponding integrand by Δ. We form all the normal forests U_2, whose renormalizations parts are in τ but does not coincide with it. These renormalization parts shall be made finite by applying the usual counting rules for renormalization.

Then the other forests U_1 will either contain τ or be disjoint from it. Again, those which are disjoint will be subtracted with the renormalization rules. However, those which contain or coincide with τ will be subtracted with the new divergence counting rules. Let us denote such subtraction operations by t' and the degree of divergence d'. When, it can be shown that[11]

$$R_\Gamma = \sum_\tau \sum_{U_1} \sum_{U_2} \prod_{\gamma' \varepsilon U_1} (-t')^{d'(\gamma')} I/\Delta$$

$$(t^{d'(\tau)} - t^{d(\tau)}) \prod_{\gamma \varepsilon U_2} (-t)^{d(\gamma)} \Delta + \text{neglected terms.}$$

The 'neglected terms' are at least $O(1/\text{large scale})^2$ smaller than the terms shown.

Because of the short-circuiting operations $t^{d'(\tau)} - t^{d(\tau)}$ we have almost a factorization formula. In Example 1, after summing up all the diagrams, carrying out all the internal integrations, and taking out some polynomial momentum factors in $t^{d'(\tau)}-t^{d(\tau)}$, we can identify the upper parts to be the universal coefficient functions. The lower parts together with such momentum polynomials are operator inserted matrix elements. To be more precise, for an n light particle proper amputated Green's function, we can summarize the heavy particle effects by the formula

$$\Gamma^n_{\text{full theory}} = \Gamma^n_{\text{light theory}}$$

$$+ \frac{1}{M^2} \sum_i C_i \left(\frac{M}{m}\right) \Gamma^n_{\text{light theory}} (O_i) + O\left(\frac{1}{M^4}\right).$$

Here, 'full theory' quantities are calculated with the Lagrangian which contains the heavy, as well as the light particles. 'Light theory' quantities are calculated with the Lagrangian without the heavy particles. Note that $C_i(M/m)$ do not depend on the number of external legs and contain all the heavy mass dependence. O_i's are (integrated) local operators with are made up of the light fields alone. This result has been explicitly obtained for QED and QCD.[7]

In the inclusive annihilation process of example 2, the general result which emerges is that moments of the structure function $W(q,p)$ factorize. Thus, let

$$F_\sigma(q^2,p^2) = \int_0^1 d\omega\ \omega^{\sigma+2}\ W(q,p)/q^2$$

One obtains[12]

$$F_\sigma(q^2,p^2) = C_\sigma(q^2)\ \Gamma_\sigma\ (p^2)$$

Again, $C_\sigma(q^2)$ are universal process independent functions of q^2. $\Gamma_\sigma(p^2)$ are cut vertices, the counterparts of matrix elements of local operators.

III. RENORMALIZATION GROUP EQUATIONS

We have indicated how matrix elements of certain processes factorize into L(long distance)·S(short distance). We want to discuss in this section the mechanics of calculating S. We shall only briefly summarize results of renormalization group analysis as applied to heavy particle effects and inclusive annihilation. The details have already appeared in the literature.

Heavy Particle Effects:

In investigating affects of heavy particles on low energy light particle physics, the factorization property gives precise meaning to the notion of an effective Lagrangian. Here C_i's are identified as the effective coupling constants and O_i's are the effective vertices. It should be emphasized once again that the procedure for renormalizing O_i's is precisely given by the over-subtraction procedure. C_i's in turn are precisely determined by the underlying fundamental theory via a set of renormalization group equations. These features distinguish the present approach from the usage of effective Lagrangians in some other areas, such as soft pion physics,[13] where the relative strengths of coupling constants and the renormalization prescription of vertices are, so far, not derivable.

There are two sets of operators in the μ-e-γ quantum electrodynamics, which are naturally divided according to the mass dimension (N=5,6) of their densities. We shall write the index i = Nb, where b is a label for operators of the same dimension. Just on dimensional ground, wc can see that the coefficients C_{5b} must be proportional to the electron mass. Also, by chiral argument, it is perhaps obvious that dimension six operators are closed under renormalization and therefore the coefficient functions C_{6b} should obey a closed set of renormalization group equations. Indeed, the dimension six operators are multiplicatively renormalized

$$O_{6a}^{bare} = (Z^{-1})_{6a,6b} \, O_{Nb}$$

By considering the response to an infinitesional change in the renormalized electron mass m, with bare masses, bare charge and ultra-violet cutoff held fixed, one can derive the following equation

$$[(m \frac{\partial}{\partial m} + \beta_e \frac{\partial}{\partial e}) \delta_{6a,6b} - \gamma^t_{6a,6b}] \, C_{6b} = 0$$

where

$$\beta_e = m \frac{d}{dm} e$$

$$\gamma_{6a,6b} = (Z \, m \frac{d}{dm} Z^{-1})_{6a,6b}$$

and the superscript 't' stands for transposed matrix. Note that β and γ are to be calculated in the light theory. Heavy mass comes in because of dimensional analysis, i.e. C_{6b} are functions of M/m, where M is the muon mass. Also, the full theory enters the solution only through the integration constant.

We hasten to add that there was some flaw in making the electron mass variation, because this cannot be accomplished. Once the bare masses are fixed, the renormalized electron mass is determined! Another scale must be introduced and we choose it to effect mass insertion. Thus, we write the bare electron mass as[14]

$$m_o = a_o + \lambda\, b_o$$

where a_o and b_o are independent of λ. It turns out that this trick saves the previous equation for C_{6b}. At the same time it even helps us in calculating C_{5a}, which now has the following explicit dependence on λ

$$C_{5a} = m\, C_{5a}^{(0)} + \lambda\, C_{5a}^{(1)}.$$

The operators are renormalized with the modification

$$O_{Ma}^{bare} = (\tilde{Z}^{-1})_{Ma,Nb}\, O_{Nb}$$

where

$$\tilde{Z}^{-1}_{Ma,Nb} = \begin{pmatrix} Z^{-1}_{6a,6b} & \lambda\, Z^{-1}_{6a,5b} \\ 0 & Z^{-1}_{5a,5b} \end{pmatrix}$$

in which Z's are independent of λ. It then follows that

$$\gamma_{Ma,Nb} = (\tilde{Z}\, m \frac{d}{dm}\, \tilde{Z}^{-1})_{Ma,Nb}$$

$$= \begin{pmatrix} \gamma_{6a,6b} & \lambda\, \gamma_{6a,5b} \\ 0 & \gamma_{5a,5b} \end{pmatrix}$$

in which the λ dependence has been explicitly shown. Let us define

$$m \frac{d}{dm} \lambda = -m(1-\beta_\lambda^{(0)}) + \lambda\ \beta_\lambda^{(1)}$$

(with λ dependence again explicitly given). The renormalization group equations for C_{5a} are

$$[(m \frac{\partial}{\partial m} + \beta_e \frac{\partial}{\partial e} + 1)\ \delta_{5a,5b} - \gamma^t_{5a,5b})\ C^{(0)}_{5b}$$

$$- (1-\beta_\lambda^{(0)})\ C^{(1)}_{5a} = 0$$

and

$$(m \frac{\partial}{\partial m} + \beta_e \frac{\partial}{\partial e} + \beta_\lambda^{(1)})\ C^{(1)}_{5a} - \gamma^t_{5a,5b}\ C^{(1)}_{5b}$$

$$- \gamma^t_{5a,6b}\ C_{6b} = 0$$

Of course, only C_{6a} and $C^{(0)}_{5a}$ are of interest, as λ is only a mathematical device which should be turned off at the end. Nonetheless, $C^{(1)}_{5a}$'s appear in the coupled equations to remind us the necessity for taking mass insertion into account.

It may be said that it demands no more labor to calculate C's than the old approach in assessing muon effects for specific processes. The merit here is that once C's are determined, they can be used for <u>all</u> processes. In fact, one loop renormalization group evaluation of them has been performed.[7]

The renormalization group equation for dimension six operators in QCD has also been derived[7]. Applications are being pursued.

Inclusive Annihilation:

The renormalization of the cut vertices is well prescribed. Furthermore, the universal functions $C_\sigma(q^2)$ in inclusive annihilation are calculable via renormalization group. This we shall describe in brief.

From the procedure, it can be shown that cut vertices are multiplicatively renormalized, i.e.

$$\Gamma_\sigma = Z_\sigma^{-1} \Gamma_\sigma^{bare}$$

Let μ be the mass scale introduced in the method of dimensional regularization, which makes the coupling constant g dimensionless. We note that the inclusive annihilation amplitude is proportional to

$$\langle \overline{T}(j(x)\ \phi(y))\ T(j(0)\ \phi(z))\rangle_{1PI}$$

$$= Z_j^{-2}\ Z_\phi \langle \overline{T}(j(x)^{bare}\ \phi(y)^{bare}) T(j^{bare}(0)\ \phi^{bare}(z))\rangle$$

$$\sim \sum_\sigma C_\sigma\ (q^2/\mu^2)\ \Gamma_\sigma\ (p^2)$$

$$= \sum_\sigma C_\sigma(q^2/\mu^2)\ Z_\sigma^{-1}\Gamma_\sigma^{bare}\ (p^2)$$

Here $\overline{T}$ stands for anti-time ordering. We apply $\mu \frac{d}{d\mu}$ to the second and the fourth lines of the above and use the independence of Γ_σ. This yields

$$(\mu \frac{\partial}{\partial\mu} + \beta_g \frac{\partial}{\partial g} + 2\gamma_j - 2\gamma_\phi - \gamma_\sigma)\ C_\sigma = 0$$

where

$$\beta_g = \mu \frac{d}{d\mu} g$$

$$\gamma_j = \mu \frac{d}{d\mu} \ln Z_j$$

$$2\gamma_\phi = \mu \frac{d}{d\mu} \ln Z_\phi$$

and

$$\gamma_\sigma = \mu \frac{d}{d\mu} \ln Z_\sigma$$

The solution of this equation with various one loop anomalous dimensions as input agrees with the result of leading $\ln$ sum.[15] A point worth remarking is that the leading $\ln$ sum can be obtained through skeleton expansion. This in turn implies that certain unitarity is built into the renormalization group analysis.

CONCLUDING REMARKS

We have given two concrete examples (1) heavy particle effects and (2) inclusive annihilation, to illustrate how the Zimmermann's oversubtraction procedure can be used to organize the flow of certain large scales. This leads to factorization of the long and short distance properties. Furthermore, it has been shown how renormalization group argument can be applied to calculate the short distance universal functions.

Clearly, this approach has much wider applicability and its potential usage merits further investigation.

I have benefited much from my collaborator Dr. Yoichi Kazama on this subject.

REFERENCES

1. The kinematical argument here follows that of, for example, R. Brandt, Phys. Rev. D1, 2808 (1970).

2. K. Wilson, Phys. Rev. 179, 1499 (1969).

3. N. Christ, B. Hasslacher and A. Mueller, Phys. Rev. D6, 3543 (1972); C.G. Callan and D.J. Gross, ibid. 8, 4383 (1973); D.J. Gross and F. Wilczek, ibid. 9, 980 (1974); H. Georgi and H. Politzer, ibid. 9, 416 (1974).

4. R.P. Feynman, in Photon Hadron Interactions, (Benjamin, Reading, Mass. 1972).

5. There are quite a few articles on these. Here is a partial list: (i) leading-logarithmic approximation: Yu.L. Dokshitzer, D.I. D'Yakonov, and S.I. Troyan, SLAC Report No. SLAC Trans-183, 1978 (unpublished); C. Llewellyn-Smith, Oxford University Report No. TP 47/78 (unpublished); Yoichi Kazama and York-Peng Yao, Phys. Rev. 19, 3121 (1979); W.R. Frazer and J.F. Gunion, Phys. Rev. D19, 2447 (1979). (ii) Factorization: S. Libby and G. Sterman, ITP-SB-78-41 (unpublished), and Phys. Rev. D18, 3252 (1978); R. Ellis, H. Georgi, M. Machacek, H. Politzer, and G. Ross, Phys. Lett. 78B, 281 (1978); Caltech Report No. CALT 68-684, 1978 (unpublished); D. Amati, R. Petronzio, and G. Veneziano, Nucl. Phys. B140, 54 (1978): B146, 29 (1978); A.H. Mueller, Phys. Rev. D18, 3705 (1978); S. Gupta and A.H. Mueller, Phys. Rev. D20, 118 (1979).

6. A. Duncan and A.H. Mueller, Phys. Rev. D21, to be published; A. Efremov and A. Radyushkin, Dubna report E2-11983 (1979); S. Brodsky and G.P. Lepage, Phys. Letts. 87B, 359 (1979);

G. R. Farrar and D. R. Jackson, Phys. Rev. Letts. 43, 246 (1979); Huang Chao-shang, paper presented at this meeting; G. Parisi, LPIENS 79/7 (unpublished).

7. Yoichi Kazama and York-Peng Yao, Phys. Rev. Letts. 43, 1562 (1979); UM HE 79-16; UM HE 79-29, FERMILAB-PUB-79/64-THY; UM HE 79-40; see also T. Appelquist and J. Carazzone, Phys. Rev. D11, 2856 (1975); E. Witten, Nucl. Phys. B122, 109 (1977).
8. This property was clearly stated in the work by N. Christ, B. Hasslacher and A. Mueller, Phys. Rev. D6, 3543 (1972). It was emphasized by H.D. Politzer, Phys. Letts. 70B, 430 (1977).
9. C. Callan, Phys. Rev. D2, 1541 (1970); K. Symanzik, Commun. Math. Phys. 18, 227 (1970).
10. G. 't Hooft (unpublished); D.J. Gross and F. Wilczek, Phys. Rev. Lett. 30, 1343 (1973); H.D. Politzer, ibid 30, 1346 (1973).
11. W. Zimmermann, in Lectures on Elementary Particles and Quantum Field Theory, 1970 Brandeis University Summer Institute in Theoretical Physics, edited by S. Deser, M. Grisaru and H. Pendleton (MIT Press, Cambridge, Mass. 1970).
12. The general approach is given by A.H. Mueller, Phys. Rev. D18, 3705 (1978). For $(\phi^3)_6$, Takahiro Kubota, UT-Kmaba 79-12 (preprint); L. Baulieu, E.G. Floratos, C. Kounnas, LPTENS 79/16 (preprint); DPH-T/138 (preprint).
13. See, e.g., S. Weinberg, Physica 96A, 327 (1979).
14. C.K. Lee, Stanford Ph.D. Thesis (1975) unpublished.
15. Yoichi Kazama and York-Peng Yao, Phys. Rev. Letts. 41, 611 (1978); Phys. Rev. D19, 3111 (1979).

ON THE METHOD OF CLOSED TIME PATH GREEN'S FUNCTION

Kuang-chao Chou

Institute of Theoretical Physics,

Academia Sinica

Abstract

This article is a brief account of the work on the method of closed time path Green's function (CTPGF) and its application in nonequilibrium statistical mechanics. As an illustration, the time dependent Ginzburg-Landau equation for order parameters and the generating functional for the statistical field theory of Martin, Siggia and Rose are reproduced from the formalism of CTPGF in the low frequency limit.

In last few years, Zhao-bin Su, Lu Yu, Bai-lin Hao and I studied some problems in non-equilibrium statistical mechanics using the method of closed time-path Green's function (CTPGF) [1-5]. This report is a short account of the results obtained. [6-11].

In the last twenty years there has been great progress in the study of nonequilibrium steady states owing to the experiments done for the ordered phenomena like laser and the Benard flow. The theoretical approaches can be classified as being of a microscopical, a semi-phenomenological or a phenomenological kind. The semiphonomenological approaches are based on master equations, Fokker-Planck equations or Langevin equations which are not derived from an underlying microscopic description, but are based on a macroscopic description with the influence of fluctuations added in a plausible manner [12-16]. Both semi-phenomenological and phenomenological approaches have achieved great successes in explaining general properties of some ordered macroscopic steady states far from thermal equilibrium [12-16]. It has been found that the basic ideas developed for thermal ordered states can also be used to describe the appearance of order in other macroscopic systems in non-equilibrium steady states. Similar Lyaponoff function like the general free energy and similar equation like the time dependent Ginzburg-Landau equation are introduced in describing a non-equilibrium steady states of a macrosystem.

Through the works of Mori, Zwanzig, Nakajima and others

[17-20] , the microscopic theories of non-equilibrium steady states has also been developed using a projection operator technique. In this approach, the master equation for the diagonal elements of the density matrix is derived from the Liouville equation. The resulting equation has the form of a general Langevin equation in semi-phenomenological theory, and therefore it establishes the relation among the macroscopic and the microscopic theories.

The method of CTPGF was first suggested by Schwinger in 1961[1] , and developed subsequently by Keldysh, Craig, Hall and others[2-4] . This method was also designed for the study of non-equilibrium statistical system. Although some results like the kinetic equation and the relaxation of some processes were obtained by this method, its value as a powerful method in non-equilibrium statistical physics had not been fully recognized.

Our primary motivation was to make clear the microphysical basis of the various parameters occuring in the phenomenological theories and the range of the applicability of these macro-theories. The method of projection operator was tried firstly. However it seemed to us that the method was not only very complicated in calculation (like the old non-covariant perturbation theory in quantum mechanics), but also was unclear in physical interpretation of the individual terms obtained. When we turned to the method of CTPGF, we had found with an improvement of the technique that this method seemed to be simpler and easier to interpret. It has many advantages, such as listed below:

1). This method is quite flexible. It is possible to use the same technique to study those behaviour of the system both near thermal equilibrium and in non-equilibrium steady states. Thus many of the concepts can be used in common, and, at the same time, it is not difficult to understand their differences.

2). In every step of calculation, the causality requirement is guaranteed automatically.

3). Almost all of the techniques developed in field theory such as the Feynman path integral, the Dyson-Schwinger equation, the Lehmann spectrum representation, and the Ward-Takahashi identity etc., can be reformulated in CTPGF. With the help of these techniques, equations familiar in non-equilibrium statistical mechanics, like the Langevin, the Fokker-Planck and the Ginzburg-Landau equations, can easily be deduced within certain well defined approximations.

4). After the relation between the micro- and the macro-theories having been established, the connection of the macro-parameters with the microphysical variables can easily be found.

5). The theory can be written in covariant form, and thus is easy to be generalized to systems interested in high energy physics.

In the last two years, we have studied the method of CTPGF systematically, and have applied it to some problems in non-equilibrium statistical physics. Our main results are the following:

1). The general form for the perturbation expansion of the CTPGF is obtained using the Hori formula and the cumulant expansion for the initial correlations [6] . Our results are more compact compared with the same expansion obtained by Hall [4] . With the help of this general form of perturbative expansion. The ultraviolet divergences of CTPGF are shown to be canceled to each order of perturbation by the same counter terms that make the ordinary field theory finite, provided that the initial correlation functions have reasonable ultraviolat behavior [6] . Therefore in the high momentum region, the CTPGF satisfy the same renormalization group equation as the Green's functions of the ordinary field theory. The same result has been obtained by Kislinger and Morley [22] for temperature Green's functions by different methods. As CTPGF does not require that system under consideration should be in states of thermal equilibrium, our result is more general than theirs.

2). In[7] the Dyson-Schwinger equation is written in a very simple and compact form, and from which the kinetic equation for quasiparticles near stationary states is easily deduced. The basic points of this derivation have been given by Kadanoff, Baym, Keldysh and others [2,23] . Our improvement was technical which made the derivation particularly simple.

3). In[7,8] , the Ward-Takahashi identities for CTPGF

have been derived, and a general discussion on stability and spontaneous symmetry breaking is given. The possibility of the existence of laser like solution for the vacuum expectation value of a field $\varphi(x)$ i.e. $\langle\varphi(x)\rangle \sim e^{-ik\cdot x}$ is studied in[9] , and the model of one mode laser in the saturation region is used as an illustration. Such kind of solution can exist in classical theory without fluctuation. Once fluctuation is switched on, the vacuum expectation value $\langle\varphi(x)\rangle$ must decay owing to the existence of finite energy Goldstone mode in this case. The decay constant γ can be calculated by using Ward-Takahashi identities. This is an example to show the restoration of symmetry after spontaneous symmetry breaking. Korenman [5] has studied this problem before also using the CTPGF method. He obtained a value of γ , which is larger by a factor of 2, because the Ward-Takahashi identities are not respected in his calculation. We found with the help of these identities that the Goldstone mode splits into two parts in such a way that they have equal probability of one half but different dissipations. The decay constant γ is related only to that half with smaller dissipation. In this way the discrepancy of the factor 2 can be removed.

4). The equations for order parameters and conserved quantities near equilibrium or non-equilibrium steady states are studied in[10,11] . When the potential conditions are satisfied (which are automatically fulfilled in the case of thermal equilibrium), the equations obtained are just the

Ginzburg-Landau equations which in the tree approximation become the Langevin equations when stochastic forces are added. In certain approximation they are also equations obtained by Kawasaki[20] using the method of projective operator. Our derivation is not only very simple, but also suitable to more general cases. The physical meaning for the parameters occured is also clearer in our formalism.

The simplest way to take the fluctuation into consideration is to use the Feynman path integral representation of the CTPGF. The field function $\varphi(x)$ defined on the closed time path consists of a positive branch $\varphi_+(x)$ and a negative branch $\varphi_-(x)$. We have proved[11] that the sum $\frac{1}{2}(\varphi_+(x)+\varphi_-(x))$ is related to the field variable used in the Langevin equation in the semi-phenomenological theory, and the difference $\Delta(x)=\varphi_+(x)-\varphi_-(x)$ is related to the stochastic force appeared in the Langevin equation, or to the response field introduced by Martin, Siggia and Rose[24] in their path integral formulation of the statistical field theory. The results of the MSR field theory can be derived from the present formalism if the path integral for $\Delta(x)$ is evaluated to the Guassian approximation. In this way the microscopic basis of the MSR statistical field theory can be established and it is possible in principle to study the consequences of the approximation involved in deriving the MSR field theory.

5). In(25) the generalized Onsager relations are obtained for those macroparameters appearing in the Ginzburg-Landau equation by using the time reversal symmetry for the CTPGF. The potential conditions for systems in non-equilibrium steady states are also discussed. Similar problems have been studied by Graham and Haken starting from the phenomenological Fokker-Planck equation and the requirement of detail balance. Our results are similar to theirs for pure dissipative systems.

6). In[10] we have studied the structure of the higher order CTPGF, which generalized the results of Keldysh[2] for second order CTPGF. The results obtained are helpful in calculating loop diagrams, and it is expected that it will be useful in problems such as non-linear response.

We have listed the main results obtained by our group in the last two years. It is not possible to explain them all in this short report. As most of the particle physists are not familiar with the method of CTPGF and the non-equilibrium statistical mechanics, we shall briefly explain the central idea involved and as an illustration, derive the time dependent Ginzburg-Landau equation and the generating functional in the MSR field theory near thermal equilibrium by the method of CTPGF.

II. A Brief Summary of Critical Dynamics

The properties of the system near critical point are described in terms of a set of order parameters $Q = \{Q_i ; i = 1, 2 \cdots n\}$. The time evolution of these stochastic variables obeys the generalized Langevin equation

$$\frac{\partial Q_i(t)}{\partial t} = K_i(Q) + \xi_i(t) \tag{1}$$

where the random force $\xi_i(t)$, reflecting the effects of all degrees of freedom not included in $\{Q_i\}$, is assumed to be Gaussian distributed, i.e.

$$\langle \xi_i(t) \rangle = 0 \quad , \tag{2}$$

$$\langle \xi_i(t)\, \xi_j(t') \rangle = \sigma_{ij}\, \delta(t-t') \quad .$$

As the aim is only to illustrate our method, we shall consider pure dissipative model near termal equilibrium in the present paper for simplicity. In this model the function $K_i(Q)$ in Eq. (1) can be expressed in the form

$$K_i(Q) = -\gamma_{ij}^{-1} \left(\frac{\delta F}{\delta Q_j} - J_j \right) \tag{3}$$

where J_j is the external force, and $F[Q]$ is the free energy functional. The static equilibrium condition $\delta F / \delta Q_i = J_i$ appears to be the Ginzburg-Landau equation. Therefore Eq. (1) without the random force ξ_i is often called the time dependent Ginzburg-Landau equation (TDGL for short). In a pure dissipative model near thermal equilibrium, σ_{ij} and γ_{ij} are symmetrical matrices and related to each other by the fluctuation-dissipation theorem

$$\sigma_{ij} = 2\, T\, \gamma_{ij}^{-1} \quad ,$$

where T is the absolute temperature.

The most compact presentation of the perturbation procedure in solving the stochastic Eq. (1) is given by the field theory of Martin, Siggia and Rose [24] , we shall briefly report some general results based on the MSR statistical field theory.

The Gaussian stochastic process $\xi_i(t)$ can be presented by a stochastic functional. Eq. (1) can be considered as a mapping of the Gaussian process $\xi_i(t)$ onto a more complicated process $Q_i(t)$. The generating functional for the stochastic variable $Q_i(\xi)$ can therefore be written as

$$Z[J] = N\int[d\xi]\, e^{-\frac{1}{2}\xi\cdot\sigma^{-1}\xi - iJ\cdot Q} \quad , \tag{4}$$

Hereafter we shall omit the integration sign and the volume element d^4x for simplicity. Multiplying the right hand side of (4) by

$$\begin{aligned} 1 &= \int[dQ]\ \Delta(Q)\,\delta\left(\frac{\partial Q}{\partial t} - K(Q) - \xi\right) \\ &= \int[dQ]\ \frac{d\hat{Q}}{2\pi}\,\Delta(Q)\, e^{i\left(\frac{\partial Q}{\partial t} - K(Q) - \xi\right)\hat{Q}} \quad , \end{aligned} \tag{5}$$

where

$$\Delta(Q) \propto \exp\left(-\frac{1}{2}\,\frac{\delta K(Q)}{\delta Q}\right) \quad , \tag{6}$$

is the Jacobian of transformation from ξ to Q [27].
Introducing the external source term for $\hat{Q}$ and integrating over ξ , we obtain

$$\begin{aligned} Z[J,\hat{J}] = N\int[dQ][d\hat{Q}]\exp\Big\{ &-\frac{1}{2}\hat{Q}\sigma\hat{Q} + \\ &+ i\hat{Q}\left(\frac{\partial Q}{\partial t} - K(Q)\right) - \frac{1}{2}\frac{\delta K(Q)}{\delta Q} - iJQ - i\hat{J}\hat{Q}\Big\} . \end{aligned} \tag{7}$$

$\hat{Q}$ is the response field introduced by MSR: It is clear from Eq. (7) that looking as a dynamical variable, $\hat{Q}$ is conjugate to Q and they do not commute with each other, The introduction of $\hat{Q}$ doubles the number of field variables, and this is a general phenomenon in statistical field theories. The Eq. (7) is the starting point of the MSR statistical field theory, and we shall derive it from CTPGF in section 4.

III. Summary of the CTPGF.

For simplicity, consider a set of real scalar fields $\varphi(x) = (\varphi_i(x);\ i=1,2,\cdots n)$. Let $\hat{\varphi}_i(x)$ and $\hat{\rho}$ denote respectively the fields and the density matrix in Heisenberg representation. The CTPGF for the field φ is defined as

$$G_p(1,\cdots\cdots n) = (-i)^{n-1}\,\mathrm{tr}\{T_p(\hat{\varphi}(x_1)\cdots\cdots\hat{\varphi}(x_n))\hat{\rho}\} \quad , \tag{8}$$

where the label p denotes a closed time path from $t = -\infty$ to $t = +\infty$ (positive branch) and then from $t = +\infty$ back to $t = -\infty$ (negative branch). The time coordinates t_i ($i = 1,\cdots,n$) can situate either on the positive or on the negative branches. Tp is an order operator along the closed path p.

The generating functional for the CTPGF is defined to be

$$Z[J(x)] = \mathrm{tr}\{T_p(\exp\{-i\int_p \hat{\varphi}(x)J(x)\})\hat{\rho}\} = e^{-iW[I]} \; , \tag{9}$$

where the integral is performed along the closed path p , and the volume element d^4x is abbreviated. In (9), the external sources on the positive and the negative branches $J_+(x)$ and $J_-(x)$ are assumed to be different. They are taken to be equal only in the final stage of calculation when comparison with observed quantities is made. As in ordinary field theory,

W[J] is the generating functional for the connected CTPGF. By differentiation with respect to J (x), we get

$$\varphi(x) = \frac{\delta W[I]}{\delta J(x)} \quad , \tag{10}$$

which is the average value of the field operator $\hat{\varphi}(x)$ when $J_+(x) = J_-(x)$.

The generating functional for vertex CTPGF is defined in the usual way

$$\Gamma[\varphi] = W[J] - \int_p \varphi(x) J(x) \quad , \tag{11}$$

from which it is easily deduced that

$$\frac{\delta \Gamma[\varphi]}{\delta \varphi(x)} = -J(x) \quad . \tag{12}$$

By the definition of the generating functional in Eqs. (9) and (11), it is easily convinced that

$$W[J_+(x), J_-(x)]\Big|_{J_+(x) = J_-(x)} = 0 \tag{13}$$

and

$$\Gamma[\varphi_+(x), \varphi_-(x)]\Big|_{\varphi_+(x) = \varphi_-(x)} = 0 \quad . \tag{14}$$

Differentiating Eq. (14) with respect to $\varphi_0(x) = \varphi_+(x) = \varphi_-(x)$, a series of relations is obtained, e.g. for the second order vertex functions, we have

$$\Gamma_{++}(x,y) - \Gamma_{+-}(x,y) = \Gamma_{-+}(x,y) - \Gamma_{--}(x,y) \quad , \tag{15}$$

where

$$\Gamma_{\alpha\beta}(x,y) = \frac{\delta^2 \Gamma}{\delta \varphi_\alpha(x) \delta \varphi_\beta(y)}\Bigg|_{\varphi_+(x) = \varphi_-(x)} ; \qquad \alpha, \beta = +.- \quad . \tag{16}$$

Similarly for connected CTPGF, we have

$$G_{++}(x,y) - G_{+-}(x,y) = G_{-+}(x,y) - G_{--}(x,y) \quad , \tag{17}$$

where

$$G_{\alpha\beta}(x,y) = \left.\frac{\delta^2 W}{\delta J_\alpha(x)\,\delta J_\beta(y)}\right|_{J_+(x)=J_-(x)} ; \qquad \alpha,\beta = +,- \quad . \tag{18}$$

We know from Eq. (17) that only three of the four second order CTPGF are independent, which are usually chosen to be the retarded CTPGF

$$\begin{aligned} G_r(x,y) &= G_{++}(x,y) - G_{+-}(x,y) \\ &= i\,\mathrm{tr}\{T(\hat{\varphi}(x)\hat{\varphi}(y))\hat{\rho}\} - i\,\mathrm{tr}\{\hat{\varphi}(y)\hat{\varphi}(x)\hat{\rho}\} \\ &= i\,\theta(x-y)\,\mathrm{tr}\{[\varphi(x), \varphi(y)]\hat{\rho}\} \quad ; \end{aligned} \tag{19}$$

the advanced CTPGF

$$\begin{aligned} G_a(x,y) &= G_{++}(x,y) - G_{-+}(x,y) \\ &= -i\,\theta(y-x)\,\mathrm{tr}\{[\hat{\varphi}(x), \hat{\varphi}(y)]\hat{\rho}\} \end{aligned} \tag{20}$$

and the correlation CTPGF

$$\begin{aligned} G_c(x,y) &= G_{-+}(x,y) + G_{+-}(x,y) \\ &= i\,\mathrm{tr}\{[\hat{\varphi}(x)\hat{\varphi}(y) + \hat{\varphi}(y)\hat{\varphi}(x)]\hat{\rho}\} \quad . \end{aligned} \tag{21}$$

The corresponding independent components of the second order vertex CTPGF are chosen to be the dispersion part

$$D = -\frac{1}{2}(\Gamma_r + \Gamma_a) = -\frac{1}{2}(\Gamma_{++} - \Gamma_{--}) \quad ; \tag{22}$$

the dissipative part

$$A = \frac{i}{2}(\Gamma_r - \Gamma_a) = \frac{i}{2}(\Gamma_{-+} - \Gamma_{+-}) \tag{22'}$$

and the quasiparticle number density

$$n = \frac{\Gamma_{+-}}{\Gamma_{-+} - \Gamma_{+-}} \tag{23}$$

From the Dyson-Schwinger equation one easily obtain

$$G_r = -\Gamma_r^{-1} = \frac{1}{D + iA} \,, \qquad G_a = -\Gamma_a^{-1} = \frac{1}{D - iA} \,, \tag{24}$$

$$G_F = G_{++} = G_r(1+n) - nG_a \,,$$

where G_F is corresponding to the Feynman amplitude in the ordinary field theory. From the above Eqs. (24), one sees easily that the energy spectra for the quasiparticles is determined by the dispersion part D and the dissipation by A. The quasiparticle number density satisfies an equation

$$n\Gamma_r - \Gamma_r n = (1+n)\Gamma_{+-} - n\Gamma_{-+} \,, \tag{25}$$

which is also a consequence of the Dyson-Schwinger equation. In the vicinity of an equilibrium system where the dependence of the number density on the space-time variables is slow, Eq. (25) is just the kinetic equation. The right hand side of Eq. (25) represents the collision term, because it can be verified from spectral decomposition that Γ_{-+} (Γ_{+-}) is proportional to the rate of absorption (emission) of the quasiparticles.

In momentum representation with respect to the relative coordinates, let

$$d + ia = \frac{d}{dk_0}(D + iA)\Big|_{k_0 = 0} \,,$$

where k_μ, $\mu = 0, 1, 2, 3$ is the momentum vector. The hermiticity condition then requires that the matrix a be real symmetrical and the matrix d be purely immaginary and antisymmetrical.

In the case of one real scalar field $\hat{\varphi}(x)$, D is an even function of k_0 and therefore d is equal to zero in this case.

Now let us turn back to the discussion of the generating functional for CTPGF, which can be written in the form of Feynman Path integral

$$
\begin{aligned}
Z[J(x)] = N\int[d\varphi(x)]\, e^{i\int_p(\mathcal{L}(x)-\varphi(x)J(x))} \\
\cdot \langle \varphi(\vec{x},t_+=-\infty)|\hat{\rho}|\varphi(\vec{x},t_-=-\infty)\rangle = e^{-iW[J(x)]}
\end{aligned}
\quad . \tag{26}
$$

Consider an order parameter Q , which is the average value of a composite operator $\hat{Q}[\hat{\varphi}(x)]$. In general, Q(x) is a slowly varying function of space-time (a macro-variable). The generating functional for the CTPGF of the composite order operator $\hat{Q}[\hat{\varphi}(x)]$ is expressed in the form

$$
Z[h(x)] = tr\{ T_p(exp[-i\int_p \hat{Q}[\hat{\varphi}(x)]h(x)])\hat{\rho}\} \quad . \tag{27}
$$

where h(x) is the external source for Q. The Eq. (27) can also be written in the form of path integral. For this purpose let us multiply the factor

$$
\begin{aligned}
1 &= \int[dQ]\,\delta(Q(x)-\hat{Q}[\hat{\varphi}(x)]) \\
&= \int[dQ][\frac{dI}{2\pi}]\exp\{i\int_p(Q(x)-\hat{Q}[\hat{\varphi}(x)])I(x)\}
\end{aligned}
\tag{28}
$$

to the right hand side of Eq. (27). With the help of the Eq. (26) we obtain the relation

$$
Z[h] = N\int[dQ]\, e^{iS_{eff}[Q]-i\int_p Q(x)h(x)} \quad , \tag{29}
$$

where

$$
e^{iS_{eff}[Q]} = \int[dI]\, e^{i\int_p Q(x)I(x)}\, e^{-iW[I]} \quad . \tag{30}
$$

and

$$e^{-iW[I]} = N\int[d\varphi]\, e^{i\int_p (\mathcal{L}(\varphi(x)) - Q[\varphi(x)]\, I(x))}$$
$$\cdot \langle \varphi(\vec{x}, t_+ = -\infty | \hat{\rho} | \varphi(\hat{x}, t_- = -\infty) \rangle \quad . \tag{31}$$

One concludes from Eq. (29) that S_{eff} $[Q]$ is the effective action for the composite field $Q[\varphi(x)]$. Integrating over I(x) to one loop approximation (Guassian approximation), we obtain from Eq. (30) that

$$e^{iS_{eff}[Q]} = \left|\det \Gamma_2\right|^{\frac{1}{2}} e^{-i\Gamma[Q]} \quad , \tag{32}$$

where $\Gamma[Q] - W[I] - \int_p I(x)Q(x)$ with I(x) determined by the equation $\frac{\delta W[I]}{\delta I(x)} = Q(x)$. The matrix

$$\Gamma_2 - \begin{pmatrix} \Gamma_{++} & \Gamma_{+-} \\ \Gamma_{-+} & \Gamma_{--} \end{pmatrix} \tag{33}$$

is the second order vertex function. It is easily verified by means of the Eq. (15) that

$$\left|\det \Gamma_2\right|^{\frac{1}{2}} = \left|\det \Gamma_r \cdot \det \Gamma_a\right|^{\frac{1}{2}} = \left|\det \Gamma_r\right| \quad , \tag{34}$$

where

$$\Gamma_r(x,y) = \frac{\delta^2 \Gamma}{\delta Q_c(y)\, \delta\Delta(x)} = \Gamma_{++} - \Gamma_{+-} \tag{35}$$

with
$$Q_c(x) = \frac{1}{2}\left(Q_+(x) + Q_-(x)\right)$$
$$\tag{36}$$
and
$$\Delta(x) = Q_+(x) - Q_-(y) \quad .$$

These relations will be useful in the following section.

IV. Time Dependent Ginzburg-Landau Equation (TGLE) and the Statistical Field Theory of MSR. (SFTMSR)

In the present section we shall derive the TGLE and the starting generating functional used in SFTMSR in the vicinity of thermal equilibrium under certain well specified approximations.

Consider, at first, the equation

$$\left.\frac{\delta \Gamma}{\delta Q_+(x)}\right|_{Q_+(x)=Q_-(x)=Q_c(x)} = -J_c(x) \tag{37}$$

satisfied by the order parameters $Q_c(x)$. We shall prove in the following that this equation is just the TGLE when the space-time variation of $Q_c(x)$ is slow.

Let the value of $Q_c(\vec{x},\tau)$ at time $t=\tau$ be known, and the Eq. (37) be expanded around $Q_c(x) = Q_c(\vec{x},\tau)$. We obtain

$$\begin{aligned}\left.\frac{\delta \Gamma}{\delta Q(x)}\right|_{Q_+(x)=Q_-(x)=Q_c(\vec{x},t)} \\ = \left.\frac{\delta \Gamma}{\delta Q_+(x)}\right|_{Q_+(x)=Q_-(x)=Q_c(\vec{x},\tau)} + \int \Gamma_r(x,y)(Q_c(y)-Q_c(\vec{y},\tau))d^4y\end{aligned} \tag{38}$$

where $\Gamma_r(x,y)$ is the retarded vertex function. To the first order of variation of the order parameter, we can write the second term on the right hand side of Eq.(38) in matrix form (with three dimensional $\vec{x}$ and $\vec{y}$ as indices of the matrix) as

$$\int \Gamma_r(t_x,t_y)\,(Q_c(t_y)-Q(\tau))\,dt_y = \int \Gamma_r(t_x\ t_y)(t_y-\tau)\,dt_y \frac{\partial Q(\tau)}{\partial \tau}.$$

Letting $\tau \longrightarrow t_x - t$ on the right hand side of Eq. (38) we have

$$-J_c(x) = \left.\frac{\delta\Gamma}{\delta Q_+(x)}\right|_{Q_+(x)=Q_-(x)=Q_c(\vec{x},t)} - \gamma \cdot \frac{\partial Q_c}{\partial t} \tag{39}$$

where γ is a matrix determined by

$$\gamma = \int \Gamma_r(t_x, t_y)(t_x - t_y)\,dt_y = i \left.\frac{\partial \Gamma_r}{\partial k_0}\right|_{k_0=0} \tag{40}$$

It should be noted that the first term on the right hand side of Eq. (39), is considered as a functional of $Q_c(\vec{x},t)$ with fixed value of time t. We shall show below that, provided certain potential conditions are satisfied, this term can be expressed as a functional derivative of a generalized free energy, which is itself a functional of the three dimensional function $Q_c(\vec{x})$ (the time t in $Q_c(\vec{x})$ should be considered as a parameter).

Let

$$I_i(Q(\vec{x})) = \left.\frac{\delta\Gamma}{\delta Q_{i+}(x)}\right|_{Q_+(x)=Q_-(x)=Q_c(\vec{x})} , \tag{41}$$

one easily verifies that

$$\frac{\delta I_i(Q(\vec{x}))}{\delta Q_j(\vec{y})} - \frac{\delta I_j(Q(\vec{y}))}{\delta Q_i(\vec{x})} = \lim_{k_0 \to 0} \left(\Gamma_{-+ij} - \Gamma_{+-ij}\right) \tag{42}$$

As in thermal equilibrium

$$\Gamma_{+-} = e^{-\beta k_0}\,\Gamma_{-+} , \tag{43}$$

so the right hand side of Eq. (42) vanishes. This is the potential condition which is sufficient for the existence of a free energy functional such that

$$I_i(Q(\vec{x})) = -\frac{\delta F}{\delta Q_i(\vec{x})} . \tag{44}$$

Substituting Eq. (44) into Eq. (39), we obtain the TGLE

$$\gamma \cdot \frac{\partial Q_c}{\partial t} = - \frac{\delta F}{\delta Q_c(\vec{x})} + J_c \quad . \tag{45}$$

In the case of non-equilibrium steady states the vanishing of the right hand side of Eq. (42) is also sufficient for the existence of a generalized free energy, and the equation satisfied by the order parameters can also be written in the form of TGLE. However, this condition is not a necessary condition, and there may be other possibilities.

Next we turn to study the fluctuations of the order parameters. For this purpose the integrating variables $Q_+(x)$ and $Q_-(x)$ occured in the positive and negative branches of the path integral (29) should be change into the variables $Q_c(x)$ and $\Delta(x)$ defined by Eq. (36). The effective action S_{eff} is expanded with respect to $\Delta(x)$ to the second order (Gaussian approximation for the variable $\Delta(x)$) and has the form

$$\begin{aligned} S_{eff}[Q_+, Q_-] = S_{eff}[Q_c, Q_c] + \frac{1}{2}\int \left(\frac{\delta S_{eff}}{\delta Q_+} + \frac{\delta S_{eff}}{\delta Q_-} \right)\Delta(x) + \\ + \frac{1}{8}\int \Delta(x)\,(S_{++} + S_{+-} + S_{-+} + S_{--})(x,y)\Delta(y) + \cdots \quad . \end{aligned} \tag{46}$$

In the limit of long wave length (i.e. to the lowest order of space-time variation for the order parameter Q_c we obtain from Eq. (38) and Eq. (45)

$$\begin{aligned} \frac{1}{2}\left(\frac{\delta S_{eff}}{\delta Q_+} + \frac{\delta S_{eff}}{\delta Q_-} \right)\Bigg|_{Q_+ = Q_- = Q_c} &= - \frac{\delta \Gamma}{\delta Q_+}\Bigg|_{Q_+ = Q_- = Q_c} \\ &= - \gamma \frac{\partial Q_c}{\partial t} - \frac{\delta F}{\delta Q_c} = - \gamma \left(\frac{\partial Q_c}{\partial t} - K(Q_c) \right) \quad , \end{aligned} \tag{47}$$

where
$$K(Q_c) = \gamma^{-1} \cdot \frac{\delta F}{\delta Q_c}$$

and
$$\begin{aligned} e^{iS_{eff}[Q_c,Q_c]} &= |\det \Gamma_r| \\ &= \det\left[\frac{\delta}{\delta Q_c(x)}\left(-\gamma \frac{\partial Q_c(y)}{\partial t} - \frac{\delta F}{\delta Q_c(y)}\right)\right] \\ &= \det\left[\frac{\delta}{\delta Q_c(x)}\left(\frac{\partial Q_c(y)}{\partial t} - K(Q_c(y))\right)\right] \cdot \text{const} \quad , \end{aligned} \tag{48}$$

provided that γ is not a functional of $Q_c(\vec{x})$ as is usually assumed in the SFTMSR. Eq. (48) is just the Jacobian factor introduced in Eq. (6). Further, we put
$$\begin{aligned} \frac{i}{4}(S_{++} + S_{-+} + S_{+-} + S_{--}) &= \frac{-i}{2}(\Gamma_{-+} + \Gamma_{+-}) \\ &= -\gamma \cdot \sigma \cdot \tilde{\gamma} \quad , \end{aligned} \tag{49}$$

In the case of thermal equilibrium, γ is a symmetrical matrix of the form
$$\begin{aligned} \gamma &= i \left.\frac{\partial \Gamma_r}{\partial k_0}\right|_{k_0=0} = \frac{i}{2} \frac{\partial}{\partial k_0}(\Gamma_{-+} - \Gamma_{+-})\Big|_{k_0=0} \\ &= i\beta(\Gamma_{-+} + \Gamma_{+-})\Big|_{k_0=0} \quad . \end{aligned} \tag{50}$$

Therefore in the long wave length limit we have from Eqs.(49) and (50) that
$$\sigma = \frac{2}{\beta}\gamma^{-1} = 2T\gamma^{-1} \quad . \tag{51}$$

This is the famous fluctuation-dissipation theorem, which has been putting in by hand in the phenomenological theory of

Langevin equation.

Consider now the generating functional by substituting Eq. (46) into Eq. (29), and obtain finally

$$Z[h_c(x), h_\Delta(x)] = N\int [dQ_c][d\Delta]\cdot \tag{52}$$

$$\cdot\exp\left\{-\frac{1}{2}\int\Delta\cdot\gamma\cdot\sigma\cdot\tilde{\gamma}\cdot\Delta + i\int\Delta\cdot\gamma\cdot\left(\frac{\partial Q_c}{\partial t} - K(Q_c)\right) - \frac{1}{2}\frac{\delta K}{\delta Q_c} - i\int(Q_c h_\Delta + \Delta h_c)\right\} .$$

Changing the variables Δ , h_Δ and h_c to

$$\hat{Q} = \tilde{\gamma}\cdot\Delta , \quad J = h_\Delta , \quad \hat{J} = \gamma^{-1}h_c ,$$

we get immediately the starting generating functional (7) in SFTMSR.

V. Discussions

Recent experimental and theoretical progresses on high energy physics indicate that the vacuum should be a complicated system with infinite many degrees of freedom. The motion of a single particle such as quark is inevitably connected with its surrounding. Many physists conjecture that the collective motion of the vacuum will be important in the understanding of the structure of hadrons. We feel that it might be helpful in solving such complicated many body problems by borrowing more concepts and methods from the non-equilibrium statistical mechanics into the particle theory, and the method of CTPGF might be an useful bridge between these two fields which are so different at first looking.

In our works we have only discussed the properties of non-equilibrium systems near their steady states. A more difficult problem is to study the evolution of the general non-equilibrium states. If the initial state is an incident scattering state, it will not become a stationary state in general. However, the time evolution of the CTPGF also determines the changes of the observables in scattering. For example the asymptotic behavior of the quasiparticle density n at time $t \longrightarrow \infty$ will be proportional to its inclusive cross section in the reaction processes. The use of CTPGF might be helpful in such multiple production processes where some kinds of ordering appear in momentum space. We hope that works on this field can be developed in the near future.

References

1. J. Schwinger, J. Math. Phys. 2, 407 (1961).
2. L.V. Keldysh, JETP. 20, 1018 (1965).
3. R.A. Craig, J. Math, Phys., 9, 605 (1968).
4. A.G. Hall, Molecular Phys. 28, 1 (1974)
   J. Phys. A8, 214 (1974).
5. V. Korenman, Ann. Phys. (N.Y.) 39, 72 (1966).
6. Kuang-chao Chou (Zhou Guangzhao) and Su Zhaobin, Physica Energiae Fortis et Physica Nuclearis (Beijing) 3, 304 (1979).
7. ibid. 3, 314 (1979).
8. Kuang-chao Chou (Zhou Guangzhao) and Su Zhaobin, Closed Time Path Green's Function and Its Applications in Non-equilibrium Statistical Physics, Ch 5 in "Progress in Statistical Physics" (Press Kexue, to be published in Chinese).

9. Kuangchao Chou (Zhou Guangzhao) and Su Zhaobin, Acta Physica Sinica (to be published)

10. Kuang-chao Chou (Zhou Guangzhao), Su Zhaobin, Yu Lu and Hao Bailin, Acta Physica Sinica (to be published).

11. ibid. Closed Time Path Green's Functions and Critical Dynamics, Preprint of Institute of Theoretical Physics, Beijing, ASITP 79-003.

12. P. Glansdorff and I. Prigogine, Thermodynamic theory of structure, stability and fluctuations, Wiley, New York 1971. G. Nicolis and I. Prigogine, Self-organization in Non-equilibrium Systems, Wiley, New York 1975.

13. R. Graham, Macroscopic theory of fluctuations and stabilitie in Optics and Hydrodynamics, in Fluctuations, Instabilities and Phase Transitions, ed. T. Riste, Plenum, New York 1975.

14. H. Haken, Rev. Mod. Phys. 47, 67 (1975). H. Haken, Synergetics, Springer, Berlin 1977.

15. F. Schlogl, Z. Physik 243, 309; 244, 199 (1971).

16. L. Onsager and S. Machlup, Phys. Rev. 91, 1505; 91, 1512 (1953) L. Tisza and I. Manning, Phys. Rev. 105, 1695 (1957)

17. S. Nakajima, Prog. Theor. Phys. 20, 948 (1958).

HADRON PHENOMENOLOGY: HARD PROCESSES

Rudolph C. Hwa

Institute of Theoretical Science and Department of Physics

University of Oregon, Eugene, Oregon 97403

Recent developments in the phenomenology of hard processes are reviewed. Particular attention is given to deep inelastic scattering, lepton-pair production, e^+e^- annihilation and large-p_T reaction. Experimental data are interpreted in the context of QCD. Because of uncertainties of non-leading order contributions, critical tests of QCD cannot be made. However, the data are found to be in general consistency with the quark-gluon model of hadron structure and in rough agreement with leading order result of hard scattering in QCD modified by renormalization group corrections.

In the quest for an understanding of the structure of hadrons at the constituent level, a large number of experiments have been done in the past ten years on hard processes involving hadrons. As is well known, a process is hard when there is a large transfer of momentum, which is necessary in order to probe short-distance behavior. In the beginning the experiments explored virgin territory where no theory ventured; the results were only interpreted in theoretical models, among which Feynman's parton model[1] emerged as the sole survivor. As theory developed, quantum chromodynamics (QCD)[2] forged ahead and took the lead over experiments. By now experiments on hard processes serve mainly as tests of QCD.

I shall review in this talk the phenomenology of hard processes. Many figures of experimental data will be shown. If you don't have an ax to grind (which is an expression meaning not having any personal involvement in any particular aspect of the problem), and if you are like me, then your absorption efficiency of what I shall present will likely be low. You might even fall asleep which would be quite understandable and quite commonly done by theorists listening to experimental talks. To help you not feel guilty, let me show you the general conclusion first: there is no experiment that has established a definitive proof that QCD is wrong. Now that you know the answer and if you feel at ease with it, you can sit back and relax.

There are two types of hard processes: those mediated by photons and those mediated by gluons. The former can be reasonably approximated by one-photon exchange since the electromagnetic coupling is weak. The latter by one-gluon exchange relies on the smallness of the running coupling constant in an asymptotically free theory and assumes that higher order QCD diagrams are unimportant which may or may not be right. The virtual vector particle (photon or gluon) can be either space-like or time-like, and the hadron can be either in the initial state (for hadron structure) or in the final state (for jet structure). Thus we have a matrix of processes for each type of mediator of large momentum transfer. They are illustrated as in Fig. 1. I shall attempt to cover most of these processes and discuss the salient points in recent advances. More specifically, I shall cover them in the following order: deep inelastic scattering, lepton-pair production, quark fragmentation, and large-p_T process.

II. DEEP INELASTIC SCATTERING

What are measured in experiments in deep inelastic scattering are structure functions F_1, F_2, etc. but what are easiest to describe in QCD are their moments. In either case it is the Q^2 dependence where theory and experiment confront each other. QCD has definite predictions about how the structure functions evolve in Q^2, The evolution being specified by the Altarelli-Parisi equation[3]

$$\frac{\partial}{\partial \ln Q^2} G^{NS}(x,Q^2) = \frac{\alpha_s(Q^2)}{\pi} \int_x^1 \frac{dy}{y} G^{NS}(y,Q^2) P_{q\to q}\left(\frac{x}{y}\right) \tag{2.1}$$

where $G^{NS}(x,Q^2)$ is the flavor nonsinglet component of the quark distribution functions [related to $F_1(x,Q^2)$ by an x^{-1} factor]. $\alpha_s(Q^2)$ is the running coupling constant and $P_{q\to q}(z)$ is the probability function for finding a quark at momentum fraction z in a quark after emitting a gluon. For the singlet component (2.1) is generalized to a matrix. Because (2.1) is a differentio-integral equation the solution depends on the boundary condition at some Q_0^2. Calculable methods in QCD do not help us to determine the distribution at Q_0^2, which must therefore be taken from experiments. Fig. 2 shows how the x dependence[4,5] changes with Q^2 over the range 1 to 30 GeV^2. Fig. 3 shows the same over a large range of Q^2.[6] The variation with Q^2 is pivotal at $x \approx 0.2$, and becomes less rapid as Q^2 increases. Eq. (2.1) itself is checked by Baulieu and Kounnas[7] in a plot of $dF_2^{p-n}(x,Q^2)/d\,\ell nQ^2$ vs. x as shown in Fig. 4 for $Q^2 = 3.2$ GeV^2. The agreement with data is good. It should, however, be recognized that (2.1) gives only leading order result which is not reliable at low Q^2. At high Q^2 where it is reliable, the dependence on Q^2 is not pronounced since scaling violation diminishes as $(\ell nQ^2)^{-1}$. At low Q^2 where the effect is more pronounced, non-leading terms in α_s as well as higher-twist contributions can be important.[8] At present ranges of Q^2 the tests of QCD in DIS are not clean enough to be definitive.

Comparison with data has also been done in terms of moments. The ordinary moments are defined by

$$M_i(n,Q^2) = \int_0^1 dx\, x^{n-2} \mathcal{F}_i(x,Q^2) \tag{2.2}$$

where $\mathcal{F}_i$ can be F_2 or xF_3. The Nachtmann moments[9] can be defined with the target mass taken into account and are more closely related to theory at low Q^2. The two moments agree at high Q^2. The Q^2 dependence of the moments is known explicitly at least in leading order:

$$M^{NS}(n,Q^2) = A_n^{NS}\left(\ell n\frac{Q^2}{\Lambda^2}\right)^{-d_n^{NS}}\left[1 + O\left(\frac{\alpha_s}{\pi}\right) + O\left(\frac{\mu^2}{Q^2}\right) + \cdots\right] \tag{2.3}$$

$$\begin{aligned} M^S(n,Q^2) = {} & A_n^+\left(\ell n\frac{Q^2}{\Lambda^2}\right)^{-d_n^+}\left[1 + \cdots\right] \\ & + A_n^-\left(\ell n\frac{Q^2}{\Lambda^2}\right)^{-d_n^-}\left[1 + \cdots\right] \end{aligned} \tag{2.4}$$

where A_n^{NS} and $A_n^{\pm}$ are hadron matrix elements that are not calculable. The anomalous dimensions d_n^{NS} and $d_n^{\pm}$ are known. Λ is a scale of strong interaction that can only be fixed by experiment and is ill-defined until the second order terms are specified. The non-leading terms depend upon the renormalization scheme; the whole result can become independent of such choices only if all terms are calculated and summed. The higher-twist terms of order $O(\mu^2/Q^2)$ are not necessarily small for $Q^2 < 100$ GeV^2 depending upon the choice of value for μ.[8] The two "classic" tests of QCD on the basis of which it is claimed that the neutrino scattering data are consistent with QCD, ignore all non-leading terms. Before we summarize them below, it is appropriate to mention the nature of moment analysis. For the ABCLOS collaboration[10] their data from BEBC cover a wide enough range in x (though not with outstanding statistics) so that moments of the structure functions can be determined. For the CDHS data,[11] on the other hand, the high x region is not covered for some Q^2 values, so the moments are determined by fitting the measured points which have high statistics with curves of the form $x^\alpha(1-x)^\beta$. A sample of the data presented in terms of moments is shown in Fig. 5.

If the higher order terms in the square brackets in (2.3) are ignored, then one obtains

$$\ln M^{NS}(n,Q^2) = \text{const.} + \frac{d_n^{NS}}{d_m^{NS}} \ln M^{NS}(m,Q^2) \tag{2.5}$$

where

$$d_n^{NS} = \frac{4}{33-2f}\left[1 - \frac{2}{n(n+1)} + 4\sum_{j=2}^{n}\frac{1}{j}\right] \tag{2.6}$$

f being the number of flavors. Thus in a log-log plot of moment versus moment of different orders at various Q^2, one should get a straight line with known slope. That indeed turns to be the case as shown in Fig. 6 where the BEBC/GGM data[10] are presented. Agreement with the theoretical slopes is evidently good. The same is also true with the CDHS data.[11] The other way of checking (2.3) is to write it in the form

$$\left[M^{NS}(n,Q^2)\right]^{-1/d_n^{NS}} = C_n \ln\frac{Q^2}{\Lambda^2} \tag{2.7}$$

where C_n is unknown theoretically. But the experimental data for the l.h.s. when plotted against $\ell n Q^2/\Lambda^2$ should be a straight line for every n, and the straight lines should intersect the x-axis at $Q^2 = \Lambda^2$. These features are indeed borne out by the data as shown in Fig. 7 for both the ordinary and Nachtmann moments. Deviation from the straight lines are expected at low Q^2 where non-leading order terms are important. In fact, such terms cause the scale parameter Λ to depend on n as well as the structure functions.[12-14] If Λ_n is defined by[14]

$$\left[M^{NS}(n,Q^2)\right]^{1/d_n^{NS}} = \left(\bar{C}_n \ln \frac{Q^2}{\Lambda_n^2}\right)^{-1} \left[1 - \frac{\beta_1}{\beta_0^2} \frac{\ln \ln Q^2/\Lambda_n^2}{\ln Q^2/\Lambda_n^2}\right] \tag{2.8}$$

where β_0 and β_1 can be calculated in QCD, then the dependence of Λ_n on n are shown in Fig. 8. The curves are theoretical predictions by Duke and Roberts[13] for F_2 and F_3, which can be shifted up or down since the absolute normalization cannot be fixed by theory. Any disagreement between theory and experiment cannot be regarded as significant enough to invalidate QCD.

It is important to note that the slopes of the straight lines in Fig. 7 are phenomenological values for C_n in (2.7) and contain information about the structure of nucleon. On the basis of a model that at low Q^2 a nucleon consists of three valence quark clusters, called valons, it is possible to determine from C_n the valon momentum distribution.[15] This distribution, rather than the quark distribution at high Q^2, is more closely related to the wave function of the nucleon constituents in the bound-state problem. A more detailed description of this problem and its applications is given elsewhere in these proceedings.

III. LEPTON-PAIR PRODUCTION (LPP)

The production of massive lepton pairs is a hard hadronic process because a highly virtual time-like photon is involved. It reveals nucleon structure almost as effectively as deep inelastic scattering, while in the case of mesons it is the only method by which the meson structure function has been extracted. There are five essential variables in LPP: M the mass of dilepton, q_T the transverse momentum of the pair, x_F the longitudinal momentum fraction of the pair in the initial c.m. system, θ and ϕ the two angles of the leptons in the c.m. frame of the dileptons. There are various ways of writing the differential cross section in terms of a general set of functions. For example, in terms of the helicity functions one would have

$$\frac{d^5\sigma}{dM^2 dq_T^2 dx_F d\cos\theta d\phi} \propto W_T(1+\cos^2\theta) + W_L \sin\theta + W_\Delta \sin 2\theta \cos\phi + W_{\Delta\Delta} \sin^2\theta \cos 2\phi \tag{3.1}$$

On general grounds it can be shown[16] that W_Δ is of order $0(q_T/M)$ and $W_{\Delta\Delta}$ of order $0(q_T^2/M^2)$. If the $q\bar{q}$ annihilation model of Drell and Yan[17] is applied, then various constraints among the helicity functions can be derived.[16] Among them, for instance, is $W_L = 2W_{\Delta\Delta}$, which implies that, for $q_T^2 \ll M^2$, W_T is the only significant function that describes the differential cross section. From (3.1) one immediately infers that the angular distribution of the lepton pair is $1 + \cos^2\theta$.

To interpret the M and x_F dependences of the differential cross section it is necessary to consider in detail the basic Drell-Yan mechanism and corrections of it. If x_i denotes the momentum fraction of a quark (or antiquark) in hadron h_i, $i = 1, 2$, and $G(x_i,Q^2)$ signifies the probability for it to occur at Q^2, then we have

$$\frac{d^2\sigma}{dM^2 dx_F} = \frac{4\pi\alpha^2}{9M^4}\frac{x_1 x_2}{\sqrt{x_F^2+4x_1x_2}} \sum_i e_i^2 \left[G_{q_i/h_1}(x_1,Q^2)G_{\bar{q}_i/h_2}(x_2,Q^2) + q_i \leftrightarrow \bar{q}_i\right] \left[1 + \frac{\alpha_s(Q^2)}{2\pi}\frac{4}{3}\left(1+\frac{4\pi^2}{3}\right) + \cdots\right] \tag{3.2}$$

where $x_F = x_1 - x_2$, and $Q^2 = M^2 = sx_1x_2$. The first line in (3.2) is the basic Drell-Yan formula while the second term in the square bracket is a correction term of order $O[(\ell nQ^2)^{-1}]$.[18] More about the latter will be discussed in the following. Multiplying the first line of (3.2) by $2M^4$ yields a scaling quantity according to theory. Experimental results for the quantity in the case $x_F = 0$ are plotted in Fig. 9 as functions of $\sqrt{\tau} = M/\sqrt{s}$; indeed, they are consistent with scaling.

The dependence on $\sqrt{\tau}$ can be compared with theory only if we know the quark and antiquark distributions in the incident hadrons. While the quark distributions in nucleon can be reasonably well extracted from deep inelastic scattering data, the antiquark distribution is not. Thus (3.2) may be used in reverse to give $G_{\bar{q}/h}(x,Q^2)$ from LPP data.[19] In Fig. 10 are shown results on the antiquark distributions as well as from neutrino scattering. The agreement should be regarded as remarkably good. The normalization appears to be off by roughly a factor of two. But even that may be accounted for by the correction term in the square bracket of (3.2). It arises from the virtual gluon corrections to the $q\bar{q}$ annihilation diagram.[18] Other contributions such as quark-quark scattering or quark-gluon scattering with virtual photon radiated are also not negligible.[20] The total effect due to the one-gluon correction is about 100% for a wide range of M^2 values. Even if it improves the agreement of LPP with neutrino data in Fig. 10, one cannot, however, take comfort in it. If $O(\alpha_s)$ correction turns out to be $O(1)$, how can we justify ignoring $O(\alpha_s^2)$ and $O(\alpha_s^3)$, etc.? Indeed, the perturbative QCD method is cast in considerable doubt.

Using pion beam in LPP is now the only way to extract the valence quark distribution in a pion. In a π^- beam the dominant Drell-Yan process is the annihilation of $\bar{u}$ antiquark in π^- with u quark in target nucleons. The longitudinal momentum distribution of the lepton pair produced is significantly different from that in pN interaction, as evidenced by the CIF data[21] shown in

Fig. 11. It signifies a grossly different valence (anti)quark distribution in a pion which is exhibited in Fig. 12 and is fitted by

$$xG_{\bar{q}/\pi}(x) \sim \sqrt{x}\,(1-x)^{1.27 \pm 0.06} \tag{3.3}$$

Other experiments done at CERN-SPS give exponents 0.9 ± 0.1 (NA-3) and 1.56 ± 0.18 (SISI).[19] These results are to be compared to the x dependence of the u-quark distribution in a nucleon, viz. $(1-x)^3$.

The problem about the transverse momentum q_T of the dilepton is very intriguing. If we believe that it is due primarily to the "primordial" transverse momentum k_T of the quarks in the initial hadron, then we would expect from the Drell-Yan annihilation mechanism

$$\langle q_T^2 \rangle \sim 2\langle k_T^2 \rangle \tag{3.4}$$

Since the dominant process in hadronic collisions at high energies involves production of pions in the central region with $\langle p_T \rangle \sim 350$ MeV, we would expect $\langle k_T \rangle$ also to be about 350 MeV. Then (3.4) would imply that $\langle q_T \rangle$ should be no more than 500 MeV. However, the experimental results on $\langle q_T \rangle$ far exceed that estimate, as evidenced by the data shown in Fig. 13. In fact, the value of $\langle q_T \rangle$ increases linearly with $\sqrt{s}$ according as[22]

$$\langle q_T \rangle = 0.6 + 0.022\sqrt{s} \quad \text{GeV} \tag{3.5}$$

up to ISR energies. There has been rejoicing in certain circles that such a behavior is as expected in perturbative QCD. But I believe that such enthusiasm is premature for the following reason.

The explanation for the linear dependence of $\langle q_T \rangle$ on $\sqrt{s}$ is based on a hard scattering non-Drell-Yan process: $q + \bar{q} \to \gamma^* + g$ where $\gamma^* \to \mu^+\mu^-$ and g stands for a gluon. With the possibility of a gluon recoil the virtual photon can, of course, acquire a large transverse momentum. However, if that is really the dominant process responsible for the large $\langle q_T \rangle$, then two-body kinematics would require that x_F must be small when $\langle q_T \rangle$ is large, and that as x_F increases $\langle q_T \rangle$ would decrease perceptibly. That is not borne out by experiment. As Fig. 14 clearly shows, $\langle q_T \rangle$ is essentially independent of x_F for $x_F < 0.9$. The simple QCD process is therefore not likely to be dominant for the q_T range examined.[23]

Finally, we mention an interesting result concerning the angular distribution of the lepton pair. If we parametrize it as

$$\frac{d\sigma}{d\cos\theta} \propto 1 + \alpha \cos^2\theta \tag{3.6}$$

then

$$\alpha = \frac{W_T - W_L}{W_T + W_L} \tag{3.7}$$

When the Drell-Yan mechanism is dominant, W_L is nearly zero, so $\alpha \approx 1$. Indeed, the angular distribution in LPP is mainly that of $1 + \cos^2\theta$. However, in the limit of $x_F \to 1$ in a $\pi^- p$ reaction the antiquark in π^- must have its momentum fraction also approaching one, causing it to become far off-shell due to kinematical constraints. The higher twist correction of order $O(m^2/Q^2)$ cannot be ignored then. Berger and Brodsky[24] have shown that an observable effect of such a correction to the simple Drell-Yan formula is $W_L \gg W_T$, i.e. $\alpha \to -1$ as $x_F \to 1$. Recent data from CIF[25] indicate the expected trend as shown in Fig. 15, although the small x_F behavior is not as expected. Whether the change in α is due solely to the higher twist effect is at present unclear.

IV. QUARK FRAGMENTATION

The fragmentation of a quark into a hadron jet appears to be independent of whether the quark is created by a time-like photon (as in e^+e^- annihilation) or knocked out of a nucleon by a space-like photon $\to$ boson (as in deep inelastic scattering). The fragmentation function, D(x), describes the probability that a hadron is detected in a quark jet with momentum fraction x of the initial quark. In Fig. 16 is shown its x dependence in various processes. Their general agreement is evidence for the universality of the jet structure.

To be more complete in the description of the fragmentation function, one should also specify the value of Q^2 that the virtual photon has. Dependence on Q^2 is scaling violation that has the same origin as for structure functions discussed in Sec. II. At high Q^2 scaling violation is logarithmic and therefore indiscernible in present data which have large errors except in the small x region where approximate scaling is being approached at PETRA energies (see Fig. 17).

Thus far there has been no way to calculate D(x) from first principles, just as it is with the structure functions F(x). In both cases the stumbling block is the confinement problem. However, if we recognize the common origin of the difficulty, phenomenological analysis of one should be sufficient as an input to facilitate a complete calculation of the other without further phenomenological input. That is precisely what V. Chang and I have done in determining D(x).[26] There are two parts to the calculation. The first part is to find the quark and antiquark joint distribution in a quark jet when Q^2 is allowed to degrade to a low Q_o^2; this can be reliably done in perturbative QCD when Q^2 is large. The second part is to calculate the probability of pion formation through the recombination of the $q\bar{q}$ pair at Q_o^2. This is intimately related to the wave function of the pion at low Q^2 and is to be extracted from the phenomenology of hadron structure.[15,27] No free parameters are involved in the calculation. The result for D(x) agrees well with experiment in both normalization and shape, as evidence in Fig. 18.

V. LARGE p_T REACTIONS

Like Rutherford scattering, large p_T reaction is a way of probing "point-like" constituents in a target. The most common experiments have been the measurement of inclusive cross sections for

$$p + p \rightarrow \pi + X$$

$$p + p \rightarrow \pi_1 + \pi_2 + X$$

where the pions are detected at large angles relative to the incident momenta in the c.m. system. In the case of the single-pion inclusive reaction at 90° there are two essential variables: p_T and s, or p_T and $x_T \equiv 2p_T/\sqrt{s}$. The inclusive cross section can be written in the general form

$$E \frac{d^3\sigma}{dp^3} \sim p_T^{-n} f(x_T) \tag{5.1}$$

where $f(x_T)$ is some scaling function. From dimensional consideration one should have n = 4 if there are no basic scales in the problem. But the value turned out to be n ≃ 8 in the beginning even at ISR energies. Fig. 19 shows the data taken at Fermilab preferring n = 8.2 - 8.5.

Various models were invented to explain this non-scaling behavior,[28-30] all of which were eliminated by a change of data as time went on. Among those models are the constituent-interchange model of Brodsky and Gunion[28] and the black-box model of Field and Feynman.[29] The former assumes the existence of a composite subsystem in a nucleon thereby introducing a scale. The latter assumes quark-quark scattering in a non-scaling way. Apart from inclusive cross sections there are various predictions on other aspects of large-p_T reactions, such as jet structure, correlations, etc. On the whole quark-quark scattering seems to be the more favorable mechanism.

The situation changed appreciably after 1978 as p_T was pushed to higher values (beyond 6 GeV) at ISR. In terms of (5.1) what was found was that the parameter n depends on x_T. For x_T between 0.3 and 0.5 three experimental groups at ISR have found n to be more in the range of 5 to 7.[31] Fig. 20 shows the p_T distributions at various energies. The fact that n has dropped to values closer to 4 definitely rules out non-scaling models that require n = 8. But the difference from n = 4 is sufficiently significant that an explanation is still needed.

The most extensive analysis that attempts to provide an explanation has been the work of Feynman, Field and Fox.[32] For $A + B \to h + X$, the basic formula in a hard-collision model is

$$E\frac{d^3\sigma}{dp^3} = \int dx_a\, dx_b\, G_{a/A}(x_a, Q^2)\, G_{b/B}(x_b, Q^2)\, D_{h/c}(z_c, Q^2)\,(z_c\pi)^{-1}\frac{d\hat{\sigma}}{d\hat{t}}(\hat{s}, \hat{t}) \tag{5.2}$$

where $d\hat{\sigma}/d\hat{t}$ describes the large-angle scattering subprocess $a + b \to c + d$, x_a is the momentum fraction of quark a in A, x_b for b in B, and z_c is the momentum fraction of hadron h in the jet associated with the quark c. A factorizable form involving the product of the probability functions G and D has been assumed. If G and D are scale invariant and $d\hat{\sigma}/d\hat{t}$ corresponds to a single gluon exchange in QCD, then (5.2) would yield the scaling result in (5.1) with n = 4. By letting the gluon interact through the running coupling constant $\alpha_s(Q^2)$, scaling violation is introduced and the effective value for n for $2 < p_T < 8$ GeV/c becomes 4.5. If the G functions are also allowed to depend on Q^2 in accordance to the prescription of QCD, as described in Sec. II, then n_{eff} is increased to 5.0. If further the D function is also given scaling violation, then n_{eff} becomes 5.8. Finally, when smearing by the parton transverse momenta k_T is taken into account, n_{eff} can be raised even higher to fit the experimental data. The fit is achieved by using $\langle k_T \rangle$ = 848 MeV/c for the G function and 439 MeV/c for the D function. One should not push for excellent fits in the small p_T region on the basis of the simple hard-scattering formula because it is most likely oversimplified. There are many unknowns that may be important in that region, such as gluon distribution, gluon k_T, cut-off at small $\hat{s}$ and $\hat{t}$, precise definition of Q^2, not to mention higher order corrections terms in QCD. Besides, there may be other processes such as the ones considered in CIM[28] that may not be negligible.

Because of the parton k_T the detection of a single particle at large angle would favor the configuration in which the large-angle scattering subprocess has partons whose initial momenta lean toward the detector; in that way the parton scattered toward the detector experiences a smaller-momentum transfer, hence larger cross section. To minimize this effect one measures the back-to-back cross section, i.e. two detectors on opposite sides triggering only on events with equally large p_T on both sides. It is found that back-to-back trigger even at FNAL energies yields $n_{eff} = 6.2 \pm 0.2$ whereas single-particle trigger would give $n_{eff} = 8.2 \pm 0.2$. It means that by minimizing k_T smearing the large-p_T regime arrives earlier.

There are many other aspects about large-p_T reactions where one must be careful in order to give proper interpretation to data. The basic difference from an experiment in e^+e^- annihilation is that in a large-p_T reaction the momenta of the hard scattering partons are not fixed by kinematics. Because the cross section is damped by such a high power of p_T (between 6 and 8), there is a built-in bias for the participating partons not to waste momentum. For example, a jet trigger should yield a much larger cross section than a single-particle trigger at the same p_T. That is because the jet associated with the detected particle must have larger momentum, hence lower cross section. Experimental evidence for this can be seen in Fig. 21.

In relating single-particle triggered to jet-triggered cross sections it is important to mention what has come to be known as the trigger bias.[33] Given a trigger momentum for a single hadron, the associated jet momentum (of the fragmenting quark) would be biased toward the low rather than the high side since minimization of the quark momentum maximizes the cross section. Thus associated with a trigger particle should be other hadrons in the same direction but carrying small momenta. This is a phenomenon that is clearly demonstrated in Fig. 22 where the momentum sum of all associated particles amounts to only about 10% of the trigger momentum.

While the single-particle trigger does not give definitive information about the jet momentum, jet trigger using a calorimeter not only determine the jet momentum but also enhances the cross section. FLPW collaboration[34] has examined the momentum balance of two oppositely directed jets at a large angle by using a double-arm calorimeter. If p_{TL} and p_{TR} denote the jet momenta measured by the left and right calorimeters their difference $p_{TX} = p_{TL} - p_{TR}$ should be zero in the special case where the two jets are due to a large-angle scattering of two partons with no transverse k_T in the initial hadrons. Thus a measurement of the p_{TX} distribution provides information about the parton k_T. Such a distribution is shown in Fig. 23 for 200 GeV pp collisions with $p_{TL} + p_{TR} = 6.0$ to 6.5 GeV/c. From the p_{TX} distribution the average parton k_T can be determined according to

$$\langle p_{TX}^2 \rangle = \langle k_{TX1}^2 \rangle + \langle k_{TX2}^2 \rangle = \langle k_{TX1}^2 \rangle + \langle k_{TY1}^2 \rangle = \langle k_{T1}^2 \rangle$$

The dependence of $\langle p_{TX}^2 \rangle^{1/2}$ on p_{TA} which is $(p_{TL} + p_{TR})/2$ is shown in Fig. 24 for both proton and pion beams.[34] Note that they are roughly independent of the type of hadron beam. This is significant in that large p_T reactions are due mainly to large-angle scattering of valence quarks, and the result implies that the k_T distributions of valence quarks in proton and pion are approximately the same. This is to be contrasted from the average q_T in dilepton production, which is greater for pion than for proton. A comparison of the various $\langle p_T \rangle$ and $\langle q_T \rangle$ is shown in Fig. 25. What one learns from this is that $\langle k_T \rangle$ is smaller for the sea quarks than for valence quarks, since sea quarks contribute significantly only to the proton initiated Drell-Yan process.

Another direction to explore the nature of large-p_T events is to examine the coplanarity of the jets. If there are two and only two jets at large (polar) angle, then they should be in a plane together with the beam direction. If one of the jets is along the azimuthal angle $\phi = 0$, then the other jet should be at $\phi = \pi$ and a minimum in multiplicity should be detected at $\pi/2$. Fig. 26 shows the distribution of charged particles correlated to a neutral trigger at $\phi = 0$. Of course, one can say that kinematics demands momentum balance at $\phi = \pi$. To minimize that effect the ABCS collaboration[35] measures the azimuthal correlation between two large-p_T π^0's for fixed values of total energy radiated transversely. A dip at $\phi = \pi/2$ is clearly seen in the data shown in Fig. 27. If one compares large-p_T events with low-p_T soft collisions, it is found (Fig. 28) that the relative frequency of observing charged particles at $\phi = \pi/2$ is about the same even when the total observed charged multiplicity is 4, but at $\phi = 0$ or π the ratio is significantly higher. Thus coplanarity is very well established.

We come finally to the transverse momentum spread in a jet. The k_T distribution with respect to the jet axis (as measured by the BS collaboration) is consistent with an exponential (see Fig. 29) giving $\langle k_T \rangle = 520 \pm 50$ MeV/c. The CERN-Saclay group gives $\langle k_T \rangle = 386 \pm 7$ MeV/c for all neutral and 589 ± 14 MeV/c for all charged secondaries. The dependence of k_T on the hadron momentum is shown in Fig. 30 and is roughly independent of the trigger momentum. It should be recognized that the corresponding angular spread of the jet is not small, so perturbative QCD method cannot be used to interpret the result for that range of jet momentum.

VI. CONCLUSION

As I have already announced at the beginning we have no definitive experimental result in any of the hard processes that can invalidate QCD. Some results agree quite well with predictions of the theory, but none so crucial as to accredit QCD with an incontroversial seal of validity. Theoretical predictions of QCD themselves are uncertain because non-leading terms have one-by-one been found recently to be not so small as initially expected. The latest development[36] (after the Guangzhou conference) is that in large-p_T reactions there are terms beyond the Born diagram for the hard-scattering subprocess which turn out to be quite sizable and completely alter the results of Ref. 32. What we have described here are mainly experimental results, which are not likely to change with time. An indisputable outcome of the experimentation is that hadrons have fractionally-charged quarks as constituents which interact through neutral fields. It is important to ascertain whether the interaction is asymptotically free as prescribed by QCD. It is unfortunate that where theoretical predictions are simple (viz. at high Q^2) experimentation is difficult, and vice-versa. Definitive tests remain to be done. Nevertheless, we are fortunate in that the only candidate theory for strong interaction is still in the running and is, in fact, doing remarkably well.

ACKNOWLEDGMENT

I am grateful to Academia Sinica for arranging my participation at the Guangzhou Conference and for its generous hospitality. The work was supported in part by the U.S. Department of Energy.

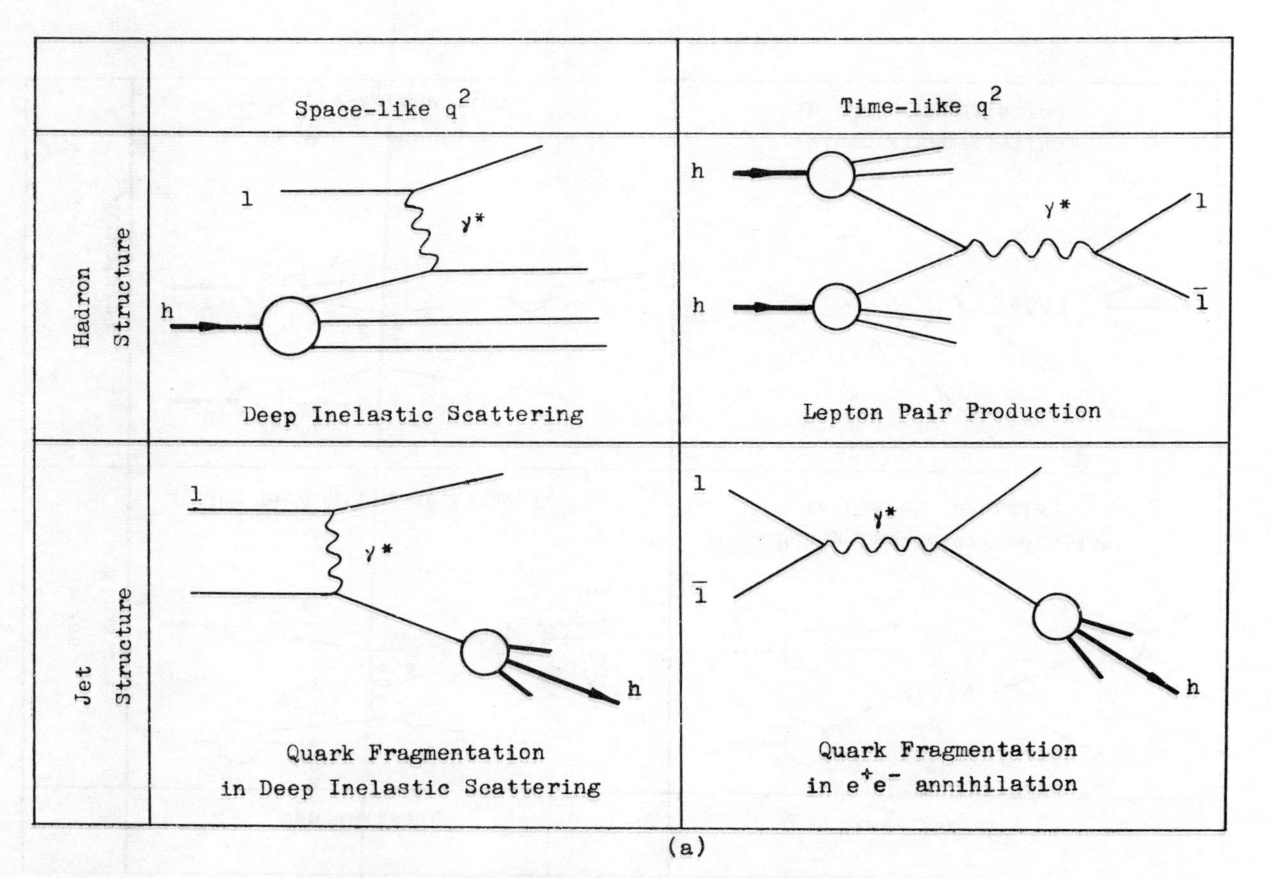

Fig. 1: Hard processes mediated by (a) photon and (b) gluon.

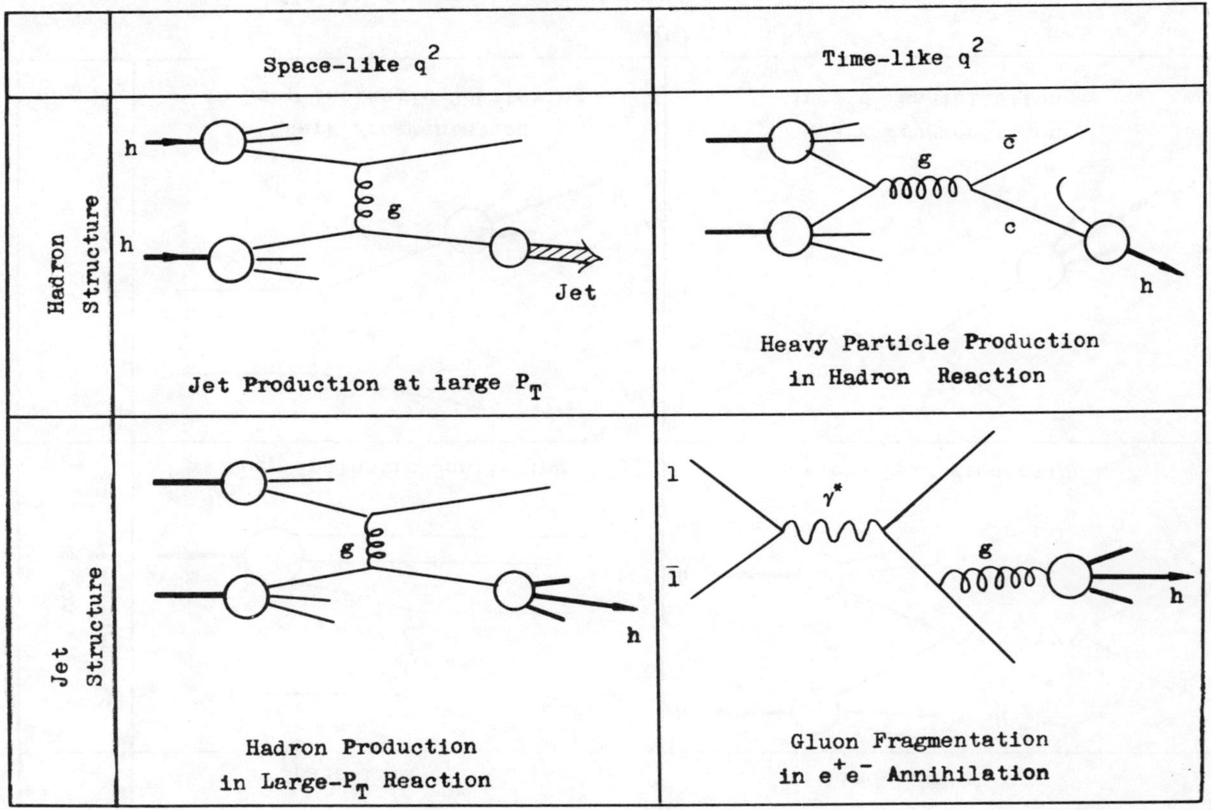

	Space-like q^2	Time-like q^2
Hadron Structure	Jet Production at large P_T	Heavy Particle Production in Hadron Reaction
Jet Structure	Hadron Production in Large-P_T Reaction	Gluon Fragmentation in e^+e^- Annihilation

Fig. 1(b)

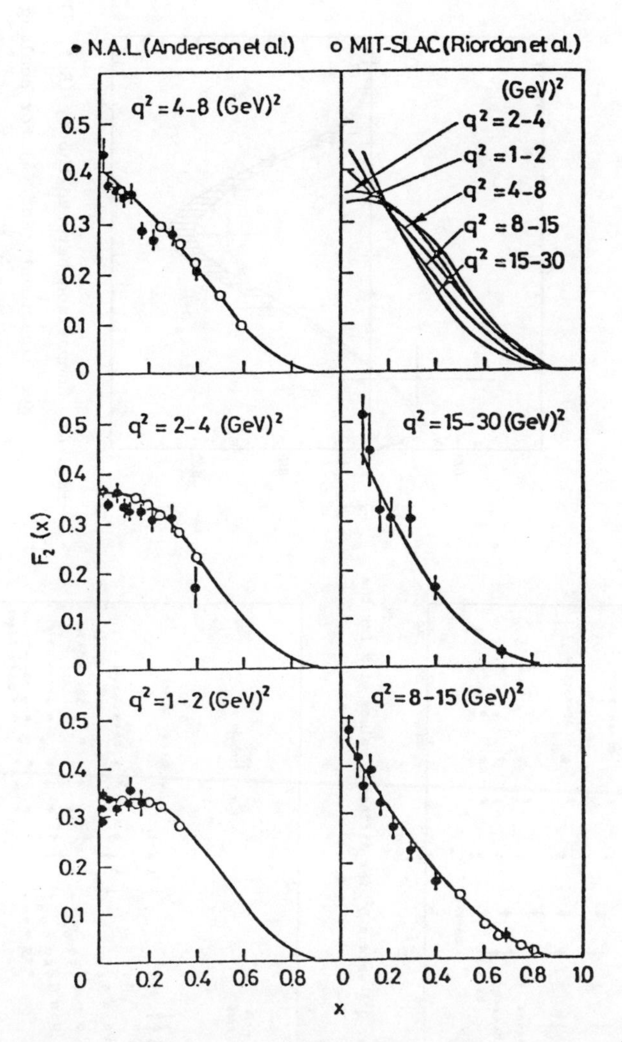

Fig. 2: Structure funcion F_2 for various values of Q^2.

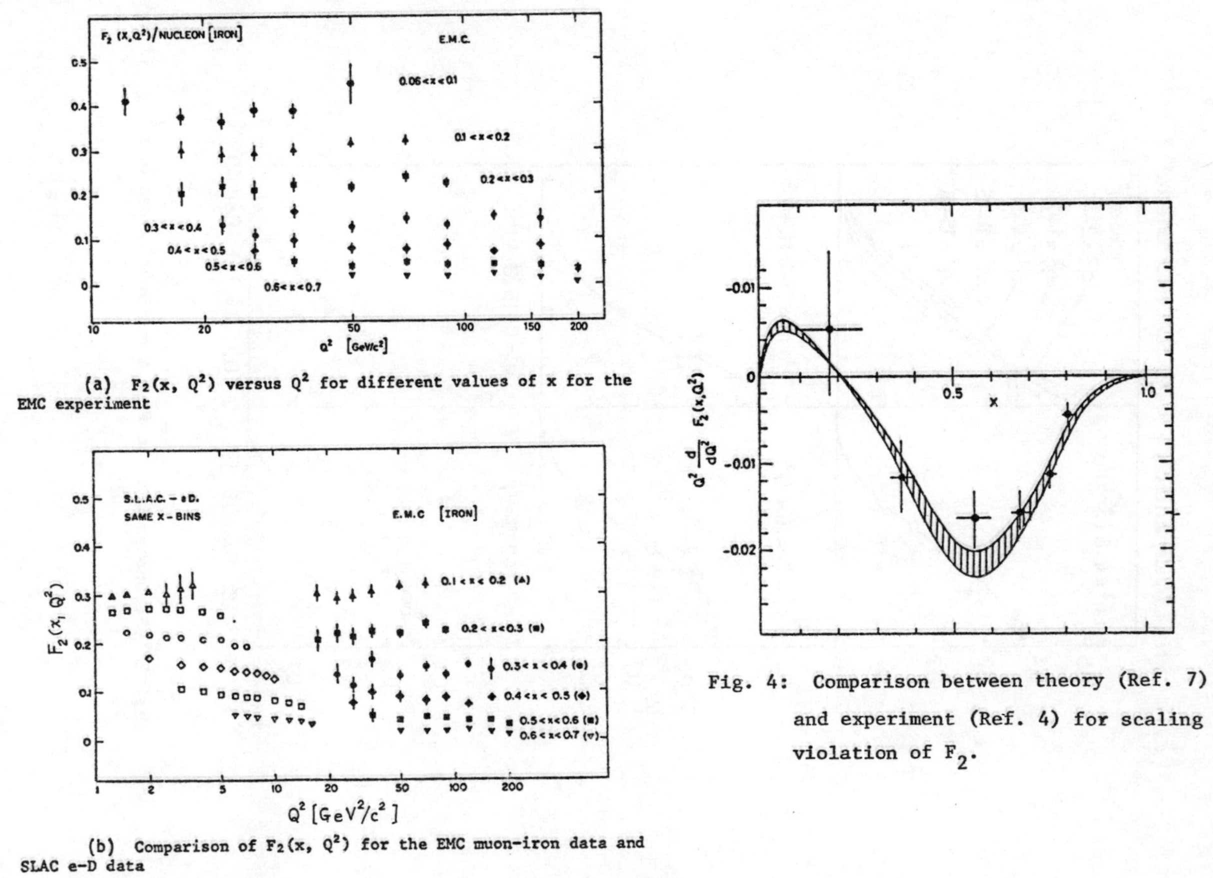

(a) $F_2(x, Q^2)$ versus Q^2 for different values of x for the EMC experiment

(b) Comparison of $F_2(x, Q^2)$ for the EMC muon-iron data and SLAC e-D data

Fig. 3: F_2 from EMC (Ref. 6) and comparison with SLAC data.

Fig. 4: Comparison between theory (Ref. 7) and experiment (Ref. 4) for scaling violation of F_2.

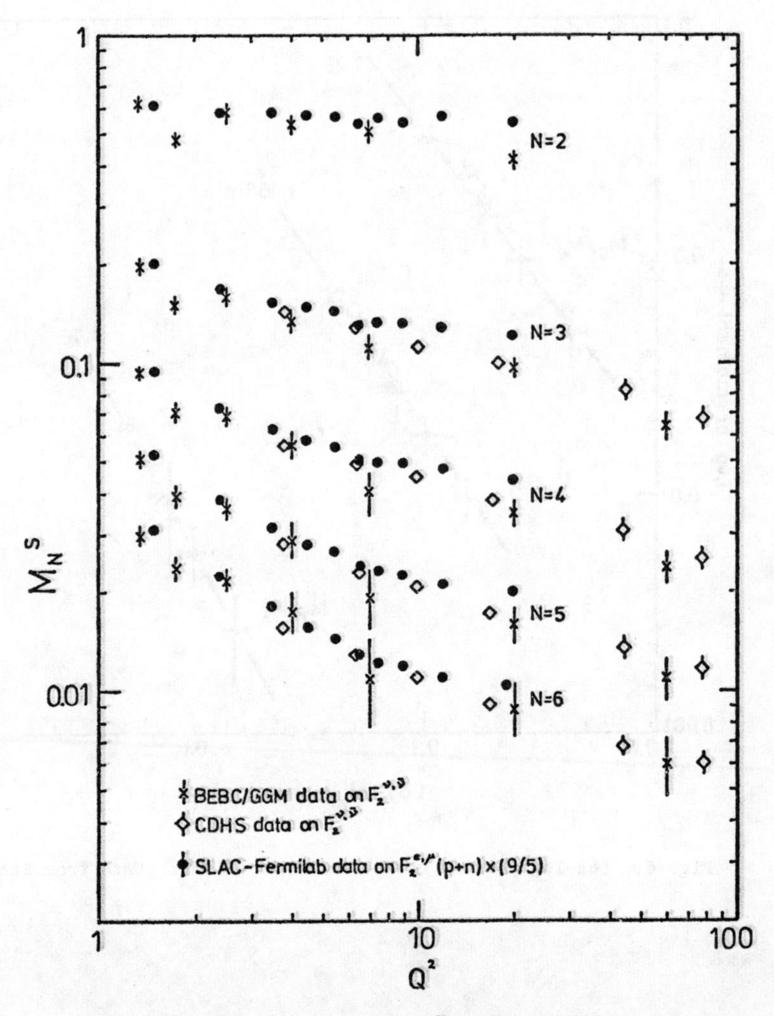

Fig. 5: Singlet moments versus Q^2. Data taken from Ref. 13.

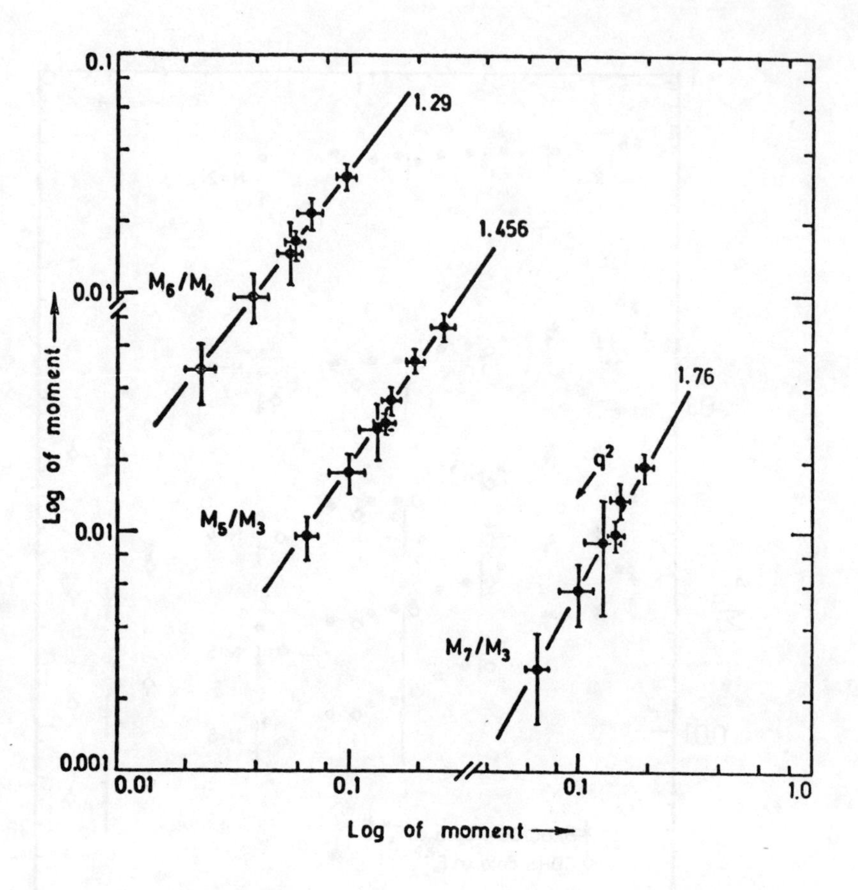

Fig. 6: Log-log plot of various moments of xF_3. Data from Ref. 10.

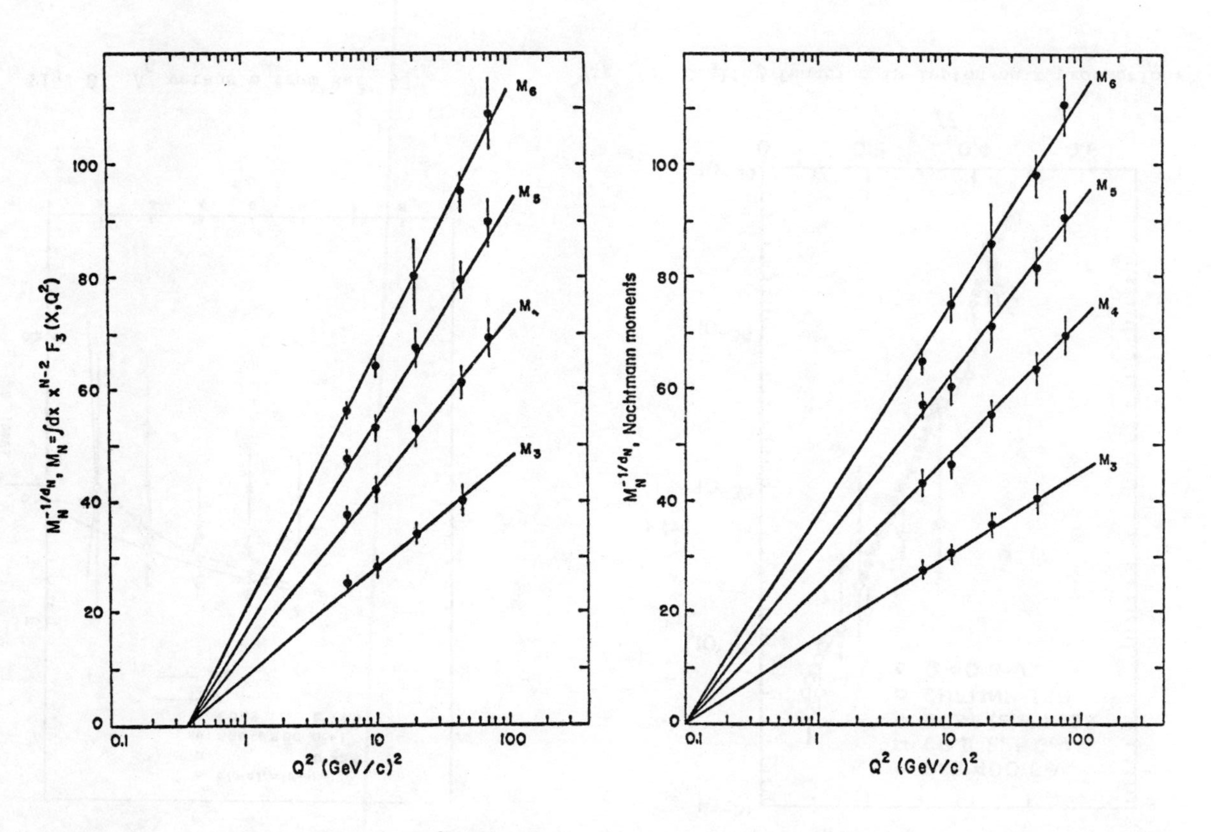

Fig. 7: Q^2 dependence of moments. Data from Ref. 11.

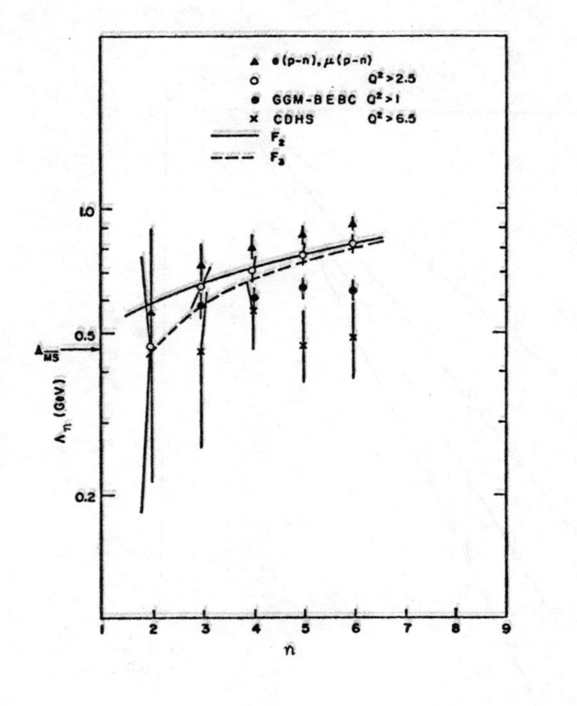

Fig. 8: Λ_n versus n from Ref. 45.

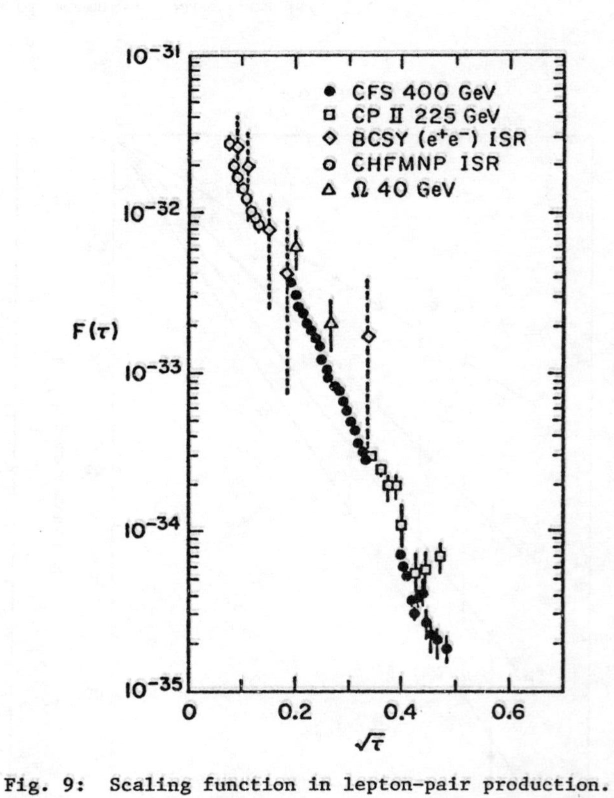

Fig. 9: Scaling function in lepton-pair production.

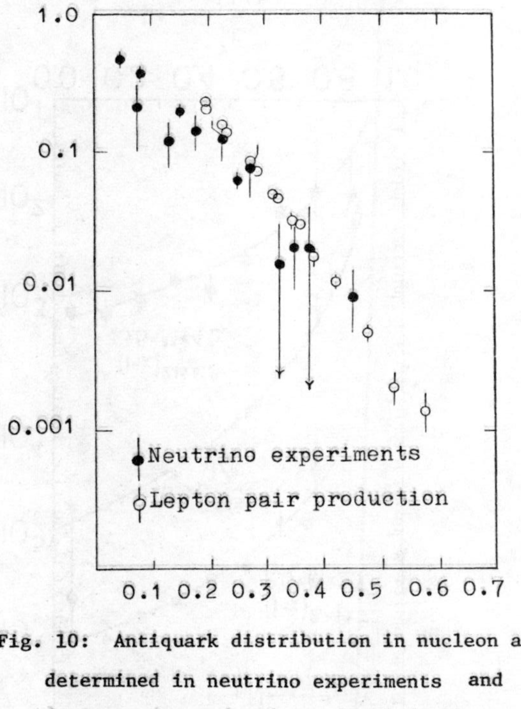

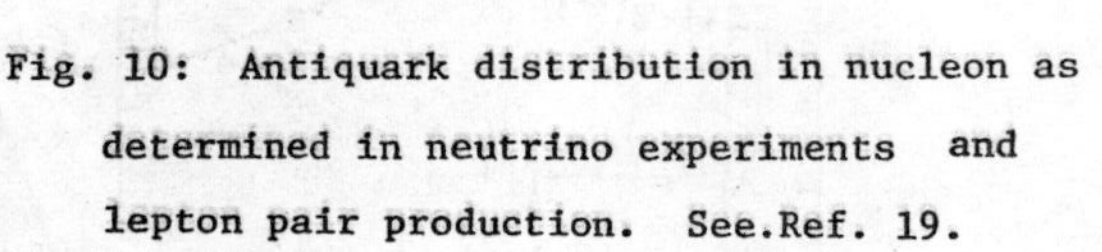

Fig. 10: Antiquark distribution in nucleon as determined in neutrino experiments and lepton pair production. See.Ref. 19.

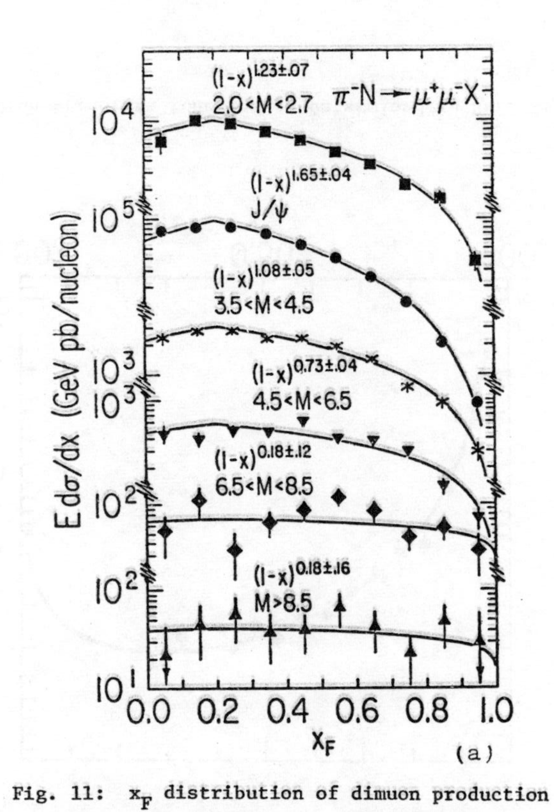

Fig. 11: x_F distribution of dimuon production cross section. Data from Ref. 21.

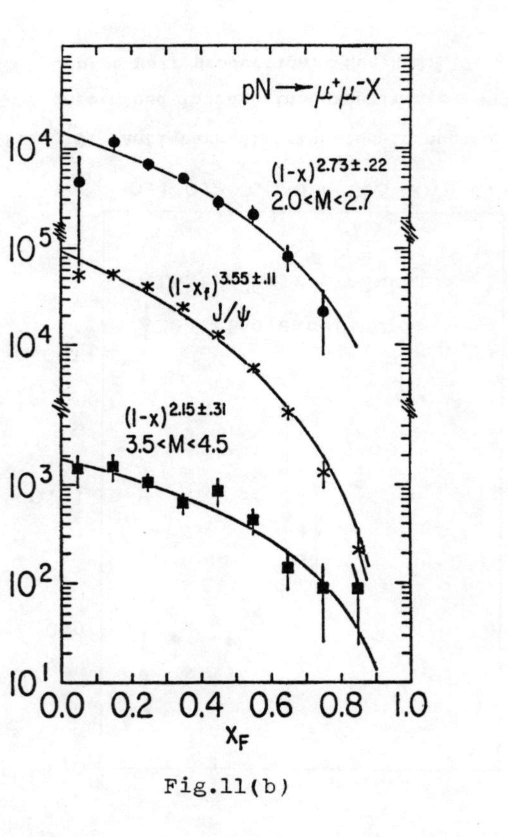

Fig.11(b)

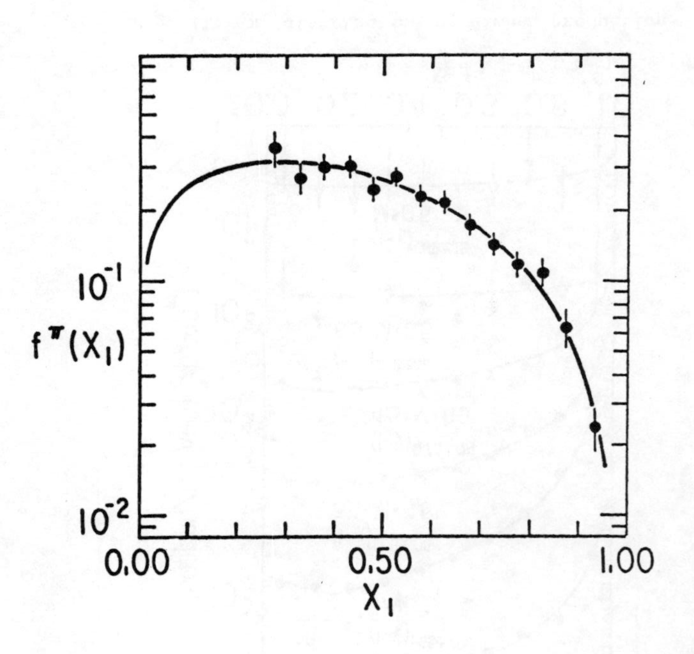

Fig. 12: Pion structure function as determined in Ref. 21.

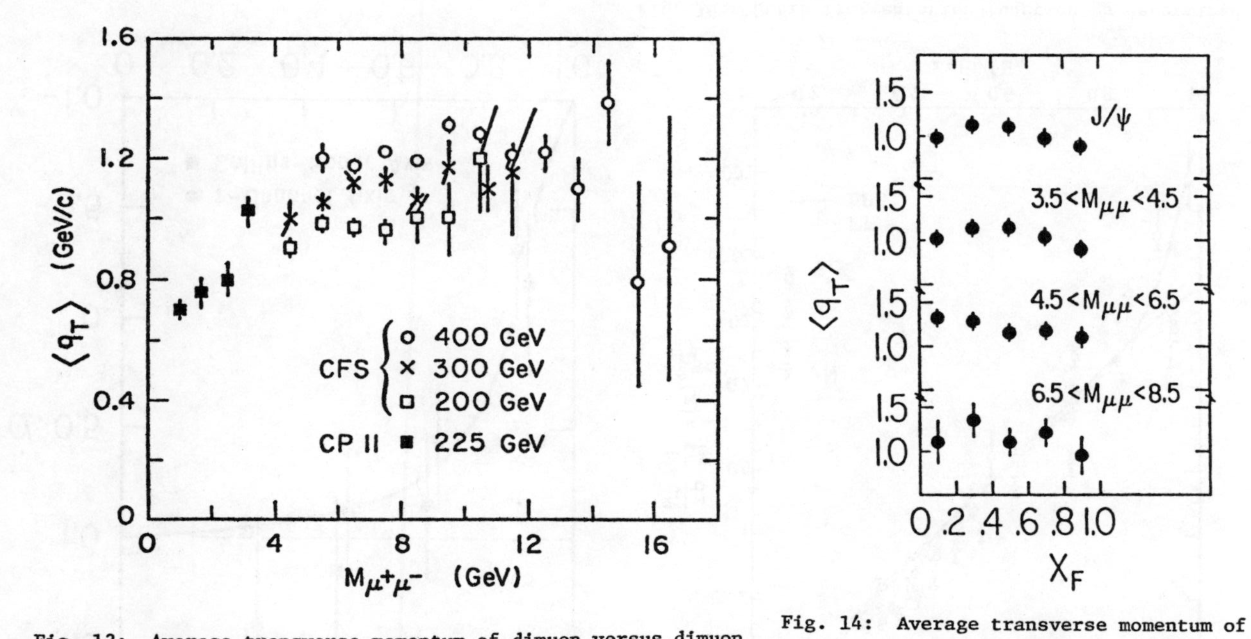

Fig. 13: Average transverse momentum of dimuon versus dimuon mass. Data from Ref. 37.

Fig. 14: Average transverse momentum of dimuon versus x_F. Data from Ref. 21.

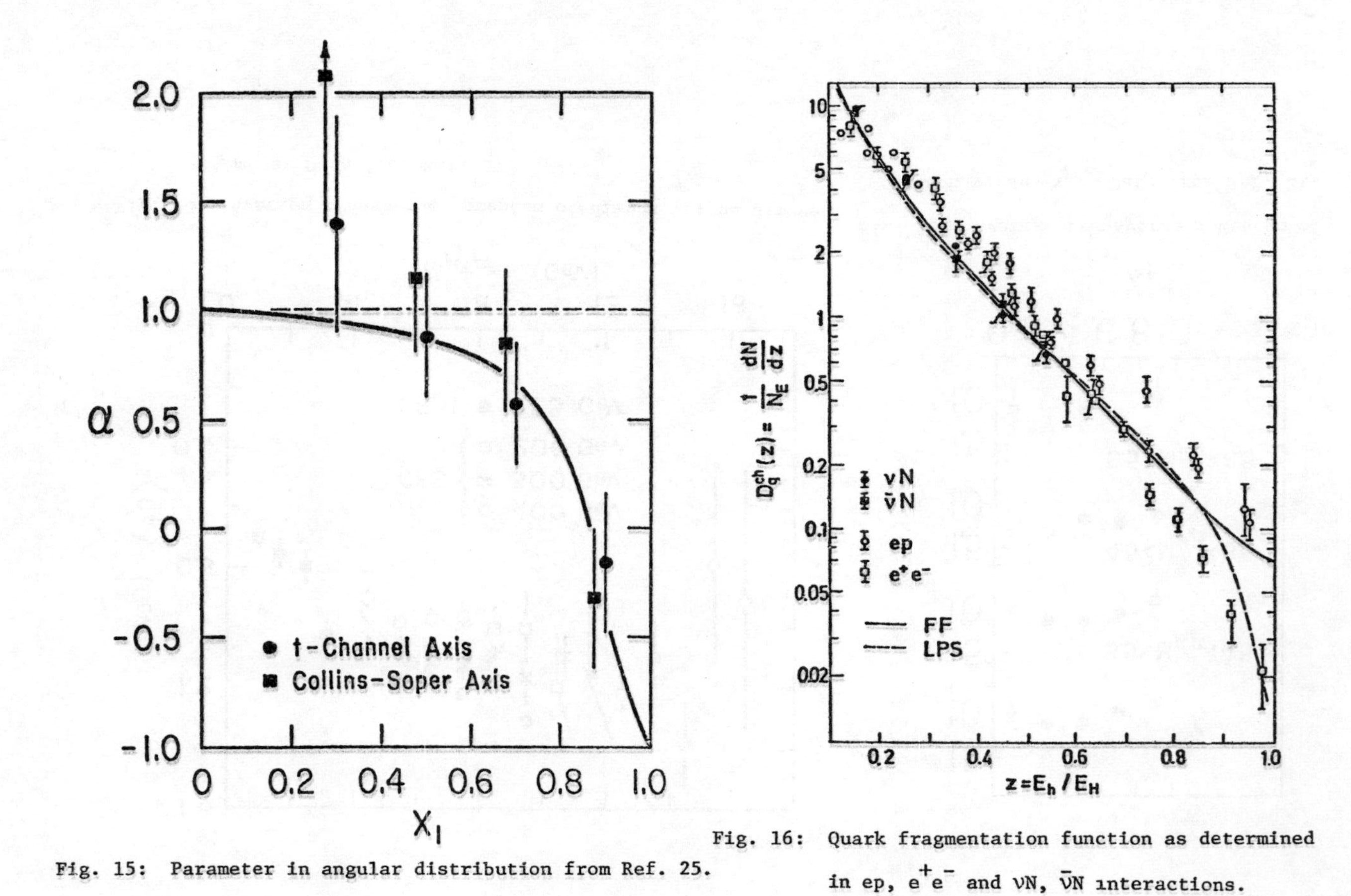

Fig. 15: Parameter in angular distribution from Ref. 25.

Fig. 16: Quark fragmentation function as determined in ep, e^+e^- and νN, $\bar{\nu}N$ interactions.

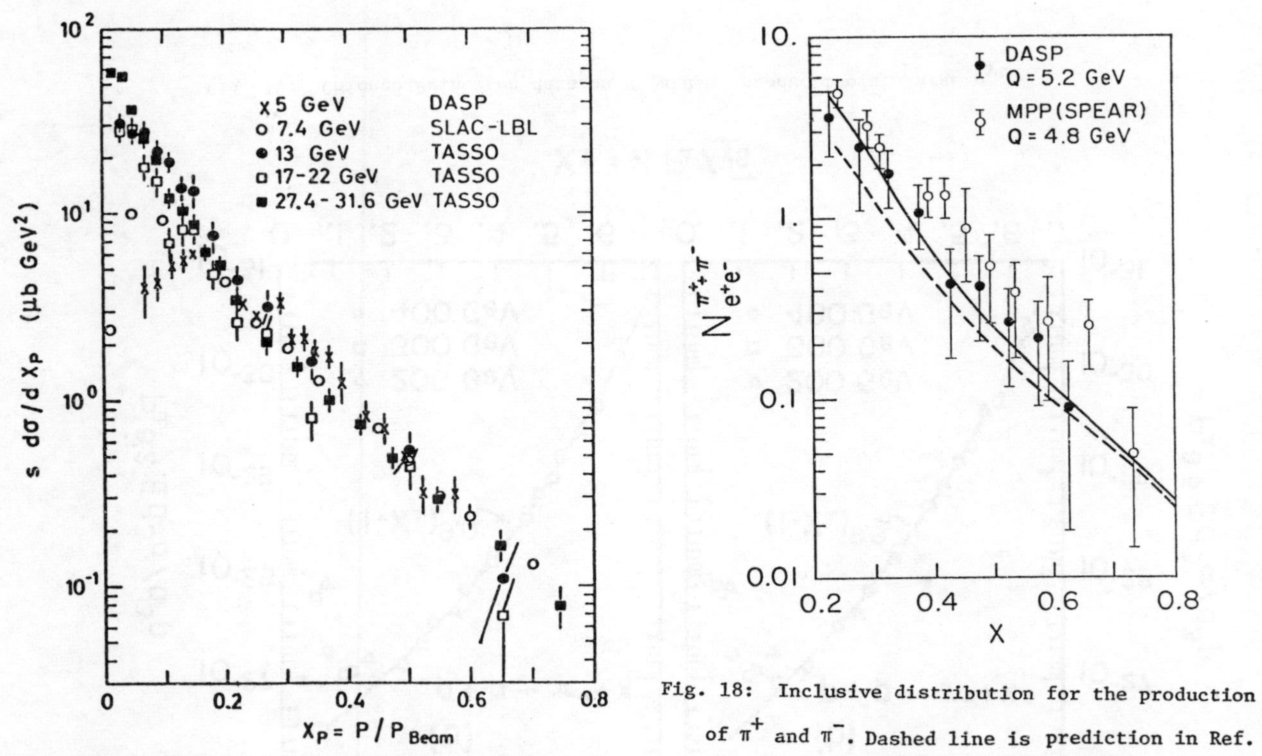

Fig. 17: Charged particle production cross section for various energies of e^+e^- annihilation process. Data from Ref. 38.

Fig. 18: Inclusive distribution for the production of π^+ and π^-. Dashed line is prediction in Ref. 26 without resonance production; solid line is with resonance production.

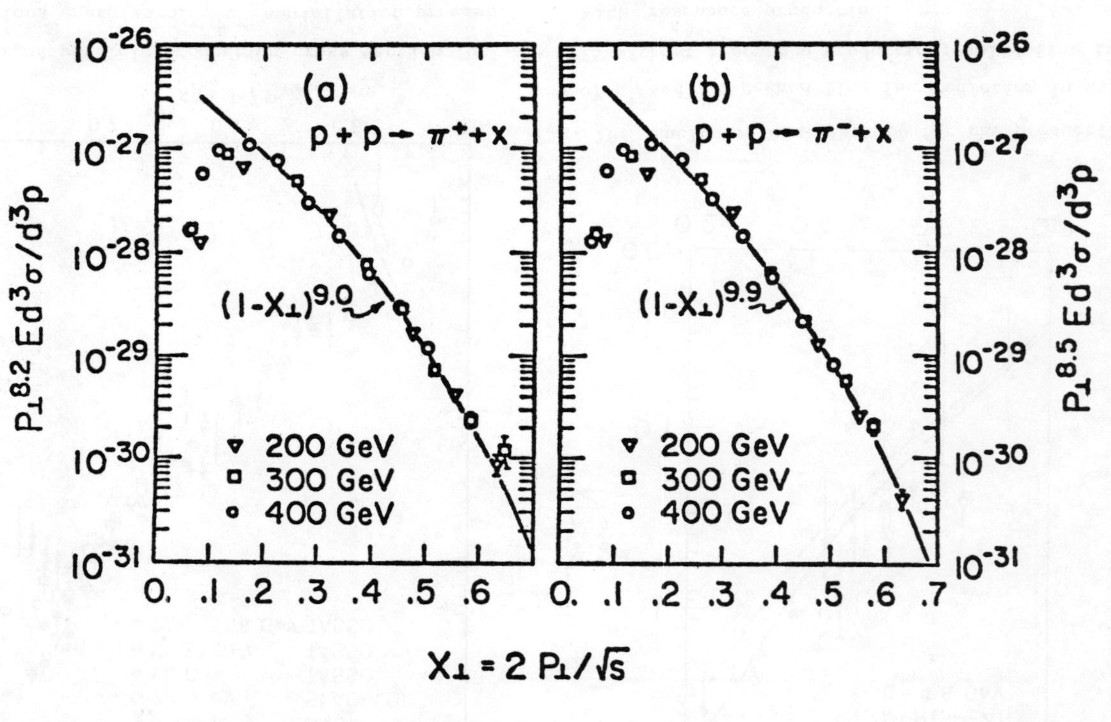

Fig. 19: Chicago Princeton data on π^+ and π^- production at large p_T.

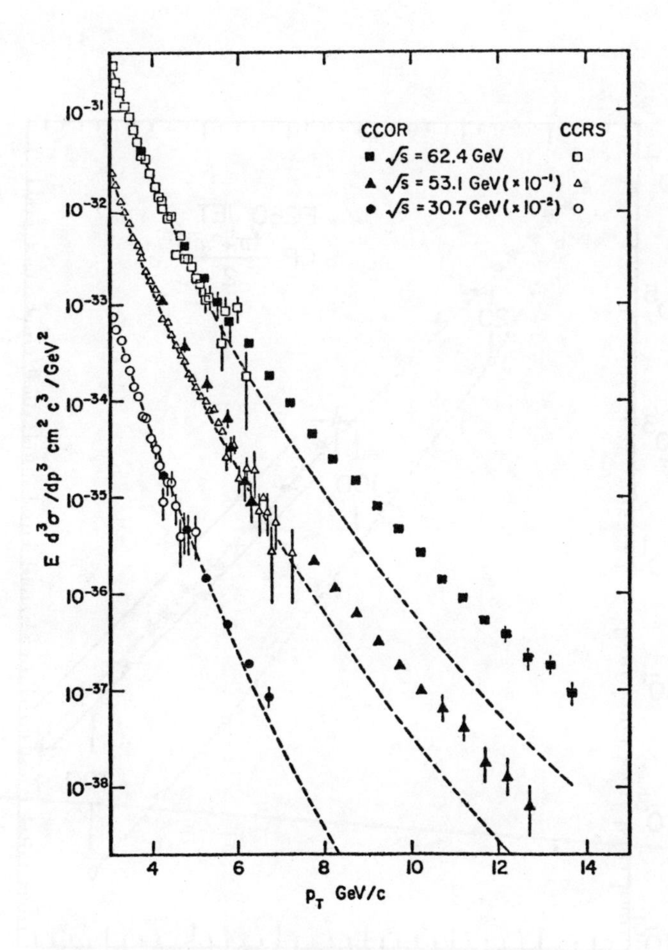

Fig. 20: Inclusive cross section at large p_T at ISR energies. Data from Ref. 39.

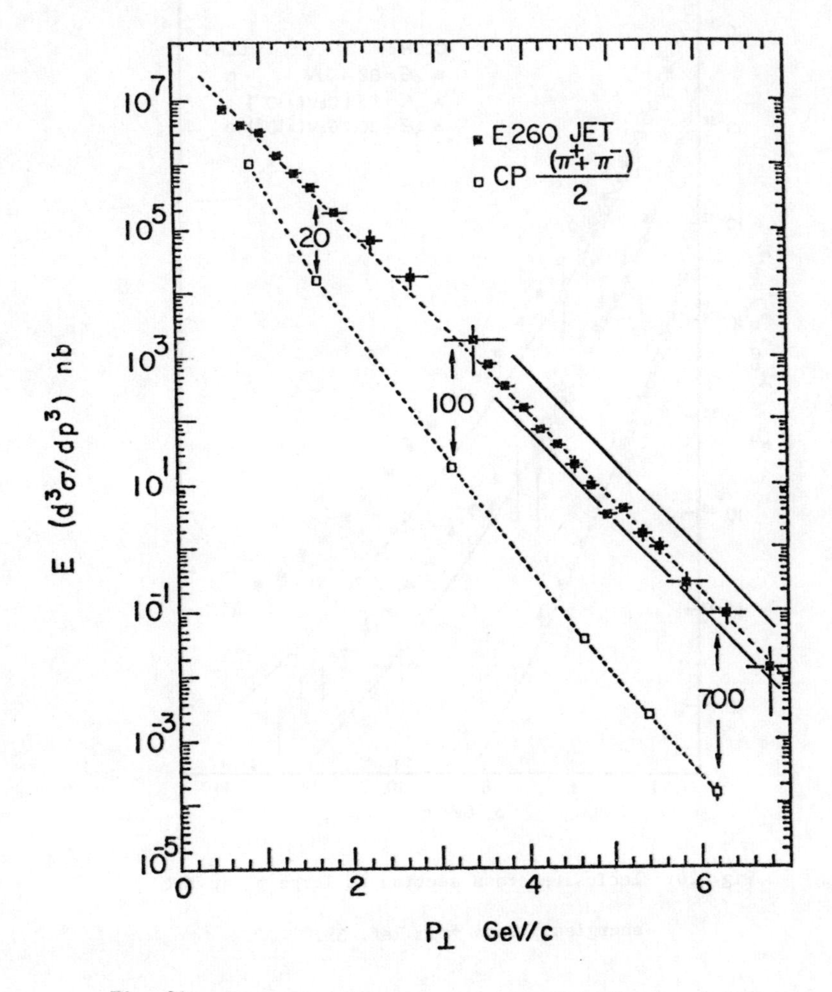

Fig. 21: Comparison between cross sections for jet and single particle production. Data are from Refs. 40 and 41.

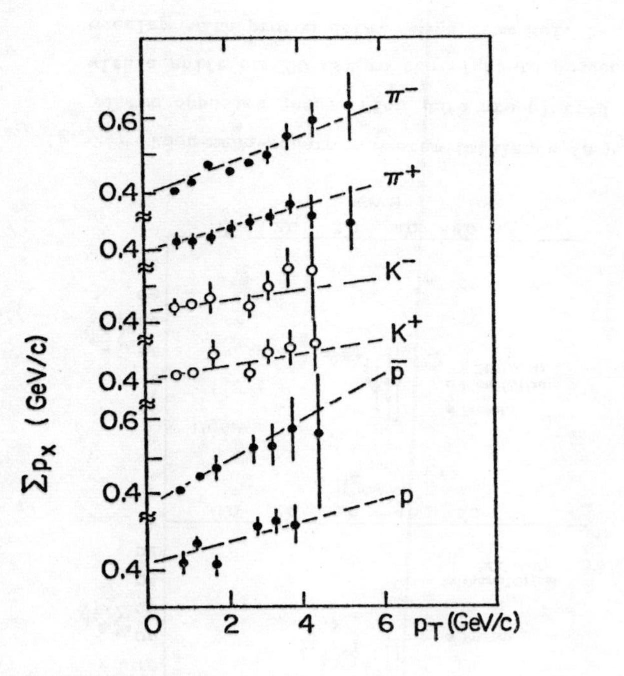

Fig. 22: Mean momentum carried by charged particles accompanying a high-p_T trigger particle. Data are from British-French-Scandinavian Collaboration, Ref. 42.

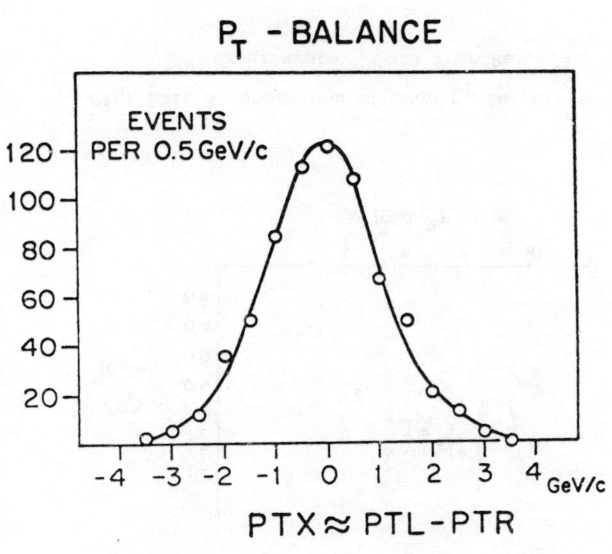

Fig. 23: Distribution in momentum imbalance in p_T of two opposite jets. Data from Ref. 34.

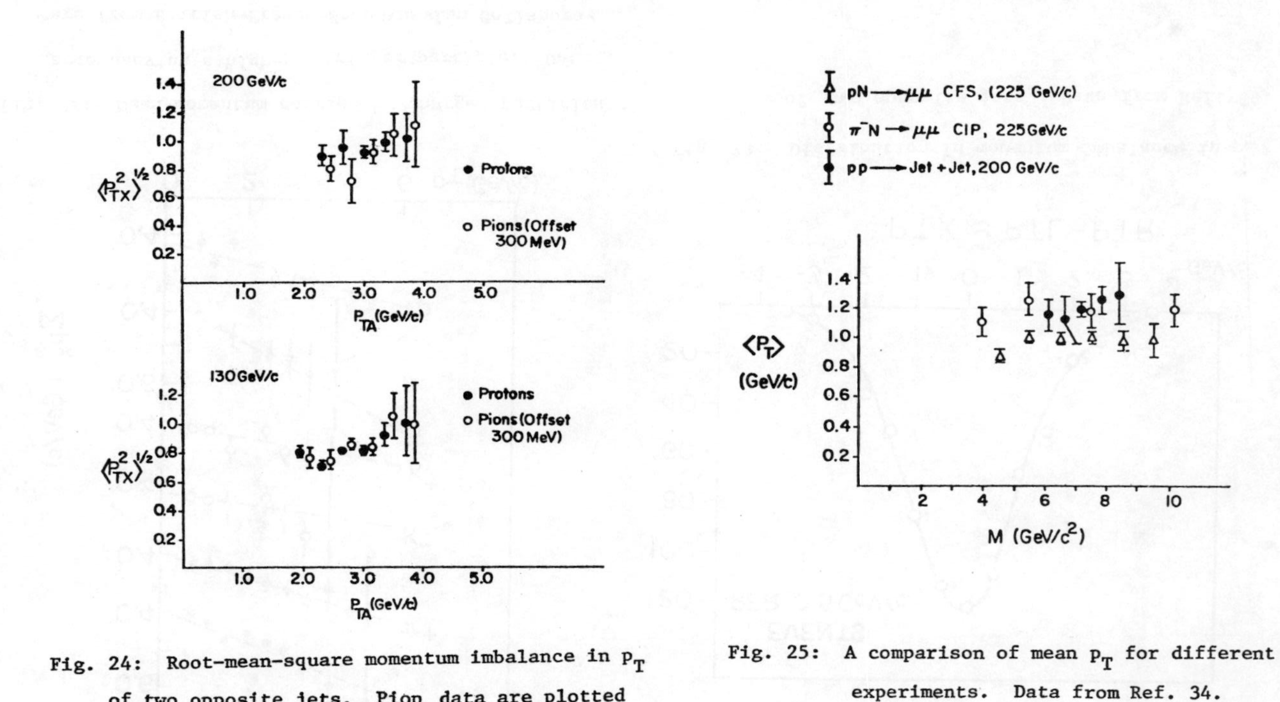

Fig. 24: Root-mean-square momentum imbalance in p_T of two opposite jets. Pion data are plotted with a shift of 300 MeV to the right to prevent overlap with proton data. Data from Ref. 34.

Fig. 25: A comparison of mean p_T for different experiments. Data from Ref. 34.

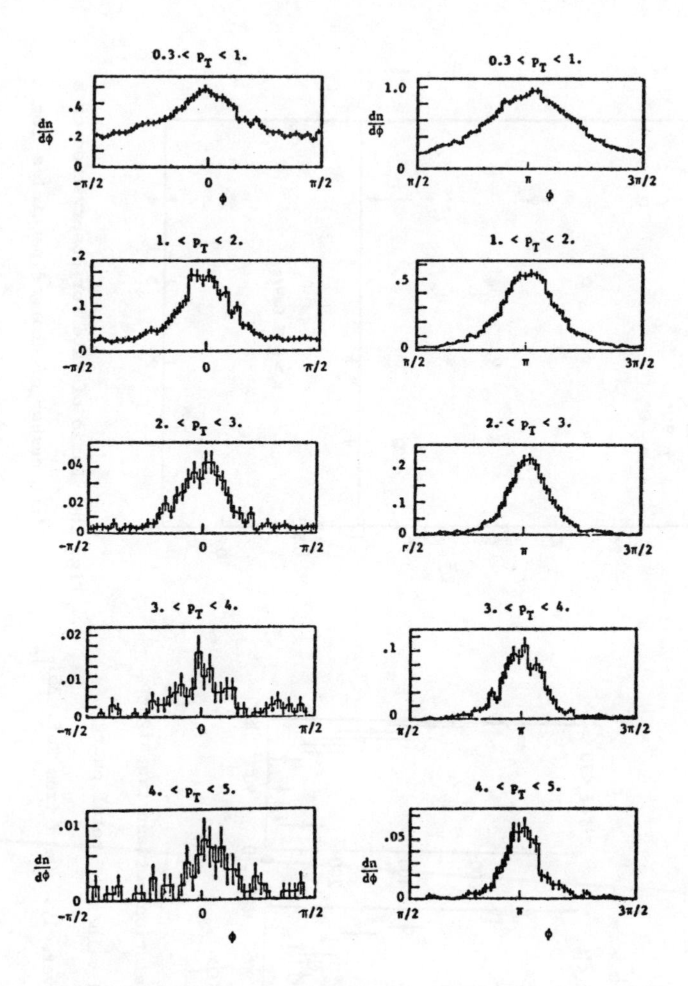

Fig. 26: Azimuthal correlation of charged particle detected relative to a neutral trigger particle. Data from Ref. 43.

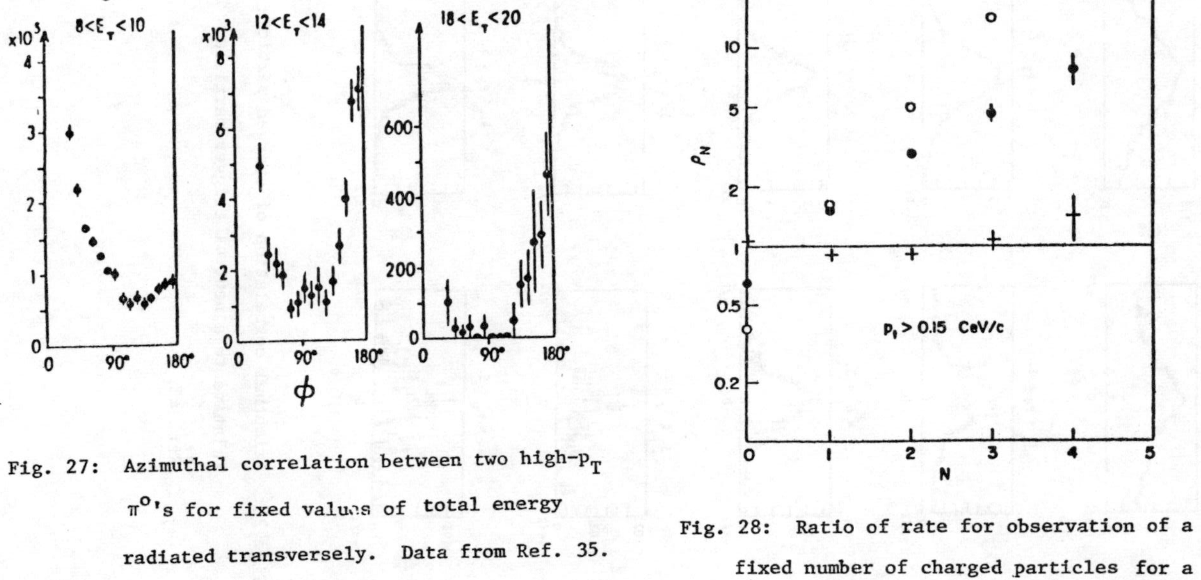

Fig. 27: Azimuthal correlation between two high-p_T π^o's for fixed values of total energy radiated transversely. Data from Ref. 35.

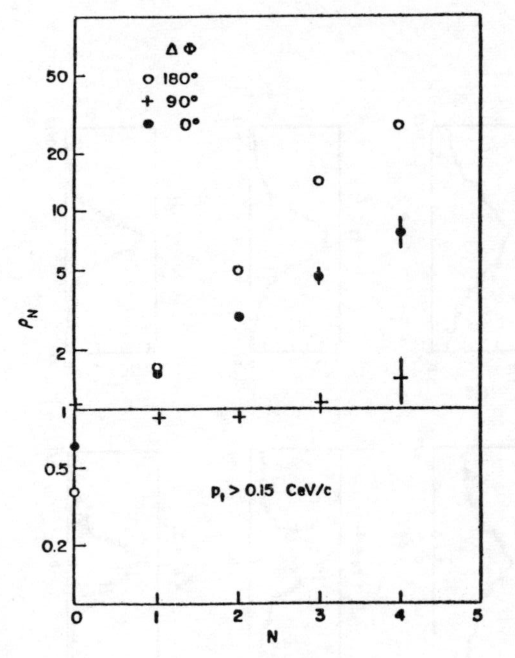

Fig. 28: Ratio of rate for observation of a fixed number of charged particles for a large-p_T (p_T > 5 Gev/c) relative to a typical low-p_T reaction. Data from Ref. 44.

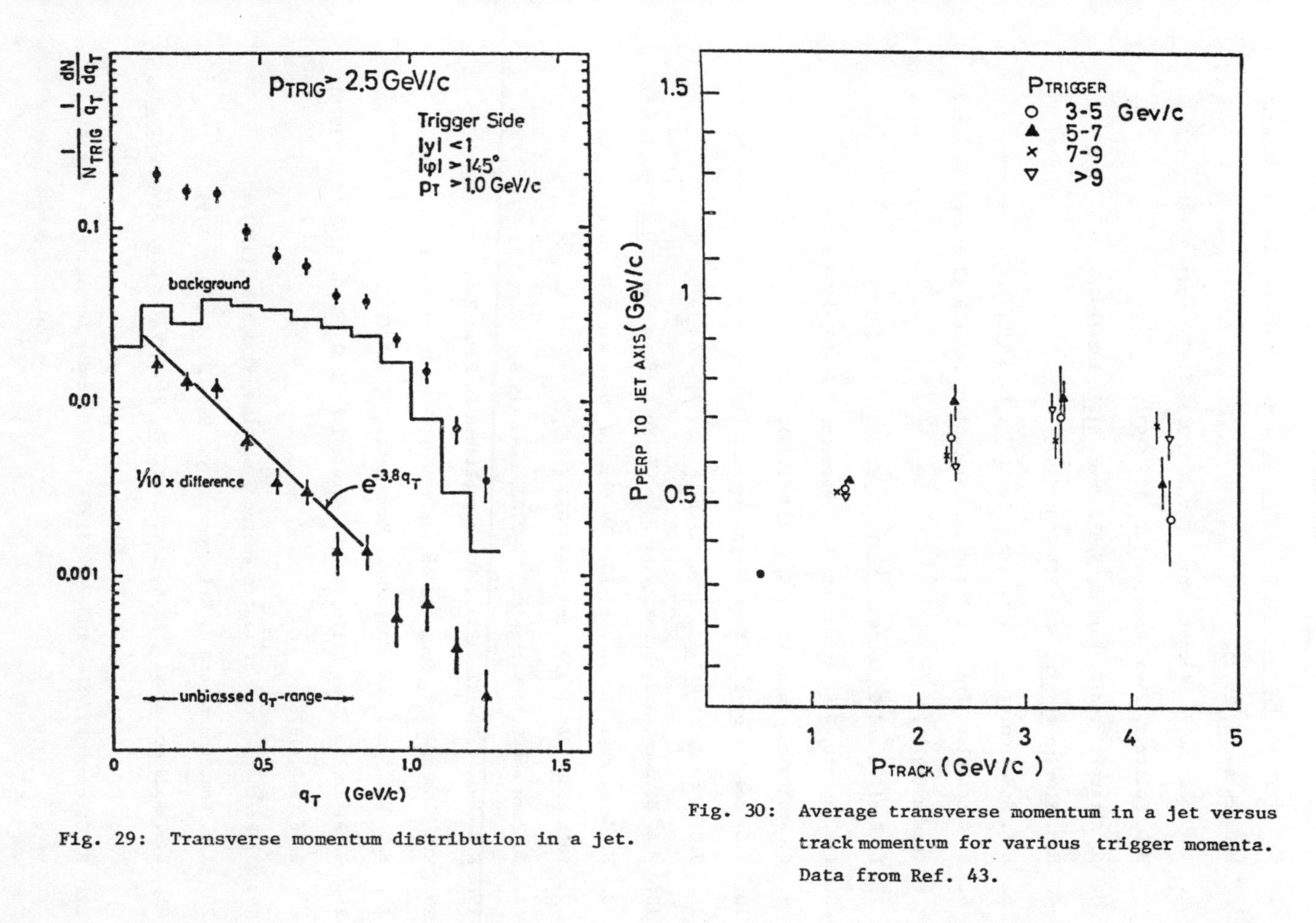

Fig. 29: Transverse momentum distribution in a jet.

Fig. 30: Average transverse momentum in a jet versus track momentum for various trigger momenta. Data from Ref. 43.

REFERENCES

1. R.P. Feynman, Photon-Hadron Interactions, W.A. Benjamin, Inc. (Reading, Massachusetts, 1972).
2. D. Gross and F. Wilczek, Phys. Rev. D 8, 3633 (1973) and D 9, 980 (1974); H.D. Politzer, Phys. Roports 14C, 129 (1974)
3. G. Altarelli and G. Parisi, Nucl. Phys. B126, 298 (1977).
4. E.M. Riordan et al., SLAC-PUB-1634 (1975).
5. H.L. Anderson et al., Phys. Rev. Lett. 38, 1450 (1977).
6. E. Gabathuler, Proc. of International Conference on High Energy Physics, Geneva (1979).
7. L. Baulieu and C. Kounnas, Nucl. Phys. B155, 429 (1979).
8. L.F. Abbott, W.B. Atwood and R.M. Barnett, SLAC-PUB-2400 (1979).
9. O. Nachtmann, Nucl. Phys. B63, 237 (1973).
10. P.C. Bosetti et al., Nucl. Phys. B142, 1 (1978).
11. J.G.H. de Groot et al., Phys. Lett. 82B, 292, 456 (1979).
12. W.A. Bardeen, A.J. Buras, D.W. Duke, T. Muta, Phys. Rev. D 18, 3998 (1978).
13. M. Bacé, Phys. Lett. 78B, 132 (1978); D.W. Duke and R.G. Roberts, Phys. Lett. 85B, 289 (1979) and RL-79-073 (1979).
14. A. Para and C.T. Sachrajda, TH-2702, CERN (1979).
15. R.C. Hwa, OITS-112 (1979), to be published in Phys. Rev.
16. C.S. Lam and W.K. Tung (to be published); W.K. Tung, these proceedings.
17. S.D. Drell and T.M. Yan, Phys. Rev. Lett. 25, 316 (1970).
18. G. Altarelli, R.K. Ellis, and G. Martinelli, Nucl. Phys. B143, 521 (1978), B146, 544 (1978).
19. J. Pilcher, Proc. of Lepton-Photon Symposium, Fermilab (1979).
20. A.P. Contogouris and J. Kripfganz, Phys. Rev. D 19, 2207 (1979).
21. K.J. Anderson et al., Phys. Rev. Lett. 42, 944 (1979); C.B. Newman et al., Phys. Rev. Lett. 42, 951 (1979).
22. R. Stroynowski, SLAC-PUB-2402 (1979), SLAC Summer Institute on Particle Physics, 1979.

23. R.C. Hwa, Proc. 13th Recontre de Moriond, Vol. 1 (1978), ed. by J. Tran Thanh Van.

24. E.L. Berger and S.J. Brodsky, Phys. Rev. Lett. 42, 940 (1979).

25. K.J. Anderson et al., phys. Rev. Lett. 43, 1219 (1979).

26. V. Chang and R.C. Hwa, Phys. Rev. Lett. 44, 439 (1980).

27. R.C. Hwa, OITS-122 (1979), to be published in Phys. Rev.

28. R. Blankenbecler, S.J. Brodsky, and J.F. Gunion, Phys. Rev. D 12, 3469 (1975).

29. R.D. Field and R.P. Feynman, Phys. Rev. D 15, 2590 (1977).

30. R.C. Hwa, A.J. Spiessbach, and M.J. Teper, Phys. Rev. Lett. 36, 1418 (1976); Phys. Rev. D 18, 1475 (1978).

31. R. Sosnowski, Proc. of the 19th International Conference on High Energy Physics, Tokyo (1978); R. Stroynowski, Ref. 22.

32. R.P. Feynman, R.D. Field, and G.C. Fox, Phys. Rev. D 18, 3320 (1978).

33. S. Ellis, M. Jacob and P.V. Landshoff, Nucl. Phys. B108, 93 (1976).

34. M.D. Corcoran et al., Wisconsid preprint COO-088-105; M.D. Corcoran et al., to be published.

35. C. Kourkoumelis et al., CERN-EP-79-36, Nucl. Phys. (to be published).

36. R.K. Ellis, M. Furman, H. Haber, and I. Hinchliffe, LBL report (to appear).

37. D.M. Kaplan et al., Phys. Rev. Lett. 40, 435 (1978); J.K. Yoh et al., Phys. Rev. Lett. 41, 684, 1083 (1978).

38. Tasso Collaboration, DESY 79/73 (1979).

39. F.W. Büsser et al., Nucl. Phys. B106, 1 (1976); A.L.S. Angelis, Phys. Lett. (1979).

40. C. Bromberg et al., Phys. Rev. Lett. 43, 565, 1058 (1979).

41. D. Antreasyan et al., Phys. Rev. D 19, 764 (1979).

42. M.G. Albrow et al., Nucl. Phys. B145, 305 (1978); Physica Scripta 19, 99 (1979).

43. A.L.S. Angelis et al., Physica Scripta 19, 116 (1979).

44. A.G. Clark et al., CERN-EP/79-74 (1979).

45. J. Ellis, Proceedings of International Symposium on Lepton and Photon Interactions, FNAL, Batavia, Illinois (August 1979).

Soft Hadron Physics as Colour Chemistry

by

CHAN Hong-Mo*

CERN, Geneva, Switzerland

Abstract.-

Soft properties, in particular the spectroscopy, of hadrons are reviewed emphasizing the analogy with ordinary chemistry, with the aim of evaluating critically the contribution to their understanding from quantum chromodynamics.

* On leave of absence from the Rutherford Laboratory, England.

In the last few years, there has been a fundamental change in our attitude towards soft hadron physics. For perhaps the first time in history, we have now a serious candidate for the theory of strong interactions, namely quantum chromodynamics. In this theory, hadrons are supposed to be made up of quark constituents interacting via the exchange of glue quanta in much the same way as atoms and molecules are constructed out of electrons and nuclei interacting via photon exchange. Hence, soft hadron physics, which is the study of the properties and interactions of these 'atoms' and 'molecules', is just the colour equivalent of chemistry, and needs not be of fundamental interest to the particle physicists which should presumably be focussed on the constituents.

Unfortunately, this change in attitude is not accompanied by a similar advance in our understanding of soft hadron properties. The reason is that, whereas QED is capable of making quantitative predictions in atomic physics which have been verified to several parts in a million, we have not yet learned to calculate with QCD anything but asymptotic quantities lacking direct confrontation with experiment. At present therefore, soft hadron physics remains a challenge and the responsibility of the particle physicist, and cannot be delegated as yet to a new department of colour chemistry. Indeed, the purpose of this talk is just to pursue this colour chemical analogy to see how much (or how little) of soft hadron physics can now be understood in this way.

The simplest (electrical) atom, H , is obtained by combining two constituents of opposite charges, namely e and p . Similarly, the simplest way to form a colour atom is to combine a quark Q with colour charge 3 and an antiquark $\bar{Q}$ with the opposite colour charge $\bar{3}$, thus $(Q^3\bar{Q}^{\bar{3}})^1$. Because of the nonabelian nature of the colour group SU(3) , another simple atom can be obtained from three quarks as follows : $((Q^3Q^3)^{\bar{3}}Q^3)^1$. These are the familiar mesons and baryons which account for nearly all entries in the Rosenfeld tables.[1)]

How much of meson and baryon spectroscopy do we understand from QCD ? Unfortunately, much less than we could wish ! The mechanism which confines quarks to form hadrons is a nonperturbative phenomenon which none has yet demonstrated to exist in QCD, let alone calculated. In current hadron spectroscopy, its effect is just imitated, either by assuming a potential of interaction between the constituents or by imposing a boundary condition enforcing the confinement of quark and gluon fields inside a bag. The so-called successes of QCD in spectroscopy are limited to the interpretation of some secondary effects such as the hyperfine splitting between members of a spin multiplet which are arrived at as follows. QCD is known to be asymptotically free which means that at small distances the effective coupling between quarks and gluons is small. Hence, if we are interested only in small distance effects among which the spin-dependent interaction between quark pairs is believed to be one, then perturbation theory may apply.[2] To leading order in the strong coupling constant α_s, we have then the one-gluon-exchange diagram, Fig. 1, which, via the Breit reduction familiar in atomic physics, leads to a spin-dependent interaction between two quark constituents of the following form :[2]

$$V_{12}^{spin} = -\alpha_s \left(\sum_a \lambda_1^a \lambda_2^a\right)\Big\{ \frac{2\pi}{3}\frac{1}{m_1 m_2}\delta^3(\underline{r})\left[\underline{\sigma}_1\cdot\underline{\sigma}_2\right] + \frac{1}{4r^3}\left[\frac{1}{m_1^2}(\underline{r}\times\underline{p}_1\cdot\underline{\sigma}_1) - \frac{1}{m_2^2}(\underline{r}\times\underline{p}_2\cdot\underline{\sigma}_2) + \frac{2}{m_1 m_2}(\underline{r}\times\underline{p}_1\cdot\underline{\sigma}_2 - \underline{r}\times\underline{p}_2\cdot\underline{\sigma}_1)\right] + \frac{1}{4r^3}\frac{1}{m_1 m_2}\left[-\underline{\sigma}_1\cdot\underline{\sigma}_2 + 3(\underline{\sigma}_1\cdot\underline{r})(\underline{\sigma}_2\cdot\underline{r})\right] \qquad (1)$$

where λ^a are 3 X 3 Gell-Mann matrices representing the colour charges of the quark constituents while σ^r are 2 X 2 Pauli matrices representing their spins, and for antiquarks, the following replacements in the formula are understood : $\lambda_a \to -\lambda_a^*$, $\sigma_r \to -\sigma_r^*$. The potential contains a spin-spin, a spin-orbit and a tensor term. With these, one is able to obtain a good description of the main features in both the mesonic and baryonic spectra.[3]

Let us now examine to what extent these successes are due indeed to QCD or colour chemsitry. For this purpose we shall ignore the question whether the application of first order perturbation theory to the hyperfine structure of hadronic spectra is in fact justified. At this level then, the only difference between QCD and electrodynamics is an additional colour degree of freedom which is reflected in the occurence in (1) of the colour matrices λ^a . We note first that a new degree of freedom in spectroscopy will in general lead to a greater complexity in the spectrum. However, for the simple systems $Q\bar{Q}$ and QQQ that we are considering, because of the requirement that all hadrons are colour singlets, this complexity is not manifested. What remains of the whole colour structure is just the factor $\sum_a \lambda_1^a \lambda_2^a$ multiplying the Breit potential. In order to investigate the colour dependence of this factor, we need to examine quark constituent pairs with different colour contents, which however are not available either in $Q\bar{Q}$ or in QQQ spectroscopy. In mesons, of course the $Q\bar{Q}$ pair has colour 1, giving

$$\sum_a \lambda_1^a \lambda_2^a = -\frac{16}{3} \tag{2}$$

for all mesons, while in baryons, any QQ pair must necessarily have colour $\bar{3}$ in order that the baryon is a total colour singlet, giving just an overall factor :

$$\sum_a \lambda_i^a \lambda_j^a = \frac{8}{3} \tag{3}$$

for all baryon states. We may of course compare the different effects of the Breit potential on mesons and baryons. For example, for the s-wave mesons π, ρ we have from the contact spin-spin interaction in (1) a mass splitting of the form :

$$m_\rho - m_\pi = -\alpha_s\left(-\frac{16}{3}\right)\cdot\frac{2\pi}{3m^2}|\psi(0)|^2\cdot 4 \tag{4}$$

while for the s-wave baryons N, Δ, we have

$$m_\Delta - m_N = -\alpha_s\left(\frac{8}{3}\right)\cdot\frac{2\pi}{3m^2}|\psi(0)|^2\cdot(-6) \tag{5}$$

To compare (4) and (5) numerically, we need to know the wave functions of the quark constituent pairs at their centres of mass, which however depend on the confining mechanism and need not even be similar for the two cases. We are therefore left with just the relative sign of the splittings as the only test for colour !

Next, a property that QCD shares with electrodynamics is the vector nature of the quanta exchanged which is reflected in the relative strengths of the spin-spin, spin-orbit and tensor terms in the Breit potential (1). In principle, from the spectrum of meson and baryon states, one can extract the contributions of the respective terms and thereby test this aspect of the theory. However, the contact spin-spin term is operative only between s-wave constituent pairs while the spin-orbit and tensor terms are not. Their comparison will thus involve both s- and p-wave states and the ratio of $\langle 1/r^3\rangle$ of the p-state to the wave function $|\psi(0)|^2$ of the s-state at the origin, which depends again on the detail of the confining mechanism. Further, although good fits were obtained for both the meson (notably charmonium) and baryon spectra with the Breit potential, the meson spectrum requires a large spin-orbit but a small tensor force,[4] while the recent fits to the baryon spectrum are obtained with dominant tensor forces but negligible spin-orbit contributions.[5] This discrepancy does not necessarily invalidate our basic tenets because of the unknown spin-dependent effects of the confining potential,[5),6)] but it can hardly be claimed as a success of QCD as it is sometimes implied in the literature.

Although the spectroscopy of ordinary mesons and baryons by itself affords no stringent test for colour chemistry, it is nonetheless not in contradiction and can be fitted into the scheme quite comfortably. We may therefore use the information here for determining some unknown quantities in our theory to bolster its inadequacy and employing them to make other predictions. For example, from (5) and the empirical value of the Δ-N mass difference, one can estimate the effective strength of the spin-spin interaction between quark pairs, say :

$$V_{ij}^{(1)} = -C \sum_{a} \sum_{r} (\lambda_i^a \sigma_i^r)(\lambda_j^a \sigma_j^r) \tag{6}$$

yielding for ordinary quarks $q = u, d$,[7),8)]

$$C \sim 20 \text{ MeV}. \tag{7}$$

Further, by studying also excited nucleon states, one can conclude that the spin-spin interaction between one constituent pair is not very dependent on the presence of other quarks in the same hadron. If so, we can use (6) and (7) to predict the hyperfine splitting in hadrons other than $Q\bar{Q}$ mesons and QQQ baryons, if such should exist.

Now in electrodynamics, very complex atoms can be obtained by combining many constituents. Can one do the same in colour chemistry ? This is not obvious. In ordinary atoms, we have strong nuclear forces which overcome the electrical repulsion between like charged protons, concentrating them in the nucleus ; this is what makes a complex atom stable. In a system of coloured quarks however, we know yet of no equivalent to nuclear forces, so that a complex system of many quarks may readily fall apart. For example, take a system of two quarks and two antiquarks $QQ\bar{Q}\bar{Q}$. This can form two $Q\bar{Q}$ pairs each in a colour singlet, and, there being now no colour forces to hold these neutral pairs together, the system can simply dissociate into mesons, thus : $(QQ\bar{Q}\bar{Q}) \rightarrow (Q\bar{Q})^1(Q\bar{Q})^1$.
In general, the more constituents a quark system contains the more channels there are into which it may dissociate.

It is not excluded of course that such a multiquark atom happens to lie below the threshold for dissociation and appears as a metastable object. For example, the colour magnetic spin-spin interaction in (6) and (7) is quite strong and may in certain cases be sufficient to shift the mass of a multiquark system below its dissociation thresholds. In order to calculate the mass of a system containing several quarks all in the s-wave state, we need to diagonalise the matrix :

$$H^{(1)} = \sum_{i>j} V_{ij}^{(1)} \tag{8}$$

where the sum runs over all pairs of constituents. Now it often happens in diagonalising a big matrix with elements of comparable size that one state gets a particularly large eigenvalue. For the matrix (8), this is indeed found to occur to the state with the maximal symmetry in colour-spin, or in other words, since quarks are fermions, to the state with the maximal asymmetry in flavour. A prime example is the dihyperon of Jaffe [9] (udsuds) with $I(J^P) = 0(\ )$, which has a colour magnetic binding estimated from (7) of the order 600 MeV. This will very likely put it below the threshold for $\Lambda\Lambda$ decay and make it therefore stable against strong interactions. Such a stable dihyperon has been looked for experimentally but was not found. However, because of its very exotic flavour, it would in any case be hard to produce, and its absence in recent experiments is not yet a strong evidence against its existence. Another interesting possibility which may be easier to observe experimentally is $(qQ\bar{q}\bar{Q})$ with $I^G(J^P) = 0^+(0^+)$ where $q = u, d$ and $Q = c, b$. [10] Such a state is likely to exist below the thresholds of dissociation into $(q\bar{Q})(\bar{q}Q)$ and $(Q\bar{Q})\eta$ and may be looked for as a narrow resonance in the radiative decays of radial excitations of J/ψ or Υ, or else in $\bar{p}p$ formation experiments with the cooled antiproton devices now being constructed where it is expected to have appreciable cross sections. Since the prediction of these metastable complex atoms depends strictly on the form of the colour magnetic interaction (6), (7), their eventual observatinn should be quite an encouragement for colour chemists.

The examples we have just quoted, however, are mere rarities. The possibility of systematically constructing metastable multiquark systems must ultimately depend on the existence of a mechanism for preventing constituents from coagulating into smaller colourless lumps. One suggestion is that this may perhaps be effected by separating the constituents by angular momentum barriers. Take again as example a $QQ\bar{Q}\bar{Q}$ system but where the quarks Q and antiquarks $\bar{Q}$ are separated by a 'high' angular momentum L, thus :

$$(QQ) \overset{L}{\text{———}} (\bar{Q}\bar{Q}) \tag{9}$$

In order now for a Q to combine with a $\bar{Q}$ in (9) to form a colour singlet $(Q\bar{Q})$ meson, it will first have to overcome the angular momentum barrier. The system may then be inhibited from dissociating into mesons and appear as a metastable resonance.

One may ask of course why the system (9) cannot simply split along its length into two high L mesons, thus :

$$\left(\frac{Q}{Q}\right) \text{———} \left(\frac{\bar{Q}}{\bar{Q}}\right) \longrightarrow \begin{matrix} (Q) \longrightarrow (\bar{Q}) \\ (Q) \longrightarrow (\bar{Q}) \end{matrix} \tag{10}$$

The argument against such a break-up goes as follows. [11] Regge trajectories of known hadron states are all approximately linear and parallel with a slope $\alpha' \sim 1\ \text{GeV}^{-2}$. The usual explanation for this behaviour is that a high L hadron is an elongated object with its constituents separated into two clusters located at the ends and connected by a colour flux tube, i.e, literally as depicted in the example (9). For L sufficiently large, both the mass and the angular momentum of the system are carried mainly in the colour flux tube. The mass M of the hadron is then proportional to the length ℓ of the flux tube, and its angular momentum L to ℓ^2, hence

$$M = \rho \ell \quad ; \quad L = \alpha' M^2. \tag{11}$$

The fact now that all known $Q\bar{Q}$ and QQQ trajectories have the same α' suggests that two flux tubes are the same so long as they carry the same colour charge, independently of the numbers and flavours of the quark constituents in the clusters at the ends. Thus, whether the hadron (9) can split into two mesons as in (10) or not depends only on whether the linear mass density ρ of the flux tube in (9) is larger or smaller than the sum of the linear mass densities of the decay products. In case the diquarks (QQ) and $(\bar{Q}\bar{Q})$ in (9) have colour 3, namely the flux tube is the same as those in the product mesons of (10), one sees that splitting is clearly impossible. In general however, we shall need an estimate for the linear mass densities of flux tubes with different colours, which cannot be made without a more specific model for confinement. In the MIT bag model, which is at present the only viable model for this purpose, the linear density ρ_x for a flux tube of colour x is proportional to $\sqrt{\mathcal{C}_x}$ where $\mathcal{C}_x$ is the value of the quadratic Casimir operator for x .[12] If so, the splitting of a flux tube into two in the manner (10) is never possible for any multiquark system because of a certain triangular inequality between Casimir operators, namely : $\sqrt{\mathcal{C}_x} < \sqrt{\mathcal{C}_{x1}} + \sqrt{\mathcal{C}_{x2}}$ for $x_1 \times x_2 \supset x$.[10),13] Dissociation modes for a system such as (9) will therefore always be inhibited by an angular momentum barrier as it was suggested above.

Now, such metastable high L hadrons, if they exist, are best pictured as colour analogues of ionic molecules[14] (e.g. Na^+Cl^-) in which two clusters of quark constituents carrying opposite (conjugate) colour charges x and $\bar{x}$ are held together by colour electric forces, thus : $(QQ)^x$--$(\bar{Q}\bar{Q})^{\bar{x}}$. We shall therefore in future refer to such hadrons as 'chromionic' molecules, the clusters of quark constituents as 'chromions', and the flux tubes connecting them as 'chromionic bonds'. The main interest for particle physicists in these chromionic molecules is that they may afford very clear tests for several basic assumptions of colour dynamics.[7] First, the colour degree of freedom as well as its nonabelian nature are reflected in the colour representations available to the chromions. For example, for the simplest ions with two or three constituents, the possible colour charges are :

$$\begin{aligned} QQ &: \quad 3 \times 3 = \bar{3} + 6, \\ Q\bar{Q} &: \quad 3 \times \bar{3} = 1 + 8, \\ QQQ &: \quad 3 \times 3 \times 3 = 1 + 8 + 8' + 10 \\ QQ\bar{Q} &: \quad 3 \times 3 \times \bar{3} = 3 + 3' + \bar{6} + 15 \end{aligned} \tag{12}$$

Second, in **contrast** to the hydrogen-like atoms $Q\bar{Q}$ and QQQ where only colour 3 occured, different molecules can now be constructed from the same quark constituents with different internal colour configurations, e.g.

$$\begin{aligned} &(QQ)^{\bar{3}} - (\bar{Q}\bar{Q})^{3}, \quad (QQ)^{6} - (\bar{Q}\bar{Q})^{\bar{6}}; \quad (Q\bar{Q})^{8} - (Q\bar{Q})^{8}; \\ &(QQ)^{\bar{3}} - (QQ\bar{Q})^{3}, \quad (QQ)^{6} - (QQ\bar{Q})^{\bar{6}}; \\ &(Q\bar{Q})^{8} - (QQQ)^{8}; \quad etc. \end{aligned} \tag{13}$$

The complexity of the observable hadronic spectrum is therefore increased directly as a consequence of the colour degree of freedom. Therefore, if one can identify these states experimentally and distinguish them from one another, it can reasonably be claimed as a triumph for the colour theory.

The best way to identify chromionic molecules is probably via their decay. [7] Take again (9) as example. Our first task is to see how we can distinguish this from an ordinary $Q\bar{Q}$ meson which may have the same total quantum numbers. From experience we know that $Q\bar{Q}$ mesons decay mainly into mesons, e.g. $f \to \pi\pi$, with widths of order 100 MeV, which may be represented as follows :

$$(Q) \text{———} (\bar{Q}) \to (Q\bar{Q}) + (Q\bar{Q}) \tag{14}$$

In our present language we may describe (14) as a rupture of the colour 3 bond accompanied by the creation of a $Q\bar{Q}$ pair to neutralise the colour charges thus exposed. Apply now the same procedure to the chromionic molecule (9) :

$$(QQ) \text{———} (\bar{Q}\bar{Q}) \to (QQQ) + (\bar{Q}\bar{Q}\bar{Q}) \tag{15}$$

We obtain instead as the dominant mode a decay into $B\bar{B}$ final states, which when combined with the previous assertion that dissociation into mesons is inhibited implies that these states have the unusual property of preferring to decay into $B\bar{B}$ instead of mesons, in spite of the obvious kinematical advantage for and the absence of any selection rule against the latter mode. This characteristic should serve as a very distinctive signature for identifying the chromionic molecules (9) from ordinary $Q\bar{Q}$ mesons.

Second, we wish also to distinguish the two different colour configurations, namely $(QQ)^{\bar{3}}$--$(\bar{Q}\bar{Q})^{3}$ and $(QQ)^{6}$--$(\bar{Q}\bar{Q})^{\bar{6}}$, often labelled as "T" and "M" respectively in the literature. This may not at all be possible since the colour configurations are mixed by gluon exchanges between the two diquark ions. However, it has been argued that since colour mixing forces are colour dipole-dipole forces, they ought to decrease with the separation between ions by some power faster than the colour electric confining potential, and if so, may lead to small mixing between colour configurations when the separation is large. The argument is strengthened by the observation that for the special case of (qq)-$(\bar{q}\bar{q})$ molecules due to the Pauli principle, colour mixing forces are also spin flipping forces, and these are known phenomenologically from ordinary $Q\bar{Q}$ spectroscopy to decrease rapidly with separation. This leads then to estimates of colour configurations mixings which are already small $(\tan\theta \lesssim .2)$ for $L \gtrsim 1$. If so, the two configurations "T" and "M" may be regarded as practically distinct for L large. Notice now that the decay mechanism (15) applies only to the configuration "T" but not to "M", since the colour 6 diquark ion in the second case cannot combine with a quark of colour 3 to form a colour singlet baryon : $(6 \times 3 = 8 + 10 \not\ni 1)$. Thus, whereas the "T" configurations are likely to be broad $B\bar{B}$ resonances with widths of order 100 MeV, the "M" configurations, with both mesonic and $B\bar{B}$ channels effectively closed, are likely to be narrow states. The easiest decay mode of a high L "M" state would appear to be the following:

$$
\begin{aligned}
(QQ)^6 \overset{L}{\text{———}} (\bar{Q}\bar{Q})^{\bar{6}} &\longrightarrow (QQ)^6 \overset{L'}{\text{———}} (\bar{Q})^{\bar{3}} \text{—/—} (\bar{Q})^{\bar{3}} \\
&\longrightarrow (QQ)^6 \overset{L'}{\text{———}} (\bar{Q}\bar{Q})^{\bar{6}} + (Q\bar{Q})^1
\end{aligned}
\tag{16}
$$

i.e. a cascade decay into a lower "M" state with smaller L via the emission of a $Q\bar{Q}$ meson. The cascade will continue until L becomes small enough for appreciable configuration mixing to occur leading then to a final break-up of the molecule into $B\bar{B}$.

Experimentally, there are indeed frequent reports in the last few years of resonances with B = 0 which decay mostly into final states containing $B\bar{B}$ pairs. They are known in the literature as "baryoniums". [15] The evidence for most of them are still quite tentative. I have quoted in Table I those examples which appear to have the best chance for survival under closer scrutiny. [15), 16)] Notice that whereas some states are broad with widths of order 100 MeV, others are narrow with widths less than the experimental resolution of $\sim$ 20 MeV even though they are high above the $N\bar{N}$ threshold. This is qualitatively similar to what is suggested in our previous analysis of QQ--$\bar{Q}\bar{Q}$ chromionic molecules. There was even a report earlier of an example at mass 2.95 GeV which was seen to cascade into the narrow states (2.020) and (2.204) of Table I via the emission of a π-meson, [17] but it was unfortunately not seen again by later experiments under somewhat different conditions. [18), 19)] If these initial findings are in future confirmed, they will lend strong support to the colour chemical interpretation of hadron spectroscopy.

A similar analysis can obviously be applied to chromionic molecules other than the diquark-antidiquark states just considered leading to the prediction of many states of unusual properties. In particular, some of the $QQQQ\bar{Q}$

baryonic states listed in (13) should soon be observable as narrow resonances in the present generation of high sensitivity experiments.[20] Indeed, several narrow baryon states have already been reported which may be candidates for chromionic molecules, among which the two most interesting are listed in Table II.[21), 22] The fact that they choose to decay into final states containing $ss\bar{s}$ instead of the kinematically much more favourable final states with only s, is indicative that they are molecules containing at least five quarks, namely $qqss\bar{s}$, and not ordinary QQQ baryons.

Assuming now the validity of our interpretation we can go further and apply the information gained in ordinary $Q\bar{Q}$ and QQQ spectroscopy to calculate the spectrum of chromionic molecules. For example, from (6) and (7) we can estimate the hyperfine splitting of chromions, and from the slope α' of $Q\bar{Q}$ Regge trajectories the linear mass densities of chromionic bonds.[7] The assertion that both the colour mixing between bonds and the spin-dependent interactions between chromions became negligible for large ionic separation, implies that ions and bonds can be treated as individual entities with their own properties as in ordinary chemistry. Following then the procedure of the chemists, we can construct our molecules from these ingredients, and infer from their properties some properties of the molecules also.[14] In Fig. 2, we quote an example of a spectrum so calculated for the molecules $(qq)^{\bar{3}}$--$(\bar{q}\bar{q})^{3}$, which depends only on one free parameter (the trajectory intercept) fitted to experiment.[7] Its correspondence with the experimental 'baryonium' states so far reported is quite encouraging.[23]

Now besides ionic molecules in ordinary chemistry one has also covalent molecules and Van der Waals molecules. Are there equivalents of such molecules also in colour chemistry ? Covalent moelcules are formed from neutral atoms sharing between them a pair of their valence electrons. To imitate this, we draw in Fig. 3 a pair of hydrogen-like QQQ atoms, sharing between them a pair of their valence quarks. The result is a quark diagram representing the exchange of a meson between two nucleons, which according to the nuclear theorists, is the source of nuclear forces at long to medium ranges ($r > .7$ fermi). Presumably then, the equivalents of covalent molecules in colour chemistry are just ordinary nuclei such as the deuteron in Fig. 3 which are bound by just these forces. As for the colour equivalents of Van der Waals molecules, we know very little. Of the two practical models for confinement now in common use, the potential model gives nonssense, namely a long ranged colour Van der Waals potential which dominates over gravitation up to a distance of one km,[24] while in the bag model, such forces are by construction zero. At present, therefore, it seems that little can yet be learned by pushing the chemical analogy in these directions.

So far, we have gone out of our way to emphasize the similarity of hadron spectroscopy to chemistry. However, QCD is not QED and there are also some important differences. For example, one big difference is that in QCD the mediating gluon field carries itself a colour charge, and since colour is supposedly confined, a natural question is whether the gluon itself can occur as a hadron constituent. In particular, the possible existence of hadron states with only gluon constituents (the so-called glue-balls or gluoniums) has been proposed as a test of chromodynamics. Unfortunately the search for gluoniums is not an easy experimental task given the lack of any detailed predictions for their spectrum and of any distinctive signature for their identification. My own opinion is that a search for pure gluoniums is not necessarily meaningful because of their probable mixing with states containing $Q\bar{Q}$ pairs to which they are connected by pair creation and annihilation. For ordinary $Q\bar{Q}$ mesons, one may argue that their mixing with gluonic states are small since the $Q\bar{Q}$ pair with colour 1 cannot annihilate into a single gluon, and all such effects must therefore be of higher order in the coupling α_s. However, in most multiquark systems, e.g. $QQ\bar{Q}\bar{Q}$, there can be $Q\bar{Q}$ pairs with colour 8 to which the annihilation diagram of Fig. 4 can contribute,[25] but this is of the same order in α_s as the one-gluon-exchange diagram of Fig. 1 held responsible for the large hyperfine splittings of $Q\bar{Q}$ mesons and QQQ baryons. The reason why we have so far been able apparently to avoid considering annihilation effects is simply because we have discussed in detail only those states in which Q's and $\bar{Q}$'s are spatially separated so that the presumably short-ranged pair annihilation forces are likely to be unimportant. However, for any system containing s-wave $Q\bar{Q}$ pairs with colour 8, a naive estimate of Fig. 4 would yield sizeable mixing not only with states containing gluons, but also with states containing $Q\bar{Q}$ pairs of other flavours as constituents. The search for gluoniums therefore may not yield anything spectacular but merely a further complexity of states embedded in, and mixed with others containing $Q\bar{Q}$ pairs. Any interpretation of the result will then require very detailed analyses, both phenomenological and theoretical. On the theory side, we shall need first of all a proper development of QCD perturbation theory for a confined system in which the occurence of gluons as constituents is properly recognised. It is encouraging that a recent approach initiated by Lee[26] and developed by others[27] is promising to be a valuable tool in this direction. On the phenomenological side also, candidate states are beginning to emerge in which the effects of gluon constituents may possibly be studied.[25] Examples are (i) the 0^+ mesons in the 1 GeV mass region which are suggested by recent analysis to contain a strong admixture of $QQ\bar{Q}\bar{Q}$ states and perhaps of $Q\bar{Q}g$, and (ii) some tentative unusual baryon states such as those in Table II, which if interpreted as chromionic molecules $(Q\bar{Q})^8$--$(QQQ)^8$ allow a study in near isolation of a $Q\bar{Q}$ system with the same quantum numbers as the gluon.

At this point, I have already spent the bulk of my alotted space, but I have discussed only spectroscopy. Whereas, soft hadron physics - the subject of my talk - should deal not merely with the spectrum of hadrons but also with their interactions. The reason that I have emphasized spectroscopy so much is simply that I yet understand very little how colour atoms and molecules interact. In passing from the spectrum to interactions, a major change in our attitude is necessary. While discussing the spectrum, we are, so to speak, operating in the imaginary world of constituents inside the hadron, but, hadronic interactions will have to be considered in the real world, namely the space of the actual hadron states, and it is not obvious how one should establish a link between the two. Recently, Jaffe and Low [28] has proposed a formalism closely related to the Wigner R-matrix which offers an interesting possibility. To see how this works, imagine first that we are interested in the scattering of a particle from a potential well $V(r)$ of radius b . The boundary condition that the wave function should vanish at $r = b$ would produce a set of stationary states, say $\psi_n(r)$ for $r < b$ with eigenmomenta k_n . These eigenstates ψ_n need not correspond at all to resonances of the system - for example, the potential may be very weak, or even repulsive so that no resonance exist. Nonetheless, one can still unambiguously identify these eigenstates ψ_n from the scattering amplitude by considering the function

$$P = k \cot (kb + \delta(k)) \tag{17}$$

of the phase shift $\delta(k)$. One sees that the poles of P occur at the zeros of $\sin(kb + \delta(k))$, i.e. exactly when the scattering wave function has a zero at $r = b$, namely when $k = k_n$. Consider now again our $QQ\bar{Q}\bar{Q}$ system. Inside a volume of hadronic size of radius b say, we believe we know to some extent how the system behaves, namely with quark constituents interacting via gluon exchange. Suppose then we impose the boundary condition that the constituent wave function should vanish on the surface of this volume in much the same way as is done in the bag model, we would obtain a set of states (called by the authors 'primitives'). Within our framework these states are calculable, but they may have little to do with the actual hadronic spectrum. The system $QQ\bar{Q}\bar{Q}$

now contains a component in $(Q\bar{Q})^1(Q\bar{Q})^1$, i.e. a two-meson state, so that outside the radius b , it has a two-meson wave function. Following (17), let us identify the primitves as poles of the P-matrix which may be constructed if we know the two-meson scattering wave function. Indeed, using the low energy meson-meson data now available, Jaffe and Low were able to locate the poles of the corresponding P-matrix and verify their close correspondence to the primitives calculated from the MIT bag model. Suppose next we are intersted say in the decay of the ρ-meson into $\pi\pi$. According to current belief, ρ is mainly a $Q\bar{Q}$ state, but because of pair creation **diagrams** such as Fig. 5, it acquires also a component in $QQ\bar{Q}\bar{Q}$ and the amount of mixing is in principle calculable, given the perturbation theory of Lee etc. mentioned above. Once the component of ρ in $QQ\bar{Q}\bar{Q}$ is known, we can then use the Jaffe-Low prescription to connect it with the 2π wave function, leading to a prediction for the $\rho \to 2\pi$ amplitude. The knowledge of resonance decay implies the knowledge of its inverse, namely resonance formation, and hence also of resonant scattering which is known to dominate low energy hadron-hadron scattering. Then, from resonant scattering, we may proceed via 'duality' to Regge exchange, and further through 'dual topological unitarisation' to diffractive scattering etc.[29] Of course, at the moment, since none of this is yet worked out, we do not know how much can in fact be achieved with the limited tools available, but it does appear to be a program worthy of some further consideration.

I am afriad my understanding of colour chemistry finishes there. Of course, the amount of vision depends critically on the point of view, and mine is so limited probably because I have in this talk chosen a particularly obtuse angle to look at the subject. Taking different approaches, others have no doubt understood more about soft hadron interactions in terms of the quark-gluon theory than what I have reported, for the omission of which I apologise. I am glad to learn however that my very restricted review will be supplemented by a talk on inclusive soft hadron reactions by Rudy Hwa later in this meeting.[30]

It is a pleasure for me to thank the Academia Sinica and the organisers of this meeting to whom I am deeply grateful for giving **me the chance** :

(a) to attend an all-Chinese physics conference,

(b) to give a conference review in my own language,

(c) to visit my birth-place Guangzhou (Canton) where this meeting is held, none of which I have done for the last thirty years.

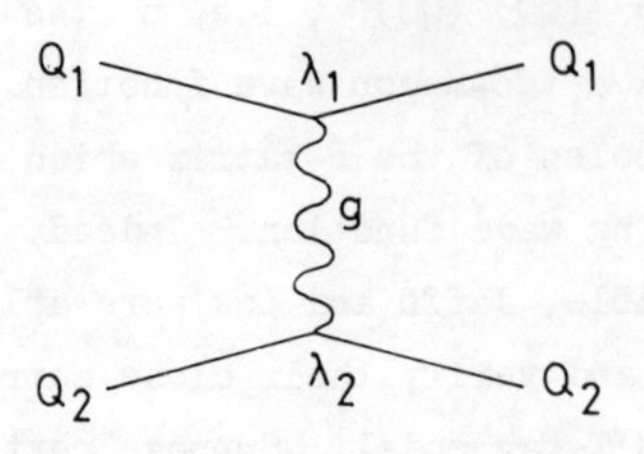

FIG 1

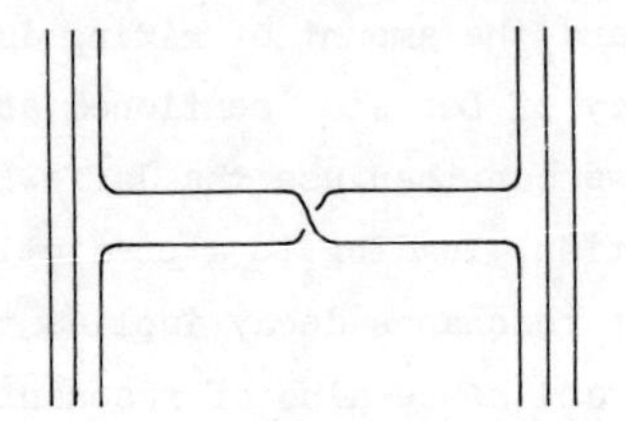

FIG 3

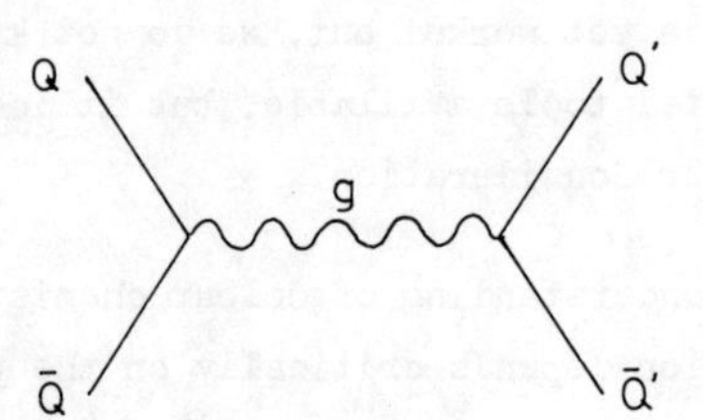

FIG 4

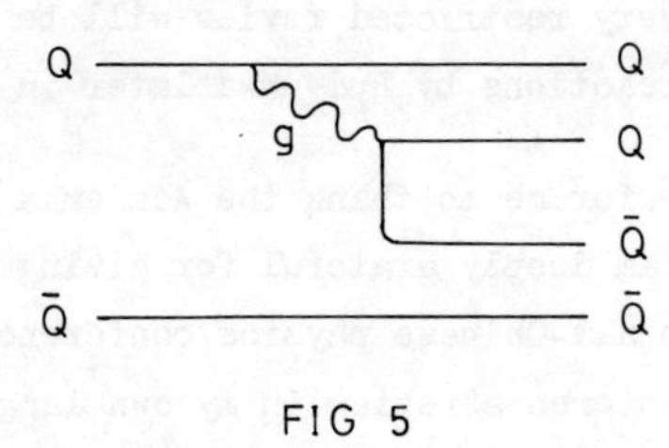

FIG 5

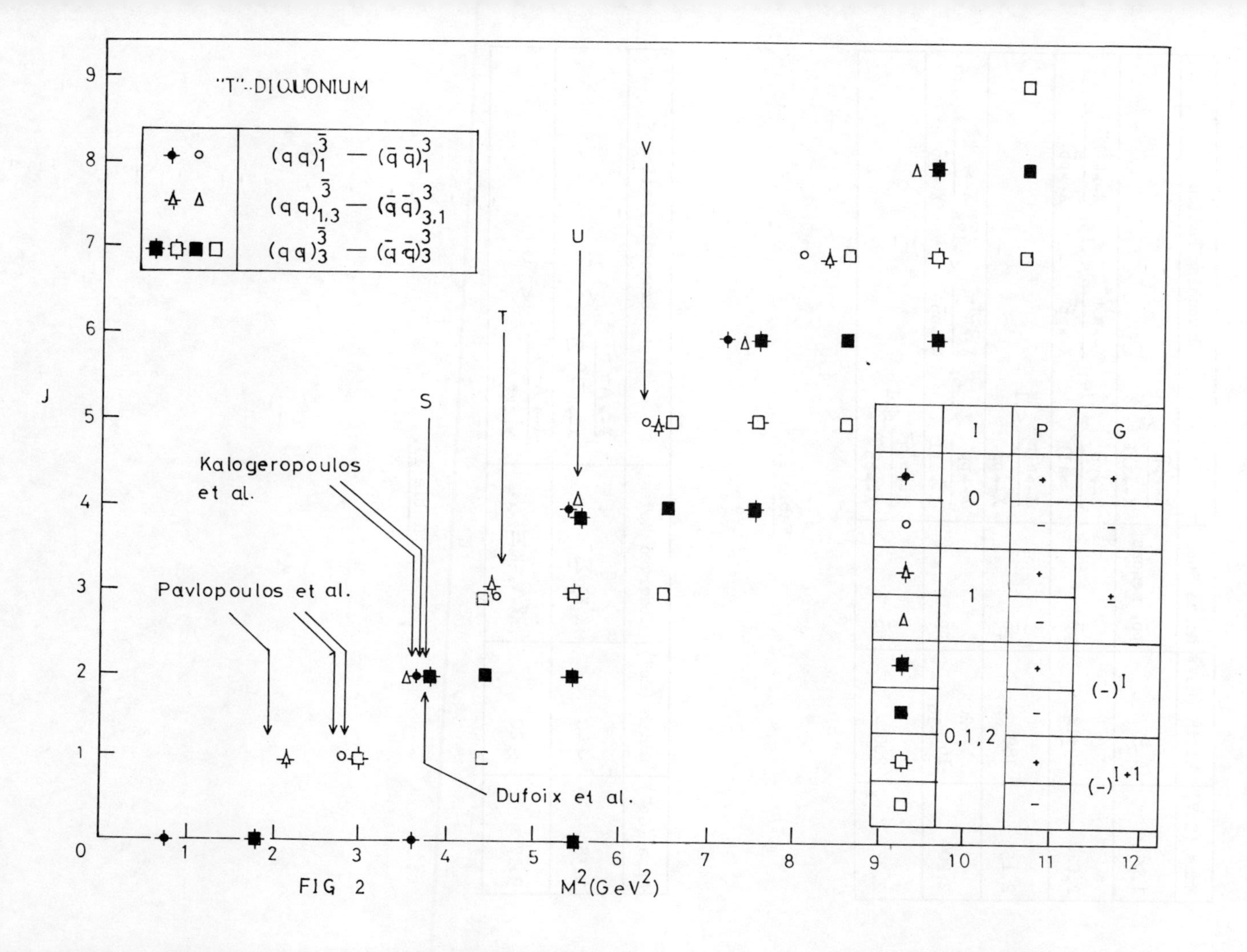
"T"-DIQUONIUM
$(qq)^{\bar{3}}_{1} - (\bar{q}\bar{q})^{3}_{1}$
$(qq)^{\bar{3}}_{1,3} - (\bar{q}\bar{q})^{3}_{3,1}$
$(qq)^{\bar{3}}_{3} - (\bar{q}\bar{q})^{3}_{3}$
J
M^2 (GeV^2)
S
T
U
V
Kalogeropoulos et al.
Pavlopoulos et al.
Dufoix et al.
I
P
G
0
1
0,1,2
$(-)^I$
$(-)^{I+1}$
FIG 2

Mass (GeV)	Width (MeV)	Modes Seen	Branching Ratios
1.936	4-8	$p\bar{p}$, mesons	$\frac{X\to p\bar{p}}{X\to all} \gtrsim .5$
2.020	24 ± 12	$p\bar{p}$	$\frac{X\to p\bar{p}}{X\to all} > .14$; $\frac{X\to \pi^+\pi^-}{X\to p\bar{p}} < .15$; $\frac{X\to K^+K^-}{X\to p\bar{p}} < .11$
2.15	180	$p\bar{p}$	$\frac{X\to \pi^+\pi^-}{X\to p\bar{p}} \lesssim .1$
2.204	16^{+20}_{-16}	$p\bar{p}$	$\frac{X\to p\bar{p}}{X\to all} > .16$; $\frac{X\to \pi^+\pi^-}{X\to p\bar{p}} < .17$; $\frac{X\to K^+K^-}{X\to p\bar{p}} < .16$
2.32	200	$p\bar{p}$	$\frac{X\to \pi^+\pi^-}{X\to p\bar{p}} \lesssim .1$

Table I.

Mass (GeV)	Width (MeV)	Observed Modes	Branching Ratios
2.58	~ 40	$K^+K^o_s\,\Sigma^-$	$\frac{X\to \Lambda\pi^+\pi^-}{X\to K^+K^o_s\,\Sigma^-} < 1.$
3.17	$\lesssim 20$	$K\bar{K}\Lambda$, $K\Xi$ ...	$\frac{X\to \Lambda+n\pi}{X\text{--}K\bar{K}\Lambda\,...} < 1.$

Table II.

References

1. Review of particle properties, Phys. Letters 75B (1978).
2. e.g.A.De Rujula, H.Georgi and S.L.Glashow, Phys. Rev. D12 (1975) 147.
3. e.g.A.G.Hey, EPS Conference, Geneva July 1979, to appear in the procedings.
4. e.g.M.Krammer and H.Kraseman, Desy preprint 79/20 (1979).
5. e.g.G.Karl, Proc. of 19th International Conf. on High Energy Phys. Tokyo 1978, 135.
6. e.g.J.D.Jackson, Proc. of 19th International Conf. on High Energy Phys. Tokyo 1978,391.
7. Chan Hong-Mo and H.Högaasen, Nucl. Phys. B136 (1978) 401.
8. Tsou Sheung Tsun, Nucl. Phys. B141 (1978) 397.
9. R.L.Jaffe, Phys. Rev. Letters 38 (1977) 195. 617.
10. G.Gelmini, to appear as Trieste ICTP preprint (1979).
11. Chan Hong-Mo and H.Högaasen, Phys. Letters 72B (1977) 400.
12. K.Johnson and C.Thorn, Phys. Rev. D13 (1976) 1934.
13. K.Konishi, private communication (1978).
14. Chan Hong-Mo et al.,Phys. Letters 76B (1978) 634.
15. L.Montanet CERN preprint EP/Phys. 77-22 (1977); 13th Rencontre de Moriond, les Arcs 1978, published as "Phenomenology of Quantum Chromodynamics", Editions Frontieres, France (1978).
16. J.Six, 4th European Antiproton Symposium, Barr (1978) 593
17. C.Evangelista et al. Phys. Letters 72B (1977) 139.
18. T.A.Armstrong et al. CERN preprint EP/79-54 (1979).

19. M.Kienzle, EPS Conference, Geneva July 1979

20. P.Sorba and H.Högaasen,Nucl. Phys. B145 (1978) 119.

21. ACNO collaboration, CERN preprint EP/Phys. 78-40 (1978).

22. T.Amirzadeh et al. CERN preprint EP/Phys. 79-101 (1979).

23. Chan Hong-Mo, 4th European Antiproton Symposium, Barr (1978) 477.

24. K.Kikkawa, 19th International Conf. on High Energy Phys. Tokyo 1978, 507.

25. H.Högaasen and Chan Hong-Mo, in preparation as CERN preprint (1980).

26. T.D.Lee, Phys. Rev. D19 (1979) 1802.

27. F.E.Close and R.R.Horgan, Rutherford Laboratory preprint, RL-79-067 (1979).

28. R.L.Jaffe and F.E.Low, Phys. Rev. D19 (1979) 2105.

29. Chan Hong-Mo and Tsou Sheung Tsun in "Many Degrees of Freedom in Particle Theory", Plenum Press N.Y.(1978) 83; G.F.Chew and C.Rosenzweig, Phys. Reports (1979).

30. R.Hwa, paper presented at this meeting.

Non-Conservation of Baryon Number*†

A. Zee
Department of Physics, University of Pennsylvania
Philadelphia, Pennsylvania 19104

ABSTRACT

The possibility that baryon number may not be conserved is examined. We discuss the experimental and cosmological consequences of this possibility, emphasizing the various implications baryon number violation would have on our understanding of the physical world.

*Research supported in part by the U.S. Department of Energy under Contract No. EY-76-C-02-3071.

†Based in part on talks delivered at the 1980 Guangzhou Conference on Theoretical Particle Physics, at the 1979 Eastern Theoretical Physics Conference, and at a number of research institutions.

I. Symmetry in Physics

The possibility of proton decay has generated a great deal of interest over the last few years. In order to understand the significance of proton decay, we must view it in the proper physical and historical context. Thus, we will begin with an account of the rise and fall of symmetry in physics.[1] One major difference between the way physics is done in the 19th century and in the 20th century perhaps lies in the use of symmetries. Maxwell and his friends experimented with coils and magnets and distilled from their observations the correct equations of motions. It was not until later that Lorentz and others realized that Maxwell's Lagrangian possesses a deep symmetry of nature. It was Minkowski, apparently, who first recognized that one could reverse this chain of reasoning. By imposing the appropriate symmetry one could arrive at Maxwell's Lagrangian, from which then flows various experimental predictions. This profound shift in physicists' point of view towards symmetries has pervaded twentieth century physics. Minkowski's view apparently made a great impression on Einstein, who constructed a theory of gravity by imposing local coordinate invariance. Were one to follow the 19th century's line of attack, one would have to start with perihelion shift of mercury and the bending of light and construct a theory of gravity by adding successive corrections to Newton's theory. This can be, and has been, done.[2] But it is clearly more laborious and less elegant than imposing symmetries from the start. In this century, as physics become more abstract and complex, physicists have come to rely more and more on imposing a conjectured symmetry in their continued gropings for the fundamental laws of physics. Minkowski's view has now come to full flowering in the development of grand unified gauge theories of the non-gravitational interactions. For instance, by imposing SU(5) symmetry, Georgi and Glashow[3] wrote down a Lagrangian which predicts proton decay, probably the most stunning prediction of our times. (An even more extreme development in this direction is represented by supersymmetry and supergravity. These theories, however, do not appear to describe Nature. Perhaps that is because supersymmetry is not strongly motivated by experiment.)

Heisenberg introduced the concept of internal symmetry into physics by formulating the notion of isospin conservation. The implications of isospin conservation were developed by Cassen, Condon, Wigner and others. Nowadays, isospin conservation has lost some of its former aura and is regarded by particle physicists as almost an accident, more due to the smallness of the up and down quark masses compared to the hadronic mass scale than due to the smallness in the difference between the up and down quark masses. Why isospin proves to be such a good symmetry in nuclear physics is thus slightly mysterious after all. This question has recently been investigated.[4]

Beginning with isospin, physicists have explored larger and larger symmetries in the continuing search for the fundamental laws of Nature. Of course, these larger and larger symmetries are more and more badly broken so that their phenomenological manifestations are more and more difficult to recognize. Nature appears to like symmetries deep down at heart but She also appears to enjoy breaking these symmetries. Perhaps the "reason" is that an exactly symmetric world is also a rather dull world. Feynman has commented that in his travels he has noticed that many architectural structures which at first glance look completely symmetrical are on closer inspection revealed to be in fact subtly asymmetrical. Architects appear to have a deep-seated need for both symmetry and asymmetry. The human face, with its slightly broken bilateral symmetry, furnishes another interesting example.

If what particle physicists have learned over the last two decades or so is to be summarized, one would probably have to say that the major lesson is that badly broken symmetries, provided they are softly broken, can be extremely useful. Understanding this point represents a tremendous advance; the older notion that only slightly broken symmetries are useful is too restrictive and not particularly illuminating. For instance, one could always speculate on the breaking of almost any symmetry by a tiny term in the Lagrangian. In contrast, the concept of soft breaking, borrowed from the profound work of Ginsburg and Landau on superconductivity and developed by Gell-Mann, Levy, Nambu, Jona-Lasinio, and others, frees us to deal with badly broken symmetries and hence to explore physics at an energy scale beyond our grasp. Thus, the tremendously successful, but badly broken, SU(2) x U(1) symmetry of Glashow, Salam, and Weinberg[5] allows us not only to correlate all known weak and electromagnetic phenomena but also to master the physics on the scale of a few hundred Gev. In other words, the method of breaking symmetries softly so as not to mar the delicate structure of the theory and hence to destroy renormalizability, now known as the Higgs mechanism, allows us to at least discuss the physics of a mass scale much beyond our presently accessible mass scales. This freedom to discuss badly broken symmetries is perhaps one reason why physics is more interesting, to some people at least, than, say, architecture.

We are now faced with the dramatic possibility that Georgi and Glashow, with their SU(5) symmetry, have liberated us (or so say the enthusiasts) from our dreary existence among the debris of symmetry breaking to dream about the (hopefully) basic physics at 10^{16} Gev.

II. The Urge to Unify

In the long march towards a physical understanding of the world, we have finally reached the point of having a theory which, while it does not explain everything, is essentially consistent with all physical phenomena. This SU(3) x SU(2) x U(1) x Einstein theory is an amazing achievement indeed. (It is perhaps sobering to recall that the closest physicists have previously come to this situation is in the late nineteenth century, just before the specific heats of various substances were carefully measured at low temperatures. However, there were already some disturbing signs. For instance, the energy output of the sun was certainly known to be inconsistent with the physics of the time. Possibly the most intrinsically disturbing sign of our times is the vanishing of the cosmological constant.)

In spite of, or perhaps because of, the success of the SU(3) x SU(2) x U(1) theory there are strong motivations to extend this theory to a grand unified theory. These include the desire (1) to understand charge quantization, (2) to have only one gauge coupling, and (3) to continue the trend towards unification started by Newton when he unified celestial physics with terrestrial physics.

We will now give the standard heuristic argument that grand unification tends (in general) to lead to proton decay. In a unified gauge theory based on a simple group the electric charge operator Q (being a generator of the simple group) must have zero trace. In other words, the electric charges of all the particles in a given representation must sum to zero. However, the charges of the known leptons (in a given family) sum up to $Q_\nu + Q_{e^-} = -1 \neq 0$, while the charges of the known quarks (in a given family) sum up to $3(Q_u + Q_d) = 3(2/3 - 1/3) = +1 \neq 0$. It is difficult, on the other hand, not to notice that $-1 + 1 = 0$. Thus, if one is not allowed to invent fermions at will then one would have to put quarks and leptons into the same representation. It follows that there are gauge bosons transforming quarks to leptons and vice versa. In general, this leads to violation of baryon and lepton numbers. (We emphasize, however, that this heuristic argument certainly does not constitute a proof that grand unification necessarily leads to proton decay.[6])

An important point to note is that the grand unification mass scale is necessarily very high. This follows because (1) the strong, weak, and electromagnetic couplings, or more accurately, the SU(3), SU(2), and U(1) couplings, are very different at low energies, (2) according to renormalization group theory these couplings change with energy logarithmically slowly, and (3) thus one has to reach very high energies before the three couplings become equal.

For the simplest SU(5) theory this mass scale turns out to be of order[7] 10^{15}Gev. This high mass scale in turns insures a very long life-time for the proton, consistent with the existing experimental limit of 10^{30} years. Thus, the first physical theory to predict an unstable proton also passes its first test: that of explaining the apparent stability of the proton. We will not discuss here all the other virtues of the simplest SU(5) theory; suffice it to say that this theory has a certain intriguing tightness and economy not shared by its competitors.

III. The Role of Exact Symmetries

Exact conservation of baryon number implies an exact symmetry in the Lagrangian. Now an exact symmetry can be either global or local.

While there is nothing in principle wrong with exact global symmetries, many people find them repugnant. For instance, if isospin is an exact global symmetry, then once the convention of defining which nucleon is the proton is fixed at one point in space and time it is fixed throughout all space and time. To put it in another way, exact isospin symmetry means that the Lagrangian is invariant uncer an arbitrary notation of the proton state into the neutron state. But if isospin is an exact global symmetry, then the performance of a rotation here in the laboratory would require the performance of exactly the same rotation behind the moon. Mach, Einstein, and others were apparently quite disturbed by this requirement. (Think of rotational invariance, say. Does it imply that to leave the Lagrangian invariant one would have to rotate the whole

Universe?) Considerations of this sort presumably led Einstein to formulate his general theory of relativity in 1915. Einstein's theory of gravity is the first theory to possess an exact local symmetry. This development inspired Weyl, in 1918-19, to try for a geometric theory of electromagnetism based on demanding invariance under a local change in the size or "gauge" of fields: $\psi \rightarrow (1+\theta)\psi$. However, he did not insert the appropriate factor of i, a faux pas for which he could hardly be blamed considering that quantum mechanics was not invented till 1925. The factor of i was inserted by Fock and by London shortly after the invention of quantum mechanics, thus making gauge symmetry something of a misnomer for a phase symmetry.[8]

The central role of exact local symmetry in physics was thus established. The next important step was taken by Yang and Mills[9] in 1954 when they wrote down a Lagrangian with nonabelian gauge symmetry. The rest is history well known to us all. Thanks to a fascinating series of theoretical and experimental developments, we now believe that the strong, electromagnetic, weak, and gravitational interactions are all based on gauge symmetries.[10]

Viewed in this context, the notion of exact baryon conservation has long been puzzling to thinking physicists. For the reasons mentioned above, many people find it repugnant to think that proton stability is guaranteed by an exact global symmetry. On the other hand, if proton stability is based on an exact local symmetry, then there would be a massless gauge boson coupled to baryon number. As Lee and Yang[11] pointed out in 1955, the rather accurately known equality between inertial mass and gravitational mass, established by a series of experimenters starting with Bessel (1830), Eötvös (1889, 1922) and ending with Dicke et al. and Braginsky et al., implies a riduculously small upper limit for the coupling of this massless gauge boson. Thus, the theoretical founcation of proton stability, already in the mid-fifties, appeared unsatisfactory to some people.

And so, when Pati and Salam[12] in 1973, Georgi and Glashow[3] in 1974, showed that the gauge principle when combined with notions on grand unification implies the decay of the proton, it meant, for many people, profound intellectual relief and satisfaction. A long-standing philosophical dilemma has been resolved.

Unfortunately, the actual calculated rate for proton disintegration is rather model-dependent and experiments designed to detect proton decay are by no means easy. Several massive experiments[13] are now either under way or being constructed and we can only hope that the rate of proton decay is within the detectible range of these experiments. In any case, those who think philosophical and aesthetic prejudices are important in physics, and this author is amongst them, are already quite convinced that protons must decay.

It might also be mentioned that Pais[14] has suggested that exact baryon number conservation may be based on a spontaneously broken abelian gauge symmetry. However, this suggestion cannot be fitted into the framework of the SU(2) x U(1) gauge theory of electroweak interactions without generating anomalies which would spoil renormalizability.

IV. Why the Universe Is Not Empty

If baryon number is not conserved, then not only can baryons disappear they could also appear. If protons will eventually die they could also have come into being in the early Universe. As pointed out by Yoshimura, Dimopolous and Susskind, Toussaint, Treiman, Wilczek, and Zee, Weinberg, and others[15], the interplay of three physical effects, (1) baryon number violation, (2) CP violation, and (3) disequilibrium in the early Universe could lead to a possible explanation of why the Universe contains matter at all.

The Universe is one vast emptiness, dotted here and there with a galaxy or two. With the discovery of the cosmic background radiation it has become possible to give a quantitative measure of this frightening[16] and almost inconceivable emptiness. By a variety of methods, it is possible to estimate, probably to within a factor of ten or so, the number density of baryons (protons and neutrons) in the Universe. The only possible dimensionless number one can consider is the ratio of the baryon number density n_B and the photon number density n_γ (which is rather accurately determined by the cosmic radiation measurements). This fundamental ratio n_B/n_γ turns out to be roughly[17] $10^{-9\pm 1}$.

It is doubly remarkable that the Universe, while almost empty, is actually not empty: n_B/n_γ is small but definitely non-zero. It would appear that the very fact that we exist establishes the non-emptiness of the Universe. However, this statement requires some amplification and clarification.

Ever since Dirac's ideas about anti-matter were experimentally confirmed, people have speculated that the Universe has an equal amount of matter and anti-matter, segregated into domains. More precisely, it was asserted that all conserved quantum numbers of the Universe should be zero. In particular, since electric charge Q is known to be zero to a high degree of accuracy, it seems "aesthetically appealing" that baryon number B and lepton number L should also be zero. Unfortunately, the weight of observational evidence is against this supposition.[18]

If baryon number is absolutely conserved the small but non-zero matter content of the Universe is then simply a matter of initial conditions and its value presumably is a question outside the domain of physics. On the other hand, the non-conservation of baryon number opens up the exciting possibility that the matter content of the Universe is actually something physicists can understand and hopefully even calculate.

The basic idea for explaining the matter content of the Universe is quite simple. It rests upon extrapolating the standard big-bang cosmology back in time, assuming the temperature becomes arbitrarily high or at least higher[19] than the grand unification mass scale of $\sim 10^{15}$ Gev. At that point, baryon number violating forces are comparable to the familiar interactions. (Thus, the proposed scenario depends on the baryon number violating interaction resulting from the exchange of heavy particles.) As the Universe cools, these forces become increasingly negligible and a net baryon number is generated, provided that a CP violation exists (at that time) which distinguishes matter from anti-matter.

It is also appealing that the small observed violation of CP in K-meson decay, which hitherto appears to have no connection with any other physical phenomena, may be after all intimately related to the non-emptiness of the Universe.

The subject is an active one, with a growing literature.[20] For further details we refer the reader to the papers cited[20]. Here we restrict ourselves to a few remarks.

(1) In the hot early Universe any net baryon number density will be quickly dissipated by baryon number violating forces. Thus, the initial condition "explanation" of the matter content of the Universe may be ruled out.

(2) Is it possible to actually calculate n_B/n_γ? Or failing that, is it possible to show, given the sign of CP violation in K-decay, that the Universe contains matter, rather than anti-matter? Unfortunately, the present state of the art does not allow us to answer these questions definitely. For instance, Barr, Segre, and Weldon[20] have shown that the simplest version of SU(5) theory with CP violation given by the Kobayashi-Maskawa mechanism[21] gives too small a value for n_B/n_γ but that the calculated value can be increased by complicating the Higgs sector. More detailed investigations have been carried out by Kolb and Wolfram.[20]

(3) Misner, Barrows, and Matzner[22] have shown that if the Universe started out in a chaotic state and subsequently smoothed itself out into a homogeneous isotropic state, the entropy per baryon would be much larger than the observed $n_\gamma/n_B \sim 10^9$. However, if n_γ/n_B is in fact determined by microphysics then one can no longer deduce from the observed value of n_γ/n_B that the Universe cannot have started in a chaotic state.[23] (It is amazing that n_B/n_γ is a number

which can be thought of as either very small or very large, and both views can be fruitful.)

(4) Since the formation of black holes may violate baryon number, one might think[24] that Hawking evaporation[25] of black holes may lead to the baryon asymmetry of the Universe, given the existence of CP violation. However, Toussaint et al.[15] showed that this is not the case unless baryon number is violated in the laws of physics.

Thus, it would appear that a rational explanation of the fact that the Universe is not empty would require baryon non-conservation. And so the enthusiasts amongst us would argue that our very existence demands the decay of protons. Cooler minds, however, will insist that while proton decay implies baryon number non-conservation, baryon number non-conservation does not imply proton decay. Indeed, Segre and Weldon[27] have constructed an SU(5) theory with a stable proton but a non-conserved baryon number. The generation of baryon asymmetry in this model differs somewhat from the standard scenario.

We are indeed fortunate that we live in that epoch of the Universe after the birth of nucleons but before their eventual death. A physicists's logarithmic history of the world,assuming an open Universe, is shown in Fig. 1. We list on the right major events in the history of the Universe, and on the left various events of particular interest to humans. Time is plotted, on a logarithmic decay, starting with the Big Bang.

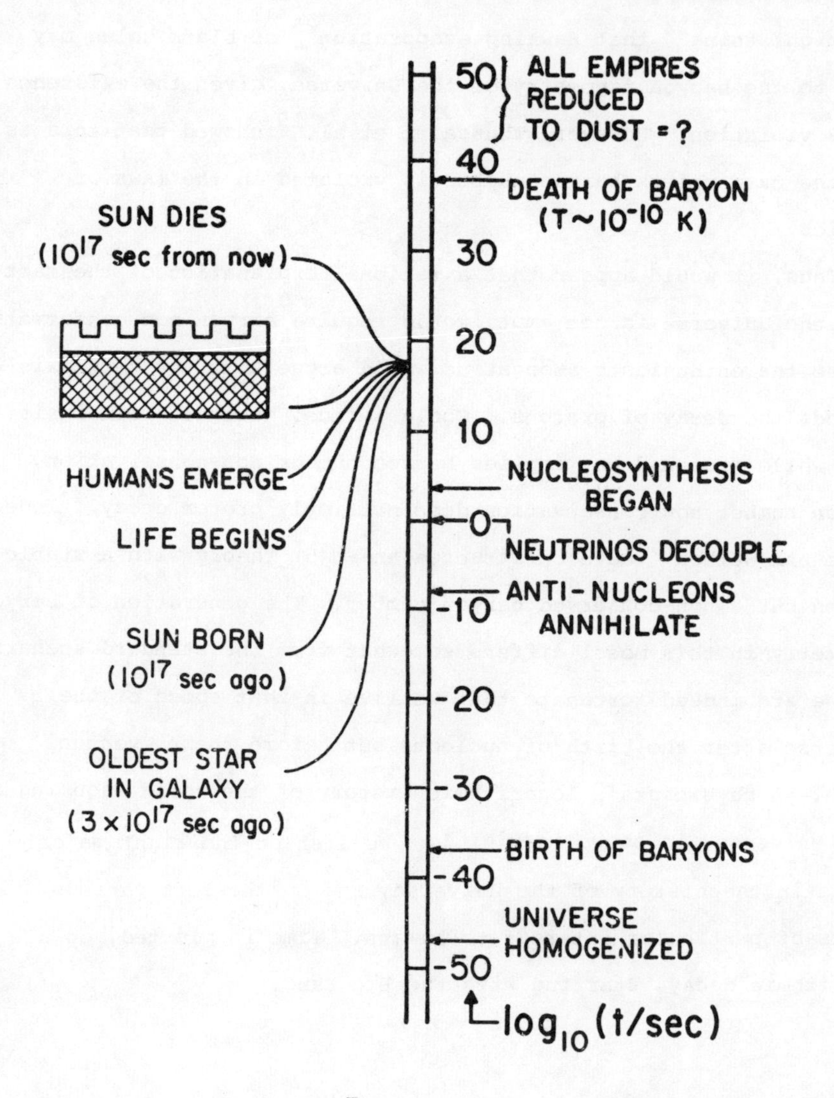

50
ALL EMPIRES
REDUCED
TO DUST = ?
40
DEATH OF BARYON
(T ~ 10^-10 K)
SUN DIES
(10^17 sec from now)
30
20
10
HUMANS EMERGE
NUCLEOSYNTHESIS
BEGAN
LIFE BEGINS
0
NEUTRINOS DECOUPLE
ANTI-NUCLEONS
ANNIHILATE
-10
SUN BORN
(10^17 sec ago)
-20
OLDEST STAR
IN GALAXY
(3 x 10^17 sec ago)
-30
BIRTH OF BARYONS
-40
UNIVERSE
HOMOGENIZED
-50
log10 (t/sec)

Fig. 1

V. Proton Decay

We are now faced with a situation perhaps unprecedented in the history of physics. We have reasons to think that the "true" underlying physics occurs at a stupendous mass scale of order 10^{15} Gev while we live and explore at an energy scale of $\sim 10^2$ Gev amidst the debris of a smashingly broken symmetry. In the simplest version of SU(5) there is no new physics between 10^2 Gev and 10^{15} Gev; between these two mass scales lies what Glashow has called a vast "desert". Many people find this scenario unpalatable; on the other hand, we must be reminded that while life may be linear, physics tends to be logarithmic, as is evidenced by Fig. 1. We will come back to this issue later.

Perhaps amazingly enough, it is not hopeless for us to learn something about the physics at 10^{15} Gev. The neutral current parameter[7] $\sin^2\theta_W$, proton instability, and to a certain extent the early Universe[27] may all offer us a glimpse of grand unification physics.

In view of the far-reaching implications of the proton-decay experiments, it is clearly important to carry out a detailed phenomenological analysis[28,29] of the process. As is often the case in physics, it is appropriate to attack the problem from both sides. On the one hand, we would like to derive model-independent results in order to test the entire framework, rather than a specific model. On the other hand, we would also like to distinguish between competing models. A particularly accessible experimental signature which may allow us to distinguish between different models concerns the question of whether B-L (baryon minus lepton number) is conserved or not, or in other words, the question of whether the proton decays into a lepton (as in the mode $p \to \mu^- \pi^+ \pi^+$) or into an anti-lepton (as in the mode $p \to \mu^+ \pi^o$).

It turns out that a simple phenomenological analysis[28,29] is possible and proceeds as follows. We wish to construct an effective interaction Lagrangian to describe proton decay. Since the physics describing proton decay presumably occurs at a very large mass scale M_x (perhaps of order 10^{15} Gev) we could to a good approximation (to the extent that $M_W/M_x << 1$) treat SU(3) x SU(2) x U(1) as an exact symmetry. In other words, we could set $M_W = 0$. Theorists certainly think nothing of setting to zero the mass of a conjectured particle too massive yet to be discovered by experimentalists. The effective Lagrangian in question is to be constructed as a sum of operators formed out of quark and lepton fields (and possibly boson fields as well). The saving point is that we can restrict ourselves to operators of the smallest possible dimension. The reason is simply that operators of higher dimensions have to be multiplied by appropriately negative powers of masses so as to appear in the Lagrangian (which has mass dimension four, of course). However, M_W has been set to zero, and the only mass left in the game is M_x and so higher dimension operators are suppressed by powers of $1/M_x$. In order to form an operator transforming as a color singlet we must have at least three quark fields. In order to have a Lorentz scalar we must combine the three quark fields with another fermion field, either a lepton field or an anti-lepton field. The homework problem we assign ourselves is thus quite finite: find all Lorentz invariant four-fermion operators, either of the form $qqq\ell$ or of the form $qqq\bar{\ell}$, which transform as

SU(3) x SU(2) x U(1) singlets and which violates baryon number. This is in fact analogous to the problem which Fermi[30] assigned himself and which he solved[31] in 1934: find all Lorentz invariant four-fermion operators which transform as singlets under the symmetry of 1934, namely the electromagnetic U(1).

The possible operators can be found readily. In the simplest case of including only one generation of quarks and leptons, there exist only four operators. The Lorentz properties of two of these are such that they could arise from the exchange of a vector boson. The other two are such that they could arise from the exchange of a scalar boson. We will refer to these operators as vector and scalar respectively, in analogy to the famous SPVAT classification.

Physical consequences follow upon inspection of the list of operators.

(1) These operators all conserve B-L. Thus, B-L is conserved to lowest order in M_W/M_X, quite independent of model. In other words, any violation of B-L must be proportional to the breaking of SU(2) x U(1).

(2) $\Delta S \geqslant 0$ in proton decay.

(3) If the two vector operators dominate (as would be the case if Higgs exchange is suppressed because Higgs are composite and/or more massive than gauge bosons) further predictions follow. This is because the two vector operators involve oppositely polarized lepton fields and because the two vector operators have definite transformation properties under strong isospin. Therefore, predictions can be made concerning the polarization of the μ^+ in proton decay (which is said to be experimentally accessible). Also, isospin-like

relations between different decay modes, such as between $O^{16} \to e^+X$ and $O^{16} \to \bar{\nu}X$, can be derived. These relations depend on the relative admixture of the two vector operators. It is at this point that model dependence comes in. The precise relative admixture can of course be determined in specific models such as SU(5), SO(10) etc. (The actual phenomenological value is subject to renormalization[32] down to 1 Gev from $\sim 10^{16}$ Gev). If the day should come when the world has witnessed hundreds of events of proton decay, these relations will help us to distinguish between competing models.

VI. Conservation or Violation of B-L

We learned in the last section that B-L is effectively conserved to a high degree of accuracy if no new physics exist between M_W and M_X. Thus, the observation of B-L violation in proton decay would have far-reaching implications bearing on the question of whether or not "oases" exist in Glashow's desert. Put another way, the theorem from the last section only states that B-L is conserved to order M_W/M where M is the next mass scale in the theory; with the existence of oases, M may be much lower than 10^{15} Gev.

Now that we have convinced ourselves that B and L are both violated, are we to let B-L stand as the last citadel upholding the notion of exact non-gauged symmetry. We have argued against, and Nature appears to abhor, exact symmetries which are not gauged; the only symmetries we know to be exact to a high degree of accuracy are color and electric charge. So, let us press onward and violate B-L.

We would do well by first examining some specific models. In the simplest and original version of SU(5), B-L turns out to be conserved exactly. This result can be traced back to the fact that with the minimal choice of Higgs fields the allowed Yukawa terms are so few that the Lagrangian admits a global U(1) symmetry. The global U(1) is broken by the Higgs vacuum expectation value which also breaks the SU(5) gauge symmetry (of course.) However, it is clear that some linear combination of the global U(1) and a U(1) subgroup of SU(5) remains unbroken; this combination turns out to be none other than B-L. Thus, exact B-L conservation in this theory appears more or less "accidental" and in general would fail if additional Higgs fields are introduced.

The simplest choice[33] is a ten-dimensional Higgs field. The additional couplings allowed by the gauge symmetries remove the global U(1) symmetry. Decays changing B-L by two units, such as $n \to \mu^- K^+$, $p \to \mu^- K^+ \pi^+$, can now proceed. A characteristic prediction is that B-L violation in nuclear decay is accompanied by a strange particle. (One can in fact show[33] that this result is independent of the specific model under discussion by extending the analysis in the previous section to include dimension seven operators.) We should emphasize that the violation of B-L discussed here is quite consistent with the theorem derived in the previous section: the violation of B-L can indeed be computed in this specific model to be proportional to M_W but the mass M_W appears not divided by the grand unification mass but by the mass of the ten-dimensional Higgs.

The actual magnitude of B-L violation is beyond our present ability to predict; however, it can be shown that, with "reasonable" choice of parameters, one can have the B-L violating amplitude overwhelm the B-L conserving amplitude in proton decay.

Thus, while the observation of B,L violating nucleon decay would be most exciting, the observation of B-L violating and B,L violating nucleon decay would be doubly exciting. It might mean that "oases" exist in Glashow's desert. In our specific model, the "oasis" is none other than the additional Higgs structure.

Should proton decay violate B-L by two units, an amusing implication for the final state of an open Universe follows. Presumably, at the high temperatures when baryons were generated SU(2) x U(1) symmetry was restored and B-L was an exact symmetry. Thus, a universe born with all quantum numbers equal to zero would have $n_B = n_L$ in maturity, but could end up in old age with a net lepton number density if proton decay violates B-L.

With the violation of B-L all sorts of exotic processes, such as neutrino Majorana mass, and neutrinos oscillating into antineutrinos, all become possible.[35] A simple phenomenological way of incorporating these processes is to complicate the Higgs sector of the theory. Observation of these processes would also shed important light on the questions discussed here.

VII. Unsolved Problems in Grand Unification

To conclude, the idea of grand unification liberates us from our dreary existence amidst the debris of symmetry breaking, freeing us to dream about the "true" physics at 10^{16} Gev in a magnificent leap of imagination probably unprecedented in the history of physics. Over the last few years, the successful calculation of $\sin^2\theta_W$ and the possibility of explaining the content of the Universe have bolstered the beliefs of grand unification in their faith. Viewed in this context, the observation of proton decay would be most significant.

However, in our enthusiasm we should not be blind to the problems. To begin with, the present theory is only consistent with all physical phenomena; much remains to be explained. In particular, almost none of the physically interesting quantities can be calculated from first principles. There is of course also the matter of finding the correct grand unifying group and theory. We list here some outstanding conceptual problems, whose solution might help us in our search for the correct grand unifying theory.

(1) The breaking of grand unification symmetry and of SU(2) x U(1) symmetry apparently occur at vastly different mass scales. In the context of the Higgs mechanism, this implies vastly different vacuum expectation values, a situation which cannot be arranged "naturally" and which is unstable with respect to radiative correction. This is the so-called hierarchy problem.[36]

(2) Perhaps not unrelated to the hierarchy problem is the problem of dynamical symmetry breaking, about which nobody has anything substantial to say so far. A very interesting approach

involves the concept of technicolor[37], which unfortunately generates additional problems. Technicolor may be an oasis in Glashow's desert. It should be noted that a theory without explicit Higgs fields typically possesses a number of global symmetries if fermions are assigned to several irreducible representations. Thus, any dynamical symmetry breaking scheme must address the problem of what to do with the multitude of Nambu-Goldstone bosons which would appear upon symmetry breaking.

(3) With spontaneous symmetry breaking, be it dynamical or due to explicit Higgs fields, there will in general be an energy (density) difference between the symmetric vacuum and the broken vacuum. This energy (density) of condensation appears as a cosmological constant in the presence of gravity.[38] If we normalize the energy scale so that the symmetric vacuum has zero energy then the cosmological constant would be enormous, of the order of the fourth power of a typical Higgs mass. This is surely the largest discrepancy between an observed value and a calculated value in the history of physics. We are of course faced with this problem even without grand unification, but grand unification exacerbates the situation. Does grand unification have anything to do with gravity?[39] How can we insure a phase transition with no latent heat?[40]

(4) The hierarchy of fermion masses is also difficult to explain.[41]

(5) Perhaps closely related to the above is the so-called family problem. Rabi's famous question, "Who ordered the muon?" has been escalated in our times to the puzzling question "Why does Nature repeat Herself, so needlessly and tediously?" To construct the physical world as we know it, only the first family of quarks and

leptons are apparently necessary. Cosmologists claim to have an upper bound on the number of different massless neutrinos. But perhaps neutrinos have small masses, increasing with each generation. Recently, people have begun to try to comprehend Nature's extravagance.

(5a) The simplest approach is to merely incorporate the repetitive structure phenomenologically, without any attempt to understand its origin. The idea is to introduce a so-called horizontal gauge symmetry under which the families transform into each other. One hopes to obtain relations between masses and mixing angles.[42] Some of the results obtained so far appear to be encouraging, but the approach suffers from a great deal of arbitrariness.

(5b) A more dynamical approach, proposed by Georgi[43], starts with an SU(N) gauge theory with N greater than five and with fermions assigned to some anomaly-free set of representations. The SU(N) gauge symmetry is supposed to break down to an SU(5) symmetry with the actual breaking mechanism unspecified. A plausible dynamical assumption is made that all fermions allowed to have SU(5)-invariant masses get masses of the order of 10^{15} GeV. The observation that there are no bare fermion mass terms in the standard SU(3) x SU(2) x U(1) and SU(5) theories provides the empirical basis for this assumption. In this way, one can, for a given initial choice of group and representations, determine how many generations of light fermions emerge at low energies.

(5c) A rather promising approach utilizes the spinorial representations of orthogonal groups. One aesthetically displeasing feature of the SU(5) theory is that fermions in each generation are assigned to two irreducible representations, a $\bar{5}$ and a 10. It seems reasonable to expect that in a truly unified theory all fermions should belong to a single irreducible representation. The SO(10) theory represents a slight improvement over SU(5) in this respect: at the cost of introducing a neutral lepton field for each generation, one can put all the fermions of a given generation in a single 16-dimensional spinorial representation. (In general, neutrino Majorana masses are induced as a result.[44]) However, the existence of families is still accommodated by merely having the 16-dimensional representation repeated. Now the spinorial representation of SO(2n) have dimension 2^{n-1}, in contrast to the tensor representation (of unitary and orthogonal groups) whose dimensions only increase like a power, rather than exponentially, with the rank of the group. Thus, we might imagine putting[45,46] all fermions into a single spinorial representation of some SO(2m+10) gauge theory which breaks down into an SO(10) theory. The 2^{m+4} dimensional representation of fermions would then decompose into 2^m spinors of SO(10). The outstanding virtue of this proposal is of course that the repetitive structure of fermion families emerges guaranteed by the mathematics. The number of families is determined to be a power of two: if we believe the cosmologists, there are four families. Unfortunately for this scheme, it turns out that only half of the 2^m spinorial representations have V-A weak interactions, the other half having V+A weak interactions. Perhaps the theory is right after all: these V+A fermions may just have been too heavy to have been seen. One (rather unnatural) approach[4] is to simply push up the masses of the unwanted fermions by the Higgs mechanism. A more interesting approach[45] invokes technicolor to confine the unwanted fermions.

This concludes our list of difficult problems. One wonders how many of these problems will have been solved, or at least will have been argued to be irrelevant, by the end of this decade just dawning on us. Most likely, the real problems are not even on this list. With this standard piety we come to the end of our brief survey of some of the issues debated by particle physicists today. It is remarkable and exciting that we are now talking about energy scales fourteen orders of magnitude beyond our experience and discussing issues which only a few years ago were thought to be outside the domain of physics.

Acknowledgment

We are grateful to many colleagues for discussions over the years, but most particularly to our frequent collaborator, Frank Wilczek. Much of what is described here is based on work done with him.

References

(This list of references is obviously not meant to be complete.)

1. Symmetries in physics are discussed in books by Weyl, Yang, and Sakurai. Our account follows closely a recent article of C. N. Yang (CERN preprint 1979).
2. D. Boulware and S. Deser, Ann. of Phys. 89, 193 (1975).
3. H. Georgi and S. L. Glashow, Phys. Rev. Letters 32, 438 (1974).
4. See e.g., P. Langacker and D. Sparrow, Phys. Rev. Letters 43, 1101 (1979).
5. S. Glashow, Nucl. Phys. 22, 579 (1961); S. Weinberg, Phys. Rev. Lett. 19, 1264 (1967); A. Salam and J. Ward, Phys. Lett. 13, 168 (1964).
6. In fact, for examples of grand unified theories which admit a stable proton, see P. Langacker, G. Segre, A. Weldon, Phys. Lett. 73B, 87 (1978(; Phys. Rev. D18, 552 (1978); G. Segre and A. Weldon, Penn preprint 1980. However, these models always involve an imposed global symmetry.
7. H. Georgi, H. Quinn, S. Weinberg, Phys. Rev. Lett. 33, 451 (1974).
8. The brief account, which follows Yang (ref. 1) is certainly not meant to be historically complete.
9. C. N. Yang and R. Mills, Phys. Rev. 95, 631 (1954); Phys. Rev. 96, 191 (1954).
10. For a review see for example E. Abers and B. Lee. Phys. Reports 9C, 1 (1973); D. Gross, in Methods in Field Theory, ed. by R. Balian.

11. T. D. Lee and C. N. Yang, Phys. Rev. 98, 1501 (1955).

12. J. Pati and A. Salam, Phys. Rev. D8, 1240 (1973);
D10, 275 (1974).

13. In the U.S., a University of Pennsylvania experiment is now under way in the Homestake Gold Mine in South Dakota, and a Brookhaven-Irvine-Michigan experiment is planned for the Morton Salt Mine outside of Cleveland. There are also experiments planned in Europe. These are notably inexpensive experiments, and may be well-suited for countries such as China. The total integrated cost of the Pennsylvania experiment will run to a few x 10^5 U.S.$ (which incidentally is of the same order as the profit of the Homestake Mining Company per day with the present price of gold.)

14. A. Pais, Phys. Rev. (1973).

15. A. D. Sakharov, ZhETF Pis'ma 5, 32 (1967);
M. Yoshimura, Phys. Rev. Lett. 41, 281 (1978), 42, 746(E) (1979);
S. Dimopolous and L. Susskind, Phys. Rev. D18, 4500 (1978);
D. Toussaint, S. Treiman, F. Wilczek, A. Zee, Phys. Rev. D19, 1036 (1979); S. Weinberg, Phys. Rev. Lett. 42, 859 (1979).

16. "Le silence éternel de ces espaces infinis m'effraie" - Pascal.

17. For a discussion and references, see e.g., S. Weinberg, Gravitation and Cosmology, Wiley.

18. G. Steigman, Ann. Rev. of Astron. and Astrophys. 14, 339 (1976).

19. It has been noted that 10^{15} Gev is roughly the kinetic energy of a typical bus. Thus, during this epoch of the Universe, to be hit by even a single photon is no laughing matter.

20. The literature includes M. Yoshimura, preprints.
J. Ellis, M. K. Gaillard, and D. V. Nanopoulos, Phys. Lett. 86B, 360 (1979), 82B, 464(E) (1979); S. Barr, G. Segre and H. Weldon, to be published; D. Nanopoulos and S. Weinberg, to be published; A. Yildiz and P. Cox, to be published; A. Yu. Ignatiev et al., Phys. Lett. 87B, 114 (1979); M. S. Turner and D. Schramm, Nature 279, 303 (1979); E. W. Kolb and S. Wolfram, Caltech preprint. We recommend especially the paper by Kolb and Wolfram which exposes the weak points of some previous analyses.

21. M.Kobayashi and K. Maskawa, Prog. Theor. Phys. 49, 652 (1973).

22. C. W. Misner, Astrophys. J. 151, 431 (1968); J. D. Barrow, Nature 272, 211 (1978) and references therein.

23. M. Turner, Nature 281, 549 (1979); S. Bludman, private communication.

24. S. Hawking, Nature 248, 30 (1974); Comm. Math. Phys. 43, 199 (1975).

25. S. Hawking, as cited in B. J. Carr, Astrophys. J. 206, 19 (1976).

26. G. Segre and A. Weldon, to be published.

27. Besides baryon asymmetry see the recent work on monopoles in the early Universe. J. P. Preskill, Phys. Rev. Lett. 43, 1365 (1979); A. Guth and H. Tye, SLAC preprint 1979; M. Einhorn, D. L. Stein and D. Toussaint, preprint.

28. The analysis described here is due to S. Weinberg, Phys. Rev. Lett. 43, 156C (1979); F. Wilczek and A. Zee, Phys. Rev. Letters 43, 1571 (1979).

29. Several issues of great practical importance such as that of pion propagation through nuclei; are not touched on here. They have been investiged by D. Sparrow, preprint, C. Dover and L. L. Wang, preprint. A specific SU(5) analysis was performed earlier by M. Mahacheck, preprint.
30. E. Fermi, Z. Physik 88, 161 (1934).
31. Actually, his solution is incomplete. G. Gamow and E. Teller, Phys. Rev. 49, 895 (1936).
32. F. Wilczek and A. Zee, ref. 3. J. Ellis, M. Gaillard and D. Nanopoulos, CERN preprint.
33. This section is based on F. Wilczek and A. Zee, Phys. Lett. 88B, 311 (1979). Similar remarks have been made by S. Glashow, HUTP-79 (A029).
34. H. Georgi, S. Glashow, S. Weinberg, private communication.
35. A. Zee, Penn preprint UPR-0150T.
36. E. Gildener, Phys. Rev. D14, 1667 (1976); S. Weinberg, Phys. Lett. 82B, 387 (1979); K. T. Mahanthappa, M. A. Sher, and D. G. Unger, Phys. Lett. 84B, 113 (1979). For a recent attempt to solve the problem, see T. P. Cheng and L. F. Li, to be published.
37. S. Weinberg, Phys. Rev. D13, 974 (1976); L. Susskind, Phys. Rev. D20, 2619 (1979).
38. M. Veltman, unpublished. J. Dreitlein, Phys. Rev. Lett. 33, 1243 (1974).
39. A. Zee, Phys. Rev. Letters 42, 417 (1979); L. Smolin, Nucl. Phys. B160, 253 (1979); S. L. Adler, IAS preprint 1980.

40. At the moment, the only glimmer of a hope involves supersymmetry and supergravity.

41. For a recent attempt in the context of grand unification, see S. Barr, to be published.

42. For a construction involving SU(3) x SU(2) x U(1) x SU(2) see F. Wilczek and A. Zee, Phys. Rev. Letters 42, 421 (1979). For a discussion at the grand unification level involving SU(5) x SU(2) see M. Gell-Mann, P. Ramond, and R. Slansky, Caltech preprint.

43. H. Georgi, Nucl. Phys. B156, 126 (1979); P. Frampton and S. Nandi, Phys. Rev. Lett. 43, 1460 (1979); J. Chakrabarti, M. Popovic, R. Mohapatra, CCNY preprint 1979.

44. M. Gell-Mann, P. Ramond, R. Slansky, unpublished. E. Witten, Harvard preprint.

45. M. Gell-Mann, P. Ramond, R. Slansky, unpublished.

46. F. Wilczek and A. Zee, preprint 1979.

Geometrical Model of Hadron Collisions

T. T. Chou
Department of Physics
University of Georgia
Athens, Georgia 30602, U.S.A.

Chen Ning Yang
Institute for Theoretical Physics
State University of New York
Stony Brook, New York 11794, U.S.A.

Abstract

Results obtained from the geometrical model for high energy hadron-hadron and hadron-nucleus collisions are summarized. The discussion is focused on elastic scattering, diffraction dissociation process and the concept of matter current distributi inside a polarized hadron or nucleus.

We wish to review some of the main results of the geometrical model for high energy collisions obtained since 1967. The study of this model has led to a number of specific predictions. These are:

(1) For elastic pp scattering,

   a. the values for the angular distribution $d\sigma/dt$ in the region for $|t| < 1.2(\text{GeV}/c)^2$,

   b. the existence in $d\sigma/dt$ of a first minimum and a second maximum,

   c. the slow inward movement of the positions for these minimum and maximum as the incident energy increases.

(2) The matter distribution inside hadrons, and the values of the rms radii for π and K mesons.

(3) The dip and kink structures for the diffraction dissociation processes.

(4) The hypothesis of limiting fragmentation in particle production processes.

(5) The existence of a second minimum in elastic pp differential cross sections.

(6) The matter current distribution inside polarized hadrons and nuclei.

While predictions (1) through (4) have all been verified, the confirmation of (5) and (6) will have to await future experiments. It should be emphasized that no adjustable parameters has been used in comparing most of these predictions with experiments.

In this talk the work on three general areas, the elastic scattering, the diffraction dissociation process, and the concept of velocity profile will be summarized. We shall present here only the main physical ideas and important experimental evidences. Detailed discussions which can be found in the literatures will be omitted.

I. Elastic Scattering

The geometrical description[1] of elastic hadron-hadron scattering at high energy is based on the fundamental assumption that all hadrons are extended objects with structures. This description provides a direct relationship between the hadronic matter density functions and the elastic differential cross section. The main physical concepts invoked in this picture are the following three:

1. The eikonal approximation.
2. The exponentiation for the transmission coefficient in impact-parameter space.
3. The convolution integral formalism for the opaqueness of the colliding system.

Combining the above ideas, one obtains the elastic scattering amplitude for hadrons A and B:

$$a_{AB} = \text{two dimensional Fourier transform of}$$
$$\{1 - \exp -[(\text{constant})\ D_A \otimes D_B]\} \qquad (1)$$

where D_A and D_B are the compressed matter densities for hadrons A and B, and $\otimes$ denotes convolution in impact-parameter space. The quantity in the square bracket represents the opaqueness.

Some results are readily derivable from this model:

(A) The hadronic matter distributions can be computed by using the experimentally measured differential cross section.

The charge form factors of hadrons may then be inferred if one makes an additional assumption that hadronic matter and charge distributions are proportional to each other. The proton form factor thus computed agrees amazingly well with the experimental G_E form factor of the proton. (See Fig. 3 of Ref. [2].) Application of the geometrical model to πp and Kp scatterings has also been used to determine the pion and kaon radii.[3] Theoretical predictions seem to be in good agreement with values obtained from direct experimental measurements.[4]

(B) With the proton form factor and the observed total cross section used as input, the model predicted[5] the pp differential cross section which shows a dip near $|t| \cong 1.4 (GeV/c)^2$, and a maximum at a slightly higher-$|t|$ value. The calculated positions of the minimum and the maximum were in good agreement with the experimental results.[6] Similar structures are also expected to exist in np elastic scattering, since it is reasonable to assume that neutron and proton have approximately the same matter form factor. This conjecture has indeed been borne out by a recent experiment.[7]

(C) The constant appearing in the opaqueness expression in equation (1) is related to the total cross section σ_T. With the discovery[8] of rising σ_T with increasing energy, this "constant" becomes dependent on the incoming energy. Because of increasing opaqueness, the model predicted[9] an inward shift of the minimum and the maximum positions with energy, and a concurrent rise of the second maximum. These characteristics are also in general agreement with the newest experiments.[10]

We now make a few remarks:

(1) We have tacitly assumed that only the spin-independent amplitude contributes predominantly to pp scatterings at high energies and that it is imaginary. The observed energy dependence of the cross section necessarily requires the existence of a real part in the scattering amplitude. This small real part is neglected in our consideration. These approximations are in accordance with the experimental fact[11] that at high energies the polarization parameter P in pp scattering is approximately zero except near the dip region.

(2) Extensive work[12] on the high energy limit of a large class of Feynman diagrams for certain couplings in field theory has been carried out by Cheng and Wu. It is remarkable that the general validity of the three basic physical concepts used in the elastic model has been confirmed, at least partly, by their studies.

(3) As the opaqueness increases with the total cross section, the geometrical model predicts[13] the existence of many dips in pp elastic scattering at very high energies. Although none of the dips other than the first one has been observed so far, the development of many dips is inevitable in view of the fact that increasing opaqueness would produce an effective black disc as the scattering center at sufficiently high energies.

(4) In describing high energy elastic scattering, the geometrical picture envisages the passage of one hadron through another with attenuation. The attenuation factor depends on

the impact parameter b, and not very much on the energy. Since the elastic and inelastic channels are strongly coupled, we argued that in high energy inelastic processes the probability amplitude for the target or the projectile to be excited into any specific mode must also be a function of b alone, and would be largely independent of incident energy. This hypothesis [14] was referred to as the hypothesis of limiting fragmentation. Experimental verification [15] of the hypothesis has been carried out at the ISR at CERN.

II. Diffraction Dissociation

In recent years there has been experimental data [16] concerning the angular distribution of hadron-nucleus and hadron-hadron diffraction dissociation. One of the conspicuous features of all these experiments is the existence of dip or kink structures similar to that observed in pp elastic scattering. The dip in diffraction dissociation occurs generally at a smaller $|t|$ value than the dip in elastic scattering. We pointed out that the geometrical model can offer a natural explanation [17] of these dip structures.

Consider the passage of an incoming hadron through an extended target. At an impact parameter b the dissociation can take place at any point along its path during its traversal. The probability for the process to occur is approximately proportional to the thickness of the material traversed, or $\Omega(b)$. There is also absorption of the incoming wave before dissociation, and of the outgoing wave after dissociation. Assuming equal mean free path for incoming and outgoing waves, the total absorption factor can be written as $\exp[-\Omega(b)]$.

Thus the source function for the outgoing hadron in diffraction dissociation may be approximated by $\Omega \exp(-\Omega)$. This approximation was first used in charge exchange scatterings and was given the name "coherent droplet model."[18]

With Ω determined from electron scattering experiments together with hadron-hadron total cross sections, numerical computations for differential cross section in diffraction dissociation processes have been made.[17] The computation contains no adjustable parameters. The calculated dip positions for both elastic scattering and diffraction dissociation process are in very good agreement with experimental values.

III. Hadronic Matter Current Distribution[19]

Since geometrical concepts such as sizes, matter distribution, or the opaqueness distribution of hadron are very useful in discussing high energy collisions, one may raise the interesting question whether matter current exists inside a polarized hadron. Once we have accepted the concept of an extended hadron with a matter density in it, it seems to us inevitable that we must also accept the existence of a matter current in a polarized hadron. The concepts of matter density and matter current necessarily complement each other, resulting in a four-vector. The existence of a matter current produces observable effects. Experimental test of this idea is now possible if one utilizes the rising total cross section with increasing incoming energies.

Consider a Kp scattering with the target proton polarized along the x-direction. If the motion of hadronic matter in a region of the target is toward the projectile, that region would have greater opaqueness than another region in which the motion is away from the projectile. Thus, to a kaon coming along the +y axis, the proton would appear more opaque on the +z axis side than the -z side. The result of this difference in opaqueness is a non-vanishing rotation parameter R(t). I.e., the polarization vector of the recoil proton for a collision in which the K is scattered in the x-y plane toward the +x axis has a y-component equal to R(t). This Wolfenstein parameter R is measurable and its values for different momentum transfers t will determine the hadronic matter current density in the target.

Numerical computation showed that for small t, R is negative. This sign has a natural classical interpretation.[20] The estimated value of R at $t = -0.5 (GeV/c)^2$ ranges from -0.05 to -0.2 radian for πp and Kp scatterings at 200 GeV, indicating that the matter current effect is measurable with available experimental techniques.

The given estimate for the R parameter is obtained on the assumption that the polarized target is infinitely heavy. Corrections clearly have to be made if the target has a non-vanishing recoil velocity. This effect, however, cannot be dealt with rigorously.

The above discussion is clearly applicable to elastic scattering of hadron off polarized nuclei, where the existence of nucleonic current is conceptually much easier to accept.

The work of T. T. C. was supported in part by the Department of Energy under Contract No. DE-AS09-76ER00946. The work of C. N. Y. was supported in part by the National Science Foundation under Grant No. PHY7615328.

References

[1] T. T. Chou and C. N. Yang, Phys. Rev. 170, 1591 (1968).

[2] A. W. Chao and C. N. Yang, Phys. Rev. D8, 2063 (1973).

[3] T. T. Chou, Phys. Rev. D19, 3327 (1979).

[4] E. B. Dally *et al.*, Phys. Rev. Lett. 39, 1176 (1977); A. Beretvas *et al.*, contributed paper No. 857 submitted to the XIXth International Conference on High Energy Physics, Tokyo, 1978 (unpublished).

[5] T. T. Chou and C. N. Yang, Phys. Rev. Lett. 20, 1213 (1968); L. Durand and R. Lipes, Phys. Rev. Lett. 20, 637 (1968).

[6] A. Böhm *et al.*, Phys. Lett. 49B, 491 (1974).

[7] C. E. DeHaven, Jr. *et al.*, Phys. Rev. Lett. 41, 669 (1978).

[8] U. Amaldi *et al.*, Phys. Lett. 44B, 112 (1973); S. R. Amendolia *et al.*, *ibid*. 44B, 119 (1973); A. S. Carroll *et al.*, Phys. Rev. Lett. 33, 932 (1974). The increase of total cross section was conjectured theoretically by H. Cheng and T. T. Wu, *ibid*. 24, 1456 (1970).

[9] F. Hayot and U. P. Sukhatme, Phys. Rev. D10, 2183 (1974).

[10] N. Kwak *et al.*, Phys. Lett. 58B, 233 (1975); E. Nagy *et al.*, Nucl. Phys. B150, 221 (1979).

[11] J. H. Snyder *et al.*, Phys. Rev. Lett. 41, 781 (1978).

[12] H. Cheng and T. T. Wu, Phys. Rev. 182, 1852 (1969) and later papers.

[13] T. T. Chou and C. N. Yang, Phys. Rev. D17, 1889 (1978); ibid. D19, 3268 (1979).

[14] J. Benecke, T. T. Chou, C. N. Yang, and E. Yen, Phys. Rev. 188, 2159 (1969).

[15] G. Bellettini et al., Phys. Lett. 45B, 69 (1973).

[16] G. Goggi et al., Phys. Lett, 79B, 165 (1978); T. Ferbel, lecture presented at the First Workshop on Ultra-Relativistic Nuclear Collisions, LBL (1979).

[17] T. T. Chou and C. N. Yang, Dip and kink structures in hadron-nucleus and hadron-hadron diffraction dissociation (preprint).

[18] N. Byers and C. N. Yang, Phys. Rev. 142, 976 (1966). This approximation was later utilized in T. T. Chou, Phys. Rev. 176, 2041 (1968); and H. Cheng, J. K. Walker and T. T. Wu, ibid. D9, 749 (1974).

[19] T. T. Chou and C. N. Yang, Nucl. Phys. B107, 1 (1976).

[20] T. T. Chou, in High Energy Collisions - 1973, edited by C. Quigg (AIP Conf. Proc. No. 15, New York, 1973) pp. 118-123; C. N. Yang in Proc. Int. Symposium on High Energy Physics, edited by Y. Hara et al. (University of Tokyo, Japan, 1973) pp. 629-634.

On Grand Unification

A Review

by

Ngee-Pong Chang*
Physics Department
City College of the City University of New York
New York, N.Y.

Abstract

We review the physical motivation and objectives of grand unification. The importance of renormalisation effects is pointed out, especially in connection with proton lifetime estimates. We present an asymptotically free, one coupling constant, one mass scale SU(5) model with three generations of light fermions. The low energy structure is identical to the standard SU(5) model.

* Supported in part by a grant from the National Science Foundation, NSF-PHY77-01350, and in part by a grant from the City University Research Foundation, PSC-BHE

1. Introduction

It is a great pleasure to be here in China participating in the first international Conference on Theoretical Particle Physics, organized by Academic Sinica. It is quite appropriate that a major theme of this Conference is grand unification.

Grand unification[1] is an attempt at the grand synthesis of our description of the three forces of nature: strong, electromagnetic and weak. The objective is to be able to understand all three forces as being due to the exchange of the Yang-Mills[2] quanta belonging to a single, grand gauge group, G. From this grand unified theory we hope to be able to understand, someday, all of the low energy parameters of the theory, viz.,

α_s, α_W, α

$\sin^2\theta_W$

current quark masses

generation and mixing angles

etc.

Prof. Zee[3] in his talk already summarised the cosmological implications as well as the phenomenological consequences of grand unification so that I can well afford to pass over them in this review. In this talk I shall try to give a brief introduction to the ideas of grand unification, the problems and hopes involved in this approach and some new results obtained in this area.

2. SU(3)×SU(2)×U(1) Theory

Let me begin with a quick review of our present day understanding of the three fundamental forces of nature. The leading candidate for the theory of strong interactions is QCD, the Yang-Mills field theory in its purest form:

$$L_{QCD} = -\frac{Tr}{4}\left(\partial_\mu g_\nu - \partial_\nu g_\mu - \frac{i}{\sqrt{2}} g_3 [g_\mu, g_\nu]\right)^2$$

$$- \bar{\psi}\gamma_\mu\left(\partial_\mu \psi - \frac{i}{\sqrt{2}} g_3 g_\mu \psi\right) \quad . \tag{1}$$

In this theory there is only one coupling constant, g_3. The successes of this theory in explaining a wide spectrum of data ranging from deep inelastic through e^+e^- to pp and πp scatterings are by now quite well documented. In principle, as we have learned from Professor Lee,[4] spectacular dynamical calculations of the spectroscopic masses (π, ρ, η,···,N, N*,···) are already under way using non-perturbative treatments of this one parameter QCD Lagrangian. Although QCD cannot be said to have been completely tested experimentally (in spite of the Aug. 30 '79 announcement in the New York Times), it is quite likely to be the correct theory of strong interactions. In grand unification we take this for granted and embed this SU(3) color gauge group in G.

The Lagrangian for all known "low energy" physics contains, in addition, the electroweak theory of Galshow-Salam-Weinberg.[5] The SU(2)×U(1) model has largely been confirmed by all present experimental data and is clearly the basis for any further unification with strong interactions. Although it is customary these days not to do so, let us nevertheless exhibit the complete Lagrangian as we know it:

$$\mathcal{L} = -\frac{Tr}{4}\left(\partial_\mu G_\nu - \partial_\nu G_\mu - \frac{i}{\sqrt{2}} g_3 [G_\mu, G_\nu]\right)^2$$

$$
\begin{aligned}
& - \frac{Tr}{4} \left(\partial_\mu W_\nu - \partial_\nu W_\mu - \frac{i}{\sqrt{2}} g_2 [W_\mu, W_\nu] \right)^2 \\
& - \frac{1}{4} \left(\partial_\mu B_\nu - \partial_\nu B_\mu \right)^2 \\
& - \sum_{\text{generations}} \bar{q}_L \gamma_\mu \left(\partial_\mu - \frac{i}{\sqrt{2}} g_3 G_\mu + \frac{i}{\sqrt{2}} g_2 W_\mu + \frac{i}{2\sqrt{15}} g_1 B_\mu \right) q_L \\
& - \sum_{\text{generations}} \bar{\ell}_L \gamma_\mu \left(\partial_\mu + \frac{i}{\sqrt{2}} g_2 W_\mu - \frac{3i}{2\sqrt{15}} g_1 B_\mu \right) \ell_L \\
& - \sum_{\text{generations}} \bar{q}_R \gamma_\mu \left(\partial_\mu - \frac{i}{\sqrt{2}} g_3 G_\mu + \frac{3i}{\sqrt{15}} Q g_1 B_\mu \right) q_R \\
& - \sum_{\text{generations}} \bar{\ell}_R \gamma_\mu \left(\partial_\mu - \frac{3i}{\sqrt{15}} g_1 B_\mu \right) \ell_R \\
& - \left| \partial_\mu \phi + \frac{i}{\sqrt{2}} g_2 W_\mu \phi - \frac{3i}{2\sqrt{15}} g_1 B_\mu \phi \right|^2 \\
& + \mu^2 \phi^\dagger \phi - \lambda \left(\phi^\dagger \phi \right)^2 \\
& - \sum_{\text{generations}} h_q \left(\bar{q}_L q_R \phi + h.c. \right) \\
& - \sum_{\text{generations}} h_\ell \left(\bar{\ell}_L \ell_R \phi + h.c. \right) \qquad (2)
\end{aligned}
$$

In Eq.(2) the set of quark fields, q, refers to

$$
\begin{pmatrix} u \\ d' \end{pmatrix}_L \qquad \begin{pmatrix} c \\ s' \end{pmatrix}_L \qquad \begin{pmatrix} t \\ b' \end{pmatrix}_L
$$

$$
u_R,\ d'_R \ , \qquad c_R,\ s'_R \ , \qquad t_R,\ b'_R
$$

where[6] ($c_i \equiv \cos\theta_i$, $s_i \equiv \sin\theta_i$, $i=1,2,3$)

$$
\begin{aligned}
\underline{d}' &= c_1 \underline{d} + s_1 c_3 \underline{s} + s_1 s_3 \underline{b} \\
\underline{s}' &= -s_1 c_2 \underline{d} + (c_1 c_2 c_3 - s_2 s_3 e^{i\delta}) \underline{s} + (c_1 c_2 s_3 + s_2 c_3 e^{i\delta}) \underline{b} \\
\underline{b}' &= s_1 s_2 \underline{d} + (-c_1 s_2 c_3 - c_2 s_3 e^{i\delta}) \underline{s} + (-c_1 s_2 s_3 + c_2 c_3 e^{i\delta}) \underline{b}
\end{aligned} \qquad (3)
$$

The leptons, ℓ, refer to the set

$$\begin{pmatrix} \nu_e \\ e \end{pmatrix}_L \qquad \begin{pmatrix} \nu_\mu \\ \mu \end{pmatrix}_L \qquad \begin{pmatrix} \nu_\tau \\ \tau \end{pmatrix}_L$$

$$e^-_R \; , \qquad \mu_R \; , \qquad \tau_R \; ,$$

Lagrangian (2) summarises all strong weak and electromagnetic interactions as we know them to date. A notable feature of Eq.(2) is the artificial combination of weak hypercharge assignments that are needed to make the quark and lepton charges come out right. One of the fringe benefits of grand unification will be to provide a natural though not the only group theoretical rationale for these assignments. Other ways to understand the weak hypercharge assignments include, among others, embedding the standard electroweak theory in a highly unusual super-group SU(2/1),[7] as well as in a more conventional Lie group such as SU(3)×U(1),[8] $SU_L(2)\times SU_R(2)$[9] or even $SU_L(2)\times SU_R(2)\times U_L(1)\times U_R(1)$.[10]

In these latter attempts (SU(2/1), SU(3)×U(1) and $SU_L(2)\times SU_R(2)\times U_L(1)\times U_R(1)$) the value of $\sin^2\theta_W$ in fact comes out to be exactly .25. In contrast, grand unified theories predict, at grand unification energies, a value of 3/8 for $\sin^2\theta_W$. The renormalisation of $\sin^2\theta_W$ down to present low energies is of course a central problem of grand unified field theories which I will discuss shortly.

The other equally striking feature of Eq.(2) is its total lack of simplicity and elegance, especially when you compare with the original Yang-Mills Lagrangian for QCD. Our complete Lagrangian here has

3 gauge couplings

9 Yukawa couplings (for the 3 generations)

1 Higgs quartic coupling

1 Higgs mass scale

and all these are in addition to the three mixing angles and CP violating phase.

Given this situation there are two ways to go. One is the way of subunification, in which we think of quarks and leptons as being themselves made of constituents. The fundamental theory may itself be a gauge theory, with consequent asymptotic freedom. In such a theory, Higgs fields are not elementary but are fermion-antifermion excitations.[11] Work in this direction has been reported at this Conference by Kerson Huang,[12] by Ling-Fong Li[13] and by Ni and Su.[14] Professor Huang's idea is especially interesting since it deals with a renormalisable theory in which fermion-antifermion pairing can occur. The induced couplings of the Higgs to itself as well as to other quarks and leptons can one day be calculated in that fundamental theory. A central problem in this approach is the precise formulation and understanding of the dynamical symmetry breakdown needed to generate masses. It is fair to say that as of today we do not have a complete good working theory of dynamical symmetry breakdown.[15]

The other way is the more phenomenological one of grand unification. Here we simply put the quarks and leptons in a big multiplet and study the resulting gauge theory of the grand unifying group, G. This G unifies the SU(3)×SU(2)×U(1) into a simple Lie group such as SU(5), the minimal candidate. While it is not obvious that this phenomenological grand unification will make sense, results obtained so far with this approach have been quite encouraging.

3. SU(5) Basics

Since much of what follows devolves upon this SU(5) model it is a good point to describe briefly this, simplest, unification. Let

$$\psi_R^{\alpha} = \begin{pmatrix} d_1 \\ d_2 \\ d_3 \\ e^+ \\ \nu^c \end{pmatrix}_R$$

be the fundamental multiplet. Clearly α=1,2,3 refers to the SU(3) subgroup while α=4,5 refers to the SU(2) subgroup. In addition a U(1) subgroup acts (differently) on the SU(3) and SU(2) subspaces.

The remaining quarks and lepton are assigned to the completely antisymmetric 10 multiplet, viz.,

$$\psi_L^{\alpha\beta} = \begin{bmatrix} 0 & u_3^c & -u_2^c & -u_1(\theta) & -d_1 \\ & 0 & u_1^c & -u_2(\theta) & -d_2 \\ & & 0 & -u_3(\theta) & -d_3 \\ & & & 0 & -e^+ \\ & & & & 0 \end{bmatrix}_L$$

Note that as a result of the antisymmetry in $\alpha\beta$ the SU(3) content of $\psi^{\alpha\beta}$ contains only 3 and 3* besides the SU(3) singlet, so that both U_L and U_L^c (i.e., U_R) can be accommodated, besides e_L^+.

The assignment 5_R and 10_L has two prominent features:

(i) In the simplest unification scheme, there is no room for ν_R (unless you add an SU(5) singlet ν_R field)

(ii) The theory is anomaly free.[16] (The triangular anomaly associated with 5_R cancels exactly the anomaly associated with 10_L). This is special to SU(5). For SU(N), N > 5, $\psi^{\alpha} \oplus \psi^{[\alpha\beta]}$ is not anomaly-free.

The gauge bosons $A^{\alpha}_{\mu\beta}$ naturally decompose into the bosons of the SU(3)×SU(2)×U(1) subgroup (8 gluons, 3 W-bosons, U(1) boson) plus the new leptoquarks which can mediate

$$d_{iR} \longrightarrow X_i + e^+_R \qquad\qquad Q(X_i) = -\frac{4}{3}$$

$$d_{iR} \longrightarrow Y_i + \nu^c_R \qquad\qquad Q(Y_i) = -\frac{1}{3}$$

Because of the presence of these X, Y leptoquarks a quark can now convert into a lepton. These X (and Y) bosons therefore can virtually convert a proton into a lepton plus mesons, through the effective 4-fermion matrix element

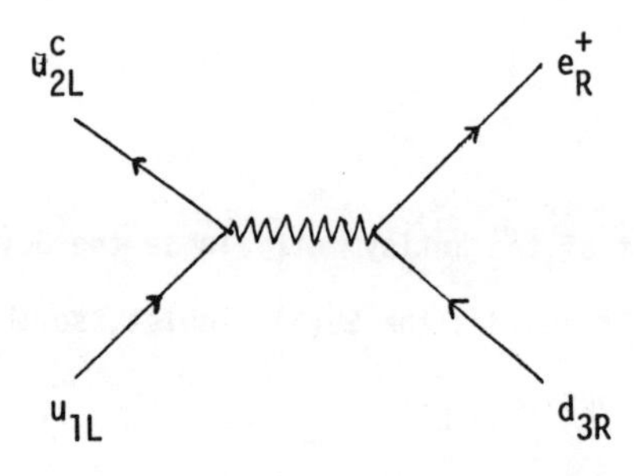

so that proton decay could proceed via

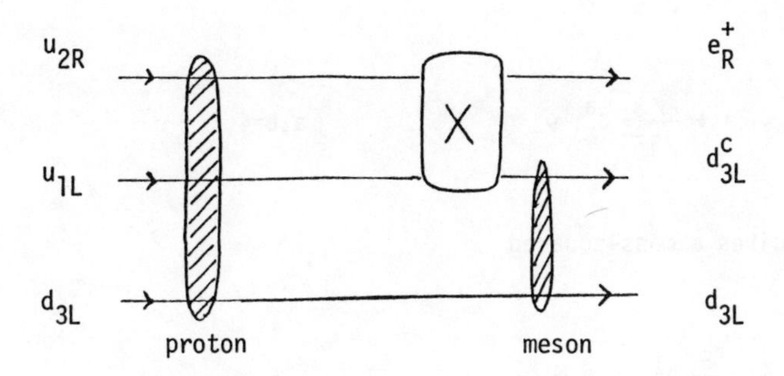

Clearly a central problem in grand unification has to do with the determination of M_x $(=M_y)$ since the proton stability depends on it.

But before we can even go about an intelligent discussion of the proton lifetime, we must already ask how X (and Y) acquired their masses. One way is to wish it on some dynamical symmetry breakdown. I believe ultimately, that would be the correct way. It is fair to say, however, that at present there is no complete theory of dynamical symmetry breakdown that would allow a detailed calculation for comparison with experiment.

The other way is the phenomenological way of the ubiquitous Higgs.[17] I will come back to the role of these Higgs in just a minute, but let me simply observe the minimum number of Higgs needed to do the job.

To generate the X-masses, a 24-Higgs (ϕ^{α}_{β}) is necessary, with

$$< \phi^i_j > = - \frac{\sqrt{2}}{\sqrt{15}} \delta^i_j \, v \qquad i,j=1,2,3$$

$$< \phi^a_b > = + \frac{\sqrt{3}}{\sqrt{10}} \delta^a_b \, v \qquad a,b=4,5 \tag{4}$$

so that X acquires a mass-squared

$$M_X{}^2 = \frac{5}{12} g^2 v^2 = M_Y{}^2 \tag{5}$$

while to generate the W-masses, a $\underline{5}$-Higgs (H^α) is also needed. To arrange for this desired pattern of symmetry breakdown

$$SU(5) \rightarrow SU(3)\times SU(2)\times U(1) \rightarrow SU(3)\times U(1)$$

the Higgs potential that is needed now must contain 2 mass scales and 5 new quartic self-couplings.[18]

$$V = - \frac{\mu^2}{2} Tr(\phi^2) - \frac{\nu^2}{2} H^+ \cdot H + \frac{\lambda_1}{4} (Tr\phi^2)^2 + \frac{\lambda_2}{2} Tr(\phi^4)$$

$$- \frac{\lambda_3}{2} (H^+ \cdot H)^2 - \frac{\lambda_4}{2} H^+ \cdot H \, Tr(\phi^2) - \frac{\lambda_5}{2} H^+ \phi^2 H \tag{6}$$

The grand unified Lagrangian, the minimum one needed for the job, now begins to look a little less comforting. It is

$$L = - \frac{Tr}{2} (\partial_\mu A_\nu - \partial_\nu A_\mu - ig[A_\mu, A_\nu])^2$$

$$
\begin{aligned}
&- \frac{Tr}{2}\left(\partial_\mu \phi - ig\,[A_\mu, \phi]\right)^2 \\
&- \left|\partial_\mu H - ig\, A_\mu H\right|^2 \\
&- \bar{\psi}_R \gamma_\mu D_\mu \psi_R \quad - \quad \bar{\psi}_L \gamma_\mu D_\mu \psi_L \\
&- V \\
&- \sqrt{2}\, h \left(\bar{\psi}_{L\alpha\beta}\, \psi_R^\alpha H^\beta + h.c.\right) \\
&- \frac{h'}{4}\, \varepsilon_{\alpha\beta\mu\nu\lambda} \left(\tilde{\psi}_L^{\alpha\beta} C^{-1} \psi_L^{\mu\nu} H^\lambda + h.c.\right)
\end{aligned}
\qquad (7)
$$

Setting aside temporarily the question of the number of arbitrary parameters in this grand Lagrangian, let me note the few immediate consequences.

(i) Grand unification fixes, <u>at grand unification energies</u>,[19]

$$\sin^2 \theta_0 = \frac{3}{8}$$

This is most simply seen from the 5_R representation

	I_{3W}	Y ($\equiv U(1)$ charge)	Q
d_1	0	$-1/\sqrt{15}$	$-1/3$
d_2	0	$-1/\sqrt{15}$	$-1/3$
d_3	0	$-1/\sqrt{15}$	$-1/3$
e^+	$1/2$	$3/(2\sqrt{15})$	1
ν^c	$-1/2$	$3/(2\sqrt{15})$	0

The normalization of Y (which transforms like $\lambda/2$) follows from the usual requirements $Tr(\lambda_i) = 0$, $Tr(\lambda_i \lambda_j) = 2\,\delta_{ij}$. From the above table it is clear that

$$Q = I_3 + \sqrt{\frac{5}{3}}\, Y \qquad (8)$$

By the definition of mixing angles, we find

$$\cot \vartheta_0 = \sqrt{\frac{5}{3}}$$

and the above result for $\sin^2 \vartheta_0$ follows. The result is a clear consequence of the usual fractional charge assignments for quarks and integral ones for the lepton.

At present energies, of course, $\sin^2 \vartheta$ differs from 3/8 as a result of renormalization effects, as we shall see.

(ii) Grand unification fixes a scale for M_x.

To see this recall that the symmetry breakdown of the grand unifying group SU(5), say, into the subgroup SU(3) x SU(2) x U(1) is entirely due to X- (and Y-) bosons having acquired a mass $M_x(=M_y)$. At low energies, of scale m, the difference between g_1 and g_2 must be attributed to this M_x. Since all renormalization effects are proportional to logarithmic scales, we expect that

$$g_1 - g_2 \propto \log \frac{M_x}{m}$$

More precisely, the typical relation reads[19] (i=1,2,3)

$$\frac{1}{g_i^2(m)} = \frac{1}{\bar{g}^2(m)} + b_i \log \frac{M_x}{m} + a_i \tag{9}$$

where $\bar{g}^{-2}(m)$ is the symmetric coupling constant at the low energy point, m, and b_i and a_i are calculable numbers.

But for the present energies, even SU(2) is broken so that the W-boson has also acquired mass and the relation (9) should properly be generalized to[20]

$$\frac{1}{g_i^2(m)} = \frac{1}{\bar{g}^2(m)} + b_i \log \frac{M_X}{m} + b_i' \log \frac{M_W}{m} + a_i \qquad (10)$$

This is a result, involving nested gauge hierarchy, which we have very recently been able to derive, based on a new renormalization group analysis.

4. Nested Hierarchies: A Theorem

The result may best be presented in the form of a theorem:[20,21]

- Let G be broken down through <u>a</u> stages of hierarchy into the remaining subgroup $g_1 \times g_2 \times g_3 \times \ldots$;
- Let the gauge bosons which acquired mass, M_a, at the <u>a</u>th stage of hierarchy be labelled X_a, Y_a ($M_1 > M_2 > M_3 \ldots$)
- Define

$$c_i = \sum_{jk} f_{ijk} f_{ijk} \text{ (summed over } g_i\text{)} \qquad (11)$$

$$c_i^a = \sum_{X_a, Y_a} f_{iX_aY_a} f_{iX_aY_a} \qquad (12)$$

so that

$$c_i + \sum_a c_i^a = C_2(G) \qquad (13)$$

Then

$$\frac{16\pi^2}{g_i^2(m)} = \frac{16\pi^2}{\bar{g}^2(m)} + \sum_a c_i^a \left[\frac{22}{3}\log\frac{M_a}{m} - \frac{1}{3}\right] + c_i\left[\frac{10s}{3\sqrt{3}} - \frac{76}{9}\right]$$
$$- \sum_H T_H(R)\left[\frac{2}{3}\log\frac{M_H}{m}\right] - \sum_h T_H(R)\left[-\frac{8}{9}\right]$$
$$- \sum_F T(R)\left[\frac{4}{3}\log\frac{M_F}{m} - \frac{1}{3}\right] - \sum_f T(R)\left[-\frac{10}{9}\right] \qquad (14)$$

$s = 2.029884\ldots$

for <u>every</u> range of m that satisfies the constraints

$m \ll M_a, M_H, M_F$

$m \gg$ remaining gauge bosons of $g_1 \times g_2 \times g_3 \times \ldots$

" light Higgs, fermion masses

In eq. (14), $T_H(R)$ and $T\ (R)$ are representation dependent group theory numbers.

For SU(5), we may use as inputs

(i) $\alpha \equiv e^2/(4\pi)$, where

$$\frac{1}{e^2} = \frac{1}{g_2^2} + \cot^2\theta_0 \frac{1}{g_1^2} \tag{15}$$

and specify = 1/137.036 at m_e scale, resulting in $\alpha(m)$ = 1/133.058 when m = 6 GeV (it being intermediate between the end of charmonium and the onslaught of the upsilon).

(ii) m_W = 85 GeV, based on our current low energy estimates.

(iii) and $\alpha_s \equiv g_3^2/(4\pi)$.

For the range of presently allowed values of $\alpha_s(m)$, the corresponding values for M_x are given by:

$\alpha_s(m)$	M_x
.2	8.17×10^{14} GeV
.25	1.75×10^{15} "
.3	2.91×10^{15} "
.35	4.18×10^{15} "

(16)

Actually eq. (14) also allows us a direct calculation of $\sin^2\theta$ at m, viz.

$$\sin^2\theta(m) = \sin^2\theta_0 \cdot \frac{e^2(m)}{16\pi^2}\left\{ \frac{16\pi^2}{e^2(m)} + c_2 \cot^2\theta_0 \left[\frac{105}{3\sqrt{3}} - \frac{76}{9}\right] + \cot^2\theta_0 \sum_a (c_2^a - c_1^a)\left[\frac{22}{3}\log\frac{M_a}{m} - \frac{1}{3}\right]\right\} \tag{17}$$

For the same input parameters, we may list the values of $\sin^2\theta$ at 6 GeV,

$\alpha_s(m)$	$\sin^2\theta$ (m)
.2	.211
.25	.207
.3	.204
.35	.202

(18)

Both (16) and (18) are independent of the Higgs and fermion content of the theory.[22]

For those of you who recall the pessimistic value of ~.19 obtained earlier by Georgi, Quinn and Weinberg[19] based on the one hierarchy formula, the more realistic two hierarchy formula, obviously gives a definite improvement in $\sin^2\theta_W$. If we follow Marciano's argument,[23] then $\sin^2\theta_W$ as measured experimentally is to be related to $\sin^2\theta(m_W)$. The relation between $\sin^2\theta(m_W)$ and $\sin^2\theta$ at 6 GeV depends very much on detailed Higgs and fermion assignments. In any case, at the very least, present data is not inconsistent with this new improved value for $\sin^2\theta$.

Let us come back to the prediction for M_x. It should be clear by now that in a sense grand unification has passed its first test. It could have happened that the M_x thus determined is so low that proton stability requirements rule out grand unification immediately. Contrary to naive expectations, M_x is not an arbitrary tunable parameter in grand unification but is completely fixed by low energy input. The only leeway comes in the present uncertainty in the low energy α_s.

5. Proton Lifetime

With the present allowed range of values for M_x we turn next to the question of the prediction of proton lifetime in a grand unified model. Here, a naive calculation may be made of the effective 4-fermion operator

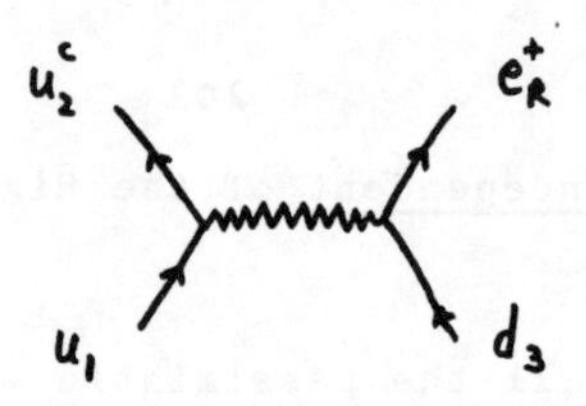

with the effective Fermi constant given by

$$\frac{"G_F"}{\sqrt{2}} = \frac{g_x^2}{8 M_x^2} \qquad (19)$$

By using for g_x the grand unified coupling g at M_x scale, eq. (19) has been used to give a naive estimate for τ_p.[19]

But eq. (19) is subject to many corrections. Some of them, involving gluon radiative corrections to g_x, have been included in earlier analyses.[18] There is however a glaring correction which must in principle be taken into account.

That is, eq. (19) involves the x- propagator, which apart from spin, is

$$\propto \frac{1}{p^2 + M_x^2 + \Sigma(p^2)} \tag{20}$$

where

$$\Sigma(p^2)\Big|_{p^2\approx -M_x^2} = O((p^2+M_x^2)^2) \tag{21}$$

At the X-pole, eq. (20) is correctly normalised. However the proton decay involves the X propagator for $p^2 \approx 1\ (\mathrm{GeV})^2$, at which point $\Sigma(p^2)$ is no longer zero but of order $g^2M^2 \log M^2/p^2 + \ldots$ Since M_x is so large there is a lot of room for the effective mass to evolve away from M_x.

The question then is whether the effective mass at low energies is appreciably <u>smaller</u> or larger than M_x. If M_{eff} is smaller, the naive estimate for τ_p may be already in trouble with present experimental limit.

To study this question requires a quantitative renormalisation group analysis of the masses. In a standard SU(5) theory, with a $\underline{5}$ and $\underline{24}$ Higgs, there are two mass scales and five quartic self-couplings. Since there are only loose constraints on these couplings,[24] no definitive statement on even the sign of the mass renormalisation effect appears possible.

For a quantitative estimate, we have turned to an asymptotically free, one coupling constant and one mass scale SU(5) model.[25] In this model, all coupling constants have been constrained by the eigenvalue conditions that preserve the asymptotic freedom of the theory.

6. Asymptotic Freedom

I will digress here for a moment and discuss a problem with the Higgs fields. As Gross and Wilczek[26] and, later, Cheng, Eichten and Li[27] have already pointed out, Higgs fields tend to destroy the fundamental beauty of the Yang-Mills field, to wit, the asymptotic freedom of the theory.

Asymptotic freedom is the unique signature of a Yang-Mills field wherein the effective coupling constant $\bar{g}(t)$ vanishes as the logarithmic energy scale ,t, increases.[28] Physically, it reflects on the remarkable anti-screening effect which occurs in vacuum polarisation of the gauge boson.[29]

In the presence of a Higgs, whose self-coupling tend to "blow-up" at some finite, positive, t, the $\bar{g}$ loses its asymptotic freedom. This is because in the renormalisation equation

$$\frac{d\bar{g}}{dt} = - b\bar{g}^3 - a\bar{g}^5 + c\bar{g}^3\lambda + \ldots$$

and the source of the threat becomes clear. Diagrammatically, the dangerous graph is

But there is a way out of this. By imposing a set of eigenvalue conditions[30]

$$\lambda \propto g^2$$

$$h \propto g$$

(h being the Yukawa coupling to the fermions) asymptotic freedom can be saved.

And what have we gained from saving this asymptotic freedom? A true unified field theory with one coupling constant! There are some who have argued that they could live with "temporary freedom", and suppose that the "blow-up" would be cured somehow upon inclusion of gravity. In such a standard approach, the grand unified theories enjoy a proliferation of constants which are a delight to phenomenologists.

Be that as it may, as I said at the beginning, an increasing number of people are converting to the point of view that Higgs are not elementary but are composite fermion-antifermion excitations. In that not-too-distant future when a precise working dynamical symmetry-breakdown theory is at hand, the fundamental Lagrangian will be simple and elegant. That theory will most likely still be a gauge theory and it will have asymptotic freedom.

In that theory all Higgs couplings will be induced couplings and are fully calculable. I suspect that the result of those calculations will reproduce the eigenvalue solutions that were necessary to preserve asymptotic freedom. A tantalising clue in this direction may be found in the work of Mahanthappa and Lemmon[31] who have found that the eigenvalue conditions are equivalent to the condition that the wave function renormalisation constant for the Higgs be zero.

7. Asymptotically Free, One Coupling Constant, One Mass Scale SU(5) Model

Ashok Das, Juan Perez-Mercader and I have recently launched into a massive program of study of SU(5). One aspect of our study was the search, along the lines that I have just described, for a truly unified field theory model, with one coupling constant.[25]

The question we asked was, could the SU(5) model be made asymptotically free? The answer is no, if only light fermions (belonging to $\underline{5}_R$ and $\underline{10}_L$) are present. To obtain asymptotic freedom, it is necessary to introduce heavy fermions.[32] (This is reminiscent of the regulator fields needed for renormalisability.) The heavy fermions will have masses of the scale of M_X.

For economy, we assumed the heavy fermions to be "supersymmetric" with the Higgs, i.e. in the $\underline{5}$ and $\underline{24}$. I am happy to report, hot off the press, a new, realistic, asymptotically free, true one coupling constant, one mass scale SU(5) model, with three generations of light fermions. The low energy content of this theory is thus identical to the standard SU(5) model.

The Lagrangian is

$$
\begin{aligned}
\mathcal{L} = & -\frac{\mathrm{Tr}}{2}\left(\partial_\mu A_\nu - \partial_\nu A_\mu - ig\,[A_\mu, A_\nu]\right)^2 \\
& -\frac{\mathrm{Tr}}{2}\left(\partial_\mu \phi - ig\,[A_\mu, \phi]\right)^2 \\
& -\left|\partial_\mu H - ig A_\mu H\right|^2 \\
& +\frac{\mu^2}{2}\mathrm{Tr}(\phi^2) + \frac{\nu^2}{2} H^\dagger \cdot H
\end{aligned}
$$

$$
\begin{aligned}
& - \frac{\lambda_1}{4}\left(\mathrm{Tr}(\phi^2)\right)^2 - \frac{\lambda_2}{2}\mathrm{Tr}(\phi^4) - \frac{\lambda_3}{2}(H^\dagger \cdot H)^2 \\
& - \frac{\lambda_4}{2} H^\dagger \cdot H\, \mathrm{Tr}(\phi^2) - \frac{\lambda_5}{2} H^\dagger \phi^2 H \\
& - \sum_{\text{generations}} \bar{\Psi}_R \gamma_\mu D_\mu \Psi_R - \sum_{\text{generations}} \bar{\Psi}_L \gamma_\mu D_\mu \Psi_L \\
& - \bar{\chi}\gamma_\mu D_\mu \chi - \bar{B}\gamma_\mu D_\mu B \\
& - \sqrt{2}\, h\left(\bar{\Psi}_{L\alpha\beta}\Psi_R^\alpha H^\beta + h.c.\right) - \frac{h'}{4}\varepsilon_{\alpha\beta\mu\nu\lambda}\left(\tilde{\Psi}_L^{\alpha\beta} C^{-1}\Psi_L^{\mu\nu} H^\lambda + h.c.\right) \\
& - k_2 \bar{\chi}_\alpha \chi^\beta \phi_\beta^\alpha - k_4\left(\bar{B}_\beta^\alpha \chi^\beta H_\alpha^\dagger + h.c.\right) \\
& - k_5 \bar{B}_\beta^\alpha B_\gamma^\beta \phi_\alpha^\gamma - k_6 \bar{B}_\beta^\alpha B_\alpha^\gamma \phi_\gamma^\beta
\end{aligned} \tag{22}
$$

where

$$
\begin{aligned}
& h = -.254406g \quad k_2 = -.635486g \quad k_4 = -.942053g \quad k_5 = -.809565g \\
& h' = -.746771g \quad k_6 = .706639g \\
& \lambda_1 = .029244g^2 \quad \lambda_2 = .457611g^2 \quad \lambda_3 = 1.196053g^2 \quad \lambda_4 = -.012374g^2 \\
& \lambda_5 = .909170g^2 \\
& n_f = 3
\end{aligned} \tag{23}
$$

and

$$
\nu^2(t) = -.927207\, \mu^2(t) \tag{24}
$$

(In our earlier version, frankly, we forgot the abnormal h' coupling. As a result, n_f = no. of generations of light fermions, with u,d,e,ν being the first generation, had to be seven before asymptotic freedom could be saved.[25] Astrophysical data suggest that the number of 2-component neutrinos with the usual weak interactions is between 3 and 4.[33] I am glad to be able to report that an $n_f = 3$ solution to the renormalisation group equation exists.)

In principle, our grand Lagrangian is now totally determined, at grand unification energies. At this stage, only the X and Y bosons have mass while the gluons, W-bosons and the U(1) boson are all massless. The study of how the parameters evolve as we scale down the energy involves a massive study of the renormalisation program. In particular a study of the quartic couplings of the H^i (i=1,2,3) and the H^a (a=4,5) would show how and where at the final stage of hierarchy breakdown the W-bosons acquire mass. This problem is under study.

As a first priority project we have used our SU(5) model to study the proton lifetime.

8. Proton Lifetime Resumed

As I have pointed out earlier, a potentially critical factor in the estimate of proton lifetime involves a study of the actual mass renormalisation effects on M_X. Now that we have a concrete calculable SU(5) model with the same low energy content as the standard SU(5), we can have confidence that our predictions are not irrelevant to experiment.

You may wonder why in this asymptotically free SU(5) model the two apparently independent mass scales μ^2 and ν^2 are related. The eigenvalue conditions alone do not demand the relation. However, by a study of the renormalisation group equation for the two masses μ^2, ν^2 we find (upon substituting in the eigenvalues to simplify the equation)

$$16\pi^2 \frac{d\mu^2}{dt} = 13.419956\, \bar{g}^2 \mu^2 + .847298\, \bar{g}^2 \nu^2$$
$$16\pi^2 \frac{d\nu^2}{dt} = 8.134063\, \bar{g}^2 \mu^2 + 21.406988\, \bar{g}^2 \nu^2 \tag{25}$$

Eq. (25) in general have solutions which for t going to infinity become proportional to each other

$$\nu^2(t) \xrightarrow{t\to\infty} 10.353678\ \mu^2(t) \tag{26}$$

Such a behavior would destroy the $SU(3)_c$ as well as the SU(2) symmetry of the vacuum at high energies. There is, however, a special solution to the coupled set of equations which relates ν^2 to μ^2 under which the $SU(3)_c$ and SU(2) symmetry of the vacuum at high energies is preserved.

With these parameters and mass scales all fixed we can now directly calculate the equation for the effective mass of the X-boson

$$16\pi^2 \frac{d\tilde{M}^2}{dt} = \left(\frac{8}{3} n_f + 4.634335\right) \bar{g}^2 \tilde{M}^2 \tag{27}$$

(n_f = 3 here) so that we find

$$\frac{\tilde{M}^2\ (1\ \ \text{GeV})}{\tilde{M}^2\ (10^{15}\text{GeV})} = \frac{1}{4.1} \qquad (28)$$

Since $\tau_p \propto \tilde{M}^4$ it is therefore clear that mass renormalisation effects can SHORTEN the naive lifetime by a factor of ~ 16. For α_s = .2, the naive lifetime was, for this model, 3.1×10^{32} years

Fortunately, a complete renormalisation group analysis of g_X finds a compensating suppression of g_X at low energies. We have

$$\frac{g_X^4\ (1\ \ \text{GeV})}{g_X^4\ (10^{15}\text{GeV})} = \frac{1}{7.2} \qquad (29)$$

This, plus gluon, W and B exchange effects[34] of the type

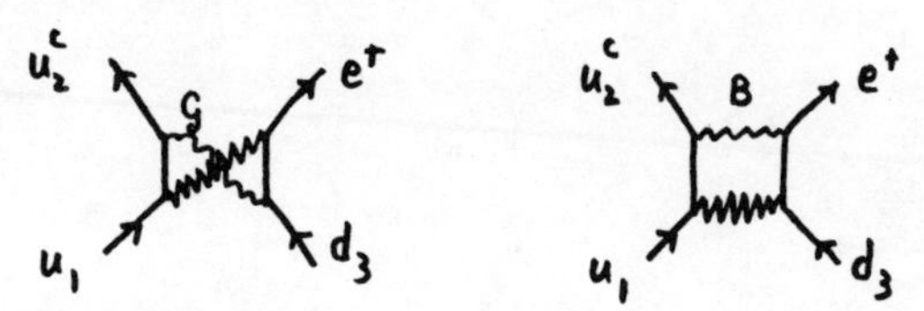

as well as fermion threshold and m_W threshold effects all lead finally[35] to a net 3.0×10^{31} years (vs. a naive 3.1×10^{32} yrs.)

To summarise our investigations, for SU(5), we have, with the input α = 1/137.036 at m_e scale, and a range of experimentally allowed values for $\alpha_s(m)$ chosen at m=6 GeV scale (chosen to be between the end of the charmonium threshold and before the onslaught of the uppsilon threshold):

	$\alpha_s(m)$	$\sin^2\theta(m)$	M_X (GeV)	τ_p (yrs.)
$n_f=3$	.2	.211	8.17×10^{14}	3.0×10^{31}
	.25	.207	1.75×10^{15}	4.7×10^{32}
	.3	.204	2.91×10^{15}	2.7×10^{33}
	.35	.202	4.18×10^{15}	7.9×10^{33}
$n_f=4$	.2	.211	8.17×10^{14}	1.9×10^{31}
	.25	.207	1.75×10^{15}	2.9×10^{32}
	.3	.204	2.91×10^{15}	1.6×10^{33}
	.35	.202	4.18×10^{15}	4.7×10^{33}

In this table I have also exhibited an n_f=4 solution that is asymptotically free, satisfying all the stability constraints. The eigenvalues are

$$h = -.0720385g \quad k_2 = -.700412g \quad k_4 = -.999376g \quad k_5 = -.838750g$$
$$h' = -.697300g \quad k_6 = .715954g \tag{30}$$

$$\lambda_1 = .030388g^2 \quad \lambda_2 = .511836g^2 \quad \lambda_3 = 1.365444g^2 \quad \lambda_4 = -.008312g^2 \tag{31}$$

$\lambda_5 = 1.083236g^2$

$n_f = 4$

$$\nu^2(t) = -1.124779\ \mu^2(t) \tag{32}$$

The renormalisation program that we have used is a new program which makes use of the minimal scheme of 't Hooft and Veltman. It takes care, very nicely, of the large broken symmetry mass, M_X, in the theory.

The calculation of τ_p reported here is but an example of the kind of calculation needed to confront SU(5) or any other grand unified model with experiment. The naive estimates must be supplemented with more definitive calculations before we can say that SU(5) is ruled out or confirmed by the anticipated measurement of the proton lifetime, hopefully soon to be reported in the Reference News as well as in the New York Times.

9. SU(N) Grand Unification

Flushed with success in SU(5), and taking note of the lack of prediction with respect to generations, Georgi[36] has taken on a higher SU(N) grand unification. He has laid down three guiding principles in the search for anomaly free theories

(i) Representations of left-handed fermions are <u>real</u> with respect to SU(3) color subgroup (i.e. equal number of 3 and 3* in the total reducible SU(N) representation)

(ii) Representations of left-handed fermions are <u>complex</u> with respect to SU(3) x SU(2) x U(1)

(iii) No irreducible representation of SU(N) should appear more than once in the total reducible, SU(N) representation.

With these rules, he finds the lowest candidate for 3 generations is SU(11), with the representation

$$\psi = [4] \oplus [8] \oplus [9] \oplus [10]$$

where $[p]$ denotes the totally antisymmetric tensor of rank p.

Under the reduction $SU(11) \supset SU(6) \times SU(5)$; the content reads

$$[4] = (1,5) + (6,10^*) + (15,10) + (20,5) + (15,1)$$

$$[8] = (15,5^*) + (6,10^*) + (1,10) + (20,1)$$

$$[9] = (6,5^*) + (1,10^*) + (15,1)$$

$$[10] = (1,5^*) + (6,1)$$

In this accounting, there are sixteen SU(5) 10 - plets and thirteen SU(5) 10*-plet. Assuming Higgs can be found to combine the 10_L 10^*_L into a heavy SU(5) 10-plet, there will thus be three SU(5) 10-plet left over. Similarly there are twenty-three SU(5) 5*-plet, and twenty SU(5) 5 plet, so that again three 5*-plet are left over.

This is the basis for the claim that 3 generations are present in SU(11). The efficiency ratio ($\epsilon \equiv$ no. of light/(total no. of fermions)) is clearly very low, being 45/561.

Frampton[37] has improved that ratio by allowing for repetitions. For SU(9), for three generations

$$\psi = [2] + [3] + [4] + [6] + \underbrace{[8] + [8] + \cdots}_{10\ \text{times}}$$

with ϵ = 45/420.

Recently Chakrabarti, Mohapatra and Popovic[38] have relaxed the working rules even further and found an SU(8) example, with n_f=3.

$$\psi = [3] \oplus [6] \oplus [7]$$

with the reduction under SU(8) $\supset$ SU(3) x SU(5), the first SU(3) being the horizontal group, and

$$[3] = (1,1) + (3^*,5) + (3,10) + (1,10^*)$$
$$[6] = (3,1) + (3^*,5^*) + (1,10^*)$$
$$[7] = (3^*,1) + (1,5^*)$$

Their claim is that the (3*,5) and (1,5*) can combine into a heavy 5-plet leaving two left-handed 5 plet which combine with the two right-handed 10-plet to form heavy fermions.

These forays into SU(N) are in my opinion to be viewed as exploratory and it is not clear what rewards have been obtained at the price of considerable proliferation in the Higgses needed to do the job.

10. B-L

In this talk I have not covered many phenomenological aspects of grand unification since they have been so vividly reviewed and portrayed by Professor Zee.[3] There is however one aspect of grand unification involving recent work with Lay-Nam Chang[39] which I will mention. The details will be discussed by him in a separate report.

This involves the issue of B-L nonconservation in physics. As Weinberg[40] and independently Wilczek and Zee[41] have already observed, the local 4-fermion operator responsible for proton decay that is invariant under SU(3)×SU(2)×U(1) necessarily conserves B-L. Our observation is simply that a mechanism exists whereby B-L violation can even occur without breaking SU(5).

An interesting consequence of B-L nonconservation is that the neutron n can, with time, evolve into an $\bar{n}$. Since $\bar{n}$ can immediately annihilate with the other nucleons in matter, neutron oscillation can contribute effectively to nuclear instability

$$
\begin{aligned}
(A,Z) &\longrightarrow (A-2,Z) + \pi^+\pi^- \\
&\longrightarrow (A-2,Z-1) + \pi^+\pi^0
\end{aligned}
$$

The lifetime for this process could be of order 10^{30} years. But I refer you to Lay-Nam's talk for further details.

11 . Conclusion

What I have reported in my talk today is but a small part of the total explosion of activities on grand unification to date. SO(10), SO(14) etc. grand unification and cosmological implications of grand unification I leave to Professor Zee.[3]

As you have already gathered, a potential embarrassment in all grand unification attempts is the concomitant proliferation of Higgs and associated coupling constants. If in trying to understand the generation problem,say, we involve fifteen new parameters then it is not so convincing to say that we have understood the generation problem.

One way out which you will find increasingly quoted in the literature is dynamical symmetry breaking. I think that is undoubtedly the correct way out, eventually. What I have tried to present is an alternative way which already allows you to make calculations in <u>anticipation</u> of that ultimate correct theory.

That a realistic (three generations) asymptotically free SU(5) model exists and has all parameters fixed is, to me, a very encouraging sign. In principle, all low energy masses and parameters can now be determined through a complete analysis of the renormalisation program.

References

1. H. Georgi and S. L. Glashow, Phys. Rev. Lett. 32. 438 (1974).
   For a review see J. C. Pati's talk presented at the Einstein Centennial Symposium held in Jerusalem, March 1979. University of Maryland Preprint #80-006.
2. C. N. Yang and R. Mills, Phys. Rev. 96, 191 (1954).
3. A. Zee, Proceedings of the 1980 Canton Conference on Theoretical Particle Physics, Academia Sinica (1980).
4. T. D. Lee, Proceedings of the 1980 Canton Conference on Theoretical Particle Physics, Academia Sinica (1980).
5. S. Weinberg, Phys. Rev. Lett. 19, 1264 (1967).
   A. Salam, Elementary Particle Theory, Nobel Symposium, N. Svartholm, ed. (Wiley-Interscience, New York, 1968).
   S. Glashow, J. Iliopoulos, L. Maiani, Phys. Rev. D2, 1285 (1970).
6. M. Kobayashi and K. Maskawa, Prog. Theor. Phys. 49, 652 (1973).
7. For an up-to-date review, see Y. Ne'eman, Proceedings of 1979 Marshak Workshop on Weak Interactions, Dec 13-15, 1979, Blacksburg, Va.
8. Chou Kuang-chao, Gao Congshou, Proceedings of the 1980 Canton Conference on Theoretical Particle Physics, Academia Sinica (1980).
9. R. E. Marshak and R. N. Mohapatra, Proceedings of 1979 Marshak Workshop on Weak Interactions, Dec 13-15, 1979, Blacksburg, Va.
10. Luo Liao-fu, Lu Tan, Yang Guo-chen, Proceedings of the 1980 Canton Conference on Theoretical Particle Physics, Academia Sinica (1980).
11. S. Dimopoulos and L. Susskind, SLAC-PUB-2126 (1978).

12. Kerson Huang, Proceedings of the 1980 Canton Conference on Theoretical Particle Physics, Academia Sinica (1980).

13. Ling-Fong Li, Proceedings of the 1980 Canton Conference on Theoretical Particle Physics, Academia Sinica (1980).

14. Ni Guang-jiong, Su Ru-keng, Proceedings of the 1980 Canton Conference on Theoretical Particle Physics, Academia Sinica (1980).

15. For a recent calculation with dynamical Higgs mechanism, see A. Carter and H. Pagels, Phys. Rev. Lett. 43, 1845 (1979).

16. S. L. Adler, Phys. Rev. 177, 2426 (1969).
J. S. Bell and R. Jackiw, Nuovo Cimento 60, 47 (1969).
W. A. Bardeen, Phys. Rev. 184, 1848 (1969).

17. P. W. Higgs, Phys. Lett. 12, 132 (1964).

18. A. Buras, J. Ellis, M. K. Gaillard, D. Nanopoulos, Nucl. Phys. B135, 66 (1978).

19. H. Georgi, H. Quinn and S. Weinberg, Phys. Rev. Lett. 33, 451 (1974).

20. N. P. Chang. A. Das and J. Perez-Mercader, "New Renormalisation Program for Broken Gauge Theories". CCNY-HEP-79-21; "Unification Formula for Nested Gauge Hierarchies", CCNY-HEP-79-22.

21. This theorem is valid only in the one loop approximation in the renormalisation group equations. Also, to simplify the presentation of the theorem, we have ignored the fine structure in the mass splittings in the heavy Higgs and heavy fermion families.

22. This is true to the extent that we have ignored the fine structures as mentioned in ref.13. The effect of including the fine structure is entirely negligible.

23. W. Marciano, Phys. Rev. D20, 274 (1979).

24. N. Cabibbo, L. Maiani, G. Parisi, R. Petronzio, CERN TH-2683, June, 1979.

25. N. P. Chang, A. Das, J. Perez-Mercader, "An Asymptotically Free, One Coupling Constant, One Mass Scale SU(5) Model", CCNY-HEP-79-23; and most recently, "An Asymptotically Free SU(5) Model with Three Generations", CCNY-HEP-80-4.

26. D. Gross and F. Wilczek, Phys. Rev. D8, 3633 (1973).

27. T. P. Cheng, E. Eichten, L. F. Li, Phys. Rev. D9, 2259 (1974).

28. D. Gross and F. Wilczek, Phys. Rev. Lett. 30, 1343 (1973); loc. cit.
H. D. Politzer, Phys. Rev. Lett. 30, 1346 (1973).

29. N. P. Chang, invited talk, Proceedings of the 1978 International Meeting on Frontier of Physics, ed. K. K. Phua, C. K. Chew, Y. K. Lim (Singapore National Academy of Science), 1978.

30. N. P. Chang, Phys. Rev. D10, 2706 (1974).
M. Suzuki, Nucl. Phys. B83, 269 (1974).
E. Ma, Phys. Rev. D11, 322 (1975); Phys. Lett. 62B, 347 (1976); Nucl. Phys. B116, 195 (1976).
E. S. Fradkin and O. K. Kalashnikov, J. Phys. A8, 1814 (1975); Phys. Lett. 59B, 159 (1975); ibid 64B, 177 (1976).
N. P. Chang and J. Perez-Mercader, Phys. Rev. D18, 4721 (1978).

31. J. Lemmon and K. T. Mahanthappa, Phys. Rev. D13, 2907 (1976).

32. E. S. Fradkin and O. K. Kalashnikov already found an asymptotically free SU(5) model in 1976 (Phys. Lett. 64B cited above in ref.22). The spectrum of heavy fermions however was more complicated than the one we chose.

33. G. Steigman, D. N. Schramm, J. E. Gunn, Phys. Lett. 66B, 202 (1977).

J. Yang, D. N. Schramm, G. Steigman, R. T. Rood, Astrophys. J. 227, 697 (1979).

For an excellent review see lectures given by G. Steigman and R. Wagoner, Summer School on Physical Cosmology, Les Houches, France, July 1979.

34. D. A. Ross, Nucl. Phys. B140, 1 (1978).

35. N. P. Chang, A. Das, J. Perez-Mercader, "Proton Stability in an Asymptotically Free SU(5) Model", CCNY-HEP-79-24; "On Proton Decay", CCNY-HEP-79-25.

36. H. Georgi, Nucl. Phys. B156, 126 (1979).

37. P. Frampton, Phys. Rev. Lett. 43, 1460 (1979).

38. J. Chakrabarti, R. N. Mohapatra, M. Popovic, "Problems of Fermion Generations in Grand Unified Theories", CCNY-HEP-79-12.

39. Lay-Nam Chang, Proceedings of the 1980 Canton Conference on Theoretical Particle Physics, Academia Sinica (1980).

40. S. Weinberg, Phys. Rev. Lett. 43, 1566 (1979).

41. F. Wilczek and A. Zee, Phys. Rev. Lett. 43, 1571 (1979).

QUANTUM FIELD THEORY OF COMPOSITE PARTICLES

Ruan Tu-nan

Department of Modern Physics,

University of Science and Technology of China

Ho Tso-hsiu

Institute of Theoretical Physics,

Academia Sinica

Abstract

The quantum field theory of composite particles is reformulated in terms of covariant equal-time two-particle operators. Because of the use of the newly derived equal-time dynamical equation of the composite particles, the question connected with the hermiticity of the destruction and creation operators has been solved. The difficulties arising from the "ghost" of B-S equation are also avoided. The theoretical studies and the applications of quantum theory of composite particles are summarized.

I. INTRODUCTION

Quantum field theory has been developed as a dynamical theory describing the elementary particles. Recently there appear even more experimental evidences indicating that hadrons are composed of more elementary parts, and there is increasing interest to extend the investigations of atoms or nuclei of atoms to high, or super-high energies. Therefore, it is necessary to develop the quantum field theory which enables us to describe various processes involving annihilation or creation of composite particles.[1-7] Obviously, a theory describing interactions between quarks and gluons is not enough in order to analyse the dynamics of hadron interactions in composite particles.

The current quantum field theory is a local theory based on the "point-like" model. Attempts to extend point-like particle to particle with finite size have ended in failure. The reason is that a nonlocal field theory would be inconsistent with microcausality condition and would lead to conflicts with the principle of special relativity, C.P.T. theorem, etc....

In the last few years, based on the Bethe-Salpeter equation and Lehman-Symanzik-Zimmermann's approach of field theory, a group of theorists in China has developed a quantum field theory of composite particles, which, as it was shown, is a local theory.[1-7] The perturbation expansion and the unperturbative method of this field theory have been formulated in [10-14], which can be applied in many cases. But the main difficulty of solving the B-S equation strongly restricts the application of this field theory. In this paper, we shall show that a quantum field theory of composite particles can be reformulated in terms of covariant equal-time operators of two particles. Because the dynamical equation of composite particles in accordance with the two particle operators used in this formulation has been derived [15-16], the problem concerning the hermiticity in the definition of destruction and creation operators of composite particles [17] and the difficulties arising from the "ghost" states in the B.S. equation are solved thoroughly, and meanwhile the advantages of the theory previously formulated remain intact. All this makes this new formulation more attractive and effective in its applications.

This paper is divided into six paragraphs. In the second paragraph the derivation of new relativistic equation of composite particles will be given concisely. In the third paragraph the postulates, on which the field theory is based, and their compatibility will be discussed. In the fourth paragraph various reduction formulas of the S-matrix and the field-current relations have been derived, and a discussion on the locality of the S-matrix is presented. In the fifth paragraph a summary on the applications of the field theory of composite particles in various problems including atoms, nuclei and structures of hadrons is given, and in the last paragraph we shall give an outlook for the development of this field theory and discuss briefly its relations with the quark confinement.

II. THE RELATIVISTICALLY COVARIANT DYNAMICAL EQUATION OF COMPOSITE PARTICLES IN C.M.SYSTEM WITH EQUAL TIME.

The advantage of the B-S equation is that the wave function of composite particle can be defined as a T-product

$$\chi^{BS}_{\vec{P}\varsigma}(x_1 x_2) \equiv \chi_{\vec{P}\varsigma}(X, x) = \langle 0 | T(\psi(X+\frac{x}{2})\bar{\psi}(X-\frac{x}{2})) | \vec{P}\varsigma \rangle \quad , \qquad (2.1)$$

which enables us to derive the dynamical equation of composite particles by using the procedures of Feynman's field theory and to preserve all the conveniences of it. But the disadvantage of the B-S equation is connected with multi-times which also arisen from a T-product. When the space-time interval $x = x_1 - x_2$ becomes time-like, it seems impossible to understand its physical meaning and the ghost state of the B-S equation does appear in this case. From the physical point of view the necessary condition for the existence of bound states may be stated as: at given coordinates the two particles must be observed at the same time. In other words, in a physical wave function of a composite particle the space-time interval of particles, of which the composite particle is composed, must be space-like. This may be expressed as:

$$\phi_{phys.}(x_1 x_2) = \theta(x^2) \langle 0 | T(\psi(X+\frac{x}{2})\bar{\psi}(X-\frac{x}{2})) | \vec{P}\varsigma \rangle \quad , \qquad (2.2)$$

or, in the c.m. system, as:

$$\phi_{\vec{P}\varsigma}(t, \vec{x}_1 \vec{x}_2) = \delta(x_{10} - x_{20}) \langle 0 \; T(\psi(x_1)\bar{\psi}(x_2)) | \vec{P}\varsigma \rangle \quad , \qquad (2.3)$$

and the covariant form of (2.3) is

$$\phi_{\vec{P}\varsigma}(x_1 x_2) = \delta(-ix\frac{\partial}{\partial X}) \langle 0 \; T(\psi(X+\frac{x}{2})\bar{\psi}(X-\frac{x}{2})) | \vec{P}\varsigma \rangle \quad . \qquad (2.4)$$

But the two-particle operator, as it is usually defined in composite particle field theory, expressed as a T-product, is not a hermitian operator:

$$T(\psi(x_2)\bar{\psi}(x_1))-\gamma_4(T(\psi(x_1)\bar{\psi}(x_2)))^{+}\gamma_4=-\varepsilon(t_1-t_2)\{\psi(x_2),\ \bar{\psi}(x_1)\}. \quad (2.5)$$

The difference between the operator and its hermitian conjugate is a time-like anti-commutator. Hence if a T-product is defined to describe the π - meson, the operator of the π -meson will not be hermitian.[17] To overcome this difficulty, one has to introduce a certain conjugate function W(p,x), and its hermitian $\bar{W}$(p,x). They should be zero when the space-time interval becomes time-like.[6] However, it is more natural and reasonable to define field operator of a composite particle as (2.2) or (2.4), and the wave function defined as (2.2) or (2.4) is but a projection of the B-S equation. Lately, we have derived an equation of this type for two particles with spins ($\frac{1}{2}$ - $\overline{\frac{1}{2}}$), ($\frac{1}{2}$ - $\frac{1}{2}$), (0 - $\frac{1}{2}$) and (0 - 0).[15-16] These results have demonstrated that an equation of space-like or equal-time wave function may be completely separated from the B-S equation. After the solution of this equation has been found, which presents the physical wave functions of composite particles, it is also shown that the time-like part of the B-S wave function will be determined uniquely by the space-like wave function. Since the time-like part has been taken away from the B-S equation, the ghost states will not appear in the equation of physical wave functions, and the corresponding two-particle operator will be hermitian. It seems that the ghost state occurs when one is going to handle the covariant Feynman's field theory by extending the time interval of integration, that is, to construct a T-product. The ghost state does appear, when the scattering amplitude of physical particles on mass shell has been continued to bound state off mass shell amplitude. Hence if the two particle operator in the field theory is defined as (2.2) (2.3) or (2.4), the ghost state will not appear at all. This means that the ghost state is not an eigenstate of the total Hamiltonian. All the advantages of the B-S equation using the Feynman's field theory and its procedure of renormalization will be preserved in the newly derived three-dimensional equation, as this equation is separated from the B-S equation. However, on the other hand, the three-dimensional equation will be more easily solved.

First of all, we shall give briefly in the following the derivation of equal-time equation of composite particles.

The physical wave function, defined as in (2.4), satisfies the following equation:

$$[\vec{\mathcal{D}} - \vec{\mathcal{V}}]\, \phi_{\vec{P}\varsigma}(x_1 x_2) \equiv [i\hat{P}(m_1 + \hat{\partial}_1) - (m_2 - \hat{\partial}_2) i\hat{P}]\phi_{\vec{P}\varsigma}(x_1 x_2) \tag{2.6}$$

$$- \int d^4(y_1 y_2)\, \delta(Px)\, V(x_1 x_2, y_2 y_1)\, \phi_{\vec{P}\varsigma}(y_1 y_2) = 0 \,,$$

where

$$V(x_1 x_2\, y_2 y_1) = \int d^4(u_1 u_2 v_1 v_2)\, S_{ff}(x_1 x_2\, u_2 u_1)\, I'(u_1 u_2 v_2 v_1) \tag{2.7}$$

$$S_{RA}(v_1 v_2, y_2 y_1) \,,$$

$$I'(x_1 x_2\, y_2 y_1) = I(x_1 x_2, y_2 y_1) + \int d^4(u_1 u_2 v_1 v_2) I'(x_1 x_2 u_2 u_1)$$

$$[G_0(u_1 u_2 v_2 v_1) - G_0'(u_1 u_2 v_2 v_1)]\, I'(v_1 v_2 y_2 y_1) \,, \tag{2.8}$$

$$G_0(x_1 x_2 y_2 y_1) = - S_f'^{(1)}(x_1 - y_1)\, S_f'^{(2)}(y_2 - x_2) \,, \tag{2.9}$$

$$G_0'(x_1 x_2 y_2 y_1) = \int d^4(u_1 u_2) S_{RA}(x_1 x_2 u_2 u_1) \delta(P_u) G_0(u_1 u_2 y_2 y_1), \tag{2.10}$$

$$S_{ff}(x_1 x_2, y_2 y_1) = [i\hat{P}(m_1 + \hat{\partial}_{x_1}) - (m_2 - \hat{\partial}_{x_2}) i\hat{P}]\, G_0(x_1 x_2, y_2 y_1), \tag{2.11}$$

$$S_{RA}(x_1 x_2, y_2 y_1) = -[i\hat{P}(m_1 + \hat{\partial}_{x_1}) - (m_2 - \hat{\partial}_{x_2}) i\hat{P}]\, i S_R^{(1)}(x_1 - y_1)$$

$$\hat{P}^{(1)} \hat{P}^{(2)} S_A^{(2)}(y_2 - x_2) \,, \tag{2.12}$$

where I denotes the kernel of integration of the B-S equation and presents the summation of all irreducible graphs. The B-S wave function can be found out from the physical wave function ϕ (x_1 x_2) as follows:

$$\chi_{\vec{P}\xi}(x_1 x_2) = \int d^4(u_1 u_2 v_1 v_2 y_1 y_2) G_0'(x_1 x_2 u_2 u_1) I'(u_1 u_2 v_2 v_1)$$

$$S_{RA}(v_1 v_2 y_2 y_1)\, \phi_{\vec{P}\xi}(y_1 y_2) \quad . \tag{2.13}$$

As can be seen, this provides a method for solving the B-S equation. Furthermore, the following normalization condition and the Green's function of (2.6) are also needed:

$$\int d^3X\, d^4(xy)\phi_{\vec{P}\xi}(X,x)Q_{\vec{P}\xi\vec{P}'\xi'}(x,y)\, i\frac{\overleftrightarrow{\partial}}{\partial X_0}\phi_{\vec{P}\xi}(X,y) \tag{2.14}$$

$$= \delta(\vec{P}-\vec{P}')\delta_{\xi\xi'}\begin{cases} \delta_{m_\xi m_{\xi'}} & \text{discrete states,} \\ \delta(m_\xi - m'_\xi) & \text{continual states,} \end{cases}$$

where

$$Q_{\vec{P}\xi\vec{P}'\xi'}(x,y) = \epsilon\frac{G^{-1}_{\vec{P}\xi}(x,y) - G^{-1}_{\vec{P}'\xi'}(x,y)}{2E_\xi(E_\xi - E_{\xi'})}\,, \quad \epsilon\delta(Px)\Big|_{PX=0} = 1\,, \tag{2.15}$$

and

$$G^{-1}_{\vec{P}\xi}(x,y) = \left[-P^2 + i\hat{P}(m_1+\hat{\partial}_x) - (m_2 - \hat{\partial}_x)i\hat{P}\right]\delta(x-y)$$

$$-\delta(Px)V_{\vec{P}\xi}(x,y)\,. \tag{2.16}$$

It should be noted that when $G^{-1}_{\vec{p}s}(x,y)$ lies on mass shell, we have $p^2 = -m_s^2$. Hence the Green's function of (2.6) will be:

$$G(x_1x_2, y_2y_1) = \frac{1}{(2\pi)^4}\int d^4P\, e^{iP(X-Y)}G_P(x,y)\,, \tag{2.17}$$

$$G_P(x,y) = \sum_\xi \frac{\phi_{\vec{P}\xi}(x)\bar{\phi}_{\vec{P}\xi}(y)}{2E_\xi(P_0 - E_\xi)} + R_P(x,y)\,. \tag{2.18}$$

But $\int d^4(u_1u_2)G(x_1x_2u_2u_1)G^{-1}(u_1u_2, y_2y_1)$

$$= \int d^4(u_1u_2)G^{-1}(x_1x_2, u_2u_1)G(u_1u_2, y_2y_1) = \delta\left(-ix\frac{\partial}{\partial X}\right)\delta(X-Y)\delta(x-y), \tag{2.19}$$

where

$$G^{-1}(x_1x_2, y_2y_1) \equiv (\vec{\mathcal{D}} - \vec{\mathcal{V}})\delta(X-Y) = \delta(X-Y)(\overleftarrow{\mathcal{D}} - \overleftarrow{\mathcal{V}})\,. \tag{2.20}$$

III. ON SOME FUNDAMENTAL ASSUMPTIONS AND THEIR COMPATIBILITY IN COMPOSITE PARTICLE QUANTUM FIELD THEORY

In order to formulate the quantum field theory of composite particles it is necessary to know how to define the basic vectors in Hilbert space. As far as the usual field theory is concerned, it is natural to choose the physical single-particle states as the basic vectors. But in cases where the physical composite particles do exist, the basic vectors must be generalized. Physically all the states with single-physical particles and composite particles are eigenstates of the total Hamiltonian. However, in the usual L-S-Z field theory which was based on Heisenberg picture, only the single-physical particle states are included in basic vectors. Therefore, a question might be raised: will the generalization to physical composite particles lead to some contradictions? As we know that, in L-S-Z field theory it has been postulated that the basic field A(x) forms a complete set, but in addition to this another assumption is needed: the corresponding "in" and "out" states of A(x), that is $A^{in}(x)$ and $A^{out}(x)$ also form a complete set respectively. The completeness of the basic fields, or the interpolating fields, has been proved in the L-S-Z field theory, provided the completeness of the "in" or "out" fields is valid, but it can not be proved vice versa. This means that the Hilbert space spanned by the basic fields will be larger or equal to the Hilbert space spanned by the "in" fields or the "out" fields. This fact is in accordance with the following physical aspect:

as in L-S-Z field theory, instead of studying the concrete Lagrangian, we emphasize the general properties of the S-matrix, which are abstracted from the Lagrangian theory. On the other hand, the existence of bound states can not be deduced from the general existence of interaction between fields without knowing some details of this interaction. For example, an interaction with repulsive or slightly attractive force would not lead to formation of bound states, but with attractive force it does probably. Therefore, as we can not determine the asymptotic state from a general existence of the basic field, it is thus quite reasonable to add in the basic vectors physical composite particle states.

The second question concerning the formulation of the composite particle field theory is the asymptotic condition of a composite particle state. In Hagg-Nishijima-Zimmermann's field theory with only a single-composite particle, as stated by Zimmermann, [18] the asymptotic condition has been expressed as

$$B(X,x) \equiv T(A(x_1)A(x_2)) = B^{\substack{in \\ out}}(X,x) - \int \Delta_{\substack{R \\ A}}(M, X-Y) I(Y x) d^4Y . \tag{3.1}$$

Corresponding to the B-S equation, however, the field operators B (X,x) have to include not only all the excited states of a composite particle, but also two-particle scattering states. Redmond and Uretski have shown explicitly that the single-particle-field operator has to be orthogonal to the two-particle's scattering states! [19] This misleading conclusion results from the incorrect description of the asymptotic condition of a composite particle. It is true that when $T \equiv X_0 \longrightarrow \pm\infty$, the composite field operators tend to all possible state of composite particle and two-particle states as a limit. Therefore, the question may be stated as: how to separate from composite field $B(X, x)$ the needed stable composite particle state. This can be done with the help of the normalization condition.

We have solved mainly two problems in formulating a field theory of composite particles. The main results will be presented in the following. For the sake of simplicity, we assume that there exist only spinor field $\psi(x)$ and its conjugate $\bar{\psi}(x)$. The composite particles are composed only of one particle and one antiparticle.

Postulate 1. There exist local, relativistically covariant and irreducible field operators $\psi(x)$ and $\bar{\psi}(x)$ with spin $\frac{1}{2}$ and the corresponding Hilbert space.

The same postulate is also needed in L-S-Z theory.

Postulate 2. Corresponding to the total energy-momentum operator P_μ there exist positively definite, nonzero discrete mass spectra m and Mi, i = 1,2,3, ... and continuous mass spectra. For single-particle state $|a\rangle$ we shall have

$$\langle 0|\psi(x)|a\rangle \neq 0, \quad \text{if } -P_\mu^2|a\rangle = m^2|a\rangle. \tag{3.2}$$

For two-particle state $|b_\xi\rangle$ we shall have

$$\left.\begin{array}{l}\langle 0|\psi(x)|b_\varsigma\rangle = 0 \\ \langle 0|B(X,x)|b_\varsigma\rangle \neq 0\end{array}\right\} \text{ if } \; -\mathbb{P}_\mu^2|b_\varsigma^i\rangle = M_i^2|b_\varsigma^i\rangle , \tag{3.3}$$

where ς denotes the spin and SU(3) indices, i = 1,2,3, ...
For the vacuum state we have

$$\mathbb{P}_\mu|0\rangle = 0, \langle 0|\psi(x)|0\rangle = 0 , \tag{3.4}$$

and

$$B(X,x) \equiv \delta(-ix\frac{\partial}{\partial X})N(\psi(x_1)\bar{\psi}(x_2)) \equiv \delta(-ix\frac{\partial}{\partial X})\cdot$$

$$T(\psi(x_1)\bar{\psi}(x_2)) - \delta(-ix\frac{\partial}{\partial X})\langle 0|T(\psi(x_1)\bar{\psi}(x_2))|0\rangle, \tag{3.5}$$

then $\langle 0|\, B\,(X,x)\,|\,0\rangle = 0.$ (3.6)

As we have seen, being a natural generalization of the field theory, postulate 2 defines the single-particle states and composite-particle states. But as we have noted, for a composite state the following relation holds:

$$\delta(-ix\frac{\partial}{\partial X})\delta(-iy\frac{\partial}{\partial Y})\langle 0|N(\psi(x_1)\bar{\psi}(x_2)\psi(y_1)\bar{\psi}(y_2))|b_\varsigma^i\rangle \neq 0. \tag{3.7}$$

This will lead to a question: which one of the two-particle state and the four-particle state does the $|b_\varsigma^i\rangle$ present? The answer is that we would define the composite-particle state as corresponding to a combination of the least number of basic fields, because the definitions (3.3) and (3.7) will give an equivalent S-matrix. [6, 20]

Some inferences would be drawn from these postulates.

From the locality of fields $\psi(x)$ we have

$$[B(X,x), B(Y,y)]=0, \text{ when } X\pm\frac{x}{2}\sim Y\pm\frac{y}{2} \quad . \tag{3.8}$$

From the invariance under inhomogeneous Lorentz transformation $x' = ax + b$ there will be

$$U(a,b)B(X,x)U^{-1}(a,b)=\Lambda(a)B(a^{-1}X-b, a^{-1}x)\Lambda^{-1}(a), \tag{3.9}$$

where U denotes the unitary transformation in infinite dimensional space, Λ the Lorentz transformation in the spinor space, a the parameter of four-dimensional rotation, b the parameter of translational transformation. For an infinitesimal translational transformation,

$$i\frac{\partial}{\partial X_\mu}B(X,x)=[\mathbb{P}_\mu, B(X,x)] \quad . \tag{3.10}$$

Postulate 3: Corresponding to basic fields ψ (x) and $\bar{\psi}$ (x), there exist "in" fields and "out" fields with the following asymptotic conditions :

$$\lim_{t\to\mp\infty}\langle\alpha|a_r(\vec{p},t)|\beta\rangle\longrightarrow\langle\alpha|a_r^{\substack{in\\out}}(\vec{p})|\beta\rangle, \tag{3.11}$$

$$\lim_{t\to\mp\infty}\langle\alpha|b_r^*(\vec{p},t)|\beta\rangle\longrightarrow\langle\alpha|b_r^{*\substack{in\\out}}(\vec{p})|\beta\rangle, \tag{3.12}$$

where

$$a_r(\vec{p},t)=\int d^3\vec{x}\,\bar{U}_r(\vec{p},x)\gamma_4\psi(x) , \tag{3.13}$$

$$b_r^*(\vec{p},t)=\int d^3\vec{x}\,\bar{V}_r(\vec{p},x)\gamma_4\psi(x) , \tag{3.14}$$

$\bar{U}$ and $\bar{V}$ are wave packets, $\langle\alpha|$ and $|\beta\rangle$ are arbitrary state vectors. The asymptotic conditions defined in this way hold in the sense of weak convergence.

The same postulate was adopted in the L-S-Z field theory. It is well-known that the Yang-Feldman equation may be deduced and that the "in" and "out" states may also be defined.

Postulate 4. Corresponding to two-particle field operators B(X,x) there exist the following asymptotic conditions: for stable bound states the asymptotic conditions are

$$\lim_{T\to\mp\infty}\langle\alpha|C_\xi(\vec{P},T)|\beta\rangle\to\langle\alpha|C_\xi^{\substack{in\\out}}(\vec{P})|\beta\rangle\ , \tag{3.15}$$

$$\lim_{T\to\mp\infty}\langle\alpha|C_\xi^*(\vec{P},T)|\beta\rangle\to\langle\alpha|C_\xi^{*\substack{in\\out}}(\vec{P})|\beta\rangle\ , \tag{3.16}$$

for unstable particles the asymptotic conditions will be

$$\lim_{T\to\mp\infty}\langle\alpha|C_\eta(\vec{P},T)|\beta\rangle\to 0, \tag{3.17}$$

$$\lim_{T\to\mp\infty}\langle\alpha|C_\eta^*(\vec{P},T)|\beta\rangle\to 0, \tag{3.18}$$

where

$$C_\xi(\vec{P},T)=\int\bar{\Phi}_{\vec{P}\xi}(X\,x)\,i\frac{\overleftrightarrow{\partial}}{\partial X_0}\,Q(x,y)B(X,y)d^3X\,d^4(x,y), \tag{3.19}$$

$$C_\xi^*(\vec{P},T)=\int B(X,-x)\,i\frac{\overleftrightarrow{\partial}}{\partial X_0}\,Q(x,y)\Phi_{\vec{P}\xi}(X,y)d^3X\,d^4(x,y), \tag{3.20}$$

and $T\equiv X_0$.

The same definitions as in (3.19) and (3.20) hold for C_η

and C_η^* only when $\varsigma \longrightarrow \eta$, and Q is the operator of normalization (see (2.15)). We emphasize that there is no ghost state in two-particle field operators defined as in (3.5), thus one does not need to introduce the assumption that all amplitudes of the ghost states vanish as $T \rightarrow \pm\infty$. As it is expected, all the solutions form a complete set, hence there will be

$$B(X,x)=\int d^3 P \sum_{\varsigma}\left(C_\varsigma(\vec{P},T)\phi_{\vec{P}\varsigma}(X,x)+C_\varsigma^*(\vec{P},T)\bar{\phi}_{\vec{P}\varsigma}(X,-x)\right), \tag{3.21}$$

in which the index ς denotes all the stable and unstable composite particles and the two particle scattering states. We also infer from postulate 4 that by using the definition of equal-time wave function and the condition of normalization, we have

$$\left[C_\varsigma^{\,\substack{in\\out}}(\vec{P})\right]^+ = C_\varsigma^{*\,\substack{in\\out}}(\vec{P}). \tag{3.22}$$

Therefore, the difficulty associated with the hermiticity of field operators has been avoided and the S-matrix will satisfy the condition of unitarity. Furthermore, from asymptotic conditions (3.15) and (3.16) and the commutation of field operators with total energy-momentum the state vectors of composite particles can be defined as the eigenstates of total Hamiltonian and the corresponding Yang-Feldman equation may also be derived. However, the same definition could not be applied to scattering states since the wave packets might have extended to infinity.[2]

Postulate 5. All the $a_r^{in}(\vec{p})$, $b_\gamma^{in}(\vec{p})$, $C_\xi^{in}(\vec{p})$ and $C^{*in}_\xi(\vec{p})$ (note that only the stable bound states are included in ξ) and the corresponding "out" operators form irreducible and complete sets in Hilbert space respectively.

It means that only the stable particles, composite particles and vacuum state present the basic vectors of Hilbert space. As it is noted that postulate 5 has extended the concept of completeness of states in quantum mechanics. In fact the in-coming waves or the out-going waves do not form a complete set only if the bound states are added.

From the completeness of operators $a_r^{in}(\vec{p})$, $b_\gamma^{*in}(\vec{p})$, $C_\xi^{in}(\vec{p})$ and $C_\xi^{*in}(\vec{p})$ the completeness of fields $\psi(x)$ and $\bar{\psi}(x)$ can be shown. However, the converse to this is not true. From the mathematical view-point a set formed by operators $a_r(\vec{p},t)$ is larger than that of $a_r^{in}(\vec{p})$, $C_\xi^{in}(\vec{p})$, ... Hence the completeness of basic fields may be drawn from the completeness of "in" operators. However, as for any concrete physical system, provided the Lagrangian is fixed, the Hilbert spaces spanned by these two sets of field operators respectively would be as large as each other. If the Lagrangian in a physical system is not given, there is ambiguity in choosing the complete set. In particular, by assuming the completeness of composite-particle states the completeness of fields $\psi(x)$ and $\bar{\psi}(x)$ may also be deduced. It is expected that the composite particle field theory may be applied as a "confined" field theory.

Having these postulates described above one easily gets the commutation relations by using the standard procedure with the following identities:

$$\left(\int d^4X d^4Y - \int d^4Y d^4X\right)\int d^4(xx'yy')\bar{\phi}_{\vec{P}\xi}(X,x')Q(xx')\delta\left(-ix\frac{\vec{\partial}}{\partial X}\right)$$

$$(-\vec{\Box}_X + m_\xi^2)T(\psi(x_1)\bar{\psi}(x_2)\psi(y_2)\bar{\psi}(y_1))(-\overleftarrow{\Box}_Y + m_{\xi'}^2)\delta(-iy\overleftarrow{\partial}Y) \tag{3.23}$$

$$\cdot Q(y,y')\phi_{\vec{P}'\xi'}(Y,y'),$$

and

$$\left(\int d^4X d^4Y - \int d^4Y d^4X\right)\int d^4(xx'yy')\bar{\phi}_{\vec{P}\xi}(X,x')Q(x'x)\delta\left(-ix\frac{\overleftrightarrow{\partial}}{\partial X}\right)$$

$$(-\vec{\Box}_X + m_\xi^2)T(\psi(x_1)\bar{\psi}(x_2)\psi(y_2)\bar{\psi}(y_1))(-\overleftarrow{\Box}_Y + m_{\xi'}^2)\delta(-iy\overleftarrow{\partial}Y)Q(yy')\phi_{\vec{P}'\xi'}(YY'). \tag{3.24}$$

It should be noted that these identities are not trivial. As we know, Zimmermann has proved these identities from the locality of ψ (x) and the mass-spectral conditions, but the result has not been published.[18] In most of the text-books or papers these identities have been used without proof. Recently a proof has been given in [21]. The commutation relations are also deduced, for example,

$$[C_\xi{}^{in}_{out}(\vec{P}), C^{*}_{\xi'}{}^{in}_{out}(\vec{P}') = \delta(\vec{P}-\vec{P}')\delta_{\xi\xi'}, \cdots . \tag{3.25}$$

Because of the wave functions entering these identities, the difficulty associated with the orthogonality of single-particle states with scattering states, which has been pointed out by Redmond and Uretski,[19] is avoided.

In deriving (3.25) one can see that it can also be obtained by using the condition of normalization. In fact, it is sufficient to establish a self-consistant composite field theory if the definitions of creation and destruction operators in postulate 4 (3.19), (3.20) are modified as

$$C_\xi(\vec{P},T)=\int \bar{W}_\xi(P,X,x)\, i\frac{\overleftrightarrow{\partial}}{\partial X_0} B(X,x)\, d^3X\, d^4x\,, \tag{3.26}$$

$$C^*_\xi(\vec{P},T)=\int B(X,-x)\, i\frac{\overleftrightarrow{\partial}}{\partial X_0} W_\xi(P,X,x)\, d^3X\, d^4x\,, \tag{3.27}$$

where $\bar{W}$ and W are arbitrary normalization functions satisfying the following conditions:

$$\int \bar{W}^i_\xi(P'X\,x)\, i\frac{\overleftrightarrow{\partial}}{\partial X_0}\phi^i_{\vec{P}\xi}(X,x)\, d^3X\, d^4x=\delta(\vec{P}-\vec{P}')\,\delta_{\xi\xi'}\,, \tag{3.28}$$

$$\int \bar{\phi}^i_{\vec{P}\xi}(X,x)\, i\frac{\overleftrightarrow{\partial}}{\partial X_0} W^i_{\xi'}(P'X,x)\, d^3X\, d^4x=\delta(\vec{P}-\vec{P}')\,\delta_{\xi\xi'}\,, \tag{3.29}$$

with index i denoting particles with the same mass $m_i, p^2 = p'^2 = m_i^2$, and that functions W^i or $\bar{W}^i$ may not be orthogonal to wave function ϕ^j or $\bar{\phi}^j$ with different mass m_j.[26] The arbitrariness reflects that in the S-matrix when it is continued of mass shell.

We shall return to this problem in the following paragraph.

IV. S - MATRIX, FIELD-CURRENT RELATIONSHIP AND LOCALITY.

After defining the state vectors, the commutation relations of destruction and creation operators are also obtained. The S-matrix can be defined as

$$\langle \text{in} \,|\, S \,|\, \text{in} \rangle = \langle \text{out} \,|\, \text{in} \rangle \tag{4.1}$$

with the help of asymptotic conditions and commutation relations. The reduction formulas of S-matrix can also be deduced. For the sake of brevity we shall consider only the following case:

$$\begin{aligned} \langle \vec{P}'\varsigma' | J_\mu(z) | \vec{P}\varsigma \rangle = & \int d^4(X Y x x' y y')\, \bar{\Phi}_{\vec{P}'\varsigma'}(X, x')\, Q(x'x)\, \delta\left(-ix\frac{\partial}{\partial X}\right) \\ & (m^2_{\varsigma'} - \vec{\Box}_X) \langle 0 | T(\psi(x_1)\bar{\psi}(x_2) J_\mu(z) \psi(y_2)\bar{\psi}(y_1)) | 0 \rangle (m^2_\varsigma - \overleftarrow{\Box}_Y) \\ & \delta\left(-iy\frac{\overleftarrow{\partial}}{\partial Y}\right) Q(y y')\, \phi_{\vec{P}\varsigma}(Y y'). \end{aligned} \tag{4.2}$$

However, for the convenience in using (4.2) for calculation, we modify (4.2) by inserting δ-functions in the T-product and using (2.19), we get

$$\begin{aligned} \langle \vec{P}'\varsigma' | J_\mu(z) | \vec{P}\varsigma \rangle = & \int d^4(X Y x x' y y' u_1 u_2 v_1 v_2 u_1' u_2' v_1' v_2')\, \bar{\Phi}_{\vec{P}'\varsigma'}(X x') \\ & Q(x'x)(m^2_{\varsigma'} - \vec{\Box}_X) G(x_1 x_2 u_2' u_1') G^{-1}(u_1' u_2' u_2 u_1) \langle 0 | T(\psi(u_1)\bar{\psi}(u_2) \\ & J_\mu(z) \psi(v_2)\bar{\psi}(v_1)) | 0 \rangle G^{-1}(v_1 v_2 v_2' v_1') G(v_1' v_2' y_2 y_1)(m^2_\varsigma - \overleftarrow{\Box}_Y) \\ & Q(y y')\, \phi_{\vec{P}\varsigma}(Y, y'). \end{aligned} \tag{4.3}$$

With (2.15) and Green's function G (2.17), (2.18), (2.20) we also get

$$\langle \vec{p}'\varsigma' | J_\mu(z) | \vec{p}\varsigma \rangle = \int d^4 (x_1 x_2 y_1 y_2)\ \bar{\Phi}_{\vec{p}\varsigma'} (X,x)$$
$$(\vec{\mathcal{D}} - \vec{\mathcal{V}}) \langle 0 | T(\psi(x_1)\bar{\psi}(x_2) J_\mu(z) \psi(y_2) \bar{\psi}(y_1)|0\rangle \tag{4.4}$$
$$(\overleftarrow{\mathcal{D}} - \overleftarrow{\mathcal{V}})\ \Phi_{\vec{p}\varsigma} (Y,y).$$

Replacing the inserted δ-function in (4.2) by Green's function of the B-S equation as

$$(\vec{O} - \vec{I})\ K(x_1 x_2 y_2 y_1) = K\ (x_1 x_2 y_2 y_1)\ (\overleftarrow{O} - \overleftarrow{I}) = \delta(x_1 - y_1)\ \delta(x_2 - y_2), \tag{4.5}$$

where $(\vec{O} - \vec{I})$ is the integro-differential operator of B-S equation, and using the expression of Green's function K at poles and (2.15) we obtain

$$\langle \vec{p}'\varsigma' | J_\mu(z) | \vec{p}\varsigma \rangle = -\int d^4 (x_1 x_2 y_1 y_2)\ \bar{X}_{\vec{p}'\varsigma'} (X,x)(\vec{O} - \vec{I})$$
$$\langle 0 | T(\psi(x_1)\bar{\psi}(x_2) J_\mu(z) \psi(y_2) \bar{\psi}(y_1))|0\rangle (\overleftarrow{O} - \overleftarrow{I})\ X_{\vec{p}\varsigma}(Y,y). \tag{4.6}$$

After using (2.15), (2.17), (2.18), (2.20) and (4.4) we have

$$\int d^4(X,Y) \frac{e^{-iP'X}}{\sqrt{2E_{\varsigma'}(2\pi)^3}} (m_{\varsigma'}^2 - \Box_X) \delta(-ix\frac{\partial}{\partial X}) \langle 0 | T(\psi(x_1)\bar{\psi}(x_2)$$
$$J_\mu(z)\psi(y_2)\bar{\psi}(y_1))|0\rangle \delta(-iy\frac{\partial}{\partial Y})(m_\varsigma^2 - \Box_Y) \frac{e^{iPY}}{\sqrt{2E_\varsigma(2\pi)^3}} \tag{4.7}$$
$$= \sum_{\varsigma\varsigma'} \Phi_{\vec{p}'\varsigma'}(x) \langle \vec{P}'\varsigma' | J_\mu(z) | \vec{P}\varsigma \rangle \bar{\Phi}_{\vec{P}\varsigma}(y).$$

An important feature of (4.7) is that the right-hand side of (4.7) allows a separation of wave function from matrix element. By multiplying any normalized conjugate wave function all the wave functions appeared on the right-hand side of (4.7) may be cancelled out. For example, by multiplying (2.15), the S-matrix of form (4.2) is obtained. By multiplying (3.28), (3.29) the S-matrix of form (4.2) is obtained. By multiplying (3.28), (3.29) the S-matrix of form defined by (3.26) (3.27) is also obtained. As is noted, the general form of $\phi_{\vec{p}\xi}(x)$ would be

$$\phi_{\vec{p}\xi}(x)=\delta(Px)\left\{\gamma_5 f_1(x)+\frac{i\hat{P}}{m_\xi}\gamma_5 f_2(x)+\frac{iP_\mu x_\nu}{m_\xi}\sigma_{\mu\nu}\gamma_5 f_4(x)\right\}, \quad (4.8)$$

and

$$\bar{\phi}_{\vec{p}\xi}(x)=\delta(Px)\left\{\gamma_5 f_1(x)-i\frac{\hat{P}}{m_\xi}\gamma_5 f_2(x)-i\frac{P_\mu x_\nu}{m_\xi}\sigma_{\mu\nu}\gamma_5 f_4(x)\right\}. \quad (4.9)$$

Multiplying on both sides by $\pm i\,\frac{\hat{P}}{m_\xi}\gamma_5$ and the SU(3) matrix A(S) and taking trace of both sides, we get

$$\begin{aligned}\langle\vec{p}'\xi'|J_\mu(z)|\vec{p}\xi\rangle = &\int d^4(XY)\, f^*_{\vec{p}'\xi'}(X)\, f_{\vec{p}\xi}(Y)\\ &(m^2_{\xi'}-\Box_X)(m^2_\xi-\Box_Y)\langle 0|T\Big(\frac{1}{4m_{\xi'}f_2'(0)}\frac{\partial}{\partial x_\alpha}\bar{\psi}(x)\gamma_\alpha\gamma_5\\ &A(S')\psi(x)J_\mu(z)\frac{1}{4m_\xi f_2(0)}\frac{\partial}{\partial y_\beta}\bar{\psi}\gamma_\beta\gamma_5 A(S)\psi(y)\Big)|0\rangle.\end{aligned} \quad (4.10)$$

Let us write the field-current relation as

$$\varphi_\xi(x)=\frac{1}{4m_\xi f_2(0)}\frac{\partial}{\partial x_\mu}\left(\bar{\psi}(x)\gamma_\mu\gamma_5 A(S)\psi(x)\right), \quad (4.11)$$

and finally we obtain the reduction formulas of S-matrix as in the usual local field theory.

$$\langle \vec{P}'\xi'|J_\mu(Z)|\vec{P}\xi\rangle = \int d^4(XY) f^*_{\vec{P}\xi'}(Z) f_{\vec{P}\xi}(Y)$$
$$(m_\xi^2 - \Box_X)(m_\xi^2 - \Box_Y)\langle 0|T(\varphi_{\xi'}(X) J_\mu(Z) \varphi_\xi(Y))|0\rangle. \tag{4.12}$$

It is obvious that (4.11) presents the PCAC, which are often used in various cases.

The results expressed in (4.11) and (4.12) are general, since for any given composite particles the similar expressions hold. One can, therefore, unify the derivations of a series of field-current relations similar to those of (4.11) and (4.12) , including field-current relations of high-spin mesons and hadrons composed of three quarks, and all the reduction formulas of L-S-Z Field theory for particles with high spins. Meanwhile in usual theories all of these could be obtained only by assuming various Lagrangians. (4.11) and (4.12) also explain why L-S-Z field theory, which has been constructed on the base of point-like model, and its associated dispersion relations, can be so nicely applied to composite particles as hadrons, nuclei or atoms. It is worthwhile to inquire why the field theory of composite particles formulated above has given the same results as the point-like theory does? In fact it is not so strange as it appears at first sight. Quantum mechanics provides a limiting case of quantum field theory, and classical mechanics presents a special case of quantum mechanics. As is well-known in classical mechanics, expression F = ma may be equally well applied to the dynamics of material point or of the center of mass of material body. It is therefore inconsistent if quantum field theory does not give as a limiting case the same results of classical mechanics.

The field-current relation described as in (4.11) is of the weak form. It means that (4.11) would hold provided the matrix element of π - mesons is on mass-shell. As it has been pointed by Brandt and Preparata from pole approximation, there are more theoretical and experimental verifications for PCAC of the weak form than of the strong form.[22] This conclusion may be regarded as a restatement of the view-point of Chou Kuang-chao [23] and Nambu [24] in 1960. In our opinion the field-current relations of the weak form are rigorous. The field-current relations of the strong form or the relations between the field operators and current operators can be derived from composite particles field theory.[10] By multiplying $\frac{i\hat{P}}{m_\varsigma}\gamma_\mu\gamma_5 A(S)$ to (3.21) and taking trace, PCAC including all excited states would be obtained when $x \longrightarrow 0$. This is just the EPCAC deduced previously by us with a doubtful assumption that all solutions of the B-S equation formed a complete set. By constructing equal-time composite-particle operator the ghost states have been excluded. But in deriving this the contributions of scattering states should be neglected. Hence as an approximation it must be considered carefully. It is possible that in a field theory of "confined quarks" the field-current relations of the strong form would hold.

In short, we have obtained the following results:

1. A comparatively complete and self-consistent field theory of composite particles, which makes possible the description of different processes including hadrons, nuclei and atoms in a unified way in terms of the S-matrix,
2. Various forms of S-matrix, among which the most important ones are:
   a. S-matrix of L-S-Z field theory and the corresponding field-current relations,
   b. Reduction formulas of S-matrix based on B-S wave functions,
   c. Reduction formulas of S-matrix based on equal-time wave functions. These unify the theory of dispersion relations, current algebra, light-cone algebra and straton model on the basis of S-matrix theory.
3. A series of field-current relations as PCAC, VMD ... and field-current relations corresponding to different particles. Combining these relations with current algebra, light-cone algebra etc. would make extensive applications possible.
4. A preliminary result indicating that in the case of "confined quarks" a theory of S-matrix may not be inconsistent.

It should also be mentioned that a quantum field theory including composite particles was reformulated by Chang

Gun-ching from Peking University[25] on the basis ofthe field theory with strong convergence due to Haag, Ruelle, Steiman and Hepp. Similarly , formulation has also been given by Huang and Weldon. [20] In particular they have first shown that the wave functions may be distinguished as with reducible or irreducible forms, and proved the equivalence between them.

V. APPLICATIONS OF QUANTUM FIELD THEORY OF COMPOSITE PARTICLES.

In this paragraph we shall summarize some applications of composite-particle field theory briefly.

1. <u>Applications of Perturbation Expansion</u>.

By assuming that the following condition is satisfied:

$$\sum_{\text{bare}} \left| \text{bare particle} \right\rangle \left\langle \text{bare particle} \right| = \sum_{\text{phys.}} \left| \text{phys. particle} \right\rangle \left\langle \text{phys.particle} \right| = 1 \tag{5.1}$$

and using the S-matrix described in the preceding paragraphs we can transform this theory from Heisenberg representation to interaction representation and obtain a series of perturbation expansions(which correspond to Feynman diagrams and also give the rules for selecting diagrams). However, some requirements must be fulfilled: the series should be renormalizable, the infra-red catastrophe should be eliminated, and the calculations must be gauge-invariant. As the solutions of the B-S equation do not represent an expansion in coupling constant, hence much complications have arisen. All of these problems have been discussed and solved thoroughly in [12-14].

The first test of the perturbation expansion of composite-particle field theory would be the study of coulombic atoms. For this purpose in [26-33] the solutions of the B-S equation with spin $(0 - 0)$, $(0 - \frac{1}{2})$, $(\frac{1}{2} - \overline{\frac{1}{2}})$, and equal or unequal masses are obtained. By using these covariant wave functions various processes involving annihilation and creations of atomic systems have been calculated. As it is well-known, with a flying atomic system of high speed, the relativistic contractions must be considered, hence the relativistically covariant wave functions should be used. Among the atoms the double exotic atom is of special interests. By the double exotic atom we mean that both proton and electron in the atom are replaced by the unstable particles respectively. For example $(\pi^{\pm} \mu^{\mp})$, etc. A summary review on this subject will be given in this conference.[33]

Another application of perturbation expansion is of the scattering of high energy hadrons on nuclei. This expansion, as is already known, can not be applied to strong interactions. Nevertheless a summation of certain infinite Feynman diagrams under some particular conditions can be performed, and as a result some reliable conclusions may be drawn from this. A group of scientists in China have derived the many-body scattering amplitude by means of eikonal approximation, which is, as Glauber's theory is , independent of the forms of interaction . Under the approximation of small scattering angles the many-body scattering amplitude can be expressed as a sum of products of two-body amplitudes.[34] As it is not well founded to apply the non-relativistic Glauber theory with energies beyond 1 GeV, the relativistic Glauber theory thus obtained may be used in many cases with ISR energies without difficulty. A report on this subject will also be given in this conference.[34] One of the purposes to construct a field theory of composite particles is to study the hadron structures and the hadron interactions. Recently the investigations of perturbative QCD have indicated that there exists asymptotic freedom in high energy collisions, it seems, therefore, that a perturbation expansion theory would also be available. In 1974 a group of theorists in China used this field theory to investigate the deep inelastic scattering of electrons on protons and neutrons and obtained the relations between the structure functions and B-S wave functions.[35] Lately, through application of this theory, the calculated structure of π-meson $F_{\pi}(x)$ with the use of the light straton model suggested by [36] approximately presented a form of a(1-x),[37] which is in excellent agreement with the recent experimental measurement of Fermi Lab. .

Another application of this theory is to improve the straton model. As it is observed, the straton model presents in fact the use of the B-S equation to study the hadron structures. A review has been given in this conference by Tzu Hung-yuan on the works of 1965-1966. The S-matrix theory based on B-S equation described above, therefore, can be regarded as an improvement of the theory of straton model. However, two points should be noted:

a. There are many works where the hadron structures were studied by means of quantum mechanics with confining potential. Hence it is also worth trying to use the B-S equation with confining potential. An attempt has been made [39-40] by using a potential of harmonic oscillators.

b. As in the perturbation expansion based on B-S wave functions the correction from the next order to the lowest has arisen by exchanging two gluons at least, which corresponds to terms of g^4 (g is the coupling constant). Therefore, the lowest order expansion would yield an adequate accuracy provided $g < 1$.

When use is made of perturbution expansion based on B-S functions, the main problem that occurs is that one would have difficulty in solving this equation. It is, therefore, very necessary to find other relativistic equation of composite particles.

2. <u>The Applications of Unperturbative Methods</u>.

By this subject we mean the application of field-current relations in combination with current algebra and light-cone algebra in the study of hadron structures, which have been derived by using the field theory of composite particles. As for the various field-current relations of weak and strong forms which have been deduced from this theory, an extensive application would be found in processes involving hadrons. In [41-45], calculations were made for $\pi \rightarrow e \nu \gamma$, k -- $e\nu\gamma$, J/ψ and transitions involving high excited states of hadrons. The theoretical results thus obtained have been in good agreement with experiments. In particular the application of the generalized PCAC, which has led to an universal correction to PCAC usually used, results in a very nice coincidence with experiments [45] On the other hand, however, in the calculations with current algebra the problem concerning the analytical continuations of transition amplitude of many energy levels off mass shell occurs. In these calculations we have adopted the method of elimination of poles corresponding to the energy levels and assumed the constancy of the resulted functions approximately in their analytical continuation. But the questions about the accuracy and the region of this method are not clear yet, therefore, we shall leave these questions to further investigation.

VI. SOME PROBLEMS CONCERNING THE FURTHER DEVELOPMENT OF COMPOSITE PARTICLE FIELD THEORY

The main aim to develop the composite particle field theory is to study the dynamics of hadrons. This problem, however, cannot be solved without the knowledge of the dynamics of quarks and gluons, in particular the knowledge of the mechanism of confinement. So far the attempt to derive the confinement from QCD has failed. Hence we try to obtain a confinement solution in classical and nonlinear QCD. The preliminary results in [46-47] indicate that it is possible to obtain the coulomb potential plus the linear potential between particle-anti-particle in two nonlinear models. The study of quantizing a nonlinear Lagrangian by means of the path integral has been carried out preliminarily.[48] There are, however, many principal problems which remain unsolved. For example, the quark confinement would lead to zero the quark propagator S'_F in the spectral representation of the usual field theory.[49] It seems that the S-matrix theory formulated above should contain the confinement theory. Nevertheless, many problems still remained.

In order to widen the various applications of the composite field theory we should recommend to use the reduction formulas of S-matrix based on the equal-time wave functions as described in §II. As it is easier to obtain such wave functions than the solutions of B-S equation, an extensive application of using this formalism is expected in the study of different problems concerning the relativistic corrections in atoms, nuclei and hadrons.

References

1. Ho Tso-hsiu, Huang Tao, Kexue Tongbao, 19 (1974) 1.
2. Ho Tso-hsiu, Huang Tao, Act. Phys. Sinica, 23 (1974) 113.
3. Ho Tso-hsiu, Huang Tao, Act. Phys. Sinica, 23 (1974), 264.
4. Ho Tso-hsiu, Huang Tao, Phys. Ener. Fort. Phys. Nucl. 1 (1977) 37.
5. Ho Tso-hsiu, Huang Tao, Scientia Sinica, 18 (1975) 502.
6. Ho Tso-hsiu, Huang Tao, Kexue Tongbao, 20 (1975) 419.
7. Ho Tso-hsiu, Chang Chao-hsi, Huang Tao, Phys. Ener. Fort. Phys. Nucl. 2 (1978), 285.
8. Ruan Tu-nan, Ho Tso-hsiu, Phys. Ener. Fort. Phys. Nucl. 2 (1978), 287.
9. Ho Tso-hsiu, Ruan Tu-nan, Liu Yao-yang, Journal of the University of Science & Technology of China, 7 (1977) 1.
10. Ho Tso-hsiu, Huang Tao, Act. Phys. Sinica, 23 (1974), 418.
11. Ho Tso-hsiu, Huang Tao, Kexue Tongbao, 21 (1976) 35.
12. Ho Tso-hsiu, Chang Chao-hsi, Huang Tao, Act. Phys. Sinica, 25 (1976), 215.
13. Ho Tso-hsiu, Chang Chao-hsi, Act. Phys. Sinica, 26 (1977) 540.
14. Chang Chao-hsi, Ho Tso-hsiu, Phys. Ener. Fort. Phys. Nucl. 2 (1978), 215.
15. Ruan Tu-nan, Chu Hsi-chuang, Ho Tso-hsiu, Phys. Ener. Fort. Phys. Nucl., to be published.
16. Ruan Tu-nan, Chu Hsi-chuan, Ching Chen-rui, Ho Tso-hsiu,

Chao Wei-chin, paper submitted to the present conference.

17. Ruan Tu-nan, Ho Tso-hsiu, Huang Tao, Phys. Ener. Fort. Phys. Nucl. 3 (1979).

18. Haag R., Phys. Rev. 112 (1958), 669 Nishijima K., Phys. Rev. 111 (1958) 995. Simmermann W., Nuovo, Cimento 10 (1958) 597.

19. Redmond P.T., Uretsky J.L., Ann. Phys. 9 (1960), 106.

20. Huang K., Weldon H.A., Phys. Rev. D15 (1975), 257.

21. Wang Rong, Phys. Ener. Fort. Phys. Nucl. 3 (1979) 645.

22. Brandt R.A., Preparata G, Ann. Phys. 61 (1970) 119.

23. Chou Kuang-chao, J. Eexpt. Theor. Phys. (USSR) 39 (1960) 703. (Soviet Phys. JETP, 12 (1961) 492.

24. Nambu Y. Phys. Rev. Lett, 4 (1960), 380.

25. Chang Gung-ching, Act. Math. Sinica 19 (1976), 13.

26. Chang Chao-hsi, Ho Jui, Ho Tso-hsiu, Scientia Sinica (1979) 572.

27. Ho Jui, Chang Chao-hsi, Ho Tso-hsiu, Phys. Ener. Fort. Phys. Nucl. 3 (1979) 117.

28. Ho Jui, Chang Chao-hsi, Ho Tso-hsiu, Phys. Ener. Fort. Phys. 3 (1979) 297.

29. Ho Jui, Chang Chao-hsi, Ho Tso-hsiu, Phys. Ener. Fort. Phys. Nucl., to be published.

30. Ching Cheng-rui, Phys. Ener. Fort. Phvs. Nucl. to be published.

31. Ching Cheng-rui, Ho Tso-hsiu, Chang Chao-hsi, material to be published.

32. Ching Cheng-rui, Ho Tso-hsiu, Chang Chao-hsi, Hojui, paper submitted to the present conference.

33. Chu Hsi-chuan, Ho Tso-hsiu, Chao Wei-chin, Bao Cheng-guang paper submitted to the present conference.

34. Ho Tso-hsiu, Chan Chao-hsi, Hse Yi-cheng, Phys. Act. Sinica, 24 (1975) 115.

35. Chu Chun-yuan, An Ying, Chan Jen-tzu, paper submitted to the present conference

36. Chao Wan-yung, paper submitted to the present conference.

37. Tzu Hung-yuan, paper submitted to the present conference.

38. Wu Yong-shi, Ho Tso-hsiu, Phys. Act. Sinica, 26 (1977) 274.

39. Ho Tso-hsiu, Chang Chao-hsi, 2 (1977) 199

40. Ho Tso-hsiu, Huang Tao, Phys. Act. Sinica, 25 (1976) 409.

41. Ho Tso-hsiu, Huang Tao, Phys. Ener. Fort. Phys. Nucl. 3 (1979) 97.

42. Ho Tso-hsiu, Huang Tao, Phys. Ener. Fort. Phys. Nucl. 2 (1978) 471.

43. Ho Tso-hsiu, Huang Tao, Phys. Ener. Fort. Phys. Nucl. 2 (1978) 665.

44. Tzao Nan-wei, Phys. Ener. Fort. Phys. Nucl. to be published.

45. Ruan Tu-nan, Cheng Shi, Ho Tso-hsiu, meterial to be published.

46. Ruan Tu-nan, Cheng Shi, Ho Tso-hsiu, paper submitted to the present conference.

47. Chou Guang-chao, Ruan Tu-nan, Yin Hung-ju, paper submitted to the present conference.

48. Ruan Tu-nan, Gao Tsung-shuo, Phys. Ener. Fort. Phys. Nucl. to be published.

HADRON STRUCTURE AND THE HADRONIZATION OF QUARKS

Rudolph C. Hwa

Institute of Theoretical Science and Department of Physics

University of Oregon, Eugene, Oregon 97403

ABSTRACT

Hadron structure is investigated in the framework in which a nucleon can be described at low Q^2 by a composite system of three valence quark clusters, called valons. Using QCD for the description of valon structure functions at high Q^2, we extract the valon distribution in a nucleon from data on deep inelastic scattering. The valon representation offers a precise definition of the recombination function which describes the hadronization of quarks. The ideas have been applied to the calculation of quark fragmentation function and inclusive distribution of low-p_T hadronic reactions. Good agreements with data have been obtained in both problems without any adjustable parameters.

I. INTRODUCTION

In the search for building blocks of matter and an understanding of the nature of their interactions, physicists have progressively sharpened their focus from molecules to atoms to nuclei to particles and now to quarks. At each stage they left behind interesting and important problems in the quest for something more fundamental. Their studies have been aided by the possibility of doing experiments that can focus on the effects being investigated. That is usually achieved by going to higher energies and using beams or targets that are made up of objects to be examined. Such a chain of favorable circumstances unfortunately ceases at the level of quarks since we do not have isolated quark beams or quark targets. Thus quark physics, such as perturbative QCD which ignores hadronic complication, cannot be divorced from hadron physics so long as contact with experiment is to be established. As long as the beam and detected particles are hadrons, it is necessary to go back and forth between hadrons and quarks. The subject of this talk is to describe a way of going between these two levels.

The wave function of quarks in a hadron describes the hadronic structure. At the same time it also specifies the hadronization of quarks. The structure functions as determined in deep inelastic scattering do not serve the purpose, since they are probed at high Q^2. We need the wave function at low Q^2 where hadronization occurs. Now, we are not proposing a solution of the confinement problem which is still untamed. Instead, we propose to get some information about the wave function phenomenologically prior to the theoretical resolution of the confinement problem. The information is to be extracted from deep inelastic scattering data using QCD. What we shall attempt to do then is effectively the construction of a bridge between the bound-state and scattering problems starting from the latter end. The connection is of interest to establish in its own right, but the result turns out to be very useful for hadron reactions involving low momentum transfers.

We shall apply the result to the calculations of the quark fragmentation function and the inclusive distributions of hadronic reactions at low p_T. In both cases we have obtained good fits of the data without any adjustable parameters. They give evidences for hope that hadron physics can now be treated at the quark level in a quantitative way.

II. VALONS

Hadron structure is a subject that has been investigated in two approaches which are nearly mutually exclusive. On the one hand, one studies the bound-state problem of a hadron in terms of two or three quarks, whether by solving Bethe-Salpeter equation[1] or considering color confinement in models such as the various bag models.[2] On the other hand, one probes the hadrons with high energy leptons, and from the structure functions deduced one infers the distribution of point-like constituents.[3] There is no communication between these two approaches. Even the naive pictures of a nucleon adopted by them are different. In the former it is the quark model that involves three quarks; in the latter it is the parton model that involves an infinite number of partons (i.e. quarks, antiquarks and gluons).

We want to construct a bridge that connects these two views of the hadron. The important point to recognize is that the quark and parton models are two views of the same object with different resolutions. At high Q^2 we see partons as point-like constituents, but at low Q^2 the quarks in the quark model need not be point-like, although for simplicity in the bound-state calculations they have been treated as if they are point constituents. The two views can be reconciled if we do not require that the constituents in the quark model be the same objects as the quarks in the parton model, but rather be clusters of quarks, antiquarks and gluons. According to the quark model there should be <u>three</u> such clusters in a nucleon with appropriate quantum numbers as usually required, i.e. those of the valence quarks in the parton model. For brevity, these valence quark clusters are called valons.[4] The fuzziness of valons should not spoil the successes of the quark model since the predictions of the model (e.g. mass spectra, $\sigma_{\pi p}/\sigma_{pp}$, etc.) do not involve large momentum transfers and consequently cannot be sensitive to the "size" of the constituents. The possibility that the valons can have internal structure containing an infinity of partons obviously allows proper contact with the usual parton model. Indeed, at high Q^2 a virtual photon interacting with a quark in one of the valons, the others being spectators, presents an eminently reasonable picture that is consistent with the naive parton model.

The picture suggested here for a nucleon is not very different in spirit from the usual picture of a deuteron. At low Q^2 we view a deuteron as a composite system of two nucleons, which in turn have internal structures that can be resolved only by high Q^2 probes. The difference is that the long range part of the nuclear force is well known, unlike the chromatic force that confines the valons.

To give content to the valon representation it is necessary to know at least either the confinement potential or the valon structure. Knowing both in a consistent way would complete the bridging of the gap between the bound-state problem and the deep inelastic scattering problem of the hadron. Our understanding of the confinement potential for light valons is at present incomplete. But for the valon structure we have fortunately QCD to lean on. Indeed, a valon is defined to be a valence quark plus its cloud of quarks, antiquarks and gluons due to virtual processes in QCD. The basic assumption of the valon picture is that the virtual processes associated with each of the three valence quarks in a nucleon are sufficiently untangled that it is meaningful to think of three distinct valons at some low Q^2. Some entanglement of very low frequency gluons is, of course, necessary to effect binding in the formation of a nucleon. However, at very high Q^2 we need only examine the structure of each valon one at a time, the effects of binding being accounted for by a smearing of the valon coordinates. That smearing is described by the valon distribution, which is the absolute square of the nucleon wave function in the valon picture. In momentum representation we denote it by $G_{v/N}(y)$, where y is the momentum fraction of the valon. Its dependence on y has been determined from deep inelastic scattering data in the framework of QCD.[4,5]

Let $\mathcal{F}^h(x,Q^2)$ be a structure function (e.g. F_2 or xF_3) of a hadron h, and $\mathcal{F}^v(z,Q^2)$ be the corresponding function of a valon v. They are related by the smearing mentioned above:

$$\mathcal{F}^h(x,Q^2) = \sum_v \int_x^1 dy\, G_{v/h}(y)\, \mathcal{F}^v(x/y, Q^2) \tag{2.1}$$

In terms of the moments

$$M^{h,v}(n,Q^2) = \int_0^1 dx\, x^{n-2}\, \mathcal{F}^{h,v}(x,Q^2) \tag{2.2}$$

$$M_{v/h}(n) = \int_0^1 dy\, y^{n-1}\, G_{v/h}(y) \tag{2.3}$$

(2.1) can be rewritten as

$$M^h(n,Q^2) = \sum_v M_{v/h}(n) M^v(n,Q^2) \tag{2.4}$$

$M^v(n,Q^2)$ is known in QCD, while $M^h(n,Q^2)$ is known from experiments. Hence, $M_{v/h}(n)$ can be determined. The details of the theoretical considerations have been described in Refs. 4 and 5, and will not be repeated here. Note that in (2.4) Q^2 dependence does not appear in $M_{v/h}(n)$. Thus the formalism would be definitely wrong if a Q^2-independent $M_{v/h}(n)$ is unable to fit the data.

In Refs. 4 and 5 only F_3 of the neutrino scattering data[6] was used as experimental input, and theoretically it was assumed (for lack of adequate muon data) that the valon distribution has no flavor dependence. Subsequently, more extensive analysis has been undertaken[7] using both ν,[8] and u data.[9] I report here the preliminary result of our new finding. We have considered singlet as well as non-singlet components of the moments, and we differentiate U- from D-valon distributions. Starting from a simple expression for the exclusive distribution for a proton

$$G_{UUD/p}(y_1, y_2, y_3) = \alpha(a,b)(y_1 y_2)^a y_3^b \delta(y_1+y_2+y_3-1) \tag{2.5}$$

we obtain by double integrations

$$G_{U/p}(y) = [B(a+1, a+b+2)]^{-1} y^a (1-y)^{a+b+1} \tag{2.6}$$

$$G_{D/p}(y) = [B(b+1, 2a+2)]^{-1} y^b (1-y)^{2a+1} \tag{2.7}$$

which satisfy the normalization conditions

$$\int_0^1 G_{U/p}(y)\,dy = \int_0^1 G_{D/p}(y)\,dy = 1 \tag{2.8}$$

and the momentum sum rule [already guaranteed by the δ-function in (2.5)]

$$\sum_v \int_0^1 G_{v/p}(y)\, y\, dy = 1 \tag{2.9}$$

Thus we have only two parameters, a and b, to fit all structure functions for all lepton deep inelastic scattering at all high Q^2. The result is that all such data in Refs. 6, 8 and 9 on $F_2^{\mu p}$, $F_2^{\mu n}$, $F_2^{\nu N}$ and $F_3^{\nu N}$ for $Q^2 > 10$ GeV^2 have been well fitted by (2.4) and the choice

$$a = 0.65 \quad \text{and} \quad b = 0.35 \tag{2.10}$$

From (2.6) and (2.7) we then have

$$G_{U/p}(y) = 8\,y^{0.65}\,(1-y)^{2} \tag{2.11}$$

$$G_{D/p}(y) = 6\,y^{0.35}\,(1-y)^{2.3} \tag{2.12}$$

The fact that these Q^2-independent valon distributions can fit all high-Q^2 lepton scattering data is the strongest support we now have for the valon picture.

From the valon distribution we can calculate the quark and gluon distributions. The large x behavior can immediately be inferred from (2.1) since the "favored" evolution functions $\mathcal{F}_{u/U}(z,Q^2)$ and $\mathcal{F}_{d/D}(z,Q^2)$ are the same and finite as $z \to 1$ at moderate Q^2, while the "unfavored" evolution functions $\mathcal{F}_{d/U}(z,Q^2)$ and $\mathcal{F}_{u/D}(z,Q^2)$ vanish at $z \to 1$. Thus the u and d quark distributions behave as

$$xu(x) \sim (1-x)^3$$
$$xd(x) \sim (1-x)^{3.3}$$

in the limit $x \to 1$. Over the whole range of x the distributions of these quarks, as well as the antiquarks and gluons, can be obtained only by more detailed calculations, which is now being pursued.[7]

The valon distribution in pions can also be obtained if we use as experimental input the pion structure function as determined in massive lepton-pair production,[10] i.e. $\mathcal{F}^{\pi}(x) \sim (1-x)^1$ as $x \to 1$. It then follows from (2.1) and pion sum rules similar to (2.8) and (2.9) that

$$G_{v/\pi}(y) = 1 \tag{2.13}$$

$$\begin{aligned} F_{v_1 v_2/\pi}(y_1, y_2) &= y_1 y_2\, G_{v_1 v_2/\pi}(y_1, y_2) \\ &= y_1 y_2\, \delta(y_1 + y_2 - 1) \end{aligned} \tag{2.14}$$

It would be of interest to relate our findings here about the valon distributions to the droplet model for hadron collisions.[11] No careful investigation on the subject has yet been made, so my remark here is tentative and reflects only a general impression. A key assumption in the droplet model is that the matter density in a hadron is proportional to its charge density; otherwise, there is no predictive power. Presumably, matter density is closely related to the valon distribution in coordinate space. If the valon distribution were not flavor dependent, it is concievable that the matter and charge densities can be proportional to each other since there would be only one distribution in terms of which both densities are to be expressed. Note that in the valon representation there are no additional gluons to upset the proportionality. However, the flavor dependence of the valon distributions in (2.11) and (2.12), though in momentum space, makes it rather unlikely for the proportionality to be true. These are minimal statements that one can make without examining the question of proportionality of matter and charge densities in each valon.

III. HADRONIZATION OF QUARKS

In the preceding section we have discussed the structure of hadrons in the valon representation. If we denote the proton wave function by $\langle U_1 U_2 D | p \rangle$, then the valon distribution in a proton is

$$F_{UUD/p}(y_1, y_2, y_3) = \left| \langle U_1 U_2 D | p \rangle \right|^2$$

Similarly, for a pion we have

$$F_{U\bar{D}/\pi^+}(y_1, y_2) = \left| \langle U\bar{D} | \pi^+ \rangle \right|^2$$

The recombination of U and $\bar{D}$ to form a pion is clearly specified by the amplitude $\langle \pi^+ | U\bar{D} \rangle$, so the recombination function which describes the probability of recombination of a valon at x_1 with another at x_2 in forming a pion at x is[5]

$$R^{\pi}(x_1, x_2, x) = F_{v_1 v_2/\pi}\left(\frac{x_1}{x}, \frac{x_2}{x}\right) \tag{3.1}$$

By virtue of (2.14) we therefore know the pion recombination function completely. The same is true for other hadrons. They describe the hadronization of quarks in the valon picture. The burden of any attempt to utilize this result for any reaction involving hadronic final states is the calculation of valon distribution before hadronization. Note that since recombination is a process that does not involve large momentum transfer, it is the valon that is relevant here.

To test our mechanism for hadroniztion we discuss first the quark fragmentation funtion. When a quark (or antiquark) is produced in, say, e^+e^- annihilation at high Q^2, it radiates gluons which in turn create $q\bar{q}$ pairs. The recurring of these virtual processes leads to a proliferation of quarks and antiquarks, which become less and less off-shell. It is referred to as Q^2 degradation, which is the exact opposite of a valon structure function, probed at high Q^2, since the latter involves Q^2 evolution. When Q^2 is sufficiently degraded in a quark jet, we have a jet of valons which recombine to form hadrons. Thus if we denote the two-valon distribution in a quark jet initiated at Q^2 by $F(x_1, x_2, Q^2)$ the quark fragmentation function is given by

$$x D(x, Q^2) = \int \frac{dx_1}{x_1} \frac{dx_2}{x_2} F(x_1, x_2, Q^2) R(x_1, x_2, x) \tag{3.2}$$

where x is the momentum fraction of the detected pion in the jet. The task is therefore mainly the determination of $F(x_1, x_2, Q^2)$.

Eq. (3.2) is very similar to (2.1). In both cases there is an implicit Q_o for the valon distributions and the degradation or evolution functions. The value of Q_o is fixed phenomenologically by fitting the large Q^2 data in deep inelastic scattering (together with the determination of the scale parameter Λ) using the leading order calculation in QCD. Its physical meaning is the effective starting point (in that approximation) of the Q^2 evolution or degradation and is therefore a measure of the mass scale at which the valons are defined and resolved. Disagreeing data from different experiments imply different values for Q_o and Λ, although the differences for Q_o/Λ are not great. Typical values are $Q_o = 0.8$ GeV and $\Lambda = 0.65$ GeV. We emphasize that Q_o is determined by using the leading log approximation, so to use it in $F(x_1,x_2,Q^2)$ as the destination of Q^2 degradation, only the leading order calculation of the two-valon distribution should be done. The general expression is complicated,[12] and will not be given here. It is, however, important to mention the hadronization problem of the gluons. If we calculate in perturbative QCD only the $q\bar{q}$ pairs (which are valon pairs at low Q^2) in a quark jet, we would miss many gluons which carry a sizable fraction of the jet momentum. To account for that leakage we must convert the gluons to $q\bar{q}$ pairs and then let them recombine. We have included such pairs in adding up the total two-valon distribution. Application to (3.2) then leads to the result shown in Fig. 1, in which the dashed line is for no resonance production while the solid line takes ρ production into account. There are no free and adjustable parameters in the calculation. The agreement with data[13] in both normalization and shape is evidently very good. This is the first application of the valon picture and the result supports our model for hadronization.

As a second application of the subject we discuss the problem of determining the inclusive distribution of pions produced in hadron-hadron collisions at low p_T in the fragmentation region. This is the probem that T.D. Lee once said, "is chemistry." Now I am able to show you a no-free-parameter prediction of the inclusive distribution having excellent agreement with data.[5] The molecules of that chemistry are just the valons discussed here. And it is not that complicated.

Consider $p + p \rightarrow \pi^+ + X$. In the fragmentation region of the projectile the produced pion is independent of the target. In the parton language this is due to short-range interaction (in rapidity) among the partons. Thus we have essentially just the fragmentation problem of $p \to \pi^+$ plus many other undetected hadrons. At the constituent level we readily say that p can be represented by three valons and the final π^+ by two valons, the representation functions being known already. Since many hadrons are produced, the final state may be represented by a many-valon state. Clearly, the problem is to understand how the valons proliferate in number. Qualitatively, one can see that high energy collision of the two incident hadrons destroys the boundaries of the static bags for the hadrons. Partons in the valons that are small in rapidity can interact, while the fast partons leave the interaction region essentially unperturbed. Among the fast partons, quarks and antiquarks dress themselves up and become valons before recombination; gluons are converted by pair creation to quarks and antiquarks and contribute further to the valon distribution. Quantitatively, it is hard to calculate from first principles the parton content in the valons. The notion of partons is supposed to be meaningful at high Q^2 where point-like partons can be resolved as constituents of valons. But for low-p_T reactions the relevant Q^2 is unknown, certainly not high enough to justify leading order approximation in a QCD calculation. Instead of abandoning any attempt to do quantitative analysis at this point, it is important to recognize one crucial phenomenological fact: that is "precocious scaling." Parton model was invented by Feynman at a time when the structure functions of nucleons were determined at SLAC for Q^2 only in the range 1 - 5 GeV^2. Approximate scaling was already evident then. Gluons are found to carry nearly half the nucleon momentum even for $Q^2 \sim 2\ GeV^2$. It means that quarks radiate gluons readily and that even at such low Q^2 values most of the evolution process has taken place already. Thus we can not only discuss parton distributions in that low Q^2 range, but go further to regard such distributions as being relevant for low-p_T reactions, as originally suggested by Feynman in discussing the implications of the parton model.

More precisely, the inclusive distribution is[14]

$$\frac{x}{\sigma}\frac{d\sigma}{dx} = \int \frac{dx_1}{x_1}\frac{dx_2}{x_2} F(x_1,x_2) R(x_1,x_2,x) \tag{3.3}$$

a formula that is very similar to (3.2) except that the jet is initiated by a hadron rather than a quark. The calculation of $F(x_1,x_2)$ can be represented[5] by two diagrams in Fig. 2 for $p \to \pi^+$. The intermediate lines are valons. The shaded blobs are parton distributions in valons; they are to be determined by fitting electroproduction data at low Q^2. The gluon conversion problem is handled in the same way as in the quark jet above. The distribution $F(x_1,x_2)$ thus obtained has no more freedom for adjustment. Insertion in (3.3) yields an unambiguous determination of the inclusive distribution which is shown by the solid line in Fig. 3. It agrees perfectly with the data.[15] This is another evidence in support of the valon picture and the recombination model for hadron production at low p_T.

IV. CONCLUSION

The valon picture for hadron structure has given us an over-constrained fit of all deep inelastic lepton scattering data at moderate to high Q^2. Only two parameters are involved. The resultant valon distributions constitute a highly economical way of summarizing all structure functions at higher Q^2. Shortly, we shall have available the quark, antiquark, and gluon distributions deduced from our valon distributions.[7] Since no guesses are involved in the initial quark and gluon distributions, our result should be more reliable than the conventional parameterization due to Field and Feynman[16] or Buras and Gaemers.

The valon distributions obtained remain to be related to the bound-state problem. Can a confinement potential be invented so that it can reproduce the known mass spectra on the one hand, and yield a momentum distribution in agreement with our result on the other? If so, then the bridge between the bound-state and scattering problems will have been established in terms of the valons.

We have described how the valon picture can be applied to the problem of hadronization of quarks. The result offers real hope for quantitative treatment of hadronic reactions. So far the success is limited to the fragmentation problems. We have not yet attempted to treat problems that involve valon-valon interaction. Our picture of valons differs from that of quarks in that they do not undergo hard collisions; that is, it is far more difficult to have a 90° scattering of two valons than of two quarks. To have a quantitative description of valon-valon interaction would be a major step toward understanding hadron interactions.

ACKNOWLEDGMENT

The hospitality of Academia Sinica in arranging my participation of the Guangzhou Conference is deeply appreciated. To experience a warm atmosphere of sharing a collective knowledge rather than staking individual claims was unique and gratifying. The work was supported in part by U.S. Department of Energy.

REFERENCES

1. See, for example, the review by H.Y.Tzu, Proc. of the Guangzhou Conference (1980)
2. See, for example, T.D. Lee, these proceedings.
3. See, for example, a review of the subject by R.C. Hwa, these proceedings.

4. R.C. Hwa, OITS-112 (1979), to be published in Phys. Rev.

5. R.C. Hwa, OITS-122 (1979), to be published in Phys. Rev.

6. P.C. Bosetti _et al._, Nucl. Phys. _B142_, 1 (1978); J.G.H. de Groot _et al._, Phys. Lett. _82B_, 292, 456 (1979).

7. R.C. Hwa and M. Zahir, work in progress.

8. A. Para (private communication).

9. W.S.C. Williams (private communication).

10. C.B. Newman _et al._, Phys. Rev. Lett. _42_, 951 (1979).

11. T.T. Chou and C.N. Yang, _High Energy Physics and Nuclear Structure_, edited by G. Alexander (North Holland, Amsterdam, 1967) p. 348; Phys. Rev. _170_, 1591 (1968).

12. V. Chang and R.C. Hwa, Phys. Rev. Lett. (to be published); OITS preprint to appear.

13. R. Brandelik _et al._, Nucl. Phys. _B148_, 189 (1979).

14. K.P. Das and R.C. Hwa, Phys. Lett. _68B_, 459 (1977).

15. G.W. Brandenburg (private communication); the permission to use the preliminary data of Fermilab Experiment No. E118 is appreciated.

16. R.D. Field and R.P. Feynman, Phys. Rev. D _15_, 2590 (1977).

17. A.J. Buras and K.J.F. Gaemers, Nucl. Phys. _B132_, 249 (1979).

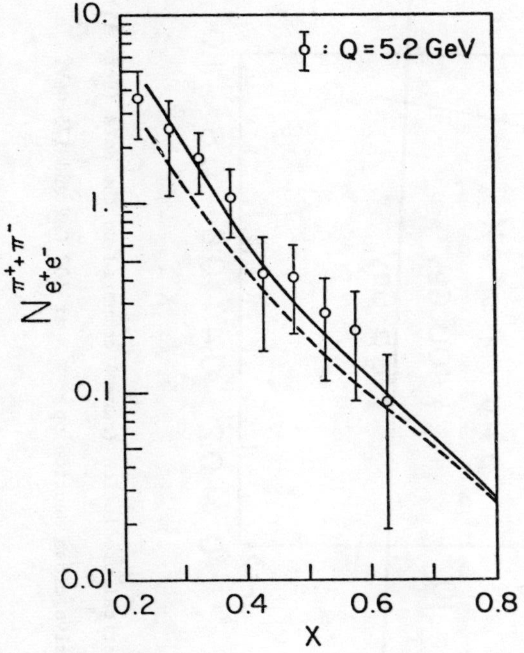

Fig. 1: Quark fragmentation function versus momentum fraction of pions. Curves are theoretical predictions with (solid) or without (dashed) resonance contribution. Data are from Ref. 13.

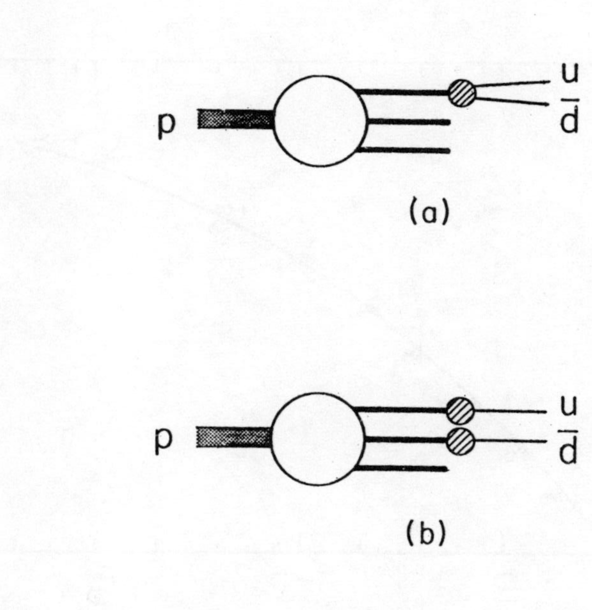

Fig. 2: Schematic diagrams for u and $\bar{d}$ quarks coming from (a) the same valon and (b) different valons.

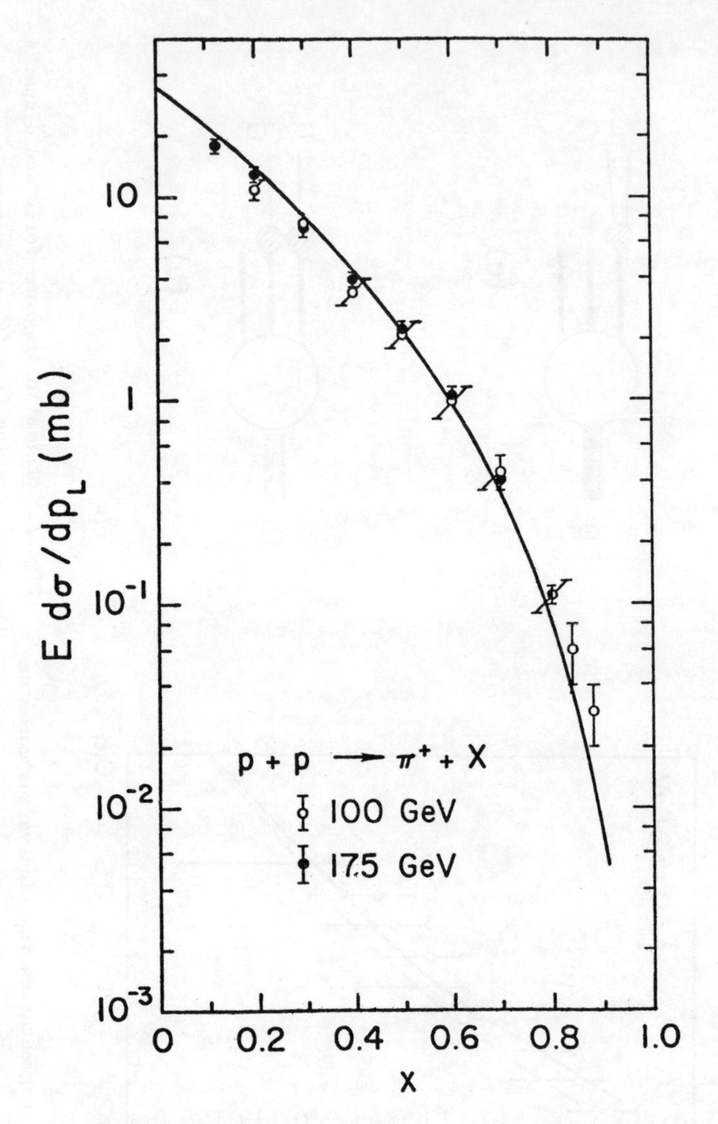

Fig. 3: Theoretical prediction (curve) compared to the data (Ref. 15) for the inclusive cross section $pp \rightarrow \pi^{+}X$ at 100 GeV and 175 GeV.

THE DYNAMICS AND PHENOMENOLOGY OF $Q\bar{Q}q\bar{q}$ STATES IN THE ADIABATIC APPROXIMATION

Chao Kuang-ta

Department of Physics, Beijing University

Abstract

The static potential energy for a $Q\bar{Q}q\bar{q}$ system is discussed in the adiabatic (Born-Oppenheimer) approximation. Using this potential and taking account of the color electric and magnetic forces the mass spectrum of $c\bar{c}q\bar{q}$ states is estimated. A possible assignment of the resonances in e^+e^- annihilation above 4-GeV as the $L=1$ $J^{PC}=1^{--}$ $(cq)_{\underline{3}^*}-(\bar{c}\bar{q})_{\underline{3}}$ states is investigated.

I. INTRODUCTION

In recent years, the theoretical study of the $qq\bar{q}\bar{q}$ systems has made significant progress.[1,2] The $c\bar{c}q\bar{q}$ states have also been proposed in order to explain the resonances in e^+e^- annihilation above 4-GeV.[3,4] De Rujula and Jaffe [5] have studied the S-wave $c\bar{c}q\bar{q}$ states in the spherical cavity approximation of the M.I.T. bag model, in which both the heavy quarks and the light quarks are treated in the same way as free Dirac particles confined within the bag. However, since the heavy quarks (Q, including c,b,t) are much more massive than the light quarks (q, including u,d,s) and are presumably moving slowly, we should probably study the problem in an adiabatic (Born-Oppenheimer) approximation.[6,7,8] In fact the successes of the non-relativistic quark model for charmonium may suggest the validity of the adiabatic approximation.

We have studied the dymamics and phenomenology of the $Q\bar{Q}q\bar{q}$ states [9,10] in the adiabatic approximation. Treating the heavy quarks as the static sources of the gluon fields, a static potential can be obtained in the framework of the M.I.T. bag model. When the separation of the heavy quark pair is small, apart from the outward pressure of the static color electric field lines, the light quarks would also exert a certain pressure on the surface and consequently the confining potential would deviate from the linear potential. When the separation is large, in the approximation of free light quarks, a linear flux tube could still be expected. However, since the interaction between the heavy quarks and the light quarks can not be ignored, the light quarks would screen the heavy quarks and hence the color configuration of the flux tube would be changed when the separation is very large. Therefore a molecule-like structure $(Q\bar{q})_{\underline{1}} - (\bar{Q}q)_{\underline{1}}$ and an ion-ion structure $(Qq)_{\underline{3}^*} - (\bar{Q}\bar{q})_{\underline{3}}$, $(Qq)_{\underline{6}} - (\bar{Q}\bar{q})_{\underline{6}^*}$, and $(Q\bar{q})_{\underline{8}} - (\bar{Q}q)_{\underline{8}}$ may be expected. Using the static potential obtained and taking into account the color electric and magnetic forces, we have estimated the spectra of the $c\bar{c}q\bar{q}$ states with various color configurations. We have found that it is possible to assign the observed resonances in the 4.1 and 4.4 GeV regions in e^+e^- annihilation as the P-wave $(cq)_{\underline{3}^*} - (\bar{c}\bar{q})_{\underline{3}}$ states, and that the

S-wave $c\bar{c}q\bar{q}$ states could be important sources of the J/ψ hadronic production.

II. THE STATIC POTENTIAL IN A SPHERICAL $Q\bar{Q}q\bar{q}$ BAG

When the separation $\vec{r}$ of $Q\bar{Q}$ is much smaller than the size of the bag, a sphere may be a good approximation to the bag's shape. To zeroth order in the quark-gluon coupling $g(\alpha_s \equiv \frac{g^2}{4\pi})$, the light quark fields confined to a spherical cavity with radius R satisfy the free Dirac equation and the frequency of the lowest mode is [11]

$$\omega(R) = \frac{2.04}{R} \tag{1}$$

for the massless light quarks. For a static heavy quark pair Q and $\bar{Q}$ located at $\frac{\vec{r}}{2}$ and $-\frac{\vec{r}}{2}$, the color charge density and current density are

$$\rho^a(\vec{x}) = g\frac{1}{2}\lambda_1^a\delta(\vec{x}-\frac{\vec{r}}{2}) + g\frac{1}{2}\lambda_2^a\delta(\vec{x}+\frac{\vec{r}}{2}) \tag{2a}$$

$$\vec{j}^a(\vec{x}) = 0 \tag{2b}$$

To lowest order in α_s the non-Abelian gluon self-coupling does not contribute, so we get eight sets of Abelian equations:

$$\nabla\cdot\vec{E}^a = \rho^a + \rho_3^a + \rho_4^a \tag{3a}$$

$$\nabla\times\vec{E}^a = 0 \tag{3b}$$

$$\nabla\cdot\vec{B}^a = 0 \tag{3c}$$

$$\nabla\times\vec{B}^a = \vec{j}_3^a + \vec{j}_4^a \tag{3d}$$

inside the bag and

$$\vec{n}\cdot\vec{E}^a = 0 \tag{3e}$$

$$\vec{n}\times\vec{B}^a = 0 \tag{3f}$$

on the surface. Here ρ_3^a (ρ_4^a) and $\vec{j}_3^a$ ($\vec{j}_4^a$) are the color charge density and current of the light quark (antiquark) respectively. In the following we shall concentrate on states in which both $Q\bar{Q}$ and $q\bar{q}$ are color singlets, i.e. $\lambda_1^a + \lambda_2^a = 0$, $\lambda_3^a + \lambda_4^a = 0$. Then we have $\rho_3^a + \rho_4^a \propto \lambda_3^a + \lambda_4^a = 0$, and

$$\nabla^2 \phi^a(\vec{x}) = -g\left(\frac{\lambda^a}{2}\right)\left[\delta\left(\vec{x} - \frac{\vec{r}}{2}\right) - \delta\left(\vec{x} + \frac{\vec{r}}{2}\right)\right] \tag{4}$$

where $\vec{E}^a = -\nabla\phi^a$. Solving the above equations,[7,11] the total energy stored in the light quarks, the gluon fields, and the bag (including the volume energy, the color electric energy, the color magnetic energy, the kinetic energy of the light quarks and the zero point energy) is found to be[10]

$$U(R) = \frac{4}{3}\pi R^3 B - \frac{4}{3}\frac{\alpha_s}{r} + \frac{4}{3}\frac{\alpha_s}{2R}\left[\frac{(r/2R)^2}{1-(r/2R)^4} + \frac{1}{2}\ln\frac{1+(r/2R)^2}{1-(r/2R)^2}\right] + \frac{A}{R} \tag{5a}$$

with

$$A = 2 \times 2.04 + \frac{4}{3}\alpha_s[2S(S+1)-3] \times 0.176 - 1.84 \tag{5b}$$

We see that the contributions of the light quarks, the color magnetic energy and the zero point energy take the form of A/R (A = 2.24 when ignoring the magnetic energy). The quadratic boundary condition is equivalent to minimizing U(R) with respect to R: $\partial U(R)/\partial R = 0$. The radius R which minimizes U as a function of r is plotted in Fig. 1. The confining potential $\bar{U}(r)$ (non-Coulombic part), the minimized energy, is shown in Fig. 2. For comparison, the results for a $Q\bar{Q}$ bag (A = 0) are also shown in Fig. 1 and Fig. 2. We see immediately that when $r \to 0$ for the $Q\bar{Q}$ bag $R \to 0$ and $\bar{U}(r)$ tends to zero linearly[7], whereas for the $Q\bar{Q}q\bar{q}$ bag

$$R(0) = \left(\frac{A}{4\pi B}\right)^{1/4} \sim 4.5\ \text{GeV}^{-1}, \quad \bar{U}(0) = \frac{4}{3}\frac{A}{R(0)} \sim 0.67\ \text{GeV}. \tag{6}$$

Here we use $\alpha_s \sim 2.2$ and $B^{\frac{1}{4}} \sim 0.145$ GeV which are determined from the spectra of the ordinary mesons and baryons.[11] The differences between $Q\bar{Q}$ and $Q\bar{Q}q\bar{q}$ are obviously caused by the fact that the light quark fields confined to a cavity also exert a certain pressure on the surface in addition to the outward pressure of the colour electric field lines and when the bag's size is small the light quarks become important. In fact, when $r \to 0$, $\frac{3}{4}$ of the total energy, (apart from the Coulombic part) comes from the light quarks. The potential obtained for small r can be used to estimate the masses of S-wave $Q\bar{Q}q\bar{q}$ states. For large r the spherical cavity can not be a good approximation and we should therefore study the problem in an arbitrary bag.

III. THE STATIC POTENTIAL IN AN ARBITRARY BAG

Next we shall deal with an arbitrary $Q\bar{Q}q\bar{q}$ bag using an energy variational method employed by Johnson.[8] The total energy of a bag is (ignoring the color indices for simplicity)

$$U = \int_{Bag} d\vec{x}\left[\frac{1}{2}(\nabla\phi)^2 + B\right] + E_\ell \tag{7}$$

with the field satisfying

$$-\nabla^2\phi = g'\left[\delta(\vec{x} - \frac{\vec{r}}{2}) - \delta(\vec{x} + \frac{\vec{r}}{2})\right] \tag{8}$$

$$\vec{n}\cdot\nabla\phi = 0 \tag{9}$$

and g' being the color charge of the static sources. E_ℓ stands for the energy of the light quarks plus the zero point energy and depends primarily on the size of the bag. If the light quarks are free inside the bag, the frequency of the lowest mode will be determined by the surface. In this case we may assume a general form of E_ℓ to be

$$E_\ell \sim \frac{C}{V^\alpha}, \quad 0<\alpha<1 \tag{10}$$

where V is the volume of the bag, and C is a constant. Let the surface which minimizes (7) be given by

$$|\vec{x}| = R(\hat{x}, r) \tag{11}$$

and the corresponding field be $\phi_0(\vec{x},r)$. For the scaled surface

$$|\vec{x}| = \frac{1}{\lambda} R(\hat{x}, \lambda r) \tag{12}$$

the corresponding field is $\phi_\lambda(\vec{x},r) = \lambda\phi_0(\lambda\vec{x},\lambda r)$ and the total energy is

$$U_\lambda = \lambda S(\lambda r) + \frac{1}{\lambda^3} BV(\lambda r) + \lambda^{3\alpha}\frac{C}{V(\lambda r)^3}$$

where

$$S(r) = \int_{B_0} d\vec{x}\, \frac{1}{2}(\nabla\phi_0)^2, \quad V(r) = \int_{B_0} d\vec{x}$$

and B_0 stands for the domain of space enclosed by the surface (11). Then $\left.\frac{\partial U_\lambda}{\partial\lambda}\right|_{\lambda=1} = 0$ gives

$$U_0(r) = S(r) + BV(r) + \frac{C}{V(r)^\alpha} = \frac{4B}{r}\int_0^r V(r')dr' + C(1-3\alpha)\frac{1}{r}\int_0^r \frac{dr'}{V(r')^\alpha} - \frac{g'^2}{4\pi r} \tag{13}$$

when $r \to 0$, using Eq. (13) we find

$$V(0) = \left(\frac{\alpha C}{B}\right)^{\frac{1}{1+\alpha}}, \quad \bar{U}(0) = (1+\alpha)\frac{C}{V(0)^\alpha} \tag{14}$$

We see that V(0) and $\bar{U}(0)$ are only related to the light quarks through C and α. Noting $E_\ell = \frac{A}{R}$ for the spherical surface, we then get $\alpha = \frac{1}{3}$, $C = A(\frac{4\pi}{3})^{1/3}$ from (10) and hence (14) will give the same result as in Section II.

When $r \to \infty$, the second term (due to the effect of the light quarks) on the right hand side of Eq. (13) vanishes for any $\alpha > 0$. The third term (Coulombic term) also vanishes. We then find that a linear flux tube with

$$V(r) \sim \frac{g'}{\sqrt{2B}} r \,, \qquad U_0(r) \sim g'\sqrt{2B}\, r \tag{15}$$

is still a consistent solution of Eq. (13), as in the case of a $Q\bar{Q}$ bag.[8]

IV. THE SCREENING EFFECTS OF THE LIGHT QUARKS

In Section III the linear flux tube is obtained only under the condition that the light quarks are treated as free Dirac particles confined in the bag. However, when r becomes large the interaction between the heavy quarks and the light quarks in fact can not be ignored. The light quarks may revolve around and screen the heavy quarks and therefore the colour configuration of the sources would be changed. In the fully screened case, $q(\bar{q})$ may neutralize $\bar{Q}(Q)$ in a color singlet and a molecule-like structure $(Q\bar{q})_{\underline{1}}$ - $(\bar{Q}q)_{\underline{1}}$ is expected. In the "partially screened" case, $Qq(\bar{Q}\bar{q})$ may form an ion in color $\underline{3}^*(\underline{3})$ and an ion-ion like structure $(Qq)_{\underline{3}^*}$ - $(\bar{Q}\bar{q})_{\underline{3}}$ (T-baryonium) may be expected. The $(Qq)_{\underline{6}}$ - $(\bar{Q}\bar{q})_{\underline{6}^*}$ (M-baryonium) and $(Q\bar{q})_{\underline{8}}$ - $(\bar{Q}q)_{\underline{8}}$ (P-mesonium) structures could also be possible but they may be difficult to excite since there could be a repulsive force between the two quarks in the ion such as $(Qq)_{\underline{6}}$ and $(Q\bar{q})_{\underline{8}}$.

In the case of the ion-ion like states, when the separation r between the two ions is large the sources of the gluon fields may be approximated as point-like sources and Eq. (8) can still be used with g' representing the effective color charge of the ion. It is easy to see $g' \propto \sqrt{C}$ where C is the eigenvalue of the color SU(3) Casimir operator for the ion. Following the procedure mentioned in Section III, we then find the static potential between the two ions to be

$$U(r) \sim \sqrt{8\pi\alpha_s BC}\; r \tag{16}$$

This linear potential can be used to estimate the spectra of high angular momentum $Q\bar{Q}q\bar{q}$ states with various color configurations.

V. PHENOMENOLOGY

We have discussed the phenomenology of the $c\bar{c}q\bar{q}$ states,[9] including their mass spectra, decay, production in hadron-hadron collision and e^+e^- annihilation, as well as photoproduction. Here we shall only concentrate on the high angular momentum ($L \gtrsim 1$) T-baryonium $(cq)_{\underline{3}^*} - (\bar{c}\bar{q})_{\underline{3}}$ states because they may offer the optimal experimental situation. For these states, the energy are mainly determined by the linear potential $U(r) = \lambda r$ between the two ions (see (16)). For states having angular momentum L the energy E_L is found to be

$$E_L = E_0 + \left(\frac{\lambda^2}{M_0}\right)^{1/3}(\zeta_L - \zeta_0) \tag{17}$$

where M_0 is the effective mass of the ion, $E_0 \sim 2M_0$, $\zeta_0 = 2.338$, $\zeta_1 = 3.361$, ... Considering that the color electric interaction may take the form

$$\frac{G^2}{4}\sum_{a=1}^{8}\lambda_i^a\lambda_j^a \tag{18}$$

the mass of the $(cq)_{\underline{3}^*}$ ion is then given by

$$M_0 = \mu_c + \mu_q + \frac{2}{3}G^2 \tag{19}$$

where $\mu_c(\mu_q)$ stands for the effective mass of the c(q) quark and can be determined by taking account of the color magnetic force:

$$\mu_u = \mu_d = \frac{1}{8}(3m_\rho + m_\pi) \sim 0.30\,\text{GeV},$$

$$\mu_s - \mu_u = \frac{1}{4}(3m_{K^*} + m_K - 3m_\rho - m_\pi) \sim 0.18\,\text{GeV}, \quad \mu_c = \frac{1}{8}(3m_\psi + m_{\eta_c}) \simeq \frac{1}{2}m_\psi \sim 1.55\,\text{GeV}.$$

The effective color electric coupling G^2 may be determined by the mass difference between a possible T-baryonium T(2.15 GeV) ($L = 1$, $J^P = 3^-$) and a possible M-baryonium X(2.20 GeV) ($L = 1$, $J^P = 3^-$): $G^2 \sim 0.084$ GeV.

From (16), we see that the linear potential used here for $(cq)_{\underset{\sim}{3}^*} - (\bar{c}\bar{q})_{\underline{3}}$ should be the same as in $c\bar{c}$, since they have the same colour flux tube, so using $m_{\psi'} - m_\psi \sim 0.59$ GeV with $\mu_c \sim 1.55$ GeV, we get $\lambda \sim 0.24\ \text{GeV}^2$. Substituting all these parameters into (17) and (19), E_L can then be obtained. In particular, when $L = 1$ we get $E_1 \sim 4.13$ GeV for q = u,d and $E_1 \sim 4.47$ GeV for q = s. Furthermore, considering that the color magnetic interaction

$$-\beta \sum_{a=1}^{8} \lambda_i^a \lambda_j^a \vec{\sigma}_i \cdot \vec{\sigma}_j \tag{20}$$

causes hyperfine splittings (h.f.s.) the mass for the ion having spin S=0 or S=1 would respectively become

$$S = 0: \quad M_0 \longrightarrow M_0 - 8\beta \tag{21a}$$

$$S = 1: \quad M_1 \longrightarrow M_0 + \frac{8}{3}\beta \tag{21b}$$

where β is determined by the h.f.s. of the charmed mesons:

$$\beta_{cu} = \frac{3}{64}(m_{D^*} - m_D) \sim 6.6\,\text{MeV}, \qquad \beta_{cs} = \frac{3}{64}(m_{F^*} - m_F) \sim 5.2\,\text{MeV}.$$

We then get the mass spacings among states having the same L but different S_i (spin of the i-th ion, i = 1,2). For L = 1 states, we have also considered the contribution of the $\vec{L}\cdot(\vec{s}_1 + \vec{s}_2)$ and the tensor forces. The predicted spectrum for L = 1 $(cq)_{\underline{3}*}$ - $(\bar{c}\bar{q})_{\underline{3}}$ states is shown in Table I.

Next we discuss the possibility of assigning the observed resonances [12] in the 4.1 and 4.4 GeV regions in e^+e^- annihilation as the L = 1, $J^{pc} = 1^{--}$ $(cq)_{\underline{3}*}$ - $(\bar{c}\bar{q})_{\underline{3}}$ states. (1) The predicted L = 1, $J^{pc} = 1^{--}$ $(cq)_{\underline{3}*}$ - $(\bar{c}\bar{q})_{\underline{3}}$ states just lie in the same region as those resonances: 4.03 - 4.17 GeV (q = u,d), 4.39 - 4.5 GeV (q = s). Although there may be some uncertainties related to the choice of the parameters involved, the relative spacings, which are essentially determined by $m_{D*} - m_D$ (q=u,d), $m_{F*} - m_F$ (q=s), and $\mu_s - \mu_u$ (determining the spacing between the resonances in 4.1 GeV region and the resonances in 4.4 GeV region), should be reliable. (2) These $(cq)_{\underline{3}*}$ - $(\bar{c}\bar{q})_{\underline{3}}$ states can decay into a charmed meson pair via light quark rearrangement (see Fig. 3). The states at 4.03 and 4.17 GeV (q = u,d) would decay into a D meson pair, whereas the states in 4.4 GeV region (q = s) may mainly decay into a F meson pair. This is compatible with the DESY experiments.[14] But these decays should be suppressed by the angular momentum barrier and their widths could be of the order of a few tens of Mev. Furthermore, according to the picture assumed in the adiabatic approximation, during the time that the light quark transition (rearrangement or creation) takes place the heavy quarks are essentially static, therefore the probability for heavy quark rearrangement, which will lead to $\psi(\eta_c)$ production, should be extremely small. The suppression factor of heavy quark rearrangement could be of order of $\left(\frac{m_q}{m_c}\right)^2 - \left(\frac{m_q}{m_c}\right)^4$, [9] i.e. $\frac{1}{25} - \frac{1}{625}$ (using $\frac{m_u}{m_c} \sim \frac{1}{5}$). Hence our assignment may not be incompatible with the ψ inclusive production experiments.[15]

(3) More importantly, these L = 1 $c\bar{c}q\bar{q}$ states can also decay, by a light quark pair creation, into a charmed meson and an L = 0 $c\bar{q}q\bar{q}$ state which will in turn readily dissociate into a charmed meson plus an ordinary meson (see Fig. 4). For q = u, d the following decays are expected:

$$c\bar{c}q\bar{q}\,(L=1,\ J^P=1^-) \rightarrow c\bar{q}q\bar{q}\,(L=0,\ J^P=0^+) + \bar{D}^*$$
$$\quad\quad\quad\quad\quad\quad\quad\quad \hookrightarrow D + \pi$$

$$c\bar{c}q\bar{q}\,(L=1,\ J^P=1^-) \rightarrow c\bar{q}q\bar{q}\,(L=0,\ J^P=1^+) + \bar{D}$$
$$\quad\quad\quad\quad\quad\quad\quad\quad \hookrightarrow D^* + \pi$$

Since these cascade decays proceed via the S-wave whereas decays into $D\bar{D}$, $D\bar{D}^*$, $D^*\bar{D}^*$ proceed via the P-wave, at least the phase space will favour former. At 4.03 GeV, in particular, part of the observed $D^*\bar{D}^*$ enhancement which led to proposing D^*-$\bar{D}^*$ type charmonium molecule [4] might actually be caused by the $D^*\bar{D}\pi$ events via these cascade decays.

(4) Assuming that the quark content scatters independently at high energies, in photoproduction the cross section of these $c\bar{c}q\bar{q}$ states could be even larger than that of the $c\bar{c}$ state ψ. Using vector dominance, we find [9]

$$\sigma(\gamma p \to c\bar{c}q\bar{q} + X) \sim 1.5 - 3.5\,\mu b$$

This suggests the possibility of partly ascribing the difference of the order $\sim 2 - 5\ \mu b$ [16] between the data and the vector dominance model calculation in $\sigma_{Tot}(\gamma p)$ in the energy range 30 - 185 GeV to the production of $c\bar{c}q\bar{q}$ states. Consequently copious charmed meson pair production via these $c\bar{c}q\bar{q}$ states may be expected in photoproduction.

To sum up, our model for $c\bar{c}q\bar{q}$ states may offer a possible explanation for the observed resonances in e^+e^- annihilation and may be a viable alternative to the standard charmonium model proposed by the Cornell group.[13]

In addition, we also argue that in hadronic production, the S-wave $c\bar{c}q\bar{q}$ states and $(c\bar{c})_{\underline{8}} - (q\bar{q})_{\underline{8}}$ states could be important sources of ψ production.[9]

References

1. R.L. Jaffe, Phys. Rev. D13, 267, 281 (1977), D17, 1444 (1978).
2. H.M. Chan and H. Høgaasen, Phys. Lett. 72B, 121, 400 (1977). Nucl. Phys. B136, 401 (1978).
3. G.L. Shaw et al, Phys. Rev. Lett. 36, 695 (1976), C. Rosenzweig, Phys. Rev. Lett. 36, 697 (1976).
4. A. De Rujula et al. Phys. Rev. Lett. 38, 317 (1977).
5. A. De Rujula and R.L. Jaffe, M.I.T. preprint CTP-658 (1977).
6. P. Hasenfratz and J. Kuti, Phys. Rep. 40C, 75 (1978).
7. L. Heller and K. Johnson, Phys. Lett. 84B, 501 (1979).
8. K. Johnson, in AIP Conf. Proc. No. 48, P112 (1978).
9. K.T. Chao, Oxford University preprint TP 27/79.
10. K.T. Chao, Oxford University preprint TP 53/79.
11. T. De Grand et. al, Phys. Rev. D12, 2060 (1975).
12. G. Feldman, in 1978 Proceedings of the Tokyo Conference on high energy physics pp. 777-789.
13. E. Eichten et al, Phys. Rev. Lett. 36, 500 (1976), Phys. Rev. D17, 3090 (1978). Cornell University preprint, CLNS-425 (1979).
14. R. Brandelik et al, Phys. Lett. 80B, 412 (1979).
15. J. Burmester et al, Phys. Lett. 68B, 283 (1977).
16. Wonyong Lee, in 1977 Proceedings of the Hamburg Conference on lepton and photon interaction

Table I. The masses and quantum numbers for L = 1 $(cq)_{\underline{3}^*}$ - $(\bar{c}\bar{q})_{\underline{3}}$ states. Here $S_1(S_2)$ is the spin of ion 1(2), $S = S_1 + S_2$.

$S_1 \times S_2$	S	J^{PC_n}	q = u, d		q = s	
			I^G	mass (GeV)	I^G	mass (GeV)
0 × 0	0	1^{--}	0^-, 1^+	4.03	0^-	4.39
(1×0) ±(0×1)	1	$0^{-\mp}$	$0^\mp$, $1^\pm$	4.05	$0^\mp$	4.40
		$1^{-\mp}$		4.08		4.42
		$2^{-\mp}$		4.12		4.46
1 × 1	0	1^{--}	0^-, 1^+	4.17	0^-	4.50
	1	0^{-+}	0^+, 1^-	4.06	0^+	4.41
		1^{-+}		4.17		4.50
		2^{-+}		4.19		4.51
	2	1^{--}	0^-, 1^+	4.06	0^-	4.41
		2^{--}		4.19		4.51
		3^{--}		4.21		4.53

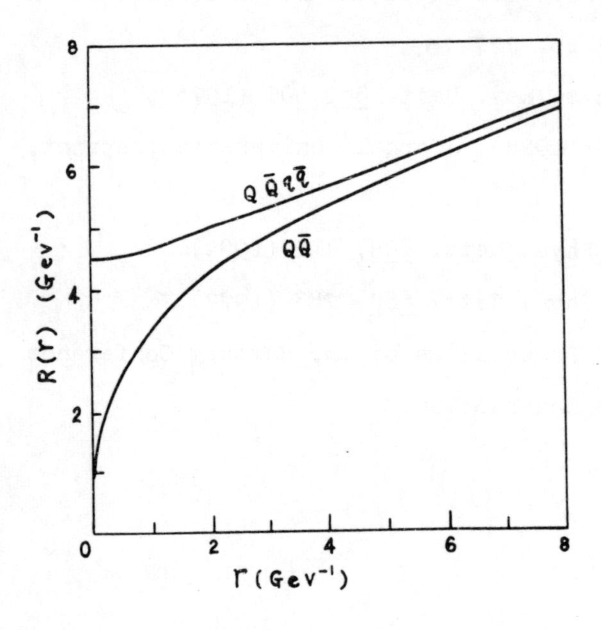

Fig. 1 The radius R of a spherical bag as a function of the separation r of the heavy quark pair. Here A= 2.24 for $Q\bar{Q}q\bar{q}$ states and A=0 for $Q\bar{Q}$ states.

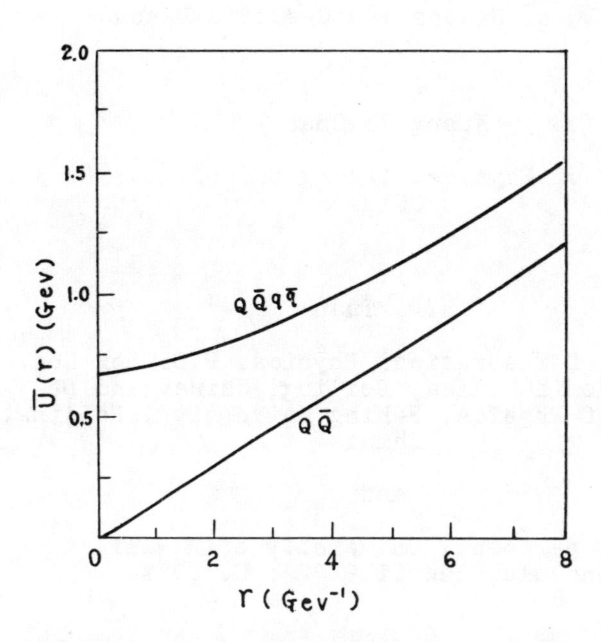

Fig. 2 The confining potential $\overline{U}$ as a function of the separation r of the heavy quark pair. Here A=2.24 for $Q\bar{Q}q\bar{q}$ and A=0 for $Q\bar{Q}$ states.

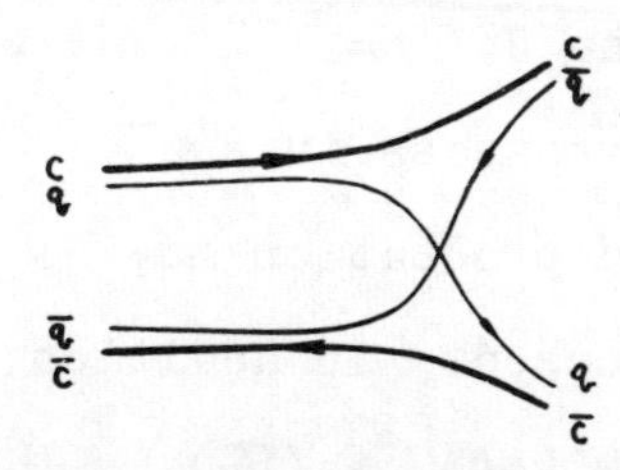

Fig. 3 Decay of an L=1 $c\bar{c}q\bar{q}$ state into a charmed meson pair via light quark rearrangement.

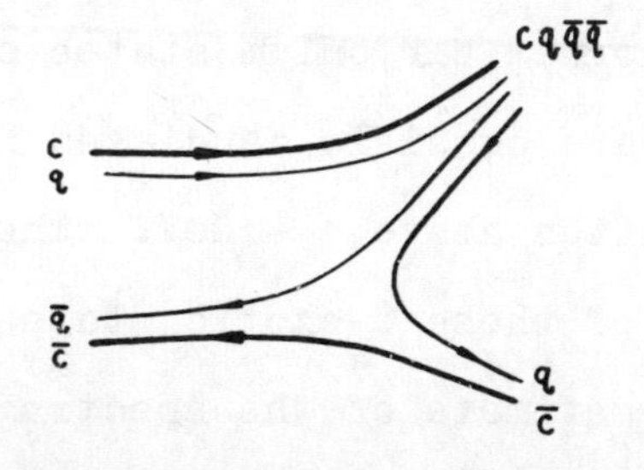

Fig. 4 Decay of an L=1 $c\bar{c}q\bar{q}$ state into an L=0 $c\bar{q}q\bar{q}$ state with a charmed meson via quark pair creation.

P-Wave $Q\bar{Q}$ $q\bar{q}$ States and C-exotic Mesons

Kuang-Ta Chao

Department of Physics, Peking University, Beijing, China

and

S.F. Tuan

Institute of Theoretical Physics, P.O. Box Nos. 2735, Academia Sinica, Beijing, China, and Department of Physics, Peking University, Beijing, China

and

Physics Department, University of Hawaii at Manoa, Honolulu, Hawaii 96822, U. S. A.

Abstract

Based on a quark-gluon model, a special type of C-exotic states, namely P-wave $(cq)_{\underline{3}*}$- $(\bar{c}\bar{q})_{\underline{3}}$ baryonium states are suggested. Their masses are nearly degenerate with C-normal baryonium states of which the $J^{PC} = 1^{--}$ members are possible candiates for the resonances in e^+e^- annihilation above 4 - GeV. The decay and production mechanism of these C-exotic states are discussed. In addition, a rough estimate of the spectrum of P-wave $(bq)_{\underline{3}*}$- $(\bar{b}\bar{q})_{\underline{3}}$ baryonium states is made in anticipation of the experimental possibilities opened up by the CESR machine. Finally it is speculated that some of the striking mass degeneracies noted between C-exotic and C-normal members of the $Q\bar{Q}$ $q\bar{q}$ model with the same spin-parity, may be manifestations of hadronic level classification under both broken SU(3) and nonchiral SU(2) symmetries suggested many years back.

I. INTRODUCTION

Multi-quark hadron states ($q\bar{q}q\bar{q}$, $c\bar{c}q\bar{q}$, $b\bar{b}q\bar{q}$, etc.) are, given the basic physics principles we work with, an unavoidable feature of hadron systems. The question is really one of how to identify these states, in as unambiguous a way as possible, given the reality that they must in general be sorted out from a background of perhaps dominant $q\bar{q}$, $c\bar{c}$, $b\bar{b}$ etc. type states. It has been known for sometimes [1] that the C-exotic mesons first proposed by Gell-Mann[2], though not coupled to the quark-antiquark system with L-excitation nor with the S-wave states formed out of say the $q\bar{q}q\bar{q}$ configuration, can nevertheless by formed as P-wave states from a four quark system. The observation of such a state will be of the utmost importance towards establishing the reality of multi-quark hadrons, if its physical parameters (e.g. mass value) are sufficiently distinctive. In this connection, four quark system involving the heavier quarks, e.g. $c\bar{c}q\bar{q}$, $b\bar{b}q\bar{q}$, are particularly useful in that the predicted C-exotic mesons are of sufficiently high mass, and hence can be differentiated from C-exotic gluonium (glue-balls) generally expected in the low mass region around 1300 MeV.[3] A particular model of $c\bar{c}q\bar{q}$ states has been proposed by one of us[4] in which the P-wave states for both the C-normal and C-exotic varieties of given spin-parity are expected to be (very nearly) degenerate and with mass values and especially mass spacings rather clearly delineated in the region between 4 to 4.5 GeV. The present paper extends the earlier work[4] by (i) making a fairly detailed analysis (to the extent possible) of the properties of these C-exotic four quark system to guide experimental search, and (ii) to map out at least roughly the spectrum of $b\bar{b}q\bar{q}$ states in anticipation of the experimental possibilities opened up by the CESR machine.

In order to explain the structures in the e^+e^- annihilation cross section above 4 GeV, states of the form $c\bar{c}q\bar{q}$ have been proposed by several groups of authors.[5,6] In particular De Rujula et al.[6] suggested that the 4.03 GeV structure is associated with a P-wave $D^*\bar{D}^*$ molecule and the 4.4 GeV structure with an S-wave molecule composed of a D (or D^*) and a P-wave $\bar{D}^{**}$. However, the predicted "charm burning" at 4.4 GeV is not compatible with the observed inclusive production of J/ψ [7]. Furthermore, their assignment of 4.4 GeV would make it difficult to explain the copious η (presumably via F meson) production in that energy region.[8]

One of us has studied the $c\bar{c}q\bar{q}$ states [4] based on a quark-gluon model in the adiabatic (Born-Oppenheimer) approximation. Taking account of the colored magnetic and electric forces, the predicted P-wave T-baryonium $(cq)_{\underline{3}^*} - (\bar{c}\bar{q})_{\underline{3}}$ spectra bear close resemblance to the experimental data. In the adiabatic approximation the slow moving heavy quarks (c and $\bar{c}$) may be treated as static sources and gluon fields and light quarks (q and $\bar{q}$) have time to continually readjust to the sources. These states can decay into charmed mesons mainly via light quark transition (rearrangement or creation). During the time that the light quark transitions take place, the heavy quarks are essentially static - they hardly change their positions, therefore the probability for heavy quark rearrangement, i.e. the so called "charm burning" should be very small. Assuming the suppression of heavy quark rearrangement to be of the the same order as that of creating heavy quark pair, we may expect this suppression to be of order of $(m_q/m_c)^2 - (m_q/m_c)^4$, (9,10,11) i.e. 1/25 - 1/625, (using $m_u/m_c \sim 1/5$) relative to light quark rearrangment. Therefore "charm burning" may not be a serious problem for our model.

The charmonium molecules proposed by De Rujula et al. (6) are essentially the $(c\bar{q})_{\underline{1}} - (\bar{c}q)_{\underline{1}}$ states and are expected to have fairly large decay widths since they can easily dissociate into charmed mesons unless they lie very close to the charmed meson pair threshold (as may be the case for the 4.03 GeV e^+e^- structure). Therefore it seems unlikely that <u>all</u> the sharp structures above 4 GeV can be ascribed to these molecules. In contrast , the P-wave T-baryonium $(cq)_{\underline{3}*}$ - $(\bar{c}\bar{q})_{\underline{3}}$ states are estimated to be below the charm baryon pair threshold and they would decay into charmed mesons only through (light) quark rearrangement and creation. These decays as discussed earlier[4] would therefore be suppressed and hence they should be fairly narrow resonances say, a few tens MeV. This is compatible with the data.

An interesting question is how to differentiate our $c\bar{c}q\bar{q}$ states from that of standard charmonium. The Cornell group[10] has worked out a sophisticated model in which the structures in e^+e^- annihilation above 4 GeV are assigned as 3^3S_1, 2^3D_1, 4^3S_1 $c\bar{c}$ states. The model seems to be very successful in accounting for many aspects of these states. However some residual problems remain: the 4.16 GeV bump has a rather large leptonic decay width ($\Gamma_{e^+e^-} \sim 0.7$ KeV) experimentally[12] and can hardly be a D-wave $c\bar{c}$ states; the $c\bar{c}$ model can not accomodate a level between 3.77 GeV and 4.03 GeV, nor one between 4.03 GeV and 4.16 GeV, both such structures seem to be suggested by experiments.[13] Therefore it is well worth examining our $\bar{c}c\bar{q}q$ model in the context of the 4 GeV situation at least as a competing if not alternative model to standard charmonium. An unambiguous differentiation between our model and the standard charmonium model would be the <u>prediction of C-exotic states, the masses of which are predicted to be degenerate with C-normal</u>

type states in our $c\bar{c}q\bar{q}$ model. The C-normal $c\bar{c}q\bar{q}$ states may be affected by mixing with nearby $c\bar{c}$ states of the same quantum numbers, and hence the degeneracy mentioned above may in practice be only an approximate one. Nevertheless with sharply predicted masses for the C-exotic members, they in principle offer the most unambiguous differentiation from standard charmonium since the $c\bar{c}$ system can not couple to C-exotic $J^{PC} = 0^{--}, 1^{-+}, 2^{+-}$, etc. states because of C-conservation in strong interactions.

In section II the spectra, decay, and production characteristics for the T-baryonium $(cq)_{\underline{3}*} - (\bar{c}\bar{q})_{\underline{3}}$ C-exotic states in the 4 GeV region are delineated. Section III extends the model (despite some theoretical uncertainties) to the $b\bar{b}q\bar{q}$ system for both C-normal and C-exotic states - in anticipation of the experimental possibilities opened up by the CESR machine. We conclude the paper in section IV with some summary comments including the possible implications of our model for earlier work on C-exotic states[14].

II. C-EXOTIC MESONS: SPECTRUM, DECAY AND PRODUCTION

Our model for $Q\bar{Q}q\bar{q}$ system[4] is based on the adiabatic (Born-Oppenheimer) approximation in which the heavy quark pair Q, $\bar{Q}$ is treated as static sources whose motion is instantaneously followed by the gluon fields and the light quarks. When the separation of the heavy quark pair is large, owing to the interaction between heavy and light quarks, the light quarks may revolve around and screen the heavy quarks and therefore the color configuration of the sources would be changed. In the fully screened case, a molecule-like structure $(Q\bar{q})_{\underline{1}} - (\bar{Q}q)_{\underline{1}}$ is expected. In the "partially screened" case, ion-ion like structures such as T-baryonium $(Qq)_{\underline{3}*} - (\bar{Q}\bar{q})_{\underline{3}}$ or M-baryonium $(Qq)_{\underline{6}} - (\bar{Q}\bar{q})_{\underline{6}*}$, P – mesonium $(Q\bar{q})_{\underline{8}} - (\bar{Q}q)_{\underline{8}}$ are expected. <u>There would be a color flux tube between the two ions</u>.

We have estimated the masses of $c\bar{c}q\bar{q}$ states(4) with various color configurations by taking account of the confining potential between two ions, the color electric and magnetic forces between the heavy quarks and the light quarks and solving the Schrödinger equation for the ion-ion system. Input into the calculation involves largely non controversial masses of $(\pi, \rho, K, K^*, J/\psi)$ and D*-D, F*-F mass differences, though some uncertainty is introduced in the estimate of color-electric force because the input here involves a <u>choice</u> of possible candidates for $q\bar{q}q\bar{q}$ states.(4) We believe the mass estimates are valid up to 100 MeV uncertainty at most, in particular the relative mass spacing between J^P states of a given L (orbital angular momentum between the two ions) should be especially well predicted. The P-wave (L=1) T-baryonium $(cq)_{\underline{3}*} - (\bar{c}\bar{q})_{\underline{3}}$ states are predicted to lie in the region 4.03 to 4.17 GeV (q = u,d) and 4.39 - 4.5 GeV (q=s). These are just in the same region as the observed $J^{PC} = 1^{--}$ resonances in e^+e^- annihilation, and we discussed the possibility of assigning these $(cq)_{\underline{3}*}-(\bar{c}\bar{q})_{\underline{3}}$ states as the observed resonances. If this assignment makes sense, then one striking consequence of the model is the prediction of <u>mass degenerate</u> states of given J^P but of opposite C-parity. For instance in the P-wave sector of the $c\bar{c}q\bar{q}$ system, $J^{PC} = 0^{-\mp}$ states are predicted at 4.05 GeV and $J^{PC} = 1^{-\mp}$ states are predicted at 4.08 GeV, to name a few.

The complete roster of P-wave C-exotic T-baryonium $c\bar{c}q\bar{q}$ states in the region 4.0 - 4.5 GeV can be summarized as follows. For q = u,d, they are $J^{PC} = 0^{--}$, $I^G = 0^-$ and/or 1^+ at 4.05 GeV and $J^{PC} = 1^{-+}$, $I^G = 0^+$ and/or 1^- at 4.08 GeV and 4.17 GeV. For q = s, they are $J^{PC} = 0^{--}$, $I^G = 0^-$ at 4.40 GeV and $J^{PC} = 1^{-+}$, $I^G = 0^+$ at 4.42 GeV and 4.50 GeV.

These states have important strong decays (via the P-wave) into charmed meson pairs through light quark rearrangement. For q = u,d, $0^{--} \rightarrow D\bar{D}^* + D^*\bar{D}$ and $1^{-+} \rightarrow D\bar{D}^* - D^*\bar{D}$, $D^*\bar{D}^*$. For q = s, $0^{--} \rightarrow F\bar{F}^* + F^*\bar{F}$ and $1^{-+} \rightarrow F\bar{F}^* - F^*\bar{F}$, $F^*\bar{F}^*$. More importantly they can also decay via the S-wave into an S-wave $cq\bar{q}\bar{q}$ state and a charmed meson by a light quark pair creation. For instance the L = 1 $c\bar{c}q\bar{q}$ states

$$\begin{aligned} 0^{--}(q=u,d) &\rightarrow (cq\bar{q}\bar{q})(L=0, J^P=0^+) + \bar{D} \rightarrow D + \pi + \bar{D} \\ 1^{-+}(q=u,d) &\rightarrow (cq\bar{q}\bar{q})(L=0, J^P=1^+) + \bar{D} \rightarrow D^* + \pi + \bar{D} \end{aligned} \quad (2.1)$$

These decay modes are sometimes called cascade decays[15] or molecular transitions.[6] The other type of cascade decay such as $c\bar{c}q\bar{q}$ (L=1) $\longrightarrow$ $c\bar{c}q\bar{q}$ (L=0) + π should be greatly suppressed according to the adiabatic approximation picture since here heavy quark transition (from P-wave to S-wave) is needed.

The strong decays of C-exotic T-baryonium states into normal hadrons will also be suppressed since a heavy quark pair annihilation is involved. The suppression factor may be comparable to charm burning (i.e. heavy quark rearrangement) say of order $(m_q/m_c)^2$ - $(m_q/m_c)^4$ or 1/25 - 1/625 of a total width of a few tens MeV, for instance. Many of these states have rather undistinguished 3π, 4π, 5π, etc. normal hadron channels open to them. A partial list of decay transitions for states (J^{PC}; I^G) is:-

$$
\begin{aligned}
&(0^{--};\ 0^-) \longrightarrow \rho\pi,\ 3\pi,\ \rho'(1600)\pi,\ etc.\\
&(0^{--};\ 1^+) \longrightarrow \omega\pi,\ \omega'\pi\ (\omega'\text{: radial excitation of } \omega),\text{etc.}\\
&(1^{-+};\ 0^+) \longrightarrow \eta+\eta',\ 2\rho,\ 2\omega,\ 2\phi,\ \pi A_1,\ KQ,\ K^*\bar{K}^*,\ K^*\bar{K}-\bar{K}^*K,\ \text{etc.}\\
&(1^{-+};\ 1^-) \longrightarrow 3\pi,\ \rho\omega,\ K^*\bar{K}^*,\ K^*\bar{K}-\bar{K}^*K,\ \text{etc.}
\end{aligned}
\qquad (2.2)
$$

Production of C-exotics poses considerable problems. In addition to the absence of $q\bar{q}$ coupling, search for $J^{PC} = 1^{-+}$ states is also inhibited by the absence of coupling to two real photons according to Yang's theorem(16)- hence we are deprived of the Weizsäcker-Williams type production mechanism(17) in e^+e^- two photon physics. In hadron physics, half the momentum of an energetic proton is not in quark-momentum and is presumably carried by gluon degrees of freedom. It has been suggested(18) that sizable production cross-section for gluonia (bound off-mass shell gluons) should be possible in hadron-hadron collision. Since $(J^{PC}; I^{G}) = (1^{-+}; 0^{+})$ states can be formed from two gluons, this might in principle be an effective mechanism for production. In practice however estimates of two gluon production mechanism [19] leading to spin 1 massive states should be taken with caution since $\Gamma(J=1 \rightarrow gg) = 0$ in lowest order also.[16]

The P-wave C-exotic states can be explored as the end product of a molecular transition from a normal T-baryonium vector state $(1^{--}; 0^{-}$ or $1^{+})$ reached in $e^{+}e^{-}$ annihilation,[20] e.g.

$$(1^{--}; 0^{-} \text{ or } 1^{+}) \rightarrow (1^{-+}; 0^{+} \text{ or } 1^{-}) + \gamma \qquad (2.3)$$

The $J^{PC} = 1^{-+}$ states could for instance be detected by the observation of peaks in the momentum distribution of low energy photons (a spin-parity analysis of the dominant decay products is however needed to establish $J^{P} = 1^{-}$!). We note that some of the $(1^{--}; 0^{-}$ or $1^{+})$ and $(1^{--}; 0^{-})$ states are predicted[4] at 4.17 and 4.5 GeV respectively, these could decay via (2.3) to lower mass $(1^{-+}; 0^{+}$ or $1^{-})$ and $(1^{-+}; 0^{+})$ states at the respective masses 4.08 and 4.42 GeV.

Finally Bjorken[18,21] has stressed production of gluonia via a virtual γgg channel in $J/\psi \rightarrow \gamma$ + gluonium (and $\Upsilon(9.46) \rightarrow \gamma$ + gluonium if gluonia prove to be unreasonably massive ~ 3 GeV). For the even heavier $(1^{-+}; 0^{+})$ T-baryonium states (which couple to the appropriate gluonium), one could in principle utilize also $\Upsilon(9.46) \rightarrow \gamma + (1^{-+}; 0^{+})$

though it would be difficult to hazard a guess on the net signal. Another possibility is that clean gluon-jets can be isolated in Υ or more massive onia decays, it is plausible that the "leading particle" in a two-gluon jet will often be (massive) gluonium which can then couple to $(1^{-+}; 0^{+})$ $c\bar{c}q\bar{q}$ states.

III. $b\bar{b}q\bar{q}$ SYSTEM

Our model for $c\bar{c}q\bar{q}$ states can be easily generalized to the $b\bar{b}q\bar{q}$ system . In fact, because b is heavier than c, the adiabatic approximation picture could at first sight be even better for the $b\bar{b}q\bar{q}$ case. The spectra of various $b\bar{b}q\bar{q}$ states can be obtained using the method of ref. (4). However, our formulation for $b\bar{b}q\bar{q}$ is in fact <u>less</u> reliable for low angular momentum states e.g. the P-wave states under discussion. This is because the separation of b and $\bar{b}$ in the P-wave $b\bar{b}q\bar{q}$ system may be too small to form an ion-ion like structure. For the P-wave $b\bar{b}$ bound states the separation r between b and $\bar{b}$ is estimated to be $r \sim 0.4$ fm[(10)], whereas we believe our classification for $Q\bar{Q}q\bar{q}$ states will make sense only when $r \gtrsim 2R$ (here R is the radius of the ideal vortex tube which is estimated to be ~ 0.5 fm[(22)]). In otherwords, there may be too much overlap between the two ions, and hence our mass formula may loose its meaning here. However it cannot be ruled out that (for unknown dynamical reasons) a situation may arise in which the light quark clouds form dense clusters around the heavy quarks and hence our classification and mass estimates would remain useful. For completeness and in anticipation of CESR's experimental capabilities, we sketch below the spectra for the $b\bar{b}q\bar{q}$ system.

As in ref. (4), the color magnetic force between two quarks takes the form

$$-\beta \sum_{a=1}^{8} \lambda_1^a \lambda_2^a \vec{\sigma}_1 \cdot \vec{\sigma}_2 \tag{3.1}$$

Then the effective mass of the b-quark is given by

$$\mu_b = \frac{1}{8}(3 m_\Upsilon + m_{\eta_b}) \tag{3.2}$$

If the mass spacing between $\Upsilon(9.46)$ and the expected $(0^{-+}; 0^+)$ η_b state $(=(64/3)\beta)$ is inversely proportional to μ_b^2, and consequently very small as expected from the non-relativistic quark model[23], we have

$$\mu_b \sim \frac{1}{2} m_\Upsilon \sim 4.73 \text{ GeV} \tag{3.3}$$

Alternatively, we can use $m_B \sim 5.3$ GeV[24], $m_D \sim 1.87$ GeV, $m_{D*} \sim 2.01$ GeV, and $\mu_c \sim 1.55$ GeV[4] as inputs to determine μ_b. In this case, using (3.1) and $m_{B*} - m_B = \frac{\mu_c}{\mu_b}(m_{D*} - m_D)$, we find

$$m_B - m_D = \mu_b - \mu_c + \frac{3}{4}\left(1 - \frac{\mu_c}{\mu_b}\right)(m_{D*} - m_D) \tag{3.4}$$

and

$$\mu_b \sim 4.92 \text{ GeV} \tag{3.5}$$

We can use the linear potential to calculate the mass spectra of $b\bar{b}q\bar{q}$ states, however here we shall follow a simpler procedure. According to our model of $c\bar{c}q\bar{q}$ states, the mass of P-wave $(cq)_{\underline{3}*}$- $(\bar{c}\bar{q})_{\underline{3}}$ states averages around 4.135 GeV for q=u,d and around 4.47 GeV for q=s (before introducing the hyperfine splittings).[4] The masses for $(bq)_{\underline{3}*}$ - $(\bar{b}\bar{q})_{\underline{3}}$ states can be easily obtained by just adding mass difference $\mu_b - \mu_c$ to the mass of corresponding $(cq)_{\underline{3}*}$ - $(\bar{c}\bar{q})_{\underline{3}}$ states. Using (3.1) we can then calculate the hyperfine splitting (h.f.s.) in a $(bq)_{\underline{3}*}$ (or $(\bar{b}\bar{q})_{\underline{3}}$) system. It is easy to find that the maxium spacing (ignoring difficult to calculate but perhaps small $\vec{L}.\vec{S}$ and tensor forces between the two ions) is equal to $m_{B*} - m_B$ amongst L=1 $(bq)_{\underline{3}*}$-$(\bar{b}\bar{q})_{\underline{3}}$ states when q=u,d, and equal to $m_{\tilde{F}*} - m_{\tilde{F}}$ when q=s (here we use the notation $\tilde{F}$ for $J^P = 0^-$ $(b\bar{s})$ meson, and $\tilde{F}*$ for $J^P = 1^-$ $(b\bar{s})$ meson). The spacings are:

$$m_{B*} - m_B = \frac{\mu_c}{\mu_b}(m_{D*} - m_D) \sim 0.05 \text{ GeV} \qquad (3.6)$$

$$m_{\tilde{F}*} - m_{\tilde{F}} = \frac{\mu_c}{\mu_b}(m_{F*} - m_F) \sim 0.04 \text{ GeV} \qquad (3.7)$$

Using $\mu_c \sim 1.55$ GeV[4], (3.3) (or (3.5)) and (3.1), (3.6), (3.7), we obtain a rough extimate of the masses of P-wave $(bq)_{\underline{3}*}$ - $(\bar{b}\bar{q})_{\underline{3}}$ states. This is shown in Table I.

Note that the mass spectrum of L=1 $(bq)_{\underline{3}*}$ - $(\bar{b}\bar{q})_{\underline{3}}$ is essentially determined by the mass difference between the b and c quarks, and the h.f.s. between B*, B and $\tilde{F}*$, $\tilde{F}$. Depending on the choice of value for $\mu_b - \mu_c$, the mass spectrum will shift, but the relative spacing will remain the same. It is readily seen that Table I has entries for both C-exotic and C-normal members of degenerate mass. Because of the present lack of detailed knowledge concerning the basic mass parameters of the b-system, we feel it is premature to make a detailed study of $b\bar{b}q\bar{q}$ decays other than the expectation that decays to $B\bar{B}$, $\tilde{F}\bar{\tilde{F}}$ will be important.

The presence of L=1 $J^{PC} = 1^{--}$ $b\bar{b}q\bar{q}$ T-baryonium states can be most appropriately studied in e^+e^- annihilation at CESR. Their coupling to the photon is expected to be 10 to 20 times weaker than the $\gamma - \Upsilon$ (9.46) coupling since the production of $b\bar{b}q\bar{q}$ by a photon involves a vertex forbidden by the Freund-Walz-Rosner rule.[25,26] However, in photoproduction, the cross section of these baryonium states could be considerably larger than that of $b\bar{b}$ states Υ (9.46). Assuming that the quark content scatters independently at high energy[4] and

$$\sigma_{tot}(\Upsilon p)/\sigma_{tot}(\psi p) \sim (m_\psi/m_\Upsilon)^2 \approx \frac{1}{10} \tag{3.8}$$

using vector dominance, we have for each $b\bar{b}q\bar{q}$ state

$$\sigma(\gamma p \longrightarrow b\bar{b}q\bar{q}+X)/\sigma(\gamma p \longrightarrow \Upsilon + X)$$
$$\approx \left(\frac{\Gamma_{e^+e^-}(b\bar{b}q\bar{q})}{\Gamma_{e^+e^-}(\Upsilon)}\right)\left(\frac{\sigma_{tot}(b\bar{b}q\bar{q}-p)}{\sigma_{tot}(\Upsilon p)}\right) \sim \frac{1}{20}\left(\frac{30\text{ mb}}{0.15\text{ mb}}\right) \simeq 10 \tag{3.9}$$

Considering that there are quite a few predicted $J^{PC} = 1^{--}$ $b\bar{b}q\bar{q}$ states, anomalously large photoproduction of B (and $\tilde{F}$) meson pairs (via decays of these $b\bar{b}q\bar{q}$ states), relative to photoproduction of Υ, could be expected.

IV. SUMMARY COMMENTS

From the above analysis, we see a very definite and

clear out pattern for the mass spectra of C-exotic states from the T-baryonium $c\bar{c}q\bar{q}$ system. We have concentrated here in particular on the T-baryonium ($\underline{3}^*$ - $\underline{3}$) system over M-baryonium ($\underline{6}$ - $\underline{6}^*$) and P-mesonium ($\underline{8}$ - $\underline{8}$), for definiteness and also because the T-baryonium system offers perhaps the optimal experimental situation. To be even more specific, let us illustrate the possible mass degeneracy complexity for a given J^P, say 1^-. Take $c\bar{c}q\bar{q}$ states with q = u,d, clearly for a given mass m a quartet of states is in principle possible, $J^{PC} = 1^{--}$, $I^G = 0^-$, 1^+ and C-exotic $J^{PC} = 1^{-+}$, $I^G = 0^+$, 1^-. Our model dynamics[4] is silent on isospin assignments for the multiquark structures, it is to be given by experiment and hopefully will be explained in the future by a more complete dynamical theory. For the T-baryonium $c\bar{c}s\bar{s}$ states, the situation is clearly simpler since the mass degeneracy is just between the duet $(J^{PC}; I^G) = (1^{--};0^-)$, $(1^{-+};0^+)$.

An interesting question to ask is the possible I^G inter-relation between C-normal and C-exotic states of a given mass. Many years back, T.D. Lee made the intuitively plausible conjecture that when hadronic mass levels become sufficiently clustered and overlapping, level classification by one scheme can by a recoupling procedure be made to look like another scheme. Actually in nuclear physics we have both collective and shell-model descriptions. In atomic physics we are familiar with electron levels associated with the approximate LS and jj coupling schemes where the general case, when both "approximate symmetries" are in evidence, belongs to the intermediate-coupling domain. A clearly pertinent question, one raised by J.R. Oppenheimer sometimes ago, is to search for the "intermediate couplings" of particle physics, and their relevance to hadronic mass-level <u>degeneracies.</u>

We are aware that the internal-symmetry group SU(3) is a good broken symmetry for the hadrons; in particular the more specific quark version $q\bar{q}$, $c\bar{c}$ with radial and L-excitations. It seems reasonable that evidence for a competing (even more broken) symmetry must at least requre us to complicate the underlying quark structure, e.g. through consideration of multi-quark states of which the $c\bar{c}q\bar{q}$ and $b\bar{b}q\bar{q}$ systems are the simplest **examples**. Our model[4] is particularly interesting in that the low lying T-baryonium $c\bar{c}q\bar{q}$ states with q=u,d offer the following possible pattern of mass degeneracies for pairs of states of opposite C-parity.

$$(J^{PC}\,;\, I^{G}) = (0^{-+};\, 0^{+}),\ (0^{--};\, 1^{+}) \qquad (4.1a)$$

and/or

$$(J^{PC}\,;\, I^{G}) = (0^{-+};\, 1^{-}),\ (0^{--};\, 0^{-}) \qquad (4.1b)$$

at degenerate mass 4.05 GeV, and

$$(J^{PC}\,;\, I^{G}) = (1^{--};\, 0^{-}),\ (1^{-+};\, 1^{-}) \qquad (4.2a)$$

and/or

$$(J^{PC}\,;\, I^{G}) = (1^{--};\, 1^{+}),\ (1^{-+};\, 0^{+}) \qquad (4.2b)$$

at degenerate mass 4.08 GeV. <u>The examples (4.1) and (4.2)</u>

do satisfy the G-parity condition[14] for $(I_1, I_2) = (½,½)$ (where the total isospin $\vec{I} = \vec{I}_1 + \vec{I}_2$) classification under the broken nonchiral SU(2) x SU(2) group (not contained in SU(3)) as the competing symmetry. Of course in practice we expect the degeneracies in each pairing to be only approximate since the C-normal members of $c\bar{c}q\bar{q}$ can mix with the $c\bar{c}$ states leading to some mass-shifts from the predicted values. There is also the possibility of mixing with $c\bar{c}g$, $q\bar{q}g$, but this is expected to be small.[18]

In this paper we have presented a rather definite pattern of C-exotic $c\bar{c}q\bar{q}$ T-baryonium states with reasonably well delineated masses (especially level splittings) and quantum numbers. Like the C-exotic gluonium case[18] the decay and production of these states appear to be rather undistinguished so far as we can ascertain. However such C-exotics (inherent in both multi-quark states and gluonia) appear to be essential ingredients in a successful theory of strong interactions. They should serve therefore as an incentive for both a more vigorous experimental search and increased effort on the part of theorists to develop novel search methods.

<table>
<tr><th rowspan="2">$S_1 \times S_2$</th><th rowspan="2">S</th><th rowspan="2">J^{PC_n}</th><th colspan="3">q = u, d</th><th colspan="3">q = s</th></tr>
<tr><th>I^G</th><th>mass* (GeV)</th><th>mass** (GeV)</th><th>I^G</th><th>mass* (GeV)</th><th>mass** (GeV)</th></tr>
<tr><td>0×0</td><td>0</td><td>1^{--}</td><td>$0^-, 1^+$</td><td>10.46</td><td>10.84</td><td>0^-</td><td>10.80</td><td>11.18</td></tr>
<tr><td>(1×0)
$\pm(0 \times 1)$</td><td>1</td><td>$0^{-\mp}$
$1^{-\mp}$
$2^{-\mp}$</td><td>$0^\mp, 1^\pm$</td><td>10.49</td><td>10.87</td><td>$0^\mp$</td><td>10.82</td><td>11.20</td></tr>
<tr><td rowspan="3">1×1</td><td>0</td><td>1^{--}</td><td>$0^-, 1^+$</td><td rowspan="3">10.51</td><td rowspan="3">10.89</td><td>0^-</td><td rowspan="3">10.84</td><td rowspan="3">11.22</td></tr>
<tr><td>1</td><td>0^{-+}
1^{-+}
2^{-+}</td><td>$0^+, 1^-$</td><td>0^+</td></tr>
<tr><td>2</td><td>1^{--}
2^{--}
3^{--}</td><td>$0^- \; 1^+$</td><td>0^-</td></tr>
</table>

Table I. The masses and quantum numbers for L = 1 $(bq)_{\underline{3}^*}$ - $(\bar{b}\bar{q})_{\underline{3}}$ states. Here S_1 (S_2) is spin of ion 1(2), $S = S_1 + S_2$, mass* is obtained by using (3.3), mass** is obtained by using (3.5).

References and Footnotes

(1) S. Pakvasa and S.F. Tuan, Phys. Rev. Lett. 20, 632 (1968)

(2) See, for instance, M. Gell-Mann and Y. Ne'eman, The Eightfold Way (W.A. Benjamin, Inc., New York, 1964), p. 98.

(3) D. Robson, Nucl. Phys. B130, 328 (1977).

(4) Kuang-Ta Chao, Oxford University Preprint Ref: 27/79, Neclear Physics B, to be published.

(5) G.L. Shaw, P.Thomas and S. Meshkov, Phys. Rev. Lett. 36, 695 (1976); C. Rosenzweig, Phys. Rev. Lett. 36 697 (1976).

(6) A. De Rujula, H. Georgi and S.L. Glashow, Phys. Rev. Lett. 38 317 (1977).

(7) J. Burmester et al., Phys. Lett. 68B, 293 (1977).

(8) R. Brandelik et al., Phys. Lett. 80B, 412 (1979).

(9) In the model proposed by the Cornell group, see ref. (10) below, the decay rate for $\psi_n(c\bar{c}) \to (c\bar{q}) + (\bar{c}q)$ (via quark pair creation) is proportional to m_q^{-4}, while A. Le Yaouanc et al. argued that this rate should be proportional to m_q^{-2} (see ref. (11) below).

(10) E. Eichten, K. Gottfried, T. Kinoshita, K.D. Lane and T. M. Yan, Phys. Rev. Lett. 36, 500 (1976), Phys. Rev. D17, 3090 (1978), also Cornell University Preprint, CLNS - 425, June 1979.

(11) A. Le Yaouanc, L. Oliver, O. Pene and J. C. Raynal, Phys. Lett. 72B, 57 (1977).

(12) R. Brandelik et al., DESY report 78/18 (1978).

(13) G. Feldman, Proc. Int. Conf. on High-Energy Physics, Tokyo, 1978 (Physical Society of Japan, Tokyo, 1979), p. 777.

(14) S.F. Tuan and T. T. Wu, Phys. Rev. Lett. 18, 349 (1967).

(15) H.M. Chan and H. Høgaasen, Phys. Lett. 72B, 121, 400 (1977), Nucl. Phys. B136, 401 (1978).

(16) See K. Nishijima, Fundamental Particles, W. A. Benjamin Inc. (1964), p. 56.

(17) F.E. Low, Phys. Rev. 120, 582 (1960); S.J. Brodsky et al., Phys. Rev. Lett. 25 972 (1970), Phys. Rev. D4, 1532 (1971).

(18) J. D. Bjorken, in Proceedings of Summer Institute on Particle Physics, Stanford, California (1979), to be published.

(19) C.E. Carlson and R. Suaya, Phys. Rev. D18, 760 (1978).

(20) S.F. Tuan, Phys. Rev. D15, 3478 (1977).

(21) P. Roy and T. F. Walsh, Phys. Lett. 78B, 62 (1978).

(22) P. Hasenfratz and J. Kuti, Phys. Rep. 40C, 75 (1978).

(23) J.D. Jackson, in Proceedings of Summer Institute on Particle Physics, Stanford, California (1976), p. 147.

(24) D. Treille, in Proceedings of the EPS High Energy Physics Conference, Geneva, Switzerland (1979), to be published; see also R. Barate et al., SISI Collaboration preprint, July (1979).

(25) M. Fukugita, Proceedings of XIII Rencontre de Moriond, Vol. I, p. 383 (1978).

(26) F.E. Close, Rutherford Lab. Report, RL-79-037.

MESON SPECTRA*

K.F. LIU AND C.W. WONG,

University of California, Los Angeles, California 90024 U.S.A.

Abstract

We discuss the possibility that there are quark-antiquark-gluon and four-quark-bag states with $J^{PC} = 0^{++}$, 1^{++} and 2^{++} among the low-lying mesons of different flavor families in addition to the more common quark-antiquark states.

Introduction

Hadron physics at intermediate energies may involve certain aspects of the internal structure of mesons, including their spatial extensions and internal composition. As a first step in providing an overall perspective on low-lying mesons of different flavor families, we discuss their descriptions as quark-antiquark ($Q\bar{Q}$), quark-antiquark-gluon ($Q\bar{Q}g$) and "four-quark bag" states of the constituent quark-

$Q\bar{Q}$ states have been discussed in terms of potential models,[1] bag models[2] and string models.[3] Of these the simplest is perhaps the non-relativistic (NR) potential model. This NR potential model is well justified for a meson made up of massive quarks, such as the charmonium meson ψ (3.10 GeV), where the kinetic energy of relative motion is only about 13% of the total rest energy of the quarks. A recent review of NR potentials for charmonium mesons has been made by Eichten.[4] A similar NR potential description for the light mesons is not realistic because the kinetic energy can exceed the rest energy of the light quarks. Nevertheless such potentials fitting light meson spectra have also been obtained.[5]

In hadron physics at intermediate energies one has to deal mostly with light hadrons. Since a simple relativistic potential

*This work was supported in part by the National Science Foundation.

model of the $Q\bar{Q}$ interaction is not available and the MIT bag model, which provides a relativistic description, may not be suitable for certain applications, one is forced to consider a NR potential model. In Sect. II, we shall show that a simple NR potential can be obtained which gives reasonable estimates of meson sizes as well as a useful orientation concerning the nature of all meson spectra. Applications of this NR potential should be made with a great deal of caution, because dynamical and kinematical effects are not completely separated. Preliminary results do suggest that it gives roughly correct masses for nucleons and Δ.[10]

A number of other mesons might have been seen which are probably not $Q\bar{Q}$ states.[6] The most famous are the three puzzling mesons, X(2.83),[16] χ(3.45) and χ(3.59), in the charmonium region. We argue that these are complicated states with one or more constituents besides $c\bar{c}$. Their relatively low masses imply that both the $c\bar{c}$ pair and the remaining constituents are likely to be separately a negative-parity object, i.e. a pseudoscalar or a vector (color singlet or octet) with an S-wave internal spatial motion. Hence these mesons are likely to have PC = ++. The presence of the additional constituent means that γ transitions to or from charmonium states might be small, with the transitions proceeding primarily through small admixtures of ^3P_J charmonium states in these mesons. The presence of a $c\bar{c}$ pair means that their annihilation widths are relatively small. Candidates for these states include the $c\bar{c}g$ states of Horn and Mandula[7] and the S-wave $c\bar{c}q\bar{q}$ bag states of Jaffe and De Rújula.[8,9]

In Sect. III we use the model of Horn and Mandula to study the properties of the $c\bar{c}g$ states for a number of $c\bar{c}$ potentials. We find that their center of mass is around 3.4 GeV and that they have roughly the right mass splittings if there is a strong spin-spin interaction. Estimates of their decay and transition properties suggest that they might be promising candidates for these three puzzling mesons.

In Sect. IV, bag-model results on four quark states obtained by Jaffe and De Rújula[8] and Jaffe[9] are reviewed. Their decay and transition properties are briefly discussed. These states are more massive than the $c\bar{c}g$ states, but can be mixed with them. This configuration mixing effect is studied. It appears to provide an important contribution to the mass splittings seen in the puzzling mesons.

If the puzzling mesons are these mixed $c\bar{c}g$ and $c\bar{c}q\bar{q}$ states, similar gluonic and four-quark states should also appear among the light mesons. With the help of the $Q\bar{Q}$ spectra obtained in the NR potential model, we suggest in Sect. V that the mesons $\varepsilon(0.7)$, $\varepsilon(1.3)$, $\kappa(1.2)$ and $\kappa'(1.43)$ might be regions of overlapping resonances containing some of these more complicated structures.

Section VI contains brief summary remarks.

II $Q\bar{Q}$ potential

Following previous works,[1,4,5] we describe charmonium mesons with the help of a NR potential containing a linear confinement term:

$$H_{Q\bar{Q}} = \frac{p^2}{2\mu} + kr - \frac{4}{3}\alpha_s\left[\frac{1}{r} + \frac{1}{m_1 m_2} f_{BF}(r)\right] + b . \tag{1}$$

Here μ is the reduced mass and b is an additive constant, which is taken to be independent of "flavors." A modified Breit-Fermi function of the form

$$f_{BF}(r) = S(r)\,\vec{s}_1 \cdot \vec{s}_2 + W(r)\,\vec{L} \cdot \vec{S} + V_T(r)\,S_{12} \tag{2}$$

is used, where

$$S(r) = -\frac{2}{3}\left\{\frac{4}{\pi^{\frac{1}{2}} r_0^{\,3}} \exp\left(-\left(\frac{r}{r_0}\right)^2\right)\right\} \tag{3}$$

$$W(r) = -\frac{3}{2}\left\{1 - \exp\left(-\left(\frac{r}{a_0}\right)^2\right)\right\}\Big/ r^3 \tag{4}$$

$$V_T(r) = -\frac{1}{4}\left\{1 - \exp\left(-\left(\frac{r}{a_0}\right)^2\right)\right\}\Big/ r^3 \tag{5}$$

$$S_{12} = 3\vec{\sigma}_1 \cdot \hat{r}\,\vec{\sigma}_2 \cdot \hat{r} - \vec{\sigma}_1 \cdot \vec{\sigma}_2 \tag{6}$$

We have dropped the velocity-dependent Darwin term and the spin-dependent δ-function term, since these tend to cancel to a certain extent.

We also find the following interesting results: (1) The same potential form, when used with a different set of potential parameters, also generates the masses of the light mesons of the ω/ρ family. (2) Other meson masses can be generated from the NR Hamiltonian of Eq. (1) if the potential parameters to be used are assumed to be the following functions of the r.m.s. radius r_V of the vector (1^3S_1) meson of the family:

$$\alpha_s(V) = 1.4/\left[1 + 1.295\ \ln\ (r_\rho/r_V)\right] \quad (7)$$

$$\omega(V) \equiv \left[4k^2(V)/\mu\right]^{1/3} = 0.75/\left[1 + 0.4314\ \ln\ (r_\rho/r_V)\right]\ \text{GeV} \quad (8)$$

$$r_o(V) = (r_V/r_\rho)^{2.06}\ 1.05\ \text{fm}\quad , \quad (9)$$

$$a_o(V) = 2.7\ r_o(V)\quad . \quad (10)$$

In these equations, all numerical constants are obtained by fitting the ψ and ω/ρ sets of empirical potential parameters. In addition, the masses of s and b quarks are obtained separately by fitting the masses of $\phi(1.02)$ and $\Upsilon(9.46)$. There are thus a total of only 13 adjustable parameters in our NR potential model:

(a) four quark masses (GeV): $m_{u,d} = 0.12$, $m_s = 0.334$, $m_c = 1.51$,

$m_b = 4.83$;

(b) one additive constant: $b = -0.08$ GeV;

(c) for ψ family: $\alpha_s = 0.54$, $k = 0.149\ \text{GeV}^2$, $r_o = 0.083\ \text{fm} = a_o/2.7$;

(d) for ω/ρ family: $\alpha_s = 1.4$, $k = 0.080\ \text{GeV}^2$, $r_o = 1.05\ \text{fm} = a_o/2.7$.

(11)

We should add that $m_{u,d} = 0.12$ GeV is actually chosen to fit[10] the experimental proton charge radius (0.81 $\pm$ 0.03 fm) of the Dirac form factor F_1.

With some exceptions, our potential model gives meson masses to better than $\approx$ 50 MeV, as shown in Fig. 1. In addition, the pion charge radius (0.53 fm) agrees with the experimental result of

0.56 ± 0.04 fm.[11] Conspicuous difficulties in the model reproduction of meson masses include the following: (1) All pseudoscalar meson masses do not come out right, but there are systematic trends in the discrepancies. For example, all I = 0 pseudoscalars appear 0.3 GeV too low. (2) Model $^1P_1$ and $^1D_2$ masses appear too low. We believe that this is due to our neglect of the Darwin term. (3) There are isolated problems such as the D(1.28) appearing at 1.18 GeV (if it is a $^3P_1$ state). (4) Other mesons might have been seen which do not appear to belong to these $Q\bar{Q}$ spectra.[6] The rest of this paper is concerned with possible structures for these additonal mesons.

III. $c\bar{c}g$ States and Their Mixing with $c\bar{c}$ ^3P_J States

Three mesons, X(2.83),[16] χ(3.45 or 3.33) and χ(3.59 or 3.18), might have been seen in the charmonium region with properties different from those of $c\bar{c}$ mesons.[6] Their relatively low masses suggest that they might be even-parity states, each made up of two odd-parity components. The small γ transition branching ratios associated with them can be understood in terms of small admixtures of charmonium ^3P_J states. All known properties concerning the puzzling mesons can then be accounted for if these composite structures themselves have very small hadronic and 2γ annihilation widths and γ transition branching ratios. We first consider the possibility that they are the $c\bar{c}g$ states studied by Horn and Mandula.[7]

The Horn-Mandula $c\bar{c}g$ states may be denoted by the composite spectroscopic symbol $(^3\boxed{L}^8_{J'},\ ^3\boxed{\ell}^g_j)\ J^{PC}$, of which the first symbol refers to a color octet $c\bar{c}$ pair. The lowest states are $(^3S_1{}^8,\ ^3s_1{}^g)J^{++}$, with J = 0,1 and 2. In estimating masses, Horn and Mandula use the trial wave function

$$\Phi\,(Q\bar{Q}g) = \phi_{Q\bar{Q}}(\underset{\sim}{d})\,\phi_g(\underset{\sim}{r}) \tag{12}$$

Here

$$\phi_{Q\bar{Q}}(\underset{\sim}{d}) = (\frac{\mu^3}{8\pi})^{\frac{1}{2}} \exp(-\frac{1}{2}\mu d),\ \phi_g(\underset{\sim}{r}) = (\frac{\lambda^3}{8\pi})^{\frac{1}{2}} \exp(-\frac{1}{2}\lambda r)\ , \qquad (13)$$

and $\underset{\sim}{d}$ and $\underset{\sim}{r}$ are the quark and gluon coordinates, respectively, as measured from the center of mass of the $Q\bar{Q}$ pair.

For the $c\bar{c}g$ Hamiltonian, we follow the Horn-Mandula additive or universality assumption under which the gluon is considered dynamically equivalent to a color-octet $q\bar{q}$ pair:

$$H = \left\{ 2m_c + p^2_{c\bar{c}}/m_c - \frac{1}{8}\left[2kd - \frac{4}{3}\alpha_s \frac{1}{2d} + b \right] \right\} + p_g$$

$$+ \frac{9}{8}\left\{ k\,(|\underset{\sim}{r}-\underset{\sim}{d}| + |\underset{\sim}{r}+\underset{\sim}{d}|) - \frac{4}{3}\alpha_s\,(\frac{1}{|\underset{\sim}{r}-\underset{\sim}{d}|} + \frac{1}{|\underset{\sim}{r}+\underset{\sim}{d}|}) + 2b \right\} \quad . \qquad (14)$$

Here $p_{c\bar{c}}$ is the relative momentum of the $c\bar{c}$ pair and p_g is the gluon momentum. The potential parameters are supposed to be those for charmonium states. Since the Hamiltonian is spin-independent, only the center of mass $\bar{M}$ of the three $({}^3S_1{}^8,\ {}^3S_1{}^g)J^{++}$ states can be calculated.

The results of this calculation are shown in Table I for three NR charmonium potentials: pot. A from Ref.[1], pot. B from Ref.[6] and pot. C, which is the spin-independent part of our potential in Sect. II. We see that all three potentials give roughly the same result of $\bar{M} \simeq 3.4$ GeV, in the neighborhood of the puzzling mesons.

The experimental meson masses actually show a very strong J dependence, which for S-wave $c\bar{c}g$ states arises from the spin-spin interaction. If the gluon is dynamically equivalent to an isoscalar, color-octet $q\bar{q}$ pair, it is possible to relate the total mass splitting Y_2-Y_0 of these states to the hyperfine mass splitting D^*-D = 0.14 GeV of the D = $c\bar{q}$ mesons. The result is

$$Y_2 - Y_o \simeq \frac{9}{4} \times \frac{3}{4}\,(D^* - D)\,(\frac{r_D}{r_{cg}})^{p} \qquad (15)$$

for an inverse p-power spin-spin potential. Here $\frac{9}{4}$ is a color factor, $\frac{3}{4}$ is a spin factor, r_D is the average separation between c and $\bar{q}$ in the D*/D mesons, and r_{cg} is the average separation between c and g in $c\bar{c}g$. Using p = 2 and calculated $r_D \simeq 0.6$ fm, $r_{cg} \simeq 0.3$ fm, we find $Y_2 - Y_0 \simeq 0.5$ GeV. This is roughly consistent with the observed mass splitting of the puzzling mesons.

We have also examined the annihilation and γ transition widths of these $c\bar{c}g$ states, and the strengths of their mixing with $c\bar{c}$ ^3P_J states. The results are as follows: (1) The 2γ annihilation and γ transition widths appear to be consistent with those required by our ^3P_J mixing model. (2) The QED-like contributions to the hadronic width are small, but the QCD contributions have not been calculated. (3) These $c\bar{c}g$ states appear to mix into $c\bar{c}$ ^3P_J states with strengths (i.e. off-diagonal mass matrix elements) of 50-250 MeV in a nonrelativistic approximation. This estimate is larger than the strengths of $\simeq$ 30 MeV needed to account for γ transitions in the ^3P_J mixing model. It is not clear if the discrepancy might not be due to our NR treatment.

IV. $c\bar{c}q\bar{q}$ States and Their Mixing with $c\bar{c}g$ States

The S-wave four-quark ($c\bar{c}q\bar{q}$) bag states (with q = u,d) have been studied by De Rújula and Jaffe.[8,9] They form a group of low-lying even parity states of which twelve are isoscalars: four 0^{++}, two 1^{++}, four 1^{+-} and two 2^{++} states. They are predicted by the MIT bag model[8] to appear at or above the J = 0 isoscalar threshold of $\eta_c\eta$ at $\simeq$ 3.6 GeV. Most of these states are also above the breakup thresholds $\psi\eta \simeq 3.7$ GeV (J = 1), $\psi\omega \simeq 3.9$ GeV (J = 0,1,2), and the charm threshold $D\bar{D}$ = 3.7 GeV. Consequently, these mesons are expected to have large hadronic widths at these predicted masses, in disagreement with the expected small hadronic widths of the puzzling mesons.

However, the hadronic widths are greatly reduced, once the masses are moved below the lowest meson production threshold. The required mass reductions do not have to be very large for isoscalar bag states (but much larger if states are isovector). On the other hand, if X(2.83)[16] is to be interpreted as the lowest $(c\bar{c}q\bar{q})0^{++}$ bag states, as the isovector δ_c and the isoscalar S_c*, as suggested by Lipkin, Rubinstein and Isgur,[12] a downward shift of about 0.8 GeV from the bag-model prediction is required. This large shift is probably unlikely unless there is a serious flaw in the bag-model description of hadron masses.

It is clear, however, that these $c\bar{c}q\bar{q}$ bag states can mix with the $c\bar{c}g$ states of Sect. III via that component in which both $c\bar{c}$ and $q\bar{q}$ are vector, color octet structures $\underline{V}$. This mixing turns out to be quite strong, and it contributes significantly to the mass splittings of the $c\bar{c}g$ states.

We study this mixing effect by first estimating the mixing strength for g (gluon) $\leftrightarrow$ $\underline{V}$ (for $q\bar{q}$). The result is

$$h(g \leftrightarrow \underline{V}) \simeq 0.16\,\mu(\mu/\lambda)^{1/2} \quad , \tag{16}$$

where λ and μ are the inverse size parameters in the $c\bar{c}g$ wave function of Eq. (13). The calculated values of $h(g \leftrightarrow \underline{V})$, also shown in Table I, are $\simeq$ 0.4 - 0.5 GeV. On allowing for the J dependence of the variational parameters λ and μ, we find $h_{J=2}(g \leftrightarrow \underline{V}) \simeq 0.4$ GeV.

The mixing problem is interesting because it gives information on both the mass splittings and the $c\bar{c}g$ probability $p(c\bar{c}g)$ of our mixed states. The answer is unfortunately very sensitive to the mass difference between the two types of states being mixed. It is quite plausible that the dynamical content of the $c\bar{c}g$ model of Sect. III is less reliable than that of the four quark states. We therefore decided to re-adjust the average mass $\bar{M}$ of the $c\bar{c}g$ states to fit the mass of X at 2.83 GeV,[16] after mixing, thereby adding a measure of realism to the results. This is done in the following way. For each choice of the strength V_{ss} of the spin-spin gluon-quark interaction

$$H_{ss} = V_{ss} \sum_{i=q,\bar{q}} (\underset{\sim}{\lambda}_i \cdot \underset{\sim}{\Gamma}_g)\,(\underset{\sim}{s}_i \cdot \underset{\sim}{S}_g)\,\frac{1}{|\underset{\sim}{r} - \underset{\sim}{r}_i|^2} \tag{17}$$

(where $\underset{\sim}{\lambda}$, $\underset{\sim}{\Gamma}$ are color SU(3) operators and $\underset{\sim}{s}$, $\underset{\sim}{S}$ are spin operators) an additive constant is added to Eq. (14) and so adjusted that the lowest J = 0 state has the observed mass 2.83 GeV at the end of the complete calculation with J-dependent parameters. The final masses M_1 and M_2 of the lowest J = 1 and J = 2 states and the value of V_{ss} used are shown in Fig. 2 as functions of the adjusted center of mass $\overline{M}$ of the $c\bar{c}g$ states for the spin-independent Hamiltonian H of Eq. (14) plus the additive constant.

It is reassuring to find in Fig. 2 that reasonable masses for the puzzling mesons are obtained when $\overline{M}$ (≃3.4-3.7 GeV) is not far from the model value of ≃ 3.4 GeV. The mixed states are made up of 73-92% of the $c\bar{c}g$ component. The remaining 27-8% is a $c\bar{c}q\bar{q}$ component which is pure $\underline{V}\cdot\underline{V}$ in J = 2, 99% pure $\underline{V}\cdot\underline{V}$ in J = 0, and 50% $\underline{V}\cdot\underline{V}$ in J = 1, in the notation of Ref. [9]. The "theoretical" meson masses shown for the two cases indicated by the arrows a and b in Fig. 2 agree roughly with the two possible sets of observed masses of the puzzling mesons.

V. Other $Q\bar{Q}g$ and $Q\bar{Q}q\bar{q}$ States in Meson Spectra

From the broader perspective of meson spectra in general, it is clear that if the puzzling mesons in the charmonium region are primarily $c\bar{c}g$ states, other $Q\bar{Q}g$ states should also appear with masses near those of $Q\bar{Q}$ mesons. Using the Horn-Mandula model of Sect. III with the spin-independent Hamiltonian (14), and the potential parameters of the $Q\bar{Q}$ potentials obtained in Sect. II, we find the results of Table II for the centers of mass $\overline{M}$ of $Q\bar{Q}g$ states.

There are, in addition, effects due to the spin-spin interaction and the mixing with four-quark bag states. When these effects are approximately included, we find that the lightest $b\bar{b}g$ meson in the upsilon region has the theoretical mass of 9.34 (9.56, 9.73) GeV for $J^{PC} = 0^{++}(1^{++},2^{++})$ and a $b\bar{b}g$ probability of ≃ 60%.

The results of Table II for light mesons are obviously unsatisfactory (too small masses), even without including the large spin-spin mass splittings expected here. It is likely that the models used are just too crude for these light mesons. A more realistic expectation might be that a $(q\bar{q}g)0^{++}$ state X coexists with a low-lying four-quark bag state under the broad structure at ε(0.7). The interpretation of ε(0.7) as a four-quark state has been given by Holmgren and Pennington.[13]

Of the eleven mesons of the K* family tabulated by Leith,[14] nine can be understood as $Q\bar{Q}$ structures, as shown in Fig. 1. The remaining two mesons are the broad structures κ(1.2) and κ'(1.43) with widths 0.45 and 0.25 GeV, respectively. We interpret the large width of κ(1.2) as an indication of the presence of three overlapping scalars – the $Q\bar{Q}$ $^3P_0$ state, the gluon state X_K and the four-quark state C_K(0.9 GeV, Ref. 9). κ'(1.43) may be Jaffe's C_K (36) and/or a gluon state. It would be useful to look experimentally for the other C_K bag states. The possibility of making more definite identification of these bag states by measuring branching ratios into two distinct decay modes should also be pointed out.

Similar analyses for mesons of the ρ, η and η' families suggest that the simple $Q\bar{Q}$ $^3P_0$ scalars are δ(0.98) I = 1, S* (0.98) I = 0, and part of ε(1.3) I = 0, respectively. It is not clear where $s\bar{s}g$ should be. If $u\bar{u}g$ really appears at ε(0.7), then $s\bar{s}g$ should also appear ≃ 0.4 GeV higher than expected, namely at ε(1.3). In this mass region, one C_η^s and a number of C^0 bag states are also expected.[9] The large experimental width (0.2-0.4 GeV) of ε(1.3) appears to be consistent with this possibility.

VI. Conclusions

We have constructed a NR potential model for $Q\bar{Q}$ mesons containing only 13 parameters. It appears useful for the study of meson sizes and for certain meson mass extrapolations. It has also been used for describing baryon masses and sizes.[10] In spite of the obvious limitation of being a NR model, it may well be useful in studies of short-range nuclear force and in hadron-structure problems at intermediate energies.

There are mesons which do not appear to be $Q\bar{Q}$ states. Examples are the three puzzling mexons X(2.83),[16] χ(3.45 or 3.33) and χ(3.59 or 3.18) in the charmonium region. We interpret these as the lowest 0^{++}, 1^{++} and 2^{++} states made up primarily of a $c\bar{c}g$ structure (63-77%). They also contain small admixtures of the $c\bar{c}$ ^3P_J states (3-29%) and the $c\bar{c}q\bar{q}$ bag states (25-6%). These admixtures appear to be important in understanding the γ transition widths and the mass splittings of these states.

The general interpretation of meson spectra given here may be summarized by the statement that in mesons the $Q\bar{Q}$ states are most readily seen. Next come the $Q\bar{Q}g$ and four-quark states.

In conclusion, we note that this interpretation of meson spectra is necessarily very speculative and incomplete. In spite of this, our studies do suggest that significant progress can be expected when more is known about puzzling mesons, past or future, of the ψ, T and also the K* families. Of these the K* family is perhaps the most promising because of its accessibility.

A more detailed report of this research has been given elsewhere.[15]

References:

[1] E. Eichten, et al., Phys. Rev. Lett. 34, 369 (1975); A. De Rujula, H. Georgi and S. L. Glashow, Phys. Rev. D12, 147 (1975).

[2] A. Chodos et al., Phys. Rev. D9, 3471 (1974); ibid, D10, 2599 (1974).

[3] Y. Nambu, Copenhagen Lectures, 1970 (unpublished); L. Susskind, Nuovo Cimento 69A, 457 (1970).

[4] E. Eichten, in New Results in High Energy Physics - 1978 (Venderbilt Conference), edited by R. S. Panvini and S. E. Csorna (AIP Conference Proceedings, No. 45, 1978), p. 252.

[5] W. Celmaster, H. Georgi and M. Machacek, Phys. Rev. D17, 879 (1978).

[6] K. Gottfried, in Proc. 1977 International Symposium on Lepton and Photon Interactions at High Energies, edited by F. Gutbrod (DESY, Hamburg, Germany, 1977), p. 667.

[7] D. Horn and J. Mandula, Phys. Rev. D17, 898 (1978).

[8] A. De Rújula and R. L. Jaffe, in Proc. Fifth International Conference on Experimental Meson Spectroscopy, 1977, edited by E. Von Goeler and R. Weinstein (Northeastern Univ. Press, Boston, 1977), p. 83.

[9] R. L. Jaffe, Phys. Rev. D15, 267,281 (1977).

[10] J. F. Rondinone, Ph.D. Thesis, UCLA, 1978, unpublished; J. F. Rondinone and C. W. Wong, unpublished.

[11] E. B. Dally, et al., Phys. Rev. Lett. 39, 1176 (1977).

[12] H. J. Lipkin, H. R. Rubinstein, and N. Isgur, Phys. Lett. 78B, 295 (1978).

[13] S.-O. Holmgren and M. R. Pennington, Phys. Lett. 77B, 304 (1978)

[14] D. W. G. S. Leith, in Proc. Fifth International Conf. on Experimental Meson Spectroscopy, 1977, edited by E. Von Goeler and R. Weinstein (Northeastern Univ. Press, Boston, 1977), p. 207.

[15] K. F. Liu and C. W. Wong, UCLA preprint, 1979, unpublished.

[16] It is communicated to us by E. Bloom that in the recent Crystal Ball experiment at SLAC, a faint signal is seen in $\psi \rightarrow \gamma X$ with a branching ratio (BR) of 0.1-0.5%. However, the mass of this X is 2.98 ± 0.02 GeV. The old X(2.83) is not seen. Throughout this paper, we have taken X(2.83) as our reference for discussion. If we were to change to X(2.98) with a mass shift of 0.15 GeV, no significant modification of our present result is expected.

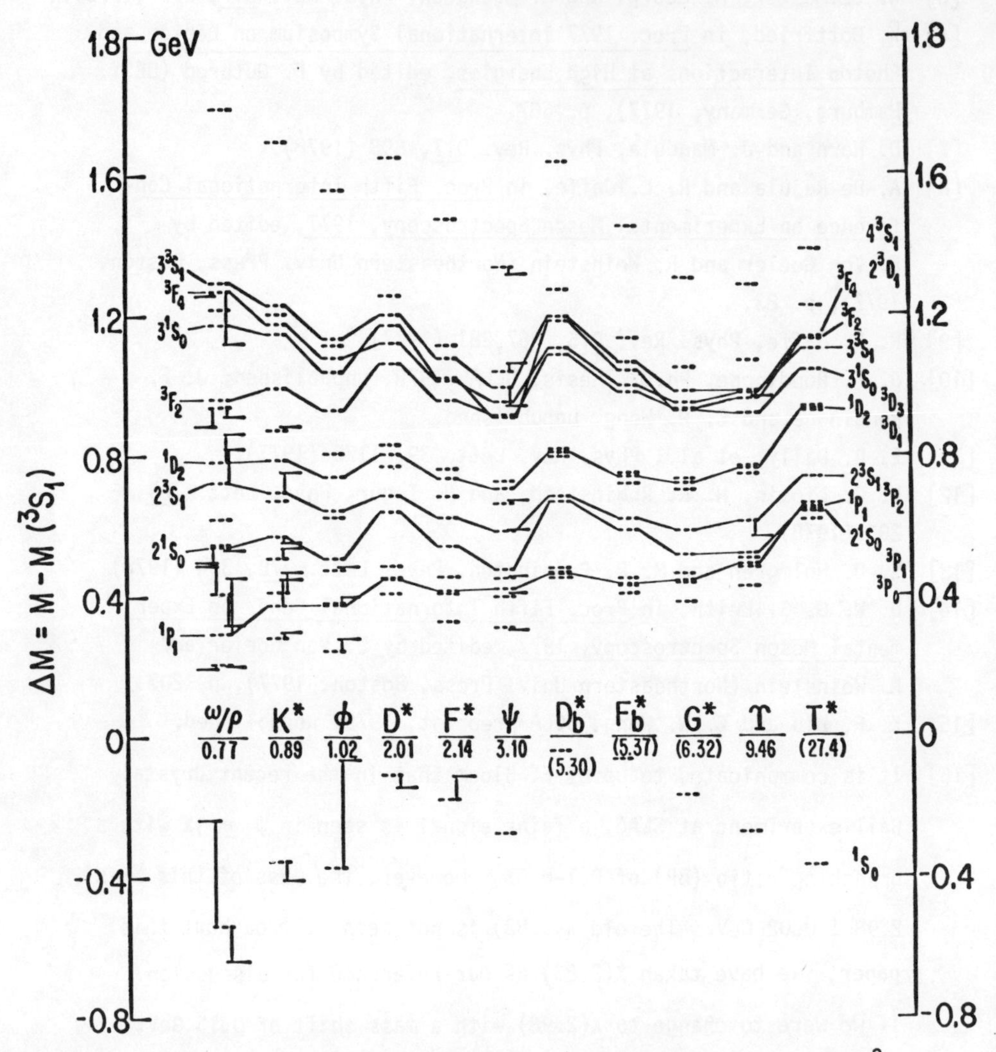

Fig. 1 Spectra of quark-antiquark mesons relative to the vector ($^3S_1$) meson of each family. The predictions of the nonrelativistic potential model of this paper are shown as broken horizontal lines. They are identified by the spectroscopic symbols and connecting lines which join the same states in different meson families. The observed mesons are s hown as thick horizontal lines. They can be identified by thin vertical lines which join them to the theoretical states to which they correspond.

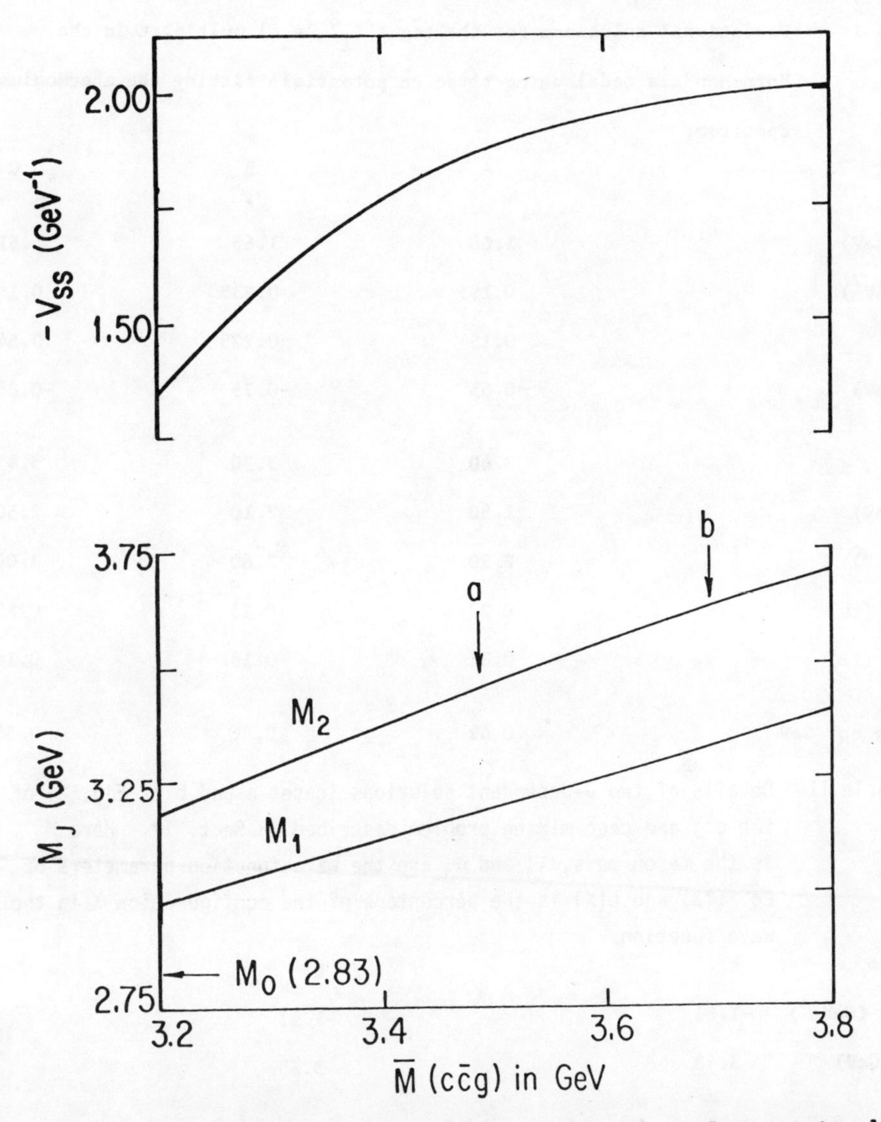

Fig. 2 Masses M_J (GeV) of the lightest meson of angular momentum J obtained in the $c\bar{c}g$ and $c\bar{c}q\bar{q}$ configuration mixing calculation described in the text plotted as a function of the assumed center of mass $\overline{M}$ of the three $c\bar{c}g$ states. The assumed strength V_{ss} (GeV^{-1}) of the spin-spin interaction in Eq. (17) used in the calculation is also shown.

Table I Variational solutions for the $c\bar{c}g$ $({}^3S_1{}^8\ {}^3s_1{}^g)$ multiplet in the Horn-Mandula model using three $c\bar{c}$ potentials fitting the charmonium spectrum.

Model	A	B	C
m_c (GeV)	1.60	1.65	1.51
k (GeV^2)	0.195	0.233	0.149
α_s	0.15	0.225	0.54
b (GeV)	-0.65	-0.75	-0.08
$\bar{M}$	3.40	3.30	3.47
λ (GeV)	1.80	2.10	2.50
μ (GeV)	2.30	2.60	3.00
$\sqrt{6}/\lambda$ (fm)	0.27	0.23	0.19
$\sqrt{6}/\mu$ (fm)	0.21	0.18	0.16
$h(g \to q\bar{q})$ GeV	0.42	0.46	0.53

Table II Details of two J-dependent solutions (cases a and b of Fig. 2) of the $c\bar{c}g$ and $c\bar{c}q\bar{q}$ mixing problem described in Sect. IV. Here M_J is the meson mass, λ_J and μ_J are the wave-function parameters of Eq. (13) and p(X) is the percentage of the configuration X in the wave function.

Case	a			b		
V_{ss} (GeV^{-1})	-1.86			-1.97		
$\bar{M}$ (GeV)	3.48			3.65		
	J = 2	J = 1	J = 0	J = 2	J = 1	J = 0
M_J (GeV)	3.48	3.18	(2.83)	(3.59)	(3.30)	(2.83)
λ_J (GeV)	2.21	3.17	5.63	2.20	3.22	7.14
μ_J (GeV)	2.73	3.50	5.44	2.72	3.55	6.57
p ($c\bar{c}g$)	0.80	0.80	0.79	0.92	0.75	0.73
p ($\underline{V}\cdot\underline{V}$)	0.20	0.10	0.20	0.08	0.12	0.26

PION EXCHANGE EFFECTS IN THE MIT BAG

M. V. Barnhill, W. K. Cheng, and A. Halprin

Physics Department, University of Delaware, Newark, DE 19711, U.S.A.

ABSTRACT

We present results of calculations based on the MIT bag model which include pion exchange effects in a manner consistent with the idea of partially-conserved-axial-current (PCAC).

I. INTRODUCTION

The remarkable success of the MIT bag model[1] in explaining the static properties of low-lying hadrons is well-known and needs no special introduction. We shall only remark that the model is invented to describe the internal structure of hadrons as composite systems of light quarks (u-, d- and s-quarks) and gluons which are confined in a static bag of radius R. The confinement mechanism is, however, an *input* but not a *derived* result. Moreover, quark-gluon interactions inside the bag are treated perturbatively according to quantum chromodynamics (QCD). The fundamental equations of the MIT bag are given below.

(a) Inside the bag ($r < R$), quarks and gluons obey the standard Yang-Mills equation:[2]

$$(i\not{\partial} - M + gA_{\mu a}\lambda_a\gamma^\mu)q = 0; \tag{1}$$

$$D_{\mu ab}F_b^{\mu\nu} = g\bar{q}\gamma^\nu\lambda_a q; \tag{2}$$

(b) outside the bag $q = 0$, $F_{\mu\nu} = 0$ identically,where

$$F_a^{\mu\nu} = \partial^\mu A_{a\mu} - \partial^\nu A_{a\mu} + g f_{abc} A_a^\mu A_c^\nu .$$

Boundary conditions at the bag surface (r = R) are

$$i n_\mu \gamma^\mu q = q \tag{3}$$

$$n_\mu F_a^{\mu\nu} = 0 \tag{4}$$

$$n_\mu \partial^\mu (\bar{q}q) - \tfrac{1}{2} \sum_a F_{\mu\nu a} F_a^{\mu\nu} = 2B . \tag{5}$$

Eqs. (3) and (4) ensure that physical hadrons are color singlets, and Eq. (5) follows from the energy-momentum conservation. (B is the confinement pressure, positive and constant throughout the bag volume). Thus we have the physical picture of hadron as shown in Fig. 1. However, to obtain the mass spectroscopy of low-lying hadrons, the third (and non-linear) boundary condition is replaced by the extremal condition (with respect to the bag radius R) of the mass formula: $\partial M(R)/\partial R = 0$, where

$$M(R) = E_V + E_o + E_q + E_m + E_e ; \tag{6}$$

and[3]

E_V = volume energy = $4\pi R^3 B/3$;

E_o = zero-point energy = $-Z_o/R$, $(Z_o > 0)$;

E_q = quark kinetic energy = $(N_q + N_{\bar{q}})\sigma$, $(E \sim 2.04/R)$;

E_m = color magnetic energy $\sim -(\alpha_c/R) \sum_{i\neq j} \vec{\lambda}_i \cdot \vec{\lambda}_j \vec{\sigma}_i \cdot \vec{\sigma}_j$

E_e = color electric energy $\sim -(\alpha_c/R) \sum_{i,j} \vec{\lambda}_i \cdot \vec{\lambda}_j$: $(\alpha_c = g^2/4\pi)$

where $\vec{\lambda}$ and $\vec{\sigma}$ are, respectively, the color spin and spin vectors, the sum is over quark constituents. We remark that:

(a) the mass formula M(R) is semi-empirical; E_V and E_o are put in by hand, E_q, E_m and E_e are derived. E_e is usually neglected in comparison with the other terms; E_m is essential for the mass splitting between the

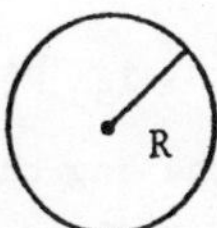

Fig. 1. The MIT bag.

observed N and $\Delta(M_\Delta - M_N \sim \alpha_c/R)$.

(b) The third boundary condition and the extremal condition of M(R) have been proven to be equivalent in the absence of gluons. The latter condition is normally used in obtaining the "best" fit to hadron masses by choosing the parameters: B, Z_o, m_s, α_c, and R. (See Table 1).

(c) As can be seen from Table 1, the mass of pion-like bag (∿280MeV) is about twice the observed value. This is intimately related to the fact that in the static MIT bag model chiral symmetry and PCAC have not been treated properly. We have therefore made a forced marriage between the MIT bag and PCAC. A discussion of the model, together with our results on hadron mass-spectroscopy, mean square charge radius and magnetic moment, is given below.[4]

II. PCAC-BAG MODEL

1. PCAC Pion

To the standard MIT bag model, we now add a PCAC pion cloud a lá the suggestion of Callan, Dashen and Gross.[5] Namely, inside the bag the pion field is zero identically, and outside the bag, a pion cloud exists (see Fig. 2 below).

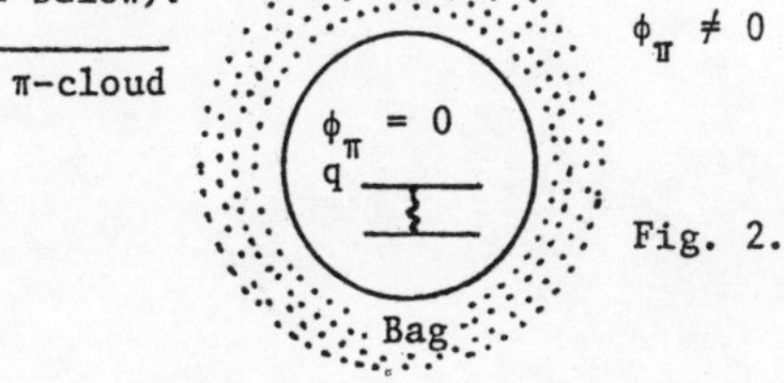

π-cloud $\phi_\pi \neq 0$ $\phi_\pi = 0$ q Bag (ϕ_π = pion field)

Fig. 2. PCAC bag: A marriage between MIT bag and PCAC.

In terms of the vacuum states inside and outside the bag (hereafter designated by $|0\rangle_{in}$ and $|0\rangle_{out}$, respectively), we may say that the model implies the following conditions: $Q_5|0\rangle_{in} = 0$ and $Q_5|0\rangle_{out} \neq 0$, where Q_5 is the chiral charge operator ($= \int j_{50} d^3x$, $j_{5\mu}$ being the axial current). The spontaneous symmetry breaking outside the bag then leads to the realization of pion as a Nambu-Goldstone boson.[6] It is interesting to note that (a) Prof. T. D. Lee[7] has described the QCD vaccua in terms of a long-range order as characterized by the (color) dielectric constant k: for $|0\rangle_{in}$, $k \cong 1$("normal" color dielectric) and for $|0\rangle_{out}$, $k \cong 0$ ("perfect" color dielectric),and (b) Callan et al[5] have characterized the vaccua in terms of two separate phases (dilute vs dense) of instanton gas:

The field equation for pion outside the bag is given by

$$(D_\pi^2 - m_\pi^2)\vec{\phi}_\pi = 0 \tag{7}$$

where $(D_\mu)_\pi = (1 + \vec{\phi}_\pi^2/f_\pi^2)^{-1}\partial_\mu$ is the chiral covariant differential operator. The boundary conditions at $r = R$ and $r \to \infty$ are, respectively:

$$r = R \qquad f_\pi n_\mu D_\pi^\mu \vec{\phi}_\pi = i\bar{q}\gamma_5 \frac{\tau}{2} q; \tag{8}$$

$$r \to \infty \qquad \vec{\phi}_\pi \to 0. \tag{9}$$

However, we make a linear approximation so that

$$(D_\mu)_\pi \to \partial_\mu \quad \text{and all terms of } O(\vec{\phi}_\pi^2) \to 0.$$

In the zero (pion)-mass limit, the model is equivalent to the linearized approximation of Callan et al.[5] The first boundary condition follows from the requirement that the normal component of the axial-vector current,

$$\vec{a}_\mu = \bar{q}\gamma_\mu\gamma_5 \frac{\vec{\tau}}{2} q + f_\pi \partial_\mu \vec{\phi}_\pi \tag{10}$$

be continuous across the bag surface r = R. Note that when $m_\pi = 0$, $\partial_\mu \vec{a}^\mu = 0$ (conserved axial current (CAC)), and when $m_\pi \neq 0$, $\partial_\mu \vec{a}^\mu = m_\pi^2 f_\pi \vec{\phi}_\pi$, (partially conserved axial current (PCAC)).

2. Calculation of Pion-Exchange Energy

To obtain the solution for the pion field, we need only the wave functions of the u- and d-quarks given below in order to satisfy the boundary condition (Eq. (8)) at r = R.

$$q(r,t=0) = \frac{N}{\sqrt{4\pi}} \begin{pmatrix} i j_o(\chi) \\ -j_1(\chi)\vec{\sigma}\cdot\hat{r} \end{pmatrix} u \tag{11}$$

where the quantity χ/R ($\chi = 2.04$) gives the lowest eigen energy of a single quark. It is easily shown that the pion field satisfying the linearized version of Eqs. (7)-(9) is given by

$$\vec{\phi}_j(\vec{r},0) = -\frac{m_\pi^2}{4\pi f_\pi} \frac{\chi}{2(\chi-1)} \frac{(1+\rho)e^{(\rho_o-\rho)}}{(2+2\rho_o+\rho_o^2)\rho^2} u_j^\dagger \vec{\sigma}\cdot\hat{r}\vec{\tau} u_j \tag{12}$$

where $\rho = m_\pi r$ and $\rho_o = m_\pi R$. For a given hadron bag, $\vec{\phi}_\pi = \sum_j \vec{\phi}_j$.

The pion field energy is

$$E_\pi = \tfrac{1}{2}\int_{r \geq R} d^3x[(\dot{\vec{\phi}}_\pi)^2 + (\nabla\vec{\phi}_\pi)^2 + m_\pi^2\phi_\pi^2] \tag{13}$$

from which we obtain the operator expression for the pion exchange energy $\hat{E}_{\pi,exch}$

$$\hat{E}_{\pi,exch} = \sum_{i\neq j} \tfrac{1}{2} \int_{r \geq R} d^3x(\nabla\vec{\phi}_i\cdot\nabla\vec{\phi}_j + m_\pi^2\vec{\phi}_i\cdot\vec{\phi}_j) \qquad (14)$$

$$= \sum_{i\neq j} u_i^+ u_j^+ \hat{E}_\pi(ij) u_i u_j ;$$

$$\hat{E}_\pi(ij) \alpha \frac{1}{f_\pi^2 R^3} \; \vec{\sigma}_i\cdot\vec{\sigma}_j \vec{\tau}_i\cdot\vec{\tau}_j . \qquad (15)$$

Notice that in E_π, there are two types of terms, namely, $\phi_i\phi_j$ $(i\neq j)$ and $\phi_i\phi_i$. We have identified the former as being responsible for the exchange energy and the latter for the self-energy. Since self-energy terms leads to the quark-mass renormalization and should be absorbed into E_q in the mass formula M(R), therefore all $\phi_i\phi_i$-terms are dropped.

The exchange energy due to the pion cloud outside a hadron bag is therefore calculated via

$$E_{\pi,exch}(R) = \langle hadron| \sum_{i\neq j} \hat{E}_\pi(ij)|hadron\rangle . \qquad (16)$$

This gives effectively a mass-renomalization for hadrons due to pion-exchange (see Fig. 3).

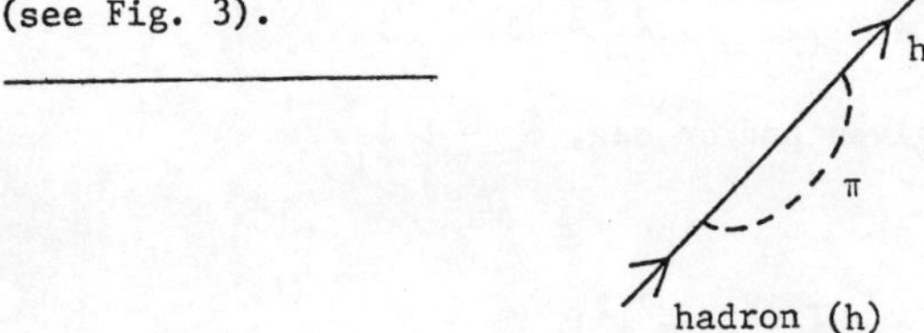

Fig. 3. Mass renomalization of hadron due to pion-exchange.

The resulting $E_{\pi,exch}$ is then incorporated into the mass formula, giving

$$M_{\pi}(R) \equiv M(R) + E_{\pi,exch}(R), \tag{17}$$

from which the new hadron mass spectroscopy is obtained by imposing the condition $(\partial M_{\pi}(R))/\partial R = 0$. This leads to new values for parameters B, Z_o, m_s, α_c and R. Our results for the hadron masses and bag-radii are listed in Table 1, together with their MIT-bag counterparts, so that a comparision can be easily made.

We note that (i) the new hadron mass spectroscopy is quite satisfactory and, with the exception of kaon and pion, there is in general a few percent change relative to the MIT-bag values, suggesting that the pion exchange energy term may be treated as a perturbation. Indeed a simple calculation of $E_{\pi,exch}$ using MIT-bag radius substantiates this point.— Hadrons with nonvanishing strangeness quantum numbers, however, receive no pion-exchange energy contribution. (ii) Again with the exception of kaon and pion, the new hadron bag-radii are larger than the corresponding MIT-bag radii. As we shall discuss below, our new bag-radii lead to values of mean square charge radius and magnetic moment which are overall in better agreement with the experimentally observed values. (iii) Our bag-radius result also suggests that the presence of PCAC pions causes kaon- and pion-bags to contract and the other low-lying hadron to expand. In the case of pion-bag, the contraction leads to its destabilization. Thus in our model, the PCAC pion and pion-like bag cannot coexist, indicating that

Table 1. Masses and Radii of low-lying hadrons with and without the inclusion of pion exchange effects. (The masses of p, Δ, Ω^- and ω are used as input in both cases.)

Baryons:	Exp. mass (GeV)	Bag-Model (MIT)	Bag-Model (PCAC)	Bag-Radius (GeV^{-1}) (MIT)	Bag-Radius (GeV^{-1}) (PCAC)
p	0.938	0.938	0.938	5.00	6.08
Λ	1.116	1.105	1.077	4.95	5.84
Σ^+	1.189	1.144	1.081	4.95	5.51
Ξ^o	1.321	1.289	1.219	4.91	5.38
Δ	1.236	1.236	1.236	5.48	6.32
Σ^*	1.385	1.382	1.373	5.43	6.21
Ξ^*	1.533	1.529	1.518	5.39	6.12
Ω^-	1.672	1.672	1.672	5.35	6.05
Mesons:					
ρ	0.77	0.783	0.816	4.71	5.49
κ^*	0.892	0.928	0.951	4.65	5.37
ω	0.783	0.783	0.783	4.71	5.24
φ	1.019	1.068	1.100	4.61	5.29
κ	0.495	0.497	0.386	3.26	2.85
π	0.139	0.280	---	3.34	--

Parameters	$B(GeV)^4$	Z_o	m_s(GeV)	α_c	f_π(GeV)
(MIT)	$4.42 x 10^{-4}$	1.84	0.279	0.55	---
(PCAC)	$3.00 x 10^{-4}$	1.53	0.291	0.79	0.092

at some deeper level, the idea of a static MIT-bag may be in conflict with chiral symmetry and PCAC. Moreover, the fact that kaon-like bag responses to the PCAC pion in a way similar to the pion-like bag leads to the possibility that if SU(3) (instead of SU(2)) PCAC condition is imposed, the kaon-like bag might become unstable as well.

3. Calculation of Mean Square Charge Radius and Magnetic Moments

The calculation of mean square charge radius, $\langle r^2 \rangle$, and magnetic moment, μ, is straightforward. The presence of pion cloud outside the bag implies that the appropriate expressions for $\langle r^2 \rangle$ and μ are, respectively, given by the following formulae:

$$\langle r^2 \rangle = \langle r^2 \rangle_{r \leq R} + \langle r^2 \rangle_{r \geq R} \tag{18}$$

$$\mu = (\mu)_{r \leq R} + (\mu)_{r \geq R} \tag{19}$$

where the second term in the above equations represents the pion-cloud contribution. However, since the pion field is static in our model, and the pion charge density ρ_π is proportional to the time-derivative of the pion field ($\rho_\pi \propto \phi \partial\phi/\partial t$), therefore $\langle r^2 \rangle_{r \geq R} = 0$. The only manifestation of pion exchange effects is therefore through the bag-radius R, for which reason our results for $\langle r^2 \rangle$ differ from those of MIT-bag by essentially a scale factor proportional to R. In contrast, the pion contribution to the magnetic moment is non-vanishing and is given by the following expression:

$$(\mu)_{r \geq R} \equiv \int_{r \geq R} d^3x \phi_\pi T_3 L_3 \phi_\pi$$

$$= \frac{R}{6\pi} \left(\frac{m_\pi}{f_\pi} \right)^2 \left(\frac{\chi}{2(\chi-1)} \right)^2 \frac{(2+\rho_o)}{(2+2\rho_o+\rho_o^2)\rho_o^2} (A + B) \qquad (20)$$

where

$$A = \sum_j (\sigma_3 \tau_3)_j ;$$

$$B = 2 \sum_{i \neq j} [(\sigma_+\tau_+)_i (\sigma_-\tau_-)_j - (\sigma_+\tau_-)_i (\sigma_-\tau_+)_j];$$

$$(i,j = u,d).$$

We remind the reader that the corresponding expressions for $\langle r^2 \rangle$ and μ for $r \leq R$ are the same as the standard MIT-bag expressions.[1] We present the results in Table 2. As easily seen, our results are in somewhat better agreement with the observed values.

Table 2. Mean square charge radii and magnetic moments of low-lying baryons.

STATE	$\langle r^2 \rangle$ (MIT)	$\langle r^2 \rangle$ (PCAC)	$\langle r^2 \rangle$ EXP	μ (MIT)	μ (PCAC)	μ EXP
p	13.2	19.6	20±2	1.90	3.30	2.79
n	0.00	0.00	-0.61±0.05	-1.27	-2.53	-1.91
Λ	1.14	1.72	---	-0.49	-0.54	-0.606±0.34
Σ^+	14.1	17.6	---	1.84	2.47	2.33±0.13
Σ°	1.14	1.49	---	0.59	0.64	---
Σ^-	11.9	14.7	---	-0.68	-1.20	-1.48±0.37
Ξ°	2.25	2.82	---	-1.06	-1.25	-1.20±0.06
Ξ^-	10.6	12.6	---	-0.44	-0.34	-1.85±0.75

III. DISCUSSION

The inclusion of the pion exchange energy in the manner described in Sec. II leads to the destabilization of a bag-pion. We note that the exchange energy $E_{\pi,exch}$ as calculated from Eq. (16) has a R^{-3} dependence and for a bag-pion it represents a large negative contribution to the mass formula $M_{\pi}(R)$, rendering it impossible to satisfy the extremal condition $\partial M_{\pi}(R)/\partial R|_{R=R_{pion}} = 0$, for values of (B, Z_o, m_s, α_c) parameters which generate the best fit to the hadron mass spectrum. In this regard it is interesting to note that in a different context, Horn and Yankielowicz have found that the instanton-induced interaction energy (E_I) has the same R^{-3} dependence and for a bag-pion E_I is also negative, leading to similar destabilization of the bag-pion. However, there is the important difference that E_I depends on spin and color spin, thus affecting only p, Λ, π and η, while E_{π} depends on spin and isospin, thus affecting all hadrons with two or more nonstrange quark constituents. Since both E_{π} and E_I are fundamentally related to the implementation of chiral symmetry and/or PCAC and the intricate properties of QCD vacuum states, it seems premature to attempt obtaining any deeper connection between PCAC-induced and instanton-induced effects. Nevertheless, it is an interesting question whether a heavy (and perhaps metastable) bag-pion may exist.

The bag model we considered in this paper represents an example of forced marriage between the MIT bag and PCAC, albeit within the framework of a linearized approximation. The results we obtained are encouraging. We are therefore motivated to investigate the possible theoretical scheme (procedure) for computing the non-linear effects due to PCAC pions. In this connection we should note that independent of us, Jaffe[8] has recently proposed a "hybrid chiral" bag model in which he has considered questions similar to ours. His pion exchange results are, however, qualitatively different from ours.

REFERENCES

1. A Chodos, R. J. Jaffe, K. Johnson, C. B. Thorn, and V. F. Weisskopf, Phys. Rev. D9, 3471 (1974); T. DeGrand, R. J. Jaffe, K. Johnson, and J. Kiskis, ibid, D12, 2060 (1975).
2. C. N. Yang and R. L. Mills, Phys. Rev. 96, 191 (1954).
3. In the second paper of Ref. 1, it is argued that Z_o is positive. However, it is worth pointing out that in QED it has been shown that $Z_o > 0$ for a slab but < 0 for a spherical cavity. No proof for the sign of Z_o for a spherical cavity in QCD exists in the literature. In our numerical work we have allowed either sign for Z_o and find that a positive Z_o does indeed lead to a better fit with and without the inclusion of the pion-exchange energy.
4. M. V. Barnhill, W. K. Cheng, and A. Halprin, Phys. Rev. D20, 727 (1979); M. V. Barnhill and A. Halprin, Phys. Rev. (in press).
5. C. Callan, Jr., R. F. Dashen, D. J. Gross, Phys. Lett. 73B, 307 (1978); Phys. Rev. D19, 1826 (1979).
6. Y. Nambu, Phys. Rev. Lett. 4, 380 (1960); J. Goldstone, Nuovo Cimento 19, 154 (1960).
7. T. D. Lee, Phys. Rev. D19 1802 (1979); Also, Professor Lee's paper in this Proceedings.
8. R. J. Jaffe, in Pionlike Structures Inside and Outside Hadrons, Proceedings of the 1979 Eric Summer School, A. Zichichi, ed., Plenum Press. Also, R. J. Jaffe, "The Sigma Term Revisted", MIT preprint CTP #822 (1979).

A MODEL OF MESON STRUCTURE WITH RELATIVISTIC INTERNAL MOTION AND THE PROBLEM OF STRATON MASSES

Chu Chung-yuan   An Ing
Institute of Theoretical Physics, Academia Sinica
Zhang Jian-zu
Shanxi University

Abstract

In this paper we have proposed a new type of the meson Bethe-Salpeter wave function with the following properties:

(1) The straton (quark) may possess small masses.

(2) The spinor structure of the wave function looks like a relativistic quark-pair state. Then we discuss the reasonable value of the u.d. straton masses, and find that it is about several hundred MeV.

How large are the u.d. straton "masses" which characterize the inertia of straton motion inside hadrons? This is a very important problem which has received much attention. During the past several years there have been a lot of different hypotheses, in which the straton mass range from very heavy to very light ones. At present it seems that the theory of light stratons is rather popular.

To investigate the possibility of a very light mass for the straton, we have done two things in this paper: 1) we have

established a model of meson under the assumption that the straton mass may be small and that they behave like relativistic free particles; a) using this model we have found some meson reactions sensitive to the straton mass and compared the theoretical estimates with the experimental results in order to determine the order of magnitude of the u.d. masses.

In the first section we define the meson Bethe-Salpeter amplitude and introduce some necessary symbols and basic requirements. In the second section we consider a system which consists of a free fermion and an anti-fermion and derive the Bethe-Salpeter amplitudes for the $J^{pc}=0^{-+}$ and 1^{--} state of this system, and in the third section we use the Bethe-Salpeter equation under the potential approximation to obtain a solution with the same form as the one derived in section two. Finally, we apply the above results to certain concrete processes and show that the u.d. straton masses cannot be too small; the reasonable values which we find are about several hundred MeV.

I. In this paper we will use the method proposed by Bethe-Salpeter to describe the relativistic bound state. [1] First we define the two-body Bethe-Salpeter (B-S) amplitude as

$$\begin{aligned}\phi^{(ij)}_{\alpha\beta}(x_1,x_2) &= \langle 0|T\psi^{(i)}_{\alpha}(x_1)\bar{\psi}^{(j)}_{\beta}(x_2)|\text{介子}\rangle \\ &= e^{iPX}[\phi_P^{(ij)}(x)]_{\alpha\beta} = e^{iPX}\int d^4p\, e^{ipx}[\phi_P^{(ij)}(p)]_{\alpha\beta}\end{aligned} \tag{1}$$

where $\psi^{(i)}(x)$ are the straton Heisenberg local field operators, the superscript i is the flavor index, $X=\frac{x_1+x_2}{2}$, $x=x_1-x_2$ and P is the meson 4-momentum. Here we would like to emphasize that the electromagnetic and weak currents consist of the local fields $\psi^{(i)}(x)$.

Suppose the whole theory is invariant under Lorentz transformation and space-time inversion, it can then be proved that [2] the 0^{-+} and 1^{--} meson B-S amplitude (1) can be written as:

$$0^{-+}:\quad \phi_P(p)=\sqrt{\frac{\mu}{E}}\gamma_5\{f_1+i\gamma_\mu[P_\mu f_2+p_\mu f_3]+i\sigma_{\mu\nu}P_\mu p_\nu f_4\}$$

$$1^{--}:\quad \phi_P(p)=\sqrt{\frac{\mu}{E}}\{i(ep)\phi^S-\epsilon_{\mu\nu\rho\sigma}e_\mu\gamma_\nu\gamma_5P_\rho p_\sigma\phi^A$$

$$+\gamma_\mu[e_\mu\phi_1^V+P_\mu(ep)\phi_2^V+p_\mu(ep)\phi_3^V] \tag{2}$$

$$+\sigma_{\mu\nu}[P_\mu e_\nu\phi_1^T+p_\mu e_\nu\phi_2^T+P_\mu p_\nu(ep)\phi_3^T]\}$$

which in the center-of-mass system is reduced to

$$0^{-+}:\quad \phi_P(p)=\gamma_5\{f_1+\hat{A}+i\sigma_{4i}f_{4i}\} \tag{3}$$

$$1^{--}:\quad \phi_P(p)=S+\hat{B}\gamma_5+\hat{\phi}^V+2ig_{4i}\sigma_{4i}+\sigma_{ij}\epsilon_{ijk}g_k$$

where μ is the meson mass, e_μ is the polarization vector of the 1^{--} meson and f_i and ϕ^I are Lorentz scalars. If we further assume that the interaction between stratons is invariant under the charge conjugate transformation, these functions must also satisfy the following conditions:

$$f_i(p^2,P\cdot p,m_i,m_j)=\begin{cases}f_i(p^2,-P\cdot p,m_j,m_i), & \text{if } i=1,2,4,\\ -f_i(p^2,-P\cdot p,m_j,m_i), & \text{if } i=3.\end{cases}$$

$$\phi^I(p^2, P\cdot p, m_i, m_j) = \begin{cases} \phi^I(p^2, -P\cdot p, m_j, m_i), & \text{if } \phi^I = \phi^S, \phi^A, \phi_1^V, \phi_2^V, \phi_1^T, \\ -\phi^I(p^2, -P\cdot p, m_j, m_i), & \text{if } \phi^I = \phi_2^T, \phi_3^T, \phi_3^V . \end{cases} \tag{4}$$

To determine the relations between all f_i and ϕ^I one must know the straton behavior inside the mesons. In Ref. [1], it is assumed that the straton masses are very large and their velocities very small, so the Bargman-Wigner type B-S amplitude is obtained. In Ref. [3] different points of view lead to other types of B-S amplitude structure. In this paper we suppose that the stratons are like relativistic free particles and obtain a new type of B-S amplitude.

II. Before evaluating the B-S amplitude for the meson bound state, we consider a system consisting of one free fermion and one antifermion, taking $\vec{p}$ as the fermion momentum and $-\vec{p}$ as the antifermion momentum in the c.m. system. According to the helicity formalism of Jacob and Wick [4], the state with definite total angular momentum in this system is:

$$|JM\mu_1\mu_2\rangle \sim \int d\Omega\, D^{J*}_{M,\mu}(\varphi, \theta, -\varphi)|\theta\varphi\mu_1\mu_2\rangle \tag{5}$$

where the angles (θ, φ) refer to the direction of $\vec{p}$, μ_1, μ_2 are, respectively, the helicities of the fermion and antifermion, $\mu = \mu_1 - \mu_2$, and $|\theta\varphi\mu_1\mu_2\rangle$ is a direct product of $|\vec{p}\,\mu_1\rangle$ and $|-\vec{p}\,\mu_2\rangle$. The phase convention follows that of Ref. [4].

Using the properties of the helicity states under space

inversion and charge conjugation, we obtain

for the 0^{-+} state

$$|0^{-+}\rangle \sim |0,0;\tfrac{1}{2},\tfrac{1}{2}\rangle - |0,0;-\tfrac{1}{2},-\tfrac{1}{2}\rangle \tag{6}$$

and for the 1^{--} state the two possible forms

$$|1^{--}\rangle \sim |1M;\tfrac{1}{2},-\tfrac{1}{2}\rangle + |1M;-\tfrac{1}{2},\tfrac{1}{2}\rangle$$

and

$$|1^{--}\rangle \sim |1M;\tfrac{1}{2},\tfrac{1}{2}\rangle + |1M;-\tfrac{1}{2},-\tfrac{1}{2}\rangle \tag{7}$$

Substituting (5)-(7) into the definition of the B-S amplitude in the c.m. system and some further calculations lead to:

$$0^{-+}:\ \phi_p(p) \sim \gamma_5\Big\{[(E_1+m_1)(E_2+m_2)+\vec{p}^{\,2}] - \gamma_4[(E_1+m_1)(E_2+m_2)-\vec{p}^{\,2}]$$
$$-i\vec{p}\cdot\vec{\gamma}(E_1+m_1-E_2-m_2) - \sigma_{4i}p_i(E_1+m_1+E_2+m_2)\Big\} \tag{8}$$

$$1^{--}:\ \phi_p(p) \sim [(E_1+m_1)(E_2+m_2)+\vec{p}^{\,2}][e_i-(\vec{e}\cdot\vec{n})n_i]\gamma_i$$
$$-i(E_1+m_1+E_2+m_2)(\vec{e}\wedge\vec{p})\cdot\vec{\gamma}\gamma_5 - [E_2+m_2-E_1-m_1](\vec{e}\wedge\vec{p})\cdot\vec{\Sigma}$$
$$+i[(E_1+m_1)(E_2+m_2)-\vec{p}^{\,2}][e_i-(\vec{e}\cdot\vec{n})n_i]\sigma_{4i}$$

and

$$\sim -i(E_1+m_1+E_2+m_2)(\vec{e}\cdot\vec{p}) + [(E_1+m_1)(E_2+m_2)-\vec{p}^{\,2}](\vec{e}\cdot\vec{n})\vec{n}\cdot\vec{\gamma}$$
$$+i(E_2+m_2-E_1-m_1)(\vec{e}\cdot\vec{p})\gamma_4 + i[(E_1+m_1)(E_2+m_2)+\vec{p}^{\,2}](\vec{e}\cdot\vec{n})\sigma_{4i}n_i \tag{9}$$

where $\vec{n} = \vec{p}/|\vec{p}|$. By use of the mass-energy relation, these two formulas can be rewritten in other forms. For example, we can rewrite Eqs. (8) and (9) as:

0^{-+} **state** $\phi_P(p) \sim \gamma_5 \left\{ (m_1 m_2 + p^2 + \frac{1}{4}\mu^2) - \gamma_4 \frac{(m_1+m_2)\mu}{2} \right.$

$$\left. + (m_2 - m_1) i \hat{p} - \mu p_i \sigma_{4i} \right\} \qquad (10)$$

1^{--} state: $\phi_P(p) \sim (m_1 m_2 + p^2 + \frac{1}{4}\mu^2)[e_i - (\vec{e}\cdot\vec{n})n_i]\gamma_i$

$$- i\mu(\vec{e}_\wedge \vec{p})\cdot\vec{\gamma}\gamma_5 - (m_2 - m_1)(\vec{e}_\wedge \vec{p})\cdot\vec{\Sigma}$$

$$+ i\left[\frac{\mu}{2}(m_1+m_2) + (m_2 - m_1)p_0\right][e_i - (\vec{e}\cdot\vec{n})n_i]\sigma_{4i}$$

and

$$\sim -i\mu(\vec{e}\cdot\vec{p}) + \left[\frac{\mu}{2}(m_1+m_2) + (m_2-m_1)p_0\right](\vec{e}\cdot\vec{n})\vec{n}\cdot\vec{\gamma}$$

$$+ i(m_2 - m_1)(\vec{e}\cdot\vec{p})\gamma_4 + i(m_1 m_2 + p^2 + \frac{1}{4}\mu^2)(\vec{e}\cdot\vec{n})\sigma_{4i} n_i \qquad (11)$$

These forms will be used in the next **section.**

III. Now we search for the B-S amplitudes of the meson bound states which have the same form as formulas (10) and (11).

To do this we use the B-S equation under the potential approximation:

$$(\hat{\partial} + \frac{i}{2}\hat{P} + m_1)\phi_P(x)(\hat{\partial} - \frac{i}{2}\hat{P} + m_2) = - \sum_{I = S,P,V,T,A} V^{(I)}(x,P)[\mathcal{P}^{(I)}\phi_P(x)] \qquad (12)$$

where $\mathcal{P}^{(I)}$ are γ -matrix projection operators satisfying

$$\mathcal{P}^{(I)}\mathcal{P}^{(J)} = \delta^{IJ}\mathcal{P}^{(J)}, \qquad \mathcal{P}^{(I)}\Gamma^{(J)} = \delta^{IJ}\Gamma^{(J)},$$

$$\Gamma^{(J)} \equiv (1, \gamma_\mu, \sigma_{\mu\nu}, \gamma_5, \gamma_\mu\gamma_5). \qquad (13)$$

and $V^{(I)}$ are the "potentials". Substituting the general expressions (1)-(3) of 0^{-+} and 1^{--} meson B-S amplitude into Eq. (12), we obtain in the c.m. system for 0^{-+} meson:

$$
\begin{aligned}
&(m_1 m_2 + V^{(P)} - \partial^2 + \tfrac{1}{4}\mu^2) f_1 + \tfrac{1}{2}(m_1+m_2)\mu A_4 + \mu \vec{\partial}\cdot\vec{f}_4 - (m_2-m_1)\vec{\partial}\cdot\vec{A} \\
&\qquad -(m_2-m_1)\partial_4 A_4 = 0 , \\
&(m_1 m_2 + V^{(A)} + \partial_4^2 - \vec{\partial}^2 - \tfrac{1}{4}\mu^2)\vec{A} - 2\partial_4\vec{\partial} A_4 - (m_1+m_2)\partial_4 \vec{f}_4 \\
&\qquad -(m_2-m_1)\vec{\partial} f_1 + \tfrac{1}{2}(m_2-m_1)\mu\vec{f}_4 = 0 , \\
&\tfrac{1}{2}(m_1+m_2)\mu f_1 + (m_1 m_2 + V^{(A)} + \vec{\partial}^2 - \vec{\partial}_4^2 + \tfrac{1}{4}\mu^2) A_4 - 2(\partial_4(\vec{\partial}\cdot\vec{A}) \\
&\qquad +(m_1+m_2)\vec{\partial}\cdot\vec{f}_4 - (m_2-m_1)\partial_4 f_1 = 0 , \\
&\mu\vec{\partial} f_1 - (m_1+m_2)\partial_4\vec{A} + (m_1+m_2)\vec{\partial} A_4 + (m_1 m_2 + V^{(T)} + \partial^2 - \tfrac{1}{4}\mu^2)\vec{f}_4 \\
&\qquad + \tfrac{1}{2}(m_2-m_1)\mu\vec{A} = 0
\end{aligned}
\tag{14}
$$

and for 1^{--} meson:

$$
\begin{aligned}
&(m_1 m_2 + V^{(S)} + \partial^2 - \tfrac{1}{4}\mu^2) S + (m_1+m_2)\partial_\mu \phi_\mu^V - 2\mu\vec{\partial}\cdot\vec{g}_4 \\
&\qquad -\tfrac{1}{2}(m_2-m_1)\mu\phi_4^V = 0 , \\
&(m_1 m_2 + V^{(A)} + \partial^2 - \tfrac{1}{4}\mu^2)\vec{B} + \mu(\vec{\partial}\wedge\vec{\phi}^V) - 2i(m_1+m_2)\partial_4\vec{g} \\
&\qquad -2(m_1+m_2)(\vec{\partial}\wedge\vec{g}_4) + i(m_2-m_1)\mu\vec{g} = 0 , \\
&(m_1+m_2)\partial_4 S + (m_1 m_2 + V^{(V)} - \vec{\partial}^2 + \partial_4^2 - \tfrac{1}{4}\mu^2)\phi_4^V + 2\partial_4(\vec{\partial}\cdot\vec{\phi}^V) \\
&\qquad -2(m_2-m_1)\vec{\partial}\cdot\vec{g}_4 - \tfrac{1}{2}(m_2-m_1)\mu S = 0 , \\
&(m_1+m_2)\vec{\partial} S - \mu(\vec{\partial}\wedge\vec{B}) + (m_1 m_2 + V^{(V)} - \partial^2 + \tfrac{1}{4}\mu^2)\vec{\phi}^V \\
&\qquad + 2\vec{\partial}(\partial_4\phi_4^V) + 2\vec{\partial}(\vec{\partial}\cdot\vec{\phi}^V) - \mu(m_1+m_2) g_4 \\
&\qquad + 2(m_2-m_1)\partial_4\vec{g}_4 + 2i(m_2-m_1)(\vec{\partial}\wedge\vec{g}) = 0 ,
\end{aligned}
\tag{15}
$$

$$-\mu\vec{\partial}S+(m_1+m_2)\vec{\partial}_\wedge\vec{B}-\frac{1}{2}(m_1+m_2)\mu\vec{\phi}^V+2(m_1m_2+V^{(T)}+\vec{\partial}^2-\partial_4^2+\frac{1}{4}\mu^2)\vec{g}_4$$

$$-4\vec{\partial}(\vec{\partial}\cdot\vec{g}_4)-4i\partial_4(\vec{\partial}_\wedge\vec{g})+(m_2-m_1)\partial_4\vec{\phi}^V-(m_2-m_1)\vec{\partial}\phi_4^V=0\ ,$$

$$\frac{i}{2}(m_1+m_2)\partial_4\vec{B}+(m_1m_2+V^{(T)}+\partial_4^2-\vec{\partial}^2-\frac{1}{4}\mu^2)\vec{g}-2i\partial_4(\vec{\partial}_\wedge\vec{g}_4)$$

$$+2i(m_2-m_1)(\vec{\partial}_\wedge\vec{\phi}^V)-\frac{i}{4}(m_2-m_1)\mu\vec{B}=0\ .$$

Generally it is difficult to solve these equations. But we find that by taking the projection potentials

$$V^{(A)}=V^{(T)}=0, \tag{16}$$

we can obtain an exact solution with the following form in the momentum space

$$\text{for } 0^{-+}:\quad \phi_p(p)=\gamma_5\Big\{(m_1m_2+p^2+\frac{1}{4}\mu^2)-\gamma_4\frac{(m_1+m_2)\mu}{2}+(m_2-m_1)i\hat{p}-\mu p_i\sigma_{4i}\Big\}f\ ,$$

$$[(m_1m_2+p^2+V^{(P)}+\frac{1}{4}\mu^2)(m_1m_2+p^2+\frac{1}{4}\mu^2)-\mu^2\vec{p}^2+(m_2-m_1)^2p^2-\mu p_0(m_2^2-m_1^2)-\frac{1}{4}\mu^2(m_1+m_2)^2]f=0\ . \tag{17}$$

and for 1^{--}:

$$\phi_p(p)=\Big\{-i\mu(\vec{e}_\wedge\vec{p})\cdot\vec{\gamma}\gamma_5+(m_1m_2+p^2+\frac{1}{4}\mu^2)[e_i-(\vec{e}\cdot\vec{n})n_i]\gamma_i+i[\frac{\mu}{2}(m_1+m_2)+(m_2-m_1)p_0][e_i-(\vec{e}\cdot\vec{n})n_i]\sigma_{4i}+(m_2-m_1)p_ie_j\sigma_{ij}\Big\}g\ ,$$

$$[(m_1m_2+p^2+V^{(V)}+\frac{1}{4}\mu^2)(m_1m_2+p^2+\frac{1}{4}\mu^2)-\mu^2\vec{p}^2+(m_2-m_1)^2p^2-\mu p_0(m_2^2-m_1^2)-\frac{1}{4}\mu^2(m_1+m_2)^2]g=0\ . \tag{18}$$

Obviously, these structures correspond to (10) and (11) respectively.[5]

Here we would like to point out that if $V^{(V)}=V^{(P)}$ then

Eqs. (17) and (18) are the same. Furthermore, it can be proved that the normalized conditions for f and g are the same also. Thus these two functions are completely identical. Of course, evaluating the concrete form of this function is related to the space-time structure of the potentials. Besides, it is difficult to solve a fourth order differential equation, so we will not discuss it further.

IV. Now we turn to the problem of the u.d. masses. First we must know which processes depend sensitively on the masses. It is easy to see that the annihilation processes are only related to a single term in the meson B-S amplitude, whereas in the "transition processes" in which a meson changes into another meson and emits other particles, all terms in the B-S amplitude will play important roles, and it would be quite difficult to single out the effect of each individual term. Furthermore, we can see that although the $\sigma_{\mu\nu} P_\mu e_\nu$ term is proportional to the straton masses in the 1^{--} meson B-S amplitude, yet this term does not appear in the electromagnetic and weak interaction, this is therefore not applicable for the 1^{--} meson. We now turn to the 0^{-+} meson. From (17) the $\gamma_4 \gamma_5$ term of the 0^{-+} B-S amplitude is proportional to the straton masses, and in the weak interaction Hamiltonian we have a $\gamma_\mu \gamma_5$ term, so the $\pi \to \mu\nu$ decay amplitude is related sensitively to the straton masses. In fact, a simple calculation gives

$$f_\pi = 2(m_u + m_\alpha)\sqrt{2\mu_\pi}\, f(x=0),$$

$$f(x=0)=\int d^4p f(p^2, P\cdot p). \tag{19}$$

Obviously if the masses of the current u.d. straton are zero, the $\pi \rightarrow \mu\nu$ decay would be forbidden. This conclusion can also be immediately obtained from kinematics. The weak interaction Hamiltonian tells us that if the u.d. masses are zero only left-handed straton and right-handed **antistraton** can annihilate through weak interaction, but from angular momentum conservation the left-handed straton and the right-handed antistraton cannot form a spin—0 state, so π is forbidden to decay into $\mu\nu$, which is in contradiction with experiments.

Is it reasonable to suppose that the u.d. masses are a few Mev? From Eq. (19), we can see that we must calculate the value of the B-S amplitude at the relative space-time coordinates x = 0 (this amplitude is usually called the zero-point wave-function). Of course, its exact value depends on the form of the "space wave-function $f(x)$" and we cannot calculate it exactly at present. However, as an estimate, we can try to use the Gaussian-type function used in Ref. (3):

$$f(x)=N e^{-\frac{\omega}{2}(\vec{x}^2+t^2)} \tag{20}$$

in the c.m system. From the normalization condition of the B-S amplitude we can evaluate the normalization constant N. Then putting this value into (19) we obtain

$$f_\pi \approx 2.5 \text{ Mev} \tag{21}$$

where we have taken $m_u + m_d$ = 11.7 Mev.[6] Comparing (21) with the experimental value $f_\pi \approx 0.97\, m_\pi$, we find that the value in Eq. (21) is 50 **times smaller.**

Another point is about the Kawarahayashi-Suzuki identity:

$$f_\pi = \frac{\mu_P}{2\gamma_P}, \qquad (22)$$

which was derived in SU(6) symmetry. This **identity** when compared with experiments has been established as correct to within the same order of magnitude. However, if the u.d. masses are very small, then f_π on the left-hand side of (22) would be very small, but the term on the right-hand side of (22) depends on the γ_μ term in the 1^{--} meson B-S amplitude, and its value varies slowly as we change the straton masses. In fact, if we use Eq. (21), the difference between the magnitude of the left- and right-hand side in (22) will be of the order of two. That is, the identity would be seriously violated.

One may ask: Do these results depend strongly on the concrete form of the space wave-function? The answer is "no". Actually if the space distribution of the meson B-S amplitude is restricted to within about 0.7 fermi [7], the value of the zero-point wave function cannot be changed significantly. Besides, even if the zero-point wave-function increases considerably, the violation of Eq. (22) will still be there, because both sides of Eq. (22) are related to the zero-point wave-function. In summary, our conclusion is that the u.d. straton masses in our consideration cannot be as small as a few Mev.

If we require our estimate of f_π to be comparable with experiment, then the u.d. masses must be several hundred **MeV.**

Then both sides of (22) would also be of the same order of magnitude.

We have also looked into some other 0^{-+}, 1^{--} meson reaction processes. We find that they are not sensitive to the straton masses, so they will not be further discussed here.

We are very grateful to Professors Zhu Hong-yuan, Hu Ning, Dai Yuan-ben and He Zuo-xiu for fruitful discussions and kind help.

References

1. E.E. Salpeter and H.A. Bethe, Phys. Rev., 84 (1951), 1232.
   Peking Theory Group of Elementary Particle, The 1966 Summer Physics Colloquium of the Peking Symposium.
2. Division of Elementary Particles, Laboratory of Theoretical Physics, Peking University, Laboratory of Theoretical Physics, Institute of Mathematics, Academia Sinica, Beijing Daxue Xuebao (Ziran kexue) 12 (1966), 209.
3. C.H. Llewellyn Smith, Ann. Phys., 53 (1969), 521.
   Chu Chung-yuan and An Ing, Scientia Sinica, 4 (1978), 387.
   M. Bohm, et al., Nucl. Phys., 51B (1973), 397.
4. M. Jacob and G.C. Wick, Ann. Phys., (N.Y.) 7 (1959), 404.
   A.D. Martin and T.D. Speanrman, Elementary Particle Physics (North-Holland, Amsterdam, 1970).
5. The solution corresponding to the second expression in Eq. (11) exists also. But, with careful examination it is shown that this is not the solution of ground state.

Because in this paper we are concentrating our attention on the mesons in the ground state only, therefore we have not discussed it.

6. S. Weinberg, A Festschrift for I.I. Rabi, edited by L. Motz (New York Academy of Sciences, N.Y. 1977) P185.
7. E. Gabathuler, Proc. 9th Int. Conf. High Energy Physics, Tokyo (1978), P841.

SOLUTION FOR THE BOUND STATE EQUATION AND THE ELECTROMAGNETIC FORM FACTOR OF THE MESON

Wang Ming-zhong, Zheng Xi-te

Chengtu University of Science and Technology

Wang Ke-lin

University of Science and Technology of China

Hsien Ting-chang

Institute of High Energy Physics, Academia Sinica

Zhang Zheng-gang

Chengtu Institute of Geology

Abstract

By proposing a relativistically covariant phenomenological potential -- the nearly -flat-bottom potential -- between the straton and anti-stration pair, we solved the Bethe-Salpeter equation for meson bound states. General properties of the solutions, the relative magnitudes of various components of the wave function, and the conditions under which the lowest partial wave approximation and the Bargman-Wigner approximation can be used are discussed. Methods for computing the space-like electromagnetic form factor and wave functions at the origin are given. It is shown that the interpretation of the B-S wave functions with regard to the probability amplitudes is reasonable.

I. INTRODUCTION

In the straton model [1] [2] [3], mesons are treated as bound states of the straton-antistraton system. The main tool for the relativistic approach to bound states is the Bethe-Salpeter equation [4]. In general, it is necessary to introduce a phenomenological potential. Many authors have investigated various different phenomenological potentials, such as: the Yukawa potential (its singularity at the origin yields a meson bound state too "slender"[2]), the difference of two Yukawa potentials, the Yukawa potential with a regularization term[5] or a cut-off[6] (all aiming at removing the singularity at the origin), the ν -potential[7] [8] obtained by differentiating the Yukawa potential ν times with respect to μ^2 (but still, they suffer from the defect of not yielding a radius for the meson "fat" enough to be compared with experimental data), the three and four dimensional harmonic oscillator potentials[10 - 14], etc. Starting with physical requirements, in §2 we propose a so called nearly-flat-bottom potential[15][16], which could yield a bigger radius for mesons.

It is needless to enumerate the difficulties in solving the B-S equation. In essence all studies in this subject had to be done under various approximations. Our aim is to investigate the meson bound states more systematically from the B-S equation under as few assumptions as possible. Apart from the form of the kernel, our only assumptions are :

(i) the masses of the straton and anti-straton are equal, and (ii) only the lowest partial waves are retained after the O(4) expansion of the wave function (However, the correction from the partial waves of one order higher is taken into consideration to examine the validity of this approximation).

The results and the general behaviours of the numerical solutions for the 0^- mesons are given in §3. In §4 we have shown how to compute[17] the space-like form factor of the 0^- meson by using directly the wave functions in the Euclidean space. An argument regarding the reasonableness of the probability-amplitude interpretation of the four dimensional Euclidean B-S wave function is given in § 5.

II. THE B-S EQUATION AND THE NEARLY-FLAT-BOTTOM POTENTIAL

The B-S equation for the meson bound state after the Wick rotation is

$$\chi_Q(P) = i\int d^4K\,[(M^2+P^2-Q^2)^2+4(P\cdot Q)^2]^{-1}(i\hat{P}-\hat{Q}-M) \cdot U(Q,P-K)\,\chi_Q(K)(i\hat{P}+\hat{Q}-M) \tag{2.1}$$

where M represents the mass of the straton and the anti-straton, P and K - the Wick rotated relative momentum, Q = (0,0,0,m/2), m-the meson mass and U-the potential function.

In a model with heavy straton and superstrong interactions, in order to prevent the energy level spacing between hadron states with different orbital angular mementum from being too large, it is necessary to assume that the potential between the straton and anti-straton is a deep flat-bottom potential[18] [19]. Experiments suggest that the superstrong interactions possess the SU(6) symmetry. However, the SU(6) symmetry is not relativistically invariant, thus it is reasonable to assume that the internal motion of the meson possesses nonrelativistic properties[1] [2]. Greenberg [20] investigated the form of the potential which can allow a relativistic motion of the particles inside a tightly bound system in the ground state and pointed out that this is possible only in the case of a square well potential or a potential superposed by attractive and repulsive Yukawa potentials. Furthermore, in order to yield the scaling behaviour as observed in the deep inelastic scattering experiment of electron, in the parton model, partons inside the nucleon ought to be nearly free. In other words, they ought to be in a deep potential with nearly flat bottom. The difficulty is how to extend the three dimensional description of a deep potential with a flat bottom to a tractable four dimensional relativistic variant, as required by the the covariance of the B-S equation.

As an attempt, we proposed a "nearly-flat-bottom" potial

$$U(K) = -\frac{iG^2}{(2\pi)^4}\sum_{j=1}^{n}\frac{a_j}{K^2+\beta_j^2 N^2} \tag{2.2}$$

which is a superposition of n single particle exchange potentials, a_j's are the relative coupling constants of various exchange potentials. The corresponding potential in the configuration space is

$$V(\vec{r}) = -\frac{G^2}{4\pi r}\sum_{j=1}^{n} a_j e^{-\beta_j N r} \tag{2.3}$$

To reduce the number of parameters in this phenomenological potential, let $\beta_1 = 1$, $\beta_{j+1} - \beta_j = \beta$ and require that V(r) has no singularity at the origin and equal to a finite value V(0). In order that the potential has a flat bottom, we require that $\frac{d^m}{dr^m}V(r)_{r=0} = 0$ up to m = n - 2. Redefine $\tilde{a}_j = -a_j NG^2/4\pi V(0)$, the above requirements lead to (for fixed n and β) a system of linear algebraic equations to determine $\tilde{a}_j$. In general, the $\tilde{a}_j$'s thus obtained are positive and negative alternatively. So (2.2) is just a nearly-flat-bottom potential superposed by a set of attractive and repulsive Yukawa potentials. Rewrite (2.2) as

$$U(K) = -\frac{i\tilde{G}^2}{(2\pi)^4}\sum_{j=1}^{n}\frac{\tilde{a}_j}{K^2+\beta_j^2 N^2} \tag{2.4}$$

where $\tilde{G}^2$ is the redefined coupling constant. $\tilde{G}^2\tilde{a}_j = G^2 a_j$. We have $\tilde{G}^2 = -4\pi\tilde{V}(0)/E$, with $\tilde{V}(0)=V(0)/M$ (in what follows we will omit the "$\sim$" and use the dimensionless quantities E=N/m and R=Mr). Now only three parameters are left : n, β and E, G^2 will be determined as the smallest eigenvalue of the system of coupled equations. The shape of the nearly-flat-bottom potential with the normalization of V(0) = -1 is depicted in Fig. 2-1, from which it is easily seen that the potential becomes flatter and broader with decreasing β , E or increasing n.

Noticing that the argument of U in eq. (2.1) is P - K, thus it can be expanded by the four dimensional spherical harmonics. According to the requirement of Lorentz covariance, the wave functions in (2.1) can be expressed in four Lorentz invariant functions and expanded by the four dimensional spherical harmonics. Using the orthogonality property and summation relations of the Gegenbauer functions we obtain

$$\chi_n^{(i)}(P,Q) = G^2 \sum_{j=1}^{4} \sum_{n'} \int_0^\infty dK\, \mathcal{K}_{nn'}^{ij}(P,Q,K)\, \chi_{n'}^{(j)}(K,Q) \tag{2.5}$$

$$n = \begin{cases} \text{even}, & \text{for } i = 1,2,4; \\ \text{odd}, & \text{for } i = 3. \end{cases}$$

This is the integral equation for the partial wave amplitudes of the wave function of the 0^- meson.

For the sake of further calculation, we introduce the following dimensionless quantities : B = m/2M = Q/M, P' = P/M, K' = K/M, and then omit the primes, with P and K understood as the dimensionless quantities. B indicates the tightness or the looseness of the bound state. Further, introduce the following Lorentz invariant functions $F_n^{(i)}$ of the same dimension : $F_n^{(1)} \equiv \chi_n^{(1)}(P,B)$, $F_n^{(i)} \equiv M\chi_n^{(i)}(P,B)$, (i = 2,3), and $F_n^{(4)} \equiv M^2 \chi_n^{(4)}(P,B)$. After this dimensionless procedure, the kernel functions are still denoted by $\mathcal{K}_{nn'}^{ij}$ (for their expressions see [9]), hence we have

$$F_n^{(i)} = G^2 \sum_{n'} \sum_{j=1}^{4} \int_0^\infty dk\, \mathcal{K}_{nn'}^{ij}\, F_{n'}^{(j)} \alpha^j \tag{2.6}$$

The α^j's introduced in eq. (2.6) assume different values corresponding to different types of coupling, see the following Table.

Type of Coupling	α^1	α^2	α^3	α^4
Scalar (S)	1	1	1	1
Pseudoscalar (P)	1	-1	-1	1
Vector (V)	4	-2	-2	0
Axial-vector (A)	4	2	2	0

III. NUMERICAL SOLUTIONS OF 0^- MESON FUNCTION

We have solved the system of integral equations (2.6) numerically on a DJS-6 computer. By introducing $x = \frac{P-1}{P+1}$, $y = \frac{K-1}{K+1}$ and changing the variables in (2.6) we can change the integration into a Gaussian summation

$$F_n^{(i)}(x_\ell) = G^2 \sum_{j=1}^{4} \sum_{n'} \mathcal{K}_{nn'}^{ij}\, F_{n'}^{(j)}(y_\ell)\, \alpha^j \frac{2W_\ell}{(1-y_\ell)^2} \tag{3.1}$$

where x_ℓ and y_ℓ are the Gaussian quadratures in the region (-1,1), W_ℓ 's are the corresponding weights. At first we retain only the lowest partial waves n, n' = 0 (i, j = 1, 2, 4) and n, n' = 1 (i, j = 3), and as an example we take M = 10 GeV for the straton mass, then the pion corresponding to B = 0.007 is a tightly bound state. For the purpose of comparison, we compute the case of B = 0.999, corresponding to a loosely bound state.

By using the iteration method, we obtain the first eigenvalue of G^2 and the corresponding eigenfunctions $F_n^{(i)}$ simultaneously. The main results are :

(i) Both tightly bound and loosely bound solutions can be found in a wide range of E (E = 0.01 3), i.e., bound state solutions exist for the nearly-flat-bottom potential with a scalar coupling.

(ii) With the lowest partial wave approximation, in the centre-of-mass system, the meson wave function has the following form :

$$\chi(P,B) = F_0^{(1)}\gamma_5 + BF_0^{(2)}\gamma_4\gamma_5 + F_1^{(3)}\hat{P}\cdot 2\cos\theta\cdot\gamma_5 + F_0^{(4)}B\epsilon_{\nu\lambda\alpha}P_\nu\sigma_{\lambda\alpha} \quad (3.2)$$

According to Llewellyn-Smith[21] we explicitly write down its spinor components :

$$\chi(P,B) = \begin{pmatrix} \chi_{+-}(P,B) & \chi_{++}(P,B) \\ \chi_{--}(P,B) & \chi_{-+}(P,B) \end{pmatrix} \quad (3.3)$$

Noticing the form $F_n^{(i)}$'s are combined in each term of (3.3), we may roughly estimate the relative contributions from each term by evaluating their maximum values, i.e., compare the absolute values of $F_1 \equiv F_0^{(1)}$, $F_2 \equiv BF_0^{(2)}$, $F_3 \equiv 2PF_1^{(3)}$, and $F_4 \equiv 2BPF_0^{(4)}$.

Our calculation shows that in the case of a tightly

bound state (B = 0.00 7) $F_1 \gg F_2, F_3, F_4$ and the meson wave function approximately has the form :

$$\chi(P,B) \approx \begin{pmatrix} 0 & \chi_{++}(P,B) \\ \chi_{--}(P,B) & 0 \end{pmatrix} \tag{3.4}$$

Moreover, $X_{++} \approx X_{--} \approx -F_0^{(1)} I$ i.e., the term with γ_5 spinor structure dominates. This corresponds to Llewellyn-Smith's model I[21]. In the case of a loosely bound state (B = 0.999), $F_1 \approx F_2 \gg F_3$, F4 (see Fig. 3-1), then approximately we have $X_{++} \gg X_{--}$, corresponding to Llewllyn-Smith's model III, i.e., the usual non-relativistic model. When the binding varies from loose to tight, the solution varies from Llewellyn-Smith's model III to model I.

(iii) The results obtained by using interactions other than the scalar coupling turn out to have the same feature. The dependence of -V(0) on B is given in Fig. 3-2 and 3-3, which shows the potential becoming broader and deeper with decreasing E and decreasing B (tighter bound states).

(iv) In the early stage during the study of the Straton Model, the Bargman-Wigner approximation was widely adopted [1] [2], in which it is assumed that $F_0^{(1)} \approx BF_0^{(2)}$, $F_3 \approx F_4 \approx 0$. Our numerical results show that for tightly bound states the B-W approximation is not valid, however, the approximation holds for loosely bound states.

(v) The correction by higher partial waves is negligible for tightly bound states, while for loosely state the correction is considerable.

(vi) The wave function of the pion $\psi^{\pi}(0)$ at the origin is defined by the following S matrix element for the weak decay process

$$\langle \mu^+ \bar{\nu} | S | \pi^+ \rangle \approx i G_w (2\pi)^4 \delta(p_{\pi^+} - p_{\mu^+} - p_{\bar{\nu}}) \sqrt{m_{\mu^+} m_{\bar{\nu}} / E_{\mu^+} E_{\bar{\nu}}}$$

$$\psi^{\pi}(0) \cos\theta \cdot i \bar{u}_{\bar{\nu}} \gamma_4 (1+\gamma_5) v_{\mu^+} \tag{3.5}$$

where G_w is the weak interaction coupling constant, θ - the Cabibbo angle and

$$|\psi^{\pi}(0)|^2 = 8\pi^4 \left(B \int_0^{\infty} dP \cdot P^3 F_0^{(2)}(P) + \int_0^{\infty} dP \cdot P^4 F_1^{(3)}(P)\right)^2 M^3 \tag{3.6}$$

Here the normalization condition of the wave function is the same as in [5].

$\psi^{\pi}(0)$ computed from eq. (3.6) is given in Fig. 3-4, to be compared with the experimental value determined from eq. (3.5) [1] $|\psi^{\pi}(0)|^2_{exp} \sim 0.76 \times 10^{-3}\ m_N^3$. A preliminary conclusion is that by a suitable choice of parameters in the theory, the computed value can be compatible with experiment.

IV. THE SPACE-LIKE ELECTROMAGNETIC FORM FACTOR OF 0^- MESONS

The electromagnetic form factor of 0^- mesons can be written as [17]

$$F(4K^2) = \frac{1}{2(m^2+K^2)} \frac{e}{(2\pi)^4} \int d^4p \, \mathrm{Tr}\left\{ \chi_{q_f}(p-K) \hat{q}_i \chi_{q_i}(p) \left[i\left(\hat{p} - \frac{\hat{q}_i}{2}\right) + M \right] \right\} \tag{4.1}$$

where q_i, q_f represent the four momenta of the meson in the initial and the final states respectively, $Q_\mu = 2K_\mu = (q_i - q_f)_\mu$ - the momentum transfer. Eq. (4.1) is written in the Minkowski space. In order that the wave functions computed numerically in the Euclidean space be directly applicable in eq. (4.1), it is necessary to continue the integration in eq. (4.1) to the Euclidean space. The analytic property of the integrand depends on those of the wave functions $\chi(P)$ and $\bar{\chi}(P-K)$, the other factors of the integrand is regular on the p_0-plane. The singularity of $\chi(P)$ consists in the cuts $\frac{1}{p_0 \pm (\omega - i\varepsilon)}$, $\omega_{min} = M - m/2$, thus $\chi(P)$ can be continued into the upper plane for $p_0 > 0$ and into the lower plane for $p_0 < 0$. The singularity of $\bar{\chi}(P-K)$ consists in the cuts $\frac{1}{(p_0 - K_0) \pm (\omega - i\varepsilon)}$. Thus for $K_0 < 0$, $K_0 + \omega_{min} < 0$ ($K_0 > 0$, $K_0 - \omega_{min} > 0$), both cuts will start from the left (right) side of the imaginary p_0 axis. Thus in order to guarantee the Wick rotation of the integrand, K_0 should be continued $K_0 \rightarrow iK_0$ simultaneously. Nevertheless, for the space-like case, we can always choose $K_0 = 0$ then $\chi(p)$ and $\bar{\chi}(p-K)$ have the same cuts, and the integrand can easily be Wick rotated. Choosing a reference frame in which $K = (K,0,0,0)$, $q_i = (K,0,0, i\sqrt{m^2+K^2})$ and $q_f = (-K, 0,0, i\sqrt{m^2+K^2})$, and neglecting the smaller third component, from (4.1) we obtain

$$F(4K^2) = \frac{1}{N}\int d^4P\{F_1^* F_1 - 2F_1^* F_2 + B^2 F_2^* F_2 + 2(2P^2 - 2P_4^2 - P_1 K) F_1^* F_4$$

$$-2P_1KF_2^*F_4+[4B^2(P^2-P_4^2)+K^2(2P^2+P_4^2-2B^2-\frac{K^2}{2}-2P_1^2)]F_4^*F_4\} \quad (4.2)$$

where by definition, $F_i^* \equiv F_i(P-K)$, $F_i \equiv F_i(P)$, N is the normalization factor such that $F(0) = 1$. All F_i's in eq. (4.2) are just the components of the meson wave functions computed from eq. (3.1). Thus we can directly compute the space-like electromagnetic form factor for the 0^- meson from (4.2). The result thus computed in the case of scalar coupling is given in Fig. 4.1, in which 0.007/0.2 is refered to the value of B/E. From Fig. 4.1 it is seen that the computed curves decrease too rapidly to fit the experiment in the large K^2 region. The results by using other types of coupling have the same feature.

V. THE RADIUS OF THE 0^- MESON.

The wave function of 0^- meson in the configuration space $\chi(\vec{x})$ is given by first taking four dimensional Fourier transform of the wave function in the momentum space and then the equal time limit ($X_4 = 0$). Though the physical intepretation of the relative time of a bound state is somewhat ambiguous, nevertheless, in the equal time limit $\chi(\vec{x})$ may be interpreted as the probability amplitude of forming the bound state at any instant by a straton-anti-straton pair separated by a distance $\vec{X}$. Thus the mean square radius of the meson is (the index p indicates that the result is obtained according to the probability interpretation)

$$\langle r^2 \rangle_p = \frac{1}{4}\int_0^\infty R^2 w(R)\, dR \tag{5.1}$$

where $w(R)$ represents the probability that the diameter of the meson bound state is R

$$w(R) = 4\pi R^2 \mathrm{Tr}(\chi^\dagger(\vec{x})\chi(\vec{x})) \Big/ \int_0^\infty 4\pi R^2 \mathrm{Tr}(\chi^\dagger(\vec{x})\chi(\vec{x}))\, dR \tag{5.2}$$

On the other hand, the electromagnetic radius of the meson is determined from its form factor by expanding $F(Q^2)$ in terms of small momentum transfer Q^2 as $F(Q^2) \approx 1 - \frac{1}{6} Q^2 \langle r^2 \rangle + \ldots$, and since $Q^2 = 4K^2$, we have

$$\langle r^2 \rangle_f = -\frac{3}{2}\frac{\partial}{\partial K^2} F(4K^2)\Big|_{K^2=0} \tag{5.3}$$

in which the index f indicates that the mean square radius is determined from the form factor. $R_p \equiv \langle r^2 \rangle_p^{1/2}$ and $R_f \equiv \langle r^2 \rangle_f^{1/2}$ for the 0^- meson are given in the cases of scalar (s), pseudoscalar (p) and vector (v) couplings in Figs. 5-1 -- 5.3. These two definitions of the radius of the 0^- meson are roughly in coincidence, though in the $S0^-$ case there exists a 20% discrepancy. This shows that to give the B-S wave function a probability interpretation is reasonable in a sense.

VI. CONCLUSION

By introducing a nearly-flat-bottom potential phenomenologically, we have systematically solved the B-S equation on an electronic computer, obtained the wave functions for various types of coupling, the value of the wave functions at the origin of the space configuration, and the space-like electromagnetic form factor for the 0^- meson. It is shown that the lowest partial wave approximation is applicable in the case of tightly bound states, but not in the case of loosely bound states. On the contrary, the Bargman-Wigner approximation is not applicable in the case of tightly bound states, but is in the case of loosely bound states.

Generally speaking, in computing the coupling constant, it is legitimate to retain only the larger components. However, in computing the value of the wave function at the origin of the space configuration, the smaller components become important, since it is determined just by the second and third (the small) components of the 0^- meson. This shows that different components play different roles in computing different physical quantities, and it is desirable to solve the B-S equation with all its Lorentz invariant functions up to a satisfactory degree.

The nearly-flat-bottom potential proposed here, though better than those phenomenological potentials mentioned in §1, still suffers from the drawback that it yields an electromagnetic form factor decreasing too rapidly at large momentum transfer, and is subjected to modification.

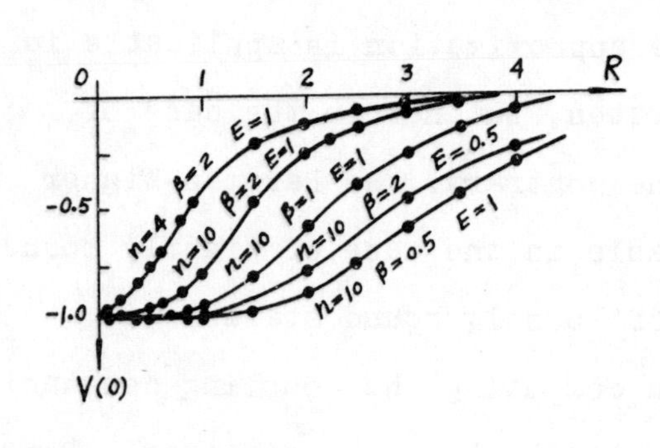

Fig. 2-1

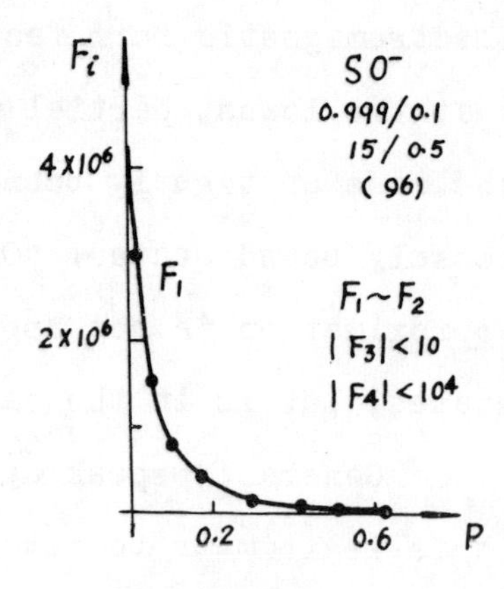

Fig. 3-1

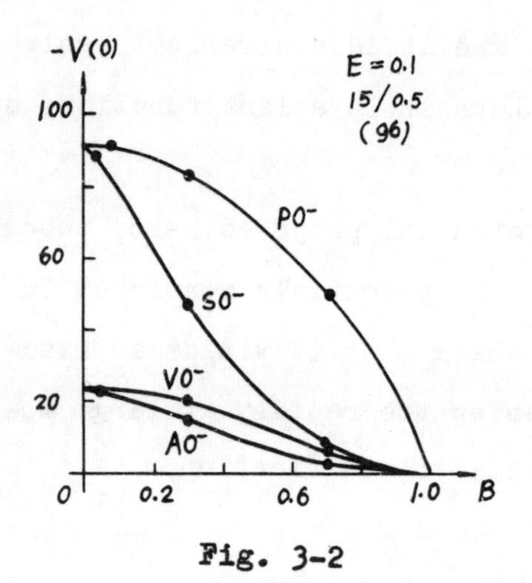

Fig. 3-2

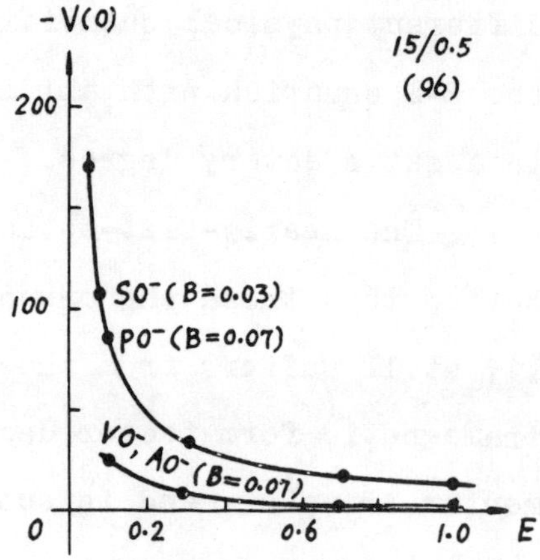

Fig. 3-3

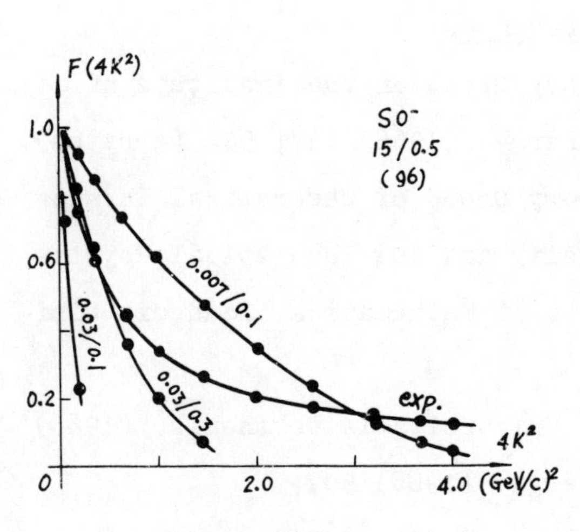

Fig. 4-1

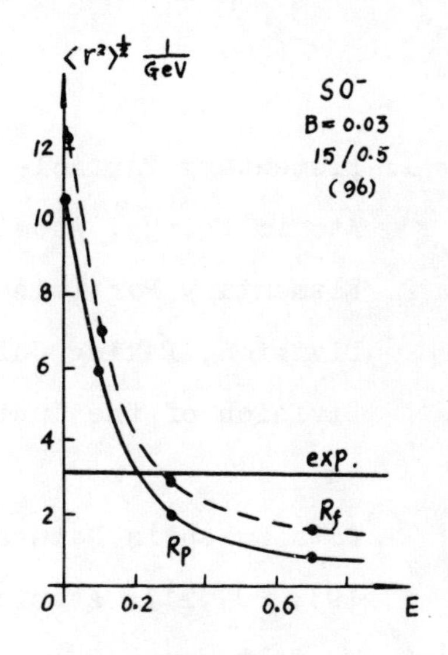

Fig. 5-1

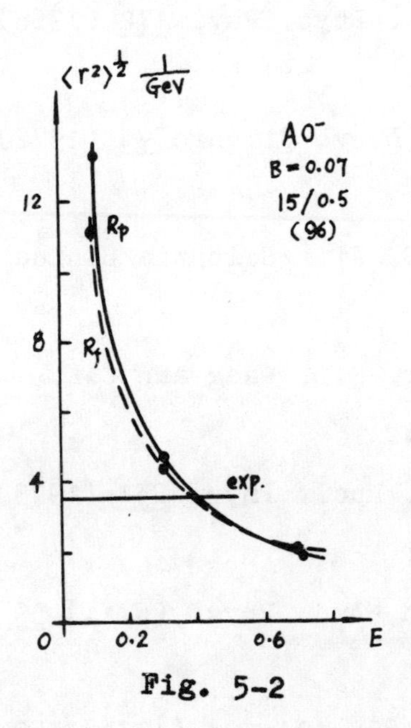

Fig. 5-2

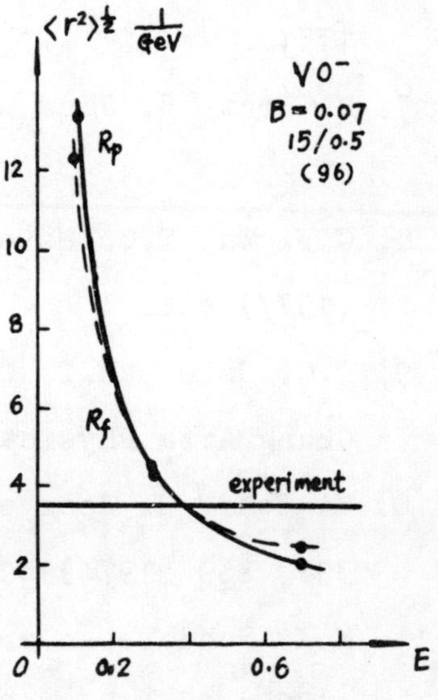

Fig. 5-3

References

1. Elementary Particle Theory Group of the Institute of Atomic Energy, Atomic Energy, (1966) 137; 514 (erratum).
2. Elementary Particle Theory Group of Theoretical Physics Division, Peking University and the Theoretical Physics Division of the Institute of Mathematics, Academia Sinica

   Acta Scientia Naturalis Universitatis Pekinenis, (1966) 103, 209,213; Atomic Energy, (1966) 507.
3. M. Gell-Mann, Phys. Letters, 8 (1964) 214.
4. E.E. Salpeter and H.A. Bethe, Phys. Rev. 82 (1974) 1232.
5. A. Guth, Ann. Phys. (N.Y.) 82 (1974) 407.
6. P. Naraynaswamy and A. Pagnamenta, Phys. Rev. 172 (1968) 1751.
7. M. Bohm, H. Joos and M. Krammer, Nuovo Cimento 7A (1972) 21.
8. C.W. Wu, T.C. Hsien, L. Gao and S. Ji, Scientia Sinica (1977) 431.
9. T.C. Hsien, H.T. Cheng, M.C. Wang, K.L. Wang and C.K. Chang Acta Physica, 27, (1978) 74.
10. M. Bohm, H. Joos and M. Krammers, Hucl. Phys. B51 (1973) 397, B69 (1974)349.
11. M.K. Sundaresan and P.J.S. Watson, Ann. Phys. (N.Y.) 59 (1970) 375, 87 (1974) 60.
12. N. Hu, Acta Physica, 25 (1976) 65, 494; 24 (1974) 458.
13. Y.S. Wu and T.H. Ho, Acta Physica, 26 (1977) 274.

14. C.Y. Chu and Y. An, Acta Physica Energiae Fortis et Physica Nuclearis, 1 (1977) 47, Scientia Sinica (1978) 387.

15. M. C. Wang, H.T. Cheng, T.C. Hsien, K.L. Wang and C.K. Chang, Acta Physica Energiae Fortis et Physica Nuclearis, 1 (1977) 7.

16. M.C. Wang, H.T. Cheng, T.C. Hsien, K.L. Wang and C.K. Chang, ibid, 2 (1978) 109.

17. M.C. Wang, H.T. Cheng, T.C. Hsien, K.L. Wang and C.K. Chang, ibid, 3 (1979) 572.

18. Xi Cheng, Bian Zheng and Li Zhong, Kexue Tongbao (1975) 28.

19. G. Morpurgo, Physics 2 (1965) 95.

20. D.W. Greenberg, Phys. Rev. 147 (1966) 1077.

21. C.H. Llewellyn-Smith, Ann. Phys. (N.Y.) 53 (1969) 521.

22. C.L. Hammer, V.S. Zidall, R.W. Reimer and T.A. Weber, Phys. Rev. D15 (1977) 696.

STRUCTURE FUNCTION OF PION IN A COVARIANT PARTON MODEL

Zhao Wan-yun
Institute of Theoretical Physics
Academia Sinica

Abstract

In this paper we formulate the parton model on the basis of a quantized field theory for composite particles. In this formalism we can compute the structure functions and the fragmentation functions of hadrons in various deep inelastic processes in terms of Feynman perturbation theory and the Bethe-Salpeter wave function of composite particles without the help of the old fashioned perturbation theory. We find that the pion structure function has the following behaviour: in the Bjorken limit, $F_{\pi}(x) \sim (1-x)$. This is in agreement with the recent Fermilab experiment.

I. INTRODUCTION

Various deep inelastic processes have been studied in the past. Theoretically, the parton model[1] itself and Quantum Chromodynamics (QCD)[2] combined with the parton model have met with a certain degree of success. But there remains to be found a reliable method for the theoretical calculation of the structure functions and the fragmentation functions in deep inelastic phenomena. Therefore, physical quantities measured in experiments have hitherto always been used as inputs for theoretical calculations.

In 1972, S.D. Drell and T.D. Lee[3] first proposed a covariant parton model and have computed the proton structure functions in terms of the Bethe-Salpeter (B-S) equation and investigated in detail relevant theoretical topics. In 1973, by using the quantized field theory of composite particles together with the light-cone expansion and the B-S wave functions of the proton, the relation between the structure functions of the proton and neutron and the B-S wave functions was obtained by He Zuo-xiu, Chang Zhao-xi and Xie Yi-cheng[4]. They pointed out that when the Bjorken variable $\omega \equiv -\frac{2(p\cdot k)}{k^2} \longrightarrow \infty$, the proton structure functions $F_{1,2}(\omega) \longrightarrow 0$. For lack of reliable three-body B-S wave functions they did not give a more concrete result. In this paper we shall discuss that on the basis of the quantized field theory of composite particles. We can give the relation between the structure functions and fragmentation functions of the pion and the B-S wave functions, and we can compute the QCD correction. Our results are independent of the old fashioned perturbation theory in the infinite momentum frame, and we introduce the assumptions of the parton model only in the perturbation expansion of the quantized field theory of composite particles. After the computation of the pion structure functions by means of the spinor structure of the pion wave function[5], we find that in the Bjorken limit, the structure function $F_{\pi}(x)$ of the pion is approximately (1-x). This result is in agreement with the recent Fermilab experiment[6].

II. COVARIANT PARTON MODEL

In our covariant parton model we assume that

1. the chosen Feynman graphs do not interfere with each other;

2. the probability that the on-shell final-state partons evolving into hadrons is one. As to the choice of the coordinate system, it is not the most important problem in the theory. Now we calculate the pion structure function to illustrate the application of our model. From the above assumptions the computation of the pion deep inelastic scattering processes will be achieved by the computation of the process

$$\gamma + \pi \longrightarrow q + \bar{q} \tag{2.1}$$

where γ is the space-like photon and q and $\bar{q}$ are the on-shell quark and anti-quark. For the process (2.1) the most important lowest order Feynman graphs are

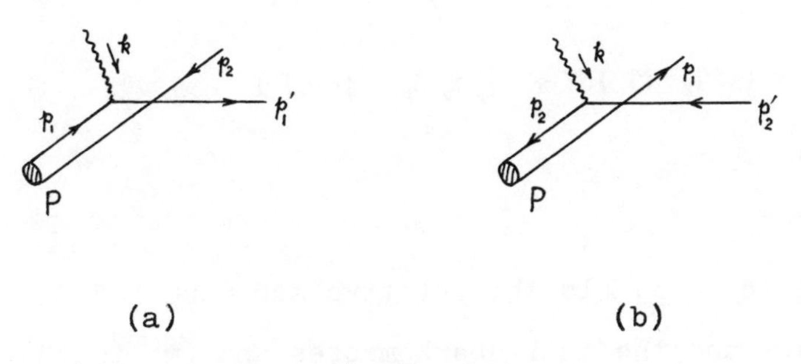

Figure 1

where P and k are the four-momenta of the pion and the photon, and p_1 , p_1' and p_2 , p_2' are the four-momenta of the quarks and the anti-quarks respectively. It should be pointed out that these two graphs (a) and (b) in figure 1 are not gauge invariant and the gauge invariance holds only when the theory takes into account all Feynman graphs of the same order in the electromagnetic coupling. As shown in Ref. [3] , in the ladder approximation and the Bjorken limit, the higher-order Feynman graphs do not contribute and only those two graphs of figure 1 contribute. According to the pertubation theory of the quantized field theory of composite particles we write the matrix element corresponding to the process (2.1):

$$M_{fi} = -\frac{i}{(2\pi)^4 k^2} \bar{u}_s(p_1') \gamma_\mu Q \chi_{(p_2+P,\, p_2)} S^{-1}_{F(p_2)} \bar{v}_r(p_2) \delta^4(P+k-p_1'-p_2) \tag{2.2}$$

where U_s and V_r are the quark and anti-quark wave functions and Q is the quark charge matrix and X is the B-S wave function of the pion. Using the spinor structure of the pion as in Ref. [5] , X can be represented by

$$\chi_{(p,P)} = \gamma_5 \left[(M_1 M_2 + p^2 + \frac{m^2}{4}) + \frac{(M_1+M_2)}{2} i \hat{P} + \right.$$

$$\left. + i(M_2 - M_1)\hat{p} + i P_\mu p_\nu \sigma_{\mu\nu} \right] f(p,P)$$

(2.3)

where $p = \frac{1}{2}(p_1 - p_2)$ is the relative momentum and M_1 and M_2 are the quark and the anti-quark masses and m is the pion mass, and $F(p,P) = \frac{g(p,P)}{p_2^2 + M_2^2}$.

We assume $M_1 = M_2 = M$ for the pion. For high energy scattering, we can neglect the quark mass in the spinor structure of the pion wave function. Actually in the process (2.1) the $\sigma_{\mu\nu}$ term does not contribute and there is only the contribution of the γ_5 term. After finishing the computation for the above M_{fi} , we can finally obtain the relation between the pion structure function and the pion B-S wave function. Choosing the laboratory system in the Bjorken limit, i.e. when $k^2 \to \infty$, $p \cdot k \to \infty$ and $\frac{P \cdot k}{k^2}$ is finite, we find that the expression of the structure function $F_\pi(x)$ is

$$F_\pi(x) = \frac{\pi m}{g} \int_{\frac{m}{2}(1-x)}^{\infty} |\vec{p}|^2 \, d|\vec{p}| \; g^2(|\vec{p}|) \cdot (1-x)$$

(2.4)

where $X = -\frac{k^2}{2(P \cdot k)}$, $0 \leqslant X \leqslant 1$, $\vec{p}$ is the relative momentum in the three dimensions.

We know from the expression (2.4) that the constraint for the wave function $g(p,P)$ is that $g(p,P)$ converges more rapidly than $|\vec{p}|^{-2}$ when $|\vec{p}| \to \infty$, and $g(p,P)$ has no singularity at the origin.

From Eq. (2.4), we find that when $x \to 1$, $F_\pi(x) \sim (1-x)$. Because the pion mass m is small, the expression (2.4) can be represented approximately by

$$F_\pi(x) \approx \frac{\pi m}{9} \int_0^\infty |\vec{p}|^2 \, d\,|\vec{p}| \; g^2(p,P) \cdot (1-x) \tag{2.5}$$

The integral in front of the (1-x) term in the expression (2.5) is constant. The constant is dependent on the wave function $g(p,P)$. Therefore $F_\pi(x) \sim (1-x)$. This is in agreement with the recent experiment worked at Fermilab (see figure 2).

With the above method, conducted at Fermilab (see Fig. 2). correction, and obtain the relation between the fragmentation function and the B-S wave function of the pion. This will be discussed in another paper.

The author wishes to express his gratitude to He Zho-xiu, Dai Yuan-ben, Zhu Chong-yuan and An Yin for their valuable help.

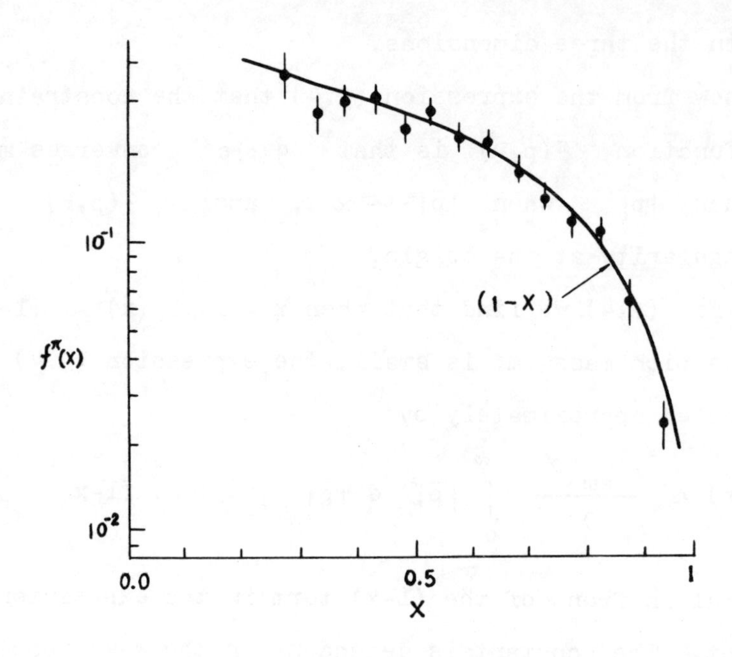

Figure 2. The pion structure function.

References

1. J.D. Bjorken, Phys. Rev. 179 (1969) , 1547; 185 (1969), 1975.
2. J.F. Owens and E. Reya, Phys. Rev. D17 (1978), 3003.
   S. Brodsky and G.R. Farrar, Phys. Rev. Lett. 31 (1973), 1153.
   G.R. Farrar, Nucl. Phys. B77 (1974), 429.
   J.F. Gunion, Phys. Rev. D10 (1974), 242.
3. S.D. Drell and T.D. Lee, Phys. Rev. D4 (1972), 1738.
4. He Zho-xiu, Chang Zhao-xi and Xie Yi-cheng, Acta Physica Sinica, 24 (1975), 115.
5. Zhu Chong-yuan, An Yin and Chang Jian-zu, paper to be published.
6. CERN COURIER, 19 (1979) 150.

MIGDAL APPROXIMATION IN GRIBOV FIELD THEORY ON A LATTICE*

By

Chung-I Tan

Brown University, Physics Department, Providence, Rhode Island, 02912 USA

ABSTRACT

The asymptotic behavior of soft hadronic collisions is studied in the framework of Gribov's field theory for interacting pomerons. An analog model of Ising-spins on a lattice with an anti-hermitian three-spin interaction is constructed. The structure of position space renormalization group analysis appropriate for this classical spin system is discussed and the existence of a desired nontrivial fixed point is demonstrated utilizing a Migdal approximation. In particular, we find that the critical index z having a "strong coupling" value $\ell n3/\ell n2 \simeq 1.58$.

---

*Supported in part by the U. S. Department of Energy under contrac No. EY-76-C-02-3130.A003 (Task A-Theoretical).

I. INTRODUCTION

Gribov's field theory (GFT) for interacting pomerons[1] has provided a systematic framework for discussing high energy behavior of hadron amplitudes at small momentum transfers where the effects of multiple-pomeron exchanges are important. Issues such as rising cross sections, which traditionally are thought to be more closely related to the direct-channel unitarity, can be discussed as infrared behavior of GFT when the renormalized pomeron mass vanishes.

In GFT, a bare pomeron is treated as a quasiparticle associated with a field $\psi(t,\underset{\sim}{x})$ in rapidity t(euclidean time), impact parameter $\underset{\sim}{x}$ (two spatial dimensions), and it satisfies a disperson relation $E = -\mu^2 + \alpha'\underset{\sim}{k}^2$ (corresponding to a trajectory $\alpha(\underset{\sim}{k}^2) = (1+\mu^2) - \alpha'\underset{\sim}{k}^2$). Renormalization group (RG) analyses using the ϵ-expansion technique strongly suggest that the infrared behavior of GFT is analogous to that of a critical phenomenon. The critical behavior can be studied by analyzing the scaling behavior of a two-point function at large t and $|\underset{\sim}{x}|$, which, upon Fourier transform in $\underset{\sim}{x}$, provides the dominant behavior of the elastic amplitude at small $\underset{\sim}{k}$, large $s(\ell ns \sim t)$,

$$F(s,\underset{\sim}{k}^2) \simeq is\,(\ell ns)^{\gamma}\ [(f(\underset{\sim}{k}^2\ (\ell ns)^{z})\ ] \tag{1.1}$$

where γ, z are critical indices.[3] However, the ϵ- expansion results are quantitatively unreliable, $\epsilon \equiv 4-D$, where D=2 for GFT. It is, therefore, desirable to search for alternative approaches which can provide further insights as well as new calculation schemes for GFT.

In this paper, an analog Ising-spin system on a lattice with an anti-hermitian three-spin interaction is constructed and its critical behavior studied. This spin system possesses identical symmetries as that of GFT; therefore, they belong to the same universality class by conventional beliefs. We discuss the structure of position space renormalization analysis appropriate for this classical spin system and demonstrate the existence of a desired nontrivial fixed point utilizing a Migdal approximation. In particular, we demonstrate that one critical index, z, naturally takes a "strong coupling" value of $\ell n3/\ell n2 = 1.58$.

II. ONE-COMPONENT FORMULATION OF GFT

GFT is defined by a Lagrangian density

$$L(\psi^{+},\psi) = -\frac{1}{2}\ \psi^{\dagger}\overleftrightarrow{\partial}_{t}\psi - \alpha'\vec{\nabla}\psi^{\dagger}\vec{\nabla}\psi - V(\psi^{\dagger},\psi) \tag{2.1}$$

where V contains an anti-hermitian triple-pomeron coupling

$$V(\psi^{\dagger},\psi) = -\mu^{2}\psi^{\dagger}\psi + \frac{ig}{2}\psi^{\dagger}(\psi^{\dagger}+\psi)\psi \tag{2.2}$$

In order to construct an analog classical spin system, we shall first reformulate GFT in a one-component form.

To simplify the discussion, we first consider the classical theory with $\alpha' = 0$, (or equivalently, D=0). Instead of working with $\psi^{\dagger}$ and ψ, which are creation and annihilation operators, we work with a single field $\phi \equiv (\psi-\psi^{\dagger})/2i$ and its conjugate momentum $\Pi = (\psi+\overline{\psi})$. We next perform a shift in Π so that the Hamiltonian does not contain a term linear in the new momentum $\overline{\Pi}$. Writing

$$H(\phi,\overline{\Pi}) = K(\phi,\overline{\Pi}) + U(\phi) \tag{2.3}$$

where $K(\phi,\overline{\Pi}) \propto \overline{\Pi}^2$ is the "kinetic energy", the new potential is

$$U(\phi) = -\frac{2}{3}\mu^2\phi^2 + (2/27\,g^2)\,[\Delta(\phi)^3 + \mu^6], \tag{2.4}$$

$$\Delta(\phi) \equiv [\mu^4 + 3g^2\phi^2]^{1/2} \tag{2.5}$$

The potential posseses (i) a unique minimum at $\phi=0$ for $\mu^2<0$, (ii) symmetric minima at $\phi=\pm\mu^2/g_o$ for $\mu^2>0$ ($\alpha_o>1$). Since $\Delta(\phi)\propto|\phi|$ as $\phi \to \pm\infty$, (2.4) is similar in structure to the Landau-Ginzburg potential, $U = -\mu^2\phi^2 + \lambda\phi^4$, in a $\lambda\phi^4$ theory.

The quantum GFT can be defined by a path integral in a Hamiltonian formalism. Alternatively, we can arrive at a Lagrangian formalism for the field by systematically integrating out the conjugate momentum $\overline{\Pi}$. This has been performed in Ref. 4, and the resulting path integral for the time evaluation operator can then be formally identified with the partition function for a classical many-body system. After returning to the case for $D\neq0$, the partition function $Z = \int \Pi_i \, d\mu(\phi_i) \exp[-\mathcal{H}]$, where,

$$d\mu(\phi) \simeq d\phi \exp[-(a_t a_s^D)U(\phi)] \tag{2.6}$$

a_t, a_s being lattice spacings, and the effective Hamiltonian is given by an expression in higher derivatives of $\phi(t,\underset{\sim}{x})$

$$\mathcal{H}[\phi] = \int dt d^D \underset{\sim}{x} \, [A(\partial_t\phi)^2 + B(\vec{\nabla}\phi)^2 + \ldots L(\partial_t\phi)(\vec{\nabla}\phi)^2 + \ldots] \tag{2.7}$$

where A,B...L..are functions of ϕ^2 only.

III. LATTICE-SPIN FORMULATION OF GFT

In the limit $(a_t a_s^D)(\mu^4/g^2) \to \infty$ with $\mu^2 > 0$, the measure $d\mu(\phi)$ approaches $\delta(\phi^2-\mu^4/g^2)$, thuo leading to an Ising spin $\sigma_i \equiv (g/\mu^2)\phi_i$ for each lattice site where σ_i takes on values $\pm$ 1. Replacing derivatives in (2.7) by finite differences, the partition function becomes $Z = \sum_{\{\sigma_i\}} e^{-\mathcal{H}}$, where, symbolically,[4]

$$\mathcal{H} = -K_1 \sum_t \sigma_i\sigma_j - K_2 \sum_x \sigma_i\sigma_j - \sum L(i,j,k)\,\sigma_i\sigma_j\sigma_j + \ldots \tag{3.1}$$

In (3.1), K_1 and K_2 terms correspond to nearest neighbor interactions in the rapidity and in the impact parameter directions respectively. The L term represents a nearest neighbor three-spin coupling whose properties will be specified shortly.

It is a common belief that the critical behavior of a many-body system exhibits universality at a critical point--all systems possessing identical symmetry properties have identical critical indices irrespective of the details of their microscopic interactions. From (2.6), our potential is invariant under $P(\underset{\sim}{x} \to -\underset{\sim}{x})$, $T(t \to -t)$, and $S(\sigma \to -\sigma)$

separately. However, the T and S symmetries are broken by the kinetic terms in (2.7) or (3.1), where the three spin term is invariant only under P and TS, (the combined time and spin reversal, not T and S separately.)

Quantum mechanically, it is readily seen that, under T, $H \rightarrow H^{\dagger}$, and $H^{\dagger} \neq H$. so that the system is not hermitian. However, our anti-hermitian three-pomeron coupling also changes sign under $S(\psi \rightarrow -\psi, \psi^{\dagger} \rightarrow -\psi^{\dagger})$, so that ST is again a quantum mechanical symmetry.

In (3.1), both K_1 and K_2 terms are invariant under P, T and S separately, which are symmetries of an ordinary Ising system. Therefore, we do not expect under a position-space RG treatment, these terms to play a crucial role for the onset of the GFT critical behavior.

For S non-invariance, we must have odd-spin interactions and, for ST invariance, the minimum number involved is three. Our nearest-neighbor three spin interaction in (3.1) is precisely what is needed, with L(i,j,k) defined by T: $L \rightarrow -L$, P: $L \rightarrow L$, and it formally follows from the L-term in (2.7).(4.6) Therefor we concentrate next on carrying out a position-space RG analysis to locate a fixed point in L.

IV. REVIEW OF POSITION-SPACE RENORMALIZATION GROUP TRANSFORMATION[7]

Given a lattice structure, we can introduce a new lattice of identical topology by grouping a neighboring sites of the former into a cell. In our 1 + 2 dimensional space, a basic nearest neighbor lattice structure is a cube; we form a new cubic cell lattice where each cell cube has a volume $(\lambda_t \lambda_s^2)$, (relative to the original cube), λ_t, λ_s being integers. (In a conventional treatment, $\lambda_t = \lambda_s = 2$ is usually chosen. For our case, relaxing the restriction $\lambda_t = \lambda_s$ turns out to be crucial.)

Associated with each cell, a new Ising-spin , σ', can be defined; its relation to the original Ising-spin is specified by a "projection matrix" $P(\sigma',\sigma)$ where

$$\sum_{\sigma'} P(\sigma',\sigma) = 1 \tag{4.1}$$

This matrix, whose specific form does not concern us for now, then defines a new Hamiltonian for our cell lattice spins by

$$e^{-H'[\sigma']+\text{const.}} = \sum_{\sigma} P(\sigma',\sigma) e^{-H[\sigma]} \tag{4.2}$$

From (4.1), we easily see that the partition function remains invariant under this transformation from $H[\sigma]$ to $H'[\sigma']$.

If we express the original Hamiltonian explicitly in terms of its couplings, collectively denoted by $\underset{\sim}{K}$ as $H(\sigma;\underset{\sim}{K})$, the transformed Hamiltonian can then be expressed as

$$H'[\sigma'] = H(\sigma';\underset{\sim}{K}') \ . \tag{4.3}$$

Eq. (4.3) in turn, defines a mapping $R(\lambda_t,\lambda_s) : \underset{\sim}{K} \to \underset{\sim}{K}'$. We say that our RG transformation has a fixed point if this transformation $R(\lambda_t,\lambda_s)$ has a fixed point $\underset{\sim}{K}^*$.

For (3.1), we single out the three-spin interaction L, and introduce an external magnetic field h, while denoting the remaining couplings by $\underset{\sim}{K}$. We now look for the fixed points of

$$R(\lambda_t,\lambda_s) : (L,K,h) \to (L',K',h') \ . \tag{4.4}$$

Our earlier symmetry consideration demands $L^* \neq 0$ and unstable. Similarly, we have $h^* = 0$ and unstable, whereas the values for K^* are unconstrained so long as they correspond to stable directions in the parameter space.

The magnetic field is introduced to facilitate the calculation of spin correlation functions. Keeping L and $\underset{\sim}{K}$ at their fixed point values, for small h, we have

$$h' = \lambda_t^{z_3} h \tag{4.5}$$

where $z_3 > 0$. It then follows by differentiations that the two-point function satisfies a scaling law at large t and $|\underset{\sim}{x}|$,

$$G(\hat{t},\hat{\underset{\sim}{x}}) = (\lambda_t)^{2z_3}(\lambda_t\lambda_s^2)^{-2}G(\hat{t}/\lambda_t,\hat{\underset{\sim}{x}}/\lambda_s) \tag{4.6}$$

where $t = \hat{t}a_t$, $\underset{\sim}{x} = \hat{\underset{\sim}{x}}\, a_s$, and the factor $(\lambda_t\lambda_s^2)$ comes from the volume of the cell cube. After repeated applications of $R(\lambda_t,\lambda_s)$ we obtain

$$G(t,\underset{\sim}{x}) = t^{-\eta} f(x^2/t^z) \tag{4.7}$$

where $-\eta = 2z_3-2-2z$ and $z \equiv 2\ell n\lambda_s/\ell n\lambda_t$. Taking Fourier transform of (4.7) with repsect to $\underset{\sim}{x}$ and comparing it with (1.1), we find $\gamma = -\eta+z$.

If the total cross section continues to rise, $(\gamma>0)$, we must have $2z_3-2-z > 0$. To satisfy the Froissart bound, we need $\gamma\leq 2$. Since G is a spin correlation function, we have $\eta\geq 0$, thus the requirement $z\leq 2$, or

$$\ell n\lambda_s/\ell n\lambda_t \leq 1 \,. \tag{4.8}$$

Of course, these conclusions presuppose the existence of a non-trivial fixed point in L.[8]

If $\underset{\sim}{K} \neq \underset{\sim}{K}^*$ and $\underset{\sim}{K}^* = 0$, above discussion must be generalized to account for the rate at which $\underset{\sim}{K}$ approaches $\underset{\sim}{K}^*$. This will modify the critical indices obtained above. We assume $\underset{\sim}{K} = \underset{\sim}{K}^*$ in what follows and refer to the critical indices so obtained as their "strong coupling" values.

V. MIGDAL APPROXIMATION

Let us begin by first considering an ordinary Ising model in two dimensions, where each lattice site is specified by $\hat{r} = (\hat{x}_1, \hat{x}_2)$, with $\hat{x}_1, \hat{x}_2$ integers. Introducing unit vectors $\hat{e}_1 = (1,0)$, $\hat{e}_2 = (0,1)$, the relevant Hamiltonian can be written as

$$H = -J_1 \sum_{\hat{r}} \sigma(\hat{r})\sigma(\hat{r}+\hat{e}_1) - J_2 \sum_{\hat{r}} \sigma(\hat{r})\sigma(\hat{r}+\hat{e}_2) \tag{5.1}$$

where we have deliberately differentiated couplings J_1 and J_2.

We now consider RG transformation by scaling $\hat{x}_1$ and $\hat{x}_2$ by λ_1 and λ_2 respectively. However, instead of performing $R(\lambda_1, \lambda_2)$ in one single step, we first perform $R(\hat{e}_1; \lambda_1)$ and then $R(\hat{e}_2; \lambda_2)$, where $R(\hat{e}_i; \lambda_i)$ is a transformation only changing the scale in $\hat{e}_i$ by λ_i. This procedure treats J_1 and J_2 asymmetrically.

A Migdal approximation[7] can be defined as

$$R(\hat{e}_i; \lambda_i) : J_j \to R(J_i)\delta_{ij} + (1 - \delta_{ij})\lambda_i J_j \tag{5.2}$$

where $R(J_i)$ is a RG transformation for a one-dimensional Ising chain with a scale change by λ_i. Note that, under $R(\hat{e}_1; \lambda_1)$, K_2 is simply multiplied by a factor λ_1 and under $R(\hat{e}_2, \lambda_2)$, K_1 by λ_2. Let $\lambda_1 = \lambda_2 = \lambda$, and after one complete cycle, we obtain

$$J_1' = \lambda R(J_1), \; J_2' = R(\lambda J_2) \; . \tag{5.3}$$

At a fixed point,

$$J_1^* = \lambda R(J_1^*), \qquad J_2^* = R(\lambda J_2^*) \ . \tag{5.4}$$

A Migdal approximation can be partly justified by noting that, in (2.6), in changing the scale for t by λ_t, with the field chosen to be dimensionless, the coupling K_2 gets multiplied by λ_t. Since K_2 does not couple spins in the $\hat{t}$ direction, we approximate the effect of this RG transformation in $\hat{t}$ by simply mapping $K_2 \to \lambda_t K_2$ and then calculating K_1' as if K_2 is absent. One next treats other directions in turn and the resulting RG transformation should be reliable, at least in locating the existence of a fixed point.

Alternatively, (5.3) can also be arrived at by a "bond-moving" technique, i.e. by doubling up every other neighboring K_2-bonds, we are left with a rectangular lattice, which can be transformed back into a square lattice by a one-dimensional RG transformation $R(\lambda_1)$. When interpreted in this manner, (5.3) corresponds to a variational calculation which leads to an upper bound for the free energy. This procedure has been used successfully in exhibiting the flow diagram of an Ising model, and it has also been used extensively in exploring critical properties of other more complicated systems. In what follows, Migdal approximations will be used in this latter sense.

We now return to (3.1) while keeping only the three-spin interaction term. Our lattice is now three dimensional, with each site labelled by

$\underset{\sim}{r} = (\hat{t}; \hat{x}_1, \hat{x}_2)$, $\hat{t}, \hat{x}_1, \hat{x}_2$ integers. Introducing unit vector $\underset{\sim}{e}_o = (1;0,0)$, $\underset{\sim}{e}_1 = (0;1,0)$, $\underset{\sim}{e}_2 = (0;0,1)$, the Hamiltonian can be written explicitly as

$$H = - \sum_{\underset{\sim}{r}} \sum_{i=1,2} [L_i][\sigma(\underset{\sim}{r}+\underset{\sim}{e}_i) + \sigma(\underset{\sim}{r}-\underset{\sim}{e}_i)][\sigma(\underset{\sim}{r})][\sigma(\underset{\sim}{r}+\underset{\sim}{e}_o) - \sigma(\underset{\sim}{r}-\underset{\sim}{e}_o)] \tag{5.5}$$

where L_1 and L_2 are couplings in $(t-x_1)$ and $(t-x_2)$ planes respectively.

We are interested in constructing a RG transformation which scales $\hat{t}$ by λ_t and $\hat{x}_1$, $\hat{x}_2$ by λ_s. Let us define this $R(\lambda_t; \lambda_s, \lambda_s)$ by first performing a transformation $R_1(\lambda_o; \lambda_1, \lambda_2)$ and then $R_2(\lambda_o; \lambda_2, \lambda_1)$; clearly, this defines $R(\lambda_t; \lambda_s, \lambda_s)$ with

$$\lambda_t = \lambda_o^2, \quad \lambda_s = \lambda_1 \lambda_2 . \tag{5.6}$$

For simplicity, we keep $\lambda_1 = \lambda_2 = \lambda$, and define $R_i(\lambda_o; \lambda, \lambda)$ by the following Migdal approximation

$$R_i(\lambda_o; \lambda, \lambda) : L_j \to \lambda R(L_i)\delta_{ij} + (1 - \delta_{ij})\lambda_o \lambda^2 L_j \tag{5.7}$$

where $R(L_i)$ is a RG transformation for a D=1 RFT spin system with scale changes by λ_o and λ in $\hat{n}_o$ and $\hat{e}_i$ respectively. Eq. (5.7) follows from a bond-moving procedure, and, applying R_2 after R_1, we obtain for our composite RG transformation $R(\lambda_t; \lambda_s, \lambda_s)$

$$L_1' = \lambda_o \lambda^3 R(L_1), \quad L_2' = \lambda R(\lambda_o \lambda^2 L_2) . \tag{5.8}$$

If a unique nontrivial fixed point (L_1^*, L_2^*) exists, it must satisfies

$$L_1^* = (\lambda_o \lambda^2) L_2^* = (\lambda_o \lambda^3) R(L_1^*) \ . \tag{5.9}$$

VI. EXISTENCE OF A NONTRIVIAL FIXED POINT

For the Ising problem, R(J) in (5.2) can be chosen to be a decimation transformation, where, for $\lambda=2$,

$$R(J) = \frac{1}{2} \ln \cosh 2J \tag{6.1}$$

and it follows that fixed point, (5.4), exists where $J_1^* = \ln \cosh 2J_1^*$. The key ingredient of (6.1) relevant to us is the existence of fixed points for R(J) itself at J = 0 and ∞, where

$$\begin{aligned} R(J) &\propto J^2 & J &\sim 0 \\ &\sim J - \ln 2 & J &\to \infty \ . \end{aligned} \tag{6.2}$$

Since R(J) does not contain a nontrivial fixed point, $J = \infty$ is unstable.

Above example indicates that if we expect (5.9) to hold as a generic property, we must construct $R(L_i)$ in (5.6) so that at a minimum $L_i = 0$ and ∞ are fixed points of $R(L_i)$, independent of whether $R(L_i)$ itself has a non-trivial fixed point or not. From a perturbative analysis, one generally expects, under any reasonably constructed RG transformation, $L_i = 0$ to be a stable fixed point, i.e., for L_i small, $R(L_i) \propto L_i^\delta$, $\delta>1$. Therefore, we must next turn to the point $L_i \to \infty$. That is, one must examine the "ground state" structure at "absolute zero". In particular, a ground state configuration must be invariant under a RG transformation.

For an Ising system, (5.1), with $J_1, J_2 \to \infty$, the ground state is doubly degenerate, with spins either pointing all up or down. The ground state degeneracy for (5.5) at D=1 is $(2^k \times 3)$ -fold[9] on a periodic square-lattice with 3k sites in $\hat{e}_o$ and even number of sites in $\hat{e}_1$. This can be seen by examining the internal energy of a plaquette of 2x2 spins; as $L \to \infty$, a minimum energy configuration is one with alternating columns ($\hat{e}_1$-direction), of down, up and alternating spins. The degeneracy is obtained by noting that spacelike shifts, (in $\hat{e}_1$-direction) of individual columns of alternating spins and timelike shifts, (in $\hat{e}_o$-direction), of the whole configuration leave the interaction energy invariant.

The RG transformation $R(L_i)$ corresponds to a scaling by factors λ_o and λ in the $\hat{e}_o$ and $\hat{e}_1$ directions respectively. A necessary and sufficient condition for $L_i = \infty$ to be a fixed point of R is the invariance of the ground state structure, which then requires $\lambda_o = 3m + 1$, $\lambda = 2n + 1$, m,n = 1,2....
The simplest possible choice being $\lambda_o = 4, \lambda = 3$. Each choice can be used as the starting point of our strong coupling RG analysis, and they are differentiated and then brought in consonant by the behavior of $\underset{\sim}{K}$ in (4.4) as they are driven to $\underset{\sim}{K}^*$. In what follows, the choice $\lambda_o = 4$ and $\lambda = 3$ is used, or from (5.6)

$$\lambda_t = 16 \qquad \text{and} \qquad \lambda_s = 9. \tag{6.3}$$

Having constructed a RG transformation $R(L_i)$ with $L_i = 0, \infty$ as fixed points, the existence of a fixed point, (5.9), can now be demonstrated. We consider the generic situation where

$$R(L_i) \sim L_i^{\delta}, \quad \delta>1 \quad \text{for } L_i \to 0$$

$$\sim \rho L_i \quad \rho \geq 1 \quad \text{for } L_i \to \infty \tag{6.4}$$

Let us differentiate two cases, (i) $\rho > 1$, and (ii) $\rho = 1$. For case (i), the fixed point of R at ∞ is stable, and a unstable fixed point at finite L_i exists. It then follows, as a generic property, a solution as (5.9) must also exist. For case (ii), $L_i = \infty$ is an unstable fixed point and it is likely that no other fixed points at finite L_i exist. However, under (5.8), it follows from (6.3) that

$$L'_i \propto L_i^{\delta}, \qquad \text{for } L_i \to 0$$

$$L'_i \propto \lambda_t^{1/2} \lambda_s L_i \quad \text{for } L_i \to \infty \tag{6.5}$$

where $\lambda_t^{1/2} \lambda_s = 36$. Clearly a solution such as (5.9) exists. (QED)

VII. AN EXPLICIT RG TRANSFORMATION

It has been proved in a formal analysis [9] that (5.5) at $D = 1$ does possess a second order phase transformation; therefore, in constructing

$R(L_i)$, if done properly, one should arrive at the situation where $\rho > 1$ in (6.4). However, in the context of a Migdal approximation, this is unncessary and we have constructed examples for both $\rho > 1$ and $\rho = 1$. We describe here one particular construct which illustrates the usefulness of a Migdal approximation in analyzing the criticality of GFT.

We now construct $R(\lambda_o, \lambda_1)$ by a further application of a Migdal approximation, in particular, in the sense of a bond-moving approximation. We first move bonds so that $L_i \rightarrow 2L_i$ and leaving the system to be a parallel rows of one-dimensional chains of plaquettes. The transformation $R_1(\lambda_o, 1)$ for this one dimensional plaquette chain can be conveniently chosen to be a decimation with $\lambda_o = 4$. Unlike the case of a one-dimensional Ising chain, however, this decimation only leads to an approximately closed mapping. After the decimation $L_1 \rightarrow 2L_1 \rightarrow D_4(2L_1)$, we complete the $R_1(4,1)$ transformation by moving half of the bonds back to form a complete square lattice, i.e.,

$$R_1(4,1) \; : \; L_1 \rightarrow 1/2 \; D_4(2L_1) \tag{7.1}$$

Without writing out a long and uninteresting expression [10], we simply note that under $R_1(4,1)$, $L_1 \rightarrow L_1^3$ for $L_1 \rightarrow 0$ and $L_1 \rightarrow L_1$ -constant, as $L_1 \rightarrow \infty$, similar in structure to the decimation transformation (6.1). We next perform $R_2(1,\lambda_1)$ by simply define

$$R_2(1,\lambda_1) \; : \; L_1 \rightarrow \lambda_1 L_1 \tag{7.2}$$

so that the combined RG transformation now read

$$R(\lambda_o, \lambda_1) \; : \; L_1 \to \frac{\lambda_1}{2} D_4 (2L_1) \tag{7.3}$$

This transformation satisfies (6.4), thus providing an explicit demonstration for fixed point, (5.9). Clearly, our crude construct, (7.2) can be greatly improved. We shall not pursue this point here further, partly because of certain intrinsic limitation of decimation transformations; therefore, we do not attempt, at this juncture to calculate all relevant critical indices. We end by pointing out that one critical index z takes on a value

$$z = 2\ell n\lambda_s/\ell n\lambda_t = \ell n3/\ell n2$$

This should be contrasted with an ε-expansion result of 1.18, (to second order in ε).

Footnotes and References

(1) V. N. Gribov, Zh. Eksp. Teor. Fiz. 53, 654 (1967) [Sov. Phys.-JETP 26, 414 (1968)].

(2) A. A. Migdal, A. M. Polyakov and K. A. TerMartirosyan, Phys. Lett. 48B, 239 (1974); H. D. I. Abarbanel and J. B. Bronzan, Phys. Rev. D9, 2397 (1974); H. D. Abarabanel, J. B. Bronzan, R. L. Sugar and A. R. White, Phys. Rep 21C, 119 (1975); M. Moshe, Phys. Rep. 37C, 255 (1978).

(3) In Ref. (2), $-\gamma$, rather than γ, is used for Eq. (1.1).

(4) B. Harms and Chung-I Tan, Phys. Rev. D16, 1186 (1977), and Phys. Lett. 67B, 435 (1977); C.-I Tan, Phys. Rev. D18, 3009 (1978).

(5) J. Cardy and R. Sugar, Phys. Rev. D12, 2514 (1975).

(6) The importance of this three-spin coupling was first emphasized in Ref. 5 in a formalism slightly different from ours in Ref. 4. We have shown in Ref. 4 that the coefficient L in (2.7) actually vanishes and a three-spin coupling comes from higher order expansion term $\dot{\phi}(\vec{\nabla}\dot{\phi})^2$. However, for a position-space RG analysis, this difference is immaterial. We also note that, in Ref. 5, the K_1 term is also missing from (3.1).

(7) For a review, see L. Kadanoff, Rev. Mod. Phys. 49, 267 (1977).

(8) In Ref. 5. $\lambda_s = \lambda_t$ is chosen; the assumption of the existence of a nontrivial fixed point in L then guarantees the Froissart bound. However, we shall shortly see that nontrivial fixed point in L cannot exist if $\lambda_s = \lambda_t$ is assumed.

(9) P. Suranyi, Phys. Rev. Lett. 37, 725 (1976).

(10) Further treatment as well as details of the present analysis will be published elsewhere.

CONSEQUENCES OF PARTON MODEL AND PERTURBATIVE QCD IN LEPTON-PAIR PRODUCTION

C.S.Lam

Stanford Linear Accelerator Center, Stanford, California, USA, and Department of Physics, McGill University, Montreal, Canada.

Wu-Ki Tung

Department of Physics, Illinois Institute of Technology, Chicago, Illinois, USA.

Abstract

Test of the parton model and perturbative QCD in lepton-pair production processes are fromulated in terms of the model-independent structure functions. Relations among the structure functions that are characteristic of the Drell-Yan parton model, as well as Quantum Chromodynamic corrections thereof, are systematically presented. These relations manifest themselves as specific predictions on the angular distribution of the outgoing lepton-pair. Special attention is directed to one relation, $W_\mu^\mu = 2W_1$, which is not modified by low order perturbative QCD diagrams -- in distinct contrast to its Callan-Gross counterpart in deep inelastic scattering, and to the integrated Drell-Yan cross-section(which is essentially W_μ^μ and known to be subject to a large correction). It is pointed out that, if QCD is correct, the apparent lepton center-of-mass angular distribution may change substantially from low to high $q_\perp$, yet the above basic relation in terms of invariant structure functions remains valid throughout. This is a unique prediction of perturbative QCD unfettered by known ambiguities and uncertainties.

I. INTRODUCTION

Lepton pair production in hadron collision processes has been studied extensively in recent years because it naturally extends and complements deep inelastic scattering in revealing the structure of hadrons. Theoretical and experimental studies of lepton pair production have relied heavily on the Drell-Yan quark-antiquark annihilation mechanism[1]. The agreement between expectations based on this model and the observed behavior of ratios such as $\sigma(\pi^+ N \to \mu^+\mu^- X)/\sigma(\pi^- N \to \mu^+\mu^- X)$ and $\sigma(\pi N \to \mu^+\mu^- X)/\sigma(NN \to \mu^+\mu^- X)$ as functions of Q^2 has provided much encouragement to this quark-parton picture[2,3]. Supporting evidence concerning the angular distribution of the final state leptons is also beginning to emerge[3]. These developments naturally lead to more detailed investigations on non-trivial dynamical corrections to the simple parton picture. Most of these activities center on the application of perturbative Quantum Chromodynamics. A significant but disturbing finding is that the next-to-leading order correction to the Drell-Yan formula is a full 100 per cent [4]. The ramifications of this result are only beginning to be sorted out. Studies of other consequences of QCD as alternative tests for its validity are clearly needed.

In this report we give a systematic account of the consequences of the parton model and perturbative QCD on the structure functions for lepton pair production which control the angular distribution of the final state leptons. The formalism is developed in parallel to that of deep inelastic scattering.[5] In addition to results analogous to those in deep inelastic scattering, new features made possible by the richer kinematics emerge. Thus, at the level of the parton model,we obtain three relations among the four independent structure functions[5]. These relations are derived without detailed assumptions on the parton distribution functions. They are generalizations of the Callan-Gross relation for deep inelastic scattering. Next, we investigate modifications to these relations due to low order QCD effects. Two features are emphasized: (i) at high transverse momenta for the virtual photon, first order QCD effects are expected to dominate, hence new structure function relations that we derive from the relevant diagrams[6] will provide useful tests of QCD; (ii) for small transverse momenta of the virtual photon, we found one of the basic structure function relation previously derived from the parton model is not modified by radiative corrections in low order QCD — in sharp contrast to the case of the Drell-Yan cross-section formula, and to its Callan-Cross counterpart in deep inelastic scattering. These results lead to rather unique Predictions of perturbative QCD on the lepton center-of-mass angular distribution as the variable $q_\perp$ increases.

II. KINEMATICS

The amplitude for lepton pair production in hadron collisions is given by:

$$f = \bar{u}(k_1)\gamma_\mu v(k_2)\frac{e^2}{q^2}\langle X|J^\mu(0)|p_1 p_2\rangle , \qquad (1)$$

where the momentum labels are shown in Fig. 1. Let $P = p_1 + p_2$, $p = p_1 - p_2$, $q = k_1 + k_2$, and $k = k_1 - k_2$.

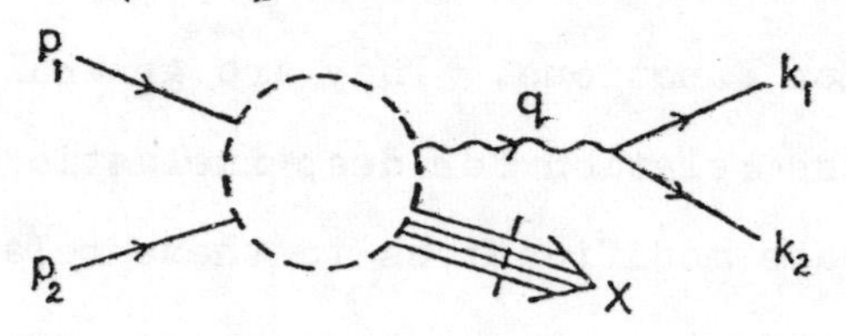

Fig. 1

If we only consider the high energy limit, then $s \equiv -P^2 = p^2$ and $Q^2 \equiv -q^2 = k^2$. The cross-section for this process is:

$$d\sigma = \left(\frac{\alpha}{SQ}\right)^2 L^{\mu\nu} W_{\mu\nu} \frac{1}{(2\pi)^4} \frac{d^3k_1\, d^3k_2}{k_1 k_2} \qquad (2)$$

where the lepton tensor is,

$$L^{\mu\nu} = g^{\mu\nu} - \frac{q^\mu q^\mu}{q^2} - \frac{k^\mu k^\nu}{k^2} \qquad (3)$$

and the hadron tensor is,

$$W_{\mu\nu} = S\int d^4z\, e^{iq\cdot z} \langle p_1 p_2 | [J_\mu(z), J_\nu(0)] | p_1 p_2 \rangle \ , \tag{4}$$

as illustrated in Fig. 2.

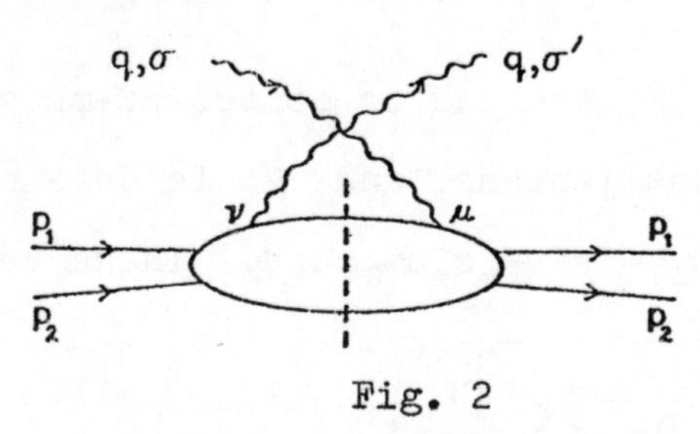

Fig. 2

The conventional Drell-Yan cross-section involves integrating over the final state lepton angles , and is given by:[1]

$$d\sigma/d^4q = \left(\frac{\alpha}{QS}\right)^2 \frac{1}{12\pi^3} W^\mu_\mu \ . \tag{5}$$

The hadron tensor amplitude $W_{\mu\nu}(q,p,P)$ must satisfy requirements of crossing symmetry, gauge invariance, and parity conservation. It can be expanded in terms of the gauge invariant projection tensor $\tilde{g}_{\mu\nu} = g_{\mu\nu} - q_\mu q_\nu / q^2$ and normalized gauge invariant vectors $\tilde{P}_\mu = \tilde{g}_{\mu\nu} P^\nu/\sqrt{S}$, $\tilde{p}_\mu = \tilde{g}_{\mu\nu} p^\nu/\sqrt{S}$ as follows:[7,5]

$$W_{\mu\nu} = W_1 \tilde{g}_{\mu\nu} + W_2 \tilde{P}_\mu \tilde{P}_\nu - W_3 (\tilde{P}_\mu \tilde{p}_\nu + \tilde{p}_\mu \tilde{P}_\nu) + W_4 \tilde{p}_\mu \tilde{p}_\nu \tag{6}$$

The invariant structure functions W_i (i = 1, 2, 3, 4) thus defined are functions of scalar variables. The cross-section formula, Eq. (2), can be expressed in terms of these functions:

$$d\sigma = \frac{\alpha^2}{32\pi^3}\frac{1}{S^2}\frac{\cos^2(\Theta/2)}{\sin^4(\Theta/2)}\, dk_1\, dk_2\, d\cos\theta_1\, d\cos\theta_2\, d\phi \tag{7}$$

$$\left[W_2 + 2\tan^2\frac{\Theta}{2} W_1 + \frac{\cos\theta_1+\cos\theta_2}{1+\cos\Theta} W_3 + \frac{1+\cos\Theta'}{1+\cos\Theta} W_4 \right]$$

where θ_1, θ_2, ϕ and Θ are hadron center-of-mass-frame angles (Fig. 3) of the leptons, and Θ' is defined by $\cos\Theta' = \cos\theta_1\cos\theta_2 - \sin\theta_1\sin\theta_2\cos\phi$. These results

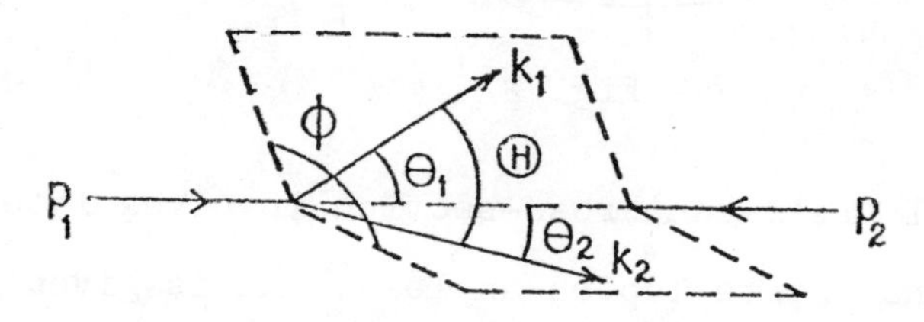

Fig. 3

closely resemble those for deep inelastic scattering, yet they have not gained wide use in the literature.

It will prove useful to introduce also helicity structure functions as follows. In the rest frame of the virtual photon (also the center-of-mass frame of the lepton pair), choose the y-axis to be normal to the hadron reaction plane. Let X^μ, Y^μ, and Z^μ be unit 4-vectors perpendicular to q^μ and parallel to the coordinate axes chosen. (X^μ and Z^μ will be linear combinations of $\tilde{P}^\mu$ and $\tilde{p}^\mu$.) Referring the polarization of the virtual photon to these axes, one can define helicity structure functions by:[5,8]

$$W_{\mu\nu}(q,p,P) = Y_\mu Y_\nu (W_T + W_{\Delta\Delta}) + X_\mu X_\nu (W_T - W_{\Delta\Delta})$$

$$+ Z_\mu Z_\nu W_L - (X_\mu Z_\nu + Z_\mu X_\nu) W_\Delta \quad , \tag{8}$$

where W_T is the transverse, W_L the longitudinal, W_Δ the single spin-flip, and $W_{\Delta\Delta}$ the double spin-flip amplitudes. Since $\tilde{g}_{\mu\nu} = X_\mu X_\nu + Y_\mu Y_\nu + Z_\mu Z_\nu$, it is not hard to see that,

$$W^\mu{}_\mu = 2W_T + W_L$$

$$W_1 = W_T + W_{\Delta\Delta} \tag{9}$$

These two relations are independent of the choice of z-axis; any further relations between the invariant and helicity structure functions, however, will depend on the specific choice.

The spin 1 nature of the virtual photon is explicitly displayed by the following expression for the differential cross-section in terms of the helicity structure functions and lepton pair center-of-mass angles Θ^* and ϕ^* :[5,8,9]

$$\frac{d\sigma}{d^4q\, d\Omega^*} = \frac{1}{2}\frac{1}{(2\pi)^4}\left(\frac{\alpha}{QS}\right)^2 [W_T(1+\cos^2\theta^*) \tag{10}$$

$$+ W_L(1-\cos^2\theta^*) + W_\Delta \sin 2\theta^* \cos\phi^* + W_{\Delta\Delta}\sin^2\theta^* \cos 2\phi^*]$$

III. FEATURES OF THE QUARK-PARTON MODEL

If we assume the Drell-Yan picture[1], as depicted in Fig. 4, the hadron tensor amplitude can be written:

$$W_{\mu\nu}(q,p,P) = \langle \omega_{\mu\nu} \rangle \tag{11}$$

where $\omega_{\mu\nu}$ is the elementary parton-parton amplitude,

$$\omega_{\mu\nu} = \tilde{g}_{\mu\nu} - r_\mu r_\nu / r^2, \tag{12}$$

with $r = r_1 - r_2$, $q = r_1 + r_2$ (cf. Fig. 4) and $\langle\ \rangle$ denoting average over parton distributions.

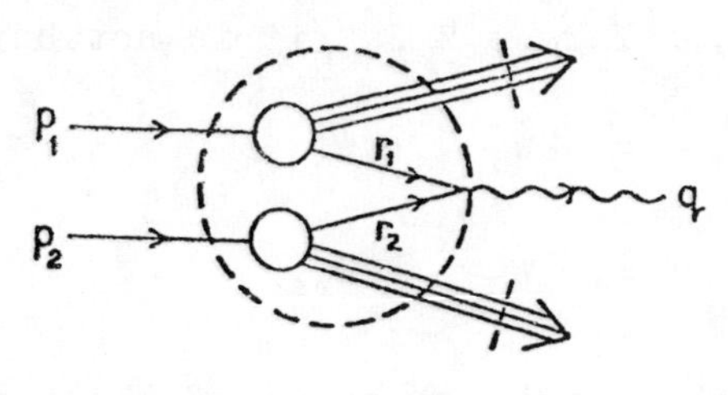

Fig. 4

Much numerical work has been done to calculate the cross-section and angular distributions using judiciously chosen (but nonunique) parton distributions. However, far more incisive tests of this picture can be obtained by studying features of the simple parton amplitude $\omega_{\mu\nu}$ (Eq. (12)) which permeate through the averaging (Eq.(11)) and manifest themselves as observable relations between the structure functions. This can be done without detailed assumptions of the parton distributions provided the transverse momenta of the partons, and hence the virtual photon, are small compared to Q. Since the observed average $q_\perp^2$ is small (and roughly consistent with expectations from confinement, i.e. uncertainty principle), we shall systematically work in expansions of the small parameters $\frac{q_\perp^2}{Q^2}$ and

$\frac{\langle r_\perp^2\rangle}{Q^2}$. In the jargon of popular QCD, terms proportional to these small parameters would be called higher twist terms.

The most interesting relation which emerges from this approach is[5].

$$W^\mu{}_\mu = 2W_1 + 2\frac{\langle r_y^2\rangle}{Q^2} \tag{13}$$

where r_y is the component of r normal to the hadron reaction plane. In terms of helicity structure functions, this relation takes the form:

$$W_L = 2W_{\Delta\Delta} + 2\frac{\langle r_y^2\rangle}{Q^2} \quad . \tag{14}$$

The 'double flip' amplitude $W_{\Delta\Delta}$ has a kinematic factor $\frac{q_\perp^2}{Q^2}$ which ensures that it vanishes in the forward direction as required by angular momentum conservation. Hence if we neglect the small correction terms,

$$\text{(a)} \quad W^\mu{}_\mu \simeq 2W_\perp \quad \text{, or (b)} \quad W_L \simeq 0. \tag{15}$$

Clearly, this is the analogue to the Callan-Gross relation for deep inelastic scattering. A remarkable fact about this relation, as will be discussed in the next section, is that it is not modified by low order QCD effects — both at high and low $q_\perp$.

More relations between the structure functions can be derived under slightly more restrictive conditions. Again, neglecting small terms involving $\frac{q_\perp^2}{Q^2}$ and assuming the root-mean-squares of the transverse momenta of the two partons are equal or, at least, $\langle r_{1\perp}^2 - r_{2\perp}^2 \rangle / q_\perp^2 \ll 1$, one can show that:[5]

$$[(q\cdot p)^2 + (q\cdot P)^2] W_1 - (q\cdot p)^2 W_2 + (q\cdot P)^2 W_4 \simeq 0 \qquad (16)$$

Likewise, provided $\langle r_x^2 - r_y^2 \rangle / q_\perp^2 \ll 1$, where r_x and r_y are transverse components of the relative parton momentum in- and out of the hadron reaction plane respectively, one obtains:[5]

$$W_1 - W_2 + W_4 \simeq 0 \qquad (17)$$

Eqs.(16) and (17) have no counterpart in deep inelastic scattering.

The three relations (15), (16), and (17) together determine all four invariant structure functions in terms of the easily measured $W^\mu{}_\mu$ (cf. Eq.(5)):

$$2W_1 = -\frac{2Q^2S}{(q\cdot p)^2} W_2 = -\frac{Q^2S}{(q\cdot p)(q\cdot P)} W_3 = \frac{2Q^2S}{(q\cdot P)^2} W_4 = W_\mu{}^\mu \qquad (18)$$

Hence, the angular distribution of the leptons in the center-of-mass frame of the hadrons is uniquely specified:

$$\frac{d\sigma}{dk_1\, dk_2\, d\cos\theta_1\, d\cos\theta_2\, d\phi} = \frac{3\pi}{8} Q^2 \frac{d\sigma}{d^4q} \cdot \frac{\cos^2(\Theta/2)}{\sin^4(\Theta/2)} \times \qquad (19)$$

$$\left[\tan^2\frac{\Theta}{2} - \frac{(q\cdot p)^2}{2Q^2S} - \frac{(q\cdot p)(q\cdot P)}{Q^2S}\,\frac{(\cos\theta_1 + \cos\theta_2)}{1+\cos\Theta} - \frac{(q\cdot P)^2}{2Q^2S}\,\frac{1+\cos\Theta'}{1+\cos\Theta}\right]$$

The transcription of Eqs. (16) and (17) into helicity structure functions depends on the choice of coordinate axes which define the latter. For the 't-channel helicity' (or

Gottfried-Jackson) frame, we obtain[5]:

$$W_T - \frac{Q}{q_\perp} W_\Delta \simeq 0, \qquad \text{and} \qquad W_T - 2\frac{Q^2}{q_\perp^2} W_{\Delta\Delta} \simeq 0 \tag{20}$$

whereas for the Collins-Soper frame[9] the results are:

$$\frac{Q}{q_\perp} W_\Delta \simeq 0 \qquad \text{and} \qquad \frac{Q^2}{q_\perp^2} W_{\Delta\Delta} \simeq 0 \tag{21}$$

Notice the spin-flip amplitudes appear in these relations with the relevant kinematic factors removed. The angular distribution of the leptons in their own center-of-mass frame takes different forms depending on the choice of coordinate axes. Eq. (21) indicates that this form is the simplest in the Collins-Soper frame[9] — if the parton picture is correct. Preliminary experimental data on the θ^*-distribution averaged over hadron variables show a behavior close to $1 + \cos^2\theta^*$, but do not clearly distinguish the various frames[3].

IV FEATURES OF PERTURBATIVE QCD EFFECTS

Part model results, such as those described in the last section, represent the impulse approximation to a non-trivial hadron interaction theory. Deviations from these zeroth-order results provide means to test the nature of that theory. In lepton pair production processes, perturbative QCD effects manifest themselves in two ways: (i) The lepton pair can have a transverse momentum $q_\perp$ large compared to the confinement scale by recoiling against a hard gluon or quark. In fact, this is expected to be the dominant mechanism for high $q_\perp$ events and it is calculable.[6,10] (ii) Also calculable are correction terms of order $\alpha_s(Q^2)$ to the Drell-Yan cross-section formula and to the parton structure function relations described in the previous section (in addition to replacing the scaling invariant parton distributions by the corresponding q^2-dependent ones from deep inelastic scattering). We describe results pertaining to each of these two aspects in turn.

A. Sturcture Function Relations and Lepton Angular Distribution at Large Transverse Momentum[6]

First order perturbative QCD diagrams responsible for large transverse momentum lepton pair production are depicted in Fig. 5: part (a) consists of quark-antiquark annihilation into a virtual photon (wavy line) with a hard gluon (dashed line) recoil; part (b) consists of quark-gluon scattering into a virtual photon with a hard quark recoil.

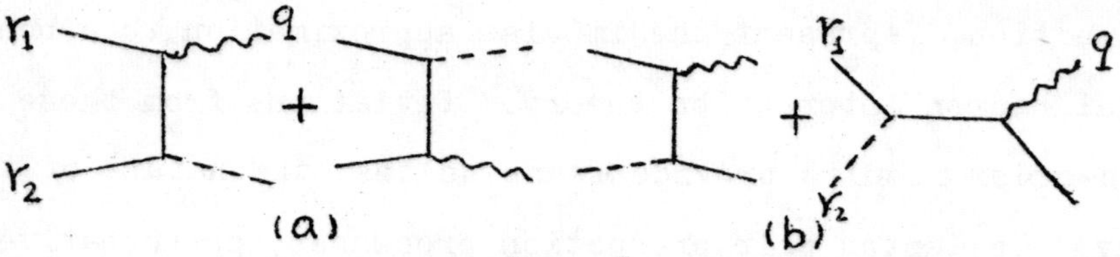

Fig. 5

Since we are only concerned with values of $q_\perp$ large compared to the intrinsic transverse momenta of the partons, it is convenient to take r_1 and r_2 (cf. Fig. 5) as purely longitudinal with respect to the incoming hadron axis in this subsection. We first calculate the parton-parton tensor amplitudes corresponding to the diagrams of Figs. 5a and 5b and denote them by $\omega^a_{\mu\nu}$ and $\omega^b_{\mu\nu}$ respectively. The results are:

$$\omega^a_{\mu\nu}(q, r_1, r_2) = (g^2/\hat{t}\,\hat{u}) \times \tag{22}$$

$$\left\{ [(\hat{t} - Q^2)^2 + (\hat{u} - Q^2)^2]\, \tilde{g}_{\mu\nu} - 4Q^2 s\, [\tilde{r}_{1\mu}\tilde{r}_{2\nu} + \tilde{r}_{2\mu}\tilde{r}_{1\nu}] \right\}$$

and

$$\omega^b_{\mu\nu}(q, r_1, r_2) = (g^2/\hat{s}\,\hat{t}) \left\{ -[(\hat{s} - Q^2)^2 + (\hat{t} - Q^2)^2]\, \tilde{g}_{\mu\nu} \right. \tag{23}$$

$$\left. + 4Q^2 s\, [2\tilde{r}_{1\mu}\tilde{r}_{1\nu} + \tilde{r}_{2\mu}\tilde{r}_{2\nu} + \tilde{r}_{1\mu}\tilde{r}_{2\nu} + \tilde{r}_{2\mu}\tilde{r}_{1\nu}] \right\}$$

where $\tilde{r}_{\alpha\mu} = \tilde{g}_{\mu\nu} r_\alpha^\nu/\sqrt{s}$, $\alpha = 1, 2$; $\hat{s}$, $\hat{t}$, $\hat{u}$ are the Mandelstam variables for the relevant parton-parton process; and g is the effective QCD coupling constant related to α_s by $\alpha_s = \frac{g^2}{4\pi}$.

The next step is to average over the parton distributions and determine which of the simple features of the above expressions remain. We find readily that both Eq. (22) and Eq.(23) lead to the familiar structure function relation (cf. Eq.(15a)):

$$W^\mu{}_\mu = 2\, W_1 \tag{24a}$$

Although this relation again implies that

$$W_L = 2\, W_{\Delta\Delta} \, , \tag{24b}$$

we can no longer infer that $W_L \simeq 0$ — because once $\frac{q_L}{Q}$ is not too small, there is no reason for $W_{\Delta\Delta}$ to be unimportant. The helicity structure functions may all be of the same order of magnitude but they **must** be constrained by Eq.(24b). (Eq.(27) below serves as an illustration to this point.) Contrasting this situation with that of last section (around Eq.(15)), we note that: <u>if QCD is correct, the apparent lepton center-of-mass angular distribution may change substantially from low to high $q_\perp$</u> (due to the shift in relative strengths of the helicity structure functions), <u>but the basic invariant structure function relation $W^\mu_\mu = 2W_1$</u> remains valid throughout. In subsection B below we shall show that the parton model results Eq.(15a,b) for small $q_\perp$, are not modified by order α_s QCD corrections. Hence the specific prediction underlined above constitutes a critical **test** of the validity of perturbative QCD. Since the relation in question follows solely from the tensor structure of the elementary parton-parton amplitudes, Eqs.(12),(22), (23), it is uniquely characteristic of the QCD picture both at low and high $q_\perp$ and, at the same time, is less intricate in its derivation than QCD corrections to cross-section formulas. We should mention also that its validity is totally independent of any uncertainties about the parton distribution functions.

In addition to Eq.(24), the 'annihilation amplitude', Eq.(22), contain other features which survive the averaging over the parton distributions. For lepton pair production dominated by this mechanism, we can show[6], in general,

$$W_2 = W_4 \tag{25}$$

and, in particular if the initial state hadrons have some symmetry under their interchange ($1 \leftrightarrow 2$) (e.g. $N\bar{N}$ and NN), then

$$2\,[W_2 - W_2(1 \leftrightarrow 2)] = W_3 + W_3(1 \leftrightarrow 2). \tag{26}$$

No corresponding relations seem to follow from the 'Compton amplitude', Eq.(23), owing to a lesser degree of symmetry both in the amplitude itself and in averaging over the quark and gluon distribution functions.

As lepton pair production in πN and $N\bar{N}$ scattering is dominated by the annihilation mechanism, the structure function relations, Eq.(24) and Eq.(25), can be used to eliminate two of the three structure function ratios in the expression for the normalized differential cross-section. The resulting expression for the lepton center-of-mass angular distribution is particularly simple with the Collins-Soper choice of axes:[10]

$$\frac{16\pi}{3\sigma}\frac{d\sigma}{d\Omega^*} = \frac{Q^2+\frac{1}{2}q_\perp^2}{Q^2+q_\perp^2}(1+\cos^2\theta^*) + \frac{q_\perp^2}{Q^2+q_\perp^2}(1-\cos^2\theta^*) + \frac{\frac{1}{2}q_\perp^2}{Q^2+q_\perp^2}\sin^2\theta^*\cos 2\phi^* + \beta\sin 2\theta^*\cos\phi^* \tag{27}$$

where β is the only remaining unspecified coefficient. If one integrates over the azimuthal angle ϕ^*, β drops out and the prediction on the θ^*-distribution is complete:[10]

$$\frac{1}{\sigma}\frac{d\sigma}{d(\cos\theta^*)} = \frac{3}{8}\,\frac{Q^2+\frac{3}{2}q_\perp^2}{Q^2+q_\perp^2}\left[1+\frac{Q^2-\frac{1}{2}q_\perp^2}{Q^2+\frac{3}{2}q_\perp^2}\cos^2\theta^*\right] . \qquad (28)$$

Preliminary data on the coefficient of $\cos^2\theta^*$ for πN scattering as a function of $q_\perp^2$ at moderately high values of $q_\perp$ are consistent with this expression but by no means prove it.[11]

B. QCD Corrections to Parton Model Relations at Low Transverse Momentum

As mentioned earlier, first order correction to the Drell-Yan cross-section formula (which, according to Eq. (5), is essentially $W^\mu{}_\mu$) has been found to be unexpectedly large[4] and takes the form of a overall normalization factor roughly equal to 2. The calculations involve delicate questions of subtraction scheme, proper regularization, and consistency with the definition of parton distribution function in deep inelastic sacttering. In this subsection we look into the related but much simpler problem of QCD corrections to the parton model structure function relations at low $q_\perp$ as presented in section III.

The first order QCD diagrams of Fig. 5, previously considered in connection with high $q_\perp$ events, are clearly also of relevance here. But in addition to squares of those amplitudes, we must also include interference terms between the zeroth order Drell-Yan vertex with second order self-energy and vertex correction terms. Self-energy corrections obviously do not affect the tensor structure of the relevant amplitude, hence we show in Fig. 6 only the vertex correction diagrams. The remarkable fact is that, as

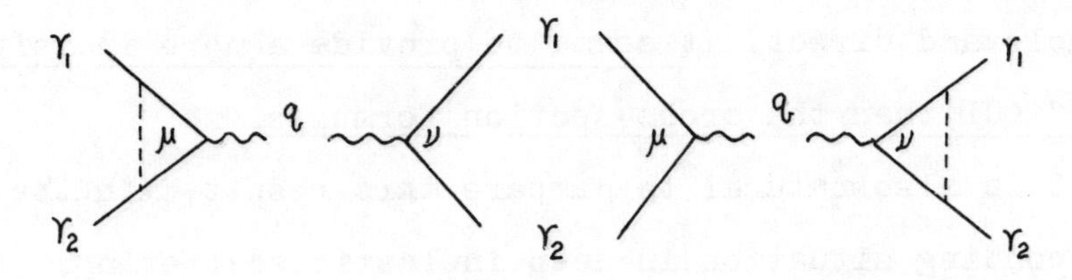

Fig. 6

a moment's reflection will reveal, even these diagrams do not modify the tensor structure of the parton-parton amplitude. The reason is that, the corrected vertex is of the form $f_1\gamma_\mu + f_2\sigma_{\mu\nu}q^\nu$ which, upon folding with the Drell-Yan vertex, yields

$$\omega_{\mu\nu} \propto \mathrm{tr}(f_1\gamma_\mu + f_2\sigma_{\mu\lambda}q^\lambda)\not{p}_1\gamma_\nu\not{p}_2 \tag{29}$$

$$\propto \mathrm{tr}\,\gamma_\mu\not{p}_1\gamma_\nu\not{p}_2 .$$

This is identical to the original Drell-Yan amplitude — except in normalization, of course. This result, together with those associated with the diagrams of Fig. 5, Eqs.(22) and (23), <u>immediately imply that the parton model relation</u>

$$W^\mu{}_\mu = 2W_1 \tag{30}$$

<u>is not subjected to any first order QCD correction at all!</u> This is in distinct contrast to the large correction found

for $W^{\mu}{}_{\mu}$ itself. So, whatever the corrections to $W^{\mu}{}_{\mu}$ are, the same must apply to W_1. As the above demonstration of the validity of the relation Eq.(30) to first order in α_s is simple and direct, it seems to provide a more incisive test of QCD than the cross-section formulas.

It is also natural to campare this result with the corresponding situation in deep inelastic scattering. How can we understand the absence of QCD corrections to this relation while it is well known that its counterpart, the Callan-Gross relation, does require first order QCD corrections? Two remarks can be made in answering this question: (i) From the phenomenological point of view, since this relation can also be cast in the form $W_L = 2W_{\Delta\Delta}$, we see that W_L is subjected to QCD corrections, just as in the case of Callan-Gross relation; but the same must apply to $W_{\Delta\Delta}$, which has no counterpart in deep inelastic scattering. (ii) From the technical point of view, although the first order QCD diagrams for the two processes are precisely the same ones (except for line-reversal), the difference arises from the fact that, with two hadrons in the initial state, the calculation in lepton pair production does not involve the final state phase space integration which gives rise to the correction to Callan-gross relation in the corresponding calculation of deep inelastic scattering.

The other parton model relations presented in section

III, Eqs. (16) and (17), <u>are</u> subjected to QCD corrections as is clear already from the results of subsection A above (applied to lwo values of $q_\perp$, of course). These corrections involve all complications that one encounters in the cross-section formulas and more. We shall not consider them here.

V. CONCLUDING REMARKS

The sturcture functions for lepton pair production provide a rich and fertile ground for testing the parton model and perturbative QCD, complementing and extending that offered by deep inelastic scattering. So far only limited aspects of this process (integrated cross-section, Q^2-distribution, average $q_\perp$, ... etc.) have been extensively studied. Can we expect to see detailed experimental results on the structure functions in order to test predictions such as those presented in this report? The first reaction to this question might be quite pessimistic, considering the difficulties of separating the structure functions in deep inelastic scattering. The actual situation, however, could turn out to be very different. The reason is, the separation of the structure functions in lepton pair production is intrinsically much easier than in deep inelastic scattering: because the time-like virtual photon gives rise to a lepton pair center-of-mass angular distribution expressable as a linear combination of trigonometric spherical harmonics whereas the space-like photon yields a lepton distribution naturally written as a linear combination of the corresponding hyperbolic functions. It is obvious that the separation of coefficients is much simpler in the first case. In practice, this advantage has to be compensated by the lower rate for lepton pair production and other difficulties, of course; but the hope for gaining relatively detailed information on these structure functions in the near future is certainly not unrealistic.

Finally, we make one comment about testing the most interesting relation $W^{\mu}_{\ \mu} = 2W_1$. Since $W^{\mu}_{\ \mu}$ is readily available, the question centers on the measurability of W_1. We observe that W_1 is just the coefficient of the term $Y_\mu Y_\nu$ in the expansion of $W_{\mu\nu}$ in terms of lepton pair center-of-mass coordinate unit vectors. Whether this observation can be translated into a simple prescription for measuring it we don't yet see.

References

1. S. Drell and T.M. Yan, Phys. Rev. Lett. 25, 316 (1970); Ann. Phys. (N.Y.) 66, 578 (1971).
2. J.G. Branson et.al., Phys. Rev. Lett. 38, 457 (1977).
3. K.J. Anderson et. al., Phys. Rev. Lett. 42, 944 (1979), and ibid. 42, 948 (1979).
4. G. Altarelli, R.K. Ellis, G. Martinelli, Nucl. Phys. B143, 521 (1978) and Erratum, B146, 544 (1978), and MIT preprint, MIT-CTP-776, 1979; K. Harada, T. Kaneko and N. Sakai, Nucl. Phys. B155, 169 (1978); B. Humpert and W.L. Van Neerven, Phys. Lett. 84B, 327 (1979), ibid. 85B, 293 (1979).
5. C.S. Lam and Wu-Ki Tung, Phys. Rev. D18, 2447 (1978).
6. C.S. Lam and Wu-Ki Tung, Phys. Lett. 80B, 228 (1979).
7. H. Terazawa, Phys. Rev. D8, 3026 (1973).
8. R.J. Oakes, Nuovo Cimento 44, 440 (1966).
9. J.C. Collins and D. E. Soper, Phys. Rev. D16, 2219 (1977).
10. J.C. Collins, Phys. Rev. Lett. 42, 291 (1979); R.L. Thews, Phys, Rev. Lett. 43, 987 (1979).
11. K.J. Anderson et. al., Phys. Rev. Lett. 43, 1219 (1979).
12. C.S. Lam and Wu-ki Tung, 'A Parton Model Relations and QCD Modifications in Lepton Pair Production', SLAC Preprint, 1980.

JET STRUCTURE IN e^+e^- ANNIHILATION

Yee Jack Ng

Institute of Field Physics, Department of Physics and Astronomy

University of North Carolina, Chapel Hill, N. C. 27514, U.S.A.

Abstract

An introduction is given to jet structures in e^+e^- annihilation, in the framework of Q.C.D. in particular. Topics discussed include 1) gluon bremsstrahlung leading to broadening of jets, and to three-jet structures; 2) quarkonium decay into three gluons; 3) the use of (only) one variable to measure the jettiness of events; 4) a specific method to extract the triple-gluon vertex; and 5) using the angular distribution of heavy quark jet production to measure the quark mass in perturbative Q.C.D. Some other topics are then briefly discussed.

I. INTRODUCTION: JET PHENOMENOLOGY

A striking discovery emerged from the study of electron-positron annihilation at SPEAR[1] is the observation that at center-of-mass energies ($\sqrt{s}$) greater than 5 GeV, hadrons are produced predominantly in back-to-back bursts called jets (Fig. 1.) These jets are characterized by three distinct signals: 1) the transverse momentum of hadrons in a jet relative to the jet axis is sharply cut off on a scale of a few hundred MeV; 2) the distribution of hadrons in the jet longitudinal to the jet axis is a function only of the fraction of jet momentum carried by the hadrons; 3) the distributions of jets relative to the axis of the incoming electron and positron momentum is, to good accuracy, $1 + \cos^2\theta$, where θ is the angle between the jet and that axis.

As the center-of-mass energy is increased it has been observed at DESY a broadening of the two jets[2] (Fig. 2.) And as $\sqrt{s}$ reaches ~30 GeV distinctive three-jet events[3] have been reported by the experimental groups at DESY (Fig. 3)

II. HOW Q.C.D. EXPLAINS THE JET STRUCTURES

The observations of two-jet structures at SPEAR[1] are most simply explained by the quark-parton model which recently has found an elegant realization in Q.C.D.: e^+e^- annihilation into hadrons proceeds via e^+e^- annihilation into a virtual photon which then decays into a quark-antiquark pair (Fig. 4.) At short distances, the quarks behave as if they were free; the observed $1 + \cos^2\theta$ angular distribution is characteristic of e^+e^- annihilation into a pair of spinors. At long distances quarks are (presumably) confined and hence must fragment into hadrons.

The next order process for $e^+e^- \rightarrow$ hadrons, in the framework of Q.C.D., is provided by the bremsstrahlung of a gluon[4,5,6] (Fig. 5) the matrix element for which is

$$\frac{1}{\sigma}\frac{d^3\sigma}{dx_1 dx_2 d\cos\phi} = \frac{\alpha_s}{8\pi}\frac{x_1^{\,2} + x_2^{\,2}}{(1 - x_1)(1 - x_2)}\,(2 + \sin^2\phi) + \phi(\alpha_s^{\,2})$$

where α_s is the color running coupling constant, and x_1 and x_2 are respectively the quark and antiquark fractional momentum ($x_i = \frac{2E_i}{\sqrt{s}}$.) The three final - state particles lie in a plane whose normal makes an angle ϕ to the e^- beam direction. At low energies the emitted gluon is relatively soft. But as the center-of-mass energy increases the emitted gluon carries more and more energy-momentum. Thus we expect the jets to be broadened due to this bremsstrahlung process. At high energies the emitted gluon forms a distinct jet giving rise to a three-jet structure as has been observed at DESY[3]. It should be pointed out that one way to test the vector nature of the gluon is to look at the angular distributions of these three-jet events: the normals to the event-planes should have a distribution[5]: $\frac{dN}{d\cos\phi} \propto 2 + \sin^2\phi$. Alternatively we may look at the angles in the planes[5,7].

Another place where gluon jets may be visible is in the decay of a bound state of heavy quarks. The decay of this state is Zweig-forbidden and proceeds via an intermediate state of three gluons[8] (Fig. 6) analogous to the electromagnetic decay of an orthopositronium state into three photons. The decay matrix element is

$$\frac{1}{\Gamma}\frac{d^2\Gamma}{dx_1 dx_2} = \frac{1}{\pi^2 - 9}\left\{\left(\frac{1 - x_1}{x_2 x_3}\right)^2 + \text{permutations}\right\}$$

where Γ is the hadronic rate and $x_i = \frac{2E_i}{\sqrt{s}}$. If the mass of the state is large enough (mass of heavy quark $\gtrsim$ 5 GeV) the gluons may form jets as they decay into ordinary hadrons. The present data for Υ decay[10] are indicative of a three-gluon decay mode. But the results are far from being conclusive because the analysis involved in the statistical reconstruction procedure[10] tends to introduce a bias in favor of a three-jet structure. Also, recent calculations[11] for quarkonium decays apparently show very large radiative corrections ($\sim 10\ \alpha_s/\pi$) for these processes. These theoretical results coupled with the possibility that gluon jets may be too wide to be separable (see below) cast doubt on the quantitative (experimental) test mentioned above, though the **qualitative feature of** a three-gluon final state in quarkonium decays remains valid.

III. PERTURBATIVE Q.C.D.: PROBLEMS AND RULES

Let us examine perturbative Q.C.D. in more rigor and detail. First of all in doing any such calculation one is encountered with two problems:

1. Infrared singularities: These singularities are due to either soft gluon emissions or collinear glon emissions. The singularities associated with the latter processes are also known as mass singularities. In general one encounters singularities if in one's calculations one treats differently physically indistinct processes, such as the emission of a quark with momentum $\vec{p}$ and that of a quark with momentum $x\vec{p}$ accompanied by a gluon with momentum $(1 - x)\vec{p}$.
2. "Hadronization" (or confinement) problem: Quarks and gluons have to **"hadronize"** to form color-singlet final states. But at present we are still ignorant of the confinement mechanism which is presumbly due to some (yet) uncalculable non-perturbative effects.

For a perturbative Q.C.D. calculation to be physically meaningful these two problems must be absent, i.e. it must be 1) infrared finite, and 2) insensitive to confinement effects. In general these two conditions imply that in calculating hadronic jet cross-sections one has to

i) choose suitable variables which are the same for physically indistinct processes (hence they must be either independent of or linear in momemtum (Fig. 7));

ii) ask suitable questions: one cannot ask questions having to do with the quarklike or gluonlike character of an event, the nature or number of particles carrying the energy-momentum. These questions involve details of the confinement mechanism. Instead one can ask for the distributions of energy flow;

iii) restrict the kinematic region in one's calculation so that the perturbative calculation is meaningful and the confinement effects are minimized.

IV. EXAMPLES OF SUITABLE VARIABLES

As mentioned above, to test the finer details of Q.C.D., one has to choose suitable variables. A popular variable used in jet calculations, that is linear in momemtum is thrust[12]:

$$T = \max \frac{\sum_i |\vec{p}^{\,i}_{||}|}{\sum_i |\vec{p}^{\,i}|} , \qquad \frac{1}{2} \le T \le 1$$

where the sum is over all hadrons. $\vec{p}_{||}$ are the momenta parallel to an axis which is chosen to maximize T and this axis is called the thrust axis.

For variables that are independent of momentum a convenient one to use is the angular radius δ first proposed by Bigi and Walsh[13], and later but independently by Grunberg, Ng and Tye[14]. The quantity that is calculated by these authors is the average hadronic energy deposited outside or inside a pair of back-to-back cones of half-radius δ around the jet axis (see below.) This quantity should be calculable at each order of perturbation theory. Perhaps it is worth mentioning that here only one variable is used to desbribe the "jettiness" of events. Hence it is a more economical variable to use than the pair of variables ε and δ introduced by Sterman and Weinberg[15] to calculate the probability that a fraction ε of the total energy lies outside two back-to-back cones of half-radius δ around the jet axis.

Just as thrust provides a measure of "two-jettiness" (multi-jet events give values of T away from 1) a variable called triplicity T_3 [16] has been proposed to give a measure of "three-jettiness". T_3 measures the fractional energy parallel to the set of three axes, which maximizes this quantity. One method of selecting three-jet events employed in the experimental analysis[3] is to pick hadronic events with small T (say $0.6 \leq T \leq 0.8$) and large T_3 (say $T_3 \geq 0.9$).

Other variables used in jet analysis include spherocity S [17] which measures the fraction of momentum transverse to the axis which minimize S, and acoplararity A [18] which measures the fraction of momentum perpendicular to the plane choosen to minimize A. The latter variable is useful in analysing four-jet events.

V. ENERGY CONE DISTRIBUTION AROUND THE JET AXIS

Having gone over some of the rudimentary aspects of perturbative Q.C.D. we are now in a position to discuss an application, namely to study the average hadronic energy deposited outside ($\sum^{out}$) or inside ($\sum^{in}$) a pair of back-to-back cones of half-radius δ around the thrust axis. We first note that $\sum^{out}$ in fact vanishes in the parton model, is of order α_s (the lowest order contributing diagrams being given by Fig. 5,) and therefore provides a measurement of Q.C.D. effects. Plotted in Fig. 8 (dashed curve) is the result $F_{quark}(\delta)$, the fraction of energy-weighted cross-section within a cone of half-radius δ with respect to the thrust axis,

$$F(\delta) = \frac{\sum^{in}}{\sigma_{total}} \simeq 1 - \frac{\sum^{out}}{\sigma_{Born}} \rightarrow \exp\left[-\frac{\sum^{out}}{\sigma_{Born}}\right] .$$

This result[14] disagrees with that found by Bigi and Walsh[13]. The exponential form of $F(\delta)$ has been used here for two reasons: first, $\frac{\Sigma^{out}}{\sigma_{Born}} \gtrsim \frac{\alpha_s}{\pi} (\ln\delta)^2$ in the limit $\delta \rightarrow 0$, therefore exponentiation ameliorates the problem at small δ. Second, exponentiation of $F(\delta)$ automatically gives the intuitively correct normalization, namely $F(\delta = 0^\circ) = 0$ and $F(\delta = 90^\circ) = 1$. But it must be pointed out that rigorous justification for this exponentiation procedure remains to be checked.

It is instructive to discuss the kinematic region where this perturbative result is applicable. From the result of $\sum^{out}$ at small δ we need $(\ln\delta)^2 \ll \frac{\pi}{\alpha_s}$ for the perturbation to be meaningful. In addition, for the result to be insensitive to the hadronization effects we have to choose $\delta^2 \gg \delta^2_{hadronization} \sim \frac{\langle p_T^2 \rangle_{hadronization}}{s}$.

Notice that the only energy dependence of $F(\delta)$ is contained in the strong coupling constant. The result of the energy-weighted cross-section as a function of the opening angle δ suggests that the typical transverse momentum p_T of quark jets grows with energy. This increase of p_T is in sharp contrast to the intrinsic limited p_T (of the order of a few hundred MeV) due to confinement effects.

The gluon jet spreading due to perturbative effects can be calculated in a similar way. Using the scalar (or pseudoscalar) source[9] to produce the gluon jets we obtain the result[19] $F_{gluon}(\delta)$ given in Fig. 8 (solid curve.) As expected at high **energies** the spread of a gluon jet is larger than that of a quark jet[21] (mainly as a result of group factors which reflect the fact that the gluon belongs to the $\underline{8}$ representation while the quark is in the $\underline{3}$ representation.) For example, for $F(\delta) = 0.7$, $\delta_{quark} \simeq 16^\circ$ while $\delta_{gluon} \simeq 33^\circ$. We expect similar results for $F_{gluon}(\delta)$ if other sources have been used. The source independence can in fact be checked by comparing different production mechanisms. For example, quarkonium decay into gluon jets has been found to give essentially the same numerical results as the scalar source[20].

VI. MEASURING THE TRIPLE-GLUON VERTEX

Basing on the above results for F_{quark} anf F_{gluon} we can now extract the triple-gluon vertex [19] whose existence is so vital to asymptotic freedom and yet has not been measured so far. First of all we need a clean source of gluon jets. Such a clean source is provided by the remarkably distinctive three-jet events [3] recently observed in e^+e^- annihilation. In each three-jet event two of the jets arise from quark fragmentation and the third one from gluon fragmentation. Though we do not know, for each event, which is the gluon jet, by measuring the spread of all the three jets and subtracting the contributions of the two quark jets, we are left with the gluon jet spread

$$F_{gluon}(\delta) = \sum_{i=1}^{3} F_i(\delta) - 2F_{quark}(\delta) \quad .$$

For F_{quark} we can use either the measurement of the two-jet events in e^+e^- annihilation or the theoretical prediction [14]. The resulting F_{gluon} can then be compared with the theoretical Q.C.D. prediction [19] to yield information on the triple-gluon vertex.

There are other suggestions to measure the effects of the triple-gluon vertex. They include asymmetries in heavy quarkonium decay,[22] scaling violation in gluon jets,[23] scaling violation in gluon distribution inside nucleons,[24] and charmed multiplicities and correlations.[25]

VII. ANGULAR DISTRIBUTIONS OF HEAVY QUARK JETS

It is reasonable to neglect the quark mass m_q at high energies. But when m_q is not negligible compared to the energies involved the quark mass effects have to be taken into account.[26] For simplicity let us adopt the "location of the pole" definition for the quark mass. Now in general the quark masses are not known to a good accuracy. It turns out that a clean place to measure the heavy quark mass in perturbative Q.C.D. is provided by the angular distribution of quark jet production in e^+e^- annihilation[27]. The idea is rather simple. Consider e^+e^- annihilation into a heavy quark pair. It suffices to consider a specific quark pair since we can separate quarks of different flavors. In the lowest order the process is described by

$$\frac{d\sigma}{d\cos\theta} \sim 1 + B_o(v)\cos^2\theta$$

where

$$B_o(v) = \frac{v^2}{2 - v^2}, \qquad v^2 = 1 - \frac{4 m_q^2}{s}$$

and θ is the angle between the quark and the electron direction. Therefore a good measurement of the angular distribution $B_o(v)$ can certainly give an accurate value of the quark mass (Fig. 9.)

At first sight it may appear that the total cross-section σ_{total} also provides a measurement of the quark mass. That this is not possible is illustrated in Fig. 9: σ_{total} approaches its asymptotic value (corresponding to $m_q = 0$) much faster than $B_o(v)$. After smearing [28], a process that renders the calculation infrared insensitive, the quark mass effects in σ_{total} are washed out whereas $B_o(v)$ still depends strongly on m_q even at energies way above the continuum threshold. It has been shown that this feature persists when perturbative Q.C.D. corrections are included. The leading Q.C.D. correction to the angular distribution of hadronic jets with the quark mass effects taken into account is given in Ref. 27.

Since the quark masses constitute, besides the color coupling, the only set of parameters of Q.C.D., it goes without saying that the determination of them is of great importance. In principle weak decays of heavy quarks also provide a means to measure m_q. It would be interesting to compare the quark masses provided by that method with those measured by the angular distribution method mentioned above.

VIII. A FEW MORE TOPICS

1. Higher Twist Effects: Hard gluon bremsstrahlung is not the only process that leads to (two-) jet broadening and to three-jet structure. To give Q.C.D. a quantitative test we have to take into account higher twist contributions[5.29] which can also lead to events that resemble three jets with one jet being a single

meson or resonance. These higher twist effects have two characteristic features: first, at high energies their cross-sections fall off faster than those for the hard gluon bremsstrahlung processes.[5] Second, the invariant mass of one of the jets is relatively low (< 2 GeV.)

2. Four-jet Events: At low thrust, to isolate the three-jet events, we also have to subtract contributions from four-jet events[5,30] which have, as final states, $q\bar{q}q\bar{q}$ or $q\bar{q}gg$ (q = quark, g = gluon.) In contrast to three-body final states four-jet events are no longer planar. A suitable variable to describe these events is acoplanarity.[30]
3. Quarkonium Decays: In addition to the three-gluon quarkonium decay mode mentioned above the two-gluon-one-photon decay mode[31,32] also provides a place to study gluon jets. The distribution of hard photons for this decay mode behaves like $1 + \cos^2\theta_\gamma$ with θ_γ being the angle between the photon and the e^+e^- direction. Detailed theoretical analyses of both P-wave and S-wave quarkonium decays[20] into gluon jets have been carried out providing additional tests of perturbative Q.C.D. On the other hand, once new states are found, jet structure analyses can be used to determine the quantum numbers of the parent and the decay states.[33]
4. Measuring α_s : The remarkable three-jet events[3] can also be used to determine the color coupling constant. The relative yields and the shape distributions of these events have enabled the MARK-J group at DESY[34] to determine $\alpha_s = 0.23 \pm 0.02$ (statistical error) with a systematic error of $\pm$ 0.04 at the center-of-mass energy $\sqrt{s}$ = 30 GeV. It would be interesting to see if at higher energies (energies beyong LEP ?) there is a decrease in α_s as required for Q.C.D.

There are many other interesting topics that we will have no chance to discuss, e.g., tests of Q.C.D. that involve energy correlation,[35] azimuthal correlation,[36] attempts to "improve" jet calculations by a resummation including multi-quanta,[37] and an interesting new source of jet structure in e^+e^- annihilation arising from photon-photon collisions,[38] to name but a few. Good review articles[39] are available to remedy the deficiency of this report.

IX. CONCLUSION

The importance of jet structure analysis is worthwhile repeating here. Due to quark and gluon confinement it is impossible to use quarks and gluons for direct experimentation. Since jets arise directly from the fragmentations of quarks and gluons it appears that jet analysis is one of the best available ways[40] to study their properties and interactions. The situation is particularly cleancut in e^+e^- annihilation because there we do not have to deal with, e.g., complications in structure functions. Also, by asking limited questions we can avoid (hopefully) the problems associated with non-perturbative effects. In short, jet analyses provide direct and meaningful tests of perturbative Q.C.D.

Acknowledgements

Materials presented in Sec. V, VI, and VII are based on work done with G. Grunberg and Sze-Hoi Henry Tye. Part of Sec. II and VIII are based on work done with T. A. DeGrand and S.-H. H. Tye. Thanks are due to all my collaborators, in particular Dr. Tye, for useful discussions. Finally it is a great pleasure for me to thank the Academia Sinica for the great hospitality extended to me and my family during our entire trip in China. This work was partly supported by the U. S. Department of Energy.

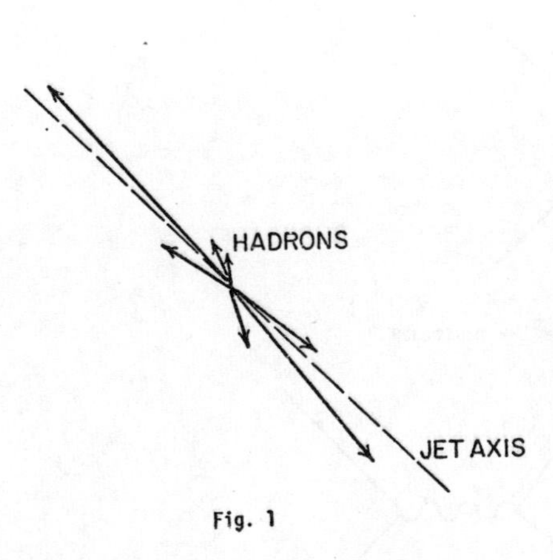

Fig. 1

A typical SPEAR two-jet event in momentum space.

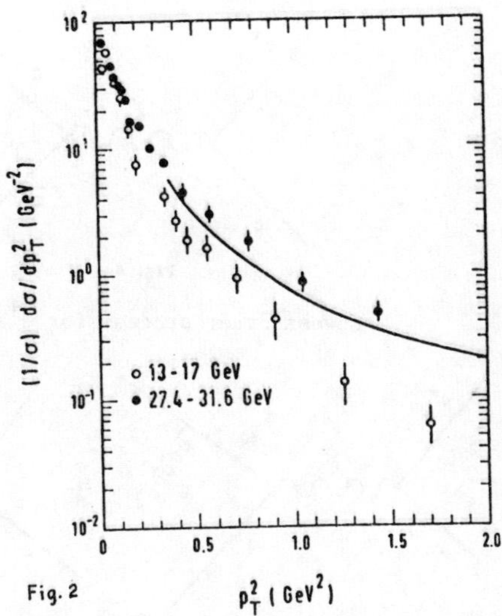

Fig. 2

The transverse momentum p_T distributions found by the TASSO collaboration [2] at different energies.

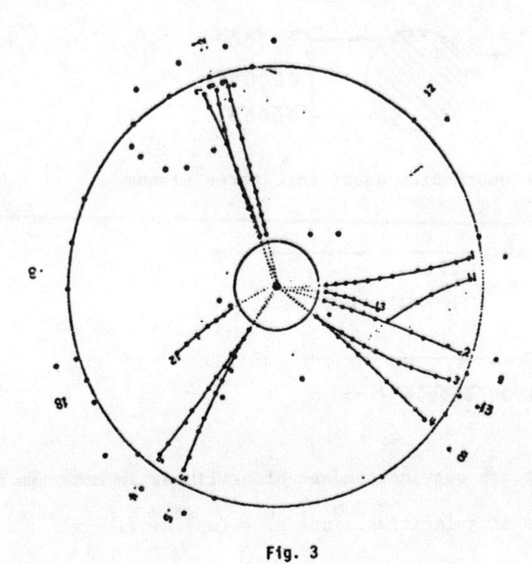

Fig. 3

A typical three-jet event reported by the PLUTO collaboration.[3]

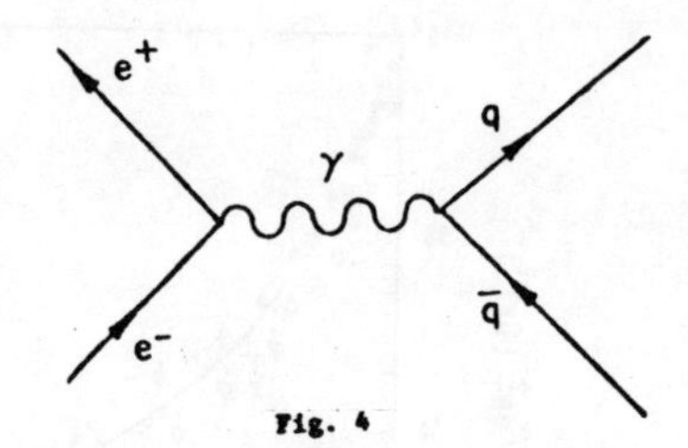

Fig. 4

Lowest order process for $e^+e^- \rightarrow$ hadrons.

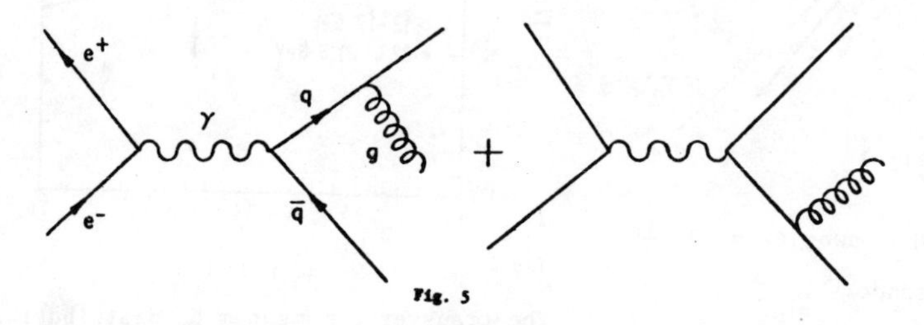

Fig. 5

Gluon bremsstrahlung diagram. Since the gluon is confined it also must fragment into hadrons.

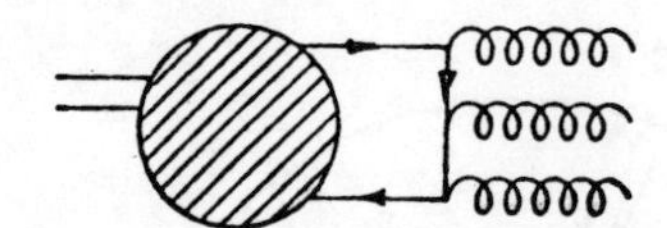

Fig. 6 Orthoquarkonium decay into three gluons.

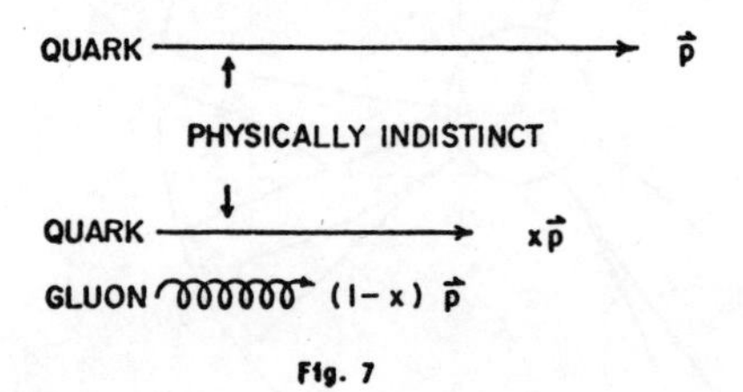

Fig. 7

Using variables that are not independent of or linear in momentum will in general lead to mass singularities since $p^n \neq (xp)^n + (1 - x)^n\ p^n$ unless $n = 1$.

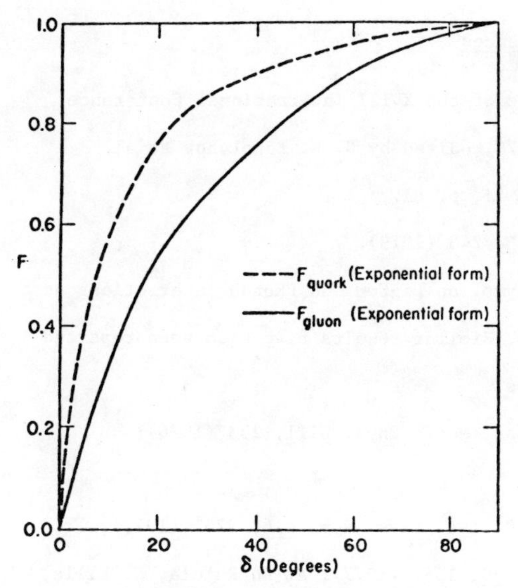

Fig. 8

The exponential forms of the functions $F_{quark}(\delta)$ (dashed curve) and $F_{gluon}(\delta)$ (solid curve) against δ (in degrees) for $\alpha_s = 0.24$ and the number of quark flavors $N_f = 5$.

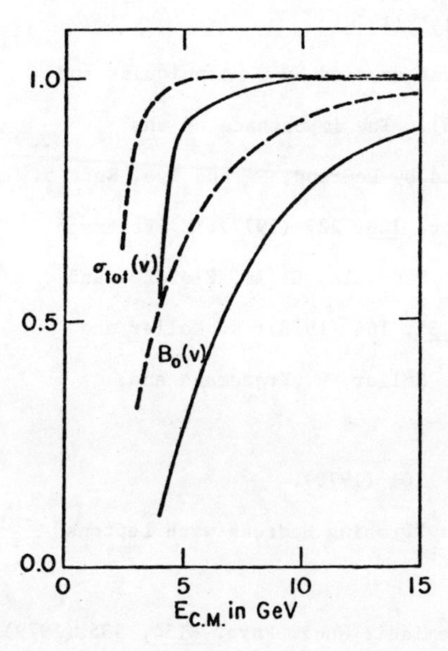

Fig. 9

The top two curves are $\sigma_{total}(m,s)/\sigma_{total}(o,s)$. The lower two curves are the coefficient B_o. Here we have considered, for illustration, the charmed quark case whose mass quoted in the literature ranges from 1.1 GeV (dashed curves) to 1.85 GeV (solid curves). The angular distribution method has been estimated to give the charmed quark mass accurate to within 0.1 - 0.2 GeV.[27]

References

[1] See, e.g., G. Hanson, in Proceedings of the XVIII International Conference on High Energy Physics, Tbilisi, 1976, edited by N. N. Bogolubov et al. (JINR, Dubna, U.S.S.R., 1977), Vol. II, p. B1.

[2] R. Brandelik et al., Phys. Lett. 86B, 243 (1979).

[3] See, e.g., Ch. Berger, Proc. Int. Symp. on Lepton and Photon Interactions at High Energies (FNAL, Batavia, 1979). Similar results have also been reported by the MARK-J, TASSO and JADE groups.

[4] J. Ellis, M. K. Gaillard, and G. Ross, Nucl. Phys. B111, 253 (1976); Erratum, B130, 516 (1977).

[5] T. A. DeGrand, Y. J. Ng, and S.-H. H. Tye, Phys. Rev. D16, 3251 (1977).

[6] Also see E. Farhi, Phys. Rev. Lett. 39, 1587 (1977); A. De Rujula, J. Ellis, E. G. Floratos and M. K. Gaillard, Nucl. Phys. B138, 387 (1978).

[7] J. Ellis and J. Karliner, Nucl. Phys. B148, 141 (1979).

[8] The three-gluon quarkonium decay mode was first studied by T. Appelquist and H. D. Politzer, Phys. Rev. Lett. 34, 43 (1975). The importance of the three-gluon jet structure was first emphasized by DeGrand, Ng and Tye, Ref. 5. See also K. Koller and T. F. Walsh, Phys. Lett. 72B, 227 (1977); H. Fritzsch and K. H. Streng, Phys. Lett. 74B, 90 (1978); DeRujula, Ellis, Floratos and Gaillard, Ref. 6; K. Hagiwara, Nucl. Phys. B137, 164 (1978); K. Koller and T. F. Walsh, Nucl. Phys. B140, 449 (1978); K. Koller, H. Krasemann and T. F. Walsh, Z. Phys. C1, 71 (1979).

[9] K. Shizuya and S.-H. H. Tye, Phys. Rev. D20, 1101 (1979).

[10] Talk by H. Meyer, Highly Specialized Seminar: "Probing Hadrons with Leptons" (Erice, 1979).

[11] R. Barbieri, E. d'Emilio, G. Curchi and E. Remiddi, Nucl. Phys. B154, 535 (1979).

[12] See, e.g., Ref. 6.

[13] I.I.Y. Bigi and T. W. Walsh, Phys. Lett. 82B, 267 (1979).

[14] G. Grunberg, Y. J. Ng and S.-H. H. Tye, to be published in Nucl. Phys. B.

[15] G. Sterman and S. Weinberg, Phys. Rev. Lett. 39, 1436 (1977).

[16] S. Brandt and H. D. Dahmen, Z. Phys., C1, 61 (1979).

[17] H. Georgi and M. Machacek, Phys. Rev. Lett. 39, 1237 (1977). A closely related variable is sphericity first introduced by J. D. Bjorken and S. J. Brodsky, Rev. D1, 1416 (1970).

[18] A. DeRujula et al., Ref. 6.

[19] G. Grunberg, Y. J. Ng and S.-H. H. Tye, Univ. of North Carolina preprint IFP 140-UNC. (1979).

[20] R. Barbieri, M. Caffo, and E. Remiddi, Max-Planck-Institut, MPI-PAE/PTh 27/79 (1979).

[21] Therefore we confirm earlier results obtained in Ref. 9, and by, e.g., M. B. Einhorn and B. G. Weeks, Nucl. Phys. B146, 445 (1978) and A. V. Smilga and M. I. Vysotsky, ITEP-93 (1978).

[22] A. DeRujula, B. Lautrup, and R. Petronzio, Nucl. Phys. B146, 50 (1978).

[23] K. Koller, T. F. Walsh, and P. M. Zerwas, Phys. Lett. B82, 263 (1979).

[24] E. Reya, Phys. Rev. Lett. 43, 8(1979).

[25] I. I. Y. Bigi, CERN preprint Ref. TH. 2707 - CERN, 1979.

[26] See, e.g., B. L. Ioffe and A. V. Smilga, preprint ITEP-69 (1978).

[27] G. Grunberg, Y. J. Ng, and S.-H. H. Tye, to be published in Phys. Rev. D.

[28] E. C. Poggio, H. R. Quinn, and S. Weinberg, Phys. Rev. D13, 1958 (1976).

[29] Of relevance here is the work by S. J. Brodsky and G. P. Lepage, SLAC preprint SLAC-PUB-2294 (1979).

[30] A. Ali, J. G. Körner, Z. Kunszt, J. Willrodt, G. Kramer, G. Schierholz, and E. Pietarinen, Phys. Lett. 82B, 285 (1979); DESY preprint DESY 79/54 (1979).

[31] M. Krammer and H. Krasemann, Phys. Lett. 73B, 58 (1978).

[32] S. J. Brodsky, D. G. Coyne, T. A. DeGrand, and R. R. Horgan, Phys. Lett. 73B, 203 (1978).

[33] H. Krasemann, Z. Phys. C1, 189 (1979).

[34] D. P. Barber et al., M. I. T. Laboratory for Nuclear Science Report 108 (1979)

[35] C. L. Basham, L. S. Brown, S. D. Ellis and S. T. Love, Phys. Rev. Lett. 41, 1585 (1978), Phys. Rev. D17, 2298 (1978).

[36] See, e.g., S.-Y. Pi, R. L. Jaffe and F. E. Low, Phys. Rev. Lett, 41, 142 (1978).

[37] See, e.g., C. H. Llewellyn Smith, Proc. XVII Internationale Universitätswochen für Kernphysik, Schadming (1978).

[38] S. J. Brodsky, T. A. DeGrand, J. F. Gunion, and J. H. Weis, Phys. Rev. Lett. 41, 672 (1978).

[39] See, e.g., M. K. Gaillard, LAPP preprint LAPP-TH-01 (1979).

[40] A complementary and useful approach is to study hadronic inclusive cross-sections. See Ref. 35.

A MODEL FOR QUARK HADRONIZATION TO JETS IN e^+e^- ANNIHILATION

Liu Lian-sou
Hua-Zhong Teachers College
Peng Hong-an
Beijing University

Abstract

A model for quark materialization into hadrons is proposed. Assuming the mechanism of quark hadronization to be closely related to quark confinement, the hadronization process is described as successive radiation of mesons under the influence of an effective confinement potential. Using this model, we find that the average multiplicity in a jet increases logarithmically with c.m.s. energy, the jet angular radius is inversely proportional to energy, and the mean transverse momentum is constant. These results are consistent with the structure of jets in e^+e^- annihilation at energies below 10GeV, and are also consistent with the structure of the quark jets in the 3-jet events appearing above 10GeV.

The jets in e^+e^- annihilation at c.m.s. energies of 3-9.5 GeV have been observed in various laboratories [1] [2]. According to current idea, the physical interpretation of these phenomena is that the virtual photon produced in e^+e^- annihilation is first transformed into quark-antiquark pair, and then the quark (antiquark) is hadronized to jets. This hadronization process, being a nonperturbative effect, cannot be studied in the framework of perturbative QCD. The usual way to get over this difficulty is to introduce an additional postulate, that the average transverse momentum $\langle k_\perp \rangle$ is constant [3] [4] [5]. Instead of making this additional postulate, we propose in this paper a model for the hadronization process. Using this model the dependence of the average multiplicity and angular radius of jet on the c.m.s. energy of e^+e^- is obtained, and the constancy of mean transverse momentum is dervied thereby.

The fundamental idea of our model is that quark hadronization should be investigated together with quark confinement. Being confined, the quark (antiquark) produced from virtual photon cannot get out, and only the physical particles produced from its hadronization can be observed experimentally. Therefore, the process of quark hadronization should be closely related to that of quark confinement.

The mechanism of quark confinement may be collective

effects of valence quarks and quark-gluon sea, or else may be some kind of vacuum phase transition, but is as yet unclear. We assume that under certain conditions the quark confinement can be described effectively by a potential, increasing indefinitely with the increasing of distance r. This confinement potential corresponds to a bag, which forbids the penetration of quarks and gluons, but is transparent for colorless hadrons. This bag is indivisibly connected with the quark and antiquark, forming a unified dynamical system. Once the quark-antiquark pair is produced from the virtual photon in e^+e^- annihilation, the bag (confinement potential) appears simultaneously. The quark and antiquark are quasi-free in the bag. But when they are projected out with high momenta, they will be retarded by the confinement potential, and will undergo "bremstrahlung" process, radiating physical particles (mainly mesons). This is the process of quark hadronization in our model.

The "bremstrahlung" concerned here is the radiation of physical mesons by quark (antiquark) under the influence of confinement potential. This is a complicated process involving nonperturbative effect. It can be regarded effectively as the excitation of a quark-antiquark pair from the quark-sea, the antiquark then combines with the original quark to form a meson, and the remaining quark continues to move along. The net effect can be viewed as "radiation" of meson M by quark q, as sketched in Fig. 1.

The momentum of the original quark (antiquark) is very high, and the first few mesons radiated have on the average

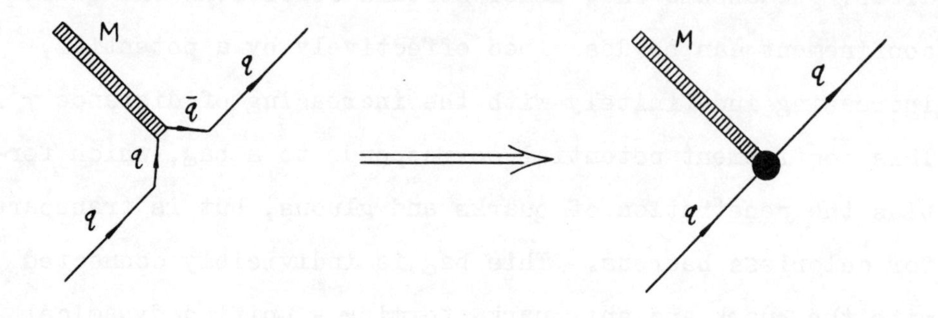

Fig. 1

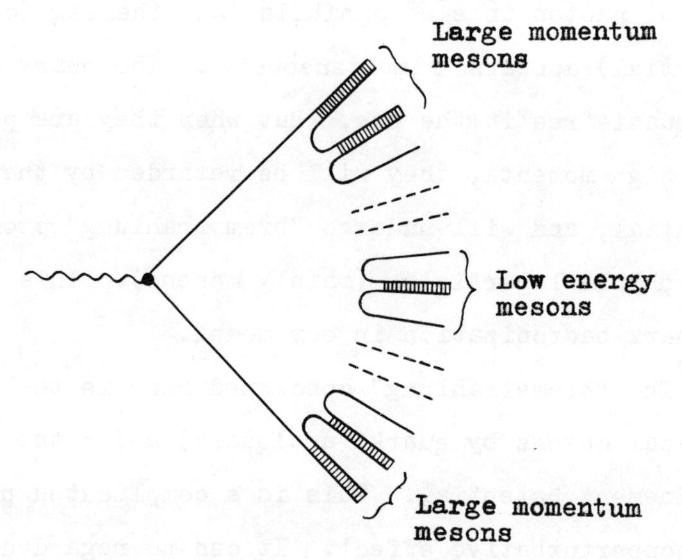

Fig. 2

relatively large momenta. After radiating many times, the quark (antiquark) exhausts its energy eventually, and when the energy decreases down to the meson mass, radiation process ceases, and the remaining quark and antiquark combine to form a low energy meson, not included in jets (see Fig. 2).

It is easily seen from this "multi-bremstrahlung" picture, that the mean multiplicity increases with energy logarithmically. Let ε be the energy of the quark (antiquark) produced from the virtual photon, and let η be the average fraction of energy lost in each radiation. The mean energy of a quark (antiquark), having radiated i mesons, is

$$\varepsilon_i = (1-\eta)\varepsilon_{i-1} = \cdots = (1-\eta)^i \varepsilon. \tag{1}$$

Having radiated $\langle n \rangle$ mesons, the quark energy decreases down to about the meson mass m :

$$(1-\eta)^{\langle n \rangle} \varepsilon \approx m ,$$

so that the mean multiplicity in a jet is

$$\langle n \rangle \approx C \ln \frac{\varepsilon}{m}, \tag{2}$$

where the constant $C = 1/\ln(1-\eta)^{-1}$.

Denoting by $A(r)$ the confinement potential, which may be, for instance, a linear potential[6] $A(r) \sim |r|$. For convenience of Fourier transformation we introduce in the potential a damping factor $e^{-r/R}$, and let $R \to \infty$ after transformation. This corresponds to regarding the perfect confinement of quark as a limiting case of imperfect confinement. It follows that

$$A(q) \sim 1/|q|^4 .$$

Let $p = (\vec{p}, \varepsilon)$ be the 4-momentum of the original quark, let $k_i = (\vec{k}_i, \omega_i)$ be that of the i-th meson radiated by it;

after the radiation of this meson, the remaining quark 4-momentum becomes $p_i=(\vec{p}_i,\varepsilon_i)$. The process is sketched in Fig. 3, where the dotted lines represent the confinement potential. If the energy of the original quark is very high, and

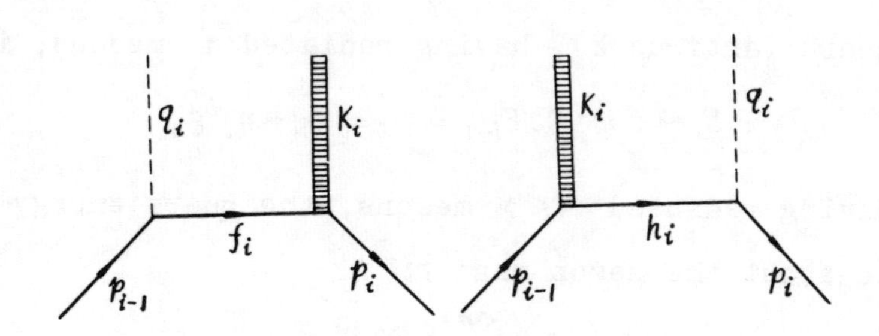

Fig. 3

if i is not too large, then $\varepsilon_i \gg M,\ \omega_i \gg m$ (M being the quark mass, which may be assumed to be of the same order of the meson mass m : $M \approx m$). In this case the dominant factors presented in the matrix elements of the processes shown in Fig. 3 will be the denominators of the propagators and that of the confinement potential $A(q)$:

$$f_i^2 + M^2 = 2p_i k_i - m^2 = -2\varepsilon_i \omega_i \left(1 - \frac{|\vec{p}_i|}{\varepsilon_i}\frac{|\vec{k}_i|}{\omega_i}\cos\theta_{ii}\right) - m^2,$$

$$h_i^2 + M^2 = -2p_{i-1}k_i - m^2 = 2\varepsilon_{i-1}\omega_i\left(1 - \frac{|\vec{p}_{i-1}|}{\varepsilon_{i-1}}\frac{|\vec{k}_i|}{\omega_i}\cos\theta_{ii-1}\right) - m^2,$$

$$q_i^2 = (p_i + k_i - p_{i-1})^2 =$$

$$= 2\varepsilon_{i-1}\omega_i\left(1-\frac{|\vec{p}_{i-1}|}{\varepsilon_{i-1}}\frac{|\vec{k}_i|}{\omega_i}\cos\theta_{i\,i-1}\right)-2\varepsilon_i\omega_i\left(1-\frac{|\vec{p}_i|}{\varepsilon_i}\frac{|\vec{k}_i|}{\omega_i}\cos\theta_{i\,i}\right)+$$

$$+\,2\varepsilon_i\varepsilon_{i-1}\left(1-\frac{|\vec{p}_i|}{\varepsilon_i}\frac{|\vec{p}_{i-1}|}{\varepsilon_{i-1}}\cos\vartheta_i\right)-2M^2-m^2,$$

where

$$\theta_{i\,i}=(\widehat{\vec{k}_i\vec{p}_i}),\quad \theta_{i\,i-1}=(\widehat{\vec{k}_i\vec{p}_{i-1}}),\quad \vartheta_i=(\widehat{\vec{p}_i\vec{p}_{i-1}}).$$

The values of the above denominators increase rapidly with the increasing of these angles, so that the radiation cross section gets large values only when these angles are small. For example:

$$h_i^2+M^2\approx\varepsilon_{i-1}\omega_i\left[\left(\frac{M}{\varepsilon_{i-1}}\right)^2+\left(\frac{m}{\omega_i}\right)^2-\frac{m^2}{\varepsilon_{i-1}\omega_i}+\theta_{i\,i-1}^2\right],$$

which is of order $O(m^2)$ when $\theta_{i\,i-1}=0$, and increases to $O(\varepsilon_{i-1}\omega_i)$ when

$$\theta_{i\,i-1}^2>\left(\frac{M}{\varepsilon_{i-1}}\right)^2+\left(\frac{m}{\omega_i}\right)^2-\frac{m^2}{\varepsilon_{i-1}\omega_i}.$$

It follows that radiation with non-vanishing probability is concentrated in the region:

$$\theta_{i\,i-1}^2\lesssim\left(\frac{M}{\varepsilon_{i-1}}\right)^2+\left(\frac{m}{\omega_i}\right)^2-\frac{m^2}{\varepsilon_{i-1}\omega_i}.$$

Using equation (1), we obtain

$$\overline{\theta}_{i\,i-1}^2\approx\left(\frac{m}{\varepsilon\eta(1-\eta)^{i-1}}\right)^2[1-\eta+\eta^2], \tag{3}$$

and analogously

$$\overline{\vartheta}_i^2\approx\left(\frac{m}{\varepsilon(1-\eta)^i}\right)^2[1-\eta+\eta^2]. \tag{4}$$

What we need in the final result is the angle θ_{i0} between

$\vec{k}_i$ and the direction of the original quark (i.e. the direction of "jet axis"). Therefore the recoil of the quark from

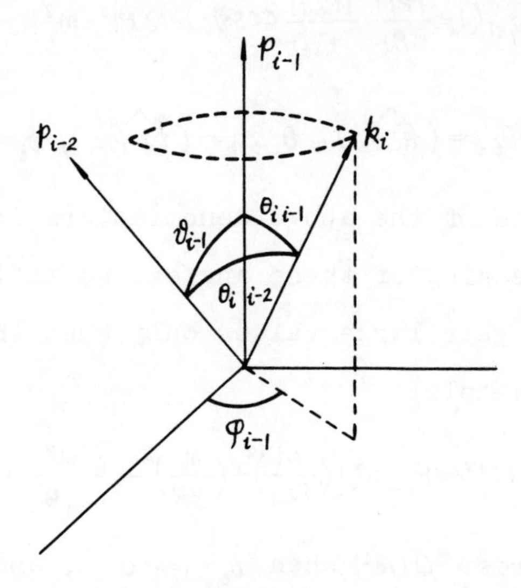

Fig. 4

meson radiation must be considered. From simple geometry (see Fig. 4) we have $\cos\theta_{i\,i-2} = \cos\vartheta_{i-1}\cos\theta_{i\,i-1} + \sin\vartheta_{i-1} \times \sin\theta_{i\,i-1}\cos\varphi_{i-1}$. Averaging over the azimuth angle φ_{i-1}, and using small angle approximation,we obtain

$$\overline{\theta^2_{i\,i-2}} \approx \overline{\vartheta^2_{i-1}} + \overline{\theta^2_{i\,i-1}} \,,$$

and so

$$\overline{\theta^2_{i0}} \approx \Big(\sum_{\alpha=1}^{i-1} \overline{\vartheta^2_\alpha}\Big) + \overline{\theta^2_{i\,i-1}} \approx \frac{m^2}{\varepsilon^2}\,\frac{1-\eta+\eta^2}{\eta^2(2-\eta)}\Big(\frac{2}{(1-\eta)^{2i-2}} - \eta\Big). \qquad (5)$$

We define the jet angular radius Θ as the energy weighted average of θ_{i0} :

$$\Theta^2 = \frac{\sum_{i=1}^{\langle n \rangle} \omega_i^2 \, \overline{\theta}_{i0}^2}{\sum_{i=1}^{\langle n \rangle} \omega_i^2} \,. \tag{6}$$

Note that this "jet angular radius" Θ is not the half angle of a cone, in which all the particles of a jet are

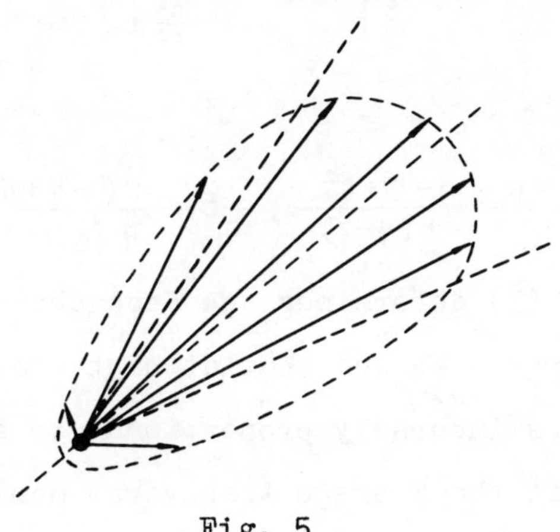

Fig. 5

included, but is only a parameter characterizing the non-isotropy of momentum distribution in a jet, as shown in Fig. 5. This is just what is observed experimentally[1]. It is easy to see that there exists an approximate relation :

$$\Theta^2 \approx \frac{2}{3} S$$

between Θ and the sphericity S defined conventionally by[1]

$$S = \frac{3}{2} \min \frac{\sum_i |p_{i\perp}|^2}{\sum_i |\vec{p}_i|^2}$$

Equation (6) is a weighted average with weights ω_i^2 , and so only the terms with large ω_i have large contributions. For this reason, substituting the above results obtained in the approximation $\omega_i \gg m$ in all the terms of equation (6) will not produce a large error. Thus from (5) and (6) we obtain

$$\Theta \approx \frac{m}{\varepsilon}\left[A + B\frac{\ln\frac{\varepsilon}{m}}{1-\frac{m^2}{\varepsilon^2}}\right]^{\frac{1}{2}}, \tag{7}$$

where

$$A = \frac{1-\eta+\eta^2}{\eta(\eta-2)}, \quad B = \frac{2(1-\eta+\eta^2)}{\eta \ln(1-\eta)^{-1}}.$$

Equation (7) determines the dependence of the jet angular radius on energy. We see that at high energies the jet angular radius is inversely proportional to energy, and the conclusion that the average transverse momentum $\langle k_\perp \rangle$ is constant follows immediately.

The dependence of the jet angular radius on the energy obtained above is different from that of perturbative QCD (with jet angular radius inversely proportional to the logarithm of energy). The reason is that here we are concerned with nonperturbative effect. Our picture should not be considered as contradictory to the QCD perturbative process of gluon radiation, but is just complementary to it. When the energy is not very high (e.g. at c.m.s. energies $W = 2\varepsilon \lesssim$ 10GeV[4][7]), the nonperturbative effect dominates, and our model can well describe the structure of the 2 jets observed experimentally. At energies exceeding 10 GeV, especially, as experimental data has shown, at energies of 27.4-31.6 GeV, the jet angular width is considerably larger than that predicated in our model. This means that the process of gluon radiation should begin to dominate at these energies. On the other hand, the data indicate that only one-sided jet is fattened and that the fattened jet appears to have an oblate structure. If this phenomenon is interpreted as a hard gluon radiated by the quark or antiquark and appears as a third jet, then the mean transverse momentum relative to each jet axis remains constant[8]. So we can conclude that, even at energies up to 31.6 GeV, our model for the hadronization process of a single quark (antiquark) is still valid.

References

1. G.G. Hanson, SLAC-PUB-2118(1978).
2. PLUTO Collaboration, DESY 78/39(1978).
3. G. Sterman and S. Weinberg, Phys. Rev. Lett. 39(1977) 1436.
4. A. De Rujula, J. Ellis, E.G. Floratos and M.K. Gaillard, Nucl. Phys. B138(1978) 387.
5. C.L. Basham, L.S. Brown, S.D. Ellis and S.T. Love, Phys. Rev. D17(1978)2298.
6. W. Marciano and H. Pagels, Phys. Reports 36C(1978)137.
7. A.De Rujula, Proceedings of the 19th International Conference on High Energy Physics, Tokyo(1978)236.
8. Mark-J Collaboration, Report on the 9th International Symposium on Lepton and Photon Interactions at High Energies, Batavia (1979), unpublished.
   PLUTO Collaboration, ibid.
   TASSO Collaboration, ibid.

GENERALIZED MOMENTS OF THREE-JET PROCESSES INDUCED BY THE MUON FOR THE TEST OF QCD

Chang Chao-hsi

Institute of High Energy Physics,

Academia Sinica

Abstract

New types of generalized moments are proposed for the analysis of a three-jet semi-leptonic process of muon. These moments consist of contributions from processes with three or more jets only. Hence it is convenient to use them for testing a high order QCD process. Numerical values of these moments for the three-jet process induced by the muon are presented. It shows that it is already feasible to test QCD by using semi-inclusive or exclusive processes of muon.

I. INTRODUCTION

QCD, in particular the so-called perturbative QCD approach has been successful in understanding phenomena such as: scaling-violation of the deep inelastic scattering, two jet structures,etc[1]. (We shall call these processes the lowest QCD processes later on). Since the perturbative QCD seems to work for the lowest processes, one would like to know whether higher order-processes, for example, those due to a hard gluon bremsstrahlung or hard quark pair production are also useful. This has become an increasingly interesting problem. Recently several papers 2)-7) in this connection have been published. The usefulness of moments and generalized moments to test QCD for the high order processes has been emphasized in Ref. 7)

Wherever possible moment-analyses are prefered to structure function and fragmentation function analyses. The reasons are: First the low values of X and Z for the structure function and fragmentation function, respectively, which are experimentally inaccessible, do not contribute significantly to high moments (N>2). Secondly the Q^2 -dependence of the moments has an analytically simple asymptotic form. So the double moments for two-jet processes [8] and the generalized moments for three-jet processes, double moment defined in terms of a new fragmentation variable Z_{out} ($Z_{out} = 2 |P_{out}| / \sqrt{Q^2}$, P_{out} is the so-called out-momentum of an outgoing hadron) instead of usual Z and a "triple" moment defined in terms of both Z and Z_{out} [7], have been introduced in order to analyze data and test QCD.

The component, perpendicular to the leptonic plane, of the momentum of an outgoing hadron, the so-called out-momentum, P_{out} (Fig. 1), has contributions partly from the transverse momentum of incoming quark (or gluon) inside a target hadron and partly from the deviation of the fragment from the jet axis (both above we call non-perturbative effects), mainly from the third jet effect if we consider a three-jet event. The former two are of Q^2-independence and we neglect them; the latter is what we are interested in. We can neglect the non-perturbative effects when either Q^2 is large enough, if the quantities that we are interested in are insensitive to the small Z_{out} region (it is so when $n \geqslant 2+m/2$, $m \geqslant 2$ for generalized moments), or they can be cancelled experimentally. The generalized moments have contribution only from processes with three or more jets in the idealcase that we neglect the non-perturbative effects, it is one of the main reasons why we introduce the generalized moments.

One more advantage of using generalized moments is that they depend on the moments of the structure function and fragmentation function only, while now the measurements for these moments are better than those for the functions directly so we can have a better input when we calculate the generalized moments.

As pointed out in the reference 7), using the generalized moments to test QCD for three-jet processes, the most accessible experiments are muon experiments, either semi-inclusive or exclusive, as the energy of the incoming muon is high enough and fixed, in contrast with neutrino's. So we will discuss the generalized moments of three-jet processes induced by muon in section II, as well as calculate them by taking the incoming energy of muon E=150 GeV and do some discussions in section III.

Now we are interested in the semi-inclusive process

$$\mu(k) + N(P) \longrightarrow \mu(k') + \pi^{\pm}(P') + \cdots \tag{1}$$

The differential cross-section can be written as follows:

$$\frac{d\sigma}{dx\,dZ\,dQ^2 dP_T' d\Phi} = \sum_{ij} \int dx_p\, dZ_p\, dp_T\, d\eta\, d\eta'\, \delta(x - \eta x_p)\, \delta(Z - \eta' Z_p)$$
$$\cdot\, \delta(P_T' - \eta' p_T)\, f_i(\eta, Q^2)\, \frac{d\sigma_{ij}}{dx_p dZ_p dQ^2 dp_T d\phi}\, D_j^{\pi^{\pm}}(\eta', Q^2)\,. \tag{2}$$

where we choose the definitions (Fig. 1):

$$q = k - k',\ Q^2 = -q^2,\ x = \frac{Q^2}{2(Pq)}\ ,\ Z = \frac{(P\cdot P')}{(Pq)}\ ,\ P_T',\ \Phi,\ P_{out} = P_T' \sin\Phi. \tag{3}$$

$f_i(\eta, Q^2)$ is the momentum distribution function of the i parton (quark, antiquark or gluon) in the target hadron N, and $D_j^{\pi^{\pm}}(\eta', Q^2)$ in the fragmentation function of the j parton to $\pi^{\pm}$. The subprocess differential cross-section is:

$$\frac{d\sigma_{ij}}{dx_p dZ_p dQ^2 dp_T d\phi} = \frac{\alpha^2}{2\pi}\, \frac{1}{Q^4}\, \delta(x_p - 1)\, \delta(Z_p - 1)\, \delta(p_T)\, (2 - 2y + y^2)\,, \tag{4}$$

for the two-jet process (Fig. 2), and

$$\frac{d\sigma_{ij}}{dx_p dZ_p dQ^2 dp_T d\phi} = \frac{\alpha^2}{2\pi}\, \frac{\alpha_s(Q^2)}{Q^4}\, \delta\left(p_T - \left[\frac{Q^2 Z_p (1-Z_p)(1-x_p)}{x_p}\right]^{\frac{1}{2}}\right) \left\{ \frac{4}{3}(A_1 + B\cos\phi + C_1 \cos^2\phi)\right.$$
$$\left. + \frac{4}{3}(A_2 + B_2\cos\phi + C_2\cos 2\phi) + \frac{1}{2}(A_3 + B_3\cos\phi + C_3\cos 2\phi) \right\} \tag{5}$$

for the three-jet process (Fig. 3), where $\alpha(Q^2)$ is the running coupling constant given as

$$\alpha(Q^2) = \frac{12\pi}{(33-2f)\ln(Q^2/\Lambda^2)} \tag{6}$$

and

$$x_p = \frac{Q^2}{2(p_a q)}, \quad z_p = \frac{(p_a p_b)}{(p_a q)}, \quad y = \frac{(p_a q)}{(p_a k)}. \tag{7}$$

We have $\quad p_a = \eta P, \quad P' = \eta' p_b, \quad \phi = \Phi, \qquad (8)$

neglecting the non-perturbative effects. A_i, B_i and C_i (i=1, 2, 3, corresponding to the contribution from Fig. 3a, 3b and 3c respectively) are functions of x_p, z_p and Y*:

$$A_1 = 8(1-y)x_p z_p + (2-2y+y^2)\left[(1-x_p)(1-z_p) + \frac{1+x_p^2 z_p^2}{(1-x_p)(1-z_p)}\right], \tag{9.1}$$

$$A_2 = 8(1-y)x_p(1-z_p) + (2-2y+y^2)\left[(1-x_p)z_p + \frac{1+x_p^2(1-z_p)^2}{(1-x_p)z_p}\right], \tag{9.2}$$

$$A_3 = 16(1-y)x_p(1-z_p) + (2-2y+y^2)\left[x_p^2 + (1-x_p)^2\right]\frac{z_p^2+(1-z_p)^2}{z_p(1-z_p)}, \tag{9.3}$$

$$B_1 = -4(2-y)(1-y)^{\frac{1}{2}}\left[\frac{x_p z_p}{(1-x_p)(1-z_p)}\right]^{\frac{1}{2}}\left[x_p z_p + (1-x_p)(1-z_p)\right], \tag{10.1}$$

$$B_2 = 4(2-y)(1-y)^{\frac{1}{2}}\left[\frac{x_p(1-z_p)}{(1-x_p)z_p}\right]^{\frac{1}{2}}\left[x_p(1-z_p) + (1-x_p)z_p\right], \tag{10.2}$$

$$B_3 = -4(2-y)(1-y)^{\frac{1}{2}}\left[\frac{x_p(1-x_p)}{z_p(1-z_p)}\right]^{\frac{1}{2}}(1-2x_p)(1-2z_p), \tag{10.3}$$

---

* We can see that these functions are consistent with those of Ref. 6) but differ from those of Ref. 2). This is the reason why we give the formulas in detail here.

$$C_1 = 4(1-y)x_p Z_p \quad , \tag{11.1}$$

$$C_2 = 4(1-y)x_p(1-Z_p) \ , \tag{11.2}$$

$$C_3 = 8(1-y)x_p(1-x_p) \ , \tag{11.3}$$

Following Ref. 7), we consider the case of non-singlet targets and fragments, namely:

$$\frac{d\sigma(ns)}{dx\,dz\,dQ^2\,dP_T'\,d\Phi} = \frac{d\sigma}{dx\,dz\,dQ^2\,dP_T'\,d\Phi}\Big[(\mu p \longrightarrow \mu\pi^+ \ldots) - (\mu p \longrightarrow \mu\pi^-) - (\mu n \longrightarrow \mu\pi^+ \ldots) + (\mu n \longrightarrow \mu\pi^- \ldots)\Big] , \tag{12}$$

where p = porton and n = neutron. Thus we have

$$\frac{d\sigma(ns)}{dx\,dz\,dQ^2\,dP_T'\,d\Phi} = \frac{\alpha^2}{2\pi}\frac{\alpha_s(Q^2)}{Q^4}\int dx_p\,dZ_p\,dp_T\,d\eta\,d\eta'\,\delta(x-\eta x_p)\,\delta(Z-\eta' Z_p) \cdot \delta(P_T'-\eta' p_T)\,F_v(\eta,Q^2)\,\delta\Big(p_T - \Big[\frac{Q^2 Z_p(1-Z_p)(1-x_p)}{x_p}\Big]^{\frac{1}{2}}\Big)\,U\,D_{\langle u\rangle}^{(\pi^+-\pi^-)}(\eta',Q^2), \tag{13}$$

where $F_v(\eta, Q^2)$ is the distribution function of valence quark in a nucleon,

$$U = \frac{5}{27}(A_1 + B_1\cos\Phi + C_1\cos 2\Phi) \ . \tag{14}$$

and

$$D_{\langle u\rangle}^{(\pi^+-\pi^-)}(\eta', Q^2) = D_{\langle u\rangle}^{\pi^+}(\eta', Q^2) - D_{\langle u\rangle}^{\pi^-}(\eta', Q^2) \ . \tag{15}$$

According to Ref.7) we have the generalized double moment

$$E^{(ns)}(n, m, Q^2) \equiv \frac{Q^4}{\alpha^2}\int_0^1 dx\int_0^1 dz\int dP_T'\int d\Phi\, x^{n-1} Z_{out}^{m-1}\,\frac{d\sigma(ns)}{dx\,dz\,dQ^2\,dP_T'\,d\Phi} \ . \tag{16}$$

where $Z_{out} \equiv \frac{2\,|P_{out}|}{\sqrt{Q^2}}$. Substituting (13) into (16), we obtain

$$E^{(ns)}(n, m, Q^2) = \frac{\alpha_s(Q^2)}{2\pi}\int_0^1 dx \int_0^1 dZ\, 2^{m-1}\left[Z(1-Z)(1-x)\right]^{\frac{m-1}{2}} x^{n-\frac{m+1}{2}} \tag{17}$$

$$\cdot \left[\alpha_{m-1} Y_1 + \beta_{m-1} Y_2\right] \hat{D}^{(\pi^+\pi^-)}_{\langle\mu\rangle}(m, Q^2) \quad ,$$

where

$$\alpha_{m-1} = \int_0^{2\pi} d\Phi\, |\sin\Phi|^{m-1} \quad , \tag{18}$$

$$\beta_{m-1} = \int_0^{2\pi} d\Phi\, |\sin\Phi|^{m-1} \cos 2\Phi \quad , \tag{19}$$

$$Y_1 = \frac{5}{27}\Big\{\Big[\hat{F}_v(n, Q^2) - \frac{Q^2}{2xEM}\hat{F}_v(n-1, Q^2)\Big]\Big[8xZ + 2(1-x)(1-Z) \tag{20}$$

$$+ 2\frac{1+x^2Z^2}{(1-x)(1-Z)}\Big] + \Big(\frac{Q^2}{2xEM}\Big)^2 \hat{F}_v(n-2, Q^2)\Big[(1-x)(1-Z) + \frac{1+x^2Z^2}{(1-x)(1-Z)}\Big]\Big\} \, ,$$

$$Y_2 = \frac{20}{27}\, x Z \left[\hat{F}_v(n, Q^2) - \frac{Q^2}{2xEM}\hat{F}_v(n-1, Q^2)\right] , \tag{21}$$

E is the energy of the incoming μ , M the mass of nucleon and the moments are:

$$\hat{F}_v(n, Q^2) = \int_0^1 dx\, x^{n-1} F_v(x, Q^2) \quad , \tag{22}$$

$$\hat{D}^{(\pi^+\pi^-)}_{\langle\mu\rangle}(m, Q^2) = \int_0^1 dZ\, Z^{m-1} D^{(\pi^+\pi^-)}_{\langle\mu\rangle}(Z, Q^2) \, , \tag{23}$$

We can see that only moments of structure function and fragmentation function occur in the double generalized moment and only Fig. 3a) contributes to the generalized moment because we consider non-singlet only. Similarly, we have the "**triple**" moments (If $m > 1$):

$$H^{(ns)}(n, m, k, Q^2) = \frac{Q^4}{\alpha^2} \int_0^1 dx \int_0^1 dz \int dp'_T \int d\Phi\, x^{n-1} Z_{out}^{m-1} Z^k \frac{d\sigma(ns)}{dx\,dz\,dQ^2 dp'_T d\Phi}$$

$$= \frac{\alpha_s(Q^2)}{2\pi} \int_0^1 dx \int_0^1 dz\, 2^{m-1} [z(1-Z)(1-x)]^{\frac{m-1}{2}} x^{n-\frac{m+1}{2}} \tag{24}$$

$$\cdot [\alpha_{m-1} Y_1 + \beta_{m-1} Y_2]\, \hat{D}^{(\pi^+-\pi^-)}_{\langle u \rangle}(m+k, Q^2)$$

There is no contribution from two-jet processes in the generalized moments because of the factor $\delta(p_T)$ in Eq. (4). From the definition of the generalized moments, we notice that, only when $n \geqslant 2 + \frac{m}{2}$, they are calculable in QCD and are not sensitive to the region of small Z_{out}.

III. DISCUSSION

For E and Q^2 given, the mean value of $x^{n-1} Z_{out}^{m-1}$ for an exclusive process, $\langle x^{n-1} Z_{out}^{m-1} \rangle_{(ns)}$, is

$$\int_{x_{min}}^1 dx \int_0^1 dz \int dp'_T \int d\Phi\, x^{n-1} Z_{out}^{m-1} \frac{d\sigma(ns)}{dx\,dz\,dQ^2 dp'_T d\Phi} ,$$

where $x_{min} = Q^2/2ME$. It can be measured directly if an exclusive process is measured. Considering the differential corss-section property in the $x \simeq 0$ region, $\langle x^{n-1} Z_{out}^{m-1} \rangle_{(ns)}$ is insensitive to x_{min} when $x_{min} \ll 1$ and $n > 2 + m/2$ $(m > 1)$. So we have

$$E^{(ns)}(n, m, Q^2) \simeq \frac{Q^2}{\alpha^2} \langle x^{n-1} Z_{out}^{m-1} \rangle_{(ns)} \tag{25}$$

if $n > 2 + m/2$ $(m > 1)$ and also E is very large, say E = 150 GeV when $1 GeV^2 \leqslant Q^2 \leqslant 20\ GeV^2$, that we can measure the generalized moments directly by means of an exclusive measurement.

Similarly, we also have

$$H^{(ns)}(n,m,k,Q^2) \simeq \frac{Q^4}{\alpha^2} \langle x^{n-1} Z_{out}^{m-1} Z^k \rangle_{(ns)} \tag{26}$$

We can also measure the generalized moments by a semi-inclusive process: First measure the differential cross section $\frac{d\sigma}{dx\,dz\,dZ_{out}\,dQ^2}$ for E and Q^2 given and then calculate the generalized moments based on Eqs. (16) and (24).

Now, in order to test QCD, all we have to do is to compare generalized moments measured directly with those calculated in the framework of QCD, namely from (16) and (24), by inserting the moments of structure function and fragmentation function as inputs.

To show the behaviour of the generalized moments and compare with experiments, as an example choosing E = 150 GeV and using the latest experimental data about the moments of structure function and fragmentation function as input, which were used in Ref. 7), we calculate and obtain the curves for $E^{(ns)}(n,m,Q^2)$ and $H^{(ns)}(n,m,Q^2)$ (Figs. 4,5).

As pointed out in reference 7), because the differential cross section (5) is of y-dependence, we must know the energy of incoming lepton when using the generalized moments to test QCD for three-jet processes. So it is more reasonable to test QCD using the generalized moments for muon experiments than for neutrino or antineutrino experiments because one can precisely know the muon energy. Especially the exclusive one will give a thorough test of QCD for the three-jet processes with changing n, m and k, and both of the Q^2-dependence and normalization of the curves for the generalized moments are crucial when testing QCD.

ACKNOWLEDGEMENT

Most part of this work was done at CERN. The author would like to thank the Theory Division of CERN for their hospitality.

REFERENCES

1. Yu L. Dokshitzer et al, Proc. XIII Winter School at Leningrad, B.P. Konstantinov, Institute of Nuclear Physics (Leningrad, 1978).
   C.H. Llewellyn Smith, Acta, Physica. Suppl. XIX (1978) 331,
   G. Altarelli and G. Parisi. Nucl. Phys. B126 (1977) 298,
   G. Sterman and S. Weinberg, Phys. Rev. Lett. 39(1977) 1436.
2. H. Georgi and H.D. Politzer, Phys. Rev. Lett, 40 (1978)3.
3. E.G. Florator, Nuovo Cimento, 43A (1978) 241.
4. G. Alterelli and G. Martinelli, Phys. Lett., 76B (1978) 89.
5. P. Bine'truy and G. Girard, CERN Preprint TH.2611 (1979).
6. A. Méndez, Nucl. Phys., B145 (1978) 199.
7. Chang Chao-hsi, CERN Preprint TH. 2736 (1979).
8. J. Ellis, M.K. Gaillard and W.J. Zakrzewski, Phys. Lett., 81B (1979) 224.

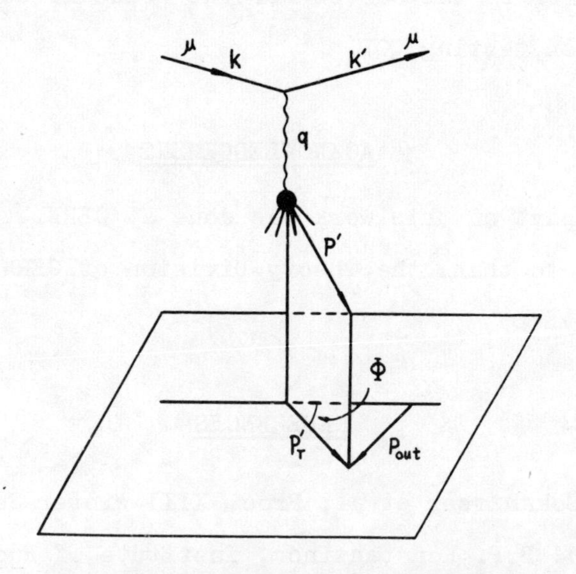

Fig. 1: Configuration of the semi-inclusive process μ N and the definitions of momenta in Lab. system.

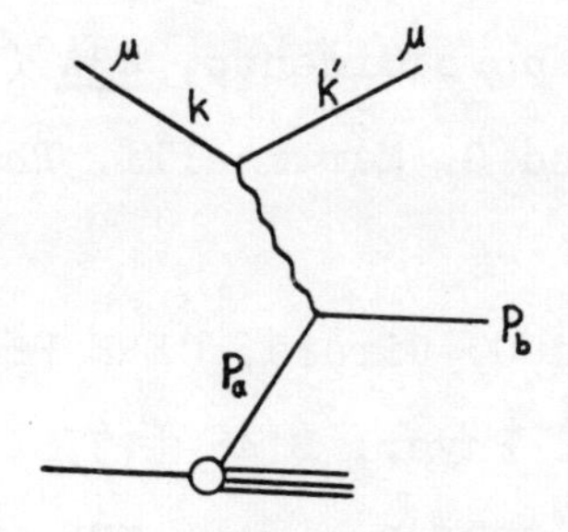

Fig. 2: The lowest perturbative QCD diagram.

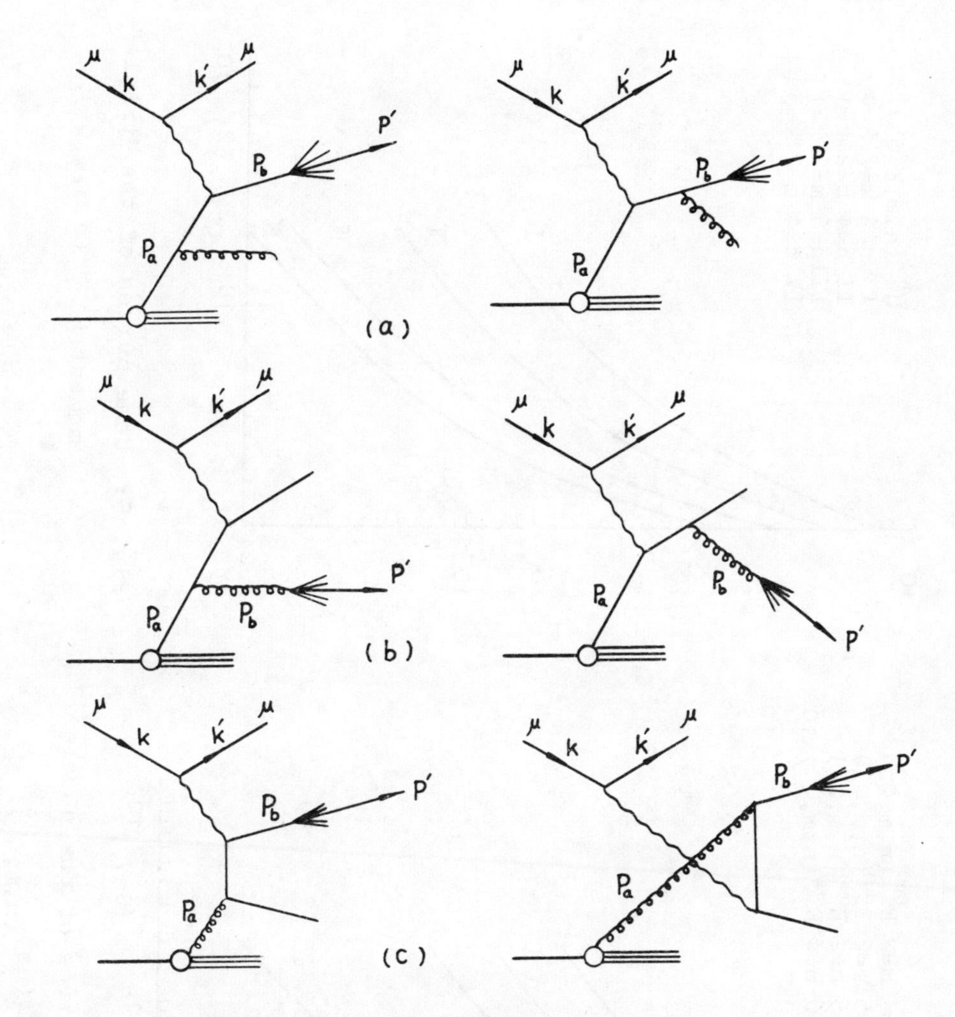

Fig. 3: Perturbative QCD diagrams:

a) a hard gluon bremsstrahlung and the hadron fragmentation from the quark;

b) a hard gluon bremsstrahlung and the hadron fragmentation from the gluon;

c) a hard pair production from a gluon inside N and the hadron fragmentation from the quarks.

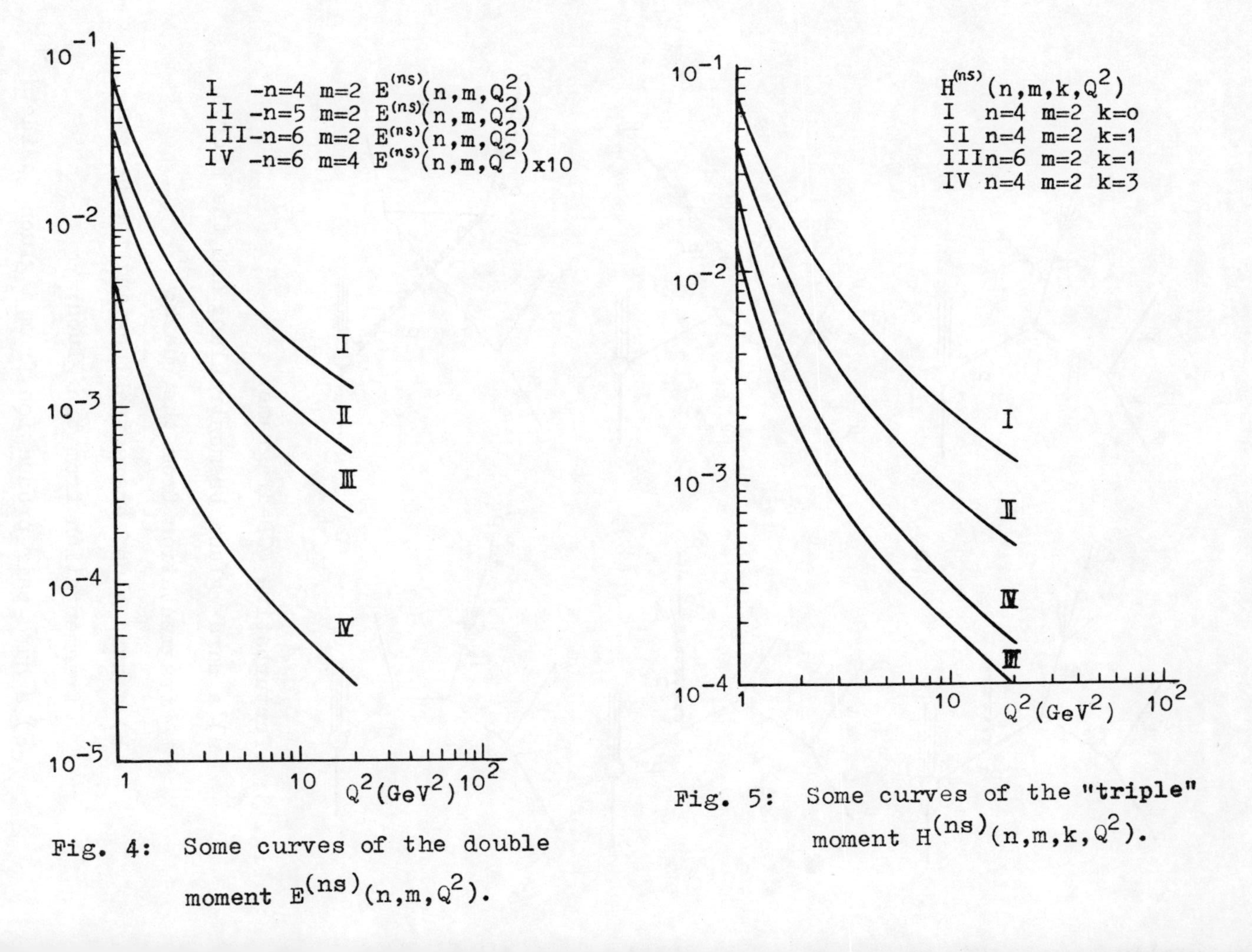

Fig. 4: Some curves of the double moment $E^{(ns)}(n,m,Q^2)$.

Fig. 5: Some curves of the "triple" moment $H^{(ns)}(n,m,k,Q^2)$.

AN INVESTIGATION OF THE EFFECTS OF CHARGED BOSONIC PARTONS (DIQUARKS) IN DEEP INELASTIC e-P SCATTERING

Peng Hong-an and Zou Guo-xing

Physics Department, Beijing University, Beijing

Abstract

We point out the possible existence of bosonic charged partons (diquarks) inside the nucleon. Certain forms of the structure functions of diquarks are assumed and the functions are suitably parametrized. Combining their effects with the usual QCD calculation, we obtain an improved theoretical fit to the experimental value of $F=\frac{\sigma_L}{\sigma_T}$ in deep inelastic e-P scattering.

I. INTRODUCTION

Scaling violation in deep inelastic lepton scattering processes(1,2) has received much attention and discussion. Perturbative QCD has led to interesting results in the interpretation of scaling violation, and the predicted structure function $F_2(x,Q^2)$ also agrees quite well with experiments(3,4). On the other hand, the QCD prediction for the structure parameter $R=\frac{\sigma_L}{\sigma_T}$ is too small, lying almost completely below the lower bound to experimental values, and it decreases rapidly with increasing Q^2, which also disagrees with experiments(3,4). Even after effects of the target mass, higher-twist(5) and parton transverse momentum(6) have been taken into account, theoretical prediction for R is still much smaller than experimental data.

In quark-parton model, R essentially vanishes because of the Callan-Gross relation. It is thus natural to suggest that, if some kind of bosonic parton is also present in the nucleon, in addition to quark partons, then R would no longer vanish and might even vary with Q^2, leading to scaling violation. When the proportion of such bosonic charged partons is small, they would only affect R but not F_2. Thus, the possible existence of bosonic partons in the nucleon might provide us with a key to solving the above difficulties.

As to the nature of bosonic charged partons, we have the following suggestion. It is well-known that, two quarks cannot be a color singlet and hence cannot form a physical particle. But inside a nucleon the $q\bar{q}$ sea, the gluons and the valence quark might generate a strong "background" field, under the action of which two other valence quarks might be very tightly bound, thus creating a dynamical diquark substructure within the nucleon. In deep inelastic e-N scattering one might expect that such diquark structùres would exist for relatively small Q^2 (say, $\lesssim$ 20GeV^2) and sufficiently large x(say, $\gtrsim$ 0.4 or 0.5), and that they would behave as point partons of integer spin in their interaction with the photon. In the following sections a certain form of distribution functions for these charged bosons is proposed and the functions are suitably parametrized. Combining their effects with the usual QCD calculations, one indeed gets a better theoretical fit to experimental values of R, while F_2 is not appreciably affected. In this preliminary attempt to investigate the effects of bosonic partons, we shall not try to make a thorough and suitable readjustment of the parametrization of quark-parton distribution functions obtained by Buras and Geamers in Ref.3, which will be directly used in our fit to experiment, even though the effects of bosonic partons (diquarks) are now taken into account. As will be seen, the boson component turns out to be quite small (~1% only), and ought not modify the quark-parton structure functions appreciably.

II. ANALYSIS AND FORMULAS

In the e-P deep inelastic reaction $e+P \to e+X$ the amplitude for single photon exchange approximation as represented in Fig. 1 is

$$T = \frac{e^2}{q^2}\bar{u}(k')\gamma^\mu u(k)\langle X|j_\mu^{em}(0)|p,r\rangle \tag{2.1}$$

The differential cross-section is then

$$\frac{d\sigma}{d\Omega d\nu} = \frac{4\alpha^2 E'^2}{q^4}\left\{W_2\cos^2\frac{\theta}{2} + 2W_1\sin^2\frac{\theta}{2}\right\} \tag{2.2}$$

where k and k' are, respectively, the initial and final 4-momenta of the electron, E and E' are, respectively, its initial and final energy in the lab system, (p,r) is the momentum and polarization of the target proton, and Θ is the scattering angle in the lab system, W_1 and W_2 are two structure functions defined as follow:

$$W_{\mu\nu}=\frac{1}{2\pi M}\sum_r\int d^4x\, e^{iqx}\langle P,r|[j_\mu^{em}(x),j_\nu^{em}(0)]|P,r\rangle$$
$$=-(g_{\mu\nu}-\frac{q_\mu q_\nu}{q^2})W_1(\nu,q^2)+\frac{1}{M^2}(P_\mu-\frac{P\cdot q}{q^2}q_\mu)(P_\nu-\frac{P\cdot q}{q^2}q_\nu)W_2(\nu,q^2) \quad (2.3)$$

where M is the proton mass and

$$q^2=-Q^2=-4EE'\sin^2\frac{\theta}{2}$$
$$\nu=\frac{p\cdot q}{M}=E-E' \quad (2.4)$$

are the two independent kinematical parameters.

With the scaling variable x defined as $x=\frac{-q^2}{2pq}=\frac{Q^2}{2M\nu}$, the physical region of e-N scattering is $0\leqslant x\leqslant 1$, where $x\rightarrow 1$ corresponds to elastic scattering and $x\rightarrow 0$ is the high-energy ($\nu\rightarrow\infty$) Regge region.

We introduce the usual dimensionless structure functions

$$F_1(x,Q^2)=M\,W_1(\nu,q^2) \qquad F_2(x,Q^2)=\nu\,W_2(\nu,q^2) \quad (2.5)$$

Equation (2.2) can then be written as

$$\nu\frac{d\sigma}{d\Omega\,d\nu}=\frac{4\alpha^2E'^2}{Q^4}F_2\left\{\cos^2\frac{Q}{2}+\frac{2}{1+R}(1+\frac{\nu^2}{Q^2})\sin^2\frac{\theta}{2}\right\} \quad (2.6)$$

where

$$R(x,Q^2)=\frac{\frac{M}{\nu}(1+\frac{\nu^2}{Q^2})F_2(x,Q^2)-F_1(x,Q^2)}{F_1(x,Q^2)}\simeq\frac{\sigma_L}{\sigma_T} \quad (2.7)$$

is just the ratio of the absorption cross-section of the longitudinal to that of the transverse virtual photon.

For $\nu\rightarrow\infty$, $Q^2\rightarrow\infty$ but x remains finite $\frac{M}{\nu}(1+\frac{\nu^2}{Q^2})\rightarrow$

$\frac{1}{2x}$ and $R \simeq \frac{F_2 - 2xF_1}{2xF_1}$. If spin $\frac{1}{2}$ quarks are the only charged partons inside the nucleon, then naive parton model yields Callan-Gross relation $F_2 - 2xF_1 = 0$, where both σ_L and R are essentially zero. As is generally supposed, a nonvanishing R comes only from corrections to naive parton model.

However, if bosonic charged partons do exist, they can give rise to a non-vanishing R even in the approximation of a naive parton model. Their effect may be accounted for as follow. We first write down the electron-parton elastic cross-section for partons of spin S(=0, $\frac{1}{2}$, 1):

$$\frac{d\sigma^{(S)}}{d\Omega} = \frac{4\alpha^2 E'^2}{Q^4}\left\{ G_2^{(S)}(\nu, Q^2)\cos^2\frac{\theta}{2} + 2\left(1+\frac{\nu^2}{Q^2}\right)G_1^{(S)}(\nu, Q^2)\sin^2\frac{\theta}{2}\right\} \quad (2.8)$$

where the kinematical factors $G_1^{(s)}$ and $G_2^{(s)}$ are given by

$$\begin{aligned} &G_1^{(0)} = 0 && G_2^{(0)} = 1 \\ &G_1^{(\frac{1}{2})} = 1 && G_2^{(\frac{1}{2})} = 1 \\ &G_1^{(1)} = (1+K)^2 \frac{2\nu^2}{3Q^2} && G_2^{(1)} = 1 + \frac{4\nu^2}{3Q^2}\left(\frac{1+K^2}{2} + K^2\frac{\nu^2}{Q^2}\right) \end{aligned} \quad (2.9)$$

K being the anomalous magnetic moment of a vector parton. Then by usual procedure we obtain the total structure functions of deep inelastic e-N scattering:

$$F_2(x, Q^2) = \sum_S F_2^{(S)}(x, Q^2)\, G_2^{(S)}(x, Q^2) \quad (2.10)$$

$$F_1(x, Q^2) = \frac{M}{\nu}\left(1+\frac{\nu^2}{Q^2}\right)\sum_S F_2^{(S)}(x, Q^2)\, G_1^{(S)}(x, Q^2)$$

with

$$F_2^{(S)}(x, Q^2) = \sum_i Q_i^{(S)\,2}\, x\, q_i^{(S)}(x, Q^2) \quad (2.11)$$

The right-hand side is summed over partons of spin S, $Q_i^{(s)}$ is the charge of the i-th parton of spin S and $g_i^{(S)}(x,Q^2)$ its momentum distribution function, which is also assumed to be dependent on Q^2. From (2.10) one obtains

$$R(x,Q^2)=\frac{\sum_S F_2^{(S)}(x,Q^2)[G_2^{(S)}(x,Q^2)-G_1^{(S)}(x,Q^2)]}{\sum_S F_2^{(S)}(x,Q^2)\,G_1^{(S)}(x,Q^2)} \tag{2.12}$$

As supposed in I (introduction) bosonic partons in the form of diquarks have much weaker distribution functions than quark partons. So it is natural to take into account the QCD corrections to the effects of quark partons, but not to those of bosonic partons. Actually, in any case there is at present no way of calculating the correct term to the naive parton model by the application of QCD to such a complicated dynamical system. For this purpose, we shall directly use the results of QCD calculations by Buras et al (3), since they have given an analytic form of the dependence of F_2 on Q^2, which appears to yield at present the best fit to experiment on F_2. Let $F^{(\frac{1}{2})}(x,Q^2)$ denote the QCD corrected quark structure function and $R^{(\frac{1}{2})}(x,Q^2)$ the corresponding parameter R obtained by these authors. Then

$$F_2(x,Q^2)=F_2^{(\frac{1}{2})}(x,Q^2)+\Delta_2(x,Q^2)$$

where

$$\Delta_2(x,Q^2)=F_2^{(0)}(x,Q^2)+F_2^{(1)}(x,Q^2)G_2^{(1)}(x,Q^2) \tag{2.13}$$

On the other hand, equation (2.12) reads

$$R(x,Q^2)=\frac{F_2'^{(\frac{1}{2})}R^{(\frac{1}{2})}+F_2^{(0)}+F_2^{(1)}[G_2^{(1)}-G_1^{(1)}]}{F_2'^{(\frac{1}{2})}+2x F_2^{(1)}G_1^{(1)}} \tag{2.14}$$

where

$$F_2'^{(\frac{1}{2})} = \frac{F_2^{(\frac{1}{2})}}{1+R^{(\frac{1}{2})}}$$

and $F_2^{(0)}(x,Q^2)$ and $F_2^{(1)}(x,Q^2)$ are the structure functions of scalar and vector partons, respectively.

III. BOSONIC CHARGED PARTONS AND THEIR DISTRIBUTION FUNCTIONS

The need for bosonic charged partons within the nucleon may be seen in two ways. First, if the quark model is to provide a consistent dynamical picture for a continuous transition from completely incoherent deep inelastic scattering to completely coherent elastic scattering ($W^2 \rightarrow M^2$ or $x \rightarrow 1$), then very tightly bound diquark or even triquark structures should be present in the nucleon. One may imagine that under the action of the "background" field mentioned in I (introduction) a small amount of tightly bound diquark substructure would exist for small Q^2 when x increases beyond a certain value, say 0.4 or 0.5. If x 1 and Q^2 remains small, tightly bound triquark states might appear and their interaction with virtual photon could conceivably give rise to elastic scattering phenomenon. However, any such triquark structure cannot by itself be a nucleon because a physical nucleon must also comprise the $q\bar{q}$ sea and gluons.

Secondly, hadron-hadron reaction experiments can only be satisfactorily accounted for by assuming diquark substructure in hadron. Thus in meson-nucleon collisions almost all charge- and strangeness-exchange processes (e.g. $K^- + P \rightarrow M^0 +$ (Λ, Σ, Y^*_{1385}), M^0 being a neutral nonstrange meson) cannot be satisfactorily explained by SU(6) symmetry, but the introduction of some diquark component will do the job quite well[7].

Let us now consider in some detail possible diquark states in the nucleon. Since we are not interested in the region $x \ll 1$ (the contribution from the $q\bar{q}$ sea is negligible for $x \leq 0.4$ or 0.5), we shall confine ourselves to diquark substructure of valence quarks only. Thus within a proton they can only be (uu) and (ud) diquarks. Their quantum numbers (see Table1) may be determined from the following considerations:

(i) the proton is a color singlet so that, since $\underline{3} \times \underline{3} = \underline{3}^* + \underline{6}$ according to color SU(3), diquarks must be $\underline{3}^*$ (color);

(ii) diquarks being composed of two fermion must be in globally antisymmetric states.

We note from Table I that, except in the last line, all vector diquarks (S=1) are in $\underline{6}$ and all scaler diquarks (S=0) in $\underline{3}^*$ of flavor SU(3). This rule is strictly valid if all diquarks are in the orbital S state ($\ell = 0$).

After the foregoing analysis , the structure functions of vector and scalar diquarks can be written as follows:

$$F_2^{(1)}(x,Q^2) = Q_{uu}^2\, x\, g_{uu}^{(1)}(x,Q^2) + 2Q_{ud}^2\, x\, g_{ud}^{(1)}(x,Q^2)$$

$$= \frac{16}{9}\, x\, g_{uu}^{(1)}(x,Q^2) + \frac{2}{9}\, x\, g_{ud}^{(1)}(x,Q^2),$$

$$F_2^{(0)}(x,Q^2) = 2\,Q_{ud}^2\, x\, g_{ud}^{(0)}(x,Q^2) = \frac{2}{9}\, x\, g_{ud}^{(0)}(x,Q^2) \qquad (3.1)$$

The functional dependence of x and Q^2 of the distribution functions $g^{(0)}$ and $g^{(1)}$ is at present essentially a matter of conjecture, and one can only make use of intuitive ideas in carrying out a preliminary analysis. First of all, if we require continuity in the transition between incoherent and coherent scattering mechanisms, then the x distribution of diquarks should have larger weight near the x=1 end than that for the valence quarks. If we further assume the validity of the Drell-Yan argument*)[9] for diquarks, then their x distribution would be of the form $(1-x)^{\beta}$, $\beta \geq 3$. On the other hand, the following considerations suggest that it should behave like x^{α} ($\alpha \geq 0$) as $x \rightarrow 0$: at fixed Q^2 and for $x \rightarrow 0$ (i.e. $\nu \rightarrow \infty$) the contribution to quark structure functions in deep inelastic e-P scattering comes principally from sea quarks and gluons (corresponding to Pomerons in high-energy diffraction processes), whose x distribution behaves like x^{-1} as $x \rightarrow 0$, while valence quarks (corresponding to Regge poles) have distribution functions behaving like $x^{-\frac{1}{2}}$, thus making a smaller contribution than sea quarks and gluons. One naturally expects the diquark contribution at $x \rightarrow 0$ to be yet smaller with behaviour like x^{α}, $\alpha \geq 0$. The x distribution of diquarks in nucleon may therefore be assumed to have the following simple form

---

*) If nucleon elastic form factor behaves as $G_m(t) \sim t^{-n}$ at large $t=q^2$, then, according to Drell and Yan, parton distribution functions in electroproduction would behave like $F_1(x)$, $F_2(x) \sim (1-x)^{2n-1}$. Starting from other dynamical assumptions Bloom and Gilman[11] arrived at the same conclusion.

The Q^2 dependence of $g(x,Q^2)$ can only be determined from the yet unknown nonperturbative mechanism by which various diquark states are formed. One may, however, expect $g(x, Q^2)$ to tend to zero as $Q^2 \longrightarrow \infty$, since there should then be no diquark formation owing to asymptotic freedom. On the other hand, we shall assume that $g_i^{(s)}(x,Q^2)$ is a smooth function of Q^2 and that the resultant diquark structure function $\Delta_2(x,Q^2)$ is essentially independent of Q^2 within a narrow range, say $2 \lesssim Q^2 \lesssim 20$ GeV^2. A simple way of doing this is to make the S=0 scalar diquark state independent of Q^2 and dependent on x only, and to give an appropriate form factor to the S=1 vector diquark state. The actual forms of these functions in Q^2 can only be determined by fit to experiments.

IV. RESULTS AND DISCUSSION

To fit data in the range $2 \lesssim Q^2 \lesssim 20$ GeV^2, $0.5 \leq x \leq 0.8$ we shall assume the diquark structure functions to have the following phenomenological forms in which "v" stands for valence quark:

$$F_{2,v}^{(0)}(x, Q^2) = b x^3 (1-x)^3$$
$$F_{2,v}^{(1)}(x, Q^2) = \frac{2b}{1+\lambda \frac{Q^2}{M^2}} x^5 (1-x)^3 \qquad (4.1)$$

where the assumed Q^2 dependence of $F_{2,v}^{(1)}$ may be regarded as the former factor of the vector diquark state. As will be seen, the introduction of this form factor ensures the diquark structure function

$$\Delta_2(x,Q^2) = F_{2,V}^{(0)}(x,Q^2) + G_2^{(1)}(x,Q^2)\,F_{2,V}^{(1)}(x,Q^2)$$

to be essentially independent of Q^2 in the region being considered and is $\sim x^3(1-x)^3$ corresponding to $\alpha=2$, $\beta=3$.

We now try to fix the parameter b and λ . In the range $0.5 \lesssim x \lesssim 0.8$ the $q\bar{q}$ sea and gluons have essentially no contribution to electroproduction, so we have to consider valence quarks only. If as usually assumed, quarks are the only charged partons in the nucleon, then the normalization condition of the distribution functions of valence quarks can be expressed in terms of the structure function $F_{2,V}$

$$\int_0^1 \frac{1}{x} F_{2,V}^{(\frac{1}{2})}(x,Q^2)\,dx = 1 \qquad (4.2)$$

In our present quark-diquark model, it is reasonable to generalize this relation to the condition

$$\int_0^1 \frac{1}{x} F_{2,V}(x,Q^2)\,dx = 1\,. \qquad (4.3)$$

Then, with (4.1), one easily obtains

$$\int_0^1 \frac{1}{x} F_{2,V}^{(\frac{1}{2})}(x,Q^2)\,dx = 1 - \frac{b}{60} - \frac{2b}{1+\lambda\frac{Q^2}{M^2}}\left\{\frac{1}{280} + \frac{1+K^2}{360}\left(\frac{Q^2}{M^2}\right) + \frac{K^2}{48}\left(\frac{Q^2}{M^2}\right)^2\right\} \qquad (4.4)$$

As was supposed in I (introduction), only when diquark substructures in the nucleon have very small probabilities it is permissible to take the QCD results for $F_2^{(\frac{1}{2})}(x,Q^2)$ from Ref. 3 directly. Thus the parameter b must be kept small and we take b=0.3 such that R can fit experimental data and Δ_2 is as small as possible. The parameter λ is then to be fixed by fitting experimental data. If $K \simeq 1$, (in units of $e\hbar/2Mc$), we choose $\lambda \simeq 20$, for which the right-hand side of (4.4) is ~ 0.991 at $Q^2=5$ GeV^2 and ~ 0.984 for $Q^2=16$ GeV^2 so diquark states contribute only 1 — 1.6%. But even such a small proprotion of bosonic partons in the nucleon can have large effects on R, as will be seen below.

Using (2.13) and (2.14), we can finally calculate the effects of diquarks in deep inelastic e-P scattering. The results are presented in Table 2 and Figures 2 and 3. It is seen that better fit to experimental data is obtained for both F_2 and than Buras et al(3) and other authors(4,5).

Finally, a few other remarks are relevant.

(1) Fig. 3 shows clearly that, even for the small proportion of 0.9 —1.6%, diquark substructure can yield quite large values of R in agreement with experiment within errors. Furthermore, in contrast to the QCD calculations of Ref. 3, our prediction of R tends to increase slowly with increasing Q^2, which is also compatible with experiment.

(2) Our values of $F_2(x,Q^2)$ as shown in Fig. 2 are greater than those of Ref. 3 by 5% and often exceed the experimental error bars, especially at large x. This apparent defect, however, ought not to be over-emphasized. In fact, we have taken the results of ref. 3 to obtain $F_2^{(\frac{1}{2})}(x,Q^2)$ and then added to them the effects of diquarks, i.e. $\Delta_2(x,Q^2)$. If one notices that in inelastic lepton scattering the values of F_2 at $x > 0.5$ are very small and approach zero rapidly as x continues to increase, one need only make a minor readjustment of the para-

metrization at $x \geqslant 0.5$ in the calculation of Ref. 3 in order to get a good fit to experiment for F_2 as well, even after inclusion of the effects of diquarks.

(3) We have chosen a large value of λ, i.e. 20. This indicates that of vector diquarks. The latter, however, is not to be neglected. Scalar diquarks only ($\lambda \to \infty$) would yield R values decreasing with Q^2 in disagreement with experiment, although, at small Q^2 theory could agree with experiment. On the other hand, if only vector diquarks were present, R would rise too rapidly with Q^2 for a good fit to experiment; R would also be too small at small Q^2.

(4) As diquarks are present in only very small amounts within the nucleon, we have not considered QCD correction to their effects, since it should not be important.

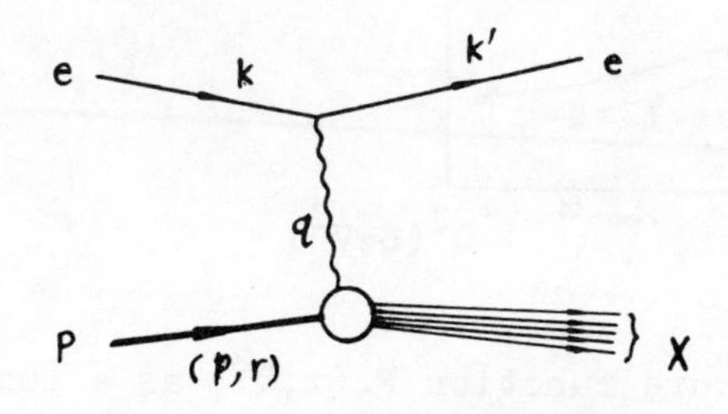

Fig. 1.

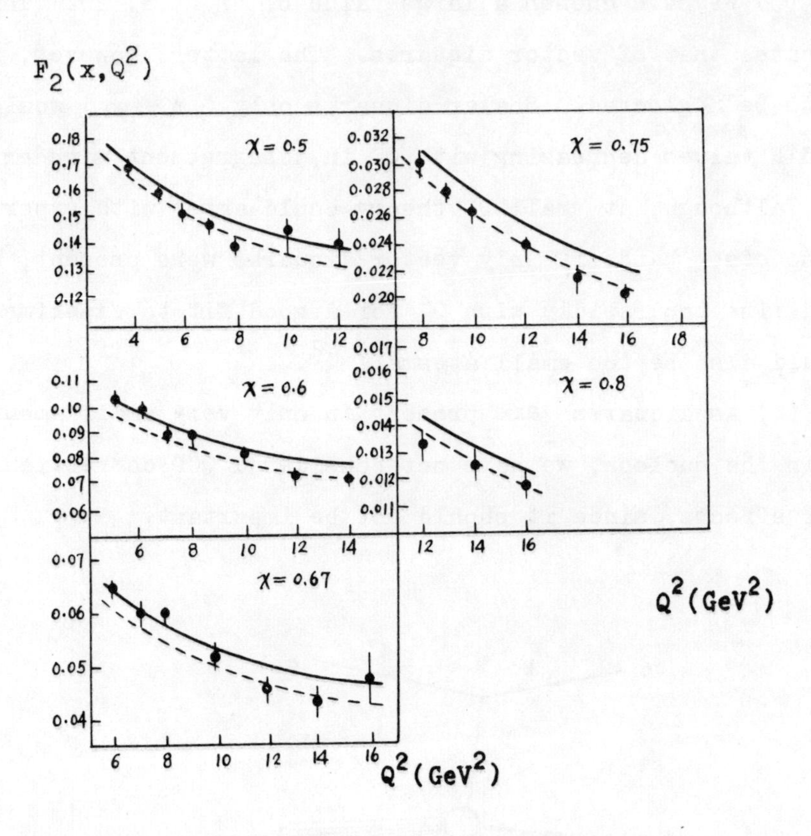

Fig. 2 Structure function $F_2(x,Q^2)$ as a function of Q^2 for five values of x. The dashed line is the fit by Buras and Geamers (3) taking into account the QCD corrections of order $\bar{g}^2$ with Λ =0.47 and target mass effect. The solid line is our fit including diquarks effects.

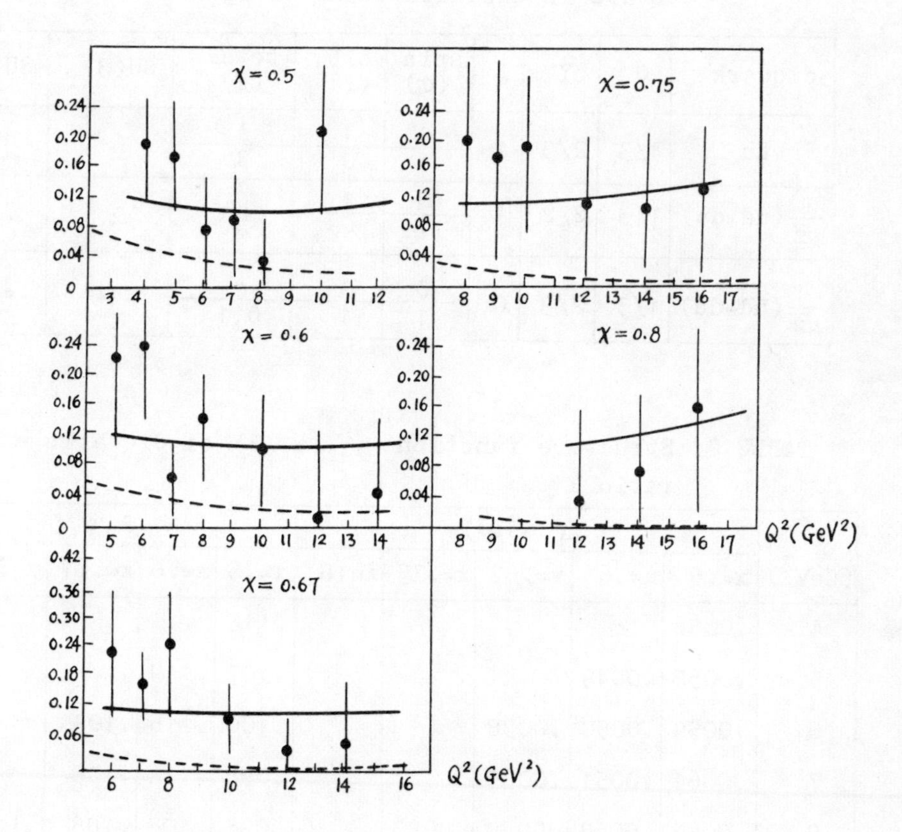

Fig. 3 $R(x,Q^2)$ as a function of Q^2 for five values of x. The dashed line is the fit of Ref. 3 taking into account QCD corrections up to $\bar{g}^2$ with Λ =0.47 and the target mass effect. The solid line is our fit including diquark effects.

TABLE I. Quantum numbers and group representations of diquark substructure in the proton. (The S=2 state is excluded owing to spin $\frac{1}{2}$ of nucleon.)

diquark	Q	Y	I	spin (σ)	orb. (ℓ)	total (S)	$SU(3)_f$	$SU(3)_c$
uu	4/3	2/3	1	0	1	1	$\underline{6}$	$\underline{3}$*
				1	0	1		
$\frac{1}{\sqrt{2}}$(ud+du)	1/3	2/2	1	0	1	1	$\underline{6}$	"
				1	0	1		
$\frac{1}{\sqrt{2}}$(ud-du)	1/3	2/3	0	0	0	0	$\underline{3}$*	"
				1	1	0,1		

TABLE 2. Structure function $\Delta_2(x,Q^2)$ of diquarks and ratio R .

Q^2	$\Delta(x,Q^2)$					$R(x,Q^2)$				
(GeV^2)	x=.5	x=.6	x=.67	x=.75	x=.8	x=.5	x=.6	x=.67	x=.75	x=.8
4	.0056					.115				
5	.0058	.0049				.107	.119			
6	.0059	.0050	.0038			.100	.116	.109		
7	.0061	.0051	.0039			.096	.112	.108		
8	.0063	.0052	.0040	.0023		.095	.108	.108	.112	
9	.0065	.0053	.0041	.0023		.094	.105	.108	.116	
10	.0067	.0054	.0042	.0024		.094	.103	.109	.121	
12	.0070	.0056	.0043	.0024	.0015	.093	.105	.110	.128	.118
14		.0058	.0044	.0025	.0015		.107	.111	.136	.133
16			.0045	.0025	.0016			.112	.144	.139

References

1. E. Riordon et al, SLAC-PUB-1634 (1975).
   W.B. Atwood et al, Phys. Lett. 64B 479 (1976).
2. H.L. Anderson et al, Phys. Rev. Lett. 37, 4 (1976); 38, 1450 (1977); 40, 1061 (1978).
3. A.J. Buras and K.J.F. Geamers, Nuci Phys. B132, 249 (1978).
   A.J. Buras et al, Nucl. Phys. B131, 308 (1977).
4. I. Hinchliffe and C.H. Llewellyn-Smith, Nucl. Phys. B128, 93 (1977).
5. G.C. Fox Nucl. Phys. B131, 197 (1977).
6. A.de Rujula, H. Georgi, and H.D. Politzer, Ann. Phys. 103, 315, (1977).
7. M. Zravek et al, Phys. Rev. D19, 820 (1979).
8. P.V. Landshoff et al, Ref-TH-2157-CERN (1976).
9. S.D. Drell and T.M.Yan, Phys. Rev. Lett. 24, 181 (1970).
10. R.McElhang et al, Phys. Rev. D8, 2267 (1973).
11. E.D. Bloom and F.J. Gilman, Phys. Rev. Lett. 25, 1140 (1970).

ON THE FORMATION OF CLUSTERS IN THE OVERLAPPING REGION OF JETS

Liu Han-chao

Physics Department, Nankai University, Tientsin

Yang Kuo-shen

Hebei College of Engineering, Tientsin

Abstract

We point out, for the first time, the formation of clusters in the rapidity-overlapping region of jets, and derive from the properties of jets and ordinary resonances, detailed structures of clusters including those which are considered in other models as basic assumptions, e.g. the average number of particles in a cluster is ~ 3, the average energy per particle ~ 0.4-0.5 GeV, $\langle K(K-1)\rangle \sim 1.1 \langle K\rangle$ and the charge of a cluster is $\sim \pm 1$ or 0, where K is the number of charged particles in a cluster. Through introducing this formation of clusters, the DTU model can give an unified interpretation of the existence of clusters of various kinds and different sizes and of almost all experimental results on clustering effects.

I. INTRODUCTION

In recent years, there are attempts to study the mechanism of soft hadronic collisions, starting from the theory of dual unitarization and topological expansion[1], where multi-particle production in soft hadronic collision is considered as simple superposition of jets. A simplified model based on this theory is called the DTU model. Applications of this model[2,3] are now still very rare.

We introduce the point of view that clusters are simply mixtures of ordinary resonances in the overlapping region of jets. The detailed structures of the clusters can then be derived, as far as you please. A unified interpretation is given for the existence of clusters of various kinds and of different sizes and for almost all experimental results on clustering effects in soft hadronic collisions.

Our point of view of the formation of clusters applies equally well to other processes (e.g. the deep inelastic scattering, the hard hadronic collisions, the annihilations of hadrons, etc.) which are superposition of several jets. The structure of the clusters produced should be the same in different kinds of processes, if their overlapping jets are the same.

II. EXPERIMENTAL RESULTS ON CLUSTERING EFFECTS IN SOFT HADRONIC COLLISIONS

The large amount of experimental results on clustering effects may be summarized[4,5] as follows:

1. In high-energy soft hadronic collisions, clusters are produced firstly (charge $\sim \pm 1, 0$); the number of final-state particles in the decay of a cluster may be different for different cluster but its average is ~ 3 and $\langle K(K-1)\rangle \sim 1.1 \langle K\rangle$, where K is the number of charged particles in a cluster; these quantities vary slowly with energy; the cluster decays isotropically, and the average energy of each final-state particle is $\sim$ 0.4-0.5 GeV.

2. For a given incident momentum P_{lab}, clusters in events with larger values of n are larger (i.e. the average number of particles in a cluster is larger), where n denotes the total number of final-state charged particles[4].

3. For low energies (e.g. $P_{lab} \lesssim 30$ GeV/c) or for high energy and low multiplicities, only ordinary mesons and baryons are produced[4,6,7].

4. Clusters begin to appear in soft pp collisions as P_{lab} increases to $\sim$ 50-60 GeV/c.

5. The rapidity half-width of the decay of clusters is similar in magnitude to that of resonances.

6. Clusters produced in $p\bar{p}$ annihilation are larger than those in soft pp collision.

7. Clusters with very large masses (e.g. several tens of GeV), are discovered in cosmic ray experiments.

III. THE FORMATION AND STRUCTURE OF CLUSTERS

According to the DTU model, the central region in the soft hadronic collision is a simple superposition in phase space of two independent hadronic jets. From experimental results and the models of jets discussed in recent years [6,7,8], hadronic jets consist mainly of mesons of broken SU_6 symmetry. The average number of particles in a meson is $\sim$ 2.0, (1.8), if the broken symmetry is described as in Ref.[7] (Ref.[8]), where the ratio of weights between the vector and the pseudo-vector mesons is 3:1 (1:1) and is refered to as I (II) in Table 1. The rapidity half-width of the decay of meson resonances is ~ 0.7-0.85.

We consider the following model for the formation of clusters:

In the overlapping region of two jets, when the rapidity interval Δ between a primary meson [8] of one jet and that of another jet satisfies $\Delta \ll 2\delta_1$ ($\sim$1.4-1.7), these two primary mesons form an ordinary cluster.

This model is reasonable, since all the final-state particles from the decay of the two primary mesons may be considered as coming from the center of the interval Δ since $\Delta \ll 2\delta_1$. Moreover, the decays of all charged (neutral) particles could be mutually independent and almost isotropic with respect to this center with a half-width similar to that of an ordinary cluster. Thus, these two primary mesons as a whole could possess all characteristic features of an ordinary cluster[12,13].

The nonexistence of such clusters in jets implies that the rapidity intervals between adjacent primary mesons in a jet are larger than Δ. This might follow from the dynamics of the production of hadronic jets, and it evidently exists in the Brodsky-Weiss model of jets[9,10].

Now let us determine the parameter Δ. The following approximations are used in the calculations:

1. The effects of baryons are neglected in the structure of jets (see Ref. [8] for explanation), the primary particles are the 36-plet mesons in the broken SU_6 symmetry[7,8], and the number of final-state particles contained in each primary meson decay is taken equal to the average number, where the ratio of the number of final-state charged mesons to the total number is put equal to 2/3.

2. The number of primary mesons in an interval Δy in the overlapping region of the two jets is taken to be the average number, and the probabilities for these primary mesons to appear at different points of Δy are equal.

3. For $\Delta y \gg \Delta$, we first calculate the probabilities for the formation of clusters of different sizes under the condition that the external effects on the primary mesons in Δy are negligible and then take into account the external effects on the clusters in the interval Δ at the left (right) end of Δy,

(the results calculated in Appendix I show that these external effects are, indeed, small).

4. For $\Delta y \gg \Delta$, the effect produced, when a primary meson is in any interval Δ, is taken to be the average of the effects when it is at the left and the right ends of the interval Δ .

Detailed calculations are given in Appendix I, taking as an example the pp collision at $\sqrt{s}$ = 60 GeV. The results are given in Table 1.

Table 1

$\sqrt{s}$	60GeV		20GeV		Experimental value
Type of SU_6 broken symmetry	(I)	(II)	(I)	(II)	
Average number of particles in a cluster	3.0_4	2.7_4	2.7	2.4	3
$\langle K(K-1)\rangle / \langle K\rangle$	1.4_2	1.2_{-1}	1.1	0.9	1.1
Probability for a cluster with charge +(-,0)	1/3	1/3	1/3	1/3	1/3
Average energy of a particle in a cluster	0.42 GeV		0.42 GeV		0.4-0.5 GeV

The constituents of the cluster in the central region of the soft hadronic collision may be calculated, as far as you please, from the average particle density $n_{1i}(n_{2i})$ of the ith kind of primary mesons in jet 1(2). Similarly, one may calculate the number distribution of primary mesons in a cluster and the constituents of the cluster in any rapidity interval (y, y+dy) of the central region.

From the above calculation, one sees that in a cluster containing 3(4) primary mesons, the rapidity intervals between the slowest and the fastest mesons may become as large as $2\Delta \sim 0.8$ ($3\Delta \sim 1.3$), which are not much smaller than the half-width ~ 1.4-1.7 of the decay of the primary meson. Hence, the effects of the decay of such clusters should be different from those of ordinary clusters. There are now two possibilities:

1. such differences are not yet discovered, since the percentages of such clusters are small and present works on clusters are too crude;
2. there exist attractive interactions between the primary mesons when their rapidity distances are less than Δ. The latter possibility is large, since all primary mesons are produced in a very short collision time of the order $\lesssim 10^{-23}$ sec. corresponding to the fragmentation time of a quark and therefore their spatial separations are $\lesssim 10^{-13}$ cm at the beginning. Since these separations are smaller than the effective range of meson forces, there must exist a minimum relative rapidity Δ in order to separate them. Such attractive interactions could reduce the rapidity intervals among the primary mesons of a large cluster and, thus, could make its difference from the ordinary clusters negligibly small. At the same time, one finds also the physical meaning of the parameter Δ.

Clusters consist of primary mesons of the overlapping jets. If the broken SU_6 symmetry is specified as in Ref.[7], each primary meson has an average mass $m \sim 0.70$, and produces, on the average, two final-state mesons. The longitudinal rapidity of the primary meson relative to the center of mass of the cluster is $y = \Delta/2$. The average transverse momentum of a primary meson in a jet is $p_T \sim 0.42$ GeV/c [8]. From $P_L = \sqrt{P_T^2 + m^2}\,\sinh y$ and $\mathcal{E} = \sqrt{P_L^2 + P_T^2 + m^2}$, one obtains the average energy $\mathcal{E}$ of each final-state particle relative to the center of mass of the cluster, $\mathcal{E} \sim 0.42$ GeV. (see Table 1).

IV. INTERPRETATION OF EXPERIMENTAL RESULTS ON CLUSTERING EFFECTS

1. Interpretation of experimental results already studied by other models.

The results in Table 1 are usually considered in other models as basic assumptions for an interpretation of a large amount of experimental, clustering phenomena.

Through introducing the formation of clusters in the framework of DTU model, we have derived in Sec. III these assumptions from more basic properties of jets and ordinary resonances and, moreover, derived the conventional independent emission cluster model in soft hadronic collisions from the independent emission cluster model of jets (see Ref. [10] for details). Thus a large amount of experimental clustering phenomena can be interpreted.

2. Interpretation of experimental results which could not be interpreted by other models.

Experimental results (2)-(7) of sec. II can be interpreted

in the framework of DTU model. Result (2) has been interpreted in Sec. III. The overlapping region of the two jets in DTU model becomes smaller as the incident momentum P_{lab} decreases (see Ref. [3]). When $P_{lab} \lesssim 50$ GeV (corresponding to $\sqrt{s} \lesssim 10$ GeV), the overlapping region becomes so small that the clusters no longer appear, and there are only ordinary mesons and baryons. The situation is similar for small values of n. These are results (3) and (4). From the relation between clusters and meson resonances described in Sec. III, one sees that their decay half-widths are similar in magnitude. This is result (5).

For a given P_{lab} , the production and the structure of clusters in events of a given number n of final-state charged particles can also be roughly calculated. Let Y_n denote the largest rapidity and assume that the final-state mesons in soft hadronic collisions are distributed uniformly along the rapidity axis. The average mass m of the final-state mesons is ~ 0.2 GeV. The average transverse momentum $P_T \sim 0.35$. From the relation given in Sec. 4 of Ref. [10], $\cosh^{-1} \dfrac{2\sqrt{s}}{3\mu\, n} = \frac{1}{2} Y_n$, one can calculate the average rapidity interval L between adjacent primary mesons of a jet. For example, for $P_{lab} \sim 2000$ GeV and $n \sim 30$ (24,12), one obtains $L \sim 0.4$(0.5, 1.3). From the last section, for $\Delta \sim 0.4$ and for large n (~ 30), larger clusters are easier to be produced. In cosmic-ray experiment, clusters with very large masses (e.g. several tens of GeV) may be produced, since P_{lab} is very large. This is result (7).

According to DTU theory, the $p\bar{p}$ annihilation is considered as a superposition of 3 mutually independent jets. Hence, the ratio of the average cluster size in $p\bar{p}$ annihilation to that in hadronic collision is 3:2. This explains result (6).

Lastly, we point out that, through introducing the formation of clusters, the DTU model can indeed interprete at least qualitatively, nearly all experimental results of soft hadronic collisions, including all basic features of these collisions as stated in Ref. [4] and those refined experimental results, directly related to the properties of quarks, as stated in Refs. [6,7] . The details will be discussed elsewhere.

Appendix 1

Derivation of the Results in Table 1

For $\sqrt{s} \sim 60$ GeV, assume $\Delta = 0.41$ and $\Delta y = 6\Delta$. From Fig. 2(b) of Ref.[3] one obtains the average rapidity interval between adjacent primary mesons ~ 0.61, since each primary meson contains two final-state particles. Thus the average rapidity interval between adjacent primary mesons in the fragmentation region of a jet is ~ 1.22. Hence there are 4 particles (i.e. 4 primary mesons) in the above interval Δy. Suppose that particles 1 and 4 (2 and 3) belong to jet 1 (jet 2). Then, the rapidity distance between 1 and 4 (2 and 3) must be $> \Delta$. Let (i.j), (i,j,k) and (1,2,3,4) (i,j,k,l=1,2,3,4) denote, respectively, the clusters containing two particles i,j, three particles i,j,k and four particles 1,2,3,4. Let $P_{(i,j)}$

$(\sum P_{(i,j)})$, $P_{(i,j,k)}$ $(\sum P_{(i,j,k)})$ $P_{(1,2,3,4)}(\sum P_{(1,2,3,4)})$ and $P_{(i,j)(k,l)}(\sum P_{(i,j)(k,l)})$ denote the probabilities (the total probabilities) of producing in the interval Δy only one two-particle cluster, one three-particle cluster, one four-particle cluster and two two-particle clusters respectively. Let P_0 denote the probability that rapidity intervals between adjacent particles in Δy are all larger than Δ, and let

$$A = \frac{1}{3} + \frac{2}{3} \cdot \frac{1}{2}(1+2), \qquad E = \frac{2}{4} \cdot \frac{1}{2}(1+2) + \frac{2}{4} \cdot 2$$

$$B = \frac{2}{3} \cdot \frac{1}{2}(1+2) + \frac{1}{3} \cdot 2 \qquad C = \frac{2}{4} \cdot \frac{1}{2}(3+2) + \frac{2}{4} \cdot 2$$

A. External effects on the formation of clusters in Δy are neglected firstly.

a) Particle 1 is produced at the center of Δy.

```
            1
  .   .   .   .   .   .   .
   (1) (2) (3) (4)  (5) (6)
```

$$P_0 = \frac{2}{6} \cdot \frac{(3+2)/2}{4} \cdot \frac{[B+(1+0)/2]/2}{4} + \frac{2}{6} \cdot \frac{(3+2)/2}{(5+4)/2} \cdot \frac{[(\frac{1}{2})+B]/2}{4} = 0.107$$

where the first factor 2/6 of the first term denotes the probability that particle 2 is produced in intervals ② and ⑤, the second factor denotes the average probability of producing particle 3 when particle 2 is in ② or ⑤, the denominator 4 of the second factor indicates that there are 4 possible intervals for the production of particle 3, since the rapidity distance between particles 3 and 2 must be $> \Delta$, the third factor denotes the average probability of producing particle 4 when particle 2 is in ② or ⑤ and particle 3 is in Δy, the factor 2/6 in the second term denotes the probability of producing 2 in ① and ⑥ , and the denominator (5+4)/2 of the second factor denotes the average possible rapidity region for the production of particle 3 when particle 2 is in ① or ⑥ , noting that ① and ⑥ are at the two ends and ② and ⑤ are in the interior of Δy.

$$\Sigma P_{(i,j)} = P_{(1,2)} + \{P_{(3,4)}\} + \{P_{(1,3)}\} + \{P_{(2,4)}\}$$

$$= \frac{2}{6}\cdot\frac{(4+3)/2}{4}\cdot\frac{[(3+2)/2+B]/2}{4} + \left\{\frac{2}{6}\cdot\frac{(3+2)/2}{4}\cdot\frac{[A+(1+2)/2]/2}{4}\right.$$

$$\left. + \frac{2}{6}\cdot\frac{(3+2)/2}{(5+4)/2}\cdot\frac{[A+(1+2)/2]/2}{4}\right\} + \left\{\frac{2}{6}\cdot\frac{(1+2)/2}{4}\cdot\frac{[(3+2)/2+E]/2}{4}\right.$$

$$\left. + \frac{2}{6}\cdot\frac{2}{(5+4)/2}\cdot\frac{[E+2(1/2)(3+2)/2]/2}{4}\right\} + \left\{\frac{2}{6}\cdot\frac{(3+2)/2}{4}\cdot\frac{[(1/3)(1/2)+(2/3)+2]/2}{4}\right.$$

$$\left. + \frac{2}{6}\cdot\frac{(3+2)/2}{(5+4)/2}\cdot\frac{[2+(2/3)+(1/3)(1/2)]/2}{4}\right\} = 0.5758$$

Similarly one obtains $\Sigma P_{(i,j,k)} = 0.0937$, $\Sigma P_{(1,2,3,4)} = 0.0179$, $\Sigma P_{(i,j)(k,l)} = 0.197$.

b) Particle 1 is produced at the left (right) end of Δy.

1
①②③④⑤⑥

$$P_0 = \frac{1}{6}\cdot\frac{(4+3)/2}{4}\cdot\frac{(A+C)/2}{5} + \frac{2}{6}\cdot\frac{3}{4}\cdot\frac{(A+B)/2}{5} + \frac{1}{6}\cdot\frac{3}{4}\cdot\frac{(B+B)/2}{5} + \frac{1}{6}\cdot\frac{(4+3)/2}{(5+4)/2}\cdot$$

$$\cdot\frac{(A+C)/2}{5} = 0.2154$$

Similarly one obtains $\Sigma P_{(i,j)}=0.5974$, $\Sigma P_{(i,j,k)}=0.0700$,

$\Sigma P_{(1,2,3,4)}=0.00394$, $\Sigma P_{(i,j)(k,l)}=0.1234$.

One takes the average values in the above two cases as the required total probabilities, i.e.

$P_o=0.1612$, $\Sigma P_{(i,j)}=0.5866$, $\Sigma P_{(i,j,k)}=0.0819$, $\Sigma P_{(1,2,3,4)}=0.0109$, $\Sigma P_{(i,j)(k,l)}=0.1602$. Their sum $=1.001$

From these, one obtains that the percentages of clusters containing 1,2,3 and 4 primary mesons are respectively 0.59_3, 0.36_{-4}, 0.041, 0.011 (e.g. the percentage of clusters containing 1 primary meson is $=\frac{.1612}{1.001}+\frac{.5866}{1.001}\left(\frac{2}{3}\right)+\frac{.0819}{1.001}\left(\frac{1}{2}\right)=.593$).

B. External effects on the formation of clusters in Δy are taken into account.

Note that Δy is in general in the interior of the overlapping region, external particles can only influence one interval of width Δ at the left (right) end of Δy, and the rapidity distance of two primary mesons in the same jet must be $>\Delta$. Let N denote

$$\longleftarrow \Delta y \longrightarrow$$

⑦ ① ② ③ ④ ⑤ ⑥ ⑧

the average number of clusters in any given interval of Δy, and N_i denote the average number of the ith clusters (each containing i primary mesons). Then

$N = .161\ (4/6) + .587\ (3/6) + .160\ (2/6) + .082\ (2/6) + .011\ (1/6) = .483,$

$N_1 = .593N, \quad N_2 = .356N, \quad N_3 = .041N, \quad N_4 = .011N.$

Let $(i|j)$ denote the probability that the ith (jth) cluster is produced in ⑦ (①) and these two resonances form a cluster containing i+j primary mesons. Let a_i denote the average probability of producing an ith cluster in interval ⑦ if there is a cluster in ① and the rapidity distance between these two clusters is $\leqslant \Delta$. Then one has $a_i = (\frac{1}{2}N_i \cdot \frac{1}{2} \cdot 2) \div 2,$

$(0|1) = (1 - \sum_{i=1}^{4} a_i) \times .593, \quad (0|2) = (1 - \sum_{i=1}^{4} a_i) \times .356,$

$(i|1) = .593 a_i\ , \quad (i|2) = .356 a_i\ , \ \ldots.$

In intervals ① and ⑥ , the percentages of clusters containing 1,2,3,4 primary mesons are .525, .357, .090, .026 respectively. In intervals ②, ③, ④ , ⑤ , the corresponding percentages are obtained above in (A). Taking the average over different intervals with correct weights, one obtains the calculated results in Table 1.

References

1. Chan, Hong Mo, et al., Nucl. Phys. B86 (1975), 479; B92 (1975), 13.

   Lee, Huan, Phys, Rev. Letters, 30 (1973), 719.

   Veneziano, G., Nucl. Phys. B74 (1974), 365; B117 (1976), 519.

   Chew, G.F. et al., Nucl. Phys., B104 (1976), 290; Phys. Reports, 41C (1978), 263.

2. Dias De Deus J. et al., Phys. Letters, 70B (1977), 73; Acta Physica Polonica B9 (1978), 249.

3. Capella, A., Sukhatme, U., Tan, Chung-I and Van J.T.T., Phys. Letters, 81B (1979), 68.

4. Dremin, I.M. and Quigg, C., Science 199 (1978), 937.

5. See, e.g. Whitmore, J., Phys. Reports, 27 C(1976), 189.

6. Kirk, H. et al., Nucl. Phys B128 (1977), 397.

   Grasslir, H. et al., Nucl. Phys. B132 (1978), 1.

7. Anisovich, V. et al., Nucl. Phys. B55 (1973), 455.

   Bjorken, J.D. et al, Phys. Rev. D9 (1974), 1449.

8. Field, R.D. and Feynman, R.P. Nucl. Phys., B136 (1978), 1.

9. Brodsky, S.J. et al., Phys. Rev., D16 (1977), 2325.

10. Liu, H.C. and Yang, K.S., "On the KNO-type scaling, inclusive fine structures in jets, local compensation of charges and mechanisms of jets". — a paper of this conference.

11. Anderson, B., et al., Phys. Letters, 69B (1977), 221; 71B (1977), 337.

    Das, K.P. and Hwa, R.C., Phys. Lett. 68B (1977), 459.

12. Bintas, A. et al., Phys. Letters 45B (1973), 337.

    Schmidt-Parzefall, W., Phys. Letters 46B (1973), 399.

    Berger, E.L., Phys. Letters 49B (1974), 369.

13. Chiu, C.B. and Wang, K.H., Phys. Rev. D8 (1973), 2929.

    Chao, A.W. et al., Phys. Rev. D9 (1974), 2016.

Relativistic Wave Equation and Heavy Quarkonium Spectra

H. Suura and W. J. Wilson

School of Physics and Astronomy

University of Minnesota

Minneapolis, Minnesota 55455

and

Bing-lin Young

Ames Laboratory-USDOE and Department of Physics

Iowa State University

Ames, Iowa 50011

Abstract

We solve an equal-time relativistic wave equation with a confinement potential for heavy quark-antiquark systems. A new pair of coupled equations, which are appropriate in the presence of a linear confinement potential, are introduced for the $J^{pc} = 1^{--}$ trajectory. With a set of standard potential parameters used in the non-relativistic analysis, we obtain a satisfactory fit to the charmonium and upsilon spectra, to the extent of neglecting magnetic interactions. In contrast to the non-relativistic treatment, we obtain the absolute mass values of the quark-antiquark systems. Furthermore, the quark masses are found to be much smaller than the corresponding non-relativistic values, which we show, cannot be identified with the physical quark mass.

I. Equal-time relativistic wave equation with a confinement potential.

In an earlier paper[1] (referred as A), one of us (H.S.) proposed that an equal-time relativistic equation of the Breit type with a linear confinement potential be used as a substitute for the Bethe-Salpeter equation to determine the spectra on quark-antiquark (q-$\bar{q}$) systems. It was pointed out there that the conventional Bethe-Salpeter amplitude is not locally gauge invariant in quantum chromodynamics, and therefore cannot be used to derive a wave equation describing hadrons. Instead, one can derive[2,3] a wave equation for a gauge invariant equal-time q-$\bar{q}$ amplitude, avoiding the use of undefinable quark and gluon propagators. This amplitude is defined by

$$\chi(1,2) = (\Psi_o, \mathrm{tr}^c \{\exp[ig \int_1^2 \vec{A}(\vec{r})\cdot d\vec{x}]\, q(1) q^+(2)\}\, \Psi) \tag{1}$$

(see Ref. 2 for notation) which satisfies in the center of mass frame the following equation:

$$[(-i\vec{\alpha}\cdot\vec{\nabla} + \beta m), \chi(\vec{r})] = [M - V(\vec{r})]\, \chi(\vec{r}) \quad , \tag{2}$$

with $\vec{r}$ being the relative distance between q and $\bar{q}$. The wave function $\chi(\vec{r})$ is a 4 x 4 matrix with elements $\chi_{ij}(\vec{r})$ where the indices i and j represent the spinor components of the quark and antiquark respectively. M is the mass eigenvalue and $V(\vec{r})$ is the q-$\bar{q}$ potential to be specified later. $\vec{\alpha}$ and β are Dirac matices, and the left-hand side of (2) represents the kinetic energy of q and $\bar{q}$. The quark mass, m, should be regarded as renormalized.

As discussed in A, Eq. (2) is singular at r = R, where M - V(R) = 0, and the equation is infested with the disease called the Klein paradox for r > R. However, it was shown that the requirement of local normalizability of the wave function imposes a unique boundary condition at r = R, which leads naturally to a bag solution confined within r = R.

The solutions of Eq. (2) for light q - $\bar{q}$ systems were systematically worked out by Geffen and Suura[4] (referred to as B) using a potential

$$V(r) = k^2 r - \frac{\kappa r}{r^2+a^2} \quad . \tag{3}$$

The pseudo-coulomb part was motivated by theoretical considerations concerning the nature of the potential near r = 0; a possible new interpretation will be given in Sec. II. A remarkably good fit of the low lying meson spectrum was obtained with $K^2 = 0.09\ GeV^2$ and $\kappa = 2.28$.[4]

In the present paper we apply Eq. (2) to heavy q - $\bar{q}$ systems. In the charmonium region we are not yet in a completely non-relativistic regime so that a relativistic treatment is certainly desirable. In addition we can determine the absolute masses of mesons in this approach. The general features of the solutions of Eq. (2) are presented in Sec. II, using just the linear confinement part of the potential (3) for the quark mass ranging between 0 and 1.2 GeV. We obtain the correct overall structure of the spectrum over this entire range of quark masses. This gives us confidence that the wave equation, (2), and the boundary condition in the presence of confinement potential are basically correct. In Sec. III, we derive for the $J^{PC} = 1^{--}$ trajectory a new set of eigenvalue equations which are appropriate for confinement. A suitable linear combination of them are shown to be almost decoupled in the limit of large quark masses. In Sec. IV, we present the results of the numerical analysis of the charmonium and the upsilon spectra with a potential of the form (3). Concluding remarks will be given in Sec. V.

II. Spectrum in a linear confinement potential.

In order to demonstrate the correctness of Eq. (2) and the boundary condition, we apply them to the $q - \bar{q}$ system bound by a linear potential, neglecting all the short range interactions,

$$V(r) = k^2 r. \tag{4}$$

The derivation of the eigenvalue equations used to obtain these spectra are deferred to the next section. The spectra obtained numerically for $k^2 = 0.2\ GeV^2$ and plotted against the quark mass are given in Fig. 1. The plot shows some remarkable features. First, in the charmonium region ($m > 1$ GeV) the gross structures of the charmonium spectrum are correctly reproduced, including the right sequence of the levels, and even the right magnitude of separation between the ground states ($^3S_1$ and $^1S_0$), P-states, D-states and the radially excited

$^3S_1$, which crosses over with $^3D_2$ at around m = 0.9 GeV. Moreover, even the absolute masses can be fitted with m = 1.18. For m > 0.6 GeV, the states with the same orbital angular momentum L but from different trajectories are clustered to form L-multiplets. Except for the radially excited S-states, different L-multiplets are almost parallel. The radially excited S-states have a smaller slope than that of the ground states, so that in the upsilon region the separation between the ground and excited S-states will be much smaller than the actual value. As we will see in Sec. IV, the introduction of the Coulomb term into the potential will remedy the situation. However it is obvious that the Coulomb potential plays a rather minor role in determining the gross structure of the spectrum.

In the light quark region (m < 0.4 GeV), where the relativistic effects are large we see large splittings among states with the same orbital angular momentum L, which is no longer a good quantum number. Especially note that the $^3S_1$ behaves more like a P-state as $m \to 0$. It was suggested in A that this is the reason for the large separation between π and ρ masses. Again the correct sequence of levels is obtained. Taking the I = 1 states and choosing m = 0.2 $\sim$ 0.3 GeV, we find in order of increasing meson masses[5] $\pi(^1S_0)$, $\rho(^3S_1)$, $\delta(^3P_0)$, $A_1(^3P_1)$, $B(^1P_1)$, $\rho'(1250)$ $(^3D_1)$, $A_2(^3P_2)$, $F_1(^3D_2)$, $\rho'(1600)$ $(^3S_1)$, $A_3(^1D_2)$, and $g(^3D_3)$. The existence and classification of the states $\rho'(1250)$[6] and $F_1(1540)$ are not clear, and no obvious candidate for the radially excited $^1S_0$ has been found.

The absolute masses of the light quarkonia bound by just the linear potential are more than 1 GeV higher than the actual values as shown in Fig. 1. As shown in B, these masses can be brought down by a strong Coulomb potential. It was also pointed out there that the wave equation is chirally symmetric in the limit $m \to 0$ but that the boundary condition is not.[7] This leads to a large separation of a pair of chiral partners π and δ mesons [see Fig. 1]. We believe that this is another indication that Eq. (2) and the boundary condition are basically correct.

Some qualifications must be made regarding interactions at short distances. It is well known[8] that the Breit equation is not exact at short distance even if we consider only the electrostatic Coulomb potential. One must use the Salpeter equation[9] which takes into account virtual pair creation. In the latter equation the Coulomb singularity at the origin is suppressed,[10] and the suppression is stronger than that due to asymptotic freedom.[11] Equivalent, we may modify the Breit equation by including a correction term, which tend to cancel the Coulomb Singularity at the origin. We must do exactly the same with Eq. (2), especially if we want to obtain sensible values for the wave function at the origin. In the present paper, however, we will merely adopt the pseudo-coulomb potential (3) as a phenomenological expression for the suppression. The result of the calculation shows that the spectrum is extremely insensitive to the details of the suppression. Namely, we can change the value of the parameter a in (3) over a wide range with no appreciable change in the spectrum. Hence, in effect we have only two potential parameters, κ and k^2. Finally, one has to keep in mind that Eq. (2) does not include magnetic interactions, which must be separately evaluated perturbatively.

III. Reduction of the wave equation to second-order differential equations.

Derivation of second-order differential equations as eigenvalue equations for certain large components of the wave function χ from Eq. (2) has been done in B. The procedure is straightforward for $J^{pc} = 0^{-+}$ and 1^{++} trajectories, but is somewhat involved for the 1^{--} trajectory because of the relativistic tensor coupling between L = J - 1 and L = J + 1 waves. The eigenvalue equations in this case are two coupled second order differential equations. The set of equations derived in B are not for pure L - J $\pm$ 1 waves. Here we will derive a new set of equations for the 1^{--} trajectory, which are simpler and more symmetric than those given in B, although the latter are still valid. By rotating among the new set, we also derive equations for another pair, one representing an "almost" pure L - J - 1 wave, the other an L = J + 1 wave. The coupling between this latter pair of amplitudes becomes very small in the large quark mass limit.

As in A and B (see B for notation), we decompose χ into four 2 x 2 spin-orbital amplitudes χ_i:

$$\chi = \rho_1\chi_1 + \rho_2\chi_2 + \rho_3\chi_3 + \chi_4 . \qquad (5)$$

For the 1^{--} trajectory (parity and charge conjugation parity ε_p, ε_c = $(-1)^J$, $(-1)^J$), the χ_i must have the following structure:

$$\chi_{1,2} = \sigma\cdot\vec{r}\ Y_{JM}\ C_{1,2}(r) + \frac{i\ D_{1,2}(r)}{\sqrt{J(J+1)}}\ r\ \vec{\sigma}\cdot\vec{\nabla}\ Y_{JM} \quad ,$$

$$\chi_3 = Y_{JM}\ F_3(r) \quad ,$$

$$\chi_4 = \vec{\sigma}\cdot\vec{L}\ Y_{JM}\ F_4(r) \quad . \tag{6}$$

Introducing (6) into (2), we obtain four linear equations for the four χ_i's. Normalizability of the χ_i's requires the following boundary condition at $r = R$:

$$C_1,\ D_2 \sim (r - R) \quad . \tag{7}$$

It develops that there are no conditions on C_2 and D_1 except that they be finite at $r = R$. Thus, we need to derive differential equations for C_1 and D_2 or for a pair related to them. After some algebra, we obtain

$$C_1'' + \left(\frac{4}{r} + \frac{N'}{N} - \frac{S'}{S}\right) C_1'$$

$$+ \left[\frac{N^2}{4} - m^2 - \frac{J(J+1)-2}{r^2} + \frac{2}{r}\frac{N'}{N} + \left(\frac{N'}{N}\right)' - \frac{2}{r}\frac{S'}{S} - \frac{S'}{S}\frac{N'}{N}\right] C_1$$

$$= -\frac{m\sqrt{J(J+1)}\ N}{S}\ D_2 \quad , \tag{8a}$$

$$D_2'' + \left(\frac{4}{N} + \frac{N'}{N} - \frac{S'}{S}\right) D_2'$$

$$+ \left[\frac{N^2}{4} - m^2 - \frac{J(J+1)-2}{r^2} + \frac{3}{r}\frac{N'}{N} + \left(\frac{N'}{N}\right)' - \frac{1}{r}\frac{S'}{S} - \frac{S'}{S}\frac{N'}{N}\right] D_2$$

$$= - \frac{m\sqrt{J(J+1)}\, N}{S} C_1 \quad , \tag{8b}$$

where prime means differentiation with respect to r,

$$N \equiv M - V(r) \quad , \tag{9}$$

and

$$S \equiv m^2 r^2 + J(J+1) \quad . \tag{10}$$

The coupling between C_1 and D_2 vanishes for $m = 0$. In this limit the C_1 equation is identical to the eigenvalue equation for the 1^{++} trajectory and the D_2 equation to that for 0^{-+}. The indicial relation at $r = R$ shows that there are two solutions for C_1 which behave like $r - R$ and $(r - R)^{-1}$ respectively. D_2 behaves similarly. Hence (7) is consistent with (8a) and (8b).

For numerical solutions to be discussed in Sec. IV., we transform the C_1 and D_2 equations into the one-dimensional Schrödinger form, using

$$C_1 = \frac{1}{r^2}\sqrt{\frac{S}{N}}\,\tilde{C}_1$$

$$D_2 = -\frac{1}{r^2}\sqrt{\frac{S}{N}}\,\tilde{D}_2 \quad . \tag{11}$$

We have

$$\tilde{C}_1'' + U\,\tilde{C}_1 = \sqrt{J(J+1)}\,U_c\,\tilde{D}_2 \tag{12a}$$

and

$$\tilde{D}_2'' + (U + \Delta U)\,\tilde{D}_2 = \sqrt{J(J+1)}\,U_c\,\tilde{C}_1 \quad , \tag{12b}$$

where

$$U = \frac{N^2}{4} - m^2 - \frac{J(J+1)}{r^2} + \frac{1}{2}\left(\frac{N''}{N} + \frac{S''}{S}\right) - \frac{3}{4}\left[\left(\frac{N'}{N}\right)^2 + \left(\frac{S'}{S}\right)^2\right]$$

$$- \frac{1}{2}\frac{S'}{S}\frac{N'}{N} \quad ,$$

$$\Delta U = \frac{1}{r}\left(\frac{N'}{N} + \frac{S'}{S}\right) \quad ,$$

and

$$U_c = \frac{mN}{S} \quad . \tag{13}$$

For a large quark mass m, the coupling between C_1 and D_2 remain finite and the coupled equations must be solved. However, we can almost decouple the equations in this limit by introducing the following two new amplitudes by rotating C_1 and D_2

$$\alpha = r^2 N \left[\sqrt{\frac{J}{2J+1}}\, C_1 + \sqrt{\frac{J+1}{2J+1}}\, D_2\right] / \sqrt{SN} ,$$

$$\beta = r^2 N \left[\sqrt{\frac{J+1}{2J+1}}\, C_1 - \sqrt{\frac{J}{2J+1}}\, D_2\right] / \sqrt{SN} . \tag{14}$$

The differential equations are, in the one dimensional Schrödinger form

$$\tilde{\alpha}'' + \left[U + \frac{J+1}{2J+1}\Delta U + \frac{2J(J+1)}{2J+1} U_c\right]\tilde{\alpha} = \frac{\sqrt{J(J+1)}}{2J+1}\left[\Delta U - U_c\right]\tilde{\beta} . \tag{15a}$$

$$\tilde{\beta}'' + \left[U + \frac{J}{2J+1}\Delta U - \frac{2J(J+1)}{2J+1} U_c\right]\tilde{\beta} = \frac{\sqrt{J(J+1)}}{2J+1}\left[\Delta U - U_c\right]\tilde{\alpha} , \tag{15b}$$

where U, ΔU, and U_c are defined in Eq. (13). In the limit $m \to \infty$ the coupling terms vanish if the ratio $M/2m$ approached one. In the charmonium region the ratio is still far from this limit, as shown in section IV. Nevertheless, the numerical calculation shows that the coupling is extremely small and the right hand sides of Eq. (15a, b) can be neglected for all practical purposes.

In the large m limit, me may put, for $J \neq 0$,

$$S \to m^2 r^2 \quad \text{and} \quad mN/S \to 2/r^2$$

and Eqs. (15a,b) reduce to

$$\tilde{\alpha}'' + [\frac{N^2}{4} - m^2 - \frac{J(J-1)}{r^2} - \frac{J}{2J+1}\frac{N'}{rN} + \frac{1}{2}\frac{N''}{N} - \frac{3}{4}(\frac{N'}{N})^2]\tilde{\alpha} = 0 \quad . \qquad (16a)$$

$$\tilde{\beta}'' + [\frac{N^2}{4} - m^2 - \frac{(J+1)(J+2)}{r^2} - \frac{J+1}{2J+1}\frac{N'}{rN} + \frac{1}{2}\frac{N''}{N} - \frac{3}{4}(\frac{N'}{N})^2]\tilde{\beta} = 0 \quad . \qquad (16b)$$

Thus we see that in this limit the $\tilde{\alpha}$-branch of the 1^{--} trajectory has the orbital angular momentum $L = J - 1$ and the $\tilde{\beta}$-branch $L = J + 1$. Hence $J/\psi(^3S_1)$ satisfies the eigenvalue equation Eq. (16a) and $\psi''(^3D_1)$ satisfies Eq. (16b).

IV. Numerical results for charmonium and upsilon spectra

The coupled differential equations $\tilde{C}_1$ and $\tilde{D}_2$, Eqs. (12a) and (12b), and the "decoupled" $\tilde{\alpha}$ and $\tilde{\beta}$ equations, Eqs. (15a) and (15b) omitting the coupling term, are solved numerically by means of difference equations.[12] For given parameters m, k^2, κ, and a, the uncoupled equations like Eqs. (16a, b) can be solved starting at $r = 0$ where the correct boundary condition is imposed as input, then changing the mass eigenvalue until the boundary condition like Eq. (7) is satisfied. In this way we obtain the unique mass eigenvalue or equivalently the unique bag radius R. In the case of the coupled equations like Eqs. (12a, b), we change the mass eigenvalue as well as the relative magnitude of the two amplitudes, independently, in order to satisfy the boundary conditions at R for both amplitudes simultaneously.

We found that the "decoupled" $\tilde{\alpha}$ and $\tilde{\beta}$ equations give solutions almost indistinguishable from the coupled $\tilde{C}_1$ and $\tilde{D}_2$, indicating that the coupling terms in the $\tilde{\alpha}$ and $\tilde{\beta}$ equation make very little contribution. We summarize the various features of our solutions:

(a) The $J = 1$, $\tilde{C}_1$ and $\tilde{D}_2$ equations possess two ground states whose wave functions do not have nodes in $0 < r < R$. The one with lower mass eigenvalue is identified with the non-relativistic $^3S_1$, the one with higher mass eigenvalue the $^3D_1$, i.e., J/ψ and ψ'' respectively. The wave functions of their first excited states will have one node in $0 < r < R$.

(b) For a given quark mass, the bulk of the meson spectrum is determined by the linear confinement potential.

(c) The Coulomb part of the potential, for $\kappa = 0.5$, contributes less than 10% to the total meson mass. It also contributes to the splitting of the $^3S_1$ and $^1S_0$, and also $^3S_1$ and its radial excitation. These splittings increase with increasing κ. The effect, however is rather small for the $^3S_1$ and $^1S_0$ splittings. But it is very important to obtain the correct spacing between the ground state $^3S_1$ and its radial excitation. We are able to obtain the correct spacing between J/ψ and ψ' and T and T', which are of the order of 550 MeV with $\kappa = 0.5$. If we use $\kappa = 0.2$ instead, we obtain $M_{T'} - m_T = 410$ MeV.

(d) The parameter a in the pseudo-coulomb part of the potential has very little effect on the mass eigenvalues as already mentioned. For example changing a = 0.1 to 0.001 results in a negligible change of the meson mass.

(e) The potential parameters used in our fit are $k^2 = 0.2$, $\kappa = 0.5$, and $a = 0.1$, which are consistent with those in the non-relativistic Schrödinger calculation.[13] Our calculated values for the J = 1 series of the charmonium and the ϒ-family are listed in Table 1. The experimental mass values and quark masses are also given there. Note that the quark masses in our fit are considerably smaller than their non-relativistic counterparts.[13] The discrepancy can be explained easily. A numerical calculation shows that using the non-relativistic Schrödinger equation we overestimate the separation between ψ/J and ψ' by more than 50 MeV, due to primarily a larger binding energy for ψ' in the non-relativistic calculation than in the relativistic one. The difference is small compared to the absolute masses of J/ψ and ψ', but fairly large compared to the separation itself. In most non-relativistic approaches, the quark mass is partially determined to fit the spacing between J/ψ and ψ'. Since the spacing changes very slowly with the changing quark mass, we must change (increase) the quark mass considerably to compensate for the difference of 50 MeV. Thus, the quark mass determined non-relativistically in this way can hardly be identified with the actual physical quark mass. The situation is reflected in a very large additive constant V_o, required to fit the absolute masses of mesons in the non-relativistic approach. There is no theoretical ground for such an additive constant. If we are in a truely non-relativistic region, we should have a vanishing V_o. Let us mention again that our predictions are the absolute mass values, not only the level spacings as in the non-relativistic approach.

(f) We can accomodate a small change in the value of k^2. For instance with $k^2 = 0.182$, which is the value used by the Cornell group in their most recent work,[13] we can maintain the fit for the charmonium state by adjusting the charm quark mass to be 1.328 GeV.

(g) We have not concerned ourselves with the splittings of the P-states. To discuss these levels, we need to consider magnetic interactions. Relativistic effects are not completely negligible as we see from the small splitting between $^3S_1$ and $^1S_0$ states.

V. Conclusion

Using a newly derived set of coupled eigenvalue equations for the 1^{--} trajectory, appropriate for dealing with confinement potentials, we are able to investigate the low lying mesons as well as heavy quarkonia in the scheme proposed in A and B. The basic correctness of the scheme, i.e., Eq. (2) plus the boundary condition, Eq. (7), are demonstrated in the resulting spectra (Sections II and IV) obtained with a potential commonly used in this type of game, modified slightly to suppress the Coulomb singularity at the origin. The spectra are essentially determined by three parameters; potential parameters k^2 and κ and the quark mass, m, similarly to the non-relativistic Schrödinger approach. However, unlike the latter, due to relativistic invariance, there does not exist in the present scheme a free parameter which allows one to define the ground state mass, we must fit the absolute masses of hadrons. [We are assuming that the non-confining part of the potential vanish at infinity.] In this sense, we have one less parameter. By considering the spectra produced by just the linear potential, Eq. (4), first and the full potential, Eq. (3) later, a clear picture of the roles played by the two parts of the potential energies; The long distance confinement part produces the bulk of mass, M, and essentially determines the spectrum. The short distance Coulomb-like part increases the level spacing so as to reproduce the data.

There are other important differences between the present scheme and the non-relativistic one: Our quark mass, $m_c = 1.32$ GeV and $m_b = 4.75$ GeV, are considerably smaller then those appearing in the latter (Eq. $m_c = 1.84$ GeV and $m_b = 5.17$ GeV as given in Ref. 13). We pointed out in Sec. IV that the discrepancy is due to the unjustified non-relativistic procedure of fitting the separation between the ground state and radially excited $^3S_1$ state. In our scheme, the threshold energy to produce a pair of particles with the bare flavor quantum number of the quark is not necessarily 2m. To find the value of this energy, one needs to calculate the mass of the bare flavor particle.

Two interesting topics are omitted in the present article, i.e., the decay properties of J/ψ, etc., and the fine structure of the spectrum. We believe that a considerable refinement of the conventional schemes is necessary to successfully account for these properties and are leaving it for future work.

Acknowledgement

We would like to thank H. C. Crawley, Kai Ming Ho and Sam S. Liu for helpful discussions of numerical solutions of differential equations. For one of us (BLY), this work was supported by the U.S. Department of Energy, contract No. EY-76-C-02-1764 and contract No. W-7405-Eng-82, Office of Basic Sciences (HK-02-01-01).

References

1. H. Suura, Phys. Rev. Lett. 38, 636 (1977).
2. H. Suura, Phys. Rev. D17, 469 (1978).
3. H. Suura, Phys. Rev. D20, 1412 (1979).
4. D. A. Geffen and H. Suura, Phys. Rev. D16, 3305 (1977).
5. Particle data group, Phys. Letts. 75B, No. 1 (1978).
6. If ρ(1250) is a pure D-wave, it will not decay into lepton pairs. But it may contain a small admixture of S-wave. Then, its decay rate to e^+e^- will be small. This offers an explanation of the difficulty in observing this state in e^+e^- colliding beam experiment.
7. According to conventional wisdom, the so-called constituents mass of the quark remains finite in the chiral limit because of the spontaneously broken chiral symmetry in this limit. Therefore, our quark mass m should not be identified with the constituent quark mass. The equal-time formulation employed here avoids the use of the quark propagator and hence the constituent mass. We believe that our assumption, that the spring constant κ remains finite in the limit $m \to 0$, is the manifestation of the spontaneously broken chiral symmetry.
8. H. W. Bethe and E. E. Salpeter, Quantum Mechanics of one- and two-electron atoms, Springer-Verlag (1957), p. 170.

9. E. E. Salpeter, Phys. Rev. 87, 328 (1952).

10. E. Haqq and H. Suura, private communication.

11. T. Appelquist and H. D. Politzer, Phys. Rev. Lett. 34, 43 (1975).

12. See, for example, William E. Milne, "Numerical Solution of Differential Equations," p. 88; John Willey and Sons, Inc. (1953).

13. E. Eichten, K. Gattfried, T. Kinoshita, K. D. Lane, and T. M. Yan "Charmonium: Comparison with Experiment," CLNS-425, June 1979; C. Quigg and J. L. Rosner, Phys. Lett. 71B, 153 (1977); D. Pignon and C. A. Picketty, Nucl. Phys. B137, 340 (1978).

14. V. Luth, talk given at the 1979 International Symposium on Lepton and Photon Interactions at High Energies, unpublished; E. Bloom, *ibid*; H. Meyer, *ibid*.

Table 1. J = 1 states of the Charmonium and Υ-family.

	Predicted	Data	m_q
J/ψ	3.096	3.097 ± 0.002	1.32
ψ'	3.679	3.686 ± 0.003	1.32
ψ"	3.749	3.764 ± 0.004[a]	1.32
η_c	3.091	2.976 (?)[b]	1.32
Υ	9.48	9.46 ± 0.01[c]	4.75
Υ'	10.01	10.01 ± 0.05[c]	4.75

a. V. Luth, Ref. 14

b. E. Bloom, Ref. 14

c. H. Meyer, Ref. 14

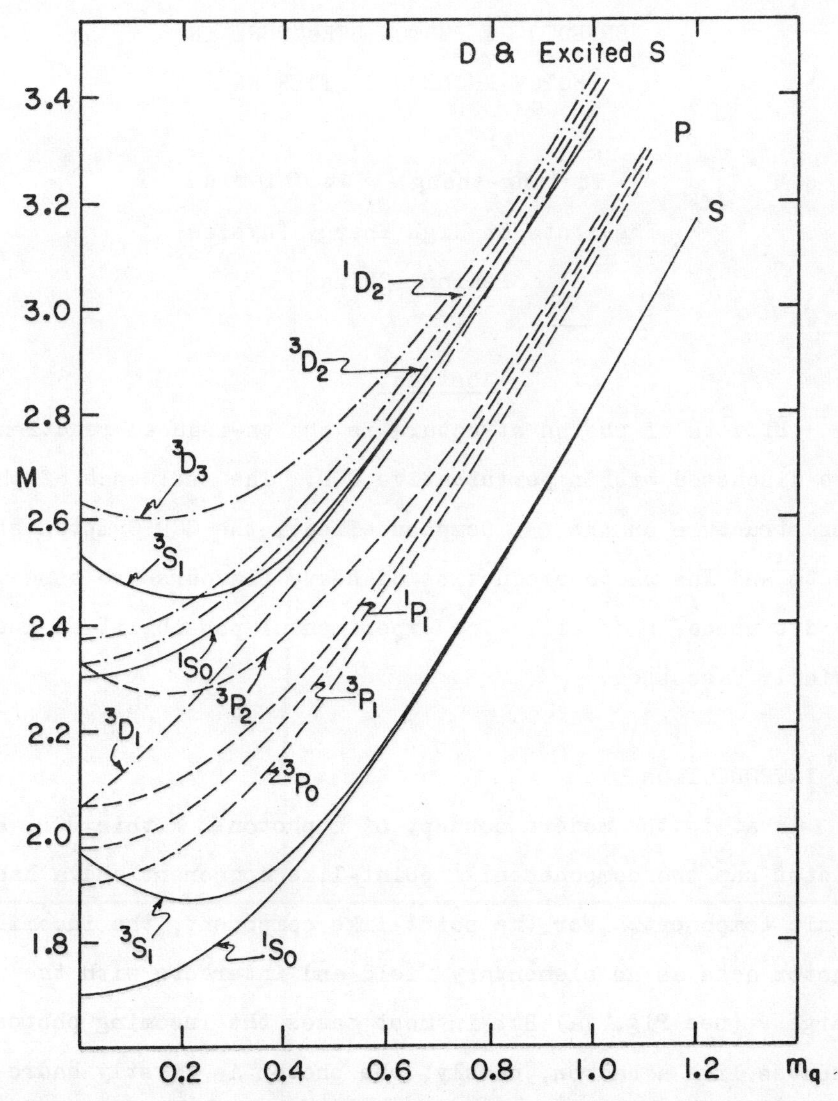

Meson mass spectrum as a function of quark mass with linear confinement potential, where $k^2 = 0.2\ GeV^2$. The solid lines are for the S-states, dotted and dash-dotted lines are for P- and D-states respectively. The S-states in the region of the D-states are the first radially excited states.

EFFECTS OF PHOTON STRUCTURE IN PHOTON-NUCLEON SCATTERING

Tu Tung-sheng Wu Chi-min
Institute of High Energy Physics,
Peking, China

Abstract

Effects of photon structure in photon-induced reactions are discussed within perturbative QCD. The influence of photon structure on the QED Compton effect, the QCD Compton effects and the photo-production of heavy flavours are especially discussed in detail. The experimental possibilities are briefly discussed.

I. INTRODUCTION

What is the modern concept of a photon? Within QCD, a photon has two components: a point-like component and a hadronic component. For the point-like component, the incoming photon acts as an elementary field and interacts with the target. (see Fig. 1a) But in most cases the incoming photon behaves like a hadron, namely, the photon is firstly hadronized and then interacts with the target.

In large transverse momentum processes, the incoming photon firstly disociated into partons and then interacts with the target. (see in Fig. 1b) In that case, perturbative

QCD is applicable. In low transverse momentum processes, the incoming photon couples directly to a bound state of quarks and gluons, for instance, $\rho, \omega, \phi, \psi, \psi', \Upsilon, \cdots$, etc. This is a non-perturbative phenomenon and can be discribed approximately by vector meson dominance. (see Fig. 1c)

In order to understand the photon structure, we need to known the parton content inside the photon. The quark distribution function $q_i^\gamma(x,t)$ and the gluon distribution function $G^\gamma(x,t)$ obey the Altarelli-Parisi equations within the leading logarithm approximation :

$$\frac{d}{dt} q_i^\gamma(x,t) = \varepsilon_i^2 \frac{2}{2\pi} a(x) + \frac{\alpha_s(1)}{2\pi} \int_x^1 \frac{dy}{y} \left[P_{qq}\left(\frac{x}{y}\right) q_i^\gamma(y,t) + P_{qG}\left(\frac{x}{y}\right) G^\gamma(y,t) \right] \quad (1)$$

$$\frac{d}{dt} G^\gamma(x,t) = \frac{\alpha_s(t)}{2\pi} \int_x^1 \frac{dy}{y} \left[P_{Gq}\left(\frac{x}{y}\right) \sum_{j=1}^{2f} q_j^\gamma(y,t) + P_{GG}\left(\frac{x}{y}\right) G^\gamma(y,t) \right] \quad (2)$$

where α and α_s (t) are the coupling constants of the electromagnetic and the strong interactions, respectively, $t = \ln Q^2/\Lambda^2$, $a(x) = X + (1 - X)^2$, ε_i is the fractional charge of quark, i, j are flavour indices, f is the number of flavours and P_{qq}, P_{qG} are the probability functions given in Ref. [2] .

From Eq.(1) we can see that the quark distribution functions have two pieces, the "anomalous" part described by $\varepsilon_i^2 \frac{\alpha}{2\pi} a(x)$ and the vector meson part described by the integration term in Eq.(1). The "anomalous" part has been calculated by several authors [3] within QCD in the leading logarithm. The vector meaon part has also been estimated.[4] Unfortunately, there is no experimental information for the

ρ structure functions up to now. The authors in Ref.[4] only used the pion structure functions to approximate the ρ structure functions. But there is some evidences in perturbation theory calculations that this neglect of spin is unwarranted.[5]

In photon-nucleon scattering, the vector meson part mainly contributes to diffractive production processes and dies out very rapidly with increasing P_T. So, in hard processes (large P_T phenomena) we can neglect the contribution of the vector meson part of the incoming photon. In the following we shall mainly concentrate on large P_T phenomena and discuss the effect of photon structure of the "anomalous" part. In Section (2), we discuss the effects of photon structure in measuring QED Compton scattering. In Section (3), we discuss QCD Compton scattering. In Section (4), we discuss photoproduction of heavy flavours. Section (5) is devoted to a short conclusion.

II. EFFECTS OF PHOTON STRUCTURE IN MEASURING QED COMPTON SCATTERING[6]

QED Compton scattering has been discussed by Bjorken and Paschos [7] using the naive parton model quite some time ago. But up to now it is still interesting both theoretically and experimentally. The dependence of the differetial cross section for this QED Compton effect is expected to be given by the proton structure function $F_2(x,Q^2)$ whose evolution at large Q^2 can be calculated using perturbative QCD. In the old-fashioned parton model the proton structure functions are only functions of x and independent of Q^2. The process is theoretically interesting because the magnitude of the cross section will determine whether the quark charge is fractional or not. So, it is interesting to measure QED Compton effect at very high energy, especially at large P_T. The typical feature of QED Compton effect is that (see Fig. 2) the incoming photon acts as an elementary field and transmits its entire energy to the final photon and a hadronic jet which can be at large P_T. In the γP c.m. frame there are no incoming photon fragments in the forward direction.

But in most cases the incoming photon behaves like a hadron, i.e., the photon is first hadronized and then interacts with the target. In Fig. 3 the corresponding Feynman diagrams are shown. When the incoming photon transmits a large fraction, say $\gtrsim$ 80%, of its energy to large P_T particles in the final state, there are only low-energy fragments of the projectile (photon). The low energy fragments are not experimentally detectable. Thus, these processes described by Fig. 3 give the same experimental results as the QED Comption effect. Their cross sections can be calculated by perturbative QCD.

There are also other processes (see Fig. 4) which give experimental results similar to the QED Compton effect. From Fig. 4 we can see that if the emitted photon or π° carries away a large fraction of the momentum of the final parton, the remaining fragments of the parton $(q,\bar{q},g)$ will have very low energy. They are also undetectable. In final states only a single γ and a counter-balancing large-P_T jet are observed.

Now let us calculate the cross-sections for all the processes mentioned above and compare them with each other.

In the γ-P c.m. frame let k, k' denote the incoming and outgoing photon momentum, resppectively, P the momentum of the proton target, xP the interacting parton momentum inside the proton and P' the outgoing parton momentum.

Define

$$x_T = k'_T/E = p'_T/E \,,\; x_L = k'_L/E \,,\; x_j = p'_L/E \tag{3}$$

where $E = \sqrt{S}/2$, $S = (p+k)^2$,

then the double differential cross section of the QED Compton effect is

$$\left.\frac{d^2\sigma}{dx_T dx_L}\right|_{\text{QEDcompton}} \tag{4}$$

$$= \frac{\pi\alpha^2}{2E^2}\frac{x_T}{a_L(a_L-x_L)}\left[\frac{2}{a_L+x_L}+\frac{a_L+x_L}{2}\right]\frac{1}{x}\sum_q e_q^4\, x\,[q^p(x,Q^2)+\bar{q}^p(x,Q^2)]$$

where

$$a_L = (x_T^2+x_L^2)^{\frac{1}{2}} \,,\quad x = (a_L - x_L)/(2 - x_L - a_L) \tag{5}$$

and $q^p(x,Q^2)$, $\bar{q}^p(x,Q^2)$ are the quark and anti-quark distribution functions inside a proton.

The contribution of diagrams (1) - (2) in Fig. 3 is:

$$\left.\frac{d^2\sigma}{dx_T dx_L}\right|_{q\bar{q}\to\gamma g}$$

$$= \int_{-1}^{1} dx_j \frac{\pi\alpha\alpha_s}{9E} \frac{x_T}{a_L a_j}\left[\frac{a_j - x_j}{a_L - x_L} + \frac{a_L - x_L}{a_j - x_j}\right]\frac{1}{x^2y^2}\sum_q e_q^2 y q^\gamma(yQ^2)\, x\left[q^p(x,Q^2) + \bar{q}^p(x,Q^2)\right] \quad (6)$$

$0.8 \leqslant y \leqslant 1$

where yk is the momentum carried by the interacting q or $\bar{q}$ from the incoming photon, q (y, Q^2) is the quark (or anti-quark) distribution function inside the photon, and

$$a_j = (x_T^2 + x_j^2)^{\frac{1}{2}}, \quad x = \frac{1}{2}(a_L + a_j - x_L - x_j), \quad y = \frac{1}{2}(a_L + a_j + x_L + x_j) \quad (7)$$

The contribution of the diagrams (3)-(4) in Fig. 3 is:

$$\left.\frac{d^2\sigma}{dx_T dx_L}\right|_{\substack{qg\to q\gamma \\ \bar{q}g\to\bar{q}\gamma}} = \int_{-1}^{1} dx_j \frac{2\pi\alpha\alpha_s}{3E^2} \frac{x_T}{a_L a_j}\left[\frac{2x}{a_L - x_L} + \frac{a_L - x_L}{2}\right]\frac{1}{16x^2y^2} \quad (8)$$

$0.8 \leqslant y \leqslant 1$

$$\times G^p(x,Q^2)\sum_q e_q^2 y\left[q^\gamma(yQ^2) + \bar{q}^\gamma(y,Q^2)\right]$$

where $G^p(x, Q^2)$ is the gluon distribution function inside a proton.

Because of the constraint $0.8 \leqslant y \leqslant 1$, the gluon contribution in the photon structure functions (see (5)-(6) in Fig. 3) is negligible. For a similar reason, the fragmentation function of $g \to \gamma$ is also negligibly small.

The contribution of the diagrams (1)-(4) in Fig. 4 is:

$$\left.\frac{d^2\sigma}{dx_T dx_L}\right|_{\gamma\ \text{emission}}$$

$$= \int_{0.8}^{1} dz \frac{\pi\alpha\alpha_s}{E^2} \frac{x_T}{z a_L(2z - x_L - a_L)}\left\{\frac{2}{3}\left[\frac{2z}{2z - x_L - a_L} + \frac{2z - x_L - a_L}{2z}\right]\frac{1}{x^2}\sum_q e_q^2 x\left(q^p(x,Q^2) + \bar{q}^p(x,Q^2)\right)\right.$$

$$\left. D_q^\gamma(zQ^2) + \frac{1}{4}\left[\frac{a_L + a}{2z - x_L - a_L} + \frac{2z - x_L - a_L}{a_L + x_L}\right]\frac{1}{x^2} G^p(x,Q^2)\sum_q e_q^2\left[D_q^\gamma(z,Q^2) + \bar{D}_q^\gamma(z,Q^2)\right]\right\}$$

(9)

The contribution of the diagrams (5)-(10) in Fig. 4 is:

$$\left.\frac{d^2\sigma}{dx_T + dx_L}\right|_{\pi^\circ \text{emission}}$$

$$= \int_{0.8}^{1} dz \, \frac{\pi \alpha \alpha_s}{E^2} \, \frac{x_T}{z a_L (2z - x_L - a_L)} \left\{ \frac{2}{3} \left[\frac{2z}{2z - x_L - a_L} + \frac{2z - x_L - a_L}{2z} \right] \frac{1}{x^2} \sum_q e_q^2 \, x \left[q^p(x, Q^2) + \bar{q}^p(x, Q^2) \right] \right.$$

$$D_q^{\pi^\circ}(z\, Q^2) + \frac{2}{3} \left[\frac{2z}{a_L + x_L} + \frac{a_L + x_L}{2z} \right] D_u^{\pi^\circ}(z, Q^2) \, \frac{1}{x^2} \sum_q e_q^2 \, x \left[q^p(x, Q^2) + \bar{q}^p(x, Q^2) \right]$$

$$+ \frac{1}{4} \left[\frac{a_L + x_L}{2z - x_L - a_L} + \frac{2z - x_L - a_L}{a_L + x_L} \right] \frac{1}{x^2} \, x \, G^p(x, Q^2) \sum_q e_q^2 \left[D_q^{\pi^\circ}(z, Q^2) + D_{\bar{q}}^{\pi^\circ}(z, Q^2) \right] \tag{10}$$

In Eqs. (9) and (10), $x = \frac{a_L - x_L}{2z - x_L - a}$, $a = \sqrt{x_T^2 + x_L^2}$ x_L, x_T are the final photon or π° variables defined as before, $D_q^\gamma(z,Q^2)$ and $D_q^{\pi^\circ}(z,Q^2)$ are the fragmentation functions of $q \to \gamma$ and $q \to \pi^\circ$, respectively. $D_q^\gamma = D_{\bar{q}}^\gamma$, $D_q^{\pi^\circ} = D_{\bar{q}}^{\pi^\circ}$.

Here, we have used $D_g^{\pi^\circ}(z, Q^2) \sim D_u^{\pi^\circ}(z, Q^2)$ [10] for the estimation. For the numerical calculations we take $E_\gamma^{lab} \sim$ 150Gev, ($s \sim 300$ Gev^2), $\alpha_s \sim 0.5$, $\Lambda = 0.5$ Gev, $Q^2 \sim$ 1.8 Gev^2, $Q^2 \sim xys$. We use the Buras-Gaemers parametrization for the proton structure functions [11] and neglect the heavy quark contributions.

For $q^\gamma(y,Q^2)$ and $D_q^\gamma(z,Q^2)$, we use the results given by Llewellyn Smith in Ref. [1] .

For $D_q^{\pi^\circ}(z,Q^2)$ we use the Field-Feynman results [12] and neglect the Q^2 dependence.

In Figs. 5,6,7,8,9 we present $\left.\frac{d^2\sigma}{dx_T dx_L}\right|_{\substack{\text{QED} \\ \text{compton.}}}$ $\left.\frac{d^2\sigma}{dx_T dx_L}\right|_{q\bar{q}\to\gamma g}$

$$\frac{d^2\sigma}{dx_T dx_L}\Big|_{\substack{qg \to q\gamma \\ \bar{q}g \to \bar{q}\gamma}}, \quad \frac{\gamma_{emitted}}{\gamma_{QED\,compton}} = \frac{d^2\sigma}{dx_T dx_L}\Big|_{\gamma emission} \Big/ \frac{d^2\sigma}{dx_T dx_L}\Big|_{QED\,compton},$$

$$\frac{\pi^\circ_{emitted}}{\gamma_{QED\,compton}} = \frac{d^2\sigma}{dx_T dx_L}\Big|_{\pi^\circ emisoion} \Big/ \frac{d^2\sigma}{dx_T dx_L}\Big|_{QED\,compton} .$$

Now let us discuss the effects of the photon structure. Perturbative QCD is meaningful only when $P_T \gtrsim$ 1-2 Gev. Because the transverse momentum of the final particles $P_T' = k_T' = x_T E \sim$ 8.7 x Gev, all the calculations in this section are only meaningful when $X_T > 0.1$. Comparing Figs. 6 and 7 with 5, we see that the the contributions of Fig. 3 is not a serious background for measuring QED Compton effect when $x_T > 0.2$. From Fig. 8, 9, we see that the emitted photon is also not important, but the emitted π° is important. At present π° can be recognized with large efficiency $\sim 80\%$. So, for $0.2 \leqslant x \leqslant 0.4$, the number of misidentified π° is comparable with the number of the final photon of QED Compton effect. But when $x_T > 0.4$, the π° background is also small. So, if we measure QED Compton effect at $x_T > 0.4$, it is possible to eliminate all the π° background caused by the effects of photon structure.

III. EFFECTS OF PHOTON STRUCTURE IN MEASURING QCD COMPTON EFFECT.[13]

It has been conjectured by many authors that large transverse momentum phenomena can be treated in perturbative QCD, and that some of those phenomena might be suited for testing QCD theory. Fritzsch and Minkowski [14] pointed out that there is a typical event in photon-hadron scattering at high energy, namely, the QCD Compton effect, which is one of the simplest processes for testing QCD.

The typical feature of the QCD Compton effect (Fig. 10) is that the incoming photon acts like a point-like particle and collides with a quark (or antiquark) inside a proton target. The imcoming photon then transmits its entire energy to the outgoing quark (anti-quark)-gluon jets which can be at large P_T. In the γ P centre of mass system there are no incoming photon fragments in the forward direction.

But there are some other processes which give similar experimental results to the QCD compton effect.

When the incoming photon acts like a point-like particle and colides with a gluon instead of a quark (antiquark) inside the proton, the incoming photon transmits its entire energy to the outgoing $q - \bar{q}$ jets which can also be at large P_T. (See Fig. 11) If we cannot distinguish a gluon jet from a quark jet experimentally, this process gives the same results as the QCD Compton effect.

In most cases the incoming photon behaves like a hadron. When the incoming photon transmits a large fraction, say $\gtrsim 80\%$, of its energy to the large P_T jets in the final states, there are also no fragments detected experimentally in the foward direction. Again, if we can not distinguish the jet type, we also have the similar experimental results to the QCD Compton effect. The corresponding Feynman diagrams are shown in Fig. 12 where the gluon contribution inside the incoming photon is neglected for the same reason mentioned in Section (2).

Now we turn to the calculation of the differential cross sections of all the processes mentioned above by using only the point-like component and the "anomalous" part of the incoming photon.

In the γ P c.m. frame let k denote the incoming photon momentum, P the proton momentum, xP the interacting parton momentum inside the proton, and k', p' the two outgoing parton momenta.

Define

$$y_T = 2k'_T/\sqrt{s} = 2P'_T/\sqrt{s}\ , \quad y_L = 2k'_L/\sqrt{s}\ , \quad y_j = 2P'_L/\sqrt{s}\ , \tag{11}$$

where $s = (p+k)^2$ is total energy squared in the γ P c.m.s.

If we can not distinguish a gluon jet from a quark jet, the double differential cross section for the QCD Compton effect is

$$\left.\frac{d^2\sigma}{dy_T dy_L}\right|_{\substack{\text{QCD compton}\\ \text{2 large P jets}}}$$
$$= \frac{8\pi\alpha\alpha_s(Q^2)}{3s}\frac{y_T}{a_L(2-y_L-a_L)}\left\{\left(\frac{2}{2-y_L-a_L}+\frac{2-y_L-a_L}{2}\right)\right.$$
$$\left.+\left(\frac{2}{a_L+y_L}+\frac{a_L+y_L}{2}\right)\ \frac{1}{x^2}\sum_q e_q^2\, x\,[q^P(x,Q^2)+\bar{q}^P(x,Q^2)]\right\} \tag{12}$$

where

$$x = (a_L-y_L)/(2-y_L-a_L)\ , \quad a_L = (y_T^2+y_L^2)^{\frac{1}{2}}$$

and (y_T, y_L) denote jet variables of a quark jet or a gluon jet.

For the contribution of Fig. 11 without distinguishing a quark jet and a gluon jet, the two large P_T jet cross section is

$$\left.\frac{d^2\sigma}{dy_T dy_L}\right|_{\substack{\gamma g \to q\bar{q} \\ \text{2large Pjets}}} = \frac{2\pi\alpha\,\alpha_s(Q^2)}{S}\sum_q e_q^2\,\frac{y_T}{a_L(2-y_L-a_L)}\left(\frac{a_L+y_L}{2-y_L-a_L}+\frac{2-y_L-a_L}{a_L+y_L}\right)\frac{1}{x^2}\times G^P(x,Q^2) \quad (13)$$

where

$$x = (a_L - y_L)/(2-y_L-a_L)$$

and (y_T, y_L) represent the quark jet variables or the antiquark jet variables.

For the contribution of Fig. 12, without distinguishing jet type

$$\left.\frac{d^2\sigma}{dy_T dy_L}\right|_{\text{2large P jets}} = \int_{-1}^{1} dy_j\,\frac{2\pi\alpha_s^2(Q^2)}{9S}\,\frac{y_T}{a_L a_j}\,\frac{1}{x^2y^2}\left[y\,q^{\gamma}_{2/3}(y\,Q^2)\right]\cdot\left\{\left[\frac{x^2}{y^2}\,\frac{(a_L+y_L)^2+4y^2}{(a_L-y_L)^2}\right.\right.$$

$$\left.+\frac{y^2}{x^2}\,\frac{(a_L-y_L)^2+4x^2}{(a_L+y_L)^2}\right]\cdot\left[x\,u_v^P(x,Q^2)+\frac{5}{2}x\,d_v^P(x,Q^2)+2x\,S^P(x,Q^2)\right]$$

$$+\left[3\frac{x^2}{y^2}\,\frac{(a_L+y_L)^2+4y^2}{(a_L-y_L)^2}+3\,\frac{y^2}{x^2}\,\frac{(a_L-y_L)^2+4x^2}{(a_L-y_L)^2}-\frac{x^2(a_L+y_L)^2+y^2(a_L-y_L)^2}{x^2y^2}\left(2-\frac{16}{3}\,\frac{xy}{y_T^2}\right)\right.$$

$$\left.-2\left(1+\frac{4}{3}\,\frac{xy}{y_T^2}\right)\right]\cdot\left[x\,u_v^P(x,Q^2)+\frac{1}{4}x\,d_v^P(x,Q^2)+\frac{1}{2}x\,S^P(x,Q^2)\right]$$

$$+\left[\left(x^2(a_L+y_L)^2+4x^2y^2\right)\left(\frac{2}{9x^2y(a_L+y_L)}+\frac{1}{y^2(a_L-y_L)^2}\right)+\left(y^2(a_L-y_L)^2+4x^2y^2\right)\right.$$

$$\left.\left.\cdot\left(\frac{2}{9x\,y^2(a_L-y_L)}+\frac{1}{x^2\,(a_L+y_L)^2}\right)\right]\frac{27}{4}\times G^P(x,Q^2)\right\} \quad (14)$$

where $u_v^P(x,Q^2)$, $d_v^P(x,Q^2)$, $S^P(x,Q^2)$ are the valence up-quark, down-quark and the sea quark distribution functions of the proton, and we have neglected the charm quark contributions. $q^{\gamma}_{2/3}(y, Q^2)$ is the up-quark distribution function of the photon.

For the numerical results we take $E_{\gamma}^{lab} \sim 150$ Gev, $\alpha_s \sim 0.5$ *). We use the Buras and Gaemers parametrization [11] for proton structure functions and neglect the charm quark contribution. Also we take $\Lambda \sim 0.5$ Gev, $Q_0^2 \sim 1.8$ Gev^2, $Q^2 \sim xyS$. For the photon structure functions we use the results given by Llewellyn Smith [1] and neglect the charm quark and gluon contributions.

The cross sections in Eqs. (12)-(14) are presented in Figs. 13-15. We can see clearly from these figures that the cross section of $\gamma g \to q\bar{q}$ (see Fig. 14) is relatively small in comparison with the QCD Compton effect and fall off very rapidly when y_T increases. When $y_T \gtrsim 0.4$

$$\left.\frac{d^2\sigma}{dy_T dy_L}\right|_{\gamma g \to q\bar{q}} \Big/ \left.\frac{d^2\sigma}{dy_T dy_L}\right|_{\text{QCDcompton}} < 10\%$$

So, the process described by Fig. 11 is not a serious backbround for measuring QCD Compton effect. But the contribution from the "anomalous" part of the hadronized photon is of the same order as the QCD Compton effect in the whole range of y_T and y_L (see Fig. 13 and 15). Thus, without

---

*) It might be suitable to take $\alpha_s \sim 0.4$ or smaller, but it does not affect our conclusion.

distinguishing a gluon jet from a quark (antiquark) jet it is difficult to pick up QCD Compton effects from so large a background. That means we can not use calorimeters to measure QCD Compton effects.

If we can distinguish different jet types, we should only consider the diagrams (5)-(7) of Fig. 12 which give the same experimental results as the QCD Compton effects when $y \gtrsim 0.8$. The contribution of (5)-(7) of Fig. 12 depends strongly on the gluon distribution inside the proton and falls off rapidly when y_T increases. It is shown that [13] when $y_T \gtrsim 0.4$ the background induced by the "anomalous" part of the photon is very small and it is quite feasible to measure QCD Compton effects.

At moderate P_T, the "constituent interchange" mechanism [15] ($qM \to qM$, $q\bar{q} \to MM$, $qB \to qB$, etc.) will be important. But it does not affect our conclusions at large P_T because these contributions fall off faster than the QCD contribution when P_T increases.

For e-P collisions, please see referance [13] . We shall not discuss it here.

IV. EFFECTS OF PHOTON STRUCTURE ON PHOTO-PRODUCTION OF HEAVY FLAVOURS [16]

Recently a bump at $M \sim 5.3$ Gev has been observed in the $\Psi K\pi$ channel in the π -P collision experiment at CERN.[17] The result might not be conclusive, but it is promising. There is a good reason to consider this bump as a meson containing the bottom quark [18] ($b\bar{u}$, $\bar{b}u$, $b\bar{d}$, $\bar{b}d$,...). Interest in searching for "bottomed" mesons in photo-production processes is increasing. The photo-production of heavy quark flavours has been discussed within QCD by several authors [19-21] but they only concentrate on the photon-gluon amalgamation mechanism and diffractive production. There is another mechanism, namely, $q\bar{q} \rightarrow b\bar{b}$ (where $\bar{q}$ comes from the hadronized incoming photon and b stands for the bottom quark), which also contributes to the photo-production of "bottomed" mesons.

The two mechanisms, γ-g amalgamation and $q\bar{q} \rightarrow b\bar{b}$, are illustrated in Fig. 16. Although the $q\bar{q} \rightarrow b\bar{b}$ mechanism is proportional to $\alpha\alpha_s^2$ instead of $\alpha\alpha_s$ like γ-g amalgamation, m_b is quite large. The lower bound for the Feynman variable $x_{min} \sim 4m_b^2/S$. When $E_\gamma^L \sim 150$ Gev (SPS photon beams) $x \gtrsim 0.31$. In that case, the gluon distribution is seriously supressed, while the valence quark distribution is still large. Therefore for SPS photon beams the $q\bar{q} \rightarrow b\bar{b}$ mechanism may be important.

Now let us compare it with γ-g amalgamation. The contribution of γ-g amalgamation (diagram (a) in Fig. 16) is [19, 21, 22]

$$\sigma(\gamma p \rightarrow B\bar{B}X) = \int_{4m_b^2/s}^{1} dx\, \hat{\sigma}(xS)\, G(x, Q^2) \tag{15}$$

where the capital B stands for a meson containing the bottom quark and

$$\hat{\sigma}(xS) = \frac{\pi \alpha \alpha_s e_b^2}{xS} \left\{ (3-\beta^4) \ln \frac{1+\beta}{1-\beta} - 2\beta(2-\beta^2) \right\}$$

$$\beta = (1 - 4m_b^2/xS)^{\frac{1}{2}}, \quad e_b = -\frac{1}{3},$$

$G(x, Q^2)$ is the gluon distribution inside the proton.

S is the total energy squared in the γ P c.m. system. With the constraint $\int_0^1 xG(x, Q^2)dx \sim 0.5$, many authors take

$$G(x, Q^2) \sim \frac{n+1}{2}(1-x)^n / x \tag{16}$$

According to QCD, the value of n should depend on the momentum transfer Q^2. For $B\bar{B}$ production it is reasonable to take $Q^2 \gtrsim 4m_b^2 \sim 100$ Gev^2 for the Buras-Gaemers parametrization.

$$\begin{aligned} &\text{If } G(x, Q_0^2) \sim 3(1-x)^5/x \text{ then } G(x, Q^2) \sim 8.5(1-x)^{16}/x \\ &\text{If } G(x, Q_0^2) \sim 5.5(1-x)^{10}/x \text{ then } G(x, Q^2) \sim 11.5(1-x)^{22}/x \end{aligned} \tag{17}$$

where $Q_0^2 \sim 1.8$ Gev^2, $Q^2 \sim 100$ Gev^2, and we take $\Lambda \sim 0.5$ Gev. The shape of $G(x, Q^2)$ is an open question. But the powers in Eq. (17) might be reasonable.

In Ref. [21] only the case n=5 is considered; and in Ref 20 both n = 5,9 are calculated but with $S \to \infty$. Hence, for comparison, we give all the cases n = 5,7,10,16,22, for SPS photon beams.

The total cross section of the $q\bar{q} \to b\bar{b}$ mechanism (diagram (b) of Fig. 16) is

$$\sigma(\gamma p \to B\bar{B}X) \sim \iint_{xyS \geq 4m_b^2} dx\,dy\,\hat{\sigma}(xyS) \sum_q \left[\bar{q}^\gamma(y, Q^2) q^p(x, Q^2) + q^\gamma(y, Q^2)\bar{q}^p(x, Q^2)\right] \tag{18}$$

$$\hat{\sigma}(xyS) = \frac{8\pi\alpha_s^2}{27(xyS)^3}[xyS(xyS-4m_b^2)]^{\frac{1}{2}}(xyS+2m_b^2) \quad (19)$$

$$\sum_q \left(\bar{q}^\gamma(y,Q^2)q^p(x,Q^2)+q^\gamma(y,Q^2)\bar{q}^p(x,Q^2)\right)$$
$$= \frac{1}{xy}\, y\, q^\gamma_{2/3}(y,Q^2)\left(xu_v^p(x,Q^2)+\frac{1}{4}xd_v^p(x,Q^2)+\frac{1}{2}xS^p(x,Q^2)\right) \quad (20)$$

and $q^\gamma_{2/3}(y, Q^2)$ is the $\frac{2}{3}$ e charged quark distribution function of the photon, and u_v^p, d_v^p, S^p, are respectively the valence up, down, and sea quark distribution functions in the proton.

For the numerical calculations, we use

$$\alpha_s(4m_b^2) \sim 0.3 \quad (21)$$

$$\alpha_s(Q^2) \sim 0.3/[1+0.3\frac{7}{4\pi}\ln(\frac{Q^2}{4m_b^2})] \quad (22)$$

For γ-g amalgamations we simply take $\alpha_s(Q^2) \sim 0.3$. For the $q\bar{q} \to b\bar{b}$ calculation, we use $\alpha_s(Q^2)$ of Eq.(22) with $Q^2 \sim xyS$. We use $q^\gamma_{2/3}(y, Q^2)$ given by Llewellyn Smith [1] and neglect the gluon distribution of the photon. This is reasonable for SPS photon beams because of the constraint $xyS \geqslant 4m_b^2$, e.g. y cannot be smaller than $4m_b^2/S$. For the proton structure functions, we use the Buras-Gaemers parametrization.

All the results are presented in Fig. 17. From Fig. 17 we can see that for SPS photon beams the $q\bar{q} \to b\bar{b}$ contribution is much larger than γ-g amalgamation for Buras-Gaemers gluon distributions (n = 16,22). The largest value of $\sigma(\gamma p \to B\bar{B}X)$ is $\sim$ 1 n b at $E_\gamma^L \sim 200$ Gev and n = 5 ($G(x,Q^2) \sim 3(1-X)^5/X$). But n=5 is not the favoured gluon distribution at $Q^2 \sim 100$ Gev^2/c^2. So, the value 1 nb is an overestimate

Now we will say a few words about diffractive production. Following Refs. [20] and [21]

$$\frac{\sigma(\gamma p \to B\bar{B}X)}{\sigma(\gamma p \to D\bar{D}X)} \sim \frac{M_\psi}{M_\Upsilon} \cdot \frac{\Gamma_{\Upsilon \to e^+e^-}}{\Gamma_{\psi \to e^+e^-}} \cdot \frac{\sigma(\Upsilon N)}{\sigma(\psi N)} \tag{23}$$

If we assume

$$\sigma(\Upsilon N)/\sigma(\psi N) \sim M_\psi^2/M_\Upsilon^2 \tag{24}$$

then

$$\sigma(\gamma p \to B\bar{B}X)/\sigma(\gamma p \to D\bar{D}X) \sim 9\times 10^{-3} \tag{25}$$

Using

$$\sigma(\gamma p \to D\bar{D}X)_{s\to\infty} \sim 1.2\ \mu b \quad \text{(see Ref.20)} \tag{26}$$

we get

$$\sigma(\gamma p \to B\bar{B}X) \sim 10\ nb \tag{27}$$

This is more than a factor 10 larger then the QCD perturbative calculation. But the perturbative QCD calculations are probably more reliable when heavy flavours are produced. In Eq. (24) the energy dependence is not specified. Also the ratio might be m_ψ^n/m_Υ^n with $n \geqslant 3$ or $e^{-M_\Upsilon^2}/e^{-M_\psi^2}$ etc. So, the estimation of (27) is not reliable. Therefore we may expect

$$\sigma_{tot}(\gamma p \to B\bar{B}X) \lesssim \text{a few nb}$$

So, it seems difficult to search for "bottomed" mesons with the present experimental limitation. But if $\sigma(\gamma p \to B\bar{B}X) \sim$ 10 nb, it is possible to discover "bottomed" mesons in photoproduction processes.

For charm production and e-P colliding beams, please see Ref. [16] we will not discuss them here.

V. CONCLUSION

From the discussion above we conclude that at large P_T the effects of photon structure are important in measuring QCD Compton effects and in the processes of photo-production of heavy flavours. But they are not important in measuring QED Compton effects. At moderate and small P_T, it in reasonable to speculate that the effects of photon structure are very important although we do not know how to calculate these effects. In diffractive processes induced by the photon, the vector meson part of the photon structure will play an important role through vector meson dominance. Obviously, in $\gamma-\gamma$ collision proceases the effects of photon structure are more important than the examples discussed above. So the investigation of the properties of the photon is still interesting field.

Acknowledgement

We would like to thank D.Amati, A.De Rujula, J.Ellis, H. Fritzsch, D. Treille, P.Weilhammer for interesting discussions.

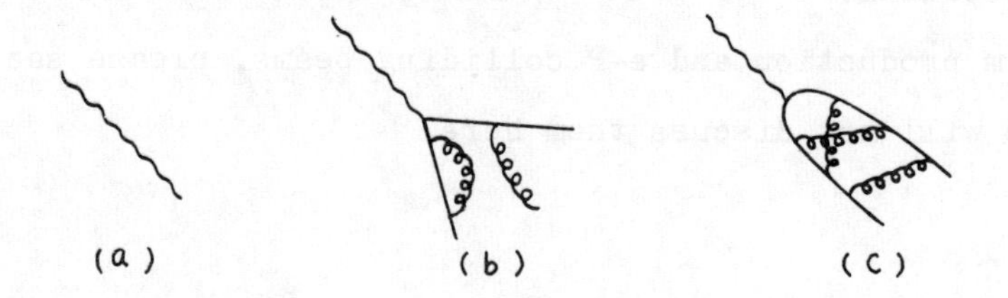

Fig. 1 : The modern concept of photon:

a) point-like component

b) photon dissociates into partons, perturbative.

c) photon couples directly to bound state, non-perturbative.

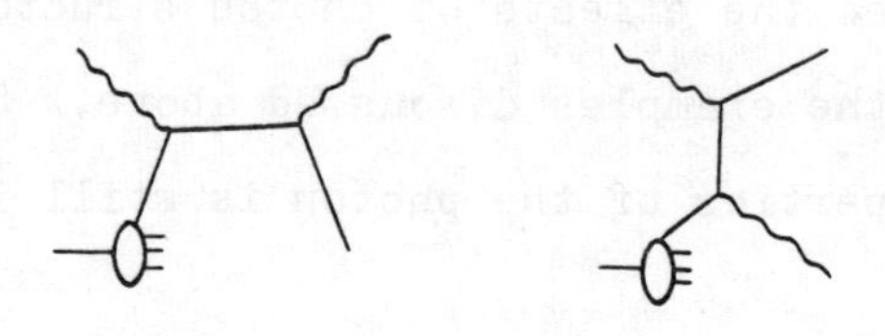

Fig. 2 : Feynman diagrams for QED Compton effect

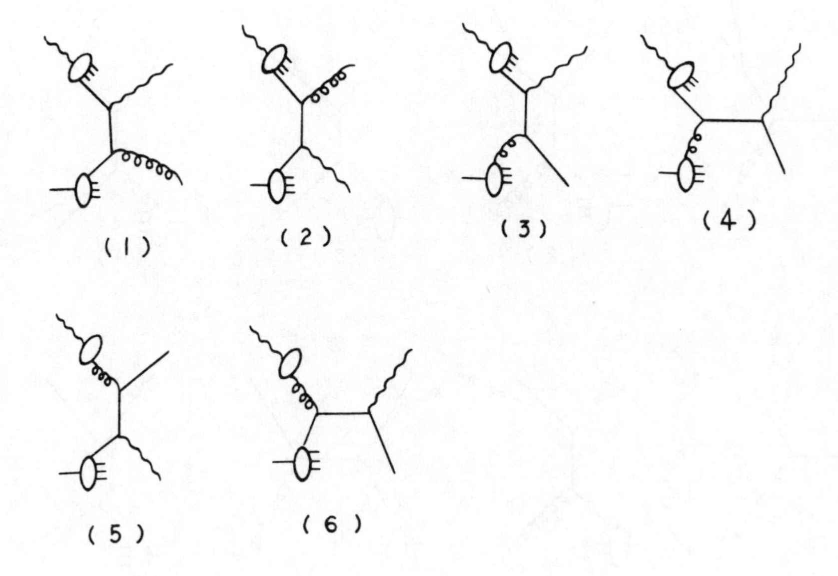

Fig. 3 : The lowest order Feynman diagrams of the hadronized incoming photon colliding with a proton.

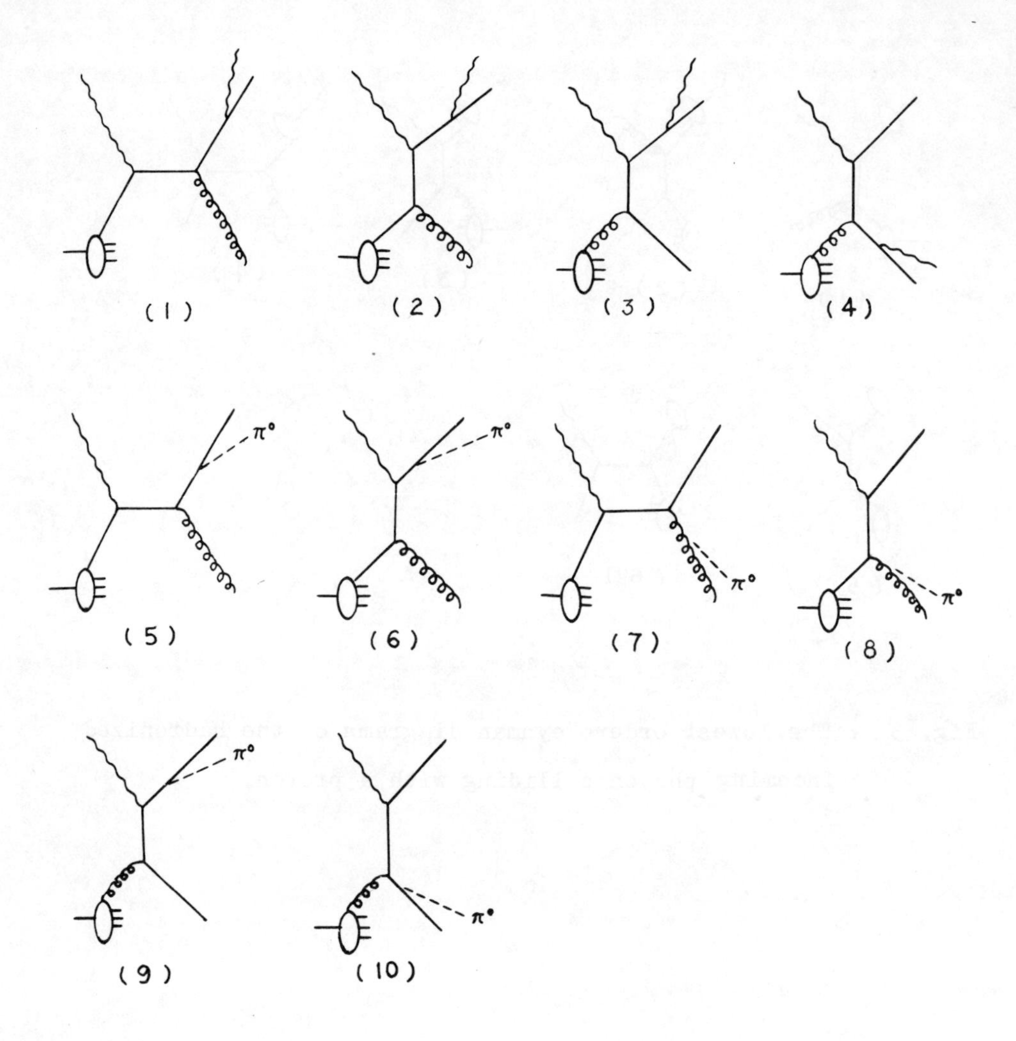

Fig. 4 : The lowest order Feynman diagrams which give important contributions to γ or π^0 emission.

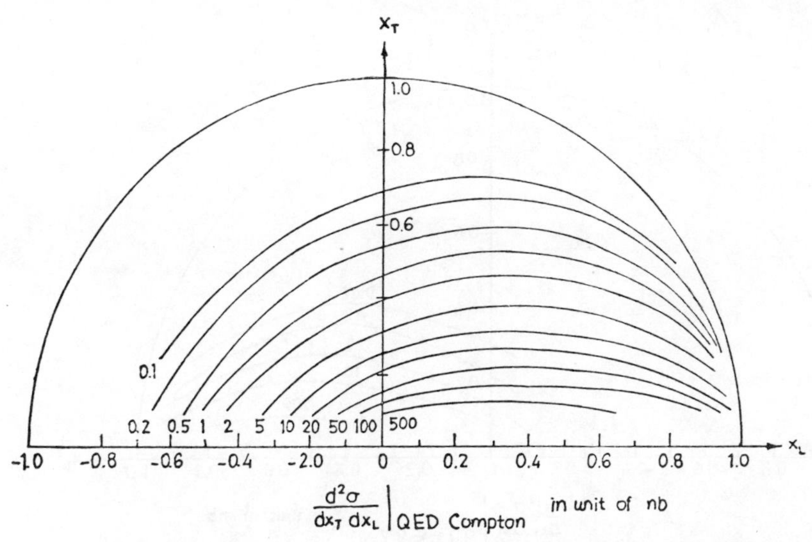

Fig. 5 : The double differential cross section of QED Compton effect. The figures by the curves are the numerical values of the cross-sections in units of nanobaens.

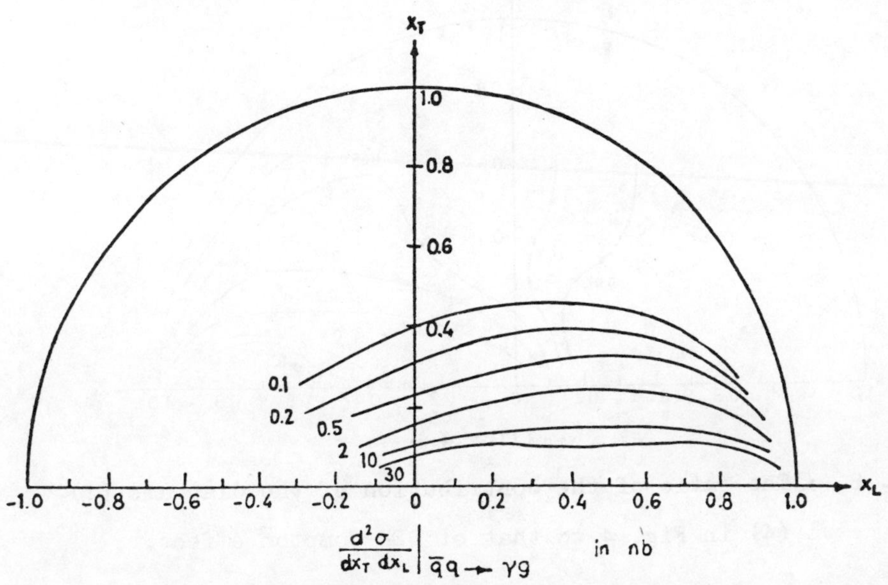

Fig. 6 : The double differential cross-section which comes from the diagrams (1)-(2) in Fig. 3.

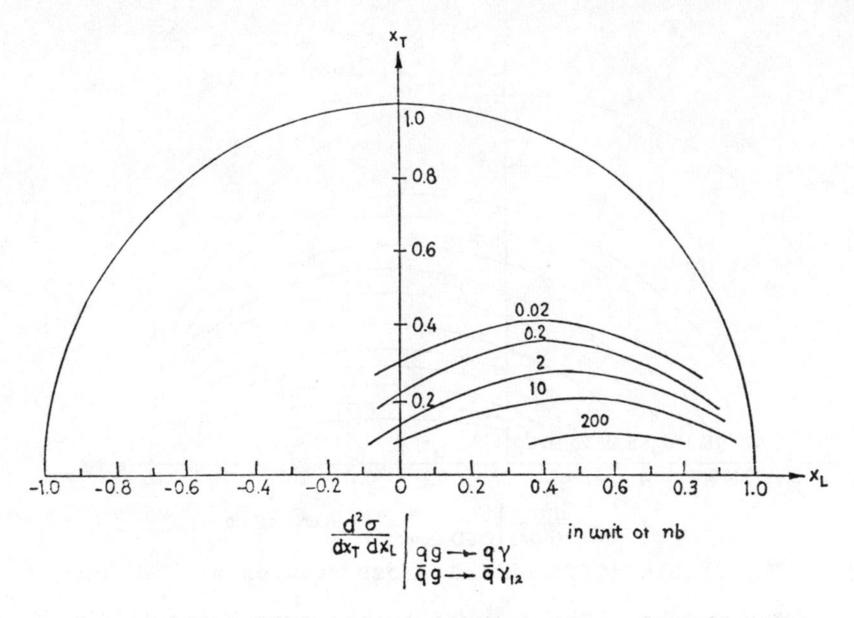

Fig. 7 : The contribution of the diagrams (3)-(4) in Fig. 3

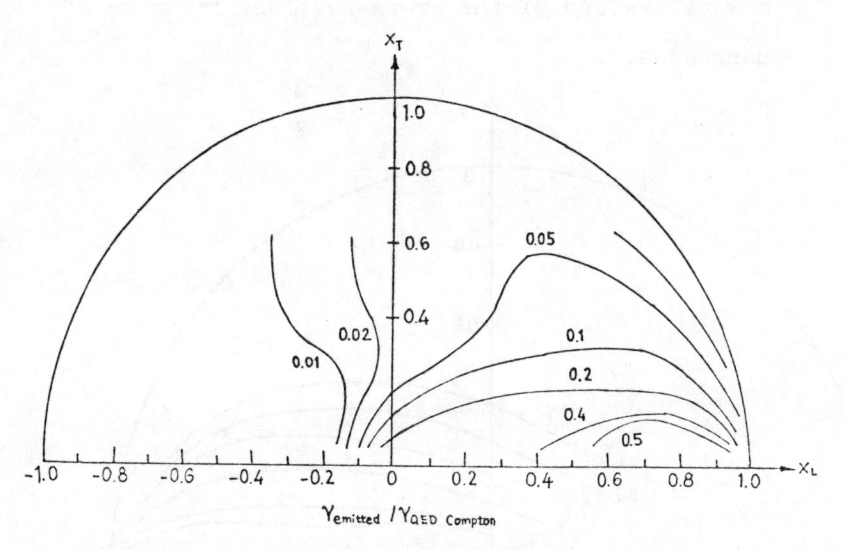

Fig. 8 : The ratio of the contribution of the diagrams (1)-(4) in Fig. 4 to that of QED Compton effect.

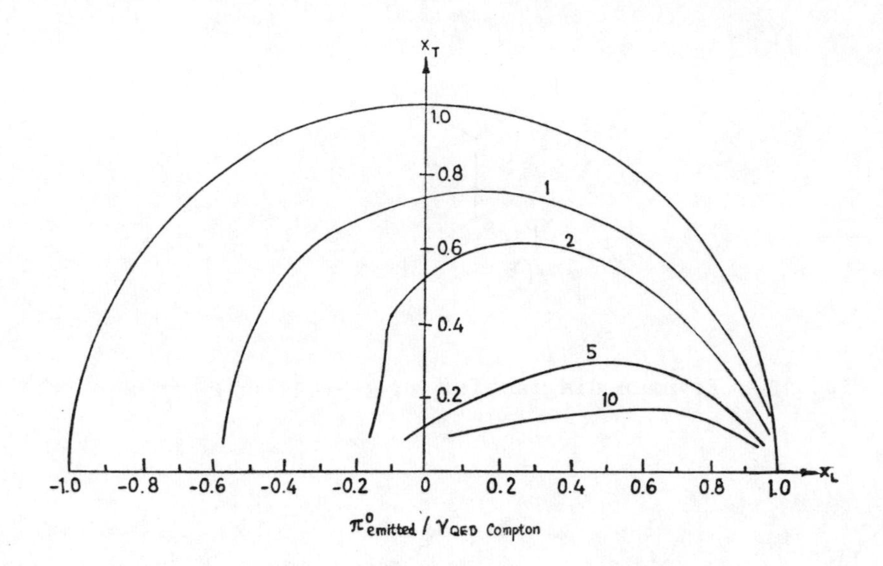

Fig. 9 : The ratio of the contribution of the diagrams (5)-(10) in Fig. 4 to that of QED Compton effect.

Fig. 10 : Feynman diagrams for QCD Compton effect.

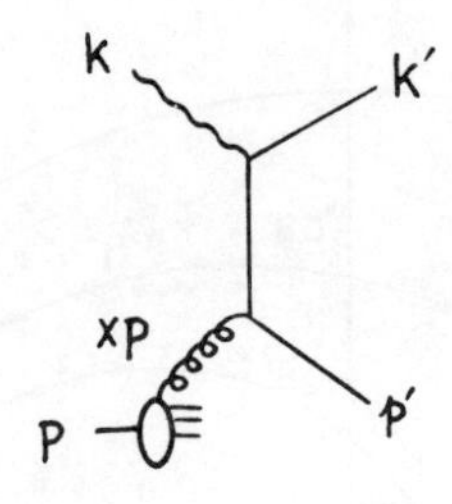

Fig. 11 : Feynman diagram for $\gamma g \rightarrow q\bar{q}$ in $\gamma P \rightarrow q\bar{q}X$.

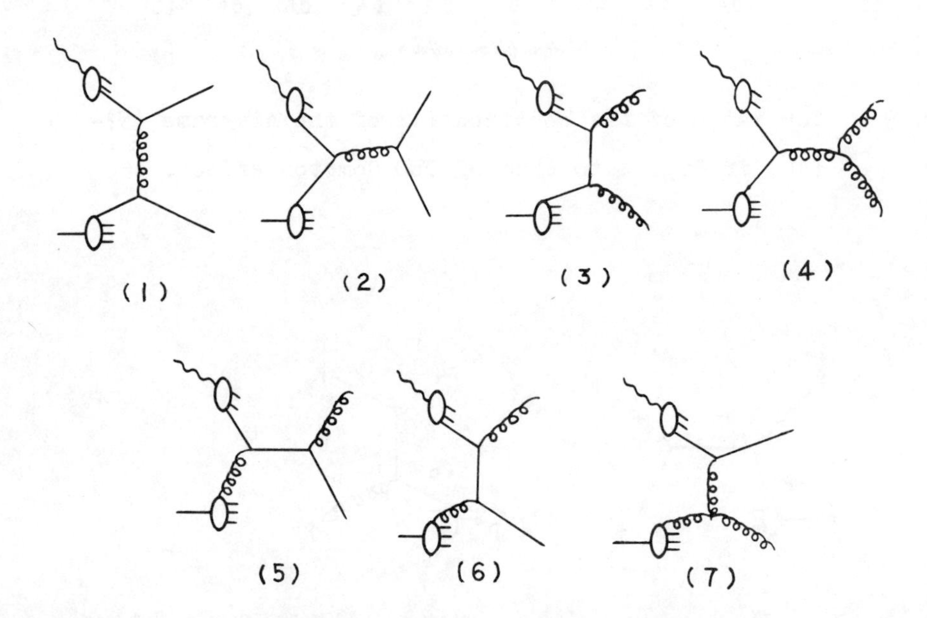

Fig. 12 : Feynman diagrams for the processes $\gamma + P \rightarrow 2$ large P_T jets.

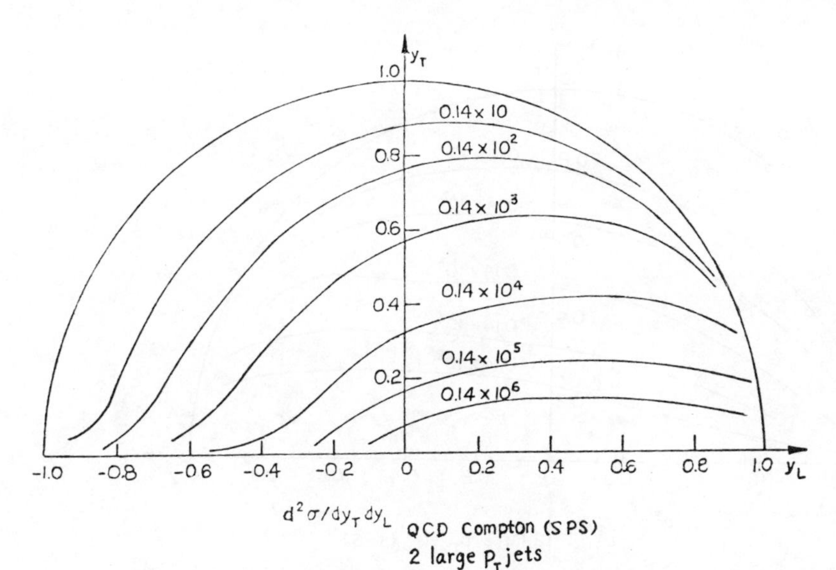

Fig. 13 : The double differential cross section for QCD Compton effect without distingushing the jet type for the SPS photon beams.

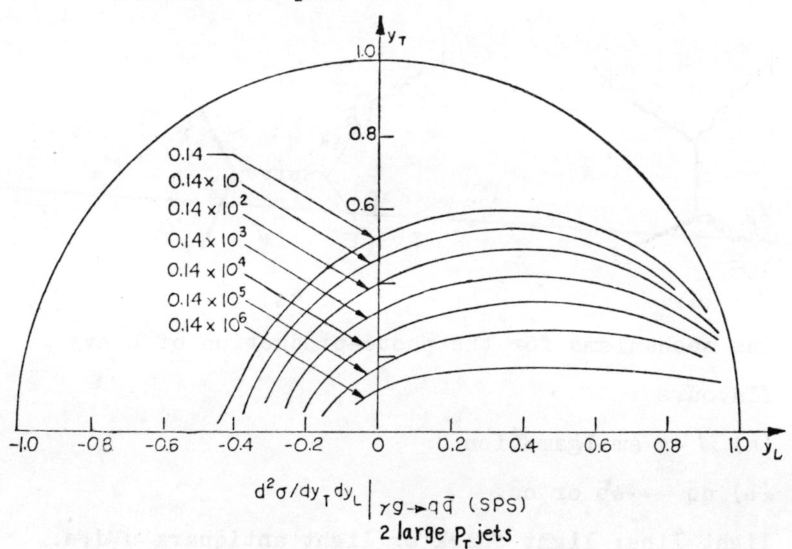

Fig. 14 : The double diffierential cross-section for the diagram in Fig. 11 without distingushing the jet type for SPS photon beams.

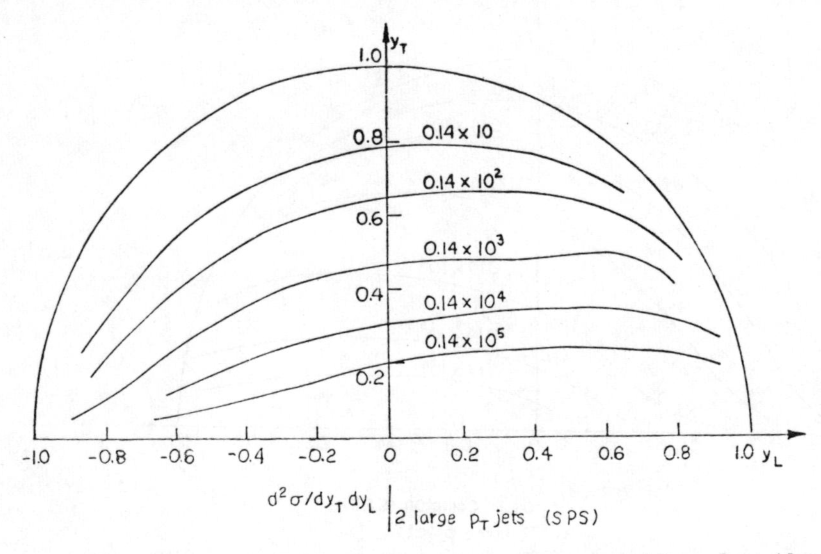

Fig. 15 : The double differential cross-section for the diagrams in Fig. 12 without distinguishing the jet type.

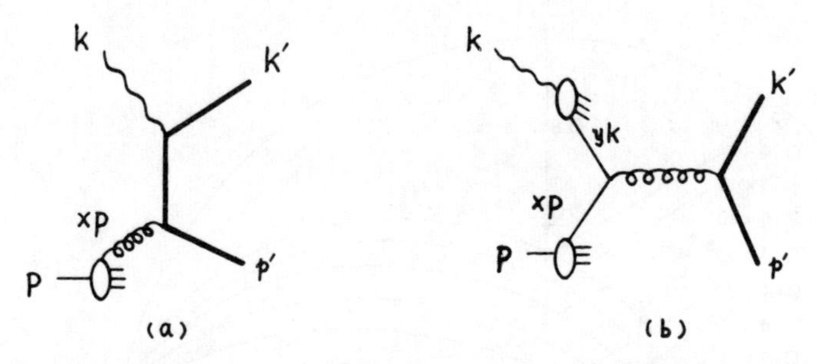

Fig. 16 : The mechanisms for the photo-production of heavy flavours :

(a) γ -g amalgamation

(b) $q\bar{q} \rightarrow b\bar{b}$ or $c\bar{c}$.

light line: light quark or light antiquark u,d,s.

bold line: heavy quark or heavy antiquark c,b,...

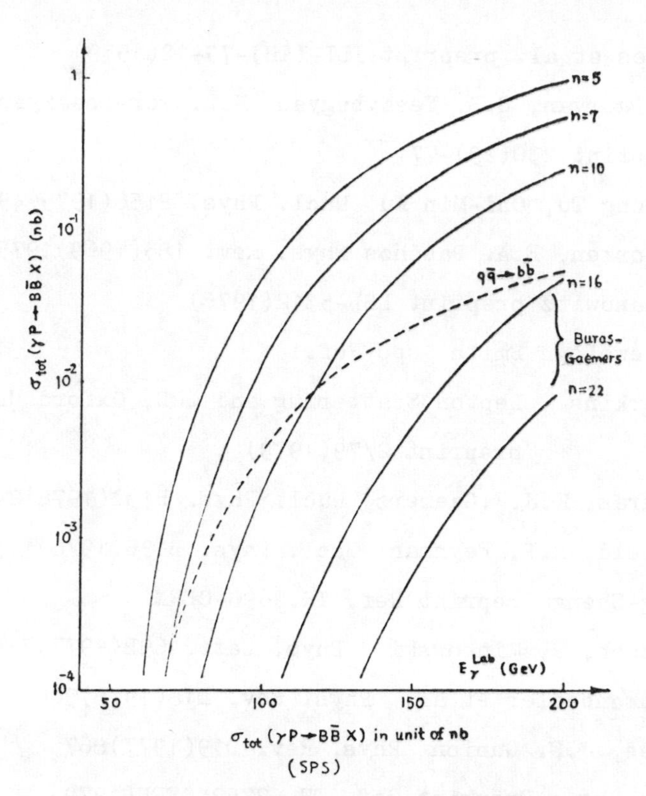

Fig. 17 : The cross section for beauty production $\sigma_{tot}(\gamma p \longrightarrow B\bar{B}X)$ for SPS photon beams. The dashed line is $q\bar{q} \longrightarrow b\bar{b}$.

References

(1) E. Witten Nucl. Phys. B120(1977)189

C.H. Llewellyn Smith Phys. Lett. 79B(1978)83

R.J. DeWitt et al. Phys. Rev. D19(1979)2046

S.J. Brodsky et al. Phys. Rev. D19(1979)1418

Phys. Rev. Lett. 41(1979)672

(2) G. Altarelli, G. Paeisi Nucl. Phys. B126(1977)298

(3) See Ref.1 C.H. Llewellyn Smith, E. Witten

W.A. Bardeen, A.J. Buras Fermilab-Pub-78/91 THY(1978)

W.R. Frazer, J.F. Gunion Phys. Rev. D20(1979)147

(4) L.M.Jones et al. preprint ILI-(TH)-79-12(1979)

(5) S.G. Grigoryan, S.B. Yesaybegyan, N.L. Ter-Isaakyan
EFI preprint 230(23)-77

(6) Tung-Sheng Tu, Chi-Min Wu Nucl. Phys. B156(1979)493

(7) J.D. Bjorken, E.A. Paschos Phys. Rev. 185(1969)1975

(8) M.S. Chanowitz preprint LBL-5312(1976)

(9) C.H. Llewellyn Smith see Ref.1

(10) D.H. Perkins Lepton Scattering and QCD, Oxford Univ.
preprint 2/79(1979)

(11) A.J. Buras, K.J.F.Gaemers Nucl. Phys. B132(1978)249

(12) R.D. Field, R.P. Feynman Nucl. Phys. B136(1978)1

(13) Tu Tung-Sheng Preprint Ref. TH.2690-CERN

(14) H.Fritzsch, P. Minkowski Phys. Lett. 69B(1977)316

(15) R. Blankenbecler et al. Phys. Rev. D18(1978)900
D. Jones, J.F. Gunion Phys. Rev. D19(1977)867

(16) Tu Tung-Sheng Preprint Ref. TH. 2752-CERN(1979)

(17) R. Baeate et al. Saclay preprint DPHPE 79-17(1979)
and CERN preprint EP/79-113(1979)

(18) H. Fritzsch, preprint Ref. TH. 2648-CERN(1979)

(19) L.M. Jones, H.W. Wyld Phys. Rev. D17(1978)759, and
reference therein

(20) H. Fritzsch, K.H. Streng Phys. Lett. 72B(1978)385

(21) H. Babcock et al. Phys. Rev. D18(1978)162

(22) J.M. Jauch, R.Rohrlich. The toeory of photons and electrons (Addison-Wesley, Reading, Mass., 1955)
p. 299

QCD PREDICTIONS FOR THE PHOTON DISTRIBUTION FUNCTIONS AND QUARK AND GLUON FRAGMENTATION FUNCTIONS USING ALTARELLI-PARISI EQUATIONS

Wu Chi-min

Institute of High Energy Physics,
Academia Sinica

Abstract

In the leading approximation, the distribution functions of the photon and the fragmentation function of quarks and gluons are investigated by using the A-P type equation. Analytic expressions for their boundary behaviours are given. Using a new method to make Mellin transformation, the equation is solved numerically. The numerical results of point-like and hadron-like parts of the photon distribution function are given separately. The results of the Q^2 dependence of fragmentation function are also given. Those results are compared with others which have been given by using other methods.

I. INTRODUCTION

Asymptotic freedom [1,2] is a remarkable property of QCD, which has made it possible to perform perturbative calculations for some experimentally measurable quantities. In the parton picture based on QCD, the quark and gluon distribution functions inside the nucleon have Q^2 dependences and explain the scaling breaking perfectly.[3] Kogut and Susskind [4] give this a physical interpretation. Altarelli-Parisi [5] have reformulated the recursion equation of Kogut and Susskind in the following integro-differential equation form which is satisfied by quark and gluon distribution functions

$$\frac{\partial q_i(x,t)}{\partial t} = \frac{\alpha_s(t)}{2\pi}\int_x^1 \frac{dy}{y}\left[P_{qq}\left(\frac{x}{y}\right)q_i(y,t) + P_{qG}\left(\frac{x}{y}\right)G(y,t)\right]$$

$$\frac{\partial G(x,t)}{\partial t} = \frac{\alpha_s(t)}{2\pi}\int_x^1 \frac{dy}{y}\left[P_{Gq}\left(\frac{x}{y}\right)\sum_{j=1}^{2f} q_j(y,t) + P_{GG}\left(\frac{x}{y}\right)G(y,t)\right] \tag{1}$$

where $t = \ln Q^2/\Lambda^2$, Λ is a scale parameter with $\Lambda \simeq 0.2$ - 0.5 GeV/c, $\alpha_s(t)$ is the running coupling constant and f is the number of flavours. It gives the same physical results.

A-P equation can be applied to discuss the photon distribution function q_γ^i, G_γ and fragmentation function, these are the subjects of this paper. A-P equation can give a clear physical picture but solving is difficult.

Now, we use a new method given by Ref. [6] to solve it. These results are compared with others which have been given using other methods.

II. PHOTON DISTRIBUTION FUNCTION

We have known for some time that the photon possesses both point-like and hadron-like behaviours. In the point-like case, it interacts with other charged particle through the electromagnetic interaction. On the other hand, when the photon acts with other particles like a hadron, we use, say, vector dominance model to describe it. Here we hope to have a unified and quantitative description in the QCD language, i.e. to reveal the behaviours of the quarks, anti-quarks and gluon inside the photon.

Recently, some authors have discussed the photon distribution functions in three different ways: (1) E. Witten[7] T. Hill et al.[8] W.A. Bardeen and Buras[9] use the operator product expansion to discuss the point-like part of photon distribution function. Witten points out that, in contrast to the structure function of the nucleon, there is a new term which comes from the point-like contribution to the photon distribution function. (2) C.H. Llewellyn Smith[10] uses a diagrammatic procedure and also gives the contribution of this point-like part. (3) R.J. DeWitt et al [11] use the A-P type equations and get a numerical solution by using an interaction procedure. Brodsky et al [12] and Frazer and Gunion [13] also discuss the photon distribution function, but emphasize the measurement of it in high energy γ-γ scattering. Now, using A-P type equation and a new method to make Mellin transformation, we discuss the behaviours of photon distribution function: point-like part, hadron-like part, boundary behaviours and so on. Finally, the numerical results for the distribution functions are given for different Q^2 values.

(II.1) <u>A-P Type Equation for Photon Distribution Function</u>

In the leading approximation, the photon distribution function obeys the following A-P type equation

$$\frac{\partial q_\gamma^i(x,t)}{\partial t} = \varepsilon_i^2 \frac{\alpha}{2\pi} a(x) + \frac{\alpha_s(t)}{2\pi}\int_x^1 \frac{dy}{y}\left[P_{qq}\left(\frac{x}{y}\right) q_\gamma^i(y,t) + P_{qG}\left(\frac{x}{y}\right) G_\gamma(y,t)\right]$$

$$\frac{\partial G_\gamma(x,t)}{\partial t} = \frac{\alpha_s(t)}{2\pi}\int_x^1 \frac{dy}{y}\left[P_{Gq}\left(\frac{x}{y}\right)\sum_{j=1}^{2f} q_\gamma^j(y,t) + P_{GG}\left(\frac{x}{y}\right) G_\gamma(y,t)\right] \quad (2)$$

where q_γ^i, G_γ are the distribution functions of the i-th quark and gluon inside the photon respectively. ε_i is the fractional charge of the i-th quark. P's are possibility functions[5]

$$\alpha_s(t) = \frac{\alpha_s(t_0)}{1 + b\alpha_s(t_0)\ln Q^2/\Lambda^2} \simeq \frac{1}{bt},$$

$$b=\frac{33-2f}{12\pi}\,, \qquad a(x)=x^2+(1-x)^2\,. \tag{3}$$

There is an "anomalous term" $\varepsilon_i^2\frac{\alpha}{2\pi}a(x)$ in Eq. (2). It comes from point-like box diagram after renormalization.

Taking the moment of each side of Eq. (2), we obtain

$$\begin{aligned}\frac{d}{dt}\langle q_\gamma^i(t)\rangle_n &= \varepsilon_i^2\frac{\alpha}{2\pi}a(n)+\frac{\alpha_s(t)}{2\pi}\frac{8}{3}A_n^{qq}\langle q_\gamma^i(t)\rangle_n+\frac{\alpha_s(t)}{2\pi}\frac{8}{3}A_n^{qG}\langle G_\gamma(t)\rangle_n\\ \frac{d}{dt}\langle G_\gamma(t)\rangle_n &= \frac{\alpha_s(t)}{2\pi}\frac{8}{3}A_n^{Gq}\sum_{j=1}^{2f}\langle q_\gamma^j(t)\rangle_n+\frac{\alpha_s(t)}{2\pi}\frac{8}{3}A_n^{GG}\langle G_\gamma(t)\rangle_n\end{aligned} \tag{4}$$

where the definitions of the moment and other quantities are

$$\langle q_\gamma^i(t)\rangle_n=\int_0^1 x^{n-1}dx\,q_\gamma^i(x,t),$$

$$\langle G_\gamma(t)\rangle_n=\int_0^1 x^{n-1}dx\,G_\gamma(x,t),$$

$$a(n)=\int_0^1 x^{n-1}dx\,a(x), \qquad A_n^{ij}=\int_0^1 x^{n-1}dx\,P_{ij}(x),$$

$$A_n^{qq}=\frac{3}{4}+\frac{1}{2n}-\frac{1}{2(n+1)}-\psi(n+1)-C$$

$$A_n^{qG}=\frac{3(2+n+n^2)}{16n(n+1)(n+2)}\,, \qquad A_n^{Gq}=\frac{2+n+n^2}{2n(n^2-1)}$$

$$A_n^{GG}=\frac{9}{4}\left[\frac{33-2f}{36}+\frac{1}{n-1}-\frac{1}{n}+\frac{1}{n+1}-\frac{1}{n+2}-\psi(n+1)-C\right] \tag{5}$$

$\psi(n+1)=\Gamma'(n+1)/\Gamma(n+1)$ is the digamma function, C is Euler's constant, C = 0.57721 ... Up to a multiplicative factor, the A's are the usual anomalous dimension in QCD.[1,2]

The solutions of Eq. (4) are

$$\langle q_\gamma^i(t)\rangle_n = \left[\langle q_\gamma^i(t_0)\rangle_n - \frac{1}{2f}\langle q_\gamma(t_0)\rangle_n\right] F_{qq1}(n,t)$$

$$+ \frac{1}{2f}\langle q_\gamma(t_0)\rangle_n F_{qq2}(n,t) + \langle G_\gamma(t_0)\rangle_n F_{qG}(n,t) + F_{qP}^i(n,t),$$

$$\langle G_\gamma(t)\rangle_n = \langle q_\gamma(t_0)\rangle_n F_{Gq}(n,t) + \langle G_\gamma(t_0)\rangle_n F_{GG}(n,t) + F_{GP}(n,t) \qquad (6)$$

The functions F(n,t) are given in Appendix A. t_0 is the initial value of t. $\langle q_\gamma(t)\rangle_n = \sum_i^{2f} \langle q_\gamma^i(t)\rangle_n$.

From Eq. (6), we see that there are contributions $F_{qP}^i(n,t)$ $F_{GP}(n,t)$ which come from the anomalous term. We call it point-like part, the rest hadron-like part. In order to obtain the contribution from the hadron-like part, we must give a initial input of the distribution functions, but for the point-like, it is not necessary.

After inverse Mellin transformation, we obtain the distribution functions themselves as

$$q_\gamma^i(x,t) = \frac{1}{2\pi i}\int_{\gamma-i\infty}^{\gamma+i\infty} dn\, \bar{x}^n \langle q_\gamma^i(t)\rangle_n$$

$$G_\gamma(x,t) = \frac{1}{2\pi i}\int_{\gamma-i\infty}^{\gamma+i\infty} dn\, \bar{x}^n \langle G_\gamma(t)\rangle_n \qquad (7)$$

where the contour ($\gamma-i\infty, \gamma+i\infty$) is to the right of all singularities of $\langle q_\gamma^i(t)\rangle_n, \langle G_\gamma(t)\rangle_n$ in the complex n plane. (Fig.1) The formulae (7) have another form which we used in our calculation:

$$q_\gamma^i(x,t)=\int_x^1\frac{dz}{z}\left[q_\gamma^i(z,t_0)-\frac{1}{2f}q_\gamma(z,t_0)\right]\tilde{F}_{qq1}\left(\frac{x}{z},t\right)$$

$$+\int_x^1\frac{dz}{z}\frac{1}{2f}q_\gamma(z,t_0)\tilde{F}_{qq2}\left(\frac{x}{z},t\right)$$

$$+\int_x^1\frac{dz}{z}G_\gamma(z,t_0)\tilde{F}_{qG}\left(\frac{x}{z},t\right)$$

$$+\tilde{F}_{qp}^i(x,t), \tag{8.a}$$

$$G_\gamma(x,t)=\int_x^1\frac{dz}{z}q_\gamma(z,t_0)\tilde{F}_{Gq}\left(\frac{x}{z},t\right)$$

$$+\int_x^1\frac{dz}{z}G_\gamma(z,t_0)\tilde{F}_{GG}\left(\frac{x}{z},t\right)$$

$$+\tilde{F}_{GP}(x,t), \tag{8.b}$$

where $\tilde{F}(x,t)$ is the inverse Mellin transformation of $F(n,t)$, $q_\gamma(z,t)=\sum_{i=1}^{2f}q_\gamma^i(z,t)$, and the s variable used later on is defined as $S=\frac{4}{3\pi b}\ln\frac{t}{t_0}$.

In order to get the inverse Mellin transformation of F(n,t), we use two different but mathematically equivalent forms of F(n,t), (see Appendix B) which are very useful when studying the behaviours of the distribution function near X = 0 and X = 1 respectively.

From Eq.(6) and Appendix A, we can also see that if we only consider the point-like contribution $F_{qp}^i(n,t)$, $F_{Gp}(n,t)$ and only leading logarithmic contribution (compared with t, we neglect the term $t_0 e^{\Lambda_n S}$), this is the results given by **Llewellyn Smith.**[10]

(II.2) Boundary Behaviours

To study the behaviours of the distributions of quark and gluon near X = 1 is equivalent to studying the limit of its Mellin transform for n going to infinity. Therefore, using Eq. (8.a) and (B2) in Appendix B, we make a 1/n expansion and take $n \rightarrow \infty$, **and calculate the** inverse Mellin transformation for the leading term.

Assuming the behaviours of parton inside photon near X = 1 for t = t_0 are

$$x q_\gamma^i (x, t_0) \simeq a_i (1-x)^{b_i}$$

$$x G_\gamma (x, t_0) \simeq a_g (1-x)^{b_g},$$

then the quark distribution function near X = 1 is

$$x q_\gamma^i (x,t) \simeq a_i e^{s(\frac{3}{4}-c)} \frac{\Gamma(b_i+1)}{\Gamma(s+b_i+1)} (1-x)^{s+b_i}$$

$$+ a_g \frac{3}{20} e^{s(\frac{3}{4}-c)} \frac{\Gamma(b_g+1)}{\Gamma(b_g+2+s)} \frac{(1-x)^{b_g+1+s}}{\left[(c-\frac{21-2f}{20}) + \ln\frac{1}{1-x} + \psi(b_g+2+s)\right]}$$

$$+ \varepsilon_i^2 \frac{\alpha}{\alpha_s} \frac{3}{8} \left[\frac{1}{\ln\frac{1}{1-x} + \frac{3\pi b}{4} - \frac{3}{4}} - \frac{e^{s(\frac{3}{4}-c-\frac{3\pi b}{4})}}{(\ln\frac{1}{1-x} + \frac{3\pi b}{4} - \frac{3}{4} + c + \psi(s+1))} \cdot \frac{(1-x)^s}{\Gamma(s+1)} \right] \quad (9)$$

The first term corresponds to a quark **having emitted a gluon,** the second one corresponds to a $q\bar{q}$ pair-produced from gluon. The third one comes from the point-like contribution.

From Eq. (9), obviously, the main contribution near X = 1 is given by the point-like part. If we only consider this part and if s is large enough, then the second term can be neglected in this part, we obtain the form: $\varepsilon_i^2 \frac{\alpha}{\alpha_s} f(x)$

where $f(x) = \dfrac{1}{\frac{8}{3}\ln\frac{1}{1-x}+2\pi b-2}$ is a scaling function. This behaviour is similar to the result given by the diagrammtic approach.[13,15]

Using Eq. (8b) and (B2) in Appendix B, we get the behaviour of gluon distribution near X = 1:

$$xG_\gamma(x,t) \simeq \frac{2}{5}\Big(\sum_i a_i\Big)e^{s(\frac{3}{4}-c)} \frac{\Gamma(b_i+1)}{\Gamma(b_i+2+s)} \frac{(1-x)^{b_i+1+s}}{\left[(c-\frac{21-2f}{20})+\ln\frac{1}{1-x}+\psi(b_i+2+s)\right]}$$

$$+a_g e^{s\frac{9}{4}(\frac{33-2f}{36}-c)} \frac{\Gamma(b_g+1)}{\Gamma(\frac{9}{4}s+b_g+1)}(1-x)^{\frac{9}{4}s+b_g}$$

$$+\frac{3}{20}\Big(\sum_i \varepsilon_i^2\Big)\frac{\alpha}{\alpha_s}(1-x)\frac{1}{\ln\frac{1}{1-x}}\frac{1}{\ln\frac{1}{1-x}+\frac{3\pi b}{4}+\frac{1}{4}} \tag{10}$$

The first term corresponds to the gluon produced by the quark through bremsstrahlung. The second one corresponds to the gluon-pair produced by gluon. The last one comes from the point-like contribution.

To study the behaviour of the parton distribution near X = 0 is equivalent to studying the behaviour of its Mellin transform near its rightmost singularity in the complex n plane. Let us assume that the behaviours of the parton near X = 0 for $t = t_0$ are

$$x q_\gamma^i(x,t_0) \simeq C_i \qquad xG_\gamma(x,t_0) \simeq C_g\,.$$

Corresponding to Eq. (9), (10), the behaviours of quark and gluon distribution function near X = 0 are:

$$\begin{aligned} x q_\gamma^i(x,t) \sim & \frac{2}{81}\Big(\sum_i c_i\Big) e^{-s\left(\frac{33}{16}+\frac{1}{72}f\right)} \Big(\frac{9s}{4\ln\frac{1}{x}}\Big)^{1/2} I_1\Big(3\sqrt{s\ln\frac{1}{x}}\Big) \\ & + \frac{1}{18} C_g e^{-s\left(\frac{33}{16}+\frac{1}{72}f\right)} \Big(\frac{9s}{4\ln\frac{1}{x}}\Big)^{1/2} I_1\Big(3\sqrt{s\ln\frac{1}{x}}\Big) \\ & + \frac{4f}{729} \varepsilon_i^2 \alpha b t_0 e^{-s\left(\frac{33}{16}+\frac{1}{72}f\right)} \Big(\frac{9s}{4\ln\frac{1}{x}}\Big)^{3/2} I_3\Big(3\sqrt{s\ln\frac{1}{x}}\Big) \\ x G_\gamma(x,t) \sim & \frac{4}{9}\Big(\sum_i c_i\Big) e^{-s\left(\frac{33}{16}+\frac{1}{72}f\right)} I_0\Big(3\sqrt{s\ln\frac{1}{x}}\Big) \\ & + C_g e^{-s\left(\frac{33}{16}+\frac{1}{72}f\right)} I_0\Big(3\sqrt{s\ln\frac{1}{x}}\Big) \\ & + \frac{4}{81}\Big(\sum_i \varepsilon_i^2\Big) \alpha b t_0 e^{-s\left(\frac{33}{16}+\frac{1}{72}f\right)} \Big(\frac{9s}{4\ln\frac{1}{x}}\Big) I_2\Big(3\sqrt{s\ln\frac{1}{x}}\Big). \end{aligned} \quad (11)$$

(II.3) Numerical Results

In order to get final result using Eq. (8), we have to input an initial value of the photon distribution function. Unfortunately, up to now, we have no data on it. However, we can estimate it by using crude, but apparently reasonable, assumptions.

In general, both the quark and gluon distribution functions contain hadron-like and point-like parts. For the hadron-like part we use vector meson dominance and SU(3) symmetry to estimate it, e.g.

$$q_\gamma^i(x,t_0) = \sum_V \frac{4\pi\alpha}{f_V^2} q_V^i(x,t_0)$$

(hadron-like part)

$$\frac{f_\rho^2}{4\pi} : \frac{f_\omega^2}{4\pi} : \frac{f_\phi^2}{4\pi} = 1 : 9 : \frac{9}{2}$$

$$\frac{f_\rho^2}{4\pi} = 2.5 \tag{12}$$

Next, we assume that the valence quark of ρ_0, ω (u,d-quark) and ϕ (s-quark) have the same distribution as that of the valence quark of π^0 and the sea quark of $\rho_0 \omega$ and ϕ have the same distribution function as that of the π_0 sea-quark if the spin has little effect on distribution function. Assuming SU(2) symmetry, π^0 and $\pi^\pm$ have the same sea-quark and gluon distribution, but the valence quark distribution in π^0 for each flavor (u, $\bar{u}$, d, $\bar{d}$) is half that in the π^+(u, $\bar{d}$). Recently, a CERN-SPS group has given the distribution function of $\pi^\pm$. [16] According to the principle mentioned above, we take the hadron-like part as

$$x q_\gamma^{u,d}(x,t_0) = \left(\frac{4\pi\alpha}{f_\rho^2} + \frac{4\pi\alpha}{f_\omega^2}\right) x q_{\pi^0(V)}(x,t_0) + \left(\frac{4\pi\alpha}{f_\rho^2} + \frac{4\pi\alpha}{f_\omega^2} + \frac{4\pi\alpha}{f_\phi^2}\right) x q_{\pi^0(S)}(x,t_0)$$

$$x q_\gamma^{s}(x,t_0) = \frac{4\pi\alpha}{f_\phi^2} 2 x q_{\pi^0(V)}(x,t_0) + \left(\frac{4\pi\alpha}{f_\rho^2} + \frac{4\pi\alpha}{f_\omega^2} + \frac{4\pi\alpha}{f_\phi^2}\right) x q_{\pi^0(S)}(x,t_0)$$

$$x G_\gamma(x,t_0) = \left(\frac{4\pi\alpha}{f_\rho^2} + \frac{4\pi\alpha}{f_\omega^2} + \frac{4\pi\alpha}{f_\phi^2}\right) x G_{\pi^0}(x,t_0) \tag{13}$$

where $$x q_{\pi^0(v)}(x, t_0) = 0.275\, x^{0.4}(1-x)^{0.9}$$

for one kind of the valence quark of π^0,

$$x q_{\pi^0(s)}(x, t_0) = 0.09(1-x)^{4.4}$$

for one kind of the sea quark of π^0

$$x G_{\pi^0}(x, t_0) = a(1-x)^5 \qquad a = 3.36$$

a is determined by momentum conservation of π^0. $Q_0^2 = 15(\text{GeV}/c)^2$.

For the point-like piece, compared with the hadron-like piece, we assume that at low Q^2 it can be neglected. Thus, using the expressions $F_{qp}(n,t)$, $F_{Gp}(n,t)$ and making Mellin transformation we can get the point-like piece at large Q^2. For calculational convenience, we assume that at $Q^2 = 4.6\ \text{GeV}^2/c^2$ it can be neglected. It corresponds to s = 0.2 from $Q^2 = 4.6\ \text{GeV}^2/c^2$ to $Q^2 = 15\ \text{GeV}^2/c^2$. After fitting the numerical output, we get the approximate expression for point-like piece at $Q^2 = 15\ \text{GeV}^2/c^2$,

$$x q_p(x) = \frac{\alpha}{\pi}(4.081)\left[(-0.12+0.3x)(1-x)^{0.1} + \frac{0.07}{x} - 0.07\right]$$

for charge =2/3

$$x q_p(x) = \frac{\alpha}{\pi}(4.081)\left[(-0.032+0.1x)(1-x)^{0.1} + \frac{0.02}{x} - 0.035\right]$$

for charge = 1/3

$$x G_p(x) = \frac{\alpha}{\pi}(4.081)\left[\frac{0.0012}{x^3} + 0.0001\, x\right] \qquad (14)$$

Then we take the sum of Eqs. (13) and (14) as initial input of Eq. (8).

For small values of x, we use the formula (B1) of Appendix B and compute the contributions of all the singularities of $F(n,t)$ in the complex n plane: n = 1, 0, -1, -2, ... Noting the fact that the inverse Mellin transform of $\frac{1}{(n+P)^r}$ is $x^P \frac{1}{\Gamma(r)}(\ln \frac{1}{x})^{r-1}$, we use the computer to sum up all the terms and finally obtain the $xq_\gamma^i(x,t)$, $xG_\gamma(x,t)$ for small x using Eq.(8).

For large values of x, we use formula (B2) of Appendix B and expand the integrand in a series in $\frac{1}{n-1}$. Noting the fact that the inverse Mellin transform of $\frac{1}{(n-1)^r}$ is $\frac{1}{x\Gamma(r)}(\ln \frac{1}{x})^{r-1}$, we obtain the inverse Mellin transform of each term of the series. After integration over v, $\widetilde{F}(x,t)$ is obtained. Also $xq_\gamma^i(x,t)$, $xG_\gamma(x,t)$ are obtained from Eqs. (8a) (8b) for large values of x.

Finally we check that the two types of calculation give the same results for $x \sim 0.6 - 0.7$.

The numerical results are shown in Figs. 2-7. If we take $\Lambda^2 = 0.25\ GeV^2/c^2$, $Q_0^2 = 15\ GeV^2/c^2$, then s = 0.2, 0.1 and 0.05 correspond to $Q^2 = 77.6$, 31.8 and 21.5 $(GeV/c)^2$ respectively.

About the hadron-like piece, we compare, as an example in Fig. 2, our up-quark distribution function with the result given in Ref.[11]. Compared with their input, we have not only a hadron-like but also a point-like piece in our input. We also take not only the ρ-meson but also ω、ϕ mesons in vector dominance. Thus our results are large than theirs. It is reasonable.

From Figs.3-5, we can see that in the general case, xq_γ^u, xq_γ^d, xq_γ^s are not equal to one another. This is due to the fact that xq_γ^d and xq_γ^s have the same point-like behaviours, but with different hadron-like behaviours. xq_γ^u and xq_γ^s have different point-like behaviours, generally speaking, they have different hadron-like behaviours too.

Our calculation shows that, at least in the region of Q^2 which we are interested in, for quark and anti-quark, the contributions given by $q \to q+g$ and $g \to q\bar{q}$ are comparable in small x region, but the contribution of $q \to q+g$ dominates in the large x region. For gluon distribution, the contributions given by $q \to q+g$ and $g \to g,g$ are comparable.

In the same leading logarithmic approximation, we compare two pieces: hadron-like and point-like.(In Fig. 7, a point-like piece is given,see Ref.[10] , we take xq_γ^u as an example of hadron-like piece). In the large x region, a point-like piece is dominant, but in the small x region, the contributions of point-like and hadron-like pieces are comparable. (in some Q^2 values)

III. FRAGMENTATION FUNCTIONS

According to the Kogut and Susskind picture, the fragmentation functions [17] of the quark and gluon are also expected to have Q^2 dependence, since the fragmentation of N-th level partons occurs through the partons of the (N-1)-th level and there are different behaviours of the fragmentation function on different levels. Following Altarelli and Parisi, there are integro-differential equations for the Q^2 dependent fragmentation functions. Several authors discuss it in this way,[18] or in field theory,[19] or in model calculation.[20] Here we shall use A-P type equations. Particular attention will be paid to getting the functions themselves, to giving boundary behaviours.

(III.1) A-P Type Equations for Fragmentation Functions

The integro-differential equations for fragmentation functions $D^h_{q_i}(z,t), D^h_G(z,t)$ are:

$$\frac{\partial}{\partial t} D^h_{q_i}(z,t) = \frac{\alpha_s(t)}{2\pi}\int_z^1 \frac{dz'}{z'}\left[P_{qq}\left(\frac{z}{z'}\right) D^h_{q_i}(z',t) + P_{Gq}\left(\frac{z}{z'}\right) D^h_G(z',t)\right]$$

$$\frac{\partial}{\partial t} D^h_G(z,t) = \frac{\alpha_s(t)}{2\pi}\int_z^1 \frac{dz'}{z'}\left[P_{qG}\left(\frac{z}{z'}\right) \sum_{j=1}^{2f} D^h_{q_j}(z',t) + P_{GG}\left(\frac{z}{z'}\right) D^h_G(z',t)\right] \quad (15)$$

where $D^h_{q_i}(z,t), D^h_G(z,t)$ are the mean number of hadrons of type h with momentum fraction z per dz in a jet initiated by a quark q_i (gluon G) at a scale t.

The physical explanation of Eq. (15) is that the Q^2 dependence of fragmentation function $D^h_{q_i}(z,t)$ is due to two processes: (1) the quark radiates a gluon and then fragments; (2) the quark radiates a gluon which fragments into the hadron. The Q^2 dependence of fragmentation function $D_G^h(z,t)$ is due to two processes: (1) the gluon can pair-produce a quark which then fragments, (2) the gluon can pair-produce gluon which fragments into the hadrons.

The solutive of Eq. (15) is similar to those in Section (II). The solutions of moment of D functions are (ignoring the mass effect):

$$
\begin{aligned}
\langle D_{q_i}^h(t)\rangle_n &= \left(\langle D_{q_i}^h(t_0)\rangle_n - \frac{1}{2f}\langle D_q^h(t_0)\rangle_n\right) F_{qq_1}(n,t) \\
&\quad + \frac{1}{2f}\langle D_q^h(t_0)\rangle_n F_{qq_2}(n,t) \\
&\quad + \langle D_G^h(t_0)\rangle_n F_{Gq}(n,t), \\
\langle D_G^h(t)\rangle_n &= \langle D_q^h(t_0)\rangle_n F_{qG}(n,t) \\
&\quad + \langle D_G^h(t_0)\rangle_n F_{GG}(n,t)
\end{aligned}
\tag{16}
$$

where some definitions are:

$$
\begin{aligned}
\langle D_{q_i}^h(t)\rangle_n &= \int_0^A z^{n-1} dz\, D_{q_i}^h(z,t) \\
\langle D_G^h(t)\rangle_n &= \int_0^1 z^{n-1} dz\, D_G^h(z,t) \\
\langle D_q^h(t_0)\rangle_n &= \sum_{i=1}^{2f} \langle D_{q_i}^h(t_0)\rangle_n \, .
\end{aligned}
$$

Making inverse Mellin transformation, we get the expressions for fragmentation functions :

$$D^h_{q_i}(z,t)=\int_z^1 \frac{dy}{y}\left[D^h_{q_i}(y,t_0)-\frac{1}{2f}D^h_q(y,t_0)\right]\tilde{F}_{qq_1}\left(\frac{z}{y},t\right)$$

$$+\int_z^1 \frac{dy}{y}\frac{1}{2f}D^h_q(y,t_0)\tilde{F}_{qq_2}\left(\frac{z}{y},t\right)$$

$$+\int_z^1 \frac{dy}{y}D^h_G(y,t_0)\tilde{F}_{Gq}\left(\frac{z}{y},t\right)\ . \qquad (17.a)$$

$$D^h_G(z,t)=\int_z^1 \frac{dy}{y}D^h_q(y,t_0)\tilde{F}_{qG}\left(\frac{z}{y},t\right)$$

$$+\int_z^1 \frac{dy}{y}D^h_G(y,t_0)\tilde{F}_{GG}\left(\frac{z}{y},t\right)\ . \qquad (17.b)$$

where $\tilde{F}(x,t)$ is the inverse Mellin transformation of $F(n,t)$, $D^h_q(z,t)=\sum_{i=1}^{2f}D^h_{q_i}(z,t)$. Similarly, we use two different, but mathematically equivalent, forms of $F(n,t)$ near $z = 0$ and $z = 1$ respectively. (see Appendix B)

(III.2) Behaviour near the Kinematic Boundary

The method which we use to discuss the boundary behaviour of photon distribution functions in Section (II.2) can be used to discuss the boundary behaviour of fragmentation functions.

If we assume that the behaviour of fragmentation functions near $z = 1$ for t_0 are:

$$z D^h_{q_i}(z,t_0)=a_i(1-z)^{b_i}$$
$$z D^h_q(z,t_0)=a(1-z)^{b}$$
$$z D^h_G(z,t_0)=a_g(1-z)^{b_g}\ . \qquad (18.a)$$

then the behaviours of fragmentation functions are:

$$z D_{q_i}^h(z,t) \simeq a_i e^{s(\frac{3}{4}-c)} \frac{\Gamma(b_i+1)}{\Gamma(b_i+1+s)} (1-z)^{b_i+s}$$

$$+ \frac{2}{5} a_g e^{s(\frac{3}{4}-c)} \frac{\Gamma(b_g+1)}{\Gamma(b_g+2+s)} \frac{(1-z)^{b_g+s+1}}{(c-\frac{21-2f}{20})+\ln\frac{1}{1-x}+\psi(b_g+2+s)},$$

$$z D_G^h(z,t) \simeq a \frac{3}{20} e^{s(\frac{3}{4}-c)} \frac{\Gamma(b+1)}{\Gamma(b+s+2)} \frac{(1-z)^{b+s+1}}{(c-\frac{21-2f}{20})+\ln\frac{1}{1-x}+\psi(b+2+s)}$$

$$+ a_g e^{s\frac{9}{4}(\frac{33-2f}{36}-c)} \frac{\Gamma(b_g+1)}{\Gamma(b_g+1+\frac{9}{4}s)} (1-z)^{b_g+\frac{9}{4}s} \tag{18.b}$$

If the behaviours of fragmentation functions near $Z = 0$ for $t = t_0$ are:

$$z D_{q_i}^h(z,t_0) \sim C_i$$

$$z D_G^h(z,t_0) \sim C_g \tag{19.a}$$

then the behaviours of $z D_{q_i}^h(z,t)$ and $z D_G^h(z,t)$ near $Z = 0$ are:

$$z D_{q_i}^h(z,t) \simeq \Big(\sum_i a_i\Big) \frac{2}{81} e^{-s(\frac{33}{16}+\frac{1}{72}f)} \Big(\frac{9s}{4\ln\frac{1}{z}}\Big)^{1/2} I_1\Big(3\sqrt{s\ln\frac{1}{z}}\Big)$$

$$+ C_g \frac{4}{9} e^{-s(\frac{33}{16}+\frac{1}{72}f)} I_0\Big(3\sqrt{s\ln\frac{1}{z}}\Big)$$

$$z D_G^h(z,t) \simeq \Big(\sum_i a_i\Big) \frac{1}{18} e^{-s(\frac{33}{16}+\frac{1}{72}f)} \Big(\frac{9s}{4\ln\frac{1}{z}}\Big)^{1/2} I_1\Big(3\sqrt{s\ln\frac{1}{z}}\Big)$$

$$+ C_g e^{-s(\frac{33}{16}+\frac{1}{72}f_2)} I_0\Big(3\sqrt{s\ln\frac{1}{z}}\Big) \tag{19.b}$$

In the lowest approximation, we obtain the boundary behaviours (9), (10), (11), (18), (19), but higher order contributions must be considered. This would be a complicated calculation.

(III.3) Numerical Results

In this section we give numerical results for the fragmentation functions of the parton into pions and kaons.

According to the assumptions in Ref. [21], there are two (three) independent fragmentation functions for pions (kaons). That is: (We only consider the light quarks)

$$
\begin{aligned}
&D_u^{\pi^+} = D_d^{\pi^-} = D_{\bar{u}}^{\pi^-} = D_{\bar{d}}^{\pi^+} \\
&D_u^{\pi^-} = D_d^{\pi^+} = D_{\bar{d}}^{\pi^-} = D_{\bar{u}}^{\pi^+} \\
&\quad \simeq D_s^{\pi^+} = D_s^{\pi^-} = D_{\bar{s}}^{\pi^+} = D_{\bar{s}}^{\pi^-} \\
&D_q^{\pi^0} = \frac{1}{2}\left(D_q^{\pi^+} + D_q^{\pi^-} \right)
\end{aligned}
\tag{20.a}
$$

and

$$
\begin{aligned}
&D_u^{K^+} = D_d^{K^0} = D_{\bar{u}}^{K^-} = D_{\bar{d}}^{\bar{K}^0} \\
&D_s^{K^-} = D_s^{\bar{K}^0} = D_{\bar{s}}^{K^+} = D_{\bar{s}}^{K^0} \\
&D_u^{K^-} = D_d^{\bar{K}^0} = D_{\bar{u}}^{K^+} = D_{\bar{d}}^{K^0} \\
&\quad \simeq D_d^{K^-} = D_u^{\bar{K}^0} = D_{\bar{d}}^{K^+} = D_{\bar{u}}^{K^0} \\
&\quad \simeq D_d^{K^+} = D_u^{K^0} = D_{\bar{d}}^{K^-} = D_{\bar{u}}^{\bar{K}^0} \\
&\quad \simeq D_s^{K^+} = D_s^{K^0} = D_{\bar{s}}^{K^-} = D_{\bar{s}}^{\bar{K}^0}
\end{aligned}
\tag{20.b}
$$

Also, following Field and Feynman[22] , we set $\alpha_S = \alpha_V$ and get the following results for the initial values of the D_q functions:

$$D_u^{\pi^+} = -3.569849 - 0.741609z + 0.089012z^2 + 0.780270\frac{1}{z} + 3.488177z^{0.46} - 0.413852\ln z$$

$$D_u^{\pi^-} = -3.153852 - 1.049605z - 0.06499z^2 + 0.780270\frac{1}{z} + 3.488177z^{0.46} + 0.263479\ln z$$

$$D_u^{K^+} = -0.338995 - 0.393414z + 0.157387z^2 + 0.184108\frac{1}{z} + 0.413914z^{0.46} + 0.041315\ln z$$

$$D_u^{K^-} = -0.477492 - 0.08541z - 0.035112z^2 + 0.184108\frac{1}{z} + 0.413914z^{0.46} + 0.125984\ln z$$

$$D_s^{K^-} = -0.828812 + 0.136393z + 0.140448z^2 + 0.184108\frac{1}{z} + 0.413914z^{0.46} - 0.503936\ln z \tag{21}$$

with $Q_0^2 = 4(\text{Gev/c})^2$ and $\Lambda = 0.4$ Gev/c.

By using the assumptions of Ref. 23 , the initial values of D_G are:

$$zD_G^{\pi} = \frac{1}{2}(1-z)^{1.5}$$

$$zD_G^{k} = \frac{1}{4}(1-z)^{1.5} \tag{22}$$

Similarly, we use the different form of F(n,t) to make inverse Mellin transformation in the small and large z regions. Using Eq. (17) and requiring those two results to be equal at z=0.6 0.7, we get the fragmentation functions. The results are shown in Figs. 8-14. For $Q_0^2 = 4(\text{Gev/c})^2$, $\Lambda = 0.4$ Gev/c, so s = 0.3, 0.2, 0.1 correspond to $Q^2 = 33.37$, 14.56, $7.22(\text{Gev/c})^2$ respectively.

From Figs. 8-9, we see that when z is close to 1, the values of $zD_u^{\pi^+}$ are larger than $zD_u^{\pi^-}$. This is due to the fact that near z = 1, the observed hadron contains the initial quark; for a u-quark it is easy to get a $\bar{d}$-quark from vacuum and form a π^+, whereas it is more difficult to form a π^-. Similar situations also happen for $zD_u^{K^+}$ and $zD_u^{K^-}$.

Compared with parton distribution functions, considerably less is known about scaling deviations in fragmentation functions. To my knowledge, this is the first time that numerical results about the Q^2 dependence of light quark and gluon fragmentation functions in leading logarithm approximation have been obtained.

I would like to thank S.D. Drell, J. Ellis, F. Martin and R. Petronzio for useful discussions. I also wish to thank F. Martin for some relevant programs.

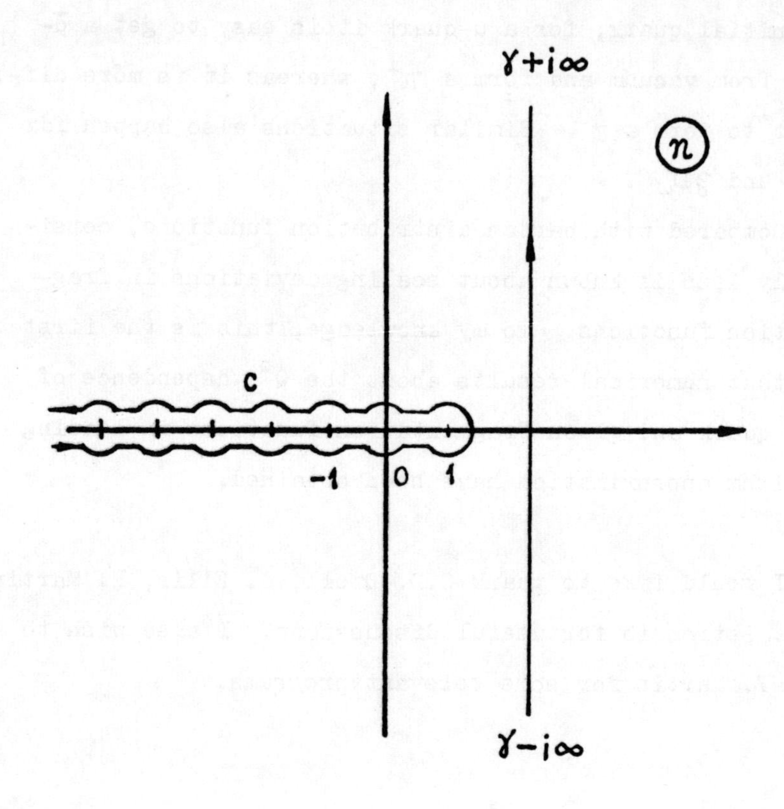

Fig. 1 The contour C in the complex n plane.

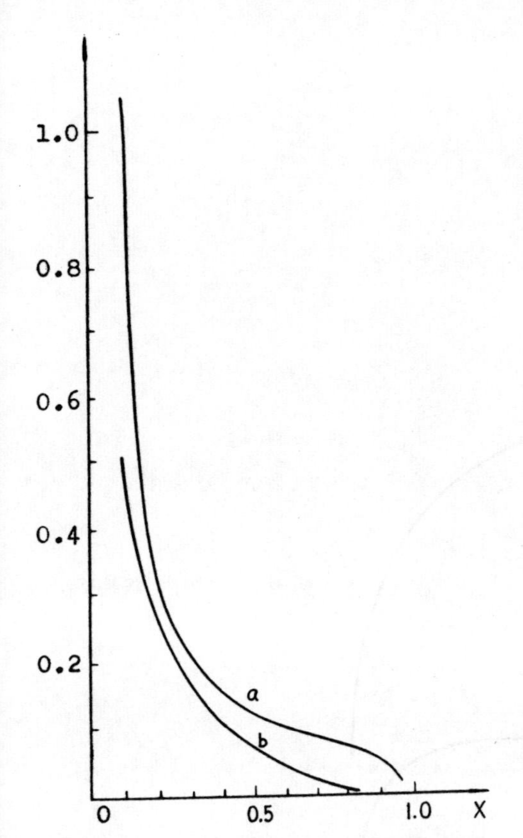

Fig. 2 Comparison of the hadron-like piece of q^u_γ (x,t) (a) with the corresponding result (b) in Ref.[11] for Q^2 =200 GeV^2/c^2.

($x10^{-2}$)

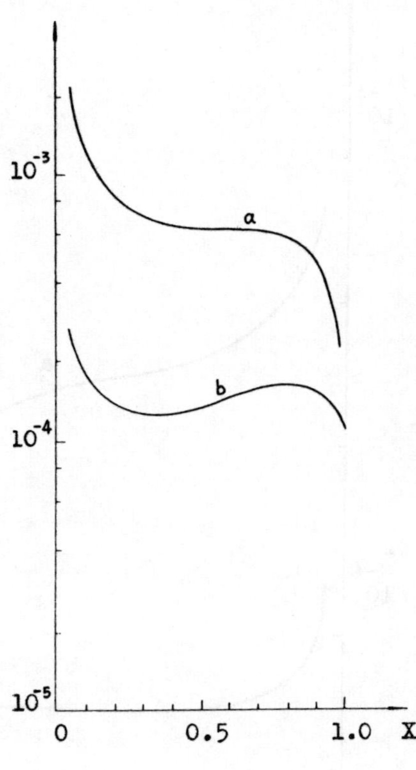

Fig. 3 The hadron-like piece of up-quark distribution function xq^u_γ (x,t) for Q^2=200 GeV^2/c^2 (a) and Q^2=20 GeV^2/c^2.(b)

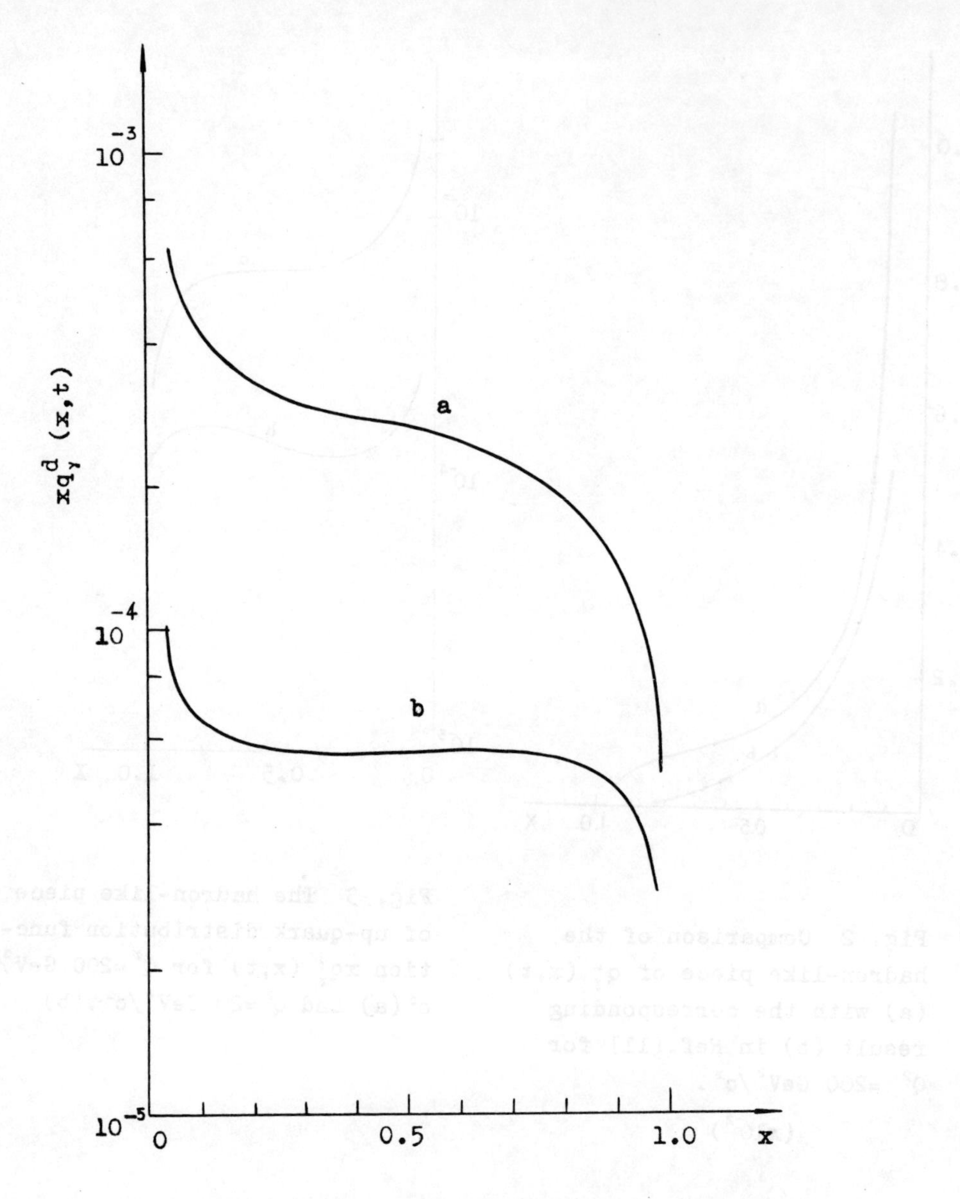

Fig. 4 The hadron-like piece of d-quark distribution function $xq_\gamma^d(x,t)$ for Q^2=200Gev^2/c^2 (a) and Q^2=20Gev^2/c^2.(b)

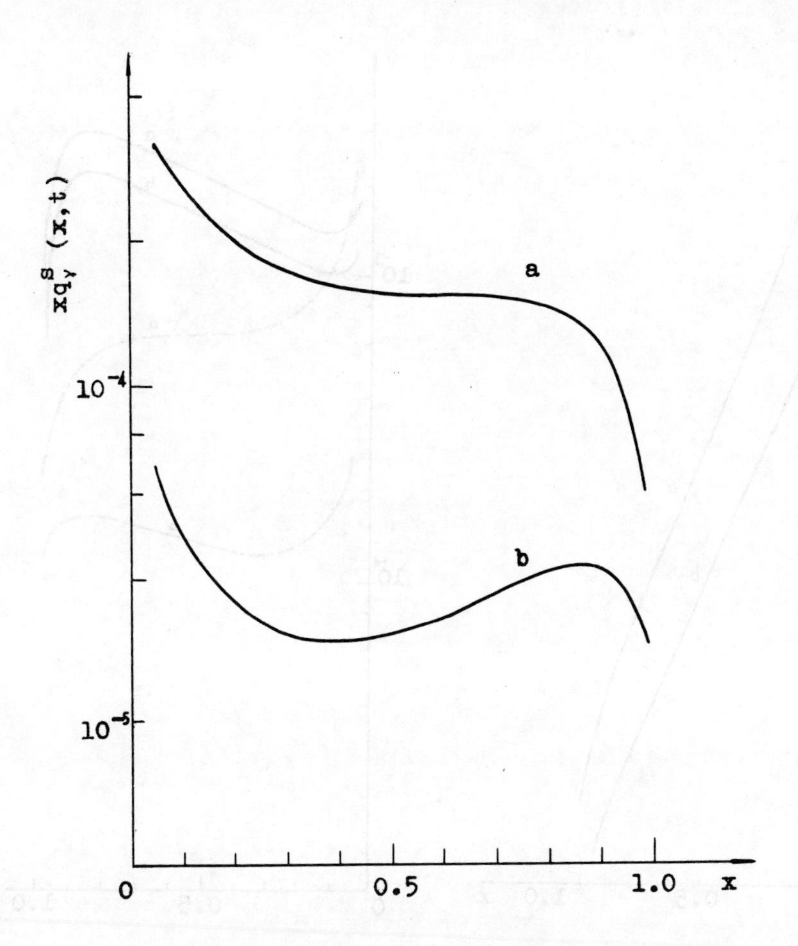

Fig. 5 The hadron-like piece of s-quark distribution function $xq_\gamma^s(x,t)$ for Q^2=200Gev^2/c^2 (a) and Q^2=20Gev^2/c^2 (b)

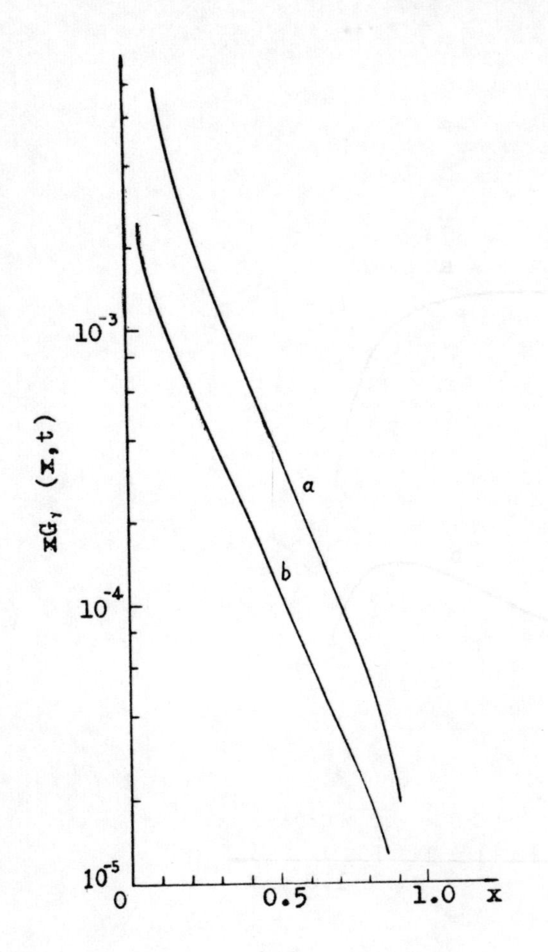

Fig. 6 The hadron-like piece of gluon distribution function $xG_\gamma(x,t)$ for Q^2 = 200 GeV^2/c^2.(a) and Q^2=20 GeV^2/c^2.(b).

Fig. 7 Comparison the point-like piece (from [10], solid line) with hadron-like piece (hadron-like piece of xq^u_γ as an example, dotted line) in the leading logarithm approximation for Q^2 =200Gev^2/c^2 (a) and Q^2=20GeV^2/c^2 (b).

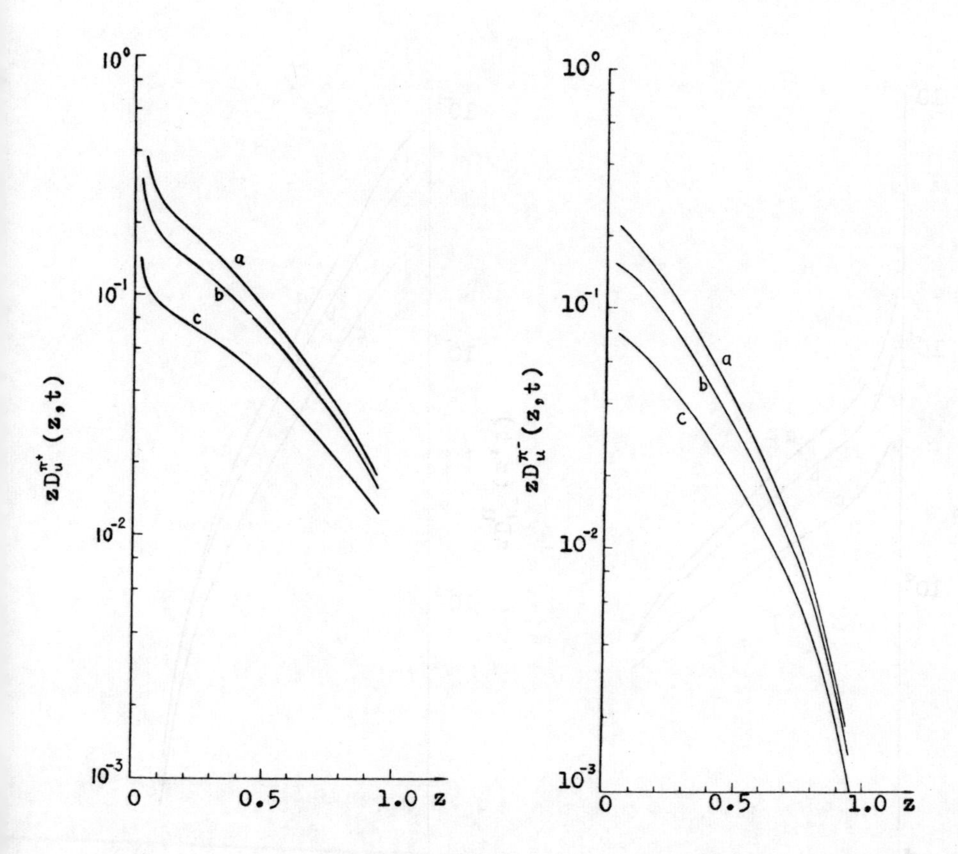

Fig. 8 Fragmentation function $zD_u^{\pi^+}(z,t)$ S=0.3(a), S=0.2(b) and S=0.1(c).

Fig. 9 Fragmentation function $zD_u^{\pi^-}(z,t)$, S=0.3(a), S=0.2(b) and S=0.1(c).

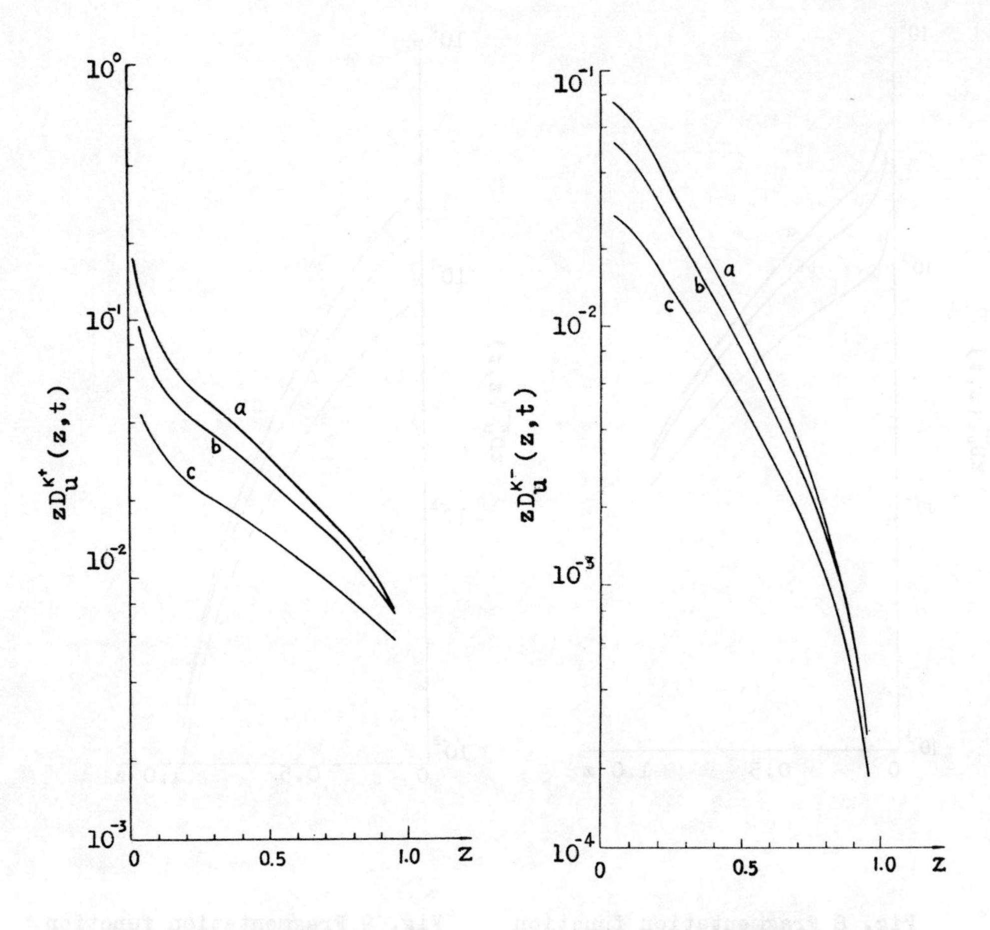

Fig. 10 Fragmentation function $zD_u^{K^+}(z,t)$. S=0.3(a), S=0.2(b) S =0.1(c).

Fig. 11 Fragmentation function $zD_u^{K^-}(z,t)$ S=0.3(a), S=0.2(b) and S=0.1(c).

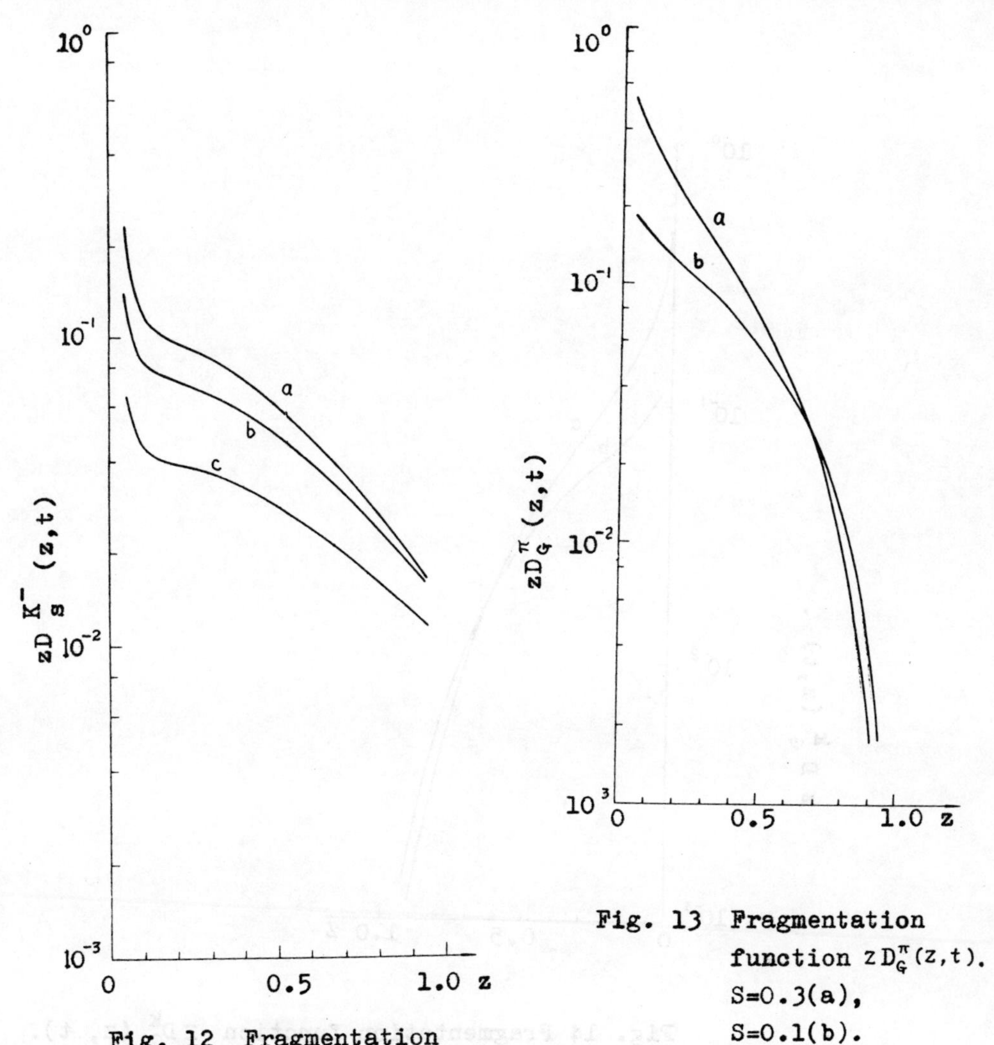

Fig. 13 Fragmentation function $zD_G^{\pi}(z,t)$. S=0.3(a), S=0.1(b).

Fig. 12 Fragmentation function $zD_s^{K^-}(z,t)$. S=0.3(a), S=0.2(b) and S=0.1(c).

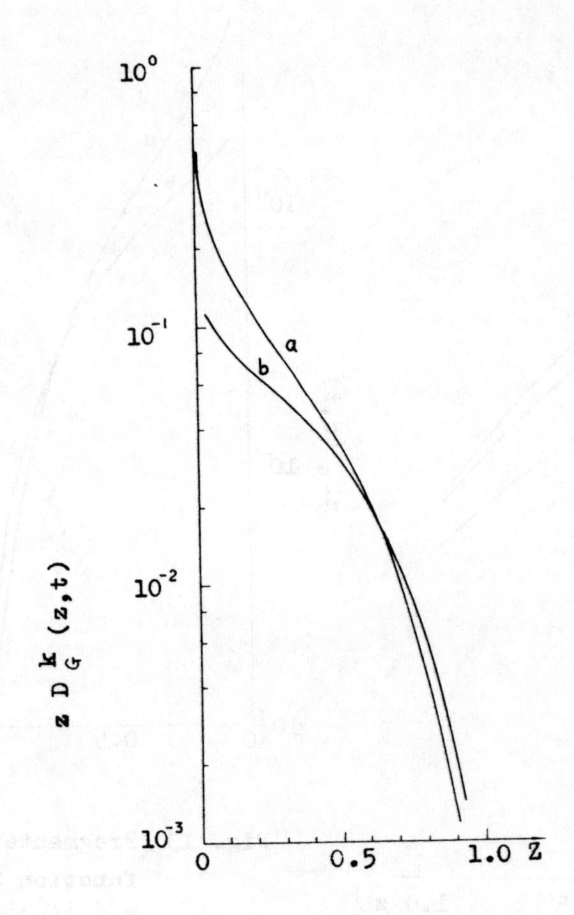

Fig. 14 Fragmentation function $z D_G^k$ (z, t). S=0.3(a), S=0.1(b).

Appendix A

$$F_{qq_1}\ (n,t) = e^{A_n^{qq}S}$$

$$F_{qq_2}\ (n,t) = \frac{1}{2}(e^{A_n^+S} + e^{A_n^-S}) - \frac{1}{2}(A_n^{GG} - A_n^{qq})\frac{e^{A_n^+S} - e^{A_n^-S}}{A_n^+ - A}$$

$$F_{qG}\ (n,t) = A_n^{qG}\frac{e^{A_n^+S} - e^{A_n^-S}}{A_n^+ - A_n^-}$$

$$F_{qp}^i\ (n,t) = (\varepsilon_i^2 - \frac{1}{2f}\sum_i \varepsilon_i^2)\frac{\alpha}{2\pi}a(n)\frac{1}{1 - \frac{4}{3\pi b}A_n^{qq}}[t - t_0 e^{A_n^{qq}S}]$$

$$+ \frac{1}{2f}(\sum_i \varepsilon_i^2)\frac{\alpha}{2\pi}a(n)\frac{1}{2}$$

$$\left\{\frac{1}{1 - \frac{4}{3\pi b}A_n^+}(t - t_0 e^{A_n^+S}) + \frac{1}{1 - \frac{4}{4\pi b}A_n^-}(t - t_0 e^{A_n^-S})\right.$$

$$\left. - \frac{A_n^{GG} - A_n^{qq}}{A_n^+ - A_n^-}\left[\frac{1}{1 - \frac{4}{3\pi b}A_n^+}(t - t_0 e^{A_n^+S}) - \frac{1}{1 - \frac{4}{3\pi b}A_n^-}(t - t_0 e^{A_n^-S})\right]\right\}$$

$$F_{Gq}\ (n,t) = A_n^{Gq}\frac{e^{A_n^+S} - e^{A_n^-S}}{A_n^+ - A_n^-}$$

$$F_{GG}\ (n,t) = \frac{1}{2}(e^{A_n^+S} + e^{A_n^-S}) + \frac{1}{2}(A_n^{GG} - A_n^{qq})\frac{e^{A_n^+S} - e^{A_n^-S}}{A_n^+ - A_n^-}$$

$$F_{GP}\ (n,t) = (\sum_i \varepsilon_i^2)\frac{\alpha}{2\pi}a(n)$$

$$\frac{A_n^{Gq}}{A_n^+ - A_n^-}\left[\frac{1}{1 - \frac{4}{3\pi b}A_n^+}(t - t_0 e^{A_n^+S}) - \frac{1}{1 - \frac{4}{3\pi b}A_n^-}(t - t_0 e^{A_n^-S})\right]$$

where

$$A_n^{\pm} = \frac{1}{2}\left[A_n^{GG} + A_n^{qq} \pm \sqrt{(A_n^{GG} - A_n^{qq})^2 + 8fA_n^{Gq}A_n^{qG}}\right]$$

$$S = \frac{4}{3\pi b}\ln\frac{t}{t_0}\ .$$

Appendix B

There are two different forms of F(n,t). Expression (B1) is convenient near x = 0 and (B2) is convenient near x = 1. The results are:

(B1)

$$F_{qq1}(n,t) = e^{A_n^{qq}S}$$

$$F_{qq2}(n,t) = \sum_{l=0}^{\infty} \frac{1}{l!} S^l (P_n)_l$$

$$F_{qG}(n,t) = A_n^{qG} \sum_{l=0}^{\infty} \frac{1}{l!} S^l (L_n)_l$$

$$F_{qP}^i(n,t) = \left(\varepsilon_i^2 - \frac{1}{2f}\sum_i \varepsilon_i^2\right)\frac{\alpha}{2\pi} t_0\, a(n) \sum_{l=0}^{\infty} \frac{1}{l!}\left(\frac{3\pi b}{4}S\right)^l (Q_n)_l$$

$$+ \frac{1}{2f}\sum_i \varepsilon_i^2 \frac{\alpha}{2\pi} t_0\, a(n) \sum_{l=0}^{\infty} \frac{1}{l!}\left(\frac{3\pi b}{4}S\right)^l (R_n)_l$$

$$F_{Gq}(n,t) = A_n^{Gq} \sum_{l=0}^{\infty} \frac{1}{l!} S^l (L_n)_l$$

$$F_{GG}(n,t) = \sum_{l=0}^{\infty} \frac{1}{l!} S^l (S_n)_l$$

$$F_{GP}(n,t) = \left(\sum_i \varepsilon_i^2\right)\frac{\alpha}{2\pi} t_0\, a(n) A_n^{Gq} \sum_{l=0}^{\infty} \frac{1}{l!}\left(\frac{3\pi b}{4}S\right)^l (H_n)_l$$

where

$$(L_n)_l = \frac{(A_n^+)^l - (A_n^-)^l}{A_n^+ - A_n^-}$$

$$(P_n)_l = \frac{1}{2}\left[(A_n^+)^l + (A_n^-)^l - (A_n^{GG} - A_n^{qq})\frac{(A_n^+)^l - (A_n^-)^l}{A_n^+ - A_n^-}\right]$$

$$(S_n)_l = \frac{1}{2}\left[(A_n^+)^l + (A_n^-)^l + (A_n^{GG} - A_n^{qq})\frac{(A_n^+)^l - (A_n^-)^l}{A_n^+ - A_n^-}\right]$$

$$(Q_n)_\ell = \frac{1-\left(\frac{4}{3\pi b}A_n^{qq}\right)^\ell}{1-\frac{4}{3\pi b}A_n^{qq}}$$

$$(R_n)_\ell = \frac{1}{2}\Bigg[\frac{1-\left(\frac{4}{3\pi b}A_n^+\right)^\ell}{1-\frac{4}{3\pi b}A_n^+}+\frac{1-\left(\frac{4}{3\pi b}A_n^-\right)^\ell}{1-\frac{4}{3\pi b}A_n^-}$$

$$-\frac{A_n^{GG}-A_n^{qq}}{A_n^+-A_n^-}\left(\frac{1-\left(\frac{4}{3\pi b}A_n^+\right)^\ell}{1-\frac{4}{3\pi b}A_n^+}-\frac{1-\left(\frac{4}{3\pi b}A_n^-\right)^\ell}{1-\frac{4}{3\pi b}A_n^-}\right)\Bigg]$$

$$(H_n)_\ell = \frac{1}{A_n^+-A_n^-}\left[\frac{1-\left(\frac{4}{3\pi b}A_n^+\right)^\ell}{1-\frac{4}{3\pi b}A_n^+}-\frac{1-\left(\frac{4}{3\pi b}A_n^-\right)^\ell}{1-\frac{4}{3\pi b}A_n^-}\right]$$

and they have following recursive formula:

$$(L_n)_\ell = (A_n^{GG}+A_n^{qq})(L_n)_{\ell-1}-(A_n^{GG}A_n^{qq}-2fA_n^{Gq}A_n^{qG})(L_n)_{\ell-2}$$

$$(P_n)_\ell = (A_n^{GG}+A_n^{qq})(P_n)_{\ell-1}-(A_n^{GG}A_n^{qq}-2fA_n^{Gq}A_n^{qG})(P_n)_{\ell-2}$$

$$(S_n)_\ell = (A_n^{GG}+A_n^{qq})(S_n)_{\ell-1}-(A_n^{GG}A_n^{qq}-2fA_n^{Gq}A_n^{qG})(S_n)_{\ell-2}$$

$$(H_n)_\ell = (A_n^{GG}+A_n^{qq})(H_n)_{\ell-1}-(A_n^{GG}A_n^{qq}-2fA_n^{Gq}A_n^{qG})(H_n)_{\ell-2}+\frac{4}{3\pi b}$$

$$(R_n)_\ell = (A_n^{GG}+A_n^{qq})(R_n)_{\ell-1}-(A_n^{GG}A_n^{qq}-2fA_n^{Gq}A_n^{qG})(R_n)_{\ell-2}+1-\frac{4}{3\pi b}A_n^{GG}$$

$$(Q_n)_\ell = 1+\frac{4}{3\pi b}A_n^{qq}(Q_n)_{\ell-1}$$

with

$$(L_n)_0 = 0,\quad (L_n)_1 = 1;\quad (P_n)_0 = 1,\quad (P_n)_1 = A_n^{qq};$$

$$(S_n)_0 = 1,\quad (S_n)_1 = A_n^{GG};\quad (H_n)_0 = 0,\quad (H_n)_1 = 0;$$

$$(R_n)_0 = 0,\quad (R_n)_1 = 1;\quad (Q_n)_0 = 0,\quad (Q_n)_1 = 1.$$

(B2)

$$F_{qq_1}(n,t) = e^{A_n^{qq}s}$$

$$F_{qq_2}(n,t) = e^{A_n^{qq}s}+\int_0^s dv\sum_{j=1}^{\infty}U^j\frac{v^j(s-v)^{j-1}}{j!(j-1)!}e^{(s-v)A_n^{GG}+vA_n^{qq}}$$

$$F_{qG}(n,t) = A_n^{qG}\int_0^s dv \sum_{j=0}^{\infty} U^j \frac{v^j(s-v)^j}{j!\,j!} e^{(s-v)A_n^{GG}+vA_n^{qq}}$$

$$F_{qP}^{i}(n,t) = \left(\varepsilon_i^2 - \frac{1}{2f}\sum_i \varepsilon_i^2\right)\frac{\alpha}{2\pi} a(n)\frac{3\pi b}{4} t \int_0^s ds' e^{\left(A_n^{qq}-\frac{3\pi b}{4}\right)s'}$$

$$+ \frac{1}{2f}\sum_i \varepsilon_i^2 \frac{\alpha}{2\pi}\frac{3\pi b}{4} t\, a(n) \cdot$$

$$\int_0^s ds' e^{-\frac{3\pi b}{4}s'}\left[e^{A_n^{qq}s'} + \int_0^{s'} dv \sum_{j=1}^{\infty} U^j \frac{v^j(s'-v)^{j-1}}{j!(j-1)!} e^{(s'-v)A_n^{GG}+vA_n^{qq}}\right]$$

$$F_{Gq}(n,t) = A_n^{Gq}\int_0^s dv \sum_{j=0}^{\infty} U^j \frac{v^j(s-v)^j}{j!\,j!} e^{(s-v)A_n^{GG}+vA_n^{qq}}$$

$$F_{GG}(n,t) = e^{A_n^{GG}s} + \int_0^s dv \sum_{j=1}^{\infty} U^j \frac{v^j(s-v)^{j-1}}{j!(j-1)!} e^{vA_n^{GG}+(s-v)A_n^{qq}}$$

$$F_{GP}(n,t) = \sum_i \varepsilon_i^2 \frac{\alpha}{2\pi}\frac{3\pi b}{4} t$$

$$a(n) A_n^{Gq}\int_0^s ds' e^{-\frac{3\pi b}{4}s'}\int_0^{s'} dv \sum_{j=0}^{\infty} U^j \frac{v^j(s'-v)^j}{j!\,j!} e^{(s'-v)A_n^{GG}+vA_n^{qq}}$$

where

$$U = 2f A_n^{qG} A_n^{Gq} .$$

References

1. D. Gross, F. Wilczek, Phys. Rev. Lett. 26 (1973) 1343
2. H.D. Politzer, Phys. Rev. Lett. 26 (1973) 1346
3. See for example

   G. Parisi, Phys. Lett. 43B (1973) 207; 50B (1974) 367;

   H. Georgi, HUTP-78/A003 (1978);

   G. Altarelli, R. Petronzio, G. Parisi, Phys. Lett. 63B (1976) 183;

   A.J. Buras, K.J.F. Gaemers, Nucl. Phys. B129 (1976) 66;

   A.J. Buras, et al. Nucl. Phys. B131 (1977) 308;

   I. Hinchiffe, C.H. Llewellyn Smith, Nucl. Phys. B128 (1977) 93;

   M. Glück, E. Reya, Phys. Rev. D14 (1976) 3034.

   H. Georgi, H.D. Politzer, Phys. Rev. D14 (1976) 1829.

   A. De Rújula, H. Georgi, H.D. Politzer Ann. Phys. (N.Y.) 103 (1977) 351.
4. J. Kogut, L. Susskind, Phys. Rev. D9 (1974) 697, 3391
5. G. Altarelli, G. Parisi, Nucl. Phys. B126 (1977) 298
6. F. Martin, Phys. Rev. D19 (1978) 1382
7. E. Witten, Nucl. Phys. B120 (1977) 189
8. T. Hill, G. Ross, Nucl. Phys. B148 (1979) 373
9. W.A. Bardeen, A. J. Buras, FNAL-Pub-78/91 THY (1978)
10. C.H. Llewellyn Smith, Phys. Lett. 79B (1978) 83
11. R.J. DeWitt et al, Phys. Rev. D19 (1979) 2046

L.M. Jones et al, ILL-(TH)-79-12 (1979)

12. S.J. Brodsky et al. Phys. Rev. D19 (1979) 1418
Phys. Rev. Lett. 41 (1979) 672

13. W.K. Frazer, J.F. Gunion Phys. Rev. D20 (1979) 147

14. R.J. DeWitt et al. see Ref. 11
M.A. Ahmed, G.G. Ross, Phys. Lett. 59B (1975) 369

15. Yu. L. Dokshitser et al. SLAC-Tr-183

16. J. Badier et al. CERN/EP 79-67 (1979)

17. R.P. Feynman, Photon-Hadron Interaction (Benjamin, New York, 1972)

18. J.F. Owens Phys. Lett. 76B (1978) 85
T. Uematsu Phys. Lett. 79B (1978) 97

19. A.M. Polyakov in Proc. of the 1975 International Symposium on Lepton and Photon Interaction at High Energies, Stanford, California, ed. W.T. Kirk (1975) P. 855

20. A.H. Mueller Phys. Rev. D9 (1974) 963
C.G. Callan, M.L. Goldberger Phys. Rev. D11 (1975) 1542, 1553
N. Coote, Phys. Rev. D11 (1975) 1611

21. R.D. Field, R.P. Feynman, Phys. Rev. D15 (1977) 2590

22. R.D. Field, R.P. Feynman, Nucl. Phys. B136 (1978) 1

23. J.F. Owens et al. Phys. Rev. D18 (1978) 1501

A DISCUSSION OF THE DECAY OF QUARKONIUM $V' \rightarrow V + 2\pi$ AS A FUNCTION OF THE QUARKONIUM MASS

Jing-yuan Wu

Institute of High Energy Physics, Beijing

Abstract

The decay rate of quarkonium $V' \rightarrow V + 2\pi$ as a function of the quarkonium mass m_V is discussed taking into full consideration that V' is the first radial excitation state of V. We find that $\Gamma(V' \rightarrow V + 2\pi) \sim m_V^{-n}$, n = 1.6 to 2.0. The main contribution comes from the class of graphs where each of the quark and antiquark lines emits a gluon, covariant gauges being assumed throughout.

I. INTRODUCTION

Although the leptonic decay rate of $\psi'(3686)$ (2.0 keV) is less than that of ψ (3097) (4.7 keV), the total width of ψ' (228 keV) is very much bigger than that of ψ (67 keV). This is because there are a few decay modes of ψ' which ψ does not possess: e.g. $\psi' \rightarrow \gamma + p$ states and $\psi' \rightarrow \psi + 2\pi$. Quigg[1] has pointed out that, with a binding potential of the form r^{ϵ}, when $\epsilon \geqslant -1$, $\Gamma(V' \rightarrow V + p$ states) decreases rapidly with the increase of the quarkonium mass m_V. If $\Gamma(V' \rightarrow V + 2\pi)$ behaves in the same way, it is quite possible that V' is narrower than V. As a result, $Br(V' \rightarrow e^+e^-)$ will be of the same order as that for V, and in $p + p \longrightarrow e^+e^- +$ anything experiments, V and V' will be produced simultaneously.

In this paper, the relation between $\Gamma(V' \longrightarrow V + 2\pi)$ and m_V is discussed with the assumption that $\Delta_V = m_{V'} - m_V \sim 590$ MeV does not depend on m_V appreciably[2]. The range of m_V considered is only up to about the mass of the quarkonium formed by t quarks ~ 30 - 40 GeV/c^2 because by cosmological reason we believe in the existence of three generations only and also quarkonium of mass much higher than this range is certainly not accessible by the present generation of accelerators. There are two factors that can affect the rate of this decay mode as m_V increases:

1. <u>The Decrease of the Relative Phase Space</u>

As m_V increases but with Δ_V fixed, the relative phase space decreases. If we take the Lagrangian

$$\mathcal{L} = g_V V_\mu V'_\mu \quad \vec{\pi}\cdot\vec{\pi} \ . \tag{1}$$

and assume the effective coupling constant g_V to be m_V independent, one gets

$$\Gamma(V' \longrightarrow V + 2\pi) \sim m_V^{-2} \ . \tag{2}$$

This result has been obtained by many authors[3], but they have neglected the second important factor concerning the special structure of the V, V' system.

2. <u>V' is the First Radial Excited State of V</u>

There are two classes of graphs (Fig. 1) governing the decay decay process $V' \longrightarrow V + 2\pi$. Although this separation is gauge dependent, each class is in fact invariant under all covariant gauges[4], because

$$K_\mu M_{\mu\nu} = 0, \quad K'_\nu M_{\mu\nu} = 0, \tag{3}$$

where $M_{\mu\nu}$ is the two-gluon-two-pion vertex and K_μ, K'_μ represent the momentum carried by each of the two internal gluons. With these gauges, the two classes of graphs give completely different m_V dependence and 2π mass distribution. As a result, these gauges are used in our analysis below.

We first make the hypothesis that graphs (a) are dominant. So we can neglect graphs (b) completely. The decay amplitude of graphs (a) contains an overlap integral

$$I_V(\vec{q}) = \int d^3x \; \phi_V^*(\vec{x}) \, \phi_{V'}(\vec{x}) \, e^{i\frac{1}{2}\vec{q}\cdot\vec{x}} \quad , \tag{4}$$

where ϕ_V and $\phi_{V'}$ are the three dimensional wave functions of V and V' respectively with normalization

$$\int d^3x \left|\phi_V(\vec{x})\right|^2 = \int d^3x \left|\phi_{V'}(\vec{x})\right|^2 = 1. \tag{5}$$

$\vec{q}$ is the momentum of the recoiling V in the rest frame of V'. Since V' is the first radial excitation of V, this overlap integral represents a process that is second order forbidden, i.e.,

$$I_V(\vec{q}) = a^2 \vec{q}^2 \quad , \tag{6}$$

where a is a parameter with the dimension of a length in the quarkonium system, whose dependence on m_V is

$$a^2 \sim m_V^{-2/(2+\epsilon)} \tag{7}$$

if the binding potential is $V(r) \sim r^\epsilon$. If Δ_V is m_V independent for all m_V up to infinity, V(r) is logarithmic, equivalent to $\epsilon = 0$. However, experimentally[5] $|\phi_V(0)|^2 \sim m_V^n$,

$n \sim 1.6$ to 1.9 corresponding to $\epsilon \sim -0.125$ to -0.42. This is not necessary a contradiction because our range of m_V is from m_ψ to $m_{t\bar{t}}$ only. If we take ϵ from 0 to -0.42, we get

$$a^2 \sim m_V^{-1} \text{ to } m_V^{-0.83} , \tag{8}$$

resulting in

$$\Gamma(V' \longrightarrow V + 2\pi) \sim m_V^{-4} \text{ to } m_V^{-3.6} \tag{9}$$

However, in Section II, we show that graphs (a) alone lead to a mass distribution of the two pions ($M_{\pi\pi}$) with maximum in the small $M_{\pi\pi}$ instead. Thus our hypothesis of the dominance of graphs (a) in covariant gauges is wrong. In Section III, we make the reverse hypothesis that graphs (b) are dominant. We discover that this hypothesis is probably correct because graphs (b) alone can give the correct mass distribution of the two pions, their contribution are indeed dominant. They lead to a mass dependence of

$$\Gamma(V' \longrightarrow V + 2\pi) \sim m_V^{-2} \text{ to } m_V^{-1.6} \tag{10}$$

II. THE DISTRIBUTION OF THE TWO PION SPECTRUM

We can approximate graphs (a) by assuming that the two gluons meet the quark or antiquark line at one point; an effective Lagrangian is

$$\mathcal{L} = g'_V m_V^{-1} \; \bar{q}_V q_V \vec{\pi}\cdot\vec{\pi} , \tag{11}$$

where q_V is the field of the quark constituenting the quarkonium V or V'. Because a quark propagator has been contracted to a point, it gives a contribution of $m_{q_V}^{-1} \sim (\frac{1}{2}m_V)^{-1}$. As a result, the dimensionless effective coupling constant g'_V is m_V independent. It is easy to work out the decay rate with graphs (6) neglected

$$\Gamma(V' \to V + 2\pi) = \frac{3 g_V'}{2(2\pi)^9} \frac{(m_V + m_{V'})\Delta_V^7}{m_V m_{V'}^2} \tag{12}$$

$$\cdot \int_{4m_\pi^2/\Delta_V^2}^{1} dZ\,(1-Z)^{\frac{1}{2}}\left(1-\frac{4m_\pi^2}{\Delta_V^2 Z}\right)^{\frac{1}{2}}\left[1-\frac{\Delta_V^2 Z}{(m_V+m_{V'})^2}\right]^{\frac{1}{2}}\left(1+\frac{E_V}{m_V}\right)^2 \left|I_V(\vec{q})\right|^2,$$

with E_V the energy of V in the rest frame of V', and

$$z = \left(\frac{M_{\pi\pi}}{\Delta_V}\right)^2 . \tag{13}$$

Using (6) and (7) and the recoil velocity of V,

$$\beta = \frac{|\vec{q}|}{m_V} = (1-z)^{\frac{1}{2}} \Delta_V (m_V m_{V'})^{-\frac{1}{2}} \tag{14}$$

neglecting $O(\Delta_V^2/m_V^2)$, one arrives at the mass relation (9), i.e., m_V^{-n}, n = 3.6 to 4.

According to Eq. (12), the mass spectrum of the two pions is

$$\frac{d\Gamma}{dM_{\pi\pi}} \propto \left(1-\frac{M_{\pi\pi}^2}{\Delta_V^2}\right)^{\frac{1}{2}} \left(\frac{M_{\pi\pi}^2}{4m_\pi} - 1\right)^{\frac{1}{2}} \left(1-\frac{M_{\pi\pi}}{\Delta_V^2}\right)^2, \tag{15}$$

the last factor comes from the β^4 factor of $|I_V(\vec{q})|^2$. This distribution is plotted in Fig. 2 (dotted), which does not agree with the experimental data at all. When $M_{\pi\pi}$ is at a maximum, the recoil of V is zero; thus the factor β^4 in (12) gives extra zeroes at maximum $M_{\pi\pi}$ shifting the distribution to the lower $M_{\pi\pi}$ side.

According the current algebra, interaction (11) needs modification. We can expand the elastic scattering amplitude $\mathcal{M}_{\pi q_v}$ of pion and q_V in powers of the momenta k_1 and k_2 of the pions[6]:

$$\mathcal{M}_{\pi q_v} = \mathcal{M}^B + \mathcal{M}^c + \mathcal{M}^\sigma + R \qquad (16)$$

with $\mathcal{M}^B$, $\mathcal{M}^c$ and $\mathcal{M}^\sigma$ representing the Born term, the current-current commutator term and the σ term respectively. R is the remainder and is second order in k_1 and k_2. $\mathcal{M}^B$ and $\mathcal{M}^c$ are first order in k_1 and k_2; however, they vanish because both q_V and the two pion system are isospin singlet. $\mathcal{M}^\sigma$ can be calculated via the σ model and is second order in k_1 and k_2. The experimental results[7] show that the two pion system has eigenvalues of the vacuum and also the recoiling ψ is not polarized (there is no angular correlation between the lepon pair decaying from ψ and the pions). Thus $\mathcal{M}_{\pi q_V}$ contains[8] only terms like $(k_1 \cdot k_2)$ but no such terms as $(k_1 \cdot \epsilon^\lambda)(k_2 \cdot \epsilon^{\lambda'})$ etc., ϵ_μ^λ and $\epsilon_\mu^{\lambda'}$ being the polarization vectors of V and V' respectively. Such amplitude satisfies Adler's self-consistency condition[9]: when one pion is on the mass shell, the amplitude approaches zero when the four momentum of the other pion approaches zero. Such amplitude can be generated by the effective Lagrangian

$$\mathcal{L} = g''_V m_V^{-1} \quad \bar{q}_V q_V \frac{\partial \vec{\pi}}{\partial x_\mu} \cdot \frac{\partial \vec{\pi}}{\partial x} \qquad (17)$$

which gives a distribution

$$\frac{d\Gamma}{dM_{\pi\pi}} \propto \left(1 - \frac{M_{\pi\pi}^2}{\Delta_V^2}\right)^{\frac{5}{2}} \left(\frac{M_{\pi\pi}^2}{4m_\pi^2} - 1\right)^{\frac{1}{2}} \left(\frac{M_{\pi\pi}^2}{2m_\pi^2} - 1\right)^2 . \qquad (18)$$

But this distribution still disagrees with the experimental one (Fig. 2, dashed curve).

However, in Eq. (18), if we delete the factor concerning the structure of V and V', i.e., letting $|I_V(\vec{q})|^2 = 1$, or deleting the contribution arriving from the factor $\beta^4 \sim (1-M^2_{\pi\pi}/\Delta^2_V)^2$, the resulting distribution (Fig. 2, solid curve) agrees very well with the experimental one. But V' is the first excitation of V, the overlap integral cannot be deleted. We note that $(\Delta^2_V - M^2_{\pi\pi})$ is just the reciprocal of a propagator with mass Δ_V. Thus, if the two pions are created through a narrow resonance with mass Δ_V, the effect of $|I_V(\vec{q})|^2$ could be cancelled away (Fig. 3). However, the only resonances with vacuum eigenvalues are[10] S*(980) $\Gamma_{S*} \sim 40$ MeV and ϵ(1200) $\Gamma_\epsilon \sim 600$ MeV. They are well away from the resonance we need and therefore have no effect at all in the cancellation of $|I_V(\vec{q})|^2$. Thus, we can draw the conclusion that for the process $V' \longrightarrow V + 2\pi$, the class of graphs (a) is not the dominant class (in covariant gauges always) and the class of graphs (b) cannot be neglected.

III. Graphs (b)

In this Section, we make the hypothesis that graphs (b) are dominant in covariant gauges. So we can neglect graphs (a).

In the rest frame of V', graphs (b) contain the factor

$$\exp(-i\vec{p}_1\cdot\vec{u}-i\vec{p}_2\cdot\vec{v}) = \exp -i\tfrac{1}{2}(\vec{p}_1-\vec{p}_2)\cdot\vec{x}\,\exp -i\tfrac{1}{2}(\vec{p}_1+\vec{p}_2)\cdot\vec{X}, \quad (19)$$

where p_1, p_2 are the momenta of the two gluons and $\vec{x}=\vec{u}-\vec{v}$, $\vec{X} = \frac{1}{2}(\vec{u}+\vec{v})$. Thus the overlap integral is

$$J_V \sim \int d^3x \ \phi_V^*(\vec{x}) \ \phi_{V'}(\vec{x}) \exp -i\tfrac{1}{2}(\vec{p}_1 - \vec{p}_2)\cdot\vec{x} \ . \tag{20}$$

If we make the ad hoc assumption that, when the two gluons annihilate into two pions, no momentum is transferred, i.e., $\vec{p}_1 = \vec{k}_1$, $\vec{p}_2 = \vec{k}_2$, when $M_{\pi\pi}$ assumes its maximum ($k_1 = -k_2$), J_V reaches its maximum. When $M_{\pi\pi}$ assumes its minimum ($\vec{k}_1 = \vec{k}_2$), J_V just vanishes. Thus the distribution of the two pions shifts to the large $M_{\pi\pi}$ side. Actually there must be a momentum transferred. Thus, when $M_{\pi\pi}$ is at maximum $|\vec{p}_1 - \vec{p}_2|$ can range from zero to infinity. But there must be a form factor at the two-gluon-two-pion vertex which limits the transfer of momentum to within a few hundred MeV, (this is so for the electromagnetic form factor). Thus $|\vec{p}_1 - \vec{p}_2|$ ranges roughly from zero to Δ_V, so that the overlap integral (20) may average to the same value for any $M_{\pi\pi}$. In other words, J_V does not affect the distribution in $M_{\pi\pi}$ very much. Other factors such as phase space and the Adler's self-consistency condition will therefore lead to a distribution in agreement with the experimental one. Thus our second hypothesis may be correct.

As for the m_V dependence of $\Gamma(V' \rightarrow V + 2\pi)$, we use the effective Lagrangian

$$\mathcal{L} = \tfrac{1}{2} i\bar{q}_V \gamma_\mu \lambda^a q_V A^a_\mu + G F^a_{\mu\nu} F^a_{\mu\nu} \vec{\pi}\cdot\vec{\pi} \ , \tag{21}$$

where A^a_μ is the gluon potential and $F^a_{\mu\nu}$ the gluon field strength, λ^a the colour SU(3) Gell-Mann 3×3 hermitian matrices. g is the coupling constant of QCD and is therefore m_V independent. G is the effective coupling for the two-gluon-two-pion vertex and is also m_V independent. The decay amplitude due to graphs (b) is

$$(\mathfrak{M}_b)^{\lambda\lambda'}_{i_1 i_2} = \frac{16 i g^2 G}{(2\pi)^4} \delta_{i_1 i_2} (m_V m_{V'})^{\frac{1}{2}}$$

$$\cdot \int \frac{d^4Q}{(2\pi)^4} \Delta_{\mu\alpha}(p_1) \Delta_{\nu\beta}(p_2) [p_{1\alpha} p_{2\beta} - \delta_{\alpha\beta}(p_1 \cdot p_2)] \chi^{\lambda\lambda'}_{\mu\nu} g \, , \tag{22}$$

where i_1 , i_2 are the isospin indices of the pions. $\Delta_{\mu\alpha}(p)$ is the propagator of the gluon carrying momentum p. $Q = p_1 - p_2$. $\chi^{\lambda\lambda'}_{\mu\nu} g$ is the four dimensional overlap integral of V and V', $\chi^{\lambda\lambda'}_{\mu\nu}$ is the colour and spin part and

$$g = \int d^4p \, \psi^*_V(p^*) \psi_{V'}(p + \tfrac{1}{2}Q) \tag{23}$$

is the space-time part, p* being the relative momentum in the rest frame of V. The four dimensional wavefunction $\psi_V(p)$ and $\psi_{V'}(p)$ can be expressed in terms of the corresponding three dimensional wavefunctions $\phi_V(p)$ and $\phi_{V'}(p)$ in the non-relativistic approximation[11] :

$$\psi_V(p) = \phi_V(\vec{p})(2\pi i)^{-1}(2W - E_V)(E_V/2 - p_0 - W + i\epsilon)^{-1}(E_V/2 + p_0 - W + i\epsilon)^{-1},$$

$$W = (\vec{p}^2 + m^2_{qV})^{\frac{1}{2}} \quad . \tag{24}$$

We need only to compute the amplitude at one point; for simplicity we take the situation when V does not recoil, i.e.,

$$-\vec{q} = \vec{k} + \vec{k} = \vec{p} + \vec{p} = 0, \qquad \vec{Q} = 2\vec{p} = -2\vec{p} \tag{25}$$

Integrating over p_0, we obtain

$$g = \int d^3 p\, \phi_V^*(\vec{p})\, \phi_{V'}(\vec{p} + \frac{\vec{Q}}{2})(2\pi i)^{-1} 4\xi [Q_0^2 - (\xi - i\epsilon)^2] , \tag{26}$$

with

$$\xi = 2(\vec{p}^2 + \vec{m}_{q_V}^2)^{\frac{1}{2}} + 2[(\vec{p} + \tfrac{1}{2}\vec{Q}) + m_{q_V}^2]^{\frac{1}{2}} - m_V - m_{V'},$$

$$\sim m_{q_V}^{-1} [\vec{p}^2 + (\vec{p} + \tfrac{1}{2}\vec{Q})^2] + (\text{binding energy of V and V'}), \tag{27}$$

since the ranges of $\vec{p}^2$ and $(\vec{p}+\frac{1}{2}\vec{Q})^2$ are limited by the wave-functions. If the average values of $\vec{p}$ and $(\vec{p}+\frac{1}{2}\vec{Q})^2$ are substituted, the first term of (27) depends on m_V weakly while the second term is of order Δ_V. As an approximation, we consider ξ as m_V independent; then

$$\begin{aligned} g &= \int d^3 x\, \phi_V^*(\vec{x})\, \phi_{V'}(\vec{x}) e^{\frac{1}{2} i \vec{Q}\cdot\vec{x}} (2\pi i)^{-1} 4\xi [Q_0^2 - (\xi - i\epsilon)^2]^{-1}, \\ &= a^2 Q^2 f(a^2 Q^2)(2\pi i)^{-1} 4\xi [Q_0^2 - (\xi - i\epsilon)^2]^{-1} , \end{aligned} \tag{28}$$

with $Q = |\vec{Q}|$. Using

$$\chi_{\mu\nu}^{\lambda\lambda'} [p_{1\mu} p_{2\nu} - \delta_{\mu\nu}(p_1 \cdot p_2)] = \frac{1}{3} \vec{\epsilon}^{\lambda} \cdot \vec{\epsilon}^{\lambda'} Q^2 , \tag{29}$$

the decay amplitude (22) becomes

$$(\mathcal{M}_b)_{i_1 i_2}^{\lambda\lambda'} = \delta_{i_1 i_2} (\vec{\epsilon}^{\lambda} \cdot \vec{\epsilon}^{\lambda'})(m_V m_{V'})^{\frac{1}{2}} \int_0^{\infty} dQ\, a^2 Q^2 f(a^2 Q^2) F(Q) , \tag{30}$$

where

$$F(Q) = \frac{i g^2 G}{12 \pi^7} \frac{Q^3 [\xi + 2Q]}{[\Delta_V^2 - (Q + \xi - i\epsilon)^2][\Delta_V^2 - (Q - i\epsilon)^2]} \tag{31}$$

is considered as m_V independent. The coupling constant G in (31) can be viewed as a form factor G(Q) which limits the integration of Q from zero to a few hundred MeV. Thus $f(a^2 Q^2)$

can be expanded in powers of a, resulting in

$$(\mathcal{M}_b)^{\lambda\lambda'}_{i_1 i_2} \sim (m_V m_{V'})^{\frac{1}{2}} \ a^2 \ . \qquad (32)$$

Comparing with Lagrangian (1) and using (8), we get

$$\Gamma(V' \longrightarrow V + 2\pi) \sim m_V^{-2} \ \text{to} \ m_V^{-1.6} \ . \qquad (33)$$

Looking at Eq. (28), the four dimensional overlap integral gives $g \sim a^2 Q^2 f(a^2 Q^2) \ \Delta_V^{-1}$, whereas for graphs (a), its contribution is $\sim a^2 \vec{q}^{\,2} f(a^2 \vec{q}^{\,2}) \ m_{q_V}^{-1}$. Thus graphs (b) should give more contribution than graphs (a), especially for large m_V. As a result, we believe Eq. (33) is the correct relation between $\Gamma(V' \longrightarrow V + 2\pi)$ and m_V.

Reference

1. C. Quigg and J.L. Rosner, Phys. Lett. 71B (1977) 153.
2. Just for ψ and Υ . $\Delta\psi/\Delta\Upsilon = (m_\psi/m_\Upsilon)^{-0.07}$. The mass dependence is indeed very small.
3. J. Ellis, M.K. Gaillard, D.V. Nanopoulos and S. Rudaz, Nuclear Physics B131 (1977) 285.
4. By covariant gauges, we mean the gluon propagator taken to be $-ik^{-2} \ \delta_{ab}[\delta_{\beta\mu} - (1-\alpha)k_\beta k_\mu/k^{-2}]$ where α is a gauge parameter. In non-cavariant gauges, for example in Coulomb gauge, the $(k^2)^{-2}$ part of the gluon propagator contains terms like $\delta_{\beta 4} k_\mu k_4$ and $k_\beta \ \delta_{\mu 4} k_4$. Although $k_\mu M_{\mu\nu} = 0$ separately for each class of graphs, both $k_4 M_{4\nu}$ and the contraction of k_β with the $q_V \bar{q}_V$ part of each class of graphs do not vanish.

5. J.D. Jackson, Lecture on the New Particles, Proceedings of Summer Institute on Particle Physics, 1976, p. 147. J.D. Jackson, C. Quigg and J.L. Rosner, Proceedings of the 19th International Conference on High Energy Physics, Tokyo, 1978, p. 391.
6. S. Weinberg, Physics Rev. Letters 17, (1966) 616.
7. G.S. Abrams et all Phys. Rev. Letters 34 (1971) 1181.
8. L.S. Brown and R.N. Cahn, Phys. Rev. Lett. 35 (1975) 1.
9. S.L. Adler, Phys. Rev. 137 (1965) B1022; 139 (1965) B1638.
10. Review of Particle Properties, Phys. Lett. 75B (1978) 1.
11. E.E. Salpeter, Phys. Rev. 87 (1952) 328.

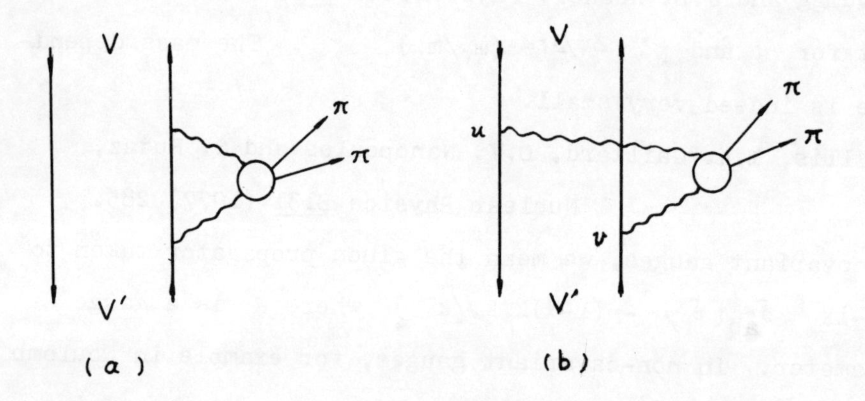

Fig. 1. Two graphs contributing to the decay V' → V + 2π.

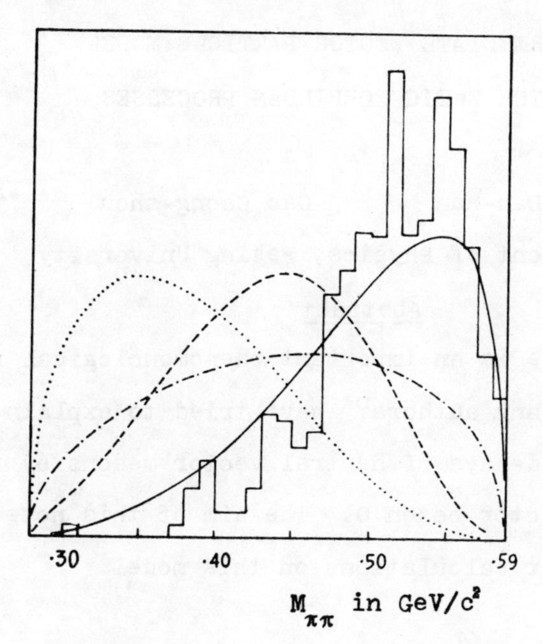

Fig. 2. Mass distribution of the two pions. Dotted-dashed curve represents pure phase space.

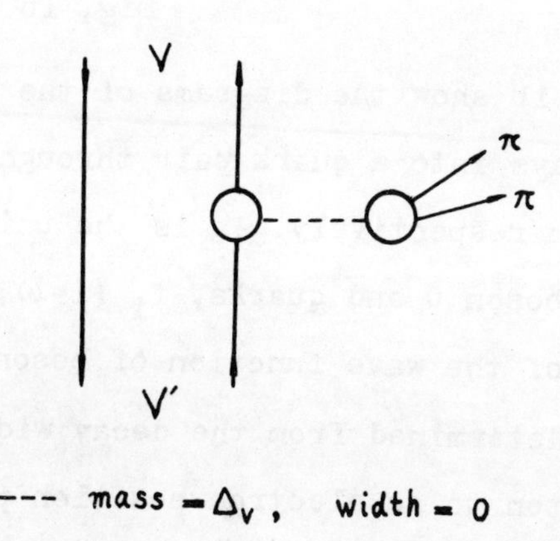

Fig. 3. Production of the two pions through a hypothetical resonance at mass Δ_v and zero width.

INTERMEDIATE VECTOR PARTICLE MODEL

OF THE ZWEIG FORBIDDEN PROCESSES

Qin Dan-hua   Gao Chong-shou

Department of Physics, Peking University

Abstract

The Zweig rule is an important phenomenological rule in hadron physics. Many authors[1] have tried to explain this rule for the forbidden decays of neutral vector mesons by introducing an intermediate vector boson O. The aim of this note is to give some results of our calculations on this model.

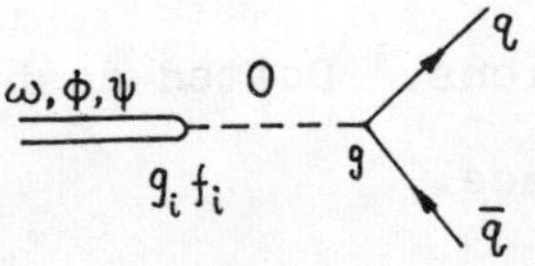

ω, ϕ, ψ O $g_i f_i$ g q $\bar{q}$

Fig. 1a

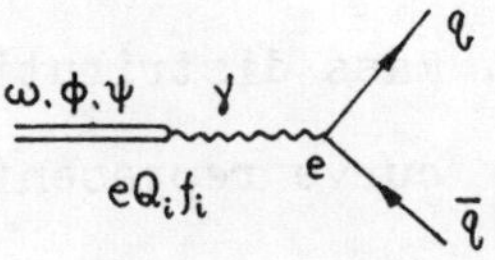

ω, ϕ, ψ γ $eQ_i f_i$ e q $\bar{q}$

Fig. 1b

Fig. 1a and Fig. 1b show the diagrams of the processes in which ω, ϕ, or ψ decays into a quark pair through the vector boson O and the photon respectively. g is the universal coupling constant between the boson O and quarks, f_i $(i=\omega, \phi, \psi)$ is related to the value of the wave function of boson i at its origin, which may be determined from the decay width of boson i through a virtual photon to an electron-position pair. The final expression for the decay width of the process shown in Fig. 1a contains only two parameters, g and the mass m_o of the boson O.

Using the experimental values $\Gamma_{\phi\, zweig}/\Gamma_{\varphi\to ee}=14.84/.031$ and $\Gamma_{\psi zweig}/\Gamma_{\psi\to ee}=86/7$, we obtain $m_o=(1.123 \pm .009)$ GeV. The corresponding width of the hadronic decay mode is given by $\Gamma_{o\to h}= 3m_o\alpha_o =(15.2 \pm 2.8)$ MeV., where $\alpha_o=g^2/4\pi$. The decay of boson O through the process $O\to\omega,\phi$ and $\psi \to$ photon $\to e^+e^-$ is also studied. The leptonic decay width of boson O is estimated to be ~ 0.57 KeV.

The relative partial widths of $\psi\to 0^- + 1^-$ are also calculated and the result is

$$\Gamma_{\psi\to x\rho} : \Gamma_{\psi\to KK^*} : \Gamma_{\psi\to\eta\phi} = 1:(0.54\pm 0.05):(0.13 \pm 0.03)$$

The following relation is also obtained

$$\Gamma_{\psi'\to e^+e^-}/\Gamma_{\psi\to e^+e^-} \approx \Gamma_{\psi'\to \text{hadrons}}/\Gamma_{\psi\to \text{hadrons}}$$

Both results are consistent with the experiment.

According to the present model the decay of Υ particle will result in two jets. This is not in accordance with the experiment which shows that the three-gluon decay mode for Υ is clearly favoured. This means that somehow the gluons must play a peculiar role at higher energy, which has not been taken into account in the present investigation.

References

1. P.G.O. Freund & Y. Nambu, P.R.L., 34 (1975), 1654; EFI (1975) 75/3; Zhao Zhi-yung, Acta Scientiarum Naturalium Universitatis Pekinensis (1979).
2. G.J. Feldman, SLAC-PUB-1851 (1976).

High Energy Hadron-Nucleus and Nucleus-Nucleus Collisions *

Meng Ta-chung

Institut für theoretische Physik der FU Berlin, Berlin, Germany

Abstract

The result of a phenomenological study of high energy hadron-nucleus and nucleus-nucleus collisions is summarized. Useful information on the basic reaction mechanism of high energy hadronic processes are obtained from a systematic analysis of the recent experimental data.

---

* This work is supported in part by the Deutsche Forschungsgemeinschaft

1. Introduction

Especially at conferences on particle physics, people usually begin a discussion on this subject with questions of the following type: "Why is it interesting to study high energy hadron-nucleus and nucleus-nucleus collisions?", "What can we learn from such complicated systems before we understand the reaction mechanism of a hadron-hadron collision at comparable energies?" I don't think I can, and I don't think I should make an exception. After all, as particle physicists, you would certainly like to know whether you want to take part in a discussion on this subject. Since the above-mentioned questions are not new, many answers to these questions can be found in the literature [1-3]. So, I shall make it short. As a theorist, I am not competent to discuss the advantages of using nuclear targets in experiments, or the advantages of using heavy ion beams in applied sciences. What I can say, is why I think studies of such reactions may lead to interesting physics.

Firstly, from high energy hadron-nucleus reactions we can obtain useful information on the space-time evolution of hadronic interactions —— a problem of some importance, but also a problem which can not be studied in a corresponding hadron-hadron process where only the asymptotic final states are observed. In fact, the nucleus in a high energy hadron-nucleus collision is used not only as a target, but also as an extremely dense "bubble chamber"!

Secondly, high-energy hadron-nucleus and nucleus-nucleus collisions can be used as a tool to study hadronic matter under extreme conditions (nuclear- and quark-matter at high temperatures and/or at high densities). In particular, if the predicted exotic objects [1], e.g. abnormal nuclear states, density isomers, pion condensates, multiquark states etc, should indeed exist, an effective way (if not the only way at present) to study them experimentally is to look at the above-mentioned collisions.

Thirdly, there exist extremely interesting experimental materials[1-8] on high-energy hadron-nucleus and nucleus-nucleus collisions, and much of them show significant deviations from the usual expectations based on extrapolations of the existing knowledge of particle and/or nuclear physics. In fact, it is not at all obvious that a theory which explains everything we observe in hadron-hadron collisions (If we have such a theory!) will also be useful in understanding the above-mentioned reactions. Let us look at some examples:

Many authors[1-3] have studied the average multiplicity $\langle n \rangle_{hA}$ of fast moving charged particles in hadron-nucleus collisions (We denote the incident hadron by h, and the target nucleus of mass number A by A) at various incident energies (E_{inc}). In comparison with the corresponding quantity $\langle n \rangle_{hp}$ where a proton (p) target is used, it is observed that the ratio between $\langle n \rangle_{hA}$ and $\langle n \rangle_{hp}$ remains to be a small number ($\lesssim 3$), and depends only very weakly on E_{inc} and on A. Furthermore, it is seen[2] that the multiplicity is independent of incident energy in two angular regions: the wide-angle region (occupies approximately 3 units of pseudorapidity η), and the forward region (occupies about 3 units for p-p and 5 units for p-Pb collisions). While the multiplicity in the very forward region ($\eta \gtrsim 4$) is independent of the target, the multiplicity-difference at wide angles is proportional to the quantity:

$$\bar{\nu} = A\, \sigma_{hp} / \sigma_{hA} \tag{1}$$

Here σ_{hp} and σ_{hA} are the total inelastic cross-section for the reactions hp and hA respectively. In fact, it is seen[2] that collisions with different hadrons (e.g.proton and pion) with the same $\bar{\nu}$ have the same behaviour!

Another striking example is the 180°- production of pions, protons, deuterons and tritons in proton-nucleus collisions. It has been observed at several laboratories[4-8] that particles can be produced beyond the kinematical limit of free nucleon-nucleon collisions. Furthermore, it is seen that the energy spectra are pure exponential functions of the kinetic energy, and that the slope in the logarithmic plots tends to a limit for increasing incident energies.[4,6,8]

The above-mentioned experimental results clearly show that Hadron-hadron collision data, and the available information on nuclear structure are not sufficient to predict the outcome of high energy hadron-nucleus processes. Indeed, taking into account the fact that at high energies, hadron-hadron collisions are predominantly inelastic, that is in general several particles will be produced when an energetic hadron interact with a nucleon inside the target nucleus, the above-mentioned experimental results strongly suggest: The time needed for the formation of multibody final states in high energy hadron-hadron collisions is long —— much longer than that for a relativistic particle to travel a distance of a few fermis. This means: In high energy hadron-nucleus collisions, the incident hadron will already be interacting with other nucleons along its path in the nucleus, before the secondaries due to the interaction between the incident hadron and the first nucleon are produced!

2. Gentle and violent h-ET collisions

In order to study the implications of this piece of information, and to study the applicability of some of the useful ideas and methods in particle physics in high energy nuclear reactions, a simple naive picture[9] has been proposed some time ago. The characteristical features of this picture are:

a) The time needed for the formation of multibody final states in high energy hadron-hadron collisions is so long that in hadron-nucleus collisions at comparable energies, the nucleons in the path of the incident hadron inside the target nucleus can be viewed as acting collectively,[10] and in the first order approximation can be considered as a single object —— an "effective target" (ET). The mass of the ET, M_{ET}, is proportional to ν_{ET}, the number of nucleons along the path of the incident hadron inside the target nucleus. That is:

$$M_{ET} = \nu_{ET} M \qquad (2)$$

M is the nucleon mass.

b) The hadron-ET process can be described by the same picture as that used to describe the collision between two hadrons. In particular, such a process is either a gentle or a violent collision. See Figure.

c) The characteristics of a gentle collision are that the energy and momentum transfers are relatively small. The colliding objects retain their identity after the interaction such that the fragments of the projectile can be differentiated from those of the target. Here, the hadron and the ET, both of which are considered as spatially extended objects, "go through each other", in general get excited and decay separately. It is obvious, that the spirit of gentle h-ET collisions is the same as that of the Chou-Yang model[11] for elastic hadron-hadron scattering, and that hadron-hadron collisions which satisfy the well-known hypothesis of limiting fragmentation[12] are typical examples of gentle collisions. Geometrically, gentle processes are predominantly peripheral collisions.

d) In violent processes, the colliding objects hit each other so hard that they arrest each other, form a conglomerate which expands and subsequently decays. That is, the colliding objects lose their identity after the interaction. The relatively large energy- and momentum-transfer in such reactions manifest themselves in producing particles with large transverse-momenta and/or in creating more particles in such collision events. Geometrically, violent processes are predominantly central collisions.

The proposed picture is obviously an idealization. In fact, it should be considered as a limiting case of a more realistic model, in which the basic ideas are not pushed into such extremes. Comparison with experiments seems to show that this idealization is not far from reality, in spite of over-simplification in many respects.

Let us first look at the gentle h-ET process in some detail: What do we expect to see in such a collision event in the laboratory? Because of the relatively small energy- and momentum-transfer (compared with the incident momentum p_{inc}) during the interaction, both the (in general) excited hadron and the excited ET will be moving in the forward directions[13], where the velocity of the ET (β_{ET}) is much lower than that of the hadron (β_h). In order to study the emission/decay processes of these objects we need to know: "What happens to the kinetic energy of the excited hadron and that of the excited ET?" Let us concentrate our attention on the ET (for h: similar, but less interesting). Viewed from the projectile frame in which the incident hadron is at rest, the fast moving excited ET may keep its kinetic energy to retain its high velocity and/or use the kinetic energy for particle-emission (bremsstrahlung analogue) in accordance with the conservation laws. The limiting cases are: i) The excited ET retains its velocity (which is approximately equal to that before the collision) until it fragments (isotopically in its rest frame into nucleons, α-particles etc.). ii) A maximum amount of this kinetic energy is used up in emitting energetic low-mass particles in the direction of the fast moving ET (i.e. 180° in lab.), while the velocity of the excited ET is reduced to a minimum. Experiments[1,4-8] show that both limiting cases exist in nature.

We now return to the experiments we discussed at the beginning of this talk. We see, without going into details[14-17], we can readily understand: a) why at $\theta_{lab}=180^{\circ}$ particles beyond the nucleon-nucleon kinematical limit are observed, b) why does the slope parameter in the (lab.) energy-distribution for a given target tend to a limiting value for increasing incident energies, c) why is the ratio $\langle n \rangle_{hA} \; / \; \langle n \rangle_{hp}$ always a small number which depends only weakly on the incident energy and on the mass-number of the nucleus, d) why is the angular distribution of the charged multiplicities independent of incident momentum in certain kinematical regions, and e) why is it independent of $\bar{\nu}$ (see Eq.1) in the very forward directions but strongly $\bar{\nu}$-dependent at wide-angles. In fact, the extremely interesting observations made by Busza et al.[2] concerning the angular distribution in proton-nucleus and pion-nucleus collisions, as well as that made by Otterlund[3] concerning the energy-dependent and energy-independent part of average multiplicities turn out to be a natural consequence of the present picture[9, 14-17].

In violent hadron-ET collisions, the energy and momentum of the incident hadron will be deposited into the ET with which a compound system (conglomerate) is formed. Compared with a corresponding gentle h-ET process, the (lab) velocity β_c of the conglomerate lies between that of the hadron and that of the ET. ($\beta_{ET} < \beta_c < \beta_h$). The products of the conglomerate formed in violent collisions will, in general, be found at all laboratory angles. The spectrum of such particles depends on the incident momentum. Experimentally, violent collisions events can be selected by high-multiplicity and/or large transverse-momentum ($p_\perp$) triggers. Large-$p_\perp$ hadron-nucleus collision experiments[18] exist. Agreement between theory and experiment has been found[9,19].

3. Gentle and violent EP-ET collisions

In the proposed picture [9], a nucleus-nucleus collision at sufficiently high energies is nothing else but the simultaneous collision between all possible pairs of effective targets and effective projectiles, where the concept "effective projectile" (EP) is introduced analogous to that of the effective target. That is, also the EP is considered in the first order approximation as a single hadron (Its mass is $\nu_{EP}M$ where ν_{EP} is the number of nucleons in the EP), and hence an EP-ET collision can be either gentle or violent. This means, a high-energy nucleus-nucleus collision is in general a mixture of both types of EP-ET processes.

Now, since it may happen that the one or the other type of EP-ET collisions dominates (For example, there will be more gentle EP-ET collisions when the two nuclei hit each other peripherally), we expect to see three different kinds of events in relativistic heavy-ion collisions. Events of the first kind predominately consist of particles with relatively small transverse momenta $p_\perp$, small emission angles in the laboratory or projectile frame, and relatively low associated multiplicities. In a $p_\perp$- y contour plot (y is the rapidity of the emitted particle) such particles are concentrated in the neighbourhood of $p_\perp/mc \ll 1$, $y=y_T$, y_P where m is the mass of the emitted particle, y_T and y_P is the rapidity of the target and that of the projectile nucleus, respectively. Particles in the second kind events have relatively large $p_\perp$ and/or higher associated multiplicities. Such particles can be found at all angles. In the corresponding $p_\perp$ - y plot, we see that they are emitted from a source with intermediate rapidity y_c: $y_T < y_c < y_P$. In events of the third category, particles of both types can be found.

Experimentally, relativistic heavy ion collision events have been classified according to their final state topologies. Precisely these three kinds of events have been observed[20].

Further immediate consequences of the present picture (EP-ET model) are the observation of the following phenomena in relativistic heavy ion reactions:

a) Particles beyond the nucleon-nucleon kinematical limit should be observed[14] not only near $\theta_{lab}=180^{o}$, but also near $\theta_{lab}=0^{o}$. This (e.g. $^{4}He+C \longrightarrow \pi^{-}+X$ at 1.05 and 2.1 GeV/A, $\theta_{lab}=2,5^{o}$) has indeed been seen[21]. The relationship between the two cases is shown[14] to be in agreement with the proposed picture.

b) There exist[15] "slowly moving sources" (SMS's) in the laboratory as well as in the projectile frame. SMS's in laboratory have been observed[22] (e.g. in Ar+Au at 0.5 GeV/A). Further experiments are in progress.

c) The rapidity (velocity) of the moving emission-system in violent EP-ET collisions depends[23] on the incident energy (per nucleon) and on the relative mass of the colliding objects. In fact, the observed average position of the highest $p_{\perp}$ point on each contour of a $p_{\perp}$ - y plot[24] (here we see events with large $p_{\perp}$ particles), as well as the position of the central peak in a $\delta - \bar{y}$ plot[25] (here we see large multiplicity events), is determined(see Eq.1 of Ref.23) simply by the kinematics of two colliding objects with average masses $\nu_{EP}M$ and $\nu_{ET}M$, where ν_{EP} and ν_{ET} is the average number of nucleons in the EP and that in the ET respectively. An experimental determination of the above-mentioned average rapidity can differentiate between two large classes of models for <u>violent (central) nucleus-nucleus collisions</u>[26], namely the class of nucleon-nucleon collision models[27] and the class of models[28] in which "collective behaviour of nucleons" is a substantial ingredient. Experimentally, collisions between equal-mass nuclei (e.g. Ne + NaF, Ar + KCl) and nuclei of different masses (e.g. Ne + Pb, Ar + Pb) at the same incident energy per nucleon (e.g. 0.8 GeV/A) show that /24/ the expected shift in $p_\perp - y$ plot indeed exists. Further experiments /29/, including a "clustering analysis " in terms of $\delta - \bar{y}$ plots are in progress.

d) Assuming that the "temperature" concept in relativistic heavy ion reactions is indeed useful, as a number of experiments and analyses suggest /30/, the density and the volume of the emission system can be determined /23,31/ from the measured spectra /24/. Unfortunately, such measurements with high-multiplicity selection (as an indicator for violent collisions) are still rare. This kind of experiment should be continued. One obvious reason for this is: Should any exotic phenomenon in connection with high-density hadronic matter indeed exist in high energy hadronic processes, this is the most probable place to see it!

4. Concluding Remarks

In contrast to the current trend in phenomenological high energy nuclear physics, the emphasis of the present investigation is _not_ to obtain _a_ good fit based on _a_ given formalism for _a_ piece of data in _a_ kinematical region, of _a_ reaction at _an_ incident energy.

The proposed picture is based on the result of a systematic analysis of all available data. Its purpose is to understand, at least qualitatively, the general, characteristic features of high energy hadron-nucleus and nucleus-nucleus collisions. Special attention has been paid to the study of the relationship between these features. Detailed comparisons with experiments —— including the most recent data —— show that the proposed picture is able to give a global, qualitative description of high energy hadron-nucleus and nucleus-nucleus collisions.

As next step, a more quantitative description of the reaction mechanism in the framework of this picture should be given. Work along this line is now under way.

References and Footnotes

/1/ See for example the papers in:

Proceedings of the Symposium on Relativistic Heavy Ion Research, GSI, Darmstadt, March 1978

Proceedings of the International Seminar on High-Energy Physics Problems, Multiparticle Production and Limiting Fragmentation of Nuclei, JINR, Dubna, June 1978

Proceedings of the 4th High-Energy Heavy Ion Summer Study, LBL, Berkeley, July 1978

A Collection of Comments on the Relevance of Relativistic Heavy Ion Physics, Editors H.H. Gubbrod et al., LBL-Report (1978)

and the references given therein

/2/ See for example:

W.Busza, in High-Energy Physics and Nuclear Structure - 1975 AIP Conference Proceedings No.26, edited by D.E.Nagle et al. (AIP, New York, 1975), p.211

J.E.Elias et al., preprint FERMILAB-PUB-79/47 ExP 7185.178 (1979)

and the papers cited therein

/3/ See for example:

I.Otterlund et al., in High-Energy Physics and Nuclear Structure - 1973, edited by G. Tibell (North Holland Amsterdam/American Elsevier, N.Y. 1974), p.427

I.Otterlund, preprint TECHNION-PH-79 (1979), and preprint LUIP-7910 (1979)

and the papers cited therein

/4/ A.M. Baldin, in High-Energy Physics and Nuclear Structure - 1975 (see Ref. 2), p.621; Dubna Report E1-11368 (1978),

and the references given therein

/5/ S. Frankel et al., Phys.Rev.Letters 36, 642 (1976)

/6/ Y.D.Bayukov et al., University of Pennsylvania preprint UPR-OOS-8E, 11/78 (1978)

/7/ G.Fujioka, in Proceedings of the 5th International Seminar on High-Energy Physics Problems (see Ref.1), p. 339

/8/ L.S.Schroeder et al., LBL-Report 9434 (1979) to be published

/9/ Meng Ta-chung, in Proceedings of the Topical Meeting on Multiparticle Production on Nuclei at Very High Energies, Trieste, Italy, 1976, edited by G. Bellini et al., (International Centre for Theoretical Physics, Trieste, 1976), p.435, and Phys.Rev. D 15, 197 (1977)

/10/ The idea of such collective behaviour has a long history References can, for example, be found in Ref.1-4, 8 and 9

/11/ T.T. Chou and C.N. Yang, Phys.Rev.170, 1591 (1968)

/12/ J. Benecke et al., Phys.Rev. 188, 2159 (1969)

/13/ We recall that in gentle collisions, $|\Delta\vec{p}_\perp| << \Delta p_{\parallel} << p_{inc}$ where $\Delta\vec{p}_\perp$ and $\Delta p_{\parallel}$ is the momentum-transfer in the transverse and in the longitudinal direction respecitvely

/14/ H.B. Mathis and Meng Ta-chung, Phys.Rev.C 18, 952 (1978)

Meng Ta-chung, in Proceedings of the Symposium on Relativistic Heavy Ion Collisions, (see Ref.1), p.238

Meng Ta-chung and E.Moeller, in Proceedings of the International Seminar on High-Energy Physics Problems (see Ref.1), p.300

/15/ Meng Ta-chung and E.Moeller, Phys.Rev.Letters 41, 1352 (1978)

/16/ Che-Ming Ko and Meng Ta-chung, Phys.Rev.Letters 43, 994 (1979)

/17/ Meng Ta-chung and E.Moeller, to be published

/18/ J.W. Cronin et al., Phys.Rev.D 11, 3105 (1975)

/19/ The conclusion "This experimental result (Ref.18) can be understood by assuming the existence of collective behaviour among the nucleons in nucleus along the path of the incident hadron" has been reached independently by three groups: S. Fredriksson (Nucl.Phys.B111, 167, 1976), A.Dar and collaborators (Y.Afek et al., in Proceedings of the Tocial Meeting in Trieste, 1976, see Ref.1, p.591) and our group (Ref.9).

A fundamental difference between their model (usually refered to as the "collective tube model", CTM) and our picture (usually refered to as the "effective target model", ETM, in the case of hadron-nucleus collisions and the "effective projectile - effective target model", the EP-ET model, in the case of nucleus-nucleus collisions) is according to our picture (see Ref.9): There exists two different types of h-ET (in the hadron-nucleus case; or EP-ET in the general nucleus-nucleus case) collisions, namely the gentle-type and the violent-type, in Nature. This point is essential in principle, and plays an important role in practice.

For example, the difficulties in the application of CTM discussed by S.A. Azimov et al., (Phys.Letters 73B, 339 (1978) are due to the neglect of this fact. (see also: Y.Afek et al., Phys.Letters 78B, 329,(1978; L.Bergström et al., Phys. Letters 78B, 337, 1978;and F. Takagi in Proceedings of the 19th International Conference on High Energy Physics, Tokyo, 1978, p.472)

/20/ H.H.Heckman et al., "An Atlas of Heavy-Ion Fragmentation Topology" (unpublished); see also: Phys.Rev.C17, 1651, (1978)

S.Y.Fung and R.T.Poe, privat communications

/21/ J.Papp et al., Phys.Rev.Letters 34, 601 (1975)

L.S.Schroeder, in High-Energy Physics and Nuclear Structure - 1975 (see Ref.2), p.642

/22/ J. Stevenson et al., Phys.Rev.Letters 38, 1125 (1977)

J. Gosset et al., Phys.Rev. C 16, 629 (1977)

P.B. Price et al., Phys.Rev.Letters 39, 177 (1977)

G.D. Westfall et al., Phys.Rev. C 17, 1368 (1978)

P.B. Price and J. Stevenson Phys.Letters 78B, 197 (1978)

/23/ Meng Ta-chung, Phys. Rev. Letters 42, 1331 (1979)

/24/ S. Nagamiya et al., J. Phys.Soc. Japan 44 (1978) Suppl. p.378 (1978);

S. Nagamiya, in Proceedings of the 4th High-Energy Heavy Ion Summer Study (see Ref.1) p.71;

S. Nagamiya, H.Steiner and I. Tanihate, private communications

/25/ For p-p collisions at the CERN Intersecting Storage Rings, see L. Foa, J. Phys. (Paris) Colloq. 10, C 1 - 317 (1973)

/26/ That is, the nucleus-nucleus collision events characterized by large multiplicity and/or large transverse-momenta (i.e. events of the second kind, see above)

/27/ References on this class of models, can, for example, be found in:
I.A. Schmidt and R. Blankenbeder, Phys.Rev. D 15, 3321 (1977)
R.L. Hatch and S.E.Koonin, Phys.Letters 81 B, 1 (1978)
and the papers cited therein.
See also Ref. 1 and Ref. 24

/28/ References on this class of models can, for example, be found in:
Y. Afek et al., Phys.Rev.Letters, 849 (1978)
Meng Ta-chung,(Ref. 23)
S. Sandoval et al., LBL-Report 8771 (1979)
and the papers cited therein.
See also Ref. 1 and Ref. 24

/29/ S. Nagamiya, private communications
K.L.Wolf, private communications

/30/ See for example:
G.D. Westfall et al., Phys.Rev.Letters 37, 1202 (1976)
Meng Ta-chung, (Ref.9 and Ref. 23)
S. Sandoval et al.,(Ref. 28)
and the references given therein as well as those given in Ref. 1 and Ref. 24

/31/ C.B. Chiu et al., to be published

gentle collision

(a)

violent collision

(b)

The two different types of hadron-nucleus collision processes (due to the two types of h-ET collisions) seen in the laboratory

What Can We Learn From High-Energy Hadron-Nucleus Interactions?

Don M. Tow

Center for Particle Theory, Department of Physics
University of Texas, Austin, Texas 78712, USA

Abstract

High-energy hadron-nucleus (hA) collisions provide the exciting possibility of giving information about the spacetime development of hadron-hadron interactions and therefore differentiating various multiparticle production models. We review some of the major developments in this field during the past decade, both experimentally and theoretically. Several general features of the data are pointed out, and several classes of models are discussed. We elaborate on a recently proposed simple spacetime model for high-energy hA collisions. We briefly comment on the extension to nucleus-nucleus interactions and the future outlook.

I. Introduction

In hadron-hadron (hh) scattering experiments, because the particle detectors are always at macroscopic distances from the interaction region,[1] one can never learn about the spacetime development of the strongly interacting process. In other words, hh scattering experiments can only measure the S-matrix. On the other hand, if one replaces the target hadron by a nuclear target, then the time of flight for the incident hadron or any produced secondary in traveling from the first nucleon to a succeeding nucleon in the nucleus is of the same order or even shorter than the typical duration of the strongly interacting process. This means the nucleons in the nucleus serve as detectors which are separated by microscopic distances. Therefore, hadron-nucleus (hA) scattering experiments offer the unique opportunity to learn about the spacetime development of a strongly interacting process. Furthermore, one of the most important areas of research in strong interaction during the past decade is to find the correct model for describing high-energy multiparticle production processes. Since different multiparticle production models for hh collisions give rise in general to different predictions for hA collisions, studying hA interactions provides a good possibility of discriminating among the various multiparticle production models and selecting the correct one. For these reasons, high-energy hA collisions have generated tremendous interest during the past few years, both experimentally and theoretically.

In this paper, we present a brief review of this subject.[2] In Section II we review the essential features of the experimental multiparticle production data. In Section III we discuss several types of hA models. Here we also elaborate on a recently proposed simple spacetime model for high-energy hA collisions. This is followed by a short section on the extension to nucleus-nucleus (AA) collisions. In the last section, we mention some other interesting effects arising from interactions with nuclear targets, and end with some conclusions and an outlook.

II. Experimental Features

There are several general features of the data.[3] One defines

$$\bar{\nu}_h \equiv \frac{A\sigma_{in}^{hN}}{\sigma_{in}^{hA}} , \qquad \text{(II.1)}$$

where σ_{in}^{hN} (σ_{in}^{hA}) is the hadron-nucleon (nucleus) inelastic cross section and A is the number of nucleons in the nucleus. The variable $\bar{\nu}_h$ can be interpreted to be the average number of inelastic collisions experienced by the hadron h as it traverses the nucleus (see Appendix A). Experimentally $\bar{\nu}_h$ is approximately energy independent in the Fermilab energy range. Table I gives the values of $\bar{\nu}$ for p and $\pi^{\pm}$ from an experiment[3b] in the energy range from 50 GeV to 200 GeV.

Table I

Element	A	$\bar{\nu}_p$	$\bar{\nu}_\pi$
H	1	1.00	1.00
C	12	1.52	1.39
Al	27	1.95	1.70
emulsion	~60	2.50	2.07
Cu	64	2.54	2.10
Ag	108	3.00	2.40
Pb	207	3.67	2.82
U	238	3.84	2.92

If one defines

$$R_A \equiv \frac{\langle N_{ch} \rangle_{hA}}{\langle N_{ch} \rangle_{hN}} , \qquad \text{(II.2)}$$

then the first feature is

(a) R_A is small (~2.5) even for the largest A (see Fig. 1) and may be approximately parametrized as

$$R_A \simeq a + b\bar{\nu} \ , \quad \text{with } a \simeq 1/2 \simeq b \ , \qquad \text{(II.3)}$$

where a and b are roughly energy independent in the Fermilab range.[4]

Let η and y be the pseudorapidity and rapidity variable, respectively, defined as $\eta \equiv -\ell n \tan(\theta_L/2)$ and $y = \frac{1}{2}\ell n \frac{E+P_z}{E-P_z}$, where θ_L is the laboratory scattering angle, E and P_z are the energy and longitudinal momentum of the particle. Experimentally it is easier to measure η, whereas theoretically it is more convenient to use y. For almost all practical purposes, they can be used interchangeably. In any case one can always change from one to another by a simple kinematical transformation. The second feature is

(b) the differential multiplicity distribution $dN/d\eta$ is approximately target independent in the projectile fragmentation region and increases roughly as $\bar{\nu}_h$ in the target fragmentation region, as shown in Fig. 2.

Different projectiles give rise to different $\bar{\nu}_h$ and result in different R_A and $dN/d\eta$ for the same A. However, if we choose A such that we have the same $\bar{\nu}$ for different projectiles, then the third feature is

(c) different projectiles but for the same $\bar{\nu}$ have almost the same R_A [3f,3h] and $dN/d\eta$.[3b]

As usual, one defines the elasticity to be the ratio of the laboratory energy of the outgoing projectile to the incident projectile's laboratory energy. For proton or neutron initiated reactions, there is little ambiguity in picking among the final hadrons the one corresponding to the projectile nucleon, because usually there is only one nucleon in the forward hemisphere. The fourth experimental feature is

(d) The elasticity for hA collisions decreases only slightly from that of hN collisions even for the largest A.[5]

Another feature is

(e) in the very forward region, i.e., for $\eta \sim \eta_{max}$, $\left(\frac{dN}{d\eta}\right)^{hA_2} < \left(\frac{dN}{d\eta}\right)^{hA_1}$ by a small amount for $A_2 > A_1$.[3e]

Finally, if one defines

$$D \equiv \sqrt{\langle N_{ch}^2 \rangle - \langle N_{ch} \rangle^2} \quad , \qquad \text{(II.4)}$$

(f) then one sees a linear dependence of the dispersion D on $\langle N_{ch} \rangle$, and the data points for various nuclei lie on the same straight line.[3a]

The above are six general features of the high-energy hA data. A successful model must explain most and eventually all of these features.

III. Theoretical Models

We discuss several types of theoretical models for high-energy hA collisions.

A. Naive Cascade Model

The most obvious model and one of the earliest models proposed is the "naive cascade model."[6] This model assumes that in hh collision, the time involved in producing physical or fully dressed secondaries[7] is very short. Therefore, in hA collision, the incident hadron h interacts with one of the nucleons near the surface of A and produces many hadrons; each of these hadrons then undergoes further interactions with the subsequent nucleons, and the process repeats until all the hadrons emerge from A. This model results in an extremely large multiplicity and is completely ruled out by the experimental result (II.3). The smallness of R_A means that for a model to have any chance of explaining the data, it must have a mechanism to freeze some of the degrees of freedom.

B. Coherent Tube Model

A high-energy incident hadron h sees the nucleus Lorentz-contracted, a second type of model[8] assumes that h interacts coherently with a tube of $\bar{\nu}_h$ nucleons. In this "coherent tube model," the net effect is that the target has a mass $\bar{\nu}_h m_N$ and gives rise to an effective incident energy of $\bar{\nu}_h E_{inc}$. This model results in an enlargement along the negative direction of the target fragmentation region. Since at present energies the inclusive distribution has not reached a plateau and its height is increasing with energy, the model does give rise to an increase in the height of $\frac{dN}{d\eta}$ for hA collisions. However, this increase is much too small to be compatible with the data.[3b,d] Therefore, we conclude that it is unlikely that the coherent tube mechanism is the dominant mechanism responsible for the bulk of the multiparticle events in hA collisions. But this does not rule out the existence of this mechanism. As a matter of fact, it is probable that it is the coherent tube mechanism which is responsible for those rare and interesting events associated with the cumulative effect.[9] Better data and perhaps also particle identification in the region $\eta \lesssim 0$ can shed light on the existence and significance of the coherent tube mechanism.

C. Long-Interaction-Time Model

From phenomenological analysis of multiparticle production in hh interaction, one concludes that a general property of hh interaction is short-range correlation in rapidity space. One of the consequences of this property is that for the incident hadron to interact with the target nucleon, it must generate a multiperipheral chain containing small rapidity particles (or partons). The rapidity difference between the latter and the target (or partons in the target) is then small and appreciable interaction can occur. Now, it takes time for a hadron to generate such a multiperipheral chain. Focusing on one link in this multiperipheral chain, one can argue from the uncertainty relation and the difference between the initial energy and the final energy that the time τ_0 (in the rest frame of the parent hadron) for the hadron to branch into two hadrons is of the order $\tau_0 \sim 1/m$, where m is the mass of the hadrons. In the laboratory frame this time is Lorentz dilated. Summing the times from all the links results in a total laboratory time given approximately by $\tau \sim \frac{E}{m} \tau_0$, where E is the energy of the incident hadron. Therefore, at high energies a long interaction time is involved. For hA collisions, h can begin to generate its multiperipheral chain while approaching A. However, for the energetic secondaries produced from the interaction of h with one of the nucleons in A, they will not be able to interact with the subsequent nucleons, because by the time they generate these small rapidity multiperipheral chains, they would have travelled a long distance and be spatially separated from A (see Fig. 3). Therefore, only the small momentum secondaries can cascade and the additional multiplicity must be concentrated in the small rapidity region, and such long-interaction-time model[10] is at least in qualitative agreement with the data. Several of these models have also been shown to be in quantitative agreement.[10f-10i] But some of these require additional assumptions. Davidenko and Nikolaev[10f] used a $\tau_0 \sim \frac{1}{150 m_\pi}$ which is much smaller than the τ_0 discussed earlier (this possibility was also discussed in Ref. [10h]). Capella and Kryzwicki[10g] suppressed all secondary cascades and assumed that several constituents of the projectile, each of which shares the incident energy roughly equally, interact with the corresponding number of nucleons in A. Bialkowski, Chiu, and Tow[10h] assumed that the probability for a produced secondary to develop a small momentum multiperipheral chain is enhanced in the presence of other hadronic matter, as in a nucleus (induced maturity concept). Valanju, Sudarshan, and Chiu[10i] generalized Feinberg's QED result[10a] and thus provided a theoretical justification for the immaturity concept of Ref. [10h]. Furthermore, by using a continuous hadronic medium for the nucleus, instead of the discrete one used in Ref. [10h], they derived naturally a factor which at present energies has the same numerical effect as the induced maturity factor used in Ref. [10h]. Therefore, to explain present hA data, it is not necessary to explicitly introduce the induced maturity concept.

Thus, this type of long-interaction-time model, perhaps with some additional assumption, has been shown to be able to explain the experimental data. This type of model has also been extended to photon- and lepton-nucleus collisions.[11]

D. Two-Sheeted Model

The previous type of model assumes that the secondary particles are produced almost instantaneously, but in some sense they are bare particles, because they have a smaller inelastic collision probability than ordinary physical particles. To dress up these particles involves a long time (the larger the momentum, the longer the time).

Another type of model[12] takes an alternative approach to freeze some of the degrees of freedom. It assumes that in hh collisions, the result of the interaction is two excited hadronic systems. Each hadronic system is consisted of a color-$\overline{3}$ diquark and a color-3 quark; in the CM system, the diquark has large momentum and the quark has small momentum. Because of the large momentum difference, the spatial separation between the diquark and the quark increases with time, and the colored confining mechanism then presumably comes into play and causes each hadronic system to evolve into a jet of hadrons after a characteristic time τ_0 (in its own rest frame) has elapsed (see Fig. 4). For incident energy in the Fermilab range, the invariant mass of each hadronic system is several GeV's; so it is reasonable to take τ_0 to be the typical lifetime of a heavy nucleon resonance, i.e., $\tau_0 \simeq 1/m_\pi$ or $1/2m_\pi$. Then at high energy, the lifetime of the excited projectile system (EPS) in the laboratory will be greatly Lorentz dilated, whereas the lifetime of the excited target system (ETS) will have a small Lorentz dilation factor. Therefore, in high energy hA collision, the first EPS does not have time to evolve into its final multi-hadron state before it reaches the next nucleon in A, whereas the ETS will have a large probability of evolving into its final multihadron state. The above process repeats itself until the EPS leaves A after $\bar{\nu}_h$ inelastic collisions. This results in one EPS and $\bar{\nu}_h$ ETS, and therefore R_A in this approximation should be given by[13]

$$R_A \simeq \frac{1}{2} + \frac{\bar{\nu}_h}{2} , \qquad \text{(III.1)}$$

in agreement with feature (a) of Section II. The model also says that $(dN/d\eta)_{hA}$ should be approximately the same as $(dN/d\eta)_{hp}$ in the large η region,[14] and $(dN/d\eta)_{hA} \simeq \bar{\nu}_h (dN/d\eta)_{hp}$ in the small η region, in agreement with feature (b).

The above hh collision model is based on the recent discovery of the two-sheeted description of the Pomeron in the Dual Topological Unitarization (DTU) approach.[15] As we have just done, it can also be rephrased in the colored quark-parton model language.[16]

Even though we use the words projectile and target to describe the two excited hadronic systems, the quark-partons from which they are formed are not the same as those of the initial projectile and target.[12] Furthermore, unlike previously proposed two-fireball models[17] which were based on the fragmentation or diffractive model,[18] this DTU two-sheeted model is consistent with short-range correlation. In addition, at asymptotic energies, the two chains overlap completely in rapidity except for two finite regions at the two ends.

There is a small probability that the ETS's also do not evolve into their final states before reaching the succeeding nucleons and thereby undergo further interactions. This results in a second generation of ETS's and therefore additional multiplicity. Because of the low energy of the ETS's, the additional multiplicity should be small and in the small η region. The final hadrons from the evolution of the ETS's can also cascade, but this contribution can be neglected because the energies involved are even smaller. Therefore, this model predicts that R_A should be slightly larger than that given by (III.1), and the excess multiplicity should be in the small η region.

The above qualitative discussion has been made quantitative[12] by choosing a specific formulation of the two-sheeted hh model. The hh model chosen is that of Capella et al.[16a] This model was chosen because it is simple, well defined, and above all separates the contribution of the ETS from that of the EPS, but the results are fairly insensitive to the specific hh model used as input (as long as it agrees with the hh data).[19] The hA model is then completely determined and there is no free parameter. Figure 5a shows the results of these zero-parameter calculations for $dN/d\eta$ for 200 GeV p-initiated reactions for $\bar{\nu} = 2, 3, 4$, as well as the $\bar{\nu} = 1$ input. The model calculations are in good agreement with the data, except in the small η bins where as discussed before we expect the model calculations to underestimate the multiplicity. Similarly, good agreements are obtained at 100 GeV and 50 GeV, as shown in Figs. 5b and 5c. Figure 5d shows the model's prediction at 400 GeV.[20] The results for R_A are shown by the solid line in Fig. 6. Again, the model calculations are approximately equal to (but as expected slightly less than) the data.

As mentioned earlier, one expects an additional multiplicity for small η. Taking $\tau_0 \simeq \frac{1}{2m}$, one obtains the mean probability of an ETS reaching the next nucleon in A before evolving into its final state; then within the same two-sheeted model, it is straightforward combinatorials to trace its subsequent interactions with the succeeding nucleons (see Appendix B). The result for R_A after adding this additional contribution is the dashed line in Fig. 6. We see that the agreement is excellent.

Inasmuch as $\langle N_{ch} \rangle$ and $dN/d\eta$ for pN, πN, and kN are nearly the same, the model predicts that these quantities for pA, πA, and kA should be approximately the same for the same $\bar{\nu}$, in agreement with feature (c). In this model, because the energy of the EPS is only slightly smaller than the incident energy, the elasticity for pA is only slightly smaller than for pp, in agreement with feature (d).

Thus, we see that this simple two-sheeted model provides a good zero-parameter explanation of the first five experimental features. The only remaining feature to be explained (which has not yet been calculated in this model) is feature (f), the linear dependence of the dispersion D on $\langle N_{ch} \rangle$. Since it is known[21] for hh collisions that a two-component model can explain the dispersion data, by including a small diffractive component in the two-sheeted description of multiparticle production, one probably can explain this last feature also.

It is an interesting question whether the two methods of suppressing the degrees of freedom in the long-interaction-time model and the two-sheeted model are equivalent or different descriptions of the same physical phenomenon.

E. Other Models

We discuss a few other theoretical models. The parton model of Brodsky et al.[22] has similarity to the model of Ref. [10g]. The model assumes that on the average $\bar{\nu}_h$ wee-partons of the projectile interact with the wee-partons of the same rapidity in $\bar{\nu}_h$ different nucleons in A, and also assumes that cascading does not occur. It gives at asymptotic energy the result

$$R_A = \frac{\bar{\nu}_h}{\bar{\nu}_h + 1} + \frac{\bar{\nu}_h}{2} . \qquad \text{(III.2)}$$

In comparison with the two-chain model of Ref. [12], this parton model corresponds to a one-chain hh process. In addition, this model is applicable only in the central region, and the inclusion of the fragmentation contributions is done by the introduction of a parameter, whereas the model of Ref. [12] is valid for the whole rapidity range.

The two-phase model of Fishbane and Trefil[23] assumes that the first collision gives rise to an excited hadronic phase which has a flat distribution over the entire rapidity range and this excited phase has a long deexcitation time, and this excited hadronic phase interacts once more within A. However, the model neglects the interaction of the second (and subsequent) excited hadronic phase formed from the interaction of the first excited hadronic phase. Even ignoring this point, this model's dN/dy is different from that of Ref. [12] (even for the same hh input), because in the latter model the EPS while traversing A does not have much hadronic matter within the rapidity interval defined by its diquark and quark.

Finally we mention that there have been interesting attempts[24] in obtaining information from hA interactions about the number of constituent quark-partons in the incident hadrons.

IV. Nucleus-Nucleus Collisions

Some of the theoretical models discussed in the previous section can be easily generalized to nucleus-nucleus (AA) collisions. For illustration purposes, we discuss only the model of Ref. [12].[25] Analogous to the definition (II.1), one defines

$$\bar{\nu}_{A_1/A_2} \equiv \frac{A_1 \sigma_{in}^{NA_2}}{\sigma_{in}^{A_1 A_2}} , \qquad \text{(IV.1)}$$

which can be interpreted to be the average number of inelastically excited nucleons in A_1 in its collision with A_2. Then the two-sheeted model gives

$$R_{A_1 A_2} \equiv \frac{\langle N_{ch} \rangle_{A_1 A_2}}{\langle N_{ch} \rangle_{NN}} \simeq \frac{\bar{\nu}_{A_1/A_2}}{2} + \frac{\bar{\nu}_{A_2/A_1}}{2} , \qquad \text{(IV.2)}$$

where the first (second) term is the contribution from the $\bar{\nu}_{A_1/A_2}$ ($\bar{\nu}_{A_2/A_1}$) excited projectile (target) systems in A_1 (A_2). Equation (IV.2) reduces to Eq. (III.1) when $A_1 = N$. The rapidity distribution $(dN/dy)_{A_1 A_2}$ is given by superposing the contributions of $\bar{\nu}_{A_1/A_2}$ EPS's and $\bar{\nu}_{A_2/A_1}$ ETS's evaluated at the incident laboratory energy per nucleon. For the model to be realistic, the latter energy must be at least 20-30 GeV per nucleon. Unfortunately, this means that in the foreseeable future the only possible source of data is from cosmic rays.

V. Conclusion

The purpose of discussing high-energy hA collisions is to learn about the spacetime development of hh interactions and to differentiate various hh multiparticle production models. From the experimental data (especially features (a) and (b) presented in Section II and the theoretical analysis discussed in Section III), one concludes that to avoid excessive cascading there must be a long time scale involved in hh interactions. In the models of Ref. [10], this is the time it takes a high-energy hadron to generate a multiperipheral chain containing small rapidity partons so that it can have appreciable interaction with an at-rest hadron. Or saying it in another way, this is the time it takes a high-energy bare particle to dress up to be a physical hadron and so to have appreciable inelastic cross section upon collision with another hadron. In the model of Ref. [12], this is the time it takes the energetic excited projectile system to evolve into its final multihadron state. This long time scale rules out any sort of instantaneous production model such as the naive cascade model.

Some of the successful hA models also tie in with recent developments for models of hh interaction. For example, the models of Refs. [10f], [10g] and [22] rely heavily on parton model ideas, and the model of Ref. [12] is based on the two-sheeted description of the Pomeron and soft multiparticle production within the Dual Topological Unitarization approach.

As discussed in Section III, the latter model can also be rephrased using the language of colored quark-parton model and color confinement. All of this leads one to speculate that one can also see glimpses of the true underlying theory governing strong interaction by looking at hA interaction.

We also remark that there may be other interesting effects, besides the ones discussed here, which arise from interactions involving nuclear targets. One is the A-dependence for the inclusive cross section for large p_T secondaries. The data[26] shows that this A-dependence for a π secondary can be parametrized as $A^{\alpha(P_T)}$, where $\alpha(P_T) \simeq 0.9$ for $P_T \lesssim 1$ GeV/c and increases to $\alpha(P_T) \simeq 1.1$ for $P_T \gtrsim 4$ GeV/c. This A-dependence is most likely due to multiple scattering[27] inside the nucleus. Another effect is the apparently large intrinsic P_T that data involving nuclear targets seem to require for perturbative QCD to explain massive lepton pair production.[28] Perhaps at least part of this large intrinsic P_T is due to an increase in the effective intrinsic P_T as a result of multiple scattering inside the nucleus. Another effect is the importance of the two-step process $p + A \rightarrow \pi + A' + \cdots \rightarrow \mu^+\mu^- + \cdots$ as compared to the one-step process $p + A \rightarrow \mu^+\mu^- + \cdots$. Since the π has a valence antiquark, the two-step process may be significant if the production process is dominated by the Drell-Yan mechanism.[29]

We see that great progress in the field of hA interactions, both experimental and theoretical, was made during the last decade; we can look forward to some more exciting progress during this decade.

Acknowledgments

I wish to thank Professor Charles B. Chiu for collaboration and for valuable discussion, and for reading the manuscript and for useful suggestions. I also want to thank the Academia Sinica for the invitation to attend this Conference. This work was supported in part by the U.S. Department of Energy.

Appendix A

Here we present a simple argument within the disk approximation that $\bar{\nu}_h$ can be interpreted to be the average number of inelastic collisions experienced by h as it traverses A. Within this approximation one can relate[30] σ_{in}^{hN} and σ_{in}^{hA} to the S-matrix and get

$$\bar{\nu}_h = \frac{A\pi r_N^2(1 - S_h^2)}{\pi r_A^2(1 - S_h^{2\bar{k}})} , \tag{A.1}$$

where $\bar{k}$ is the average (over impact parameter) number of nucleons in A lying along the path of h, r_N and r_A are the radii of a nucleon and the nucleus, respectively. But $\bar{k} \simeq A(r_N/r_A)^2$ and $S_h^{2\bar{k}} << 1$, so

$$\bar{\nu}_h \simeq \bar{k}(1 - S_h^2)$$

$$= \bar{k}P_h , \tag{A.2}$$

where P_h is the inelastic collision probability.[10h] Our desired interpretation of $\bar{\nu}_h$ follows from Eq. (A.2).

For a more refined derivation, compare the theoretically calculated values of the average number of inelastic collisions obtained by Fishbane and Trefil[17b] and the experimental values of $\bar{\nu}_h$ of Ref. [3b].

Appendix B

In this appendix we consider pA collision and calculate the additional multiplicity due to the probability Q that the ETS does not evolve into its final multihadron state before reaching the succeeding nucleon N and thereby undergoes further interactions. We parametrize this probability by $e^{-t/\gamma\tau_0} \simeq e^{-2/\gamma}$, where we set $\tau_0 \simeq 1/2m_\pi$, and $t \simeq 1/m_\pi$ to be the time to travel the average internucleon distance, and γ is the time dilation factor.[31] Now, let P be the probability that an EPS or ETS would undergo an inelastic collision in a collision with N.[32] Then the probability P' that an EPS and ETS does not evolve into its final state after traveling the average internucleon distance and then undergoes an inelastic collision with the subsequent N is given by $P'_{EPS} \simeq P$ and $P'_{ETS} \simeq Pe^{-2/\gamma_{ETS}}$.

It suffices to include only those interactions of the ETS's with those succeeding N's which had not interacted inelastically with the EPS (we call these unwounded nucleons), because the excitation of N amounts to the formation of a sheet and this provides a natural saturation mechanism for its excitation.[33] We want to derive a formula for the probability p_i that the unwounded ith nucleon experiences an inelastic collision with the preceding ETS's. It is obvious that p_2 is given by

$$p_2 \simeq P(1-P)P'_{ETS} , \quad (B.1)$$

where the first factor is the probability that the incident proton undergoes an inelastic collision with the first N to produce an ETS, the second factor is the probability that the EPS leaves the second N unwounded, and the third factor is the probability that the ETS does not evolve into its multihadron final state and then undergoes an inelastic collision with the second N. Similarly, p_3 is given by

$$p_3 \simeq [2P(1-P)](1-P)P'_{ETS} + P^2(1-P)[1 - (1 - P'_{ETS})^2] , \quad (B.2)$$

where the first (second) term is the contribution when there is one (two) ETS. The general formula is then

$$p_m \simeq \sum_{i=1}^{m-1} \phi_i^{m-1}(1-P)[1 - (1-P'_{ETS})^i] , \quad (B.3)$$

where

$$\phi_i^{\ell} = \frac{\ell!}{i!(\ell-i)!} P^i(1-P)^{\ell-i} . \quad (B.4)$$

The multiplicity from each such collision is to be evaluated at the laboratory energy of the ETS.[34] Also, only one-half of this is an additional multiplicity (as the contribution from the first generation ETS's is already included). The ratio r of this charged multiplicity to $\langle N_{ch} \rangle_{pN}$ (the latter evaluated at the incident energy) is approximately $r \simeq 0.27$.[35] Therefore, the additional contribution ΔR_A to R_A for a nucleus of $\bar{k}$ nucleons in length ($\bar{k}$ is related to $\bar{\nu}_p$ by Eq. (A.2)) is

$$\Delta R_A \simeq r \sum_{i=2}^{\bar{k}} p_i \tag{B5}$$

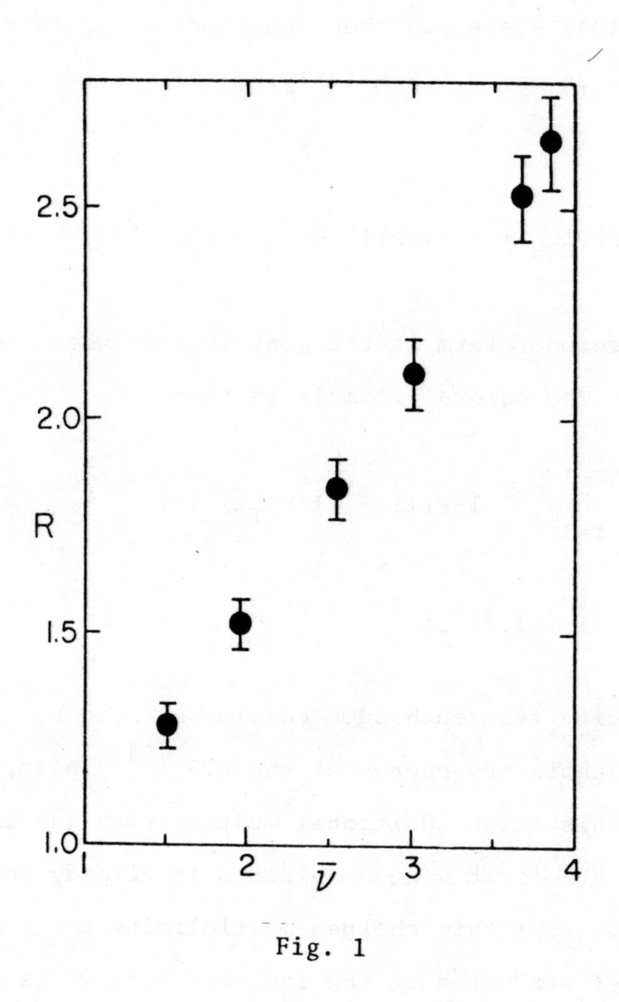

Fig. 1

The multiplicity ratio R_A versus $\bar{\nu}_p$ for 200 GeV pA collisions. Data is from Ref. [3b].

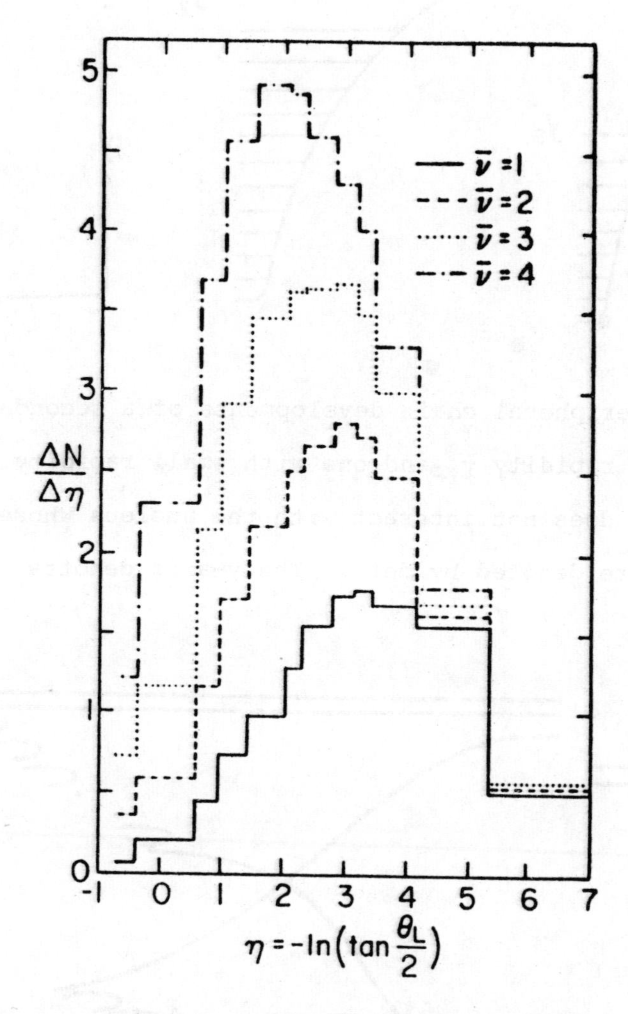

Fig. 2

Laboratory pseudorapidity distributions dN/dη for pA collisions. Data is from Ref. [3b].

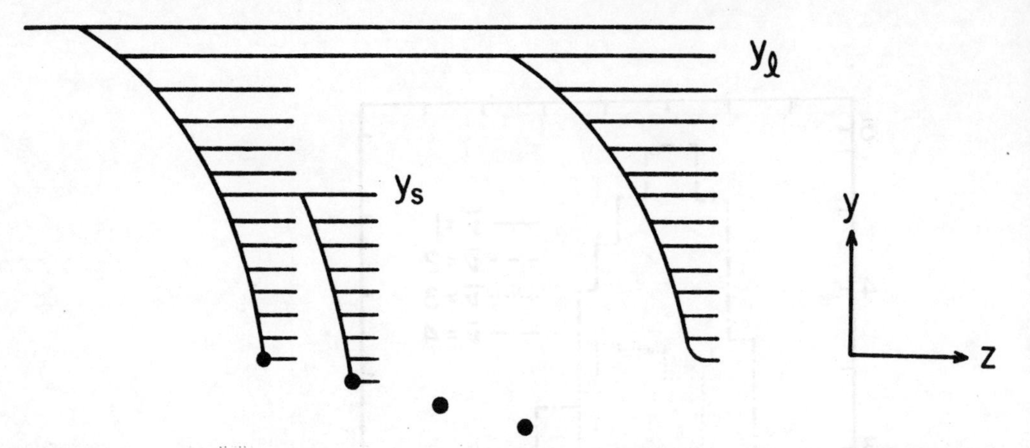

Fig. 3 The multiperipheral chain developments of a secondary with large rapidity y_ℓ and one with small rapidity y_s. The former does not interact with the nucleus whose nucleons are denoted by dots. The y-axis denotes rapidity.

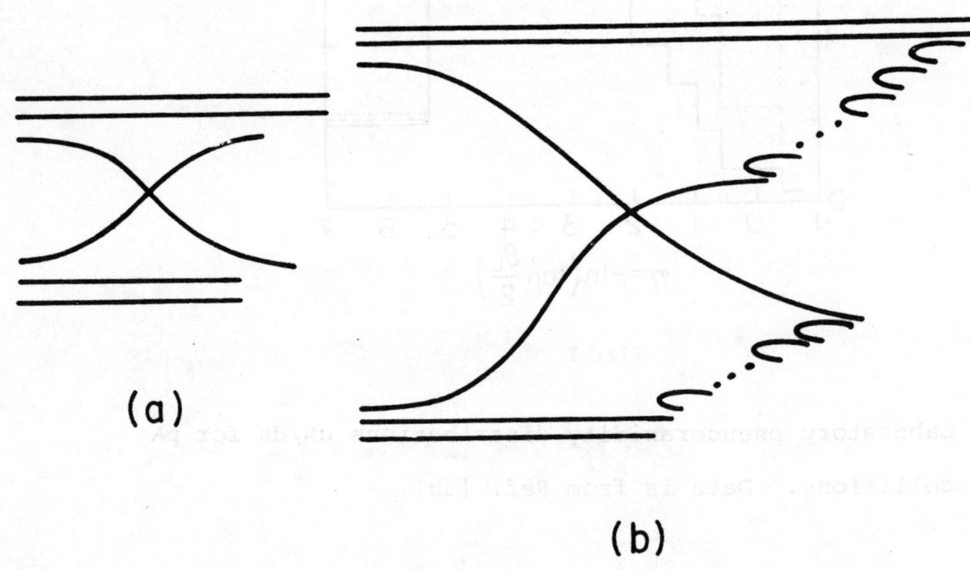

Fig. 4 The two-sheeted model for multiparticle production: (a) at short times, (b) at long times.

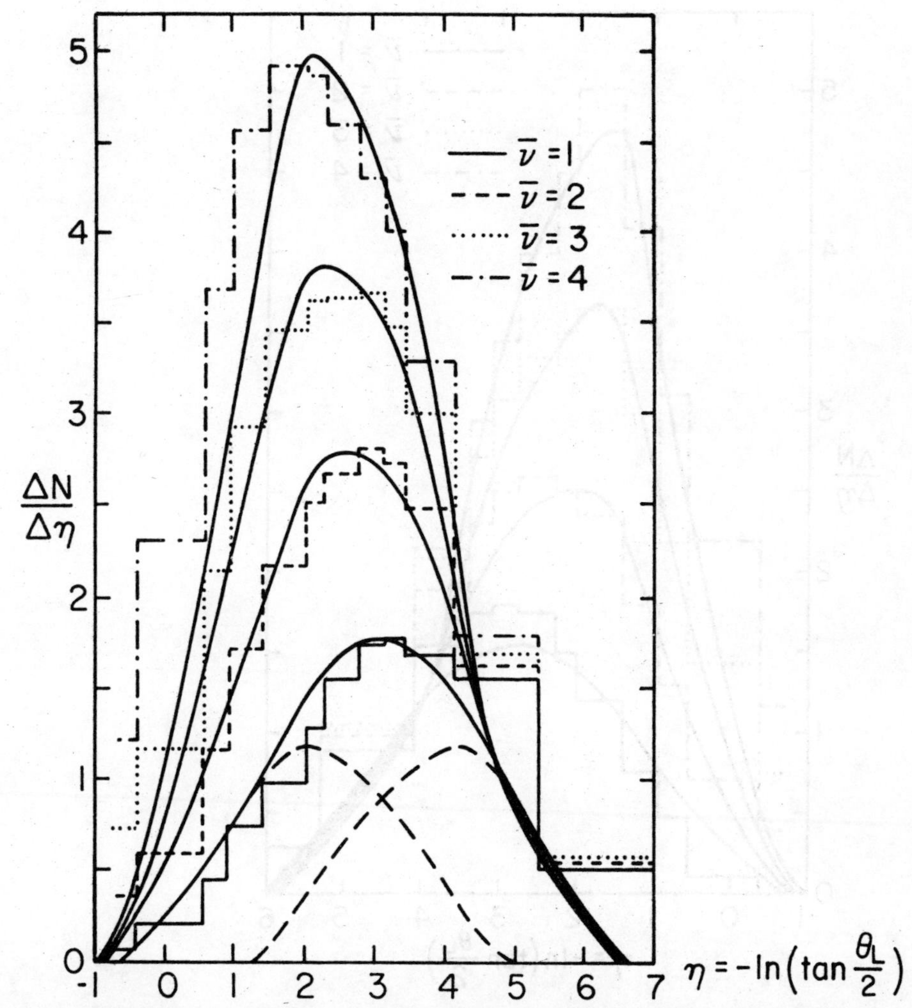

Fig. 5. Laboratory pseudorapidity distributions dN/dη for pA collisions. Solid curves are the zero-parameter results of the two-sheeted model of Ref. [12]; the $\bar{\nu} = 1$ curve is the input.

(a) At 200 GeV. Also shown are the individual contributions (dashed curves) of the EPS and ETS. The histograms are the data of Ref. [3b].

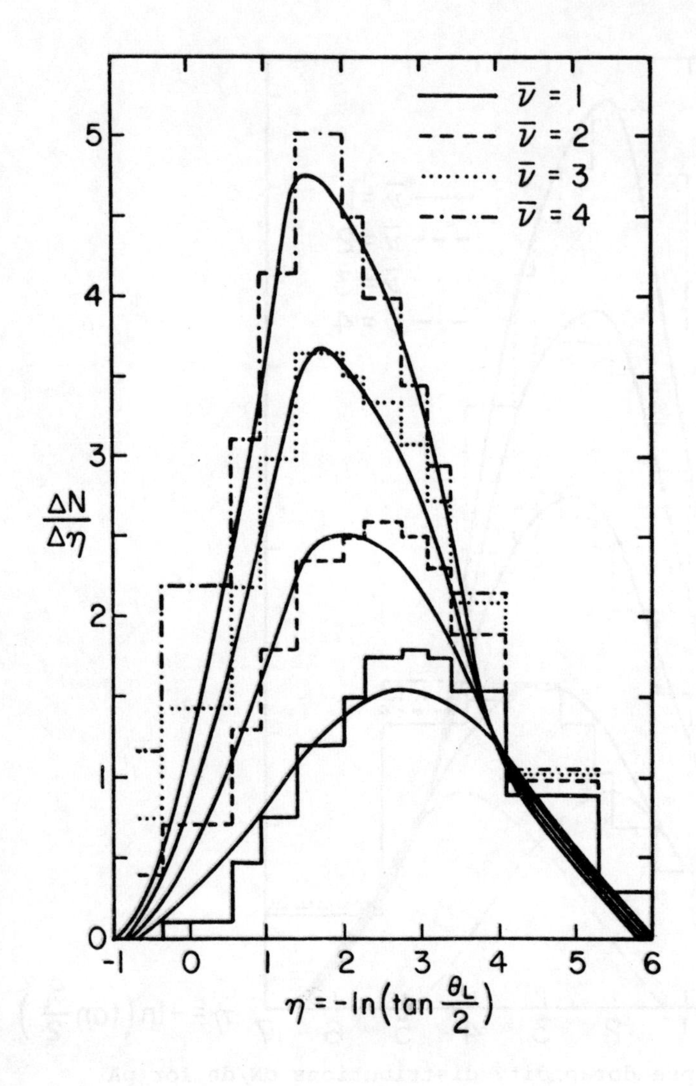

Fig. 5b

At 100 GeV. The histograms are the data of Ref. [3h].

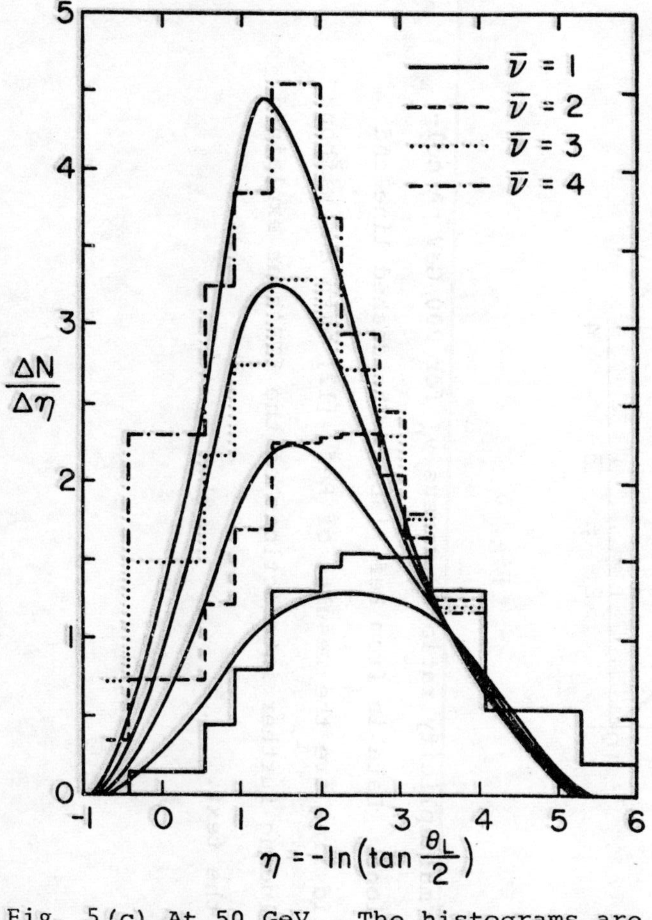

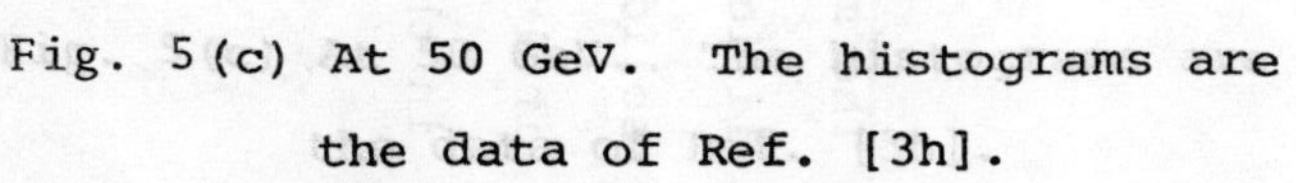

Fig. 5(c) At 50 GeV. The histograms are the data of Ref. [3h].

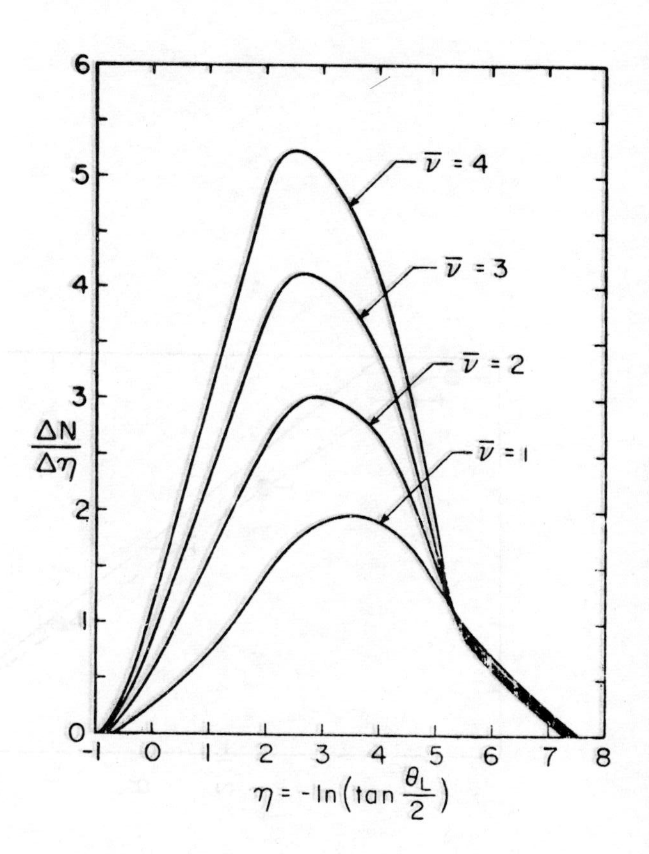

Fig. 5(d) Predictions at 400 GeV.

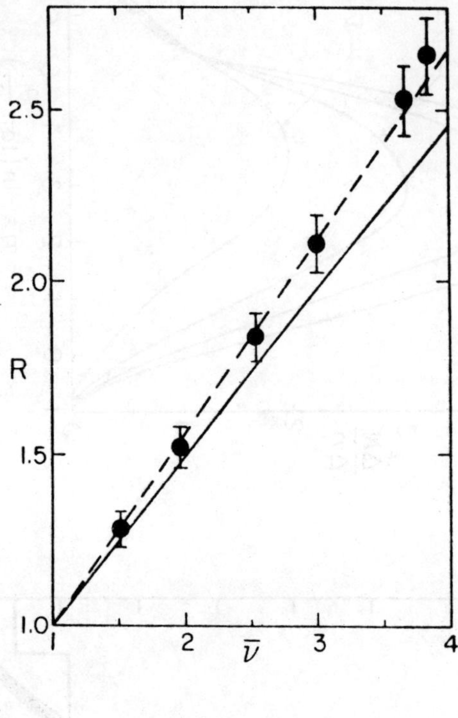

Fig. 6

The multiplicity ratio R_A versus $\overline{\nu}_p$ for 200 GeV pA collisions. Data is from Ref. [3b]. The dashed line and solid line are the results of Ref. [12] with and without including further interactions of the ETS's as explained in the text.

References and Footnotes

[1] Macroscopic distance here means any distance which is much larger than the typical range of strong interaction, i.e., 1 Fermi.

[2] For previous reviews, see

a. K. Gottfried, in Proc. of the V Intern. Conf. on High Energy Physics and Nuclear Structure-1973, Uppsala, Sweden, ed. by G. Tibell (North-Holland, Amsterdam/American Elsevier, New York, 1974).

b. W. Busza, in Proc. of the VI Intern. Conf. on High Energy Physics and Nuclear Structure-1975, Santa Fe and Los Alamos, U.S.A., ed. by D. E. Nagle et al. (American Institute of Physics Conf. Proc. No. 26, New York, 1975).

c. L. Bertocchi, ibid.

d. N. N. Nikolaev, in Proc. of the Trieste Topical Meeting on the Multiple Production on Nuclei-1976, ed. by L. Bertocchi.

e. A. Bialas, talk presented at the First Workshop on Ultra-Relativistic Nuclear Collisions, Berkeley, 1979; Fermilab preprint FERMILAB-Conf-79/35-THY (1979).

[3] a. W. Busza et al., Phys. Rev. Lett. 34, 836 (1975).

b. W. Busza et al., in Proc. of the XVIII Intern. Conf. on High Energy Physics, Tbilisi, 1976, ed. by N. N. Bogoliubov et al. (JINR, Dubna, U.S.S.R., 1977).

c. J. R. Florian et al., Phys. Rev. D13, 558 (1976).

d. C. Halliwell et al., Phys. Rev. Lett. 39, 1499 (1977).

e. D. Chaney et al., ibid. 40, 71 (1978).

f. J. E. Elias et al., ibid. 41, 285 (1978).

g. D. Chaney et al., Phys. Rev. D19, 3210 (1979).

h. J. E. Elias et al., Fermilab preprint FERMILAB-PUB-79/47-EXP 7185.178 (1979).

[4] The constant b (a) is slightly larger (smaller) than 1/2.

[5] See, e.g., Ref. [3g].

[6] a. P. M. Fishbane and J. S. Trefil, Phys. Rev. D3, 231 (1971).

b. A. Dar and J. Vary, ibid. D6, 2412 (1972).

[7] By physical or fully dressed secondary we mean that it has the same inelastic collision probability as an ordinary hadron.

[8] a. A. Z. Patashinskii, Zh. Eskp. Teor. Fiz. Pis'ma Red. 19, 654 (1974) [JETP Lett. 19, 338 (1974)].

b. F. Takagi, Nuovo Cimento Lett. 14, 559 (1975).

c. G. Berlad, A. Dar and G. Eilam, Phys. Rev. D13, 161 (1976).

d. S. Fredricksson, Nucl. Phys. B111, 167 (1976).

[9] A. M. Baldin et al., Yad. Fiz. 20, 1201 (1974) [Engl. Trans. 20, 629 (1975)]. See also A. M. Baldin, in Proc. of the VI Intern. Conf. on High Energy Physics and Nuclear Structure-1975, ibid.

[10] a. E. L. Feinberg, Zh. Eksp. Teor. Fix. 50, 202 (1966) [Sov. Phys.-JETP 23, 132 (1966)].

b. O. V. Kancheli, Zh. Eksp. Teor. Fiz. Pis'ma Red. 18, 465 (1973) [JETP Lett. 18, 274 (1974)].

c. E. S. Lehman and G. A. Winbow, Phys. Rev. D10, 2962 (1974).

d. J. Koplik and A. H. Mueller, Phys. Rev. D12, 3638 (1975).

e. L. Stodolsky, in Proc. of the VI Symp. on Multiparticle Dynamics, Oxford, 1975.

f. G. V. Davidenko and N. N. Nikolaev, Yad. Fiz. 24, 772 (1976) [Engl. Trans. 20, 402 (1976)].

g. A. Capella and A. Kryzwicki, Phys. Lett. 67B, 84 (1977), Phys. Rev. D18, 3357 (1978).

h. G. Bialkowski, C. B. Chiu, and D. M. Tow, Phys. Lett. 68B, 451 (1977), Phys. Rev. D17, 862 (1978). M. Hossain and D. M. Tow, Phys. Rev. D (to be published).

i. P. Valanju, E.C.G. Sudarshan, and C. B. Chiu, Phys. Rev. D (to be published).

[11] See the third paper of Ref. [10h].

[12] Chao W.-q., C. B. Chiu, He Z., and D. M. Tow, University of Texas preprint ORO 3992-376 (1979).

[13] Energy momentum conservation correction to this result can be shown[12] to be small.

[14] Since the EPS has a slightly smaller energy than its parent, $(dN/d\eta)_{hA}$ is actually slightly smaller than

$(dN/d\eta)_{hp}$ for $\eta \sim \eta_{max}$, in agreement with feature (e).

[15] a. H. Lee, Phys. Rev. Lett. 30, 719 (1973).

b. G. Veneziano, Phys. Lett. 43B, 413 (1973), 52B, 220 (1974), Nucl. Phys. B74, 365 (1974).

c. H.-M. Chan, J. E. Paton, and S. T. Tsou, Nucl. Phys. B86, 479 (1975); H.-M. Chan, J. E. Paton, S. T. Tsou, and S. W. Ng, Nucl. Phys. B92, 13 (1975).

d. For a review, see, e.g., G. F. Chew and C. Rosenzweig, Phys. Rep. 41C, 263 (1978).

e. For a discussion of the similarity between the DTU Pomeron and the dual resonance model Pomeron, see C. B. Chiu and S. Matsuda, Nucl. Phys. B134, 463 (1978), P. Aurenche and L. Gonzalez-Mestres, Phys. Rev. D18, 2995 (1978).

[16] a. A. Capella, U. Sukhatme, C.-I Tan, and J. Tran Thanh Van, Phys. Lett. 81B, 68 (1979).

b. A. Capella, U. Sukhatme, and J. Tran Thanh Van, Orsay preprint LPTPE 79/23 (1979). This is a refinement of the model of Ref. [16a]. It also eliminates an objection recently raised by W. Ochs and T. Shimada, Max-Planck Institute preprint MPE-PAE/PTh 34/79 (1979).

c. G. Cohen-Tannoudji et al., Saclay preprint DPh-T/79-97 (1979), and references therein.

d. For similar ideas, see also J. Dias de Deus and S. Jadach, Acta Phys. Polon. B9, 249 (1978).

[17] a. A. Dar and J. Vary, Phys. Rev. D6, 2412 (1972).

b. P. M. Fishbane and J. S. Trefil, Phys. Rev. D9, 168 (1974).

[18] a. C. Quigg, J.-M. Wang, and C. N. Yang, Phys. Rev. Lett. 1290 (1972).

b. R. Hwa, ibid. 26, 1143 (1971).

c. M. Jacob and R. Slansky, Phys. Rev. D5, 1847 (1972).

[19] For example, using the more refined version of Ref. [16b] gives essentially the same conclusions.

[20] There is one set of experimental data at 400 GeV [F. Fumuro, R. Ihara, and T. Ogata, Nucl. Phys. B152, 376 (1979)]. However, their normalization seems to be very different from those at several lower energies (50-300 GeV).[3]

[21] a. L. Van Hove, Phys. Lett. 43B, 65 (1973).

b. K. Fialkowski and H. I. Miettinen, Phys. Lett. 43B, 493 (1973).

c. C. B. Chiu and K.-H. Wang, Phys. Rev. D8, 2929 (1973).

[22] S. J. Brodsky, J. F. Gunion, and J. H. Kühn, Phys. Rev. Lett. 39, 1120 (1977).

[23] P. M. Fishbane and J. S. Trefil, Phys. Lett. 51B, 139 (1974).

[24] a. A. Bialas, W. Czyz, W. Furmanski, Acta Phys. Polon. B8, 585 (1977).

b. V. V. Anisovich, Yu. M. Shabelsky, and V. M. Shekhter, Nucl. Phys. B133, 477 (1978).

c. N. N. Nikolaev and A. Ya. Ostapchuck, CERN preprint

Ref. TH. 2575-CERN (1978).

d. See also A. S. Goldhaber, Phys. Rev. Lett. 33, 47 (1974).

[25] See also the discussion in Ref. [22].

[26] a. J. W. Cronin et al., Phys. Rev. D11, 3105 (1975).

b. L. Kluberg et al., Phys. Rev. Lett. 38, 670 (1977).

[27] J. H. Kühn, Phys. Rev. D13, 2948 (1976).

[28] See, e.g., R. C. Hwa, in Proc. of the XIII Rencontre de Morioud, Les Arcs, France, ed. by J. Tran Thanh Van, 1978.

[29] S. D. Drell and T. M. Yan, Phys. Rev. Lett. 25, 316 (1970); Ann. Phys. (N.Y.) 66, 578 (1971).

[30] See the first two papers of Ref. [10h] and Footnote 10 of the third paper of this reference.

[31] For an incident energy of 200 GeV, within the model of Ref. [12], γ_{EPS} and γ_{ETS} are respectively 34 and 3. Therefore, this probability Q for the EPS can be set approximately equal to one.

[32] In the numerical calculation we assume that the EPS or ETS has the same P as that of a proton which was estimated in Ref. [10h] to be 0.64.

[33] Another argument to support this assertion is because two ETS's have small relative momentum and so even upon a collision very little additional multiplicity will result.

[34] Within the model of Ref. [12], this energy is ~17 GeV when the incident energy is 200 GeV.

[35] See, e.g., D. M. Tow, Phys. Rev. D7, 3535 (1973).

A RELATIVISTIC GLAUBER EXPANSION FOR MANY-BODY SYSTEM

Zhu Hsi-quen and Ho Tso-hsiu

Institute of Theoretical Physics, Academia Sinica

and

Chao Wei-qin and Bao Cheng-guang

Institute of High Energy Physics, Academia Sinica

Abstract

Based on the relativistic composite field theory and its perturbative expansion, and using the eikonal approximation, an expression of the many body amplitude of a high energy hadron scattering on a nucleus or a composite particle is obtained. As in the usual Glauber theory, the many-body scattering amplitude is also expressed as a finite sum of the products of two-body amplitudes, including the necessary relativistic correction.

I. INTRODUCTION

Recently, the eikonal method developed by Glauber[1,2] has been widely applied to the problem of high energy hadron scattering on nucleus. There are two approximations in Glauber theory: the first, eikonal approximation, i.e., under the condition of small momentum transfer the propagators of the hadron and nucleons can be linearized as follows:

$$\frac{1}{(\vec{p}-\vec{q})^2-\vec{p}^2+i\epsilon} = \frac{1}{-2\vec{p}\cdot\vec{q}+\vec{q}^2+i\epsilon} \sim \frac{1}{-2\vec{p}\cdot\vec{q}+i\epsilon} \quad ; \qquad (1.1)$$

the second, the frozen approximation, i.e., the collisions between the nucleons in the nucleus can be neglected during the moment the high speed hadron passes the nucleus. Under these two approximations a very important result has been deduced by Glauber from Schrödinger eq.: the many-body amplitude of a hadron scattering on a nucleus can be expressed as a finite sum of products of two-body scattering amplitudes. This result not only avoids the very complicated intermediate progagator which exists in Watson multiple scattering expansion but also escapes from the difficulty of dealing with the strong interaction. Then the many-body scattering amplitude can be calculated directly, using the experimental two-body amplitude, or vice versa.

Glauber[2], and Chou and Yang[3] emphasize that the character in high energy many-body scattering mentioned above is essentially the result of the diffractive coherent phenomena caused by the wave property of high energy particles. Only for non-relativistic situations there is a dynamic derivation to obtain this diffractive picture,[1,4] but it is not clear how well this picture appears in relativistic situations. This restricts the application of

Glauber theory to relativistic particles, especially for inelastic scattering phenomena.

There were some papers to generalize the Glauber theory to relativistic situation, but only for two body problem.[5-9] Lately, the systematic composite particle quantum field theory[10] and its perturbative expansion provides a theoretical framework for dealing with the relativistic scattering between the composite particles. Our work starts from the composite particle quantum field theory; using the same two approximations as those in Glauber theory; extending to many body scattering the generalized ladder summation method established by Levy and Sucher for two-body problem. Finally, we obtain a relativistic expression consistent with Glauber's but substituting the three-dimensional impact parameter to the two-dimensional one in the usual Glauber theory and using the covariant transition density of nucleus. It is interesting that as in Glauber theory our result to some extent is also independent of the particular form of the interaction. Thus, this relativistic Glauber expansion may be applied to a wider area.

II. THE RELATIVISTIC GLAUBER EXPANSION

From the composite field theory the S-matrix element of a hadron a scattering on an N- body composite particle

B can be written as

$$
{}^{out}\langle a'B' \mid aB \rangle^{in} = \int d^4X_B\, d^4X'_B\, d^4(x_1^c \cdots x_N^c)\, d^4(x_1^{c'} \cdots x_N^{c'})
$$

$$
\delta\Big(\frac{1}{N}\sum_i x_i^c\Big)\, \delta\Big(\frac{1}{N}\sum_i x_i^{c'}\Big) \frac{1}{\sqrt{2\omega'}} e^{ip'_a x'_a} \frac{1}{\sqrt{2E'}} e^{iP'X'_B} \varphi^*_{P'}(x_1^c \cdots x_N^c)
$$

$$
(\vec{O}' - \vec{I}')\, \langle 0 | T(\varphi(x'_a)\varphi(x_1^{c'} + X'_B) \cdots \varphi(x_N^{c'} + X'_B)\varphi(x_a)\varphi(x_1^c + X_B)
$$

$$
\cdots \varphi(x_N^c + X_B)) | 0 \rangle (\overleftarrow{O} - \overleftarrow{I}) \varphi_P(x_1^c \cdots x_N^c) \frac{1}{\sqrt{2E}} e^{-iPX_B} \frac{1}{\sqrt{2\omega}} e^{-ip_a x_a} ,
$$

$$
(2.1)
$$

where $\varphi(x)$ is the Heisenberg field operator of a single particle, $\varphi_P(x_1{}^c \cdots x_N{}^c)$ is the B - S wave function of the N-particle system defined in the following way

$$
\frac{1}{\sqrt{2E}} \psi_P(x_1 \cdots x_N) \equiv \langle 0 | T(\varphi(x_1) \cdots \varphi(x_N)) | B \rangle
$$

$$
= \frac{1}{\sqrt{2E}} e^{-iPX_B} \varphi_P(x_1^c \cdots x_N^c) \tag{2.2}
$$

where

$$
X = \frac{1}{N}\sum_i x_i , \quad x_i^c = x_i - X , \quad i = 1 \cdots N . \tag{2.3}
$$

The wave function in the momentum space is defined as

$$
\chi_P(p_1 \cdots p_N) = \int d^4(x_1 \cdots x_N) e^{i\sum_i p_i x_i} \psi_P(x_1 \cdots x_N) . \tag{2.4}
$$

ψ_P satisfies N-body B-S eq.:

$$
(\vec{O} - \vec{I})\psi_P(x_1 \cdots x_N) = 0 . \tag{2.5}
$$

For the composite particle composed of scalar particles,

the operator $\vec{O}$ is a product of N Klein-Gorden operators and $\vec{I}$ is an integral operator representing the sum of all irreducible diagrams of the interactions between N particles.

For convenience we discuss a scalar particle a scattering on a bound state B composed of N scalar particles $(b_1 \cdots\cdots b_N)$ as shown in fig. 1, where p_a , P_a ' and P, P' represent respectively the initial and final four-dimensional momenta of a and B. $\{p_i\}^N$ and $\{p_i'\}^N$ are the initial and final momenta of $\{b_i\}^N$ in the composite paricle B. ($\{p_i\}^N$ = $(p_1 \cdots p_N)$) The shaded region in fig. 1 represents all the possible interactions. As an example, we select only all the possible generalized ladder digrams exchanging one kind of scalar particle and neglect all the radiation correction. We will discuss the more general situation later.

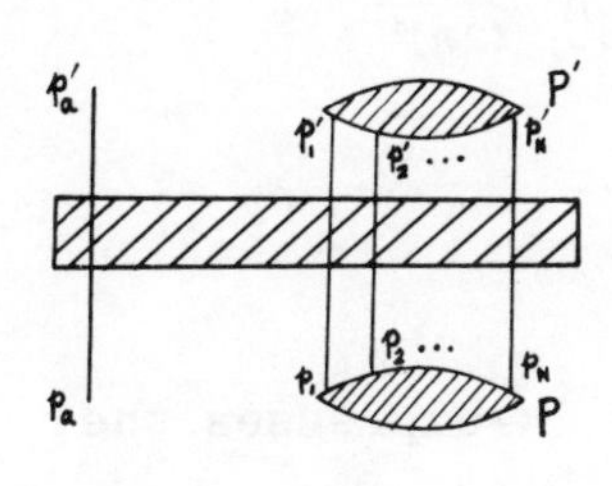

Fig. 1

Under the frozen approximation any quantum exchange between the particles in the target is neglected and only the interactions of the incident particle with the particles in the target are considered. Now the scattering amplitude T which expresses particle a interacting with the composite particle B is

$$T = \sum_{n=1}^{N} \sum_{C_N^n} T^{(n)} \qquad (2.6)$$

where $T^{(n)}$ is the T-matrix element which expresses particle a interacting with only n particles in B, for example $(b_1 \cdots\cdot b_n)$, exchanging at least one quantum with each of them; $\sum_{C_N^n}$ is the sum of all the combinations selection n particles from B. The expression of $T^{(n)}$ is

$$T^{(n)}(p_a P, p_a' P') = \frac{1}{\sqrt{16\omega\omega' E E'}} \int \prod_{i=1}^{N} \frac{d^4 p_i}{(2\pi)^4} \prod_{i=1}^{n} \frac{d^4 p_i'}{(2\pi)^4}$$

$$\chi^*_{P'}(p_1' \cdots p_n' \, p_{n+1} \cdots p_N)\, \chi_P(p_1 \cdots p_N) \quad \cdot \qquad (2.7)$$

$$M\left(p_a \{p_i\}^n, p_a' \{p_i'\}^n\right) \prod_{j=n+1}^{N} \Delta^{-1}(p_j) \quad ,$$

where $\Delta^{-1}(p_j) = -i(p_j^2 - m_b^2)$, $M\left(p_a\{p_i\}^n, p_a'\{p_i'\}^n\right)$ expresses the amplitude contributed by all the generalized ladder diagrams in which particle a exchanges at least one quantum with each of n independent particles $b_1, \cdots\cdot, b_n$.

Of course, the summation of the above diagrams is difficult. Using the eikonal approximation, for two-body scattering Levy and Sucher linearized the propagator as

$$\frac{1}{(p-q)^2 - m^2 + i\epsilon} = \frac{1}{p^2 - m^2 - 2p\cdot q + q^2 + i\epsilon} = \frac{1}{-2p\cdot q + i\epsilon} \qquad (2.8)$$

Using the identity

$$\sum_P{}' \frac{1}{a_1} \cdot \frac{1}{a_1 + a_2} \cdots\cdot \cdot \frac{1}{a_1 + \cdots + a_n} = \frac{1}{a_1} \cdot \frac{1}{a_2} \cdots\cdot \cdot \frac{1}{a_n} \qquad (2.9)$$

(where $\sum_{P}'$ is a sum of all the possible permutations), they obtained an eikonal amplitude of two-body scattering. By means of a similar method, after a complicated manipulation of permutation and combination, we finally obtained that under the eikonal approximation the scattering amplitude of particle a with n independent particles $\{b_i\}^n$ is

$$M = \int d^4x \, e^{iqx} \int \prod_{i=1}^{n} d^4x_i \, e^{iq_i x_i} \frac{e^{U_{ab_i}(x_i - x)} - 1}{U_{ab_i}(x_i - x)} \cdot \Delta_f(x_1 - x) \, U_a(x_2 - x) \cdots \cdot U_a(x_n - x) \,, \tag{2.10}$$

where

$$q = p_a - p_a' \,, \qquad q_i = p_i - p_i'$$

$$\Delta_f(x) = \int \frac{d^4k}{(2\pi)^4} e^{ikx} (-ig)^2 \Delta_f(k)$$

$$U_a(x) = \int \frac{d^4k}{(2\pi)^4} e^{ikx} (-ig)^2 \Delta_f(k) \left(\frac{i}{a_d(k)} + \frac{i}{a_u(k)} \right) \tag{2.11}$$

$$U_{ab_i}(x) = \int \frac{d^4k}{(2\pi)^4} e^{ikx} (-ig)^2 \Delta_f(k) \left(\frac{i}{a_d(k)} + \frac{i}{a_u(k)} \right) \left(\frac{i}{b_{d_i}(k)} + \frac{i}{b_{u_i}(k)} \right)$$

and

$$\Delta_f(k) = \frac{i}{k^2 - \mu^2 + i\epsilon}$$

$$a_d(k) = -2p_a \cdot k + i\epsilon \qquad b_{d_i}(k) = 2p_i \cdot k + i\epsilon$$

$$a_u(k) = 2p_a' \cdot k + i\epsilon \qquad b_{u_i}(k) = -2p_i' \cdot k + i\epsilon \,. \tag{2.12}$$

Unfortunately, the obtained many-body amplitude is not symmetric. Even if one applies some sort of artifical symmetrization device, it is impossible to express the many-body amplitude as the product of Levy's two-body amplitudes. To do this there is the need of deeper understanding about the meaning of the approximation of small momentum transfer.

The imaginary part of (2.8), i.e., the on-shell part of the propagator shows that the small momentum transfer of any physical particle is expressed as

$$p \cdot q \simeq 0 \tag{2.13}$$

This means that the small momentum transfer of an on-shell particle should be approximately perpendicular to the direction of the four-momentum of the incident particle. To show this character explicitly we choose the direction of the incident four-momentum p_a as a base vector:

$$\epsilon^0 = \hat{p}_a = \left(\frac{\omega}{m_a}, 0, 0, \frac{|\vec{p}_a|}{m_a} \right) . \tag{2.14}$$

Naturally, the other three base vectors are orientational vectors perpedicular to ϵ^0:

$$\epsilon^1 = (0,1,0,0) , \ \epsilon^2 = (0,0,1,0) , \quad \epsilon^3 = \left(\frac{|\vec{p}_a|}{m_a}, 0, 0, \frac{\omega}{m_a} \right) . \quad (2.14)$$

Under the approximation of small momentum transfer it is obvious that $p_a' \sim p_a$, and

$$\frac{i}{a_d(k)} + \frac{i}{a_u(k)} \simeq \pi \delta (p_a \cdot k) = \frac{\pi}{m_a} \delta (k^{\epsilon^0}) . \tag{2.15}$$

Substituting (2.15) into (2.11), we have

$$U_a(x) = U_a(\vec{x}_\epsilon), \qquad U_{ab_i}(x) = U_{ab_i}(\vec{x}_\epsilon) \tag{2.16}$$

where $\vec{x}_\epsilon = (x_{\epsilon_1}, x_{\epsilon_2}, x_{\epsilon_3})$ is the orientation vector of four-vector x along $\vec{\epsilon}$, and

$$U_a(\vec{x}_\epsilon) = \frac{1}{2m_a}\int d x_{\epsilon_0}\,\Delta_f(x)\,. \tag{2.17}$$

From (2.13) we know that a small momentum transfer between two physical particles means $p_a\cdot q \simeq 0$, i.e., $q^{\epsilon^0} \simeq 0$, $q\cdot x \simeq \vec{q}^{\epsilon}\cdot\vec{x}_\epsilon$ Now except for the factor $\Delta_f(x_1-x)$, all the other factors in (2.10) are independent of x_{ϵ_0}. As a result the integration of x_{ϵ_0} can be performed directly on $\Delta_f(x_1-x)$ and an entirely symmetric result can be obtained:

$$M(p_a\{p_i\}^n, p_a'\{p_i'\}^n) = 2m_a\int d\vec{x}_\epsilon\, e^{i\vec{q}^\epsilon\cdot\vec{x}_\epsilon}\prod_{i=1}^{n}\int d^4x_i\, e^{iq_ix_i}\cdot \tag{2.18}$$

$$\Gamma_{ab_i}(\vec{x}_{i\epsilon}-\vec{x}_\epsilon)\,,$$

where

$$\Gamma_{abi}(\vec{x}_{i\epsilon}-\vec{x}_\epsilon) = \frac{e^{U_{ab_i}(\vec{x}_{i\epsilon}-\vec{x}_\epsilon)}-1}{U_{ab_i}(\vec{x}_{i\epsilon}-\vec{x}_\epsilon)}\,U_a(\vec{x}_{i\epsilon}-\vec{x}_\epsilon)\,. \tag{2.19}$$

For n = 1, (2.18) returns to the usual Levy's two-body amplitude:

$$M(p_a p_b\,.\,p_a' p_b') = 2m_a\int d\vec{x}_\epsilon\, e^{i\vec{q}^\epsilon\cdot\vec{x}_\epsilon}\prod_{i=1}^{n}\int d^4x_b\, e^{iq_b\cdot x_b}\,\Gamma_{ab}(\vec{x}_{b\epsilon}-\vec{x}_\epsilon)$$

$$= (2\pi)^4\delta(p_a+p_b-p_a'-p_b')\int d\,x\, e^{iq_bx}\,\Delta_f(x)\,\frac{e^{U_{ab}(x)}-1}{U_{ab}(x)}\,. \tag{2.20}$$

If we define the three-dimensional profile function Γ_{ab_i} using the two body amplitude (2.20), the n-body amplitude is actually expressed only as the products of the two-body ones.

As in Glauber theory, assuming that the scattering amplitude of the incident particle with the particle in the target can be regarded as an on-shell one, we can substute (2.18) into (2.6) and (2.7). Neglecting the effect of Fermi energy, all the two-body profile functions are the same:

$$\Gamma_{ab_i}(\vec{x}_\epsilon) \equiv 2 m_b \Gamma_a(\vec{x}_\epsilon) \qquad (2.21)$$

Using (2.4) we obtain the final expression of the scattering amplitude of hadron a with the composite particle B composed by N particles in the coordinate representation:

$$T(p_a P, p_a' P') = \frac{2 m_a}{\sqrt{16 \omega \omega' E E'}} (2\pi)^4 \delta^{(4)}(p_a + P - p_a' - P') \cdot \int d\vec{x}_\epsilon \, e^{i\vec{q}^\epsilon \cdot \vec{x}_\epsilon} \Gamma(\vec{x}_\epsilon) \quad , \qquad (2.22)$$

where the profile function is

$$\Gamma(\vec{x}_\epsilon) = \sum_{i=1}^{N} \int \Gamma_a(\vec{x}_\epsilon - \vec{x}_{i\epsilon}) D^{(1)}_{p'p}(\vec{x}_{i\epsilon}) \, d\vec{x}_{i\epsilon}$$

$$+ \sum_{i<j}^{N} \int \Gamma_a(\vec{x}_\epsilon - \vec{x}_{i\epsilon}) \Gamma_a(\vec{x}_\epsilon - \vec{x}_{j\epsilon}) D^{(2)}_{p'p}(\vec{x}_{i\epsilon}, \vec{x}_{j\epsilon}) d\vec{x}_{i\epsilon} d\vec{x}_{j\epsilon} +$$

$$+ \cdots$$

$$\vdots \qquad (2.23)$$

$$+ \int \Gamma_a(\vec{x}_\epsilon - \vec{x}_{1\epsilon}) \cdots \Gamma_a(\vec{x}_\epsilon - \vec{x}_{N\epsilon}) D^{(N)}_{p'p}(\vec{x}_{1\epsilon} \cdots \vec{x}_{N\epsilon}) \prod_{i=1}^{N} d\vec{x}_{i\epsilon} \quad ,$$

the thickness function is

$$D^{(n)}_{p'p}(\vec{x}_{1\epsilon} \cdots \vec{x}_{n\epsilon}) = \int \rho^{(n)}_{p'p}(x_1 \cdots x_n) \prod_{i=1}^{n} dx_{i\epsilon_0} \qquad (2.24)$$

the corresponding density function $\rho^{(n)}$ is

$$\rho^{(n)}_{p'p}(x_1 \cdots x_n) = \int \varphi^*_{p'}(x_1 \cdots x_N)(2m_b)^n \prod_{j=n+1}^{N} \Delta_f^{-1}(x_j) \cdot$$

$$\cdot \varphi_p(x_1 \cdots x_N)\, \delta\left(\frac{1}{N} \sum_i^N x_i\right) d^4(x_{n+1} \cdots x_N) \quad . \qquad (2.25)$$

(2.22)-(2.25) are our final results. It is very clear that this is the relativistic generalization of the non-relativistic many-body Glauber expansion. However, the two-dimensional impact parameter is substituted by the three-dimensional one, $\vec{x}_\epsilon$. At the same time, the two-dimensional thickness function is substituted by the three-dimensional one. As a result we can consider the relativistic recoil effect of any pair of two-body collision and the Loretz boost effect of the thickness functions. It is a very interesting problem to discuss these relativistic effects concerning the scattering of a high energy hadron on a nucleus in detail.

III. THE PHYSICAL MEANING OF THE COVARIANT DENSITY FUNCTION AND THE NON-RELATIVISITIC APPROXIMATION

To show the physical meaning of the covariant transition density defined by (2.25) we discuss an electromagnetic transition process on a composite particle (fig. 2), where the vertex of the electro-magnetic transition is defined as

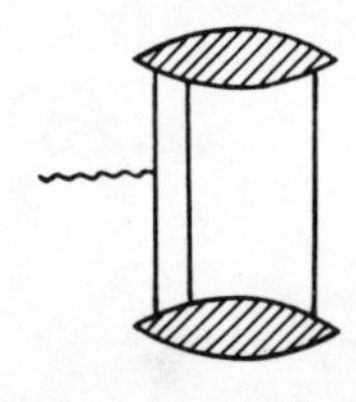

Fig. 2

$$(2\pi)^4 \delta^{(4)}(p+q-p')(p_\mu + p'_\mu) \ .$$

Under the condition of small momentum transfer, neglecting the off-shell effect of the particles in the composite particle, we obtain

$$\langle P' | J_0 | P \rangle = \frac{1}{\sqrt{4EE'}} \sum_{l=1}^{N} \int \prod_{i=1}^{N} \frac{d^4 p_i}{(2\pi)^4} \, 2m_b \prod_{j \neq l} \Delta^{-1}(p_j)$$

$$\chi^*_{P'}(p_1 \cdots p_N) \, \chi_P(p_1 \cdots p_N) \tag{3.1}$$

$$= \frac{(2\pi)^4 \delta(P-P')}{\sqrt{4EE'}} \sum_{l=1}^{N} \int \rho^{(1)}_{P'P}(x_l) d^4 x_l$$

Consequently, as in the usual Glauber many-body expansion, where the first term is related only to the single particle density, $\rho^{(1)}_{P'P}$ in the first term of (2.23) is the fourth component of the single particle current vector of a composite particle*. Similarly, calculation of the two

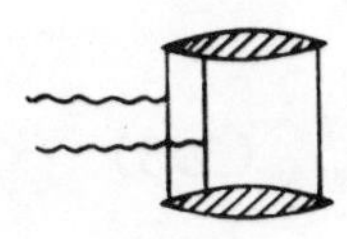

particle or many particle current (fig. 3) will give the meaning of various correlation densities.

fig. 3

To understand the physical meaning of the three-dimensional impact parameter clearly it is necessary to discuss the non-relativistic transition of our expansion, which includes two respects.

The first, neglecting the relativistic recoil of each two-body collision induces a transition of the three-dimensional profile function to the two-dimensional one. For simplicity we only discuss the situation of two-body composite particle. Neglecting the recoil of each two-body collision, i.e., $q_i^0 = 0$, we have

$$T(p_a P, p_a' P') = \frac{2P}{\sqrt{16\omega\omega' E E'}} \, 2\pi\,\delta(q^0)\int d^2x_\perp \, e^{iq_\perp x_\perp} \cdot$$

$$\int d^4x_1\, d^4x_2 \;\; \psi^*(x_1 x_2)\Big\{2m_b\Gamma(x_\perp - x_{1\perp})\,\delta(x_1^0)\,\Delta^{-1}(x_2)$$

$$+ 2m_b\Gamma(x_\perp - x_{2\perp})\;\delta(x_2^0)\,\Delta^{-1}(x_1)$$

$$+(2m_b)^2\,\Gamma(x_\perp - x_{1\perp})\,\Gamma(x_\perp - x_{2\perp})\,\delta(x_1^0)\,\delta(x_2^0)\Big\}\psi(x_1 x_2)\;, \tag{3.2}$$

*This is very similar to the idea proposed by Chou and Yang in the coherent droplet model where the distribution of the nuclear matter is made consistent with the electromagnetic one.

where

$$\Gamma(x_\perp) = e^{U(x_\perp)} - 1$$

$$U(x_\perp) = \frac{p}{m_a} \int V(\vec{x})\, d x^3 \tag{3.3}$$

$$V(\vec{x}) = \frac{1}{4 m_a m_b} \int \Delta_f(x)\, d x^0$$

The second, applying the instantaneous approximation to the kernel in the B-S eq. of the bound state, together with the non-relativistic approximation of the kinematic factor, the relativistic B-S wave function will approach the non-relativistic Schrödinger wave function.[11] In fact, the factor $\Delta^{-1}(x)$ in (3.2) will correspond to a one-time δ-function:

$$\Delta^{-1}(x_i) \longrightarrow 2\, m_b\, \delta(t_1 - t_2) \tag{3.4}$$

Substituting (3.4) into (3.2), finally a result exactly coincident with Glauber's is obtained:

$$T(p_a P, p_a' P') = \frac{2p \cdot 4\, m_b^2}{4\, m_a M} (2\pi)^4 \delta^4(p_a + P - p_a' - P') \int d^2 x_\perp e^{i q_\perp x_\perp}$$

$$\int |\Phi_S(\vec{x}_1, \vec{x}_2)|^2 \Big\{ (e^{U(x_\perp - x_{1\perp})} - 1) + (e^{U(x_\perp - x_{2\perp})} - 1) +$$

$$(e^{U(x_\perp - x_{2\perp})} - 1)(e^{U(x_\perp - x_{2\perp})} - 1) \Big\}\, d^3 x_1\, d^3 x_2 \tag{3.5}$$

IV. THE UNIVERSALITY OF THE RELATIVISTIC GLAUBER EXPANSION

The above many-body amplitude expressed as a sum of the products of two-body ones, although only using an interaction model of scalar particles, will be suitable to more general interaction forms. In fact, the whole derivation of the expansion is independent of the particular form of the propagation function of the exchanged particle. It can be an arbitrary scalar function Δ (fig. 4):

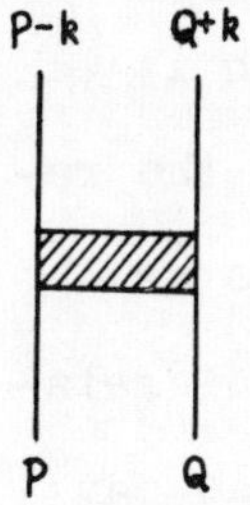

p-k Q+k

P Q

fig. 4

$$\Delta = \Delta(p^2, Q^2, k^2, p\cdot Q, p\cdot k, Q\cdot k, g_i^2) \tag{4.1}$$

$$p = p_a - \sum k_i = p_a - q$$

$$Q = p_i + \sum k_i' = p_i + q'$$

In fact, under the approximation of small momentum transfer comparing with the incident energy we have

$$\begin{aligned} p &= p_a - q \sim p_a \\ Q &= p_i + q' \sim p_i \end{aligned} \tag{4.2}$$

and

$$p^2 \sim p_a^2 = m_a{}^2 \qquad Q^2 \sim p_i^2 \sim m_b^2$$

$$p\cdot Q \sim p_a \cdot p_i$$

If the limit of the function Δ exists as $p.k, Q\cdot k \longrightarrow 0$ the form of the propagation function should be

$$\Delta = \Delta(p_a\cdot p_i, m_a^2, m_b^2, k^2, g_i^2) \quad . \tag{4.3}$$

Thus the exponentiation of the scattering amplitude is always possible, only the factor $(-ig)^2 \Delta_f(k)$ in (2.11) should be substituted by the general one (4.3). Therefor, we hope that the expansion (2.22)-(2.25) to a great extent is independent of the particular form of the interaction and that its application will not be restricted to the generalized ladder diagram, in which only the interaction of exchanging one quantum is considered, but should be able to include the "generalized" interaction, which is considered a "black blob" (as in Fig. 4), including the radiation correction and many-particle exchange. Thus as in the usuzl Glauber theory, in the eikonal theory of relativistic many-body scattering the many-body amplitude can also be calculated by using the experimental two-body amplitude.

Besides, just as Leby's generalized ladder summation method of two-body scattering can be applied to the situation of the propagation of the vector meson between two Dirac particles, therelativistic many-body Glauber expansion can also be generalized correspondingly. This however will not be discussed here.

References

1. R.J. Glauber — Lectures in Theory Physics Vol. 1 (1959) 315
2. R.J. Glauber — High Energy Physics and Nuclear Structure (1967) 311
3. T.T. Chou, C.N. Yang, — Phys. Rev. 170 (1966)1591, 175 (1968) 1832
4. J.M. Eisenberg — Ann. Phys. 71 (1972) 542
5. Bao Cheng-guang, Chao Wei-qin, — Physica Energiae Fortis et Physica Nuclearis 2 (1978) 336
6. V.A. Rizov, I.T. Todorov, Sov, Jour. Part. Nucl. 6 (1975) 269
7. M.Levy, J.Sucher, — Phys. Rev. 186 (1969) 1656
8. H.Cheng, T.T. Wu, — Phys. Rev. 186 (1969) 1611
9. H.O.I. Abarbanel, C. Itzukson, — Phys. Rev. Lett. 23 (1969) 53
10. Ho Tso-hsiu, Huang Tau, — Acta Physica Sinica, 23 (1974) 409
11. E.E. Salpeter, H.A.Bethe, Phys. Rev 84 (1951) 1232
12. M.M. Islam — Nuovo Cimento 5A (1971) 315

THE ABSOLUTE YIELD OF THE SECONDARY HADRONS IN pp COLLISIONS AND THE PRODUCTION RULE OF L=1 MESONS

Xie Qu-bing
Shantong University

Abstract

Starting from the N(Q) relation obtained previously [1] by the author based on the quark model of multiparticle production, I worked out a systematic method for calculating the absolute yields of secondaries. After analyzing the scant data about production of L=1 mesons, it is found that if the production of L=1 mesons has the same SU(6) broken symmetry as that of L=0 mesons, the ratio of their yields would be 1:3. The yields of various hadrons thus calculated are in agreement with the existing data and give a natural solution of the recent controversy on the η/ω and η'/η anomaly.

I. INTRODUCTION

The quark model of multiparticle production proposed by Anisovich et al.[2], which generalizes the quark model valid in lower energy region to high energy multiparticle production, provides the first predictions on the relative yields among the different 36-plet mesons and those among the 56-plet baryons. These predictions have recently been confirmed by a large body of experimental data, except the η/ω and η'/η anomaly.[4-11] Further predictions of absolute yields of secondary hadrons under various incident energies will evidently be important to both theoretical and experimental developments. A method of calculating the absolute yields on the basis of Anisovich model is proposed by Guman et al.[3]. However, since the existing model fails to give the dependence of the mean number N of the newborn quark pairs on the energy release Q in the collision, which is a fundamental relation and starting point for the whole procedure, one has to rely entirely on experimental results. Thus this method is semiempirical in nature. The absolute yields calculated by this method show significant deviations from the experimental results, except for $\langle\pi\rangle$ and $\langle k^-\rangle$, which are used to determine N(Q) and λ (the suppression parameter of the strange quark) respectively. This is considered to be due to the omission of the effects of the production of hadrons with relative orbital angular momentum $L > 0$.

After analyzing the difficulties of the above model, an associated production mechanism of quark - gluon is proposed in [1], from which a N(Q) dependence is deduced, thus making possible the systematic calculation of absolute yields of hadrons under various incident energies on theoretical grounds. In Sec. II, a method of calculating the absolute yields for secondary hadrons in pp collisions is developed on the basis of [2,3].

Recent experiments give the ratio $\langle f \rangle / \langle \rho^0 \rangle \sim 0.2$; it is impossible to give correct predictions about the secondary hadronic yields without taking into account the production of L=1 mesons and its contribution to that of L=0 mesons. But, up to now, production rule for L=1 mesons has not yet been formulated, either theoretically or experimentally. For example, is the SU(6) broken symmetry applicable to the L=0 mesons also valid for L=1 mesons? Does there exist a definite ratio between the L=1 and L=0 mesons? If it does, what is its value? In Sec. III, from the analysis of the yields of several L=1 mesons, it seems plausible to assume the SU(6) symmetry similar to that in the production of all L=0 mesons to hold in the production of L=1 mesons. Under this assumption and from the fact $\langle f \rangle / \langle \rho^0 \rangle \sim 0.2$, which is independent of Q, we get the universal ratio of 3:1 for the L=0 and L=1 mesons and a theoretical formula for calculating the absolute yields of all L=0 hadrons and L=1 mesons, at various energies.

In Sec. IV, we compare the yields calculated from the above formula with the existing data. The yields of long-lived charged particles $\bar{p}$, k^-, k^+, π^-, π^+ and of other L=0, 1 mesons are all in agreement with the theory, with only a few exceptions. The non-selfconsistency of the value of λ when only the production of L=0 mesons is taken into account gets also a solution. In particular, I obtain a natural explanation for the η/ω and η'/η "anomaly" through the η contributed by the highly excited states.

II. METHOD OF CALCULATION

The physical picture and calculation method of the quark-gluon model of multiparticle production developed by us based on [2,3] have been presented in [1,12]. In this model, it is assumed that the process of low P_T multihadron production at high energy may be divided into three stages. Firstly, two quarks (or antiquarks) belonging separately to the two incident hadrons collide with each other and transform that part of released energy carried by them. γQ, into the kinetic energy of N pairs newborn quarks and the potential energy required to form the N(2N-1) bonds between them. The other valence quarks in the incident hadrons act only as "spectators". Secondly, all the valence quarks in the incident hadrons and the N pairs of newborn quarks combine to form "directly" produced hadrons, according to the various possible spin statistics allowed in the usual quark model. Finally, "directly" formed short-lived particles immediately decay into long-lived particles. According to this picture, the absolute yields of various kinds of hadrons produced in pp collisions can be calculated according to the following steps:

1. Using the N(Q) relation presented by us in [1], one calculates the average number of newborn quark pairs, at c.m. energy $\sqrt{s}$:

$$N = (3.46 + 2.24Q)^{\frac{1}{2}} - 1.86$$

$$Q = \sqrt{S} - 2m_p , \tag{1}$$

where Q is the total energy released in the reaction, and m_p is the proton mass. The value of N calculated from (1) allows an error of 5%.

2. One calculates the average number $\langle n(\bar{B})\rangle$ of antibaryons $\bar{B}$ ($\bar{q}\bar{q}\bar{q}$) combined stochastically from the (N+2) quarks (N newborn quarks and two incident quarks) and N antiquarks in the central region of the collision, according to the method given in [3] :

$$\langle n(\bar{B})\rangle = \sum_{a+3b=N} b\,\omega_{ab} \tag{2}$$

where ω_{ab} is the probability for N antiquarks $\bar{q}$ to be combined as b antibaryons and the remaining a=N-3b antiquarks to be combined as mesons M.

$$\omega_{ab} = \frac{X_{ab}}{\sum\limits_{a+3b=N} X_{ab}}$$

$$X_{ab} = \frac{1}{2}\left[\frac{(a+2b)!\,3^{a-1}}{(a-1)!\,b!\,(b+1)!} + \frac{1}{b!(b+1)!}Y(a-1,2b)\right] + \frac{(a+2b-1)!\,3^{a-3}}{2(a-3)!\,b!\,(b+2)!}$$

$$+ \frac{1}{b!(b+1)!}Y(a-3,2b+1) + \frac{(a+2b)!\,3^{a}}{a!\,b!\,b!} + \frac{(a+2b-1)!\,3^{a-2}}{(a-2)!\,b!\,(b+1)!} ,$$

where $Y(p,q) = \sum\limits_{r=0}^{p} \frac{(q+r)!}{r!} 3^r$

In X_{ab} and Y (p,q), terms in negative powers of three are equal to zero.

Evidently, the total number of mesons M and baryons B directly combined is

$$\langle n(M)\rangle = N-3\langle n(\bar{B})\rangle ,$$

$$\langle n(B)\rangle = 2+\langle n(\bar{B})\rangle .$$

The M's formed by combining q with $\bar{q}$ have two types, viz. the one with q as a "spectator" is called a fragmentation meson, with its average number designated by $\langle n(M1)\rangle$, and the other without any "spectator" is directly combined in the central region, with average number designated by $\langle n(M0)\rangle$.

$$\langle n\ (M1)\rangle = \frac{11N^2 - 3N - 2}{4(N+1)\ (2N+1)} + 2D(N) , \tag{3}$$

$$\langle n\ (M0)\rangle = \langle n(M)\rangle - \langle n(M1)\rangle . \tag{4}$$

The B's formed by combining three q's have three types, viz. the fragmentation baryons $\langle n\ (B2)\rangle$ involving two "spectators", the baryons $\langle n(B1)\rangle$ involving one "spectator", and the baryons $\langle n(B0)\rangle$ formed solely by quarks in the central region without any "spectator".

$$\langle n\ (B2)\rangle = \frac{(N+2)\ (5N+2)}{2(N+1)\ (2N+1)} - D(N) , \tag{5}$$

$$\langle n\ (B1)\rangle = \frac{N+2}{4(2N+1)} , \tag{6}$$

$$\langle n\ (B0)\rangle = \langle n(B)\rangle - \langle n(B2)\rangle - \langle n(B1)\rangle . \tag{7}$$

D(N) in (3) and (5) is given by

$$D(N) = \frac{N!(N+2)!}{(2N+2)!}\sum_{p=0}^{N-3}\frac{(2p+4)!}{p!(p+4)!} .$$

3. One calculates the relative weights of the various kinds of particles in $\langle n(M0)\rangle$, $\langle n(\bar{B})\rangle$ and $\langle n(B0)\rangle$ which are directly produced in the central region, and in $\langle n(M1)\rangle$, $\langle n(B2)\rangle$ and $\langle n(B1)\rangle$ which are directly produced in the fragmentation region. Under the assumption that hadrons with L=0 satisfy SU(6) symmetry with relative suppression of strange-quarks s (or $\bar{s}$), the relative weights C_{pM_i}, $C^{(2)}_{pB_i}$ (i.e. $C^{(\frac{2}{3})}_{pB_i}$ in [2]) and $C^{(1)}_{pB_i}$ (i.e. $C^{(1/3)}_{pB_i}$, in [2]) of mesons or baryons of type i in the respective numbers $\langle n(M1)\rangle$, $\langle n(B2)\rangle$ and $\langle n(B1)\rangle$ in the direct fragmentations of the incident protons have been calculated by [2] using SU(6) wave functions. These will be employed directly. The relative weights $C_i(M0)$, $C_i(\bar{B})$ and $C_i(B0)$ of various kinds of mesons, antibaryons, and baryons in the respective numbers $\langle n(M0)\rangle$, $\langle n(\bar{B})\rangle$ and $\langle n(B0)\rangle$ are calculated directly using SU(6) wave-functions:

$$C_i(\bar{B}) = (2J_i + 1)\lambda^a \tag{8}$$

Apart from the special mixed states η and η', for which the $C_i(M0)$'s can be expressed as $C_\eta = \frac{1}{3} + \frac{2}{3}\lambda^2$ and $C_{\eta'} = \frac{2}{3} + \frac{1}{3}\lambda^2$ respectively, the other $C_i(M0)$'s can also be expressed as

$$C_i(M0) = (2J_i + 1)\lambda^a \tag{9}$$

As to $C_i(B0)$, after it has been calculated according to (8), it is necessary to consider the effects of incident quarks to the particles belonging to the same charge multiplet.

All the weights $C_i(MO)$, $C_i(\bar{B})$, $C_i(BO)$, C_{pMi}, $C^{(2)}_{pBi}$ and $C^{(1)}_{pBi}$ in hadron i, with L=0 involved a unified strange quark suppression parameter λ, determined in [3] from $\langle k^- \rangle$ to be 0.4. This value will be adopted in the following calculations where J_i is the total spin of particle i and a is the total number of strange quarks, s and $\bar{s}$, in the constituents of i. For example, in the case of $\varphi(s\bar{s})$ meson, J_φ =1, a=2, therefore from (9) $C_\varphi = 3\lambda^2 = 0.48$.

Under the assumption that all the L=1 mesons ($J^{pc} = 2^{++}$, 1^{++}, 0^{++}, 1^{+-}) obey the SU(6) symmetry similar to the L=0 mesons, the above method can be used to calculate the weights $C_{i'}$ (MO) and $C_{pMi'}$ of the i'th particle in the L=1 mesons. For example, $C_{i'}(MO)$ can be expressed as

$$C_{i'}(MO) = (2J_{i'} + 1)\lambda'^{a}, \qquad (10)$$

where λ' reflects the suppression of s or $\bar{s}$ relative to nonstrange quarks $\bar{q}$ or q in the formation of mesons having a relative orbital angular momentum L=1. Whether the value of λ' coincides with that of λ for L=0 will be decided by experiments. The production rule of L=1 mesons together with the calculation of absolute yields, taking into account the production of L=1 mesons, will be treated in the following section.

4. Calculation of absolute yields $\langle j \rangle_{direct}$.

From the above results, the yield of any hadron j produced directly in pp collisions can be obtained immediately:

meson j: $\langle j \rangle_{direct} = \frac{C_j(M0)}{\sum_i C_i(M0)} \langle n(M0) \rangle + \frac{C_{pMj}}{\sum_i C_{pMi}} \langle n(M1) \rangle$; (11)

antibaryon j: $\langle j \rangle_{direct} = \frac{C_j(\bar{B})}{\sum_i C_i(\bar{B})} \langle n(\bar{B}) \rangle$; (12)

baryon j: $\langle j \rangle_{direct} = \frac{C_j(B0)}{\sum_i C_i(B0)} \langle n(B0) \rangle + \frac{C^{(1)}_{pBj}}{\sum_i C^{(1)}_{pBi}} \langle n(B1) \rangle + \frac{C^{(2)}_{pBj}}{\sum_i C^{(2)}_{pBi}} \langle n(B2) \rangle$ (13)

5. Total absolute yields $\langle j \rangle$

The resonances produced directly will decay strongly according to definite branching ratios and the experimentally measured $\langle j \rangle$ will be the total yields including the contributions of all strong decays. The short-lived M, $\bar{B}$, and B given directly by all the six terms in (11)-(13) can product π's through strong decays and thus all these contributions to $\langle \pi \rangle$ must be taken into account. Define Br $(i \rightarrow j)$ as the "add-weighted" branching ratio of the decay of hadron from i to j : Br $(i \rightarrow j) = \sum_\alpha$ Br $(i \rightarrow \alpha) . n_j(\alpha)$. Here, α represents the decay channel and $n_j(\alpha)$ is the number of hadrons in the final state. It is obvious that all Br$(i \rightarrow j)$'s can be calculated with the aid of P.D.G. [13] data. Note that when $i = j$, Br $(i \rightarrow j) = 1$, which gives automatically the term representing direct production. Thus, the general expression for $\langle j \rangle$ can be written as

$$\langle j \rangle = \frac{\sum_i C_i(M0)\, Br(i \rightarrow j)}{\sum_i C_i(M0)} \langle n(M0) \rangle + \frac{\sum_i C_i(\bar{B})\, Br(i \rightarrow j)}{\sum_i C_i(\bar{B})} \langle n(\bar{B}) \rangle$$

$$+\frac{\sum_i C_i(B0)\,Br(i\to j)}{\sum_i C_i(B0)}\langle n(B0)\rangle+\frac{\sum_i C_{pM_i}\,Br(i\to j)}{\sum_i C_{pM_i}}\langle n(M1)\rangle$$

$$+\frac{\sum_i C^{(1)}_{pB_i}\,Br(i\to j)}{\sum_i C^{(1)}_{pB_i}}\langle n(B1)\rangle+\frac{\sum_i C^{(2)}_{pB_i}\,Br(i\to j)}{\sum_i C^{(2)}_{pB_i}}\langle n(B2)\rangle \qquad (14)$$

III. THE PRODUCTION OF L=1 MESONS

The investigation of the production of L=1 mesons is of great significance. The production characteristics of L=0 mesons predicted by [2] aroused great interest in the measurements of vector mesons. Besides the preliminary confirmation of the validity of SU(6) symmetry with a relative suppression of s ($\bar{s}$), it reveals that f has a significant effect on the yields and the momentum spectra of π [9] and thus investigation of the production of all L=1 mesons is necessary. Theoretically, only through the clarification of the production behavior of L=1 mesons can we get further examination and development of the mechanism of multiparticle production. Furthermore, by comparing the similarity and difference between the L=0 and L=1 mesons, we can also get some deeper understanding of the characteristics of $q\bar{q}$ interaction.

It is evident from Sec. II that, if the production of L=1 mesons possesses the relative broken SU(6) symmetry characterized by λ', its yields will be given by (11):

$$\langle j'\rangle=\langle j'\rangle_{\mathbf{direct}}=\frac{C_{j'}(M0)}{\sum_i C_{i'}(M0)}\langle n(M0)\rangle+\frac{C_{pMj'}}{\sum_i C_{pMi'}}\langle n(M1)\rangle.$$

At the same energy, particles with the same J_i and a and having $I_3 = 0$, such as A_2^o and f, will have the same C_j,(MO) and C_{pMj}', and thus the same yield, irrespective of the type of incident particle. In addition, particles in the same isotopic multiplet produced in particle-antiparticle annihilations, such as A_2^+, A_2^0 and A_2 must have equal yields. The values $\langle A_2^{\pm} \rangle = 0.082 \pm 0.027$, $\langle f \rangle = 0.064 \pm 0.005$ obtained recently in 9.1 GeV/c $P\bar{P}$ annihilations coincide with each other within experimental errors [11]. Therefore, we consider that the L=1 mesons do have such SU(6) symmetry and give the C_i'(MO) and C_{pMi}' values of all L=1 mesons (Table 1) according to the method given in the preceding section. The λ' value determined from the recently measured relative yields[15] of f, k^{*0} (1430) and $\bar{k}^{*0}$ (1430) are

$$\frac{\langle k^{*0}(1430) \rangle}{\langle f \rangle} = \frac{\langle \bar{k}^{*0}(1430) \rangle}{\langle f \rangle} = \lambda' \simeq 0.05$$

which is an order smaller than λ, a fact of considerable significance.

Table 1. Weights C_i'(MO), C_{pMi}' and decay branching ratios of L=1 mesons

J^{PC} i'	C'_i (MO)	$C_{PM_i'}$	decay	J^{PC} i'	C'_i (MO)	$C_{PM_i'}$	decay
A_2^+	5	20		δ^+	1	4	
A_2^0	5	15	see [13]	δ^0	1	3	$\eta\pi \sim 100$
A_2^-	5	10		δ^-	1	2	
2^{++} f	5	15	"	0^{++} ε	1	3	$\pi\pi \sim 100$
K^{*+}	$5\lambda'(0.25)$	$20\lambda'(1)$				$4\lambda'(0.2)$	
K^{*0}	"	$10\lambda'(0.5)$	"	χ ?	$\lambda'(0.05)$	$2\lambda'(0.1)$	?
K^{*-}	"	0				0	
$\bar{K}^{*0}$	"	0				0	
f'	$5\lambda^2(0.01)$	0	?	S?	$\lambda^2(\sim 0)$	0	?
A_1^+	3	12		B^+	3	12	
A_1^0	3	9	See [13]	B^0	3	9	See [13]
A_1^-	3	6		B^-	3	6	
1^{++} D	3	9	$\eta\pi\pi \sim 100$	1^{+-} ?	.		?
Q^+	$3\lambda'(0.15)$	$12\lambda'(0.6)$					
Q^0	"	$6\lambda'(0.3)$	$K\pi\pi \sim 100$	?			?
Q^-	"	0					
$\bar{Q}^0$	"	0					
E?	$3\lambda^2(<0.01)$	0	?	?			?

It is necessary to know the branching ratios in calculating the contributions of L=1 mesons to the yield of L=0 mesons. Experiments on D, Q, δ and χ indicate that there are two possible models of decay and the branching ratio can be obtained through theoretical analysis (such as [16]). It is also seen from table 1, that for certain particles, not only the decay models but also the particles themselves are left undetermined. But, fortunately, their weights are very small and can be safely neglected. It is sufficient to take the contributions of A_2 , f, $k^*(1430)$, A_1 , D, Q, δ, ε and B only. Thus we have

$$\sum_{i'} C_{i'}(MO) = 45 + 32\lambda' = 46.6$$

$$\sum_{i'} C_{pMi'} = 135 + 48\lambda' = 137.4$$

Experimental data about $\langle f(L=1)\rangle / \langle \rho^{\circ}(L=0)\rangle$ from different reactions have been reported recently

reaction	P_L (GeV/c)	Q(GeV)	$\langle f \rangle / p^{\circ}$	ref.
K^- p	10	3.03	0.18 ± 0.06	[8]
K^- p	16	4.16	0.21 ± 0.06	"
K^+ p	32	6.55	0.26 ± 0.15	[21]
π^- p	16	4.48	0.19 ± 0.03	[5]
π^+ p	16	4.48	0.21 ± 0.03	"
p p	405	27.5	0.33 ± 0.18	[22]
p p	1500	51.1	0.24 ± 0.10	[6]

which coincide within experimental errors and all lie within the range 0.18-0.22 or 0.20±0.02. Keeping in mind that both f and ρ° are particles with I_3 =0 and that their relative probability is independent of the type of reaction; this fact shows that $\langle f \rangle / \langle \rho^{\circ} \rangle = 0.20$ is independent of Q. Under the SU(6) symmetry, however, the independence of $\langle f \rangle / \langle \rho^{\circ} \rangle$ on Q leads directly to the conclusion that all L=0 and L=1 mesons are produced in a universal ratio.

Let the universal ratio of the mesons formed by q and $\bar{q}$ according to L=0 and L=1 (neglecting excited states with higher orbital angular momentum) be denoted by (1-x):x, we can determine x from $\langle f\rangle/\langle \rho^0\rangle=0.2$, given by pp reaction at Q=51.5 GeV. Since decays of L=1 meson have contributions to $\langle \rho^0\rangle$ while $\langle f\rangle$ have no contributions from higher states, from (11) and (14) it follows that

$$\langle f\rangle = x\frac{C_f(M0)}{\sum_{i'} C_{i'}(M0)}\langle n(M0)\rangle + x\frac{C_{pf}}{\sum_{i'} C_{pM_{i'}}}\langle n(M1)\rangle \tag{15}$$

$$\langle \rho^0\rangle = \left[(1-x)\frac{C_{\rho^0}(M0)}{\sum_i C_i(M0)} + x\frac{\sum_{i'} C_{i'}(M0)Br(i'\to\rho^0)}{\sum_{i'} C_{i'}(M0)}\right]\langle n(M0)\rangle$$

$$+\left[(1-x)\frac{C_{p\rho^0}}{\sum_i C_{pM_i}} + x\frac{\sum_{i'} C_{pM_{i'}}Br(i'\to\rho^0)}{\sum_{i'} C_{pM_{i'}}}\right]\langle n(M1)\rangle \tag{16}$$

All the quantities on the right-hand side of (15) and (16), except x, can be evaluated according to the formula given in the last section. We can obtain x from the experimental value $\langle f\rangle/\langle \rho^0\rangle=0.20$, giving x=0.25 and 1-x=0.75. This means that $q\bar{q}$ are always combined to form mesons with L=0 and L=1 in an approximate ratio 3:1. Therefore, the general formula (14) for secondary hadrons $\langle j\rangle$ can be further written as:

$$j = \left[0.75\frac{\sum_i C_i(M0)Br(i\to j)}{\sum_i C_i(M0)} + 0.25\frac{\sum_{i'} C_{i'}(M0)Br(i'\to j)}{\sum_{i'} C_{i'}(M0)}\right]\langle n(M0)\rangle$$

$$+\left[0.75\frac{\sum_i C_{pM_i}Br(i\to j)}{\sum_i C_{pM_i}} + 0.25\frac{\sum_{i'} C_{pM_{i'}}Br(i'\to j)}{\sum_{i'} C_{pM_{i'}}}\right]\langle n(M1)\rangle +$$

$$+\frac{\sum_i C_i(\bar{B})Br(i\to j)}{\sum_i C_i(\bar{B})}\langle n(\bar{B})\rangle+\frac{\sum_i C_i(BO)Br(i\to j)}{\sum_i C_i(BO)}\langle n(BO)\rangle+\frac{\sum_i C^{(1)}_{pBi}Br(i\to j)}{\sum_i C^{(1)}_{pBi}}\langle n(B1)\rangle+\frac{\sum_i C^{(2)}_{pBi}Br(i\to j)}{\sum_i C^{(2)}_{pBi}}\langle n(B2)\rangle \tag{17}$$

B and $\bar{B}$ in high angular momentum states and mesons with $L \geqslant 2$ must of course have some probability. But owing to the lack of any information about the production of B and $\bar{B}$ of highly excited states in the pp collisions and the rareness of data about mesons with $L \geqslant 2$ (only a few data about g^o), it is impossible to pursue any further. On the other hand, since $\langle n(\bar{B})\rangle$ and $\langle n(BO)\rangle$ are in fact an order smaller than $\langle n(MO)\rangle$ even at the highest energy of the existing accelerator and $\langle g^o\rangle$ is also an order smaller than $\langle f\rangle$, their effects may be taken as corrections of higher order and may be neglected for the time being. This approximation allows us to concentrate our attention on the test of the production rule of mesons with L=1.

IV. COMPARISON WITH EXPERIMENTS

Comparision can be made between theory and experiments by calculating the absolute yields of all secondaries with L=0 and those of mesons with L=1, by means of (17). The information now in hand includes mainly the accurate data on the long-lived charged particles π^-, π^+, k^-, k^+, $\bar{p}$, and p produced in the pp collisions at various energies. Although the data on short-lived mesons with L=0 and 1 are still limited to a few particles and energies, they can provide important tests of the theory. The quantity $\langle p \rangle_{exp.}$ includes also the contributions from the elastic diffraction, which does not belong to multi-particle production and thus cannot be compared directly with the calculated results based on (17), and will not be given here.

1. $\langle \bar{p} \rangle$

Theoretically $\langle \bar{p} \rangle$ is the simplest in that: (a) Only the contribution of $\langle n(\bar{B}) \rangle$ is present. (b) It has nothing to do with any assumptions about the production of mesons. (c) It is insensitive to the value of λ. Thus it may provide a test for the N(Q) dependence (1) used to deduce $\langle n(\bar{B}) \rangle$, and for the calculation method for quark combinations (2). From formulas (17), (8) and the decay branching ratio of Δ given in [13], we obtain

$$\langle \bar{p} \rangle = \frac{\sum_i C_i(\bar{B})\, Br(i \to \bar{p})}{\sum_i C_i\,(\bar{B})} \langle n(\bar{B}) \rangle$$

$$= \frac{5}{15.088} \langle n(\bar{B}) \rangle = 0.331 \langle n(\bar{B}) \rangle. \quad (18)$$

Fig. 1 gives a comparison between theory (solid line) and experiment (from [17]) and shows full agreement within the range of errors.

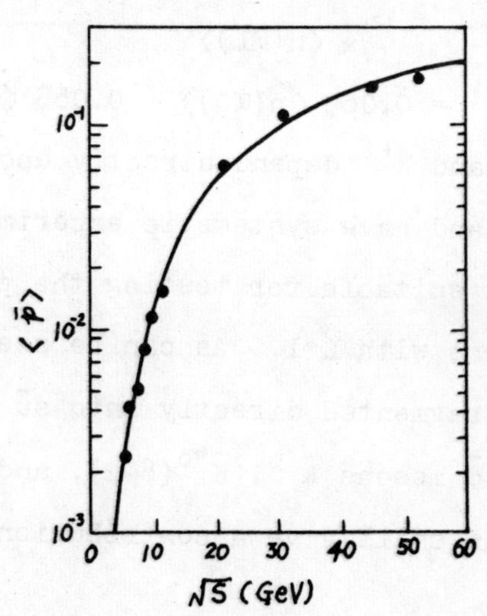

2. $\langle K^- \rangle$ and $\langle K^+ \rangle$

Due to the inability of the baryons and antibaryons in the 56 to decay into kaons, the last four terms in (17) will vanish, thus

$$\langle K^- \rangle = \left[0.75 \frac{\sum_i C_i(M0)Br(i \to K^-)}{\sum_i C_i(M0)} + 0.25 \frac{\sum_{i'} C_{i'}(M0)Br(i' \to K^-)}{\sum_{i'} C_{i'}(M0)}\right] \times \langle n(M0) \rangle + 0.25 \frac{\sum_{i'} C_{pMi'} Br(i' \to K^-)}{\sum_{i} C_{pMi'}} \langle n(M1) \rangle$$

$$= 0.063 \langle n(M0) \rangle + 0.002 \langle n(M1) \rangle \qquad (19)$$

$$\langle K^+ \rangle = \left[0.75 \frac{\sum_i C_i(M0)Br(i \to K^+)}{\sum_i C_i(M0)} + 0.25 \frac{\sum_{i'} C_{i'}(M0)Br(i' \to K^+)}{\sum_{i'} C_{i'}(M0)}\right] \times \langle n(M0) \rangle + \left[0.75 \frac{\sum_i C_{pMi} Br(i \to K^+)}{\sum_i C_{pMi}} + 0.25 \frac{\sum_{i'} C_{pMi'} Br(i' \to K^+)}{\sum_{i'} C_{pMi'}}\right] \times \langle n(M1) \rangle$$

$$= 0.063 \langle n(M0) \rangle + 0.068 \langle n(M1) \rangle . \qquad (20)$$

K^- and K^+ depend directly upon the production of mesons with L=1 and have systematic experimental data, therefore they are most suitable for testing the production characteristics of mesons with L=1. As can be seen from (19), although p cannot be fragmented directly into $s\bar{u}$ mesons K^-, $K^{*-}(892)$, and $K^{*-}(1430)$ or $s\bar{d}$ mesons $\bar{K}^0$, $\bar{K}^{*0}(892)$, and $\bar{K}^{*0}(1430)$, the fragmented A and f can still give a contribution to K^-, through the decay channel $K\bar{K}$. Therefore the contribution of fragmentation is different from zero but is still so small as to have practically no effect on $\langle K^- \rangle$. Hence $\langle K^- \rangle$ can offer a concentrated test to the production rule of L=1 mesons in $\langle n(M0) \rangle$, while $\langle K^+ \rangle$ does so in $\langle n(M1) \rangle$.

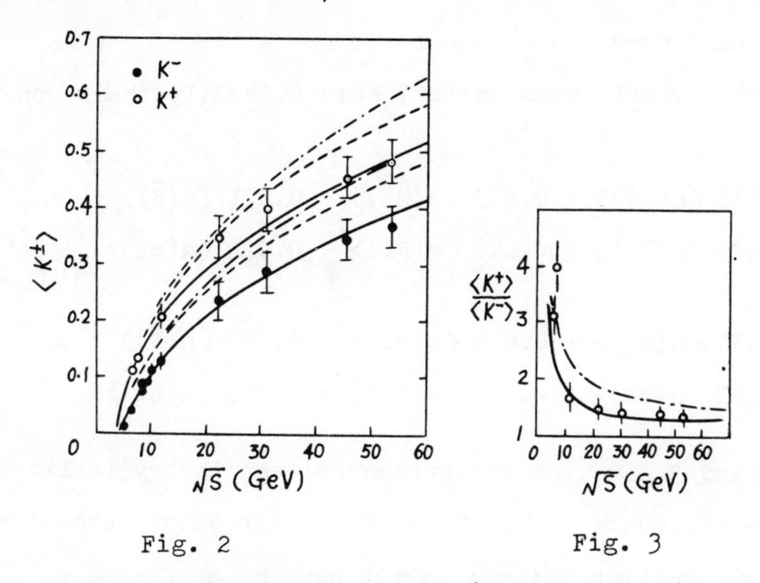

Fig. 2 Fig. 3

Fig. 2 and 3 give the comparison between theory and experiments. The solid lines are calculated according to (19) and (20) while the experimental data are taken from [17]. The theory and experiments are seen to be in agreement, except for a few points. The dotted line in Fig. 2 represents the calculated results taking $\lambda' = \lambda = 0.4$, which disagree with the experiments. The dot-dashed lines in Fig. 2 and Fig. 3 represent the culculated results neglecting the contribution of L=1 mesons. If λ is adjusted so as to make $\langle k^- \rangle$(or $\langle K^+ \rangle$) agreeing with experiment, a larger deviation in $\langle K^+ \rangle$ (or $\langle K^- \rangle$) appears.

This inconsistency of different values of λ is just the same as the situation appearing in [3]. Owing to the independence of $\langle k^+\rangle / \langle k^-\rangle$ from the value of λ, this illustrates further that without considering L=1 mesons, even $\langle K^\pm\rangle$ cannot be explained.

3. $\langle \pi^-\rangle$ and $\langle \pi^+\rangle$

For $\langle \pi^\pm\rangle$, all terms in (17) have evidently their contributions:

$$\langle \pi^-\rangle = 0.535 \langle n(M0)\rangle + 0.503 \langle n(M1)\rangle + 0.271 \langle n(\bar{B})\rangle + 0.285 \langle n(B0)\rangle + 0.148 \langle n(B1)\rangle + 0.022 \langle n(B2)\rangle \quad (21)$$

$$\langle \pi^+\rangle = 0.535 \langle n(M0)\rangle + 0.670 \langle n(M1)\rangle + 0.285 \langle n(\bar{B})\rangle + 0.271 \langle n(B0)\rangle + 0.312 \langle n(B1)\rangle + 0.265 \langle n(B2)\rangle \quad (22)$$

Fig. 4 and 5 give the comparison between theory (solid lines) and experiments [17]. They are in agreement, except at a few points. Putting $\lambda' = \lambda$, $\langle \pi^\pm\rangle$ have no appreciable change. All calculated values lie appreciably lower than experimental ones, if the contributions of mesons with L=1 are neglected.

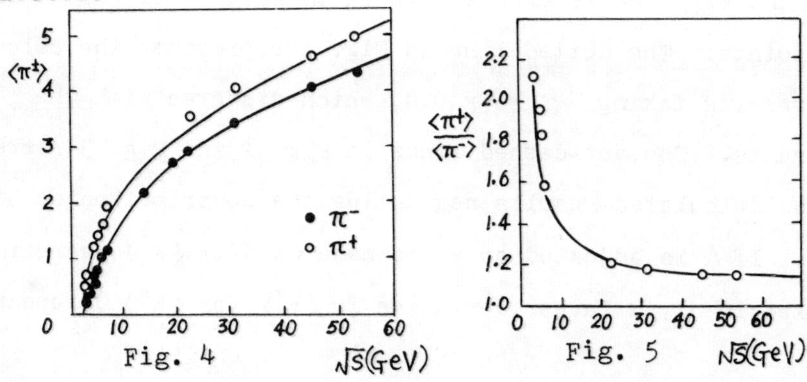

Fig. 4 Fig. 5

4. Resonance Mesons

The experimental data of the yields of resonance mesons with L=0 and 1 are still very rare, but they can provide more direct tests to the production rule presented in Sec. 3. For mesons of L=0, excluding π and K, (17) reduces to

$$\langle j \rangle = \left[0.75 \frac{C_j(M0)}{\sum_i C_i(M0)} + 0.25 \frac{\sum_{i'} C_{i'}(M0) Br(i' \to j)}{\sum_{i'} C_{i'}(M0)}\right] \langle n(M0) \rangle$$

$$+ \left[0.75 \frac{C_{pMj}}{\sum_i C_{pMi}} + 0.25 \frac{\sum_{i'} C_{pMi'} Br(i' \to j)}{\sum_{i'} C_{pMi'}}\right] \langle n(M1) \rangle \qquad (23)$$

For meson j' with L=1, this simplifies further to

$$\langle j' \rangle = 0.25 \left[\frac{C_{j'}(M0)}{\sum_{i'} C_{i'}(M0)} \langle n(M0) \rangle + \frac{C_{pMj'}}{\sum_{i'} C_{pMi'}} \langle n(M1) \rangle\right]. \qquad (24)$$

where the term $\langle n(M0) \rangle$ represents the particles in the central region while $\langle n(M1) \rangle$ represents those in the fragmentation region.

Table 3 gives comparisons of the measured yields of mesons, in pp collisions at $\sqrt{s}$ = 53 GeV, with theoretical values calculated according to (23) and (24).

Table 3

	ρ^0	ω	η	f	$K^{*0}(1430)$	$\bar{K}^{*0}(1430)$	$K^{*0}(892) + \bar{K}^{*0}(892)$
theo.	1.01 ±0.05	1.31 ±0.07	0.44 ±0.02	0.20 ±0.01	0.009	0.008	0.59 ±0.03
exp.	1.19 ±0.25	1.59 ±0.28	0.92 ±0.58	0.29 ±0.16	0.017 ±0.01	0.011 ±0.006	1.14 ±0.35
ref.	[6]				[15]		[6]

It can be seen that the first six groups of data are in agreement within the range of error. The last experimental value $\langle K^{*0}(892) + \bar{k}^{*0}(892)\rangle$, which does not agree with theory, is just the one considered to be unreliable in [19]. The measurements in [19] of pp collision at $\sqrt{s}$ = 19.6 GeV give $\langle K^{*\pm}(892)\rangle$ = 0.22, this coincides with the theoretical value $\langle K^{*+}(892)\rangle$ = 0.22.

Fig. 6 gives comparisons of the yields of ρ° in the central region measured at several energies [18] with the theoretical value of $\langle \rho^{\circ}\rangle_c$. They are in agreement within the range of errors.

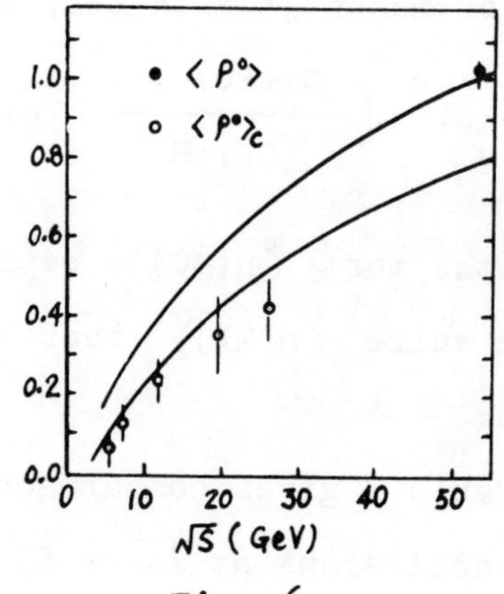

Fig. 6

5. The question of η/ω and η'/η anomaly

$\frac{\langle\eta\rangle}{\langle\omega\rangle}=0.39\pm 0.03$ and $\frac{\langle\eta'\rangle}{\langle\eta\rangle}=0.05-0.10$ measured in πp collisions deviate substantially from $\frac{\langle\eta\rangle}{\langle\omega\rangle}=0.14$, $\frac{\langle\eta'\rangle}{\langle\eta\rangle}=1.7$ predicted in [2]. The behavior of η and η' is somewhat peculiar in itself, and the difficulties just mentioned above have further given rise to a strong controversy about its properties. Some attempts have been made to explain the anomaly from various angles.[20,4,14] But this result appears naturally if we adopt (17) for unified calculations. From ref. [13] we have obtained $Br(i\to\eta')=0$, but $Br(\eta'\to\eta) = 0.66$, $Br(A_2\to\eta) =0.15$, $Br(K^*(1430)\to\eta)\approx 0.03$,

$Br(D \rightarrow \eta) \simeq 1$, $Br(\delta \rightarrow \eta) \simeq 1$, thus

$$\langle \eta' \rangle = 0.023 \langle n(M0) \rangle + 0.026 \langle n(M1) \rangle \,, \tag{25}$$

$$\langle \eta \rangle = 0.075 \left[\langle n(M0) \rangle + \langle n(M1) \rangle \right] \,, \tag{26}$$

$$\langle \omega \rangle = 0.154 \langle n(M0) \rangle + 0.286 \langle n(M1) \rangle \,. \tag{27}$$

For πp and pp reactions, while the above formulas are the same, $\langle n(M0) \rangle$ and $\langle n(M1) \rangle$ take different numerical values. Take the pp reaction as an example, at $\sqrt{s}$ = 53 GeV, calculated value $\frac{\langle \eta \rangle}{\langle \omega \rangle} = 0.43$ is in agreement with the experimental value (table 3), while $\frac{\langle \eta' \rangle}{\langle \eta \rangle} = 0.31$ is no longer 1.7 as given by [2], so the "anomaly" is due to the contributions of highly excited states not having been considered in [2], but among the mesons with L=0, such contributions for η are obvious. After we consider all these contributions, the problems can be explained naturally.

References

1. Xie Qu-bing, Phys. Ener. Fort. et Phys. Nucl. 3 (1979) 520.
2. V.V. Anisovich et al., Nucl. Phys., B55 (1973), 455.
3. V.N. Guman et al., Nucl. Phys., B99 (1975), 523.
4. H. Kirk et al., Nucl. Phys., B128 (1977), 397.
5. J. Bartke et al., Nucl. Phys., B118 (1977), 360.
6. G. Jancso et al., Nucl. Phys., B124 (1977), 1.
7. H. Grässler et al., Nucl. Phys., B132 (1978), 1.
8. P. Schmitz et al., Nucl. Phys., B137 (1978), 13.
9. P. Granet et al., Nucl. Phys. B140 (1978), 389.
10. K. Bockmann et al., Nucl. Phys., B140 (1978), 235.
11. M. Markytan et al., Nucl. Phys., B143 (1978), 263.
12. Xie Qu-bing, Journal of Shandong University (Natural Sciences) 3-4 (1978), 146.
13. Particle Data Group, Phys. Lett., 75B (1978),
14. Wang Cheng-zhi, Phys. Ener. Fort. et Phys. Nucl., 3 (1979), 122.
15. A. Böhm et al., Phys. Rev. Lett., 41 (1978), 1761.
16. Ch. De. LA Vaissiere, Nuovo Cimento., 44A(1978), 481.
17. M. Antinucci et al., Lett. Nuovo Cimento, V6 (1973), 121.
18. V. Blobel et al., Phys. Lett., 48B (1974), 73.
    R. Singer et al., Phys. Lett., 60B (1976), 385.
    A. Sheng et al., Nucl. Phys., B115 (1976) 189.
19. R. Singer et al., Nucl. Phys., B135 (1978), 265.
20. N.H. Fuchs, Lett. Nouvo Cimento, V20 (1977), 103.
21. P. Granet et al., Nucl. Phys., B140 (1978), 389.
22. A. Suzuki et al., Lett. Nuovo Cimento, V24 (1979), 449.

SPIN STRUCTURE AND DIP DEVELOPMENT IN ELASTIC PROTON–PROTON SCATTERING

S.Y. Lo, D.J. Clarke and M.M. Malone

School of Physics, University of Melbourne
Parkville, Victoria, Australia. 3052.

Abstract

A current-current interaction picture is used to obtain excellent fit for all proton proton elastic scattering above 12 GeV/c. We pay special attention to the development of first, second, and third dips as energy increases. Spin structure of proton proton elastic scattering is displayed.

1. Introduction to the Model

The nucleon-nucleon interaction is one of the most studied subjects in particle physics, both theoretically and experimentally. At low energy (< 1 GeV), the nucleon-nucleon interaction is known to be complex. However there is always a hope that at high energy the nucleon-nucleon interaction may become simple again in small angle scattering region. We propose that currents, which play a central role in electromagnetic and weak interaction may occupy an equally important position in nucleon-nucleon interaction. Specifically, the singlet current is responsible for the elastic asymptotic proton-proton scattering[1], and the flavor changing current is responsible for the decreasing energy dependent term. The S-matrix in the eikonal picture exponentiates to be:

$$S(\underset{\sim}{b},s) = \exp(-2\pi\mu_0(s) \iint d^3x' d^3x j_0^{\alpha}(x) j_0^{\alpha}(x) \Delta(x'-x) - 2\pi\mu_1(s) \iint d^3x' d^3x$$

$$j_0^{(I=1)}(x)\, j_0^{(I=1)}(x)\, \Delta^{(1)}(x'-x)]$$

where j_0^{α} is the time component of the color current, $j_0^{I=1}$ is the time component of the isovector current, and $\Delta^{(0)}, \Delta^{(1)}$ are interaction kernels.

Our theoretical motivations for proposing this particular model may be summarized as follows: (1) It is the natural quantized version of the Chou-Yang model[2]. (2) The value of the eikonal picture has been demonstrated already for QED[3], and it has recently been shown to hold also for Yang-Mills Gauge theories.[4]

Our particular formulation employs the following simplifications:

$$< p'|j_0^{\alpha}|p > = < p'|j_0^{\alpha'}|p > = < p'|j_0^{(I=0)}|p >$$

$$= \{ \frac{m^2}{p_0 p_0'} \}^{\frac{1}{2}} \bar{u}(p') \{F_1^{s}(q^2)\gamma_4 + F_2^{s}(q^2) \frac{1}{2m} \sigma_{i4} q_i\} u(p)$$

where we have identified the singlet current form factors with the isoscalar form factors F_1^{s} and F_2^{s} [5].

The I = 1 term makes use of the isovector form factors. All the form factors are well known experimentally through electromagnetic interaction.

The interaction Green's function $\Delta^{(0,1)}$ is approximated by a δ-function, $\delta^2(\underset{\sim}{b}_l - \underset{\sim}{x} + \underset{\sim}{x}')$.

The model makes no prediction as to the s-dependence, but the t-dependence is completely specified by the proton form factors. The singlet term, μ_o, is determined by the normalization point of the total cross section. The energy-dependent term, $\mu_1(s)$, which has a phase, is fitted to data and is the only adjustable parameter in the theory.

2. Data Fitting, and Development of Dips

This phenomenological model is used to fit the experimental data[5,6,7] for pp elastic scattering for Lab. momenta from 10 GeV/c up to 10^3 GeV/c. Figures 1 to 3 show the differential cross sections for 12, 19, 150, 200, 300 and 1500 GeV/c. The model provides good fits to all small angle data, which for ISR data includes the region $-t \sim 10$ GeV^2. The data extends over 12 orders of magnitude and the maximum deviation from experimental number is not more than 30%[1]. In particular it is worth noting that the model reproduces the change of slope at $-t = 1.0$ GeV^2 for $P_{Lab} = 12$ GeV/c and the positions and magnitudes of the dips at $P_{Lab} = 150$ and 300 GeV/c. For the development of the first dip, we show in fig.4 and fig.5. One can see that the first dip manifests itself as a break in the slope around $t \simeq 1$GeV^2, but the break starts to deepen to become a dip when $p < \simeq 100$ GeV. If one only measures the pp elastic scattering below $p < \lesssim 100$ GeV, one could make the announcement that diffraction picture is not valid because there is no diffraction zero. Nevertheless, such a statement is not true. Similarly one can say about the second dip in pp elastic scattering, there is no trace now, but it will show at higher energy. If we assume that the energy dependent term $\mu_1 = 0.1$ does not change from ISR energy to Isabelle energy, we show the differential cross section in fig.6. decomposing into spin flip term, spin-nonflip term, and energy dependent term. It is clear some interesting feature is to emerge.

Let us now suppress the energy dependent μ_1 term, and only concentrate on the μ_o term to discuss the third and fourth zeroes more carefully. Figure (7) shows distribution for $\frac{d\sigma}{dt}$ vs -t from 0< -t < GeV^2 at σ_{tot} = 70 mb and also σ_{tot} = 42.5 mb for comparison. Energy dependent contributions to the amplitude would turn the zeroes shown here into dips, and possibly hide some completely.

Figures (8a) and (8b) display $\frac{d\sigma}{dt}$ distributions at various values of σ_{tot}; (8a) contains the first and second dip region 0<-t< 4.5 GeV^2 and (2b) covers 6<-t< 10.5 GeV^2 which is the t-range in which the third and fourth dips develop. The latter graph shows that for σ_{tot} below about 60 mb only a shallow dip exists at about -t = 8.25 GeV^2 and it could easily be hidden by the energy dependent terms. Above 63 mb, two zeroes develop with between them a maximum which grows very quickly with σ_{tot}. Such quickly changing underlying structure would not be completely hidden by energy dependent contributions. Experimental distributions at energies corresponding to 60 - 70 mb should be rich in structure in the 7<-t< 9.5 GeV^2 region. The predicted behaviour as shown in figure (8b) comes about because of the corresponding behaviour of the amplitude $\alpha^{pp}(s,-t)$, sketched schematically in figure (9). The amplitude for σ_{tot} above 60 mb develops a well in this -t region which becomes deeper as α_{tot} (and μ') increase until the centre of the well (at about $|t|$ = 8.12 GeV^2) touches zero at σ_{tot} = 62.8 mb. Then it drops below the axis and so the cross section, which is proportional to the amplitude squared, develops a corresponding peak between the two zero points. The peak grows rapidly as σ_{tot} and μ_o increase.

The Figures discussed so far predict features which may be seen at Isabelle, in the form of dips and peaks. As well, we have also investigated how these features are expected to shift as σ_{tot} increase. The positions of the theoretical zeroes are plotted in figure(10) as a function of σ_{tot}. Corresponding dips observed in experimental distributions should move in the same way. The positions of the predicted peaks are also plotted in figure (10).

In figure (10) the first dip, which has already been seen at CERN, is predicted to move steadily closer to t = 0 as σ_{tot} increases. In contrast, it shows that the second dip - not yet conclusively identified at CERN, as discussed in Section I - is predicted firstly to move out in $|t|$ until σ_{tot} = 47.5 mb where it is located at about $|t|$= 3.8 GeV^2 and then to move back in again as σ_{tot} increases. Such behaviour may or may not be expected to be experimentally seen, depending on how soon the second dip is unambiguously observed. A specific model for the energy dependence has predicted that significant change of slope for the second dip will probably first occur at σ_{tot} = 47.7 mb which according to figure (10) is almost exactly where the second dip will start moving in towards $|t|$ = 0. On balance, initial outward movement of the second dip is unlikely to be seen due to masking of the dip by energy-dependent contributions, and like the first dip it too will eventually be observed to be moving in towards t = 0 as σ_{tot} increases.

As discussed above, an initially shallow trough in $\frac{d\sigma}{dt}$ for $6.5 \lesssim |t| \lesssim 9.5$ GeV^2 develops into two zeroes with one maximum between them as σ_{tot} increases, as shown in figure (2b). The position of the initial trough for σ_{tot} <62.8 mb stays at about -t = 8.2 GeV^2 once deep enough to be seen, but once the third and fourth dips are formed they move as σ_{tot} increases as shown in figure 4. The third dip moves steadily inwards towards smaller $|t|$ as σ_{tot} increases, in a similar fashion to the first dip but more quickly. In contrast the fourth dip moves outwards to a position $|t| \approx 9.1$ GeV^2 at σ_{tot} = 70 mb and then begins to come in again, in a similar fashion to the second dip.

The dashed line set of graphs in figure (10) display the predicted positions of the various peaks as a function of σ_{tot}. Note that the "fourth peak" only appears at $\sigma_{tot} \lesssim 63$ mb. They all move in as σ_{tot} increases except for the fifth peak which is initially "pushed out" as the fourth peak develops. Note that the estimates of peak and dip positions presented here are taken from calculated forms for $\frac{d\sigma}{dt}$ distribution, with interpolation errors of at most 2 to 3.

3. Polarization and Spin Structure

The Polarization Parameter can also be calculated from the above model. It is essentially due to the interference between the singlet current-singlet current and the flavor current-flavor current (I = 1) interaction terms. The results are shown in fig.11.

In calculating the numerical fit for the singlet current-singlet current interaction the spin non-flip term is included to 9th order but the spin flip term only to first order. We only include flavor current-flavor current interaction to the first order, so we cannot, at this moment, reliably predict the spin parameters, A and C_{nn}, especially at low energy where the flavor current with a much larger spin flip term plays a more important role.

From the success with which both differential cross section and polarization data are fitted, it is interesting to speculate on the correctness of the current-current interaction picture for proton-proton scattering through considering the significance of the spin-flip term at high energy, say that achieved by the ISABELLE accelerator.

The spin-dependent cross sections given in fig.12 may be better examined from their formulation in terms of the Wolfenstein Parameters[8].

$$\frac{d\sigma}{dt}(\uparrow\downarrow) = \pi|a - m|^2$$

$$\frac{d\sigma}{dt}(\uparrow\uparrow) = \pi|a + m + 2ic|^2$$

$$\frac{d\sigma}{dt}(\downarrow\downarrow) = \pi|a + m - 2ic|^2$$

Without a spin-flip term these cross sections are identical as implied in a pure scalar theory, but they show considerable structure in the current-current interaction picture.

The exact numerical values presented in fig.12 of course depend sensitively on the parameters we assume and the omission of higher order spin-flip terms. However these qualitative features of the three cross sections seem rather stable: the first and second dip of the mixed spin cross section $\frac{d\sigma}{dt}(\uparrow\downarrow)$, the first dip and then the first maximum for the parallel spin cross sections are insensitive to the exact value of the energy dependent term (μ_1). The three spin correlated cross sections are approximately the same for $|t|$ values smaller than the position of the first dip, but deviate after the occurence of the first dip. We hope this can be tested soon.

In conclusion, the original Chou-Yang model does not contain any spin structure. The current-current interaction picture proposed here contain spin in a natural way. It can fit the polarization data and predicts many interesting structure in spin dependent differential cross section. Spin, in future, should reveal the detailed structure of proton proton scattering.

References

1. For more detailed theoretical fits refer to "The Appearance of the Second Dip in pp Elastic Scattering", D.J. Clarke and S.Y. Lo, Phys. Rev. 20D 193 (1979).
2. T.T. Chou and C.N. Yang, Phys. Rev. 170 1591 (1968); Phys. Rev. 19D 3268 (1979).
   S.Y. Lo, Nucl. Phys. B19 286 (1970).
   D.J. Clarke and S.Y. Lo, Phys. Rev. D10 1519 (1974).
3. B.W. Lee, Phys. Rev. D1 2361 (1970).
   M. Cheng and T.T. Wu, Phys. Rev. Lett. 24 1456 (1970).
4. H. Cheng et al., Phys. Rev. Lett. 40 1681 (1978).
5. J. Allaby et al., Nucl. Phys. B52 316 (1973).
   G. Fidecaro et al., Phys. Lett. 76B 369 (1978).
   N. Kwak et al., Phys. Lett. 58B 233 (1975).
6. H. De Kerret et al., Phys. Lett. 68B 374 (1977).
   C. Rubbia et al., unpublished report in "Proceedings at the XVI International Conference on High Energy Physics, Chicago - Batavia I11, 1972".
7. M. Borghini et al., Phys. Lett. 24B 77 (1967).
   J. Snyder et al., Phys. Rev. Lett. 41 781 (1978).
8. M.J. Moravcsik, "The Two-Nucleon Interaction", (Clarendon Press, Oxford, 1963).

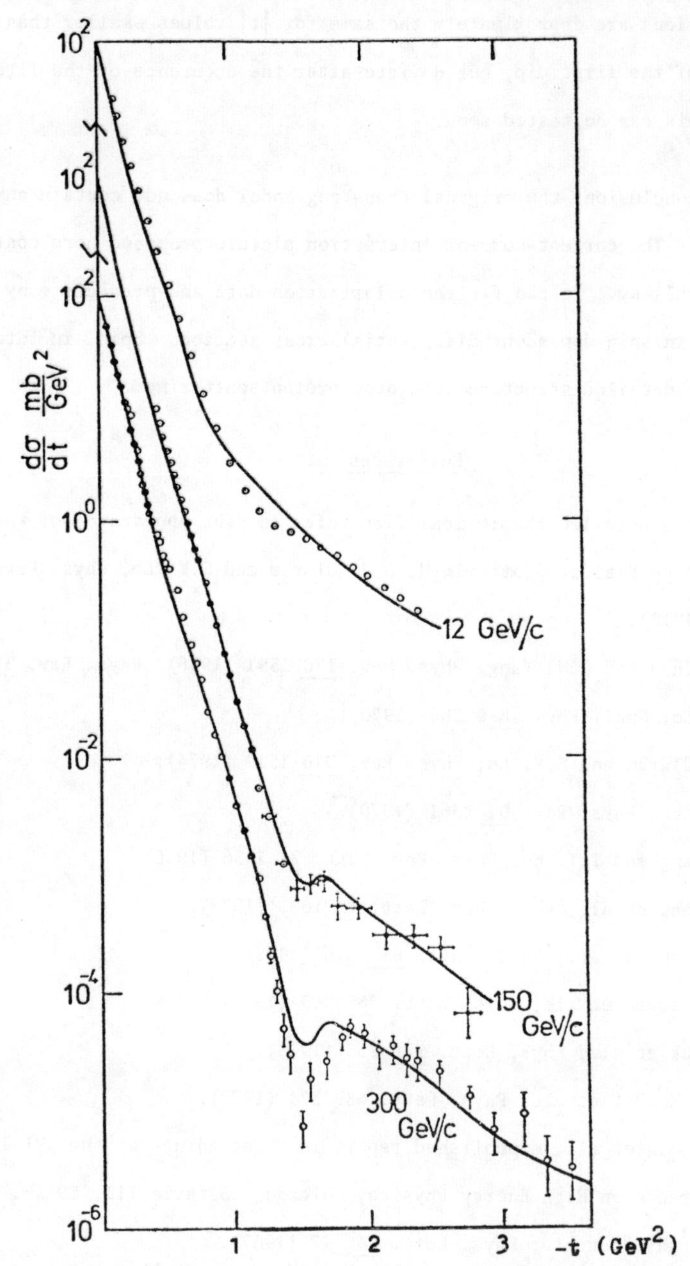

Fig.1 Theoretical curves for pp elastic scattering differential cross section data at lab momenta 12, 150 and 300 GeV/c. Data is from reference 5 and the total cross sections used to fix μ_o were 39mb and 39.1mb respectively. The values μ_1 employed were 1.15, 0.15 and 0.106 GeV^{-2}.

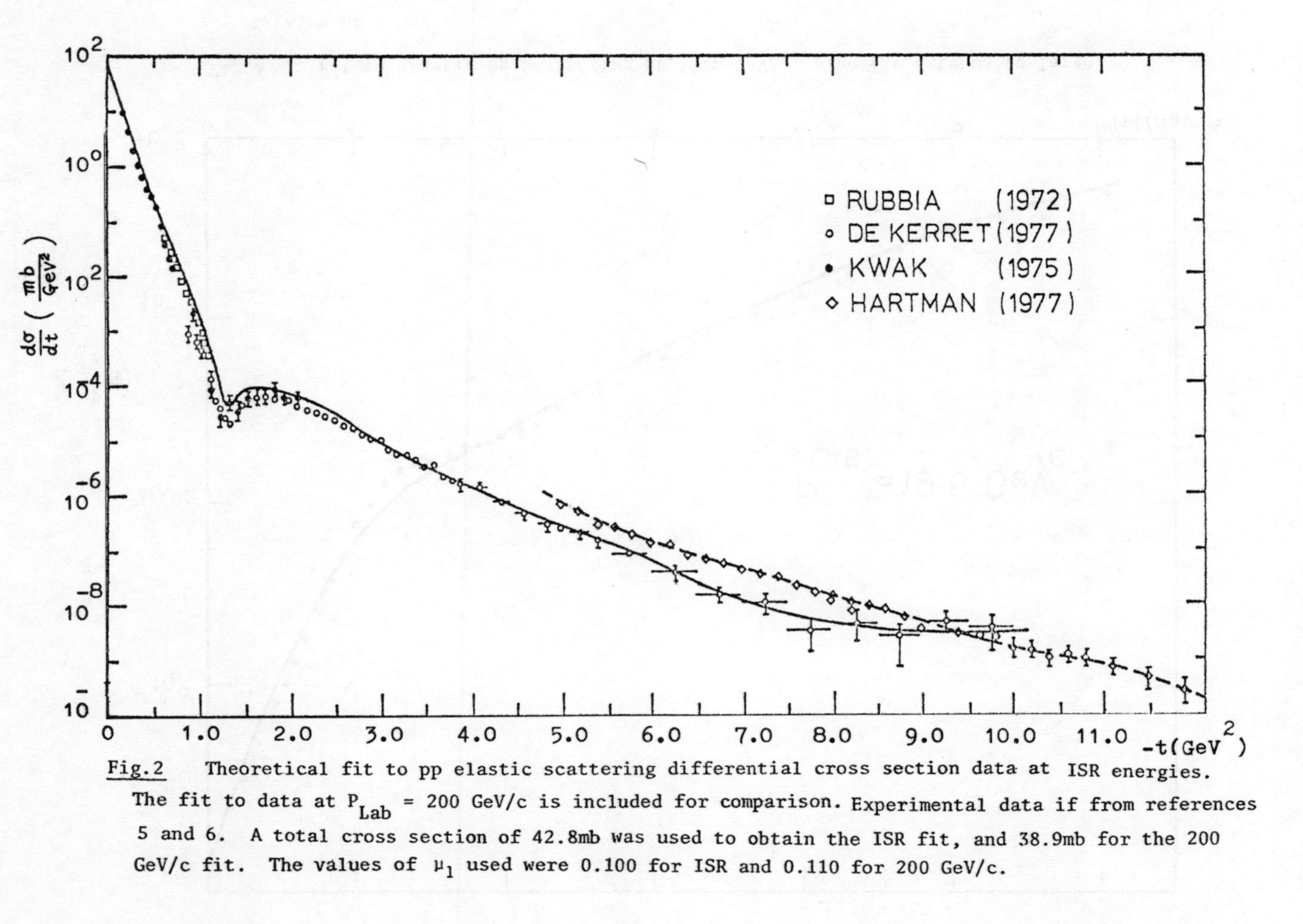

Fig.2 Theoretical fit to pp elastic scattering differential cross section data at ISR energies. The fit to data at P_{Lab} = 200 GeV/c is included for comparison. Experimental data if from references 5 and 6. A total cross section of 42.8mb was used to obtain the ISR fit, and 38.9mb for the 200 GeV/c fit. The values of μ_1 used were 0.100 for ISR and 0.110 for 200 GeV/c.

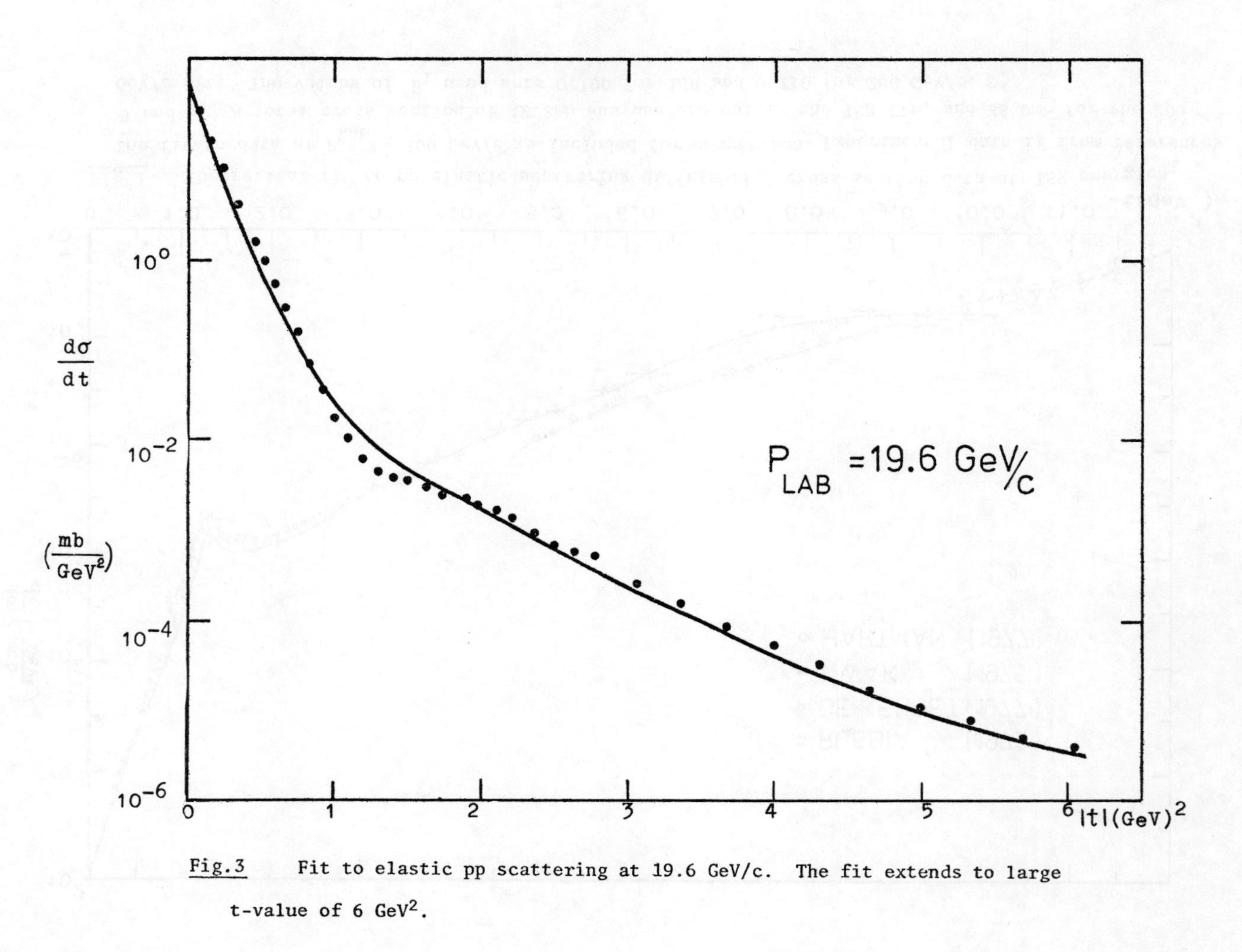

Fig.3 Fit to elastic pp scattering at 19.6 GeV/c. The fit extends to large t-value of 6 GeV^2.

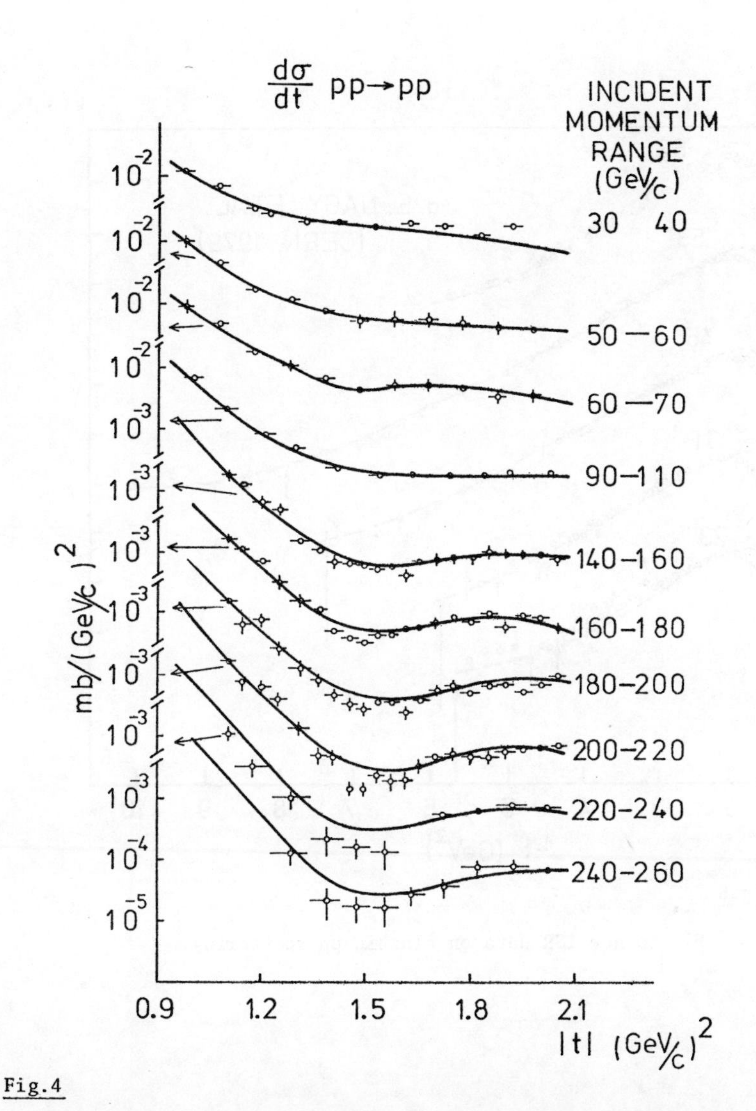

Fig.4

Differential cross sections for proton-proton elastic scattering showing the development of the first minimum. The solid lines are the theoretical curves.

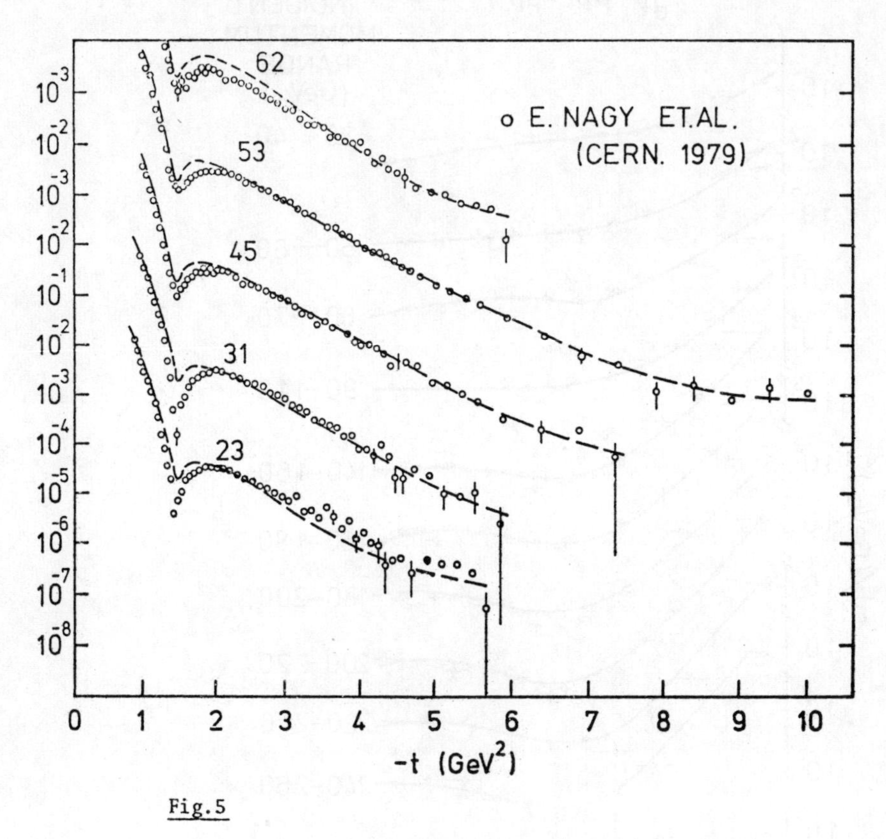

Fig.5

Fit to new ISR data on elastic pp scattering.

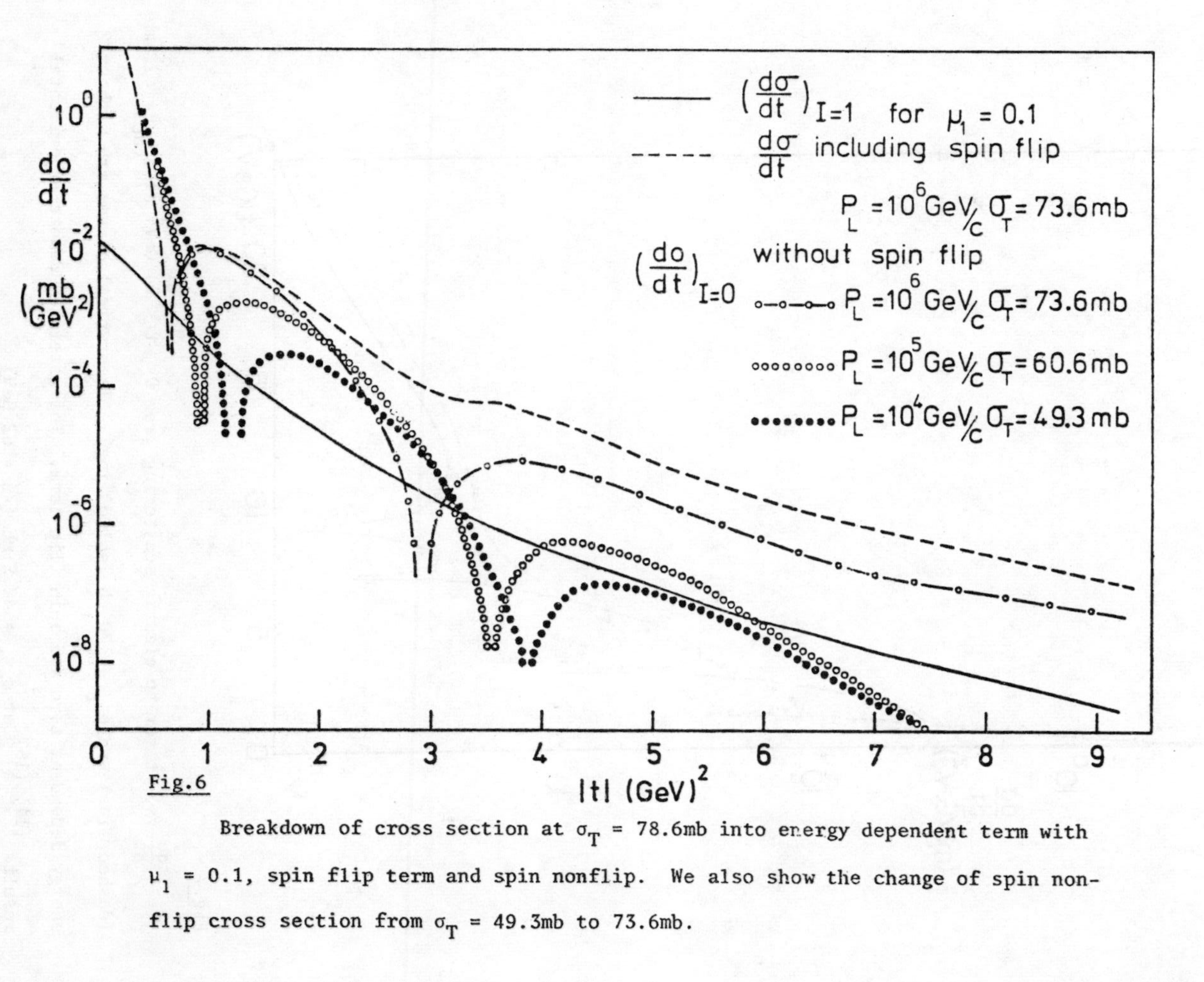

Fig.6

Breakdown of cross section at σ_T = 78.6mb into energy dependent term with μ_1 = 0.1, spin flip term and spin nonflip. We also show the change of spin non-flip cross section from σ_T = 49.3mb to 73.6mb.

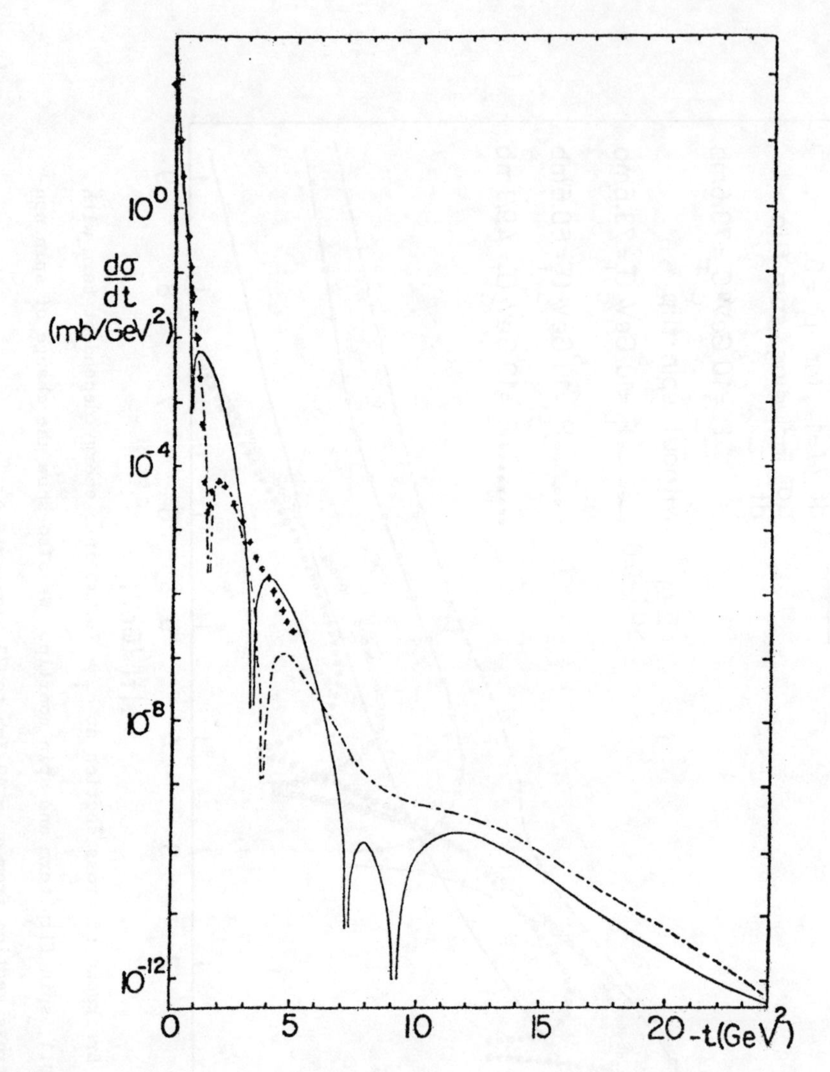

Fig.7

$\frac{d\sigma}{dt}$ vs. -t for pp elastic scattering from $0< -t < 24$ GeV^2 at σ_{tot} = 42.5mb (dashed line) and σ_{tot} = 70 mb (solid line). Theoretical curves are without energy dependent term and spin flip term. The asterisks denote experimental results (14),(15) at σ_{tot} = 42.5 mb ($\sqrt{s}$ = 62 GeV).

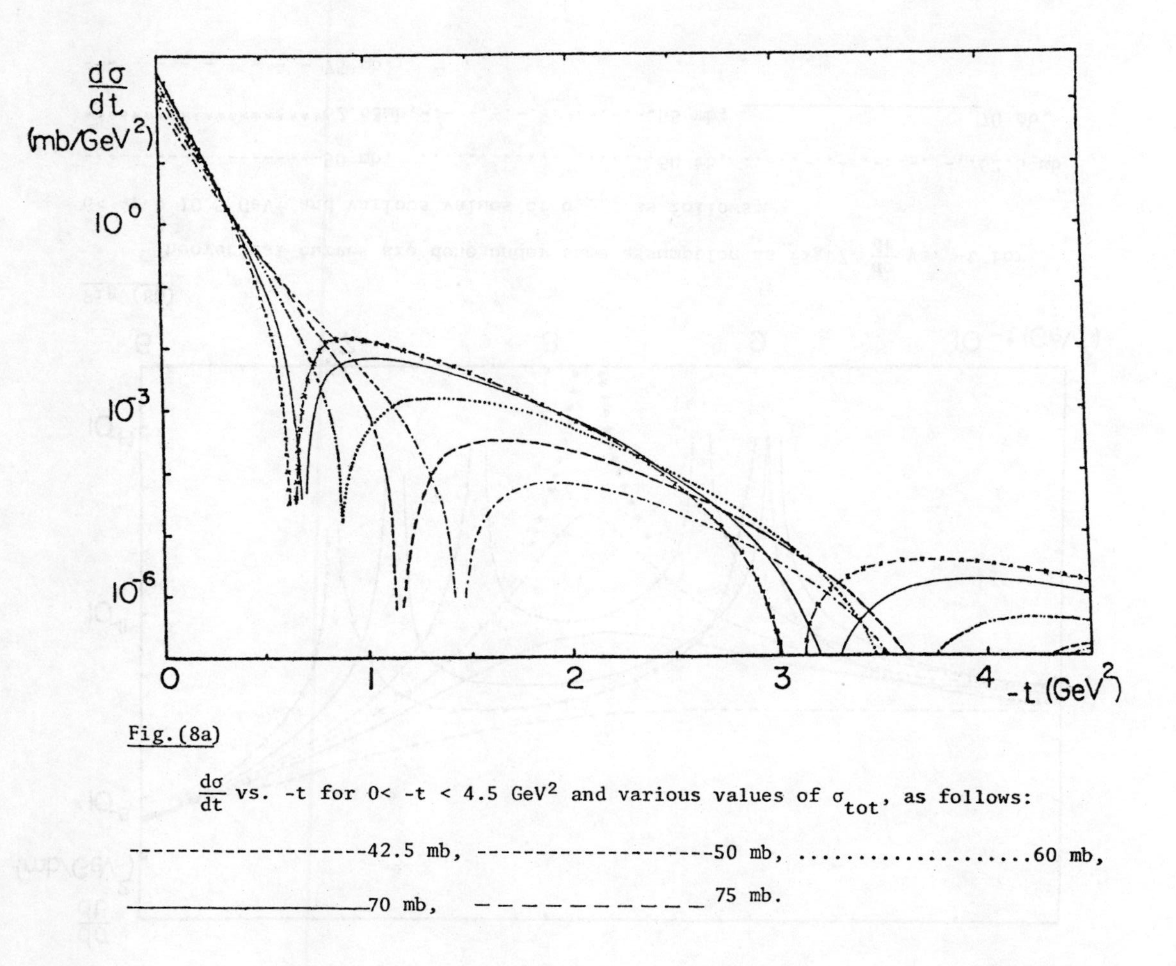

Fig. (8a)

$\frac{d\sigma}{dt}$ vs. -t for $0 < -t < 4.5$ GeV^2 and various values of σ_{tot}, as follows:

- - - - - - - - - - - - - - 42.5 mb, ——————— 50 mb, ... 60 mb,

——————— 70 mb, — — — — — — — — — 75 mb.

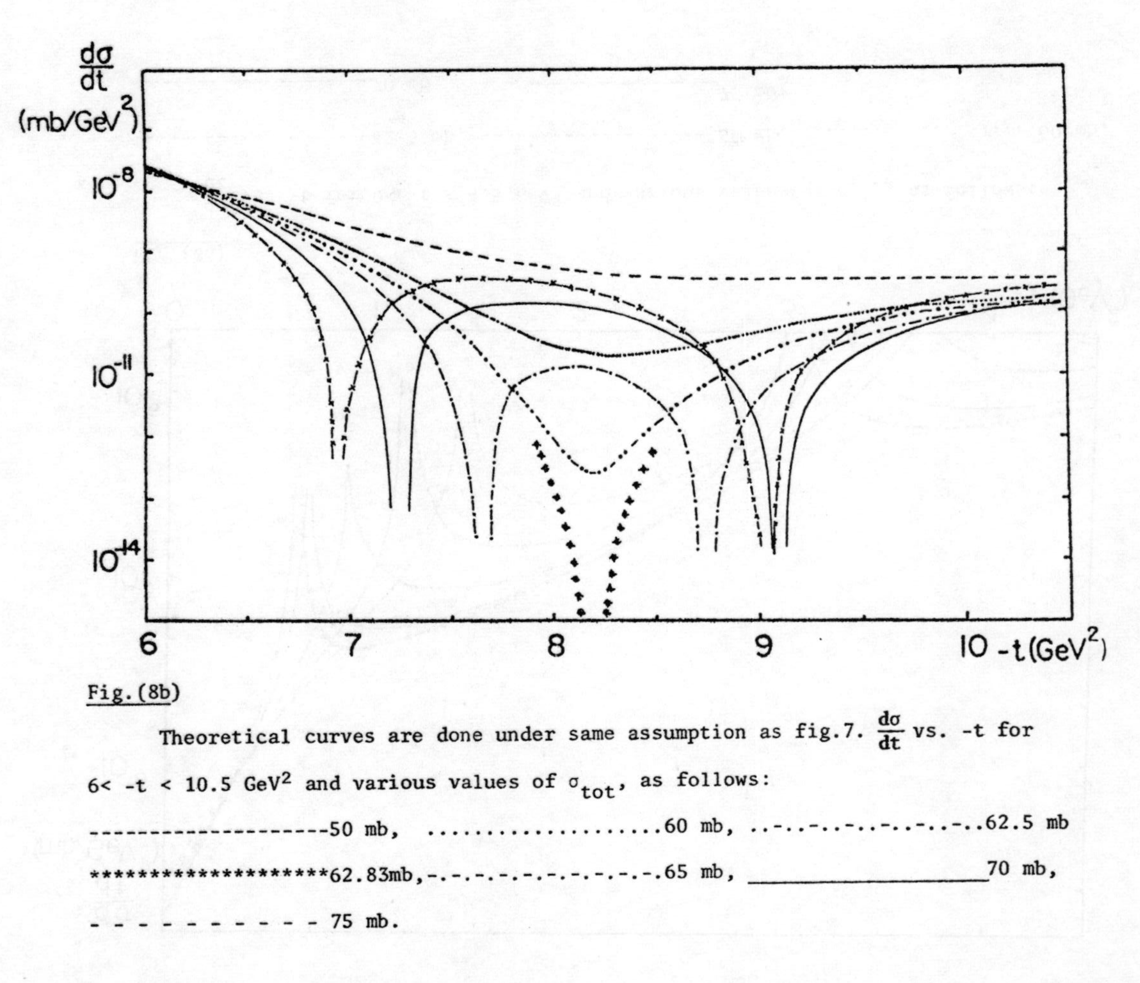

Fig.(8b)

Theoretical curves are done under same assumption as fig.7. $\frac{d\sigma}{dt}$ vs. -t for $6 < -t < 10.5$ GeV^2 and various values of σ_{tot}, as follows:

--------------------50 mb, ...60 mb, ..-..-..-..-..-..-..62.5 mb

*******************62.83mb,-.-.-.-.-.-.-.-.-.-.65 mb, ____________________70 mb,

- - - - - - - - - - - - 75 mb.

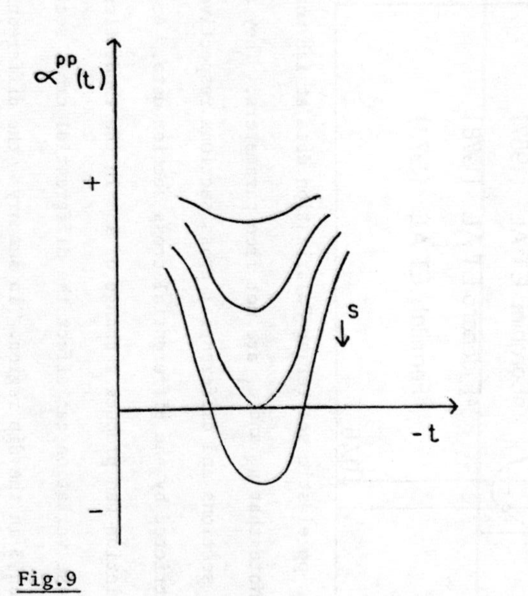

Fig.9

Schematic diagram illustrating the way in which the amplitude $\alpha^{pp}(s,-t)$ develops a trough and then two zeroes as σ_{tot} increases above 60 mb, in the region $5 < -t < 11$ GeV^2 corresponding to Figure (2b).

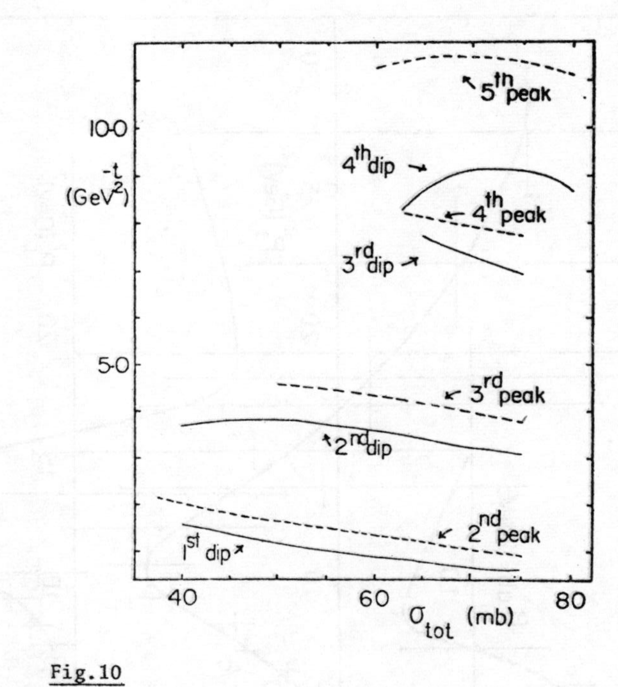

Fig.10

Change in (-t) positions of the dips and peaks as σ_{tot} increases.

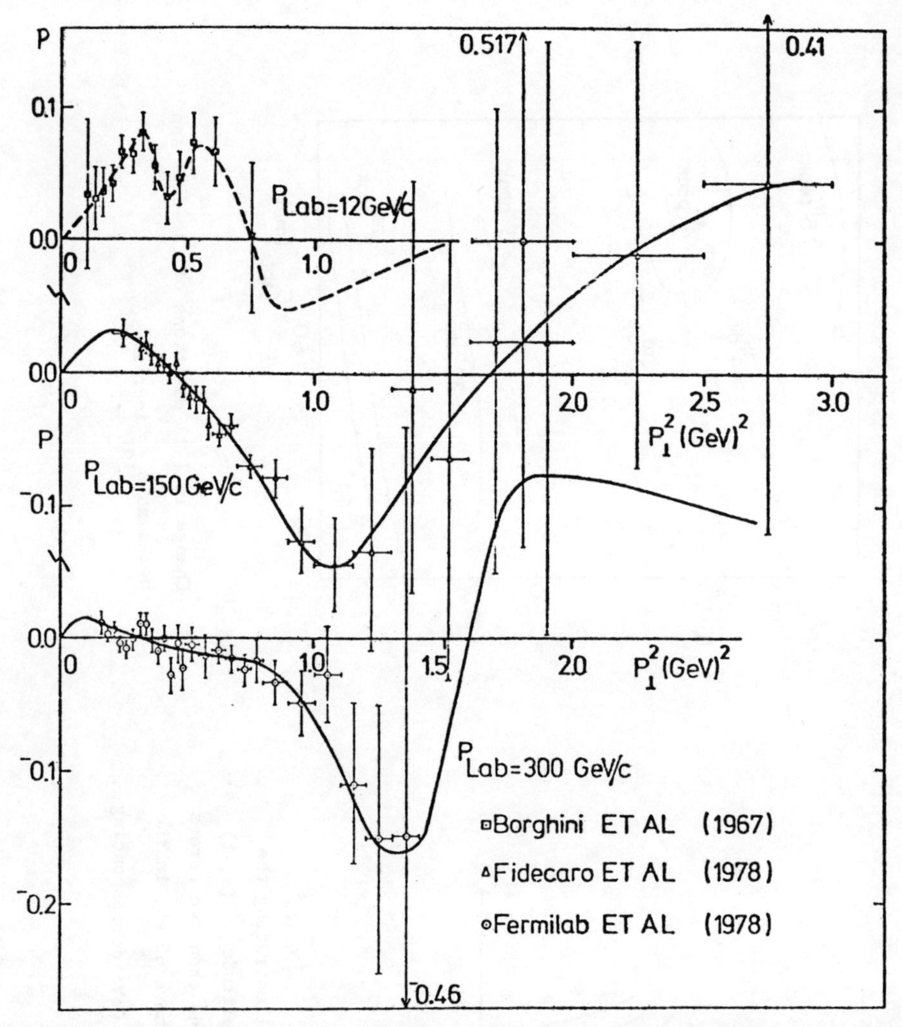

Fig.11 Theoretical fits to pp elastic scattering Polarization data at lab momenta 12, 150 and 300 GeV/c. Note that μ_0 and μ_1 are not free parameters. They are fixed by the total cross sections and differential cross sections respectively. The phase Φ, is also restricted by the differential cross section data. At small cross momentum phase variation can produce a change of 15% in the theoretical fit to $\frac{d\sigma}{dt}$. At larger $P_\perp^2$ phase variation can affect the differential cross section by a factor of 2, and even 3 in the dip region. In summary: the differential

cross section restricts the possible choice of phase and it is most encouraging that the phase values employed to fit differential cross section data are entirely consistent with the Polarization data.

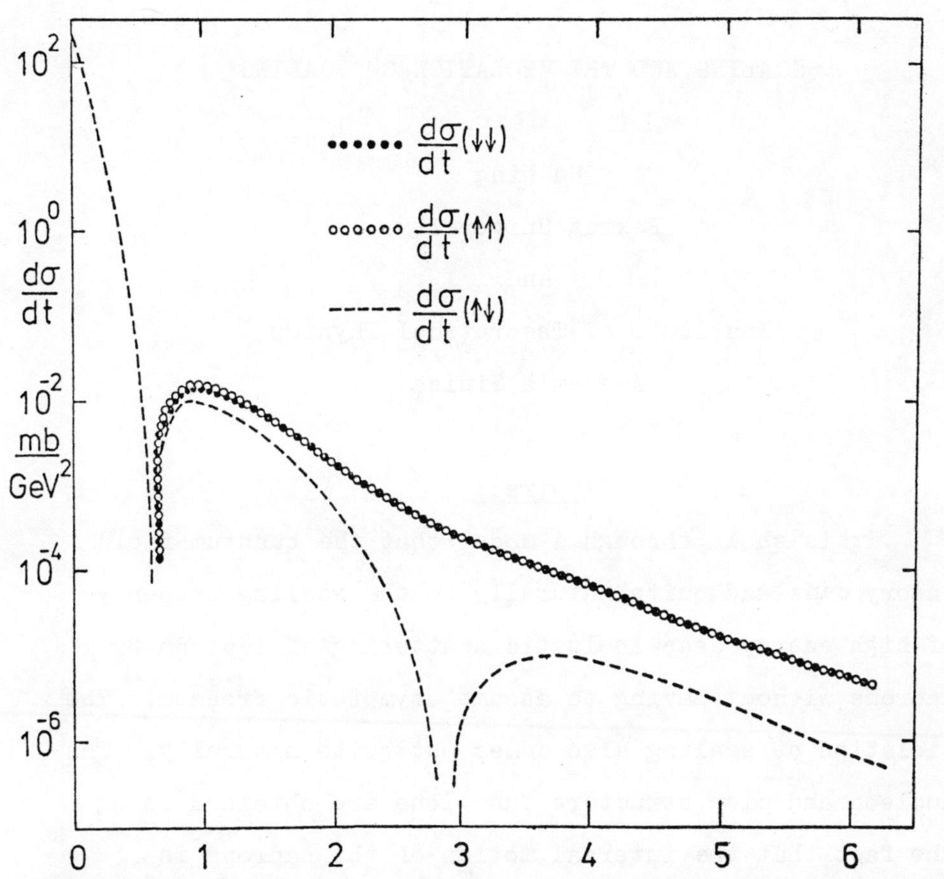

Fig. 12 Theoretical curves for spin-parallel and spin-anti-parallel differential cross sections at ISABELLE energy. A total cross section of 73.6 mb was used to fix μ_o and a value of 0.08 GeV^{-2} used in μ_1. This latter value was obtained by extrapolating the apparent trend in μ_1 to higher energy. Phase values were determined on the assumption that the same regularity observed at lower energies could be assumed for very high energies.

SCALING AND THE VIOLATION OF SCALING

Hu Ning

Peking University

and

Institute of Theoretical Physics,

Academia Sinica

Abstract

It is shown through a model that the quantum field theory can lead quite naturally to the scaling property of high energy deep inelastic scattering of leptons by hadrons without having to assume asymptotic freedom. The violation of scaling also comes out quite naturally. The nucleon and pion structure functions are obtained using the fact that the internal motion of the hadrons is simple harmonic. This fact is borne out by the Regge behavior of hadrons, which shows that the energy levels of various hadrons are indeed those of the relativistic simple harmonic oscillators. The scaling property of the Drell-Yan process is also investigated.

I. THE CONDITION OF ELASTIC SCATTERING OF A LEPTON BY STRATONS INSIDE THE HADRON

We consider the deep inelastic scattering process of the electron by a hadron. Choose the center of mass system of the hadron and the virtual photon emitted by the electron, in which q = -P, where P and q are momenta of the hadron and the virtual photon, respectively. In the following all momenta are understood as the components along the direction of P unless otherwise indicated. The 4-momentum of the photon is space-like, i.e., $Q^2 = q^2 - q_0^2 > 0$. The transition matrix element of the process is given by

$$\langle k_1 k_2 \cdots k_n | H | P, q \rangle = \sum_m \sum_{p_i} \langle k_1 k_2 \cdots k_n | V_2 | p_1 + q, p_2, \cdots p_m \rangle$$

$$\frac{1}{P_0 + q_0 - E_1' - E_2 - \cdots - E_m} \langle p_1 + q | J_\mu | p_1 \rangle \frac{1}{P_0 - E_1 - E_2 - \cdots - E_m}$$

$$\langle p_1 p_2 \cdots p_m | V_1 | P \rangle \tag{1}$$

The corresponding Feynman diagram is shown in Fig. 1. p_i and $E_i = (p_i^2 + p_{i\perp}^2 + M_i^2)^{1/2}$ are respectively momentum and energy of the i-th straton inside the hadron M_i being the mass of the i-th straton. V_1 and V_2 represent effective vertex functions which correspond to the shaded parts in Fig. 1. No creation and annihilation of straton pairs are considered since they can be shown to be small and negligible when P is very large.

Some of the internal lines p_i may be antistratons. When P is very large one may write $p_i = x_i P$ with $x_i \leqslant 1$. One has approximately

$$E_i \simeq x_i P + \frac{1}{2 x_i P}\left(p_{i\perp}^2 + M_i^2\right) \tag{2}$$

$$E_1' \simeq -(x_1 P + q) - \frac{1}{2(x_1 P + q)}\left(p_{1\perp}^2 + M_1^2\right) \tag{3}$$

The sign on the right-hand side of the second equation arises from the fact that the square root must take positive value while $x_1 P + q$ is negative. The two energy denominators in (1) are then given by

$$\frac{1}{P_0 - E_1 - E_2 - \cdots - E_m} \simeq - \frac{2P}{\sum_i x_i^{-1}\left(p_{i\perp}^2 + M_i^2\right)} \tag{4}$$

$$\frac{1}{P_0 + q_0 - E_1' - E_2 - \cdots - E_m} \simeq \Big[2x_1 P + q_0 + q - \frac{1}{P}\sum_{i=2}^{m} \frac{1}{2x_i}\left(p_{i\perp}^2 + M_i^2\right) + \frac{p_{1\perp}^2 + M_1^2}{2(x_1 P + q)} \Big]^{-1} \tag{5}$$

(4) is of the order P, (5) will also be of the order P if the following condition is satisfied:

$$2x\,P + q_0 + q = 0 \tag{6}$$

When (6) is not satisfied, (5) will be of the order $1/P$. This means that the contribution to (1) from (5) when (6) is not satisfied is a small quantity of relative order $1/P^2$ and therefore may be neglected entirely. (6) can also be written in the following form

$$2x_1 P_\mu q_\mu + Q^2 = (q - q_0)(2x\,P + q + q_0) = 0 \tag{7}$$

This means that the electron-straton scattering is elastic. The total cross-section is given by the elastic cross-section multiplied by a distribution function and integrated over all x_i. Scaling is obtained in parton model and QCD by assuming that the distribution function depends only on x_i. We shall go beyond QCD and give a derivation of this function based on our straton model.[1]

II. THE STRUCTURE FUNCTIONS OF PIONS AND NUCLEONS.

The pion may be considered as consisting of two parts. One part is the bare point particle which absorbs the photon. The other part is the rest part of the pion which may be considered as a composite particle having the quantum numbers of nucleon. The vertex functions V_1 and V_2 are given by

$$V_1 = \gamma_5 f_1$$
$$V_2 = \gamma_5 f_2 \tag{8}$$

Using the following Dirac functions

$$u(p) = \sqrt{\frac{M_1+E_1}{E_1}} \begin{pmatrix} 1 \\ \frac{\sigma p_1}{M_1+E_1} \end{pmatrix} U_1 \qquad v(p_2) = \sqrt{\frac{M_2+E_2}{E_2}} \begin{pmatrix} \frac{\sigma p_2}{M_2+E_2} \\ 1 \end{pmatrix} U_2$$

where U_1 and U_2 are two component spin wave functions, one obtains

$$\langle p_1 + q | J_\perp | p_1 \rangle \simeq Q_1 (\bar{U}_1' \sigma_\perp \sigma_3 U_3) \tag{9}$$

$$\langle p_1, p_2 | V_1 | P \rangle \simeq P^{-3/2} x_1^{-1} (1 - x_1)^{-1} (\bar{U}_1 U_2) f_1 \quad (10)$$

$$\langle X_0 | V_2 | p_1 + q, p_2 \rangle \simeq P^{-1/2} (1 - x_1)^{-1/2} (\bar{U}_2 U_1') f_2 \quad (11)$$

where Q_1 denotes the charge of the straton 1. Only the transversely polarized photon contributes to (9). The difference between (10) and (11) is due to the fact that p_1 and p_2 are in the same direction whereas $p_1 + q$ and p_2 are in opposite directions. The above results are valid only when P is very large.

The function f_1 should be closely related to the internal wave function of the hadron, which has been found to be given by [1)]

$$\varphi \sim e^{-r^2/16\lambda}, \quad \lambda \simeq 1\,\mathrm{GeV}^{-2} c^{-2} \quad (12a)$$

This represents a simple harmonic oscillator with energy levels*

$$m^2 = - m_0^2 + \lambda^{-1} (2n + \ell + \frac{3}{2}) \quad (12b)$$

The above results were obtained by us by solving the Beth-Salpeter equation with an effective pseudo-scalar potential.[1)] It is noted that m^2 instead of m appears on the left side of the above expression. The energy levels of the excited states of hadrons given by (12b) agree excellently with experiments. Therefore we shall accept (12a) as the true internal wave function describing the internal

* m_0 is different for different ground state hadrons

motion of the two parts of the pion. The function f_1 in (10) may be obtained by boosting (12a) to momentum P. One then obtains

$$f_1 \sim e^{-4\lambda p_\perp^2 - \lambda(\tau_2 x_1 - \tau_1 x_2)^2 M_\pi^2 c^2}, \quad \tau_i = M_i/(M_1+M_2) \qquad (13)$$

It is seen that (13) is a function of x_1 and x_2 only and does not contain P. When P is very large, the final state formed just after the absorption of the photon as a simple harmonic oscillator will be highly excited in the direction of P. One may consider this system to be a one dimensional harmonic oscillator in a highly excited state. Since the energy level is now very high, the system may be treated classically. One has then

$$p = A \cos \omega t \qquad (14)$$

The probability of finding the system having internal momentum p is

$$\frac{dt}{dp} \propto \frac{1}{[A^2 - p^2]^{1/2}} \qquad (15)$$

The wave function may be represented by the square root of the above expression.

$$\varphi \sim \frac{1}{[A^2 + p^2]^{1/4}} \qquad (16)$$

After the absorption of the virtual photon, the momenta of the two parts become equal and opposite and have magnitude $(1 - x_1)P$. The center of mass of the system is then at rest and the total energy is $X_0 = 2(1 - x_1)P$. The internal momentum reaches maximum just after the photon is absorbed. It follows from (16) that the probability becomes infinity at this moment as it should be for a classical oscillator. However, the concept of probability is meaningful only when the interval of measurement $dp_1 = P\,dx_1$ is specified, the observed brobability being given by the average value in this interval. Consequently (16) must be replaced by

$$\frac{dt}{dp} \propto \frac{1}{[A^2-(P_{max}-\frac{1}{2}dp_1)^2]^{1/2}} \cong \frac{1}{[(1-x)P\,dp_1]^{1/2}} \qquad (17)$$

It can easily be seen that the normalized probability is independent of $P\,dp_1$. The corresponding normalized wave function becomes

$$f_2 \sim \frac{1}{(1-x)^{1/4}} e^{-4\lambda p_\perp^2} \qquad (18)$$

It is well known that for a simple harmonic oscillator the average kinetic energy is equal to the average potential energy and each is equal to one-half the total energy. When P is very large the potential energy will become the kinetic energy of the gluon field under Lorentz transformation. This is in agreement with the experimental result that the valence stratons possess only one-half of the kinetic energy of the hadron.

The jets produced by deep inelastic scattering of neutrinos and muons on nucleons have been observed recently.[2)] The results show that the distribution of transverse momentum $P_\perp$ is given by

$$\frac{dN}{dp_\perp} \sim e^{-5.6\,p_\perp^2}, \quad (0 < p_\perp < 1\ \mathrm{GeV/c}) \tag{19}$$

This is different from the distribution of $p_\perp$ observed in high energy hadron-hadron collision, which is given by

$$\frac{dN}{dp_\perp} \sim e^{-6 p_\perp}, \quad (0.2 < p_\perp < 2\ \mathrm{GeV/c}) \tag{20}$$

The difference between the above two distributions may be explained by the fact that the gluons participate only in strong interaction. (19) shows that the distribution $p_\perp$ is indeed consistent with the wave function of the simple harmonic oscillator. The coefficient 5.6 in (19) is also consistent with the value of λ given above.

Using the above results and overlooking the difference between $p_{1\perp}^2 + M_1^2$ and $p_{2\perp}^2 + M_2^2$, one obtains

$$\langle X | H | P, q \rangle \sim Q_1 (U_1' \sigma_\perp \sigma_3 U_1)(\bar{U}_1 U_2)(\bar{U}_2 U_1') \cdot (1-x_1)^{1/4} e^{-4\lambda p_\perp^2 - \lambda(\tau_2 x_1 - \tau_1 x_2)^2 M_\pi^2 c^2} \tag{21}$$

This is not yet the matrix element (1) since the decay of the excited states into stable hadrons has not been specified. According to the present picture, the final state is a one dimensional excited simple harmonic oscillator. The two oscillating parts will move away from each other starting from the point of maximum momentum just after the absorption of the virtual photon. When the distance between these two parts increases kinetic energy will be graduately converted into potential energy. When potential energy becomes large enough that a new straton pair can be produced, the system will break into two oscillators. The two parts in each oscillator will continue to move apart until further breaking into oscillator pairs. The process of breaking apart will repeat again and again until finally two jets of un-excited hadrons are obtained. The potential of the one dimensional oscillator may be visualized as a elastic string connecting the stratons.

The structure function $F_2^{\pi}(x)$ is given by squaring (21) and then averaging over spin and charge, and finally multiplying the result by P and x:

$$F_2^{\pi}(x) \sim P \sum_i Q_i^2 x (1-x)^{1/2} e^{-2\lambda(2x-1)^2 M_\pi^2 c^2} \tag{22}$$

This shows that the condition (7) leads indeed to scaling. In the above picture M_2 should be greater than M, since M_1 represents a bare straton while M_2 includes all the rest of the pion. We take the limiting case $M_1 = M_2$. The curve for $F_2^{\pi}(x)$ is shown in Fig. 2.

We turn now to consider deep inelastic scattering of the electron by the nucleon. One may again consider the nucleon as consisting of two parts. One part is the bare straton which absorbs the virtual photon, and the other part includes the remaining part of the nucleon. This second part may be either a pseudo-scaler or a vector particle. We shall only consider the case of pseudo-scaler particle since the vector particle leads to too small a value of the structure function at small values of x. The appropriate vertex functions are that given by

$$V_i = \gamma_5 f_i \qquad (i = 1, 2)$$

with M_π in f_i now replaced by M_N. One has[3)]

$$\langle p_1 p_2 | V_1 | P \rangle \sim x_1^{-1}(1-x_1)^{1/2} P^{-3/2} \bar{U}_1 \sigma_3 U_P e^{-4\lambda p_\perp^2 - \lambda(2x_1 - 2\tau_1)^2 M_N^2 c^2} \qquad (23)$$

$$\langle X | V_2 | p_1 + q, p_2 \rangle \sim P^{-1}(1-x)^{-1/2} \bar{U}_X \sigma_3 U_1' \frac{1}{(1-x_1)^{1/4}} e^{-4\lambda p_\perp^2} \qquad (24)$$

Using similar dervation one obtains the following expression for the structure function for the nucleon:[3)]

$$F_2^N(x) \sim x(1-x)^{7/2} e^{-2\lambda(2x-2\tau_1)^2 M_N^2 c^2} \qquad (25)$$

Since the part M_2 contains at least two stratons, one expects $M_2 \approx 2M_1$, i.e., $\tau_1 = 1/3$. The curve for $F_2^N(x)$ corresponding to this value of τ_1 is shown in Fig. 2 along with the curve for $F_2^\pi(x)$. The dashed curve represents the well known empirical curve $A(1 - x)^3$, which coincides with the theoretical curve (25) in the region $0.4 < x < 1$. The maximum of the curve (25) will shift to the side of small x when τ_1 decreases. It is seen that $F_2^\pi(x)$ is more predominent in the region of large x, as was expected based on other considerations. The curve $F_2^N(x)$ becomes too small at $x \lesssim 0.2$ in comparison with the experimental one. This is due to the fact that the contribution of the wee partons has been overlooked, and also due to the deviation from scaling as discussed in the next section. It should also be stressed that the wave function for the final state (18) may not be a good approximation to the true one. The present treatment does not make any assumption on the type of interaction between the straton and the gluon field. The simple harmonic oscillator wave function is accepted only on the experimental ground. It does not excluded any theory. Our main conclusion is that scaling can be obtained without asymtotic freedom.

III. DEVIATION FROM SCALING

It has been pointed out that deviation from scaling comes from the emission of a gluon of large transverse momentum $k_\perp$ in the case of non-singlet structure functions. Using the leading logarithmic approximation, a relation between $k_\perp^2$ and Q^2 is obtained. The dependence of the structure functions on $k_\perp^2$ thus leads to their dependence on Q^2 and consequently the scaling property is violated. It will be shown in the following that a similar relation between $k_\perp^2$ and Q^2 can a lso be obtained directly when the final state interaction is not overlooked.

Fig 3 shows a diagram relevant to deviation from scaling. The corresponding transition matrix element is given by

$$\langle k_1 k_2 \cdots k_n | H | P, q \rangle = \sum_m \sum_{p_i} \langle k_1 k_2 \cdots k_n | V_2 | k, p_1 + q - k, p_2 \cdots p_m \rangle$$

$$\frac{1}{P_0 + q_0 - E_1'' - E_2 \cdots E_m - \varepsilon} \langle p_1 + q - k | J_\perp | p_1 - k \rangle \frac{1}{P_0 - E_1' - E_2 \cdots E_m - \varepsilon}$$

$$\langle p_1 - k | J_g | p_1 \rangle \frac{1}{P_0 - E_1 - E_2 - \cdots - E_m} \langle p_1 p_2 \cdots p_m | V_1 | P \rangle \tag{26}$$

where ε, E_i, E'_1 and E''_1 are energies corresponding to the momenta k, p_i, p_1 - k and p_1 - k + q, respectively. Introducing as before

$$p_i = x_i P, \quad k = (1 - z) p_1 = x_1 (1 - z) P \tag{27}$$

one obtains

$$\begin{aligned} &\varepsilon = x_1(1-z)P + \frac{k_\perp^2}{2x_1(1-z)P}, \quad E_i = x_i P + \frac{\mu_i^2}{2x_i P} \\ &E_1' = x_1 z P + \frac{k_\perp^2}{2x_1 z P}, \quad E_1'' = -(q + x_1 z P) - \frac{k_\perp^2}{2(q + x_1 z P)} \\ &\mu_i^2 = p_{i\perp}^2 + M_i^2 \end{aligned} \tag{28}$$

Inserting (28) in the denomenators of (26), one obtains further

$$\begin{aligned} &\frac{1}{P_0 + q_0 - E_1'' - E_2 - \cdots - E_m - \varepsilon} \simeq \frac{1}{q_0 + q + 2x_1 z P - [\frac{1}{x_1(1-z)} + \frac{1}{1 - x_1 z}] k_\perp^2 / 2P} \\ &\frac{1}{P - E_1' - E_2 - \cdots - E_m - \varepsilon} \simeq \frac{2P}{[\frac{1}{x_1(1-z)} + \frac{1}{x_1 z}] k_\perp^2} \\ &\frac{1}{P - E_1 - E_2 - \cdots - E_m} \simeq \frac{2P}{\sum_i \mu_i^2 / x_i} \end{aligned} \tag{29}$$

The last relation is the same as (3) given before. In the above expressions $k_\perp^2$ is considered to be much larger than μ_i^2. The right-hand side of the first expression will be of the order P instead of the order 1/P if the following condition is satisfied:

$$q_0 + q + 2x_1 z P - \left[\frac{1}{x_1(1-z)} + \frac{1}{1-x_1 z}\right]\frac{k_\perp^2}{2P} = 0 \qquad (30)$$

(30) represents a deviation from the elastic condition (7). Multiplying both sides of (30) by $q_o - q$, one obtains as before

$$Q^2 + 2x_1 z P_\mu q_\mu + \left[\frac{1}{x_1(1-z)} + \frac{1}{1-x_1 z}\right]\frac{1}{2P}k_\perp^2(q_0 - q) = 0 \quad , \qquad (31)$$

q_o in the above expression may be eliminated by using (7):

$$Q^2 - Q_0^2 = f(x_1, z)k_\perp^2 \ , \ f(x_1, z) = \left[\frac{1}{x_1(1-z)} + \frac{1}{1-x_1 z}\right](1 - x_1 z) \qquad (32)$$

where $Q_o = -2x_1 zP_\mu q_\mu$ is the value of Q when $k_\perp = 0$. Taking the logarithm and differentiating, one obtains

$$d\ln(Q^2 - Q_0^2) = d\ln k_\perp^2 \qquad (33)$$

which is different from a similar result obtained by the leading logarithmic approximation.

$$d\ln Q^2 = d\ln k_\perp^2 \qquad (34)$$

(34) is called the sub-asymptotic correction in QCD. In the present theory the correction (33) has nothing to do with the asymptotic freedom.

Since one does not consider final-state interaction in QCD, the gluon emitted by the straton is considered to be a free particle of zero mass with only transverse polarizations. Since in the present treatment final-state interaction is not overlooked, the gluon k should be considered to be a virtual particle also with longitudinal and temporal components of polarization. It can be shown that if all four components of polarization are taken into account, then the ratio of anomalous dimensions for the non-abelian vector gluon theory is given by *

$$A_n^{NS}/A_{n'}^{NS} = \left[-\frac{1}{2}+\frac{1}{n(n+1)}\right]/\left[-\frac{1}{2}+\frac{1}{n'(n'+1)}\right]$$

which is the same as that obtained in the pseudo-scalar and scalar gluon theories. Therefore the above result cannot be used to show whether QCD is better than other theories.

IV. THE DRELL-YAN PROCESS[4)]

In this section the Drell-Yan process $\pi P \rightarrow \mu^+\mu^- X$ will be studied without overlooking final-state interaction. The transition matrix element for this process is

$$\begin{aligned}\langle q, X|H|P_p, P_\pi\rangle &= \langle X|V_x|x_2' P - x_2 P\rangle\langle q|J_\mu|x_1' P, -x_1 P\rangle \\ &\quad \langle x_1' P, -x_2' P|V_\pi|P\rangle\langle -x_1 P_1, -x_2 P_2|V_p|-P\rangle \\ &\left[\left(\frac{\mu_1^2}{2x_1 P}+\frac{\mu_2^2}{2x_2 P}\right)^{-1}+\left(\frac{\mu_1'^2}{2x_1' P}+\frac{\mu_2'^2}{2x_2' P}\right)^{-1}\right]\left(\frac{\mu_1^2}{2x_1 P}+\frac{\mu_2^2}{2x_2 P}+\frac{\mu_1'^2}{2x_1' P}+\frac{\mu_2'^2}{2x_2' P}\right)^{-1} \\ &\left[\left(x_1 P + x_1' P - q_0 + \frac{\mu_1^2}{2x_1 P}+\frac{\mu_1'^2}{2x_1' P}\right)^{-1}+\left(x_2' P + x_2 P - X_0 + \frac{\mu_2'^2}{2x_2' P}+\frac{\mu_2^2}{2x_1 P}\right)^{-1}\right] \quad (35)\end{aligned}$$

---

* Hu Ning and Ma Zhong-qi, unpublished calculation.

The center of mass coordinate system of the pion and the proton is used in which $P_\pi = -P_p = P$. The following Weinberg approximation has already been used in writing down the denominators:

$$E_i = (x_i^2 P^2 + \mu_i^2)^{1/2} \simeq x_i P + \frac{\mu_i^2}{2 x_i P}, \quad \mu_i^2 = p_{i\perp}^2 + M_i^2 \tag{36}$$

$$E_i' = (x_i'^2 P^2 + \mu_i'^2)^{1/2} \simeq x_i' P + \frac{\mu_i'^2}{2 x_i' P}, \quad \mu_i'^2 = p_{i\perp}'^2 + M_i'^2$$

The complicated summations in (35) is due to contribution from different time order of the vertices. The corresponding Feynman diagram is shown in Fig. 4. As before one may consider the pion and the proton as each consisting of two parts. No concrete form of vertex functions will be introduced in this section except that they should be functions of x_i and x_i' only. The vertex functions used in foregoing sections are of such types.

The last square bracket of (35) will have contribution of the order P instead of 1/P if the following two conditions are satisfied simultaneously

$$x_1 P + x_1' P - q_0 = 0 \tag{37}$$

$$x_2 P + x_2' P - X_0 = 0 \tag{38}$$

From the condition of conservation of total energy of the whole process $2P - q_0 - x_0 = 0$ and $x_1 + x_2 = x_1' + x_2' = 1$, one sees that if one of the relations (37) and (38) is satisfied, the other will be satisfied automatically.

(35) then becomes

$$\langle qX|H|P_\pi P_p\rangle = \langle X|V_x|x_2'P_1,-x_2P\rangle\langle q|\gamma_\mu|x_1'P,-x_1P\rangle$$
$$\langle x_1'P,-x_2'P|V_\pi|P\rangle\langle -x_1P,-x_2P|V_p|-P\rangle P^3$$
$$\left(\frac{\mu_1^2}{2x_1}+\frac{\mu_2^2}{2x_2}\right)^{-1}\left(\frac{\mu_1'^2}{2x_1'}+\frac{\mu_2'^2}{2x_2'}\right)^{-1}\left[\left(\frac{\mu_1^2}{2x_1}+\frac{\mu_1'^2}{2x_1'}\right)^{-1}+\left(\frac{\mu_2^2}{2x_2}+\frac{\mu_2'^2}{2x_2'}\right)^{-1}\right] \tag{39}$$

Inserting $x_2 = 1 - x_1$ and $x'_2 = 1 - x'_1$ in (37) and (38), one obtains

$$x_1P+x_1'P-q_0=0,\ (1-x_1)P+(1-x_1')P-X_0=0 \tag{40}$$

(40) may be considered as two independent conditions, and the condition of conservation of total energy

$$2P - q_0 - X_0 = 0 \tag{41}$$

as the consequence of (40). As already pointed out, the contribution to (39) is not negligible only when both conditions of (40) are satisfied. x_1 and x_1' may then be expressed in terms of q_0/P and X_0/P. Thus x_1 and x_1' represent not only the fractions of momenta p_1/P and p_1'/P, but also the energy variables q_0/P and X_0/P. Similar Dual role is also played by x_i in the case of deep inelastic scattering.

Introducing

$$x_1=\tau e^{y},\quad x_1'=\tau e^{-y} \tag{42}$$

and noting $Q^2=q_0^2-q^2=(x_1+x_1')^2P^2-(x_1-x_1')^2P^2=4x_1x_1'P^2$

one obtains

$$\tau^2=Q^2/4P^2 \tag{43}$$

(42) then gives

$$dx_1 dx_1' = \frac{\partial(x_1' x_1)}{\partial(\tau, y)} d\tau dy = \frac{1}{2} d\tau^2 dy \tag{44}$$

The differential cross-section of the process is then given by

$$\frac{d^2\sigma}{d\tau^2 dy} \sim P^2 |\langle q X | H | P_\pi P_P \rangle|^2 \tag{45}$$

This shows that the Drell-Yan process satisfies the scaling condition, in agreement with experiments. However, it can easily be seen that the fragmentation function cannot be expressed as the product of the pion and nucleon structure functions in the present case. This point may be used as a test of asymptotic freedom.

We have shown in the above investigation through a simple model that scaling may not be related to asymptotic freedom, and that the factorizability of the fragmentation function into pion and nucleon structure functions may be utilized as a good test of asymptotic freedom. Our pion structure function rises steeply as x decreases from 1. This agrees with usual expectation. Our nucleon structure function is practically zero in the region $0.8 < x < 1$ and agrees with the experimental curve $A(1 - x)^{3/2}$ for $0.4 < x < 1$.

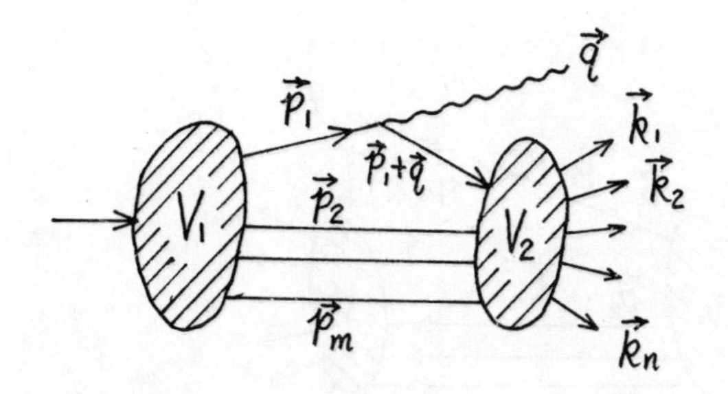

$\vec{q}$
$\vec{p}_1$
$\vec{p}_1+\vec{q}$
$\vec{p}_2$
$\vec{p}_m$
V_1
V_2
$\vec{k}_1$
$\vec{k}_2$
$\vec{k}_n$

Fig. 1.

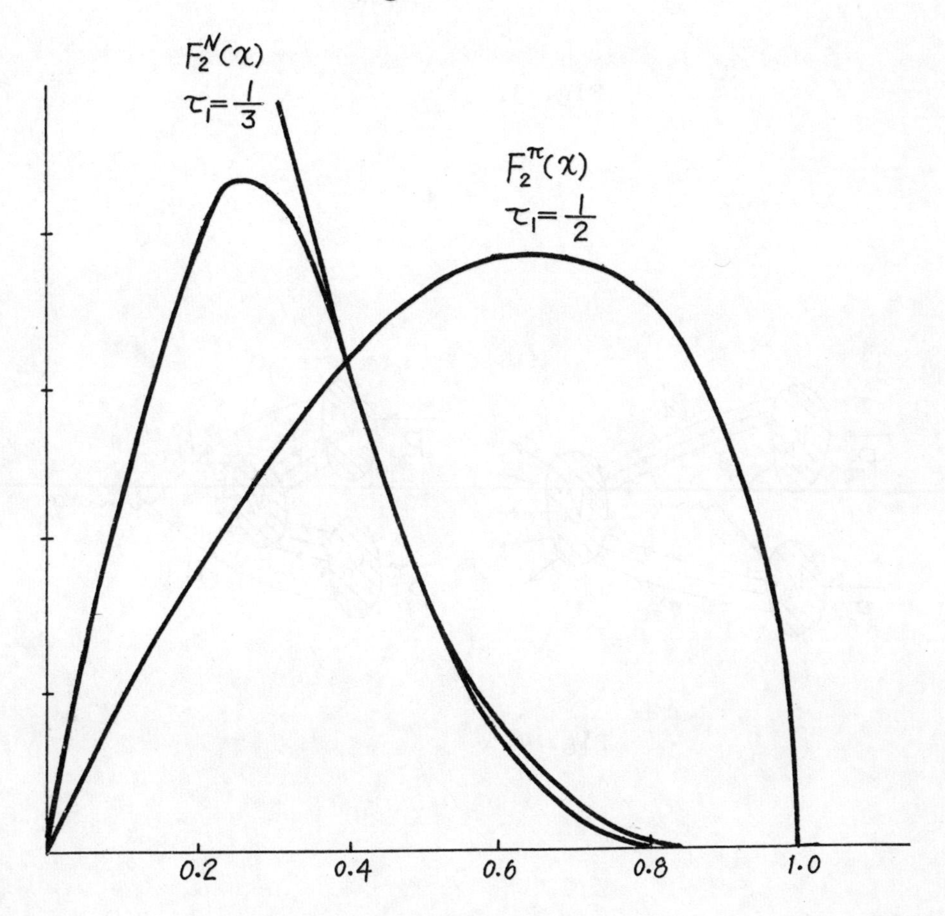

$F_2^N(x)$
$\tau_1=\frac{1}{3}$
$F_2^\pi(x)$
$\tau_1=\frac{1}{2}$
0.2
0.4
0.6
0.8
1.0

Fig. 2.

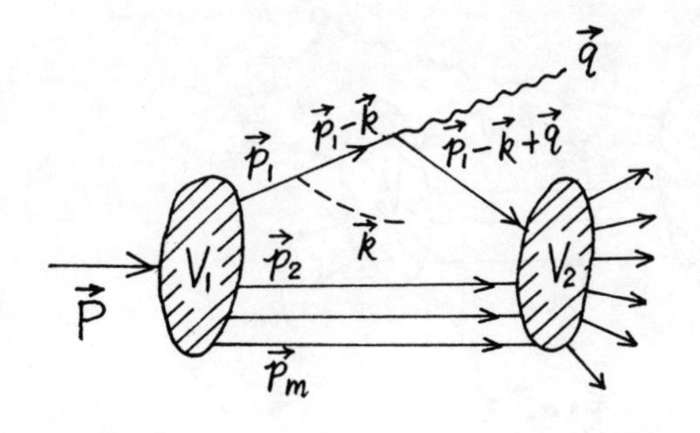

$\vec{q}$
$\vec{p}_1-\vec{k}$
$\vec{p}_1-\vec{k}+\vec{q}$
$\vec{p}_1$
$\vec{k}$
$\vec{p}_2$
V_1
V_2
$\vec{P}$
$\vec{p}_m$

Fig. 3.

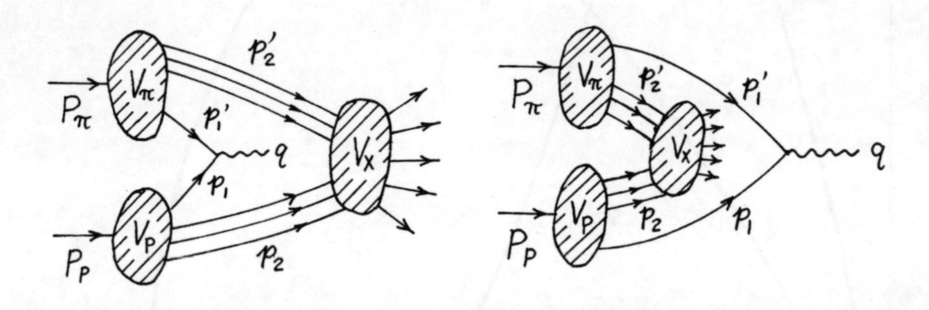

p_2'
V_π
P_π
p_1'
q
p_1
V_x
V_p
P_p
p_2

Fig. 4.

References and Notes

1. Hu Ning, Scientia Sinica, 22 296 (1979); Concluding Talk, this proceeding; similar attempt has been given by Drell, S.D. and Lee, T.D., Phys. Rev. D 5 1738 (1972).

2. Chapman et al, Phys. Rev. D 14 5 (1976).
   C. de Para et al, Phys. Rev. D 15 2425 (1977).

3. The structure function W as the transverse component of $W_{\mu\nu}$ is given by $W_1 = P\ \bar{\Sigma}_{ini.}\Sigma_{fin.}\langle X|H|P\rangle$. F_2 is defined as $\nu W_2 = 2N \times W_1$ when P is very large. Also when P is very large, the mass M in (8a) may be overlooked in comparison with E, except for the state $|X\rangle$ of (24), for which the momentum is zero and the mass is of the order P. The Dirac wave function for this particle is then

$$u \sim \begin{pmatrix} 1 \\ 0 \end{pmatrix} U_{\pm}$$

   This is responsible for the factor P^{-1} in (24). The same state $|X\rangle$ also appears in (39). Using the above expressions of Dirac wave functions, one easily varifies that the Drell-Yan process indeed satisfies the scaling condition. It will also be noted that our F (x) given by (22) differs from the expression obtained from asymptotic freedom condition by a factor P. This can also be checked by future experiments.

4. Drell, S.D. and Yan, T.M. Ann. of Phys. 66 578 (1971).

THE LARGE MOMENTUM BEHAVIOR OF THE PION WAVE FUNCTION AND THE PION ELECTROMAGNETIC FORM FACTOR

Huang Chao-shang
Institute of Theoretical Physics,
Academia Sinica

Abstract

In this paper we assume that QCD is the underlying theory of the strong interaction. Applying the methods of operator product expansion and renormalization group, we examine and obtain the large momentum behavior of the Bethe-Salpeter wave function and the electromagnetic form factor of pion.

QCD and the renormalization group method have gained conspicuous success for inclusive processes (e.g., deep inelastic electron scattering off hadrons, $e^+ e^-$ annihilation into hadrons etc.) at high energy. Recently, there have been a number of investigations[(1)] which examine fixed angle scattering of hadrons and the pion electromagnetic form factor at large momentum transfer. The field theory model that is assumed by these authors is the ϕ_6^3 scalar field theory or Yukawa theory. In this paper we assume that QCD is the underlying theory of the strong interaction. Applying the methods of the operator product expansion and the renormalization group, we derive the large momentum behavior of the Bethe-Salpeter

(B-S) wave function and the electromagnetic form factor for the 0^- meson (e.g., π meson).

I. In QCD all observed hadrons are assumed to be colour singlet states. Let $|P,\xi,\underline{1}\rangle$ represent a state vector of the 0^- meson, where P is the 4-momentum, ξ is the index of the flavor multiplet, and $\underline{1}$ represents a color singlet state. The B-S wave function of a 0^- meson is

$$\chi_{P,\xi,\underline{1}}(x_1,x_2)=\langle 0|T(\psi_\alpha(x_1)\bar{\psi}_\alpha(x_2))|P,\xi,\underline{1}\rangle\ , \quad (1)$$

where

$$\psi_\alpha(x)=\begin{pmatrix}\psi_\alpha^1(x)\\ \psi_\alpha^2(x)\\ \vdots\\ \psi_\alpha^N(x)\end{pmatrix}\ ,$$

and $\psi_\alpha^i(x)$ is the spinor field operator of the straton, i = 1, 2,... N is the flavor index, α = 1,2,3 is the color index. Hereafter for simplicity we shall suppress all color indices and write $\psi(x)$ instead of $\psi_\alpha(x)$. Then equation (1) can succinctly be written as

$$\begin{aligned}\chi_{P,\xi}(x_1,x_2)&=\langle 0|T(\psi(x_1)\bar{\psi}(x_2))|P,\xi\rangle\\ &=e^{iP\cdot X}\langle 0|T(\psi(\tfrac{x}{2})\bar{\psi}(-\tfrac{x}{2}))|P,\xi\rangle=e^{iP\cdot X}\chi_{P,\xi}(x)\end{aligned} \quad (2)$$

where $X=\frac{x_1+x_2}{2}$, $x=x_1-x_2$. In the momentum representation

$$\chi_{P,\xi}(p)=\int d^4x\, e^{-ip\cdot x}\langle 0|T(\psi(\tfrac{x}{2})\bar{\psi}(-\tfrac{x}{2}))|P,\xi\rangle\ , \quad (3)$$

where

$$P=p_1-p_2\ ,\quad p=(p_1+p_2)/2\ .$$

In order to find the large momentum ($p^2, p\cdot P\to\infty, \frac{p\cdot P}{p^2}=r$ fixed) behavior of the B-S wave function χ , we expand the operator product at $x^2\approx 0$:

$$
\begin{aligned}
T\left(\psi(\tfrac{x}{2})\bar{\psi}(-\tfrac{x}{2})\right) = \sum_{n=1}^{\infty} \sum_{i} \Big\{ & x^{S(0)} \mathcal{F}^{1}_{n,i}(x^2)\gamma_5 x_{\mu_1}\cdots x_{\mu_n} O^{n,i}_{\mu_1\cdots\mu_n} \\
& + x^{S(0)} \mathcal{F}^{2}_{n,i}(x^2)\sigma_{\mu\nu}\gamma_5 x_{\mu_1}\cdots x_{\mu_{n-1}} x_\mu O^{n,i}_{\mu_1\cdots\mu_{n-1}\nu} \\
& + x^{S(0)+1} \mathcal{F}^{3}_{n,i}(x^2)\gamma_\mu\gamma_5 x_{\mu_1}\cdots x_{\mu_{n-1}} O^{n,i}_{\mu_1\cdots\mu_{n-1}\mu} \qquad (4) \\
& + x^{S(0)-1} \mathcal{F}^{4}_{n,i}(x^2)\gamma_\mu\gamma_5 x_{\mu_1}\cdots x_{\mu_n} x_\mu O^{n,i}_{\mu_1\cdots\mu_n} \Big\} \\
& + \cdots
\end{aligned}
$$

where $x^{S(0)} = (x^2+i\epsilon)^{S(0)/2}$, $S(0) = \tau(0) - 3 = -1$ ($\tau(0)$ is the twist of $O^{n,i}_{\mu_1\cdots\mu_n}$), $i = a, \psi$ and*

$$O^{n,a}_{\mu_1\cdots\mu_n} = \frac{i^{n-1}}{n!}\left[\bar{\psi}\gamma_5\gamma_{\mu_1}\overleftrightarrow{D}_{\mu_2}\cdots\overleftrightarrow{D}_{\mu_n}\lambda^a\psi + \text{permutation}\right],$$

$$n = 1, 2, \cdots$$

$$O^{n,\psi}_{\mu_1\cdots\mu_n} = \frac{i^{n-1}}{n!}\left[\bar{\psi}\gamma_5\gamma_{\mu_1}\overleftrightarrow{D}_{\mu_2}\cdots\overleftrightarrow{D}_{\mu_n}\psi + \text{permutation}\right],$$

$$D_\mu\psi = (\partial_\mu - ig\lambda_j A_{j,\mu})\psi ,$$

---

* Footnote: More generally, $O^{n,a}$ can be written as

$$O^{n,a}_{\mu_1\cdots\mu_n} = \sum_{\substack{m,l \\ (m\geq n\geq l)}} a_{mnl}\, x_{\mu_{n+1}}\cdots x_{\mu_m}\partial_{\mu_{l+1}}\cdots\partial_{\mu_m}(\bar{\psi}\gamma_5\gamma_{\mu_1}\overleftrightarrow{D}_{\mu_2}\cdots\overleftrightarrow{D}_{\mu_l}\psi),$$

but which does not affect the ensuing discussion of the large momentum behavior of the wave function.

where λ_j $(j = 1, 2, \cdots, 8)$ is the infinitesimal operator of SU_3^c and λ^a is the infinitesimal operator of SU_N. In the expansion we have chosen here a gauge invariant combination of operators so as to extract the gauge invariant hadronic matrix element $\langle 0| O^{n,i} |P\rangle$ and to simplify our derivation. Since the form factor is gauge independent, we may expect that the form of the expansion of the wave function will be different from that of this paper but the final result about the form factor will be the same with ours if one expands with gauge dependent operators.

When ξ represents a non-singlet state, we find by inserting Eq. (4) into Eq. (3)

$$\begin{aligned}
\chi_{P,\xi}(p) = \frac{1}{(p^2)^{3/2}} m_\xi [\gamma_5 \sum_{n=2,4,\cdots} \left(\frac{p\cdot P}{p^2}\right)^n f^1_{n,a}\left(\frac{p^2}{\mu^2}\right) h^{n,a}_\xi\left(\frac{m_\xi^2}{\mu^2}\right) \\
+ \frac{p_\mu P_\nu}{p^2}\gamma_5\sigma_{\mu\nu} \sum_{n=1,3,\cdots}\left(\frac{p\cdot P}{p^2}\right)^{n-1} f^2_{n,a}\left(\frac{p^2}{\mu^2}\right) h^{n,a}_\xi\left(\frac{m_\xi^2}{\mu^2}\right)] \\
+ \frac{1}{p^4} m_\xi [\hat{P}\gamma_5 \sum_{n=1,3,\cdots}\left(\frac{p\cdot P}{p^2}\right)^{n-1} f^3_{n,a}\left(\frac{p^2}{\mu^2}\right) h^{n,a}_\xi\left(\frac{m_\xi^2}{\mu^2}\right) \\
+ \hat{p}\gamma_5 \sum_{n=1,3,\cdots}\left(\frac{p\cdot P}{p^2}\right)^{n} f^4_{n,a}\left(\frac{p^2}{\mu^2}\right) h^{n,a}_\xi\left(\frac{m_\xi^2}{\mu^2}\right)] + \cdots,
\end{aligned} \tag{5}$$

where we have taken into account space reflection and charge conjugation invariance[2].

It is easy to derive that the dimensionless Wilson coefficients $f^l_{n,a}\left(\frac{p^2}{\mu^2}\right)$ $(l = 1, 2, 3, 4)$ in Eq. (5) satisfy the following renormalization group equation[3] in the Landau gauge:

$$[\mathcal{D} + \gamma_{O^{n,a}}(g_R) - 2\gamma_\psi(g_R)] f^l_{n,a}\left(\frac{\lambda^2 p_f^2}{\mu^2}, g_R, \frac{m_R}{\mu}\right) = 0, \tag{6}$$

$$\mathcal{D} \equiv \lambda \frac{\partial}{\partial \lambda} - \beta(g_R)\frac{\partial}{\partial g_R} + \left(1+\gamma_\theta(g_R)\right) m_R \frac{\partial}{\partial m_R}, \quad \lambda p_f = p ,$$

and p_f is the fixed 4-momentum.

According to the conventional techniques the solution of Eq. (6) is given by

$$f^{\ell}_{n,a}\left(\frac{\lambda^2 p_f^2}{\mu^2}, g_R, \frac{m_R}{\mu}\right) = f^{\ell}_{n,a}\left(\frac{p_f^2}{\mu^2}, g(\lambda), \frac{m(\lambda)}{\mu}\right) \exp\left\{-\int_1^{\lambda}\left[\gamma_{0^{n,a}}(g(\lambda')) - 2\gamma_\psi(g(\lambda'))\right]\frac{d\lambda'}{\lambda'}\right\} \xrightarrow{\lambda\to\infty} f^{\ell}_{n,a}\left(\frac{p_f^2}{\mu^2}, g(\lambda), 0\right)(\ln\lambda)^{\frac{-\gamma^n+2\gamma^F}{b_0}} \tag{7}$$

where b_0 is given by the first term of the perturbation expansion of $\beta(g_R)$:

$$\beta(g_R) = -\frac{b_0}{2} g_R^3 + b_1 g_R^5 + \cdots ,$$
$$b_0 = \frac{1}{8\pi^2}\left[\frac{11}{3} C_2(G) - \frac{4}{3} T(R)\right] ; \tag{8}$$

$\gamma_{0^{n,a}}$ and γ_ψ are respectively the anomalous dimensions of the operators $0^{n,a}$ and ψ^i_α :

$$\gamma_{0^{n,a}} = \frac{g^2}{8\pi^2} C_2(R)\left[1 - \frac{2}{n(n+1)} + 4\sum_{j=2}^{n}\frac{1}{j}\right] + \cdots \equiv \gamma^n g^2 + \cdots , \tag{9}$$

and in the covariant gauge

$$\gamma_\psi = \frac{g^2}{16\pi^2}\alpha C_2(R) + \cdots \equiv \gamma^F g^2 + \cdots$$

(note that in the Landau gauge $\gamma^F = 0$).

Consequently when $p^2, p\cdot P \to \infty$ and $\frac{p\cdot P}{p^2} = \tau$ = fixed, the large momentum behavior of the wave function is

$$\chi_{P,\xi}(p) \sim \frac{1}{(p^2)^{3/2}} \text{ times powers of logarithm} + O\left(\frac{1}{p^4} \text{ times powers of logarithm}\right) \tag{10}$$

If we keep only the largest logarithm term then we have

$$\chi_{P,\xi}(p) \sim \frac{1}{(p^2)^{3/2}} \left(\ln \frac{p^2}{\mu^2}\right)^{\frac{-\gamma' + 2\gamma^F}{b_0}} m_\xi h_\xi\left(\frac{m_\xi^2}{\mu^2}\right) f\left(\frac{p_f^2}{\mu^2}, g\left(\frac{p^2}{p_f^2}\right), 0\right), \quad (11)$$

which can be rewritten as

$$\begin{aligned} \chi_{P,\xi}(p) &\equiv \chi_\xi(\lambda P_f, \lambda p_f; g_R, m_R, \mu) \\ &\xrightarrow{\lambda \to \infty} \frac{1}{\lambda^3} (\ln \lambda^2)^{\frac{-\gamma' + 2\gamma^F}{b_0}} \chi'_\xi(p_f^2, m_\xi, g(\lambda), m(\lambda), \mu) \end{aligned} \quad (12)$$

with $P^2 = (\lambda P_f)^2 = -m_\xi^2$, where χ'_ξ does not vanish when $\lambda \to \infty$. On the other hand, since $\chi_{P,\xi}(p)$ satisfies the renormalization group equation

$$[\mathcal{D} + 2 - 2\gamma_\psi(g_R)] \chi_\xi(\lambda P_f, \lambda p_f; g_R, m_R, \mu) = 0 ,$$

therefore

$$\chi_\xi(\lambda P_f, \lambda p_f; g_R, m_R, \mu) \xrightarrow{\lambda \to \infty} \frac{1}{\lambda^2} (\ln \lambda)^{\frac{2\gamma^F}{b_0}} \chi_\xi(P_f, p_f; g(\lambda), m(\lambda), \mu).$$

Comparing the above result with Eq. (12), we find when $\lambda \to \infty$

$$\chi_\xi(P_f, p_f; g(\lambda), m(\lambda), \mu) \sim \frac{1}{\lambda} (\ln \lambda)^{-\frac{\gamma'}{b_0}} \chi'_\xi(p_f^2, m_\xi, g(\lambda), m(\lambda), \mu). \quad (13)$$

That is, when $\lambda \to \infty$, $\chi_\xi(P_f, p_f; g(\lambda), m(\lambda), \mu)$ comprises a vanishing factor, implying that we can not obtain the correct large momentum behavior of the wave function from the simple dimensional consideration. We need to apply an operator product expansion which factorizes out the large momentum part responsible for the dominant asymptotic behavior.

II. We now derive the large momentum behavior of the pion electromagnetic form factor F (q^2) which is defined by the vertex function

$$V_\mu(q;P,P') = (P+P')_\mu F(q^2)$$
$$= \int \frac{d^4p}{(2\pi)^4} S_p\{\bar{\chi}_{P'}(p-\tfrac{q}{2})\Gamma_\mu(q;p+\tfrac{P}{2},p+\tfrac{P}{2}-q)\chi_P(p)\Gamma_2(p-\tfrac{P}{2})\}$$
$$+\int \frac{d^4p\,d^4p'}{(2\pi)^8} S_p\{\bar{\chi}_{P'}(p')T_\mu(q;p'+\tfrac{P'}{2},p'-\tfrac{P'}{2},p+\tfrac{P}{2},p-\tfrac{P}{2})\chi_P(p)\}, \quad (14)$$

where $\Gamma_2(p) = S_F^{-1}(p)$ and $\Gamma_\mu(q;p_1,p_1')$ and $T_\mu(q;p_1',p_2',p_1,p_2)$ are amputated Green functions (see Figs.(1a) and (1b)). Using the renormalization group equations satisfied by Green functions Γ_2, Γ_μ and T_μ and by the wave functions χ and $\bar{\chi}$, one derives readily that V_μ satisfies the following renormalization group equation:

$$[\mathcal{D}-1]V_\mu(\lambda q_f,\lambda P_f,\lambda P_f';g_R,m_R,\mu) = 0.$$

Fig.(1a)

Fig.(1b)

Therefore

$$V_\mu(\lambda q_f,\lambda P_f,\lambda P_f';g_R,m_R,\mu) \xrightarrow{\lambda\to\infty} \lambda V_\mu(q_f,P_f,P_f';g(\lambda),m(\lambda),\mu). \quad (15)$$

If $V_\mu(q_f,P_f,P_f';g(\lambda),m(\lambda),\mu)$ does not have mass singularities and did not vanish, Eq. (15) would determine the large momentum behavior of V_μ (consequently $F(q^2)$). But when $\lambda\to\infty$, $V_\mu(q_f,P_f,P_f';g(\lambda),m(\lambda),\mu)$ vanishes because of the complexity of the wave function, although $V_\mu(q_f,P_f,P_f';g(\lambda),m(\lambda),\mu)$ has no singularities. We shall argue that $V_\mu(q_f,P_f,P_f';g(\lambda),m(\lambda),\mu)$ has no singularities and then analyse how fast it vanishes when $\lambda \to \infty$.

We first investigate succinctly the mass singularities of the wave function $\chi(P_f, p; g(\lambda), m(\lambda), \mu)$ in order to examine the mass singularities of $V_\mu(q_f, P_f, P_f'; g(\lambda), m(\lambda), \mu)$. For fixed λ , simple evaluation of the one loop diagram (Fig. (2)) shows that there is no infrared divergence due to the kinematics requirement that the two straton legs can not simultaneously be on mass-shell (the order of magnitude of the off mass-shell $\geqslant O(\lambda^{-2})$). When $\lambda \to \infty$, the one loop diagram still has no infrared singularities. This conclusion can be extended to the case of multiloop diagrams. That is, when $\lambda \to \infty$ ($m(\lambda) \to 0$ and $P_f^2 \to 0$) the wave function $\chi(P_f, p; g(\lambda), m(\lambda), \mu)$ has no singularities, provided that its two straton legs are cut off.

Fig.(2)

the above discussion is valid only for the conventional wave function with $m_B < 2m$. The wave function with $m_B > 2m$ has infrared singularities when the two legs approach mass-shell and in the leading **logarithm** approximation these singularities exponentiate (For the detail see Cornwall and Tiktopoulos's paper[4].) We note that the mass singularities of the wave function are gauge dependent but the gauge invariance of V_μ allows us to work in any given gauge.

Now we argue that $V_\mu(q_f, P_f, P_f'; g(\lambda), m(\lambda), \mu)$ has no singularities when $\lambda \to \infty$. For this purpose we take the axial gauge.

Let $\Gamma = S_F^{-1} \chi S_F^{-1}$ represent the leg-amputated wave function. Then the factors of the intergrand in Eq. (14) are Γ, S_F, Γ_μ and T_μ. Firstly consider the first term in Eq. (14). The Z_2 does not contain mass and infrared singularities because the renormalization has been carried out at $m = 0$ and $p^2 = \mu^2$ [3]. Therefore when $p^2 \neq 0$ $S_F(p)$ has no singularities and when $p^2 = 0$ it contributes a damping exponential factor in the approximation of leading logarithm singularities when $m(\lambda) \to 0$ $(\lambda \to \infty)$ [5]. Since the large momentum flows through either all inner straton lines or one of them (p_1 or p_1') Γ_μ has no mass singularities in the axial gauge in either these two cases. In addition the leg-amputated wave functions Γ also have no mass singularities. Consequently the first term in Eq. (14) has no singularities when $\lambda \to \infty$. Secondly consider the second term in Eq. (14). As can be known from the Fig.(1b), T_μ is a two-particle irreducible Green function (at channels of a bound state) and so has no mass singularities in the axial gauge if $p_1^2, p_2^2 \neq 0$ or $p_1'^2, p_2'^2 \neq 0$. This conclusion can be proved by using the method of Amati et al.'s [6]. However, in the case when the large momentum q flows through p_1'(or p_1) T_μ has infrared singularities, when p_2^2 and $p_2'^2$ both approach mass-shell. But in the approximation of leading logarithm singularities the sum of all leading logarithmic divergences to all orders is a damping exponential factor. Therefore, the second term in Eq. (14) does not diverge when $\lambda \to \infty$. (Considering the momentum regime responsible for the dominant asymptotic behavior, we expect the contribution of the term to V_μ is less than the first term of Eq. (14).) Thus $V_\mu(q_f, P_f, P'_f; g(\lambda), m(\lambda), \mu)$ has no singularities when $\lambda \to \infty$.

But it comprises a vanishing factor. One knows from the above results about the large momentum behavior of the wave function that $\chi(P_f, P_f; g(\lambda), m(\lambda), \mu)$ comprises a vanishing factor when $\lambda \to \infty$ (see Eq. (13).) Using Eqs. (13) and (14), we obtain

$$V_\mu(q_f, P_f, P'_f; g(\lambda), m(\lambda), \mu) \overset{\lambda\to\infty}{\sim} \frac{1}{\lambda^2}(\ln\lambda)^{-\frac{2\gamma'}{b_0}} V'_\mu(q_f, P_f, P'_f; g(\lambda), m(\lambda), \mu), \tag{16}$$

where $V'_\mu(q_f, P_f, P'_f; g(\lambda), m(\lambda), \mu)$ is a finite (non-zero) value. Inserting Eq. (16) into Eq. (15), we find

$$V_\mu(\lambda q_f, \lambda P_f, \lambda P'_f; g_R, m_R, \mu) \xrightarrow{\lambda\to\infty} \frac{1}{\lambda}(\ln\lambda)^{-\frac{2\gamma'}{b_0}} V'_\mu(q_f, P_f, P'_f, g(\lambda), 0, \mu) \tag{17}$$

Expanding V'_μ versus $g(\lambda)$, we have

$$V'_\mu(q_f, P_f, P'_f; g(\lambda), 0, \mu) \sim g^2(\lambda) V_\mu(q_f, P_f, P'_f, \mu) + O(g^4(\lambda)).$$

(From Eq. (11) and the perturbation calculation for $f(\frac{P_f^2}{\mu^2}, g(\lambda), 0)$ we know that the expansion of V'_μ begins from the $g^2(\lambda)$ term. In ladder approximation this result can also be obtained by solving the B-S equation and computing V_μ (7).)

Therefore,

$$F(q^2) \xrightarrow{q^2\to\infty} \frac{\alpha_S(q^2)}{q^2} f_1(m_\xi) + O\left(\frac{1}{q^2}\alpha_S^2(q^2)\right) \tag{18}$$

with $\alpha_S(q^2) = g^2(q^2)/4\pi$ (Note $\gamma' = 0$, see Eq. (9)). It is evident from Eqs. (17) and (18) that the S-matrix element and the form factor are gauge independent though the wave function χ is gauge dependent (note that γ_F is gauge dependent). These are just the expected consequences.

If we take into account all terms in the expansion of the wave function, then the final result is

$$F(q^2) \xrightarrow{q^2\to\infty} \frac{\alpha_s(q^2)}{q^2}\left[f_1(m_\xi)+\sum_{n,n'}(\ln q^2)^{-\frac{\gamma^n+\gamma^{n'}}{b_0}} f_{nn'}(m_\xi)+O(\alpha_s(q^2))\right]. \quad (19)$$

This work is carried out under my supervisor Dai Yuanben's guidance. The author acknowledges his guide and help.

Note added. After this paper was submitted, I noticed some recent papers [8] in which the same results for the pion form factor have been obtained.

REFERENCES

1. T. Appolquist and E. Poggio, Phys. Rev. D10 (1974) 3280;
   C. G. Callan and D. J. Gross, Phys. Rev. D11 (1975) 2905;
   P. Menotti, Phys. Rev. D13 (1976) 1118;
   M. L. Goldberger et al., Phys. Rev. D14 (1976) 1117.
2. Division of Elementary Particles, Laboratory of Theoretical Physics, Peking University and Laboratory of Theoretical Physics, Institute of Mathematics, Academia Sinica, Acta Scientiarum Naturalium Universtatis Pekinensis, 12 (1966) 209.
3. S Weinberg, Phys. Rev. D8 (1973) 3497.
4. J. M. Cornwall and G. Tiktopoulos, Phys, Rev. D13 (1976) 3370.
5. J. Frenkel and J. C. Taylor, Nucl. Phys. B116 (1976) 185.
6. D. Amati et al., Nucl. Phys. B146 (1978) 29.
7. G. R. Farrar and D. R. Jackson, preprint CALT-68-708 (1979).
8. G. P. Lepage and S. J. Brodsky, preprint SLAC-PUB- 2343 (1979); preprint SLAC-PUB-2348 (1979);
   A. Duncan and A. H. Mueller, preprint CU-TP-162 (1979).